International Conference on Science and Applied Science (Engineering and Educational Science) 2016

Journal of Physics: Conference Series Volume 795

Solo, Indonesia
19 November 2016

Editors:

Soeparmi
Faud Anwar

ISBN: 978-1-5108-3680-8
ISSN: 1742-6588

Printed from e-media with permission by:

Curran Associates, Inc.
57 Morehouse Lane
Red Hook, NY 12571

Some format issues inherent in the e-media version may also appear in this print version.

Printed by Curran Associates, Inc. (2017)

For permission requests, please contact the Institute of Physics
at the address below.

Institute of Physics
Dirac House, Temple Back
Bristol BS1 6BE UK

Phone: 44 1 17 929 7481
Fax: 44 1 17 920 0979

techtracking@iop.org

Additional copies of this publication are available from:

Curran Associates, Inc.
57 Morehouse Lane
Red Hook, NY 12571 USA
Phone: 845-758-0400
Fax: 845-758-2633
Email: curran@proceedings.com
Web: www.proceedings.com

Table of contents

Volume 795

International Conference on Science and Applied Science (Engineering and Educational Science) 2016

19 November 2016, Solo, Indonesia

Accepted papers received: 19 December 2016
Published online: 10 February 2017

Preface

Papers

Keynote Speakers

Invited Speakers

Physics

Mathematics

Biology

Chemistry

Civil Engineering

International Conference on Science and Applied Science (Engineering and Educational Science) 2016 (ICSAS 2016)

Syariah Hotel Solo
Surakarta, Indonesia
19 November 2016

Preface

International Conference on Science and Applied Science (Engineering and Educational Science) 2016 (ICSAS 2016) was held at the Syariah Hotel Solo, Surakarta, Indonesia on 19 November 2016. The ICSAS 2016 conference is aimed to bring together scholars, leading researchers and experts from diverse backgrounds and applications areas in Science. Special emphasis is placed on promoting interaction between the science theoretical, experimental, and education sciences, engineering so that a high level exchange in new and emerging areas within Mathematics, Chemistry, Physics and Biology, all areas of sciences and applied mathematics and sciences is achieved.

In ICSAS 2016, there are five parallel sessions and five keynote speakers. It is an honour to present this volume of *Journal of Physics: Conference Series (JPCS)* and we deeply thank the authors for their enthusiastic and high-grade contribution. We would like to express our sincere gratitude to all in the Programming Committee who have reviewed the papers and developed a very interesting Conference Program, as well as thanking the invited and plenary speakers. Finally, we would like to thank the conference chairman, the members of the steering committee, the organizing committee, the organizing secretariat and the financial support from the Sebelas Maret University that allowed ICSAS 2016 to be a success.

The Editor of the ICSAS 2016 Proceedings:
Prof. Dra. Soeparmi, M.A.,Ph.D
Dr. Fuad Anwar, S.Si., M.Si

 doi:10.1088/1742-6596/795/1/011002

Organizer of International Conference on Science and Applied Science (Engineering and Educational Science) 2016

Organizer
Graduate Program, Physics Department, Sebelas Maret University, Indonesia
Jl. Ir. Sutami 36A Kentingan Jebres Surakarta 57126, Indonesia
Phone/fax : 0271-669017 psw 308
Email : icsas@mail.uns.ac.id

Chairman

1. Prof. Dra. Soeparmi, M.A., Ph.D, Sebelas Maret University, Indonesia
2. Dr. Fuad Anwar, S.Si.,M.Si, Sebelas Maret University, Indonesia

Organizing Committee

1. Prof. Drs. Cari, M.A., M.Sc., PhD, Sebelas Maret University, Indonesia
2. Ahmad Marzuki, S.Si., Ph.D., Sebelas Maret University, Indonesia
3. Dr. Eng Budi Purnama, S.Si, M.Si., Sebelas Maret University, Indonesia
4. Dr. Fahru Nurosyid, S.Si.M.Si., Sebelas Maret University, Indonesia
5. Drs. Harjana, M.Si. M.Sc., Ph.D, Sebelas Maret University, Indonesia
6. Dr. Agus Supriyanto, S.Si,M.Si. Sebelas Maret University, Indonesia
7. Dr. Yofentina Iriani, S.Si.,M.Si., Sebelas Maret University, Indonesia
8. Dr. Eng. Risa Suryana, S.Si,M.Si., Sebelas Maret University, Indonesia
9. Khairuddin, S.Si.,M.Phil, Ph.D., Sebelas Maret University, Indonesia
10. Drs. Iwan Yahya, M.Si., Sebelas Maret University, Indonesia
11. Mohtar Yunianto, S.Si,M.Si., Sebelas Maret University, Indonesia
12. Dr. Nuryani, S.Si, M.Si., Sebelas Maret University, Indonesia

ICSAS 2016

International Conference on Science and Applied Science (Engineering and Educational Science) 2016

UNS
SEBELAS MARET
UNIVERSITY

Peer review statement

All papers published in this volume of *Journal of Physics: Conference Series* have been peer reviewed through processes administered by the proceedings Editors. Reviews were conducted by expert referees to the professional and scientific standards expected of a proceedings journal published by IOP Publishing.

IOP Conf. Series: Journal of Physics: Conf. Series **795** (2017) 012001 doi:10.1088/1742-6596/795/1/012001

The Shannon entropy information for mixed Manning Rosen potential in D-dimensional Schrodinger equation

A Suparmi, C Cari, Beta Nur Pratiwi and Dewanta Arya Nugraha
Physics Department, Graduate Program, Sebelas Maret University, Indonesia

Email: soeparmi@staff.uns.ac.id

Abstract. D dimensional Schrodinger equation for the mixed Manning Rosen potential was investigated using supersymmetric quantum mechanics. We obtained the energy eigenvalues from radial part solution and wavefunctions in radial and angular parts solution. From the lowest radial wavefunctions, we evaluated the Shannon entropy information using Matlab software. Based on the entropy densities demonstrated graphically, we obtained that the wave of position information entropy density moves right when the value of potential parameter q increases, while its wave moves left with the increase of parameter α. The wave of momentum information entropy densities were expressed in graphs. We observe that its amplitude increase with increasing parameter q and α

1. Introduction

The Schrodinger equation or Dirac equation in the case of spin and pseudo spin symmetry which are reduced to Schrodinger like equation for a class of shape invariant potentials are solvable using supersymmetric quantum mechanic (SUSY QM) [1-8], Nikiforov-Uvarov method [9-10], Factorization methods [11-12] and Asymptotic Iteration Method (AIM) [13-17]. SUSY QM can be considered very similar to Factorization method. In addition, the derivation of energy eigenvalue, in AIM and SUSY QM method also very similar. Moreover, some methods are derived from generalized hypergeometric equation.

D-dimensional Schrodinger equation, Klein Gordon equation and Dirac equation in the case of spin and pseudo spin symmetry, which are reduced into one dimensional Schrodinger equation, are recently attractive for some researcher. Even the direction of this problem is not clear enough but it is worthy to be put on consideration. In this paper, we will obtain the energy eigen value and wavefunctions of single particle that is governed by separable D-dimensional mixed Manning Rosen non-central potential which are defined as

$$V(r,\theta_1,\theta_2,...,\theta_{D-1})=\frac{\hbar^2}{2M}\left\{V(r)+\frac{1}{r^2}\sum_{i=1}^{D-3}\left(\frac{V_i(\theta_i)}{\sin^2\theta_{i+1}...\sin^2\theta_{D-1}}\right)+\frac{1}{r^2}\frac{V_{D-2}(\theta_{D-2})}{\sin^2\theta_{D-1}}+V_{D-1}(\theta_{D-1})\right\} \tag{1}$$

where

$$V(r)=\alpha^2\left(\frac{v(v-1}{\sin^2\alpha r}+\frac{\gamma(\gamma-1)}{\cos^2\alpha 2}+2q\tan\alpha r+2p\cot\alpha r\right) \tag{2}$$

$$V_i(\theta_i)=\left(\frac{v_i(v_i-1)}{\sin^2\theta_i}+\frac{\gamma_i(\gamma_i-1)}{\cos^2\theta_i}+2q_i\tan\theta_i+2p_i\cot\theta_i\right) \tag{3}$$

for $i=1,2,3...$ The Manning Rosen potential is used as a mathematical model in the description of diatomic molecular vibrations and it constitutes a convenient model for other physical situations. [18-

21]. Radial wavefunction will be used to investigated the Shannon entropy information. The Shannon entropic plays an important rule in the measure of uncertainly. [22-24]

In this paper, solution of Schrodinger equation in D dimension for mixed Manning- Rosen potential was studied using SUSY QM. The Shannon entropy was calculated for explaining about the uncertainty system. The paper is organized as follows. The basic theory is presented in section 2. The results and discussion are presented in section 3 and a conclusion in section 4.

2. Basic theory

2.1. The D-dimensional Schrodinger equation

D dimensional Schrodinger is separated into 1 one -dimensional radial Schrodinger part, below [25]

$$\left\{r^2\frac{1}{r^{D-1}}\frac{\partial}{\partial r}\left(r^{D-1}\frac{\partial}{\partial r}\right)-r^2\left(V(r)-\frac{2m}{\hbar^2}E\right)\right\}R=\lambda_{D-1}R \tag{4}$$

and (D-1) one- dimensional angular Schrodinger equations as follow

$$\frac{\partial^2 P_1}{\partial\theta_1^2}-V_1(\theta_1)P_1+\lambda_1 P_1=0 \tag{5}$$

$$\left\{\frac{1}{\sin\theta_2}\frac{\partial}{\partial\theta_2}\left(\sin\theta_2\frac{\partial P_2}{\partial\theta_2}\right)\right\}-\mathrm{V}_2(\theta_2)\mathrm{P}_2-\frac{\lambda_1 P_2}{\sin^2\theta_2}+\lambda_2 P_2=0 \tag{6}$$

$$\left[\frac{1}{\sin^2\theta_3}\frac{\partial}{\partial\theta_3}\left(\sin^2\theta_3\frac{\partial \mathrm{P}_3(\theta_3)}{\partial\theta_3}\right)\right]-V_3(\theta_3)\mathrm{P}_3(\theta_3)-\frac{\lambda_2}{\sin^2\theta_3}\mathrm{P}_3(\theta_3)+\lambda_3\mathrm{P}_3(\theta_3)=0 \tag{7}$$

....

$$\frac{1}{\sin^{D-2}\theta_{D-1}}\frac{\partial}{\partial\theta_{D-1}}\left(\sin^{D-2}\theta_{D-1}\frac{\partial P_{D-1}}{\partial\theta_{D-1}}\right)-V_{D-1}(\theta_{D-1})P_{D-1}-\frac{\lambda_{D-2}P_{D-1}}{\sin^2\theta_{D-1}}+\lambda_{D-1}P_{D-1}=0 \tag{8}$$

with $\lambda_1.\lambda_2,\ \lambda_3,...,\lambda_{D-1}$ are variable separation constant.

2.2. Supersymmetric quantum mechanics

Each of the Schrodinger is solved by using SUSY QM [26] and the properties of shape invariant potential as follows the SUSY partner potentials

$$\begin{aligned}V_-(x)&=w^2(x)-\frac{\hbar}{\sqrt{2M}}w'(x)\\ V_+(x)&=w^2(x)+\frac{\hbar}{\sqrt{2M}}w'(x)\end{aligned} \tag{9}$$

where

$$V(x)=V_-(x)+E_0 \tag{10}$$

and w is super potential.

And by applying the condition of shape invariant potential which is defined as to obtain the mapping parameter:

$$V_+(a_0,x)-V_-(a_1,x)=R(a_1) \tag{11}$$

the energy eigenvalue can be obtained, following equation (12)

$$E_n=\sum_{k=1}^{k=n}R(a_k)+E_0 \tag{12}$$

Then for obtain the wavefunctions we used the lowering and raising operators,

$$A = \frac{\hbar}{\sqrt{2M}}\frac{\partial}{\partial x} + w(a_0, x);\ A^+ = -\frac{\hbar}{\sqrt{2M}}\frac{\partial}{\partial x} + w(a_0, x)$$
$$A\psi_0 = A\psi_0^{(-)} = 0\ ;\ A^+(a_0, x)\psi_0^{(-)}(a_1, x) = \psi_1^{(-)} \tag{13}$$

2.3. Shanon entropy

Shannon's information entropy which was proposed in the late 1940s has attracted many physicists [22-24] Shannon proposed the idea of probabilities that was used to measure uncertain of events and also proposed the information entropy to measure the uncertainty.

The information entropy consist of position entropy density $\rho(r)$ and momentum entropy density $\rho(p)$ are defined below

$$\rho(r) = |\Psi(r)|^2 \ln|\Psi(r)|^2 \tag{14}$$

$$\rho(p) = |\phi(p)|^2 \ln|\phi(p)|^2 \tag{15}$$

$\Psi(r)$ is the normalized eigenstate of position and $\phi(p)$ is the corresponding Fourier transform, with r is radial radius and p is momentum.

3. Solution and discussion

3.1. Solution of radial part

By using equation (2) and equation (4), the general solution for four dimensional Schrodinger equation is given as

$$\frac{1}{r^3}\frac{\partial}{\partial r}\left(r^3\frac{\partial R}{\partial r}\right) - \alpha^2\left(\frac{v(v-)}{\sin^2\alpha r} + \frac{\gamma(\gamma-1)}{\cos^2\alpha r} + 2q\tan\alpha r + 2p\cot\alpha r\right)R + \frac{2M}{\hbar^2}ER - \frac{\lambda_3}{r^2}R = 0 \tag{16}$$

By setting $R = \frac{u}{r^{3/2}}$ in equation (16) we have

$$u'' - \frac{3/4 + \lambda_3}{r^2}u - \alpha^2\left(\frac{v(v-1)}{\sin^2\alpha r} + \frac{\gamma(\gamma-1)}{\cos^2\alpha r} + 2q\tan\alpha r + 2p\tan\alpha r\right)u = -\frac{2ME}{\hbar^2}u \tag{17}$$

By applying centrifugal term approximation defined as $\frac{1}{r^2} \cong \frac{\alpha^2}{\sin^2\alpha r}$, then equation (17) becomes

$$u'' - \alpha^2\left(\frac{v(v-1) + 3/4 + \lambda_3}{\sin^2\alpha r} + \frac{\gamma(\gamma-1)}{\cos^2\alpha r} + 2q\tan\alpha r + 2p\tan\alpha r\right)u = -\frac{2ME}{\hbar^2}u \tag{18}$$

By using equations (9-13) we get the super potential, the super partner potential V_+ and V_- mapping para meter, raising and lowering operators as

$$w(r) = \gamma\alpha\tan\alpha r - v\alpha\cot\alpha r + \alpha\frac{q}{\gamma} \tag{19}$$

$$V_-(r) = \frac{\hbar^2\alpha^2}{2M}\left(\frac{v'(v'-1)}{\sin^2\alpha r} + \frac{\gamma(\gamma-1)}{\cos^2\alpha r} + 2q\tan\alpha r + 2p\cot\alpha r\right) - \frac{\hbar^2\alpha^2}{2M}(v+\gamma)^2 + \frac{\hbar^2\alpha^2}{2M}\frac{q^2}{\gamma^2} \tag{20}$$

$$V_+(r) = \frac{\hbar^2\alpha^2}{2M}\left(\frac{v'(v'+1)}{\sin^2\alpha r} + \frac{\gamma(\gamma+1)}{\cos^2\alpha r} + 2q\tan\alpha r + 2p\cot\alpha r\right) - \frac{\hbar^2\alpha^2}{2M}(v+\gamma)^2 + \frac{\hbar^2\alpha^2}{2M}\frac{q^2}{\gamma^2} \tag{21}$$

where
$$v' = \frac{1}{2} + \sqrt{v(v-1) + 1 + \lambda_3} \tag{22}$$

so we get the mapping parameters given as

$$a_0 = v', a_1 = v'+1, ..., a_n = v'+n; \quad b_0 = \gamma, b_1 = \gamma+1, ..., b_n = \gamma+n \tag{23}$$

The lowering and rising operator is determined by,

$$A = \frac{\hbar}{\sqrt{2M}}\frac{\partial}{\partial r} + \frac{\hbar}{\sqrt{2M}}\left(\gamma\alpha\tan\alpha r - v'\alpha\cot\alpha r + \alpha\frac{q}{\gamma}\right);$$
$$A^{+} = -\frac{\hbar}{\sqrt{2M}}\frac{\partial}{\partial r} + \frac{\hbar}{\sqrt{2M}}\left(\gamma\alpha\tan\alpha r - v'\alpha\cot\alpha r + \alpha\frac{q}{\gamma}\right) \tag{24}$$

By using SUSY, we obtained

$$E_0 = \frac{\hbar^2\alpha^2}{2M}(v'+\gamma)^2 - \frac{\hbar^2\alpha^2}{2M}\frac{q^2}{\gamma^2} \tag{25}$$

So the energy eigenvalue was obtained from equations (12, 23 and 25) ,

$$E_n = \frac{\hbar^2\alpha^2}{2M}(v'+\gamma+2n)^2 - \frac{\hbar^2\alpha^2}{2M}\frac{q^2}{(\gamma+n)^2} \tag{26}$$

From equation (26,) it is shown that the energy eigenvalue depends on the parameters of all components of the composed potential and also depend on the quantum number *n*.

And then from equations (19, 24) we get the ground state wave function as

$$u_0 = C_0(\sin\alpha r)^{v'}(\cos\alpha r)^{\gamma} e^{-\frac{\alpha q}{\gamma}r} \tag{27}$$

and by using raising operator, we get the first exited wave function

$$u_1 = -\frac{\hbar\alpha}{\sqrt{2M}}C_1\left\{\begin{array}{l}(2v'+1)(\sin\alpha r)^{v'}(\cos\alpha r)^{\gamma+2} - (2\gamma+1)(\sin\alpha r)^{v'+2}(\cos\alpha r)^{\gamma} \\ -\left(\frac{q\alpha}{\gamma+1}+\frac{q}{\gamma}\right)(\sin\alpha r)^{v'+1}(\cos\alpha r)^{\gamma+1}\end{array}\right\} e^{-\frac{\alpha q}{\gamma+1}r} \tag{28}$$

3.2. Solution of angular part

From equation (3) and equations (5-7), we rewrote the angular part equations below,

$$\frac{\partial^2 P_1}{\partial\theta_1^2} - \left\{\frac{v_1(v_1-1)}{\sin^2\theta_1} + \frac{\gamma_1(\gamma_1-1)}{\cos^2\theta_1} + 2q_1\tan\theta_1 + 2p_1\cot\theta_1\right\}P_1 + \lambda_1 P_1 = 0 \tag{29}$$

$$\left\{\frac{1}{\sin\theta_2}\frac{\partial}{\partial\theta_2}\left(\sin\theta_2\frac{\partial P_2}{\partial\theta_2}\right)\right\} - \left\{\frac{v_2(v_2-1)}{\sin^2\theta_2} + \frac{\gamma_2(\gamma_2-1)}{\cos^2\theta_2} + 2q_2\tan\theta_2 + 2p_2\cot\theta_2\right\}P_2$$
$$-\frac{\lambda_1 P_2}{\sin^2\theta_2} + \lambda_2 P_2 = 0 \tag{30}$$

$$\left[\frac{1}{\sin^2\theta_3}\frac{\partial}{\partial\theta_3}\left(\sin^2\theta_3\frac{\partial P_3}{\partial\theta_3}\right)\right] - \left\{\frac{v_3(v_3-1)}{\sin^2\theta_3} + \frac{\gamma_3(\gamma_3-1)}{\cos^2\theta_3} + 2q_3\tan\theta_3 + 2p_3\cot\theta_3\right\}P_3$$
$$-\frac{\lambda_2}{\sin^2\theta_3}P_3 + \lambda_3 P_3 = 0 \tag{31}$$

and by repeating the steps used to solve radial Schrodinger equation we get the angular wave functions and the constant of variable separation as follows. By solving equation (29) using SUSY QM we obtain the ground state wave function P_{10} and the constant of variable separation as

$$P_{10}(\theta_1)=C_{10}\left(\sin\theta_1\right)^{v_1}\left(\cos\theta_1\right)^{\gamma_1}e^{-\frac{q_1}{\gamma_1}\theta_1};\lambda_1=\left(\gamma_1+v_1+2n_1\right)^2-\frac{q_1^2}{\left(\gamma_1+n_1\right)^2} \quad (32)$$

By setting $P_2=\frac{Q_2}{\sqrt{\sin\theta_2}}$ and $P_3=\frac{Q_3}{\sin\theta_3}$ in equations (30) and (31) the both equations becomes

$$Q_2''-\left\{\frac{v_2(v_2-1)+\lambda_1-1/4}{\sin^2\theta_2}+\frac{\gamma_2(\gamma_2-1)}{\cos^2\theta_2}+2q_2\tan\theta_2+2p_2\cot\theta_2\right\}Q_2+\left(\lambda_2+1/4\right)Q_2=0 \quad (33)$$

$$Q_3''-\left\{\frac{v_3(v_3-1)+\lambda_2}{\sin^2\theta_3}+\frac{\gamma_3(\gamma_3-1)}{\cos^2\theta_3}+2q_3\tan\theta_3+2p_3\cot\theta_3\right\}Q_3+\left(\lambda_3+1\right)Q_3=0 \quad (34)$$

By setting $v'_2(v'_2-1)=v_2(v_2-1)+\lambda_1-1/4$ or $v'_2=1/2+\sqrt{v_2(v_2-1)+\lambda_1}$ and $v'_3=1/2+\sqrt{(v_3-1/2)^2+\lambda_2}$, these two equations are solved using SUSY QM and we get the lowest wave functions and the variable separation constants as

$$Q_{20}=C_{20}\left(\sin\theta_2\right)^{v'_2}\left(\cos\theta_2\right)^{\gamma_2}e^{-\frac{q_2}{\gamma_2}\theta_2};\lambda_2=\left(v'_2+\gamma_2+2n_2\right)^2-\frac{q_2^2}{\left(\gamma_2+n_2\right)^2} \quad (35)$$

$$Q_{30}=C_{30}\left(\sin\theta_3\right)^{v'_3}\left(\cos\theta_3\right)^{\gamma_3}e^{-\frac{q_3}{\gamma_2}\theta_3};\lambda_3=\left(v'_3+\gamma_3+2n_3\right)^2-\frac{q_3^2}{\left(\gamma_3+n_3\right)^2} \quad (36)$$

The lowest total wavefunction from equation (27), equation (32) and equations (35-36) is given as

$$\psi(r,\theta_1,\theta_2,\theta_3)=C\alpha^{3/2}\frac{\left(\sin\alpha r\right)^{v'}\left(\cos\alpha r\right)^{\gamma}e^{-\frac{\alpha q}{\gamma}r}}{\left(\sin\alpha r\right)^{3/2}}\left(\sin\theta_1\right)^{v_1}\left(\cos\theta_1\right)^{\gamma_1}e^{-\frac{q_1}{\gamma_1}\theta_1}$$
$$\left(\sin\theta_2\right)^{v'_2-1/2}\left(\cos\theta_2\right)^{\gamma_2}e^{-\frac{q_2}{\gamma_2}\theta_2}\left(\sin\theta_3\right)^{v'_3-1}\left(\cos\theta_3\right)^{\gamma_3}e^{-\frac{q_3}{\gamma_2}\theta_3} \quad (37)$$

with $C=C_0C_{10}C_{20}C_{30}$ which is normalization factor of the total lowest wave function.

3.3. *The Shannon information entropy of one-dimensional Manning Rosen potential*

For the case of $\lambda=2$ and $v'-3/2=2$, we obtain the radial wavefunction given as

$$\psi_0(x)=C_0\left(\sin\alpha r\right)^2\left(\cos\alpha r\right)^2e^{-\frac{\alpha q}{2}r} \quad (38)$$

and the momentum eigenstate is obtained by using Fourier transform on position eigenstate as follows

$$\phi_0(p)=\int e^{-ipx}C_0(\sin\alpha x)^2(\cos\alpha x)^2e^{-\frac{\alpha q}{2}x}dx \quad (39)$$

so we obtained

$$\phi_0(p)=\frac{1}{4}C_0\left\{-\frac{-ip+\frac{\alpha q}{2}}{p^2+\left(\frac{\alpha q}{2}\right)^2}+\left(\frac{(ip+\alpha q/2)\cos4\alpha x+4\alpha\sin4\alpha x}{-p^2+\alpha^2q^2/4+16\alpha^2+ip\alpha q}\right)\right\}\left(e^{-x(ip+\alpha q/2)}\right) \quad (40)$$

The Shannon position information entropy density was obtained numerically from equation (14) and equations (38) using Matlab software, which can be shown in Figure 1.

IOP Conf. Series: Journal of Physics: Conf. Series **795** (2017) 012001 doi:10.1088/1742-6596/795/1/012001

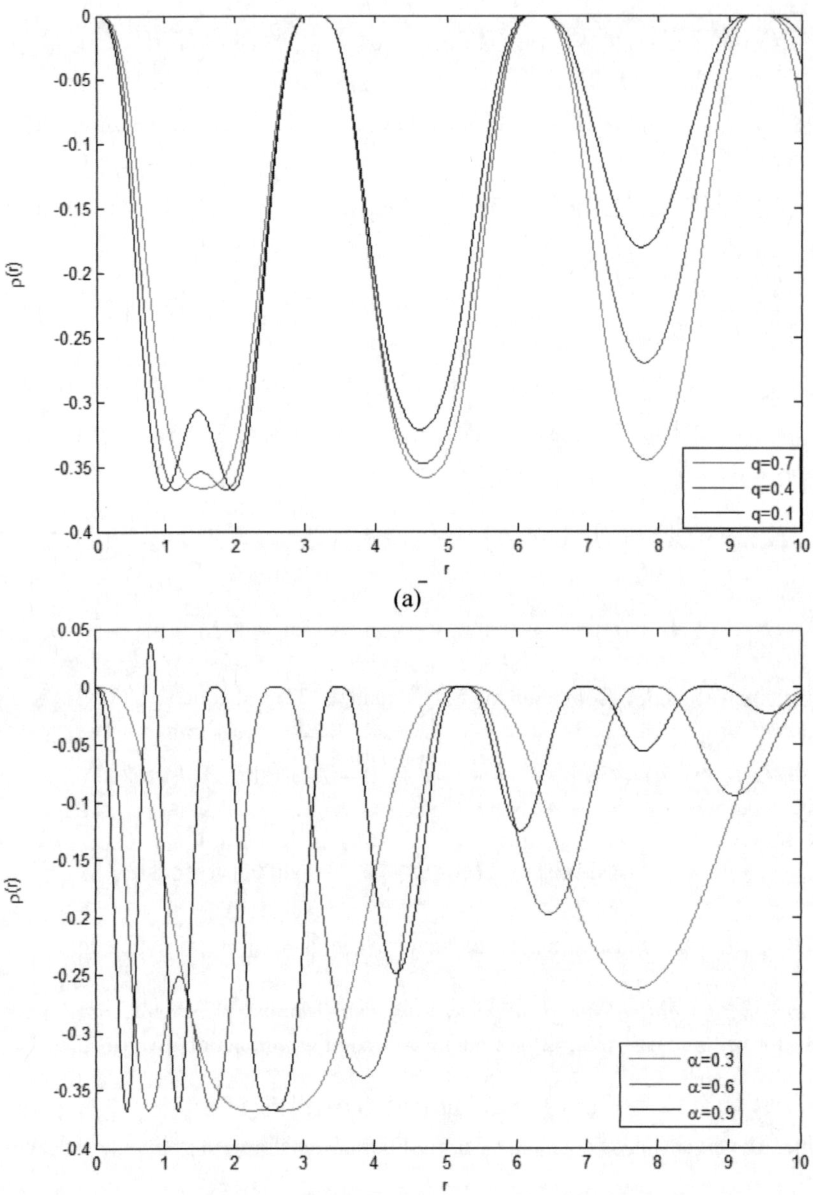

(a)

(b)

Figure 1. Plot of the position space entropy density $\rho(r)$ of the Manning Rosen potential for (a) variation of q (b) variation of α

From Figure (1), we observe that the wave of position information entropy density moves right when the value of potential parameter q increases (Figure 1.a), while its wave moves left with the increase of parameter α (Figure 1.b). And then by using equation (15) and equation (40), we found the graphically result in Figure (2) for the momentum information entropy density. It is shown that the wave amplitude increase with increasing parameter q and α

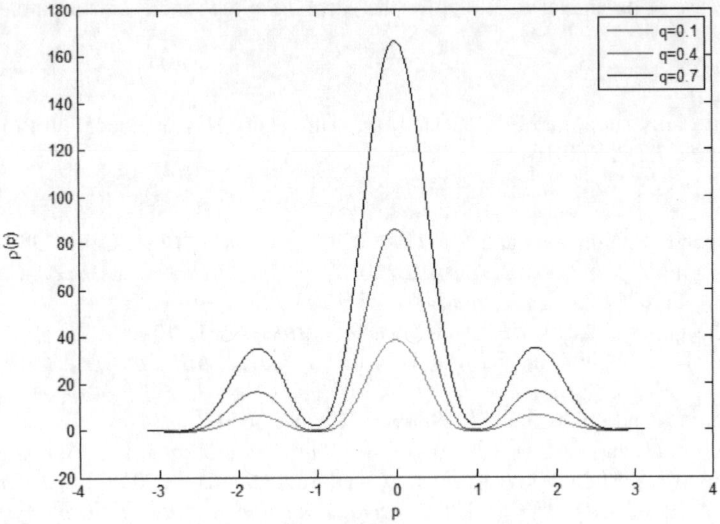

(a)

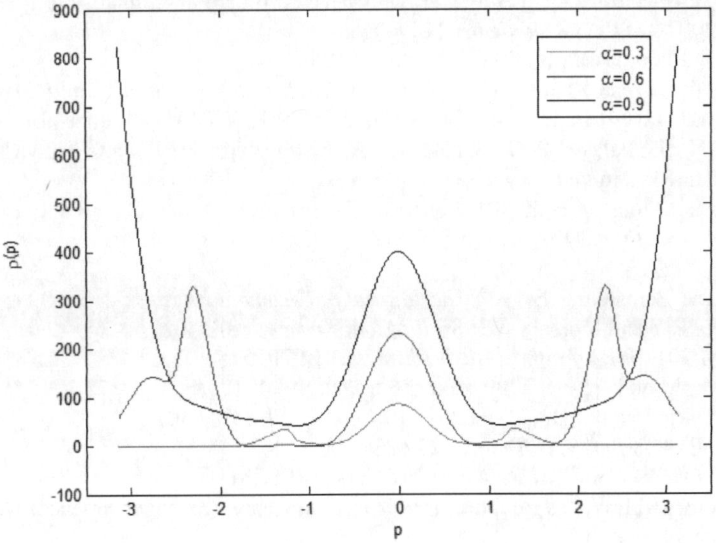

(b)

Figure 2. Plot of the momentum entropy density $\rho(p)$ of the Manning Rosen potential for (a) variation of q (b) variation of α

4. Conclusion

In this paper, we have presented the solution of D-dimensional Schrodinger for the mixed Manning Rosen equation using supersymmetric quantum mechanics. We obtained the energy eigenvalues from radial part solution, which the energy eigenvalue depends on the parameters of all components of the composed potential and also depend on the quantum number n. The radial wavefunctions and angular wavefunctions was obtained using lowering and raising operator. We presented the Shannon entropy information graphically using Matlab software. We obtained that the wave of position information entropy density moves right when the value of potential parameter q increases, while its wave moves left with the increase of parameter α. While for the wave amplitude increase with incresing parameter *q* and α

Acknowledgement
This research was partly supported by Sebelas Maret University Higher Education Project Grant with contract no. 632/UN27.21/LT/2016

References

[1] Wu Q S, Wen H B, Yun W D and Yu G Z 2002 *Comm. Theor. Phys.*(China) **38** 139
[2] Sheng J C, Ying Z, Lin Z X and Tian S L 2001 *Commun .Theor. Phys.* **36** 641-646
[3] Suparmi and Cari 2014 *J. Math. Fund. Sci.* **46** 3 205-223
[4] Cari C and Suparmi A 2014 *AIP Conference Proceedings* **1615** 101
[5] Alemgadmi K I K, Suparmi, Cari, and Deta U A 2015 *AIP Conference Proceedings* **1677** 030004
[6] Dutt R, Khare A and Sukhatme U P 1988 *Am. J. Phys.* **56** 163
[7] Suparmi A, Cari C and Pratiwi B N 2016 *Journal of Physics: Conference Series* **710** 012026
[8] Gangopadhyaya A, Mallow J V, Rasinariu C and Sukhatme U P 2001 *Chin. J. Phys.* **39**
[9] Nikiforov A E and Uvarov V B 1988 *Special Function in Mathematical Physics* (Birkhausa: Basel)
[10] Falaye B J, Oyewumi K J and Abbas M 2013 arXiv:1109.3894v4 [quant-ph]
[11] Sadeghi J 2007 *Act.Phys.Polonica A* **112** (1) 23
[12] Sadeghi J and Pourhassan B 2008 EJTP **5** (17) 193
[13] Ciftci H, Hall R L and Saad N 2003 *Journal of Physics A: Mathematical and General* **36** 11807
[14] Ikhdair S M, Falaye B J and Adepoju A G 2013 arXiv:1308.0155v1 [quant-ph]
[15] Oyewumi K J, Falaye B J, Onate C A, Oluwadare O J and Yahya W A 2014 arXiv:1402.2181v1 [quant-ph]
[16] Rostami A and Motavali H 2008 *Progress in Electromagnetic Research B* 4 171
[17] Soylu A, Bayrak O and Boztosun I 2008 *Journal of Physics A: Mathematical and Theoretical* **41** 065308
[18] Leroy R J and Bernstein R B 1970 *The Journal of Chemical Physics* **52**, 3869
[19] Cai J M, Cai P Y and Inomata A 1986 *Physical Review A* **34**, 4621
[20] Ikhdair S M 2011 *ISRN Mathematical Physics* **2012** 201525
[21] Chen Z Y, Li M and Jia C S 2009 *Mod. Phys. Lett. A* **24**, 1863
[22] Sun G H, Aoki M V and Dong S H 2013 *Chin. Phys. B* **22** 050302
[23] Shannon C E 1948 *Bell Syst. Tech. J.* **27** 623
[24] Shannon C E 1993 *Collected Papers* (New York :IEEE Press)
[25] Shi-Hai Dong 2011 Waves Equation in Higher Dimension (Springer Science: Business Media) pp 41-43, pp 101-102, p 97
[26] Witten E 1981 *Nucl. Phys. B* **188** 513

Quantum Bound States on Some Hyperbolic Surfaces

Hishamuddin Zainuddin,[1,2] Chan Kar Tim,[1,2] Nor Syazana Shamsuddin,[2] Nurisya Mohd Shah[1,2]

[1]Laboratory of Computational Sciences and Mathematical Physics, Institute for Mathematical Research, Universiti Putra Malaysia, 43400 UPM Serdang, Selangor, Malaysia
[2]Department of Physics, Faculty of Science, Universiti Putra Malaysia, 43400 UPM Serdang, Selangor, Malaysia

Abstract. The aim of this work is to highlight results of energy eigenstates on some noncompact finite hyperbolic surfaces. Such systems are known to exhibit both continuous and discrete spectra and are dependent on the subgroups of the modular group that underlie these surfaces. We study explicitly the cases of Maass cusp forms on the singly punctured two-torus and the triply punctured two-sphere for their eigenvalues. The eigenvalues for the torus system are doubly degenerate while for the sphere case, the eigenvalues are nondegenerate. We also note that the lowest eigenvalue of the sphere system is larger than that of the torus system.

1. Introduction

Systems of a particle on hyperbolic surfaces are known to exhibit chaotic properties ever since the late 19th century [1]. This led researchers to study their quantum versions to understand quantum chaos [2]. Mathematically, hyperbolic surfaces exhibit richer family of surfaces with nontrivial topology, characterised by their genus and number of cusps/punctures (points at infinity) with finite or infinite area [3]. In mathematics, there is a vast number of literature connecting the spectra of the hyperbolic laplacian with number theory [3,4].

In physics, however, there is considerably less literature on the nature of quantum states on hyperbolic surfaces and is almost exclusively limited to the discussion of scattering states on noncompact finite area hyperbolic surfaces (see [5–9]). Lesser is known about the bound states on such surfaces. In this note, we will review the results of [10, 11] for the bound states on two hyperbolic surfaces namely singly punctured two-torus and triply punctured two sphere and make comparisons between their spectra.

2. Construction of the Hyperbolic Surfaces

To construct the abovementioned hyperbolic surfaces, we begin with the upper half plane model $H = \{ z = x + iy \mid \mathrm{Im}(z) > 0 \}$ of the hyperbolic plane i.e. equipped with the hyperbolic metric $ds^2 = y^{-2}(dx^2 + dy^2)$. The geodesics on this plane are given by (portions of) vertical straight lines and semicircles meeting x-axis perpendicularly. The upper half plane can also be written as the homogeneous space $H = \mathrm{PSL}(2,\mathbb{R})/\mathrm{SO}(2)$ and there is a (projective) action of $\mathrm{SL}(2,\mathbb{R})$ on H via fractional linear transformations:

$$\begin{pmatrix} a & b \\ c & d \end{pmatrix} \cdot z = \frac{az+b}{cz+d} \quad ; \quad \begin{pmatrix} a & b \\ c & d \end{pmatrix} \in \mathrm{SL}(2,\mathbb{R}), \quad z \in H. \tag{1}$$

To construct further hyperbolic surfaces out of the hyperbolic plane, we construct tessellating polygons using discrete subgroups of $\mathrm{PSL}(2,\mathbb{R})$. A standard example is that from the modular group [12] $\mathrm{PSL}(2,\mathbb{Z}) \equiv \Gamma(1)$ whose fundamental domain is given by the hyperbolic triangle with vertices $\left\{ e^{\pm i\pi/3}, i\infty \right\}$. Elements of $\Gamma(1)$ generated by

$$S = \begin{pmatrix} 0 & -1 \\ 1 & 0 \end{pmatrix} \quad ; \quad T = \begin{pmatrix} 1 & 1 \\ 0 & 1 \end{pmatrix}, \tag{2}$$

will tessellate the whole hyperbolic plane with the equivalent hyperbolic triangles. The surface $H/\Gamma(1)$ formed by the fundamental domain is a hyperbolic surface with one cusp.

One can build further hyperbolic surfaces by choosing smaller discrete groups like subgroups of the modular group. One particular subgroup is the commutator subgroup Γ' of modular group $\Gamma(1)$. This subgroup is generated by

$$A = \begin{pmatrix} 1 & 1 \\ 1 & 2 \end{pmatrix} \quad ; \quad B = \begin{pmatrix} 1 & -1 \\ -1 & 2 \end{pmatrix}, \tag{3}$$

which are themselves commutators of the generator of the modular group i.e. $A' = ST^{-1}S^{-1}T$ and $B' = STS^{-1}T^{-1}$. The fundamental domain of this subgroup is a quadrangle whose sides are identified by the generators (3) as shown in Fig. 1. The identification is similar to the one for the parallelogram in Euclidean plane representing a two-torus but it now includes the cusp at infinity. Figure 2 graphically depicts the type of surface H/Γ' being constructed.

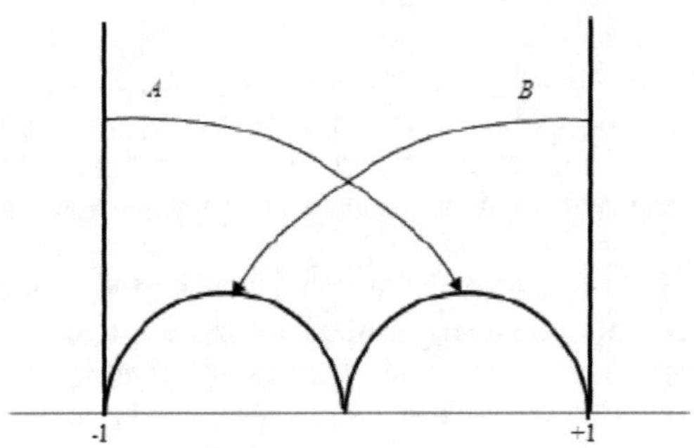

Figure 1. The fundamental domain of Γ', a parallelogram with sides identified by *A* and *B*.

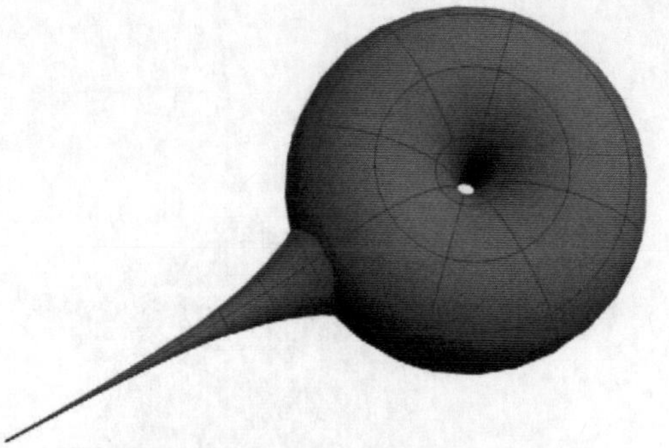

Figure 2. Singly punctured two-torus.

Next, we consider instead the principal congruence subgroup of level two of the modular group defined by

$$\Gamma(2) = \left\{ \begin{pmatrix} a & b \\ c & d \end{pmatrix} \in \mathrm{PSL}(2,\mathbb{Z}) \middle| \begin{pmatrix} a & b \\ c & d \end{pmatrix} = \begin{pmatrix} 1 & 0 \\ 0 & 1 \end{pmatrix} \bmod 2 \right\}. \tag{4}$$

This subgroup is generated by the matrices

$$A = \begin{pmatrix} 1 & 2 \\ 0 & 1 \end{pmatrix} \;\; ; \;\; B = \begin{pmatrix} 0 & 1 \\ -2 & 1 \end{pmatrix} . \tag{5}$$

Its fundamental domain is again a parallelogram with sides identified by generators (5) (see Fig. 3). Figure 4 graphically depicts the surface $H/\Gamma(2)$ formed by the fundamental domain.

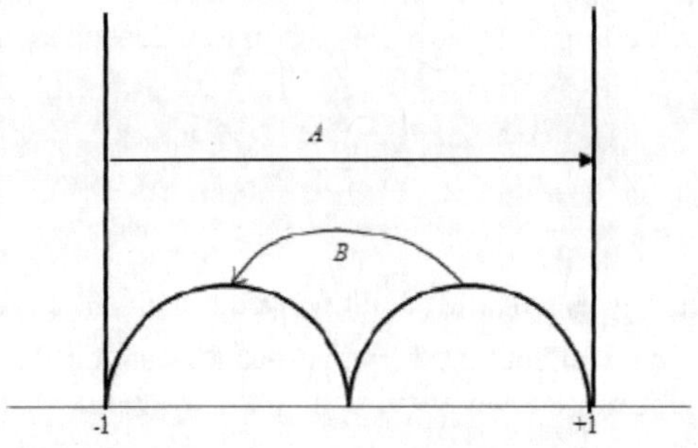

Figure 3. Fundamental domain of $\Gamma(2)$ with sides identified by A and B.

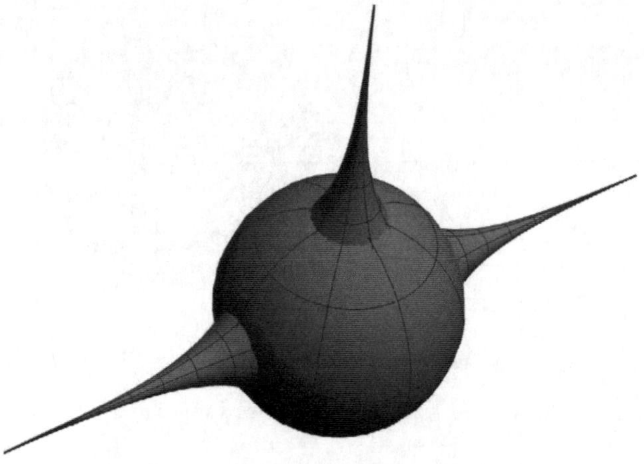

Figure 4. Triply punctured two-sphere

It is interesting to note that the fundamental domains in Figs. 1 and 3 are identical parallelograms with only different side pairings and as such have the same area of 2π. It is the different pairings that gave the resultant surfaces the different topological signature i.e. genus 1 and one cusp for H/Γ' and genus 0 and three cusps for $H/\Gamma(2)$.

3. Schrödinger wave mechanics on hyperbolic surfaces

Quantum mechanics on these hyperbolic surfaces are described by writing the Schrödinger equation with a hyperbolic Laplacian. Letting $\hbar = 1 = 2m$, the time independent Schrödinger equation is

$$H\psi = \Delta\psi = y^2\left(\frac{\partial^2}{\partial x^2} + \frac{\partial^2}{\partial y^2}\right)\psi = \lambda\psi \quad , \tag{6}$$

where $\lambda = \frac{1}{4} + r^2$ is the energy eigenvalue. Following the convention in mathematics literature, we will call r as the eigenvalue. To reflect that the wavefunctions are defined on the fundamental domain of the specified discrete group Γ , we require that the wavefunction to satisfy the automorphy condition:

$$\psi(\gamma z) = \psi(z) \quad , \qquad \gamma \in \Gamma, \quad z \in H \quad . \tag{7}$$

Note that this can be relaxed to include a phase factor, which will not be considered here. For bound states, we add a further cusp boundary condition i.e. $\psi \to 0$ at the cusps, which physically says that the quantum particle remains on the surface not shooting off to infinity. Such solutions are called Maass cusp forms in the literature [13].

The Maass cusp forms has been computed extensively for the modular group $\Gamma(1)$ initially by Hejhal [14] who found a linear stable algorithm that uses the periodicity under $z \to z+1$ on the upper half plane to compute the energy eigenfunctions, together with automorphy condition under $\Gamma(1)$. The wavefunctions are expanded as the following Fourier series:

$$\psi(x+iy) = \sum_{m=-\infty, m\neq 0}^{\infty} a_m y^{1/2} K_{ir}(2\pi m|y|) e^{2\pi imx} \quad , \tag{8}$$

where K_{ir} is the K-Bessel function, ensuring the cusp boundary condition is obeyed. This is further refined by Then [15] using adaptive search method to compute large eigenvalues $r > 40000$. To make such computations more accessible, Siddig & Zainuddin [16] have ported the calculations in Mathematica, from which the present calculations are based.

The algorithm for the modular group can be extended to the case of the commutator subgroup by first modifying its fundamental region in a way that its apparent cusp appears only at $i\infty$ as it should. This can be done using (combinations of) the generators of Γ' to transport subregions away from the x-axis. As it is done in [1, 5], the parallelogram is transformed to an 8-gon for which each side is identified with another, giving the 4-gon needed (see Fig. 5).

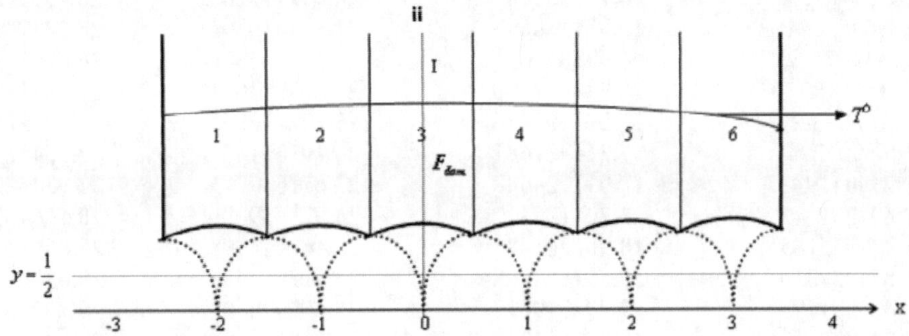

Figure 5. Transformed fundamental domain of commutator subgroup.

There are two advantages of this transformed domain, namely that there are no identified cusps appearing on the x-axis and that the domain is essentially six copies of the domain of the modular group (cusp width six). Hence one can now apply the same Hejhal-Then algorithm without much modifications apart from the periodicity in z is now $z \to z+6$. The new expansion of the wavefunction will be

$$\psi(x+iy) = \sum_{m=-\infty, m\neq 0}^{\infty} a_m y^{1/2} K_{ir}(2\pi m|y|) e^{2\pi imx/6} \quad . \tag{9}$$

For the triply punctured two-sphere, the computation is much harder since there are now three cusps to consider and one cannot simply apply the Hejhal-Then algorithm directly here. Instead, one need to divide the fundamental domain into subregions individually for each cusp as shown in Figure 6. Each subregion is further transformed in a way that maps each cusp to $i\infty$ and the linearized equations for all the three cusps are solved simultaneously, employing symmetry between the subregions of each cusp. For further details, see [17, 11]. The wavefunction expansion in this case is given by

$$\psi(x+iy) = \sum_{m=-\infty, m\neq 0}^{\infty} a_m y^{1/2} K_{ir}(2\pi m|y|) e^{2\pi imx/2} \quad , \tag{10}$$

where the cusp width is now two for each cusp.

4. Computational results

In [10, 11], we have managed to find quantum bound states on both singly punctured two-torus and triply punctured two-sphere respectively by computing their eigenvalues and eigenfunctions. We reproduce some of the results of [10, 11] and compare some of their characteristics.

For the punctured torus, we computed 104 eigenvalues (see Table 1) and they are doubly degenerate, shared between the odd and even cusp forms. Computationally, this degeneracy may be attributed to the arithmetical symmetries found in the Fourier coefficients as explained in [10], though it is desirable to seek a more physical or geometrical explanation. We also found that the lowest eigenvalue is lower than that of the modular group which could be explained to the larger area accessible to the particle in the singly punctured two-torus.

Table 1. The odd and even r-values for singly punctured two-torus.

r-values			
2.956458939	14.233387724	20.726567345	25.704632818
4.513710759	15.792883185	20.941713201	26.191978900
5.815015474	16.321150693	21.313410193	26.243239248
6.783467732	16.428861681	22.196935326	26.676103685
8.315501837	16.924117490	22.473313041	27.167179212
9.275857326	17.202297470	23.107172833	27.193225814
9.465901775	17.314337681	23.198611455	27.554581248
10.239041998	17.997724548	23.677587275	27.923882677
10.694231142	18.701198322	24.023489099	28.116403998
11.427555185	18.901261587	24.447714998	28.869327572
11.874857830	19.591854998	24.677812458	29.039011365
13.420771791	19.791709677	25.657928181	29.739983568
13.544228169	19.977886869	25.688294998	29.924027400

In Figures 6 and 7, we show the contour and density plots for the odd eigenfunction corresponding to the lowest eigenvalue $r = 2.956458939$ and a higher eigenvalue $r = 29.924027400$. They clearly show that as the energy increases, the particle's wavefunction gets more delocalised.

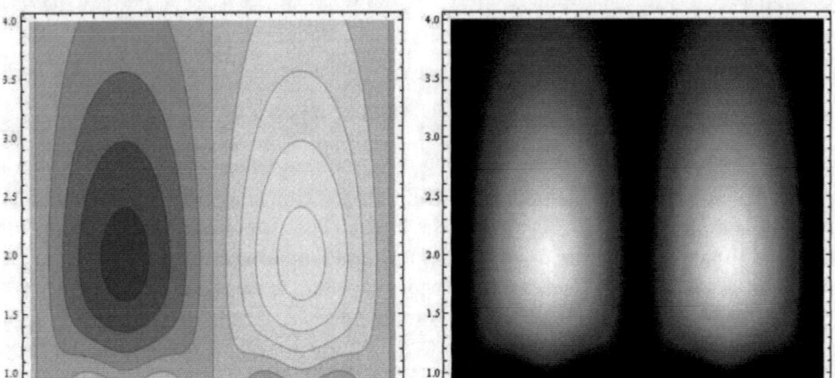

a) b)

Figure 6. The contour plot (a) and density plot (b) for odd eigenfunction corresponding to $r = 2.956458939$ of singly punctured two-torus

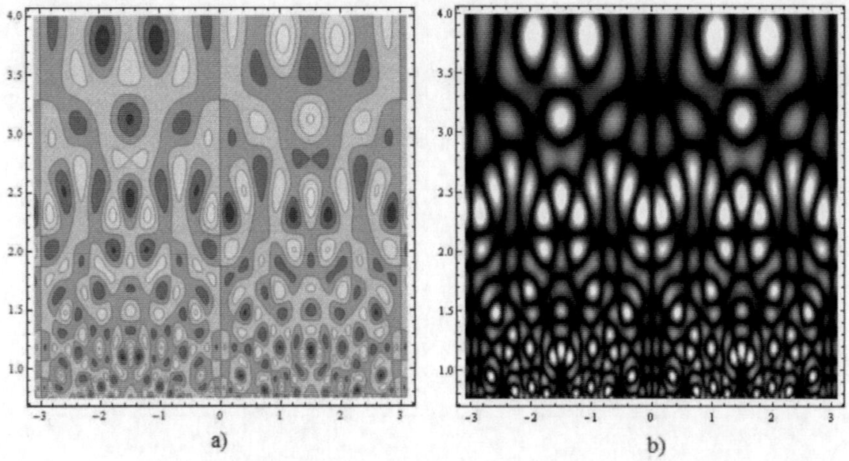

Figure 7. The contour plot (a) and density plot (b) for odd eigenfunction corresponding to $r = 29.924027400$ of singly punctured two-torus

Next, we show the case of the triply-punctured two-sphere for which we managed only to compute fewer eigenvalues due to harder computations (more simultaneous equations involved) and computational constraints. Unlike the case of the torus, the eigenvalues found are non-degenerate (see Table 2). Again, the lowest eigenvalue for the sphere is lower than that of the modular group, attributable to the larger area available for the quantum particle as in the torus case. Rather surprisingly, this lowest eigenvalue for the sphere ($r = 3.7033078$) is larger than that of the torus ($r = 2.956458939$), despite that there are three scattering channels for the sphere. This is also reflected in the plots of the eigenfunction corresponding to the lowest eigenvalue of the sphere case in Fig. 8 where one can see the eigenfunction is more localised (in comparison to Fig. 6). In Fig. 9, is the less localised higher energy eigenfunction for $r = 13.7797514$.

Table 2. The r-values for the odd and even eigenfunctions on the triply punctured two-sphere.

Odd		Even	
3.7033078	9.5336951	5.8793541	12.8776165
5.4273348	9.9349198	8.0424776	13.1720749
6.6204223	11.3176796	9.8598964	13.7797514
7.2208719	11.9727767	10.9203917	
8.2736658	12.1730084	11.4930046	
8.5225029		12.0929949	

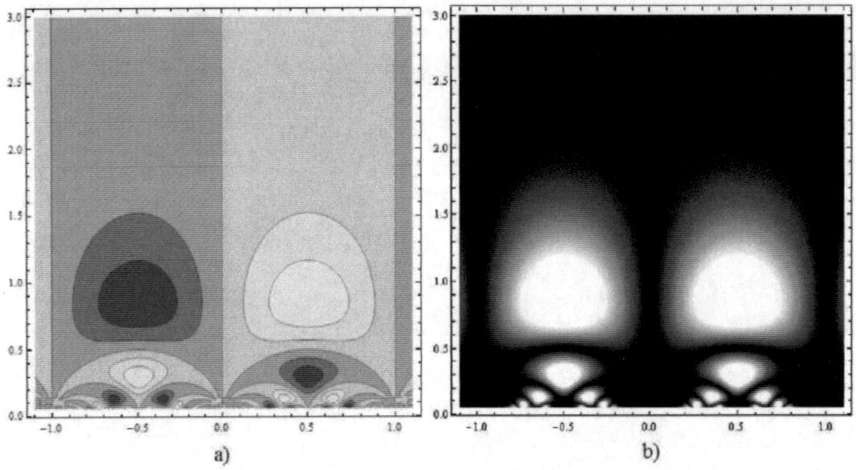

Figure 8. The contour plot (a) and density plot (b) for the eigenfunction corresponding to the lowest eigenvalue $r = 3.7033078$ of triply punctured two-sphere.

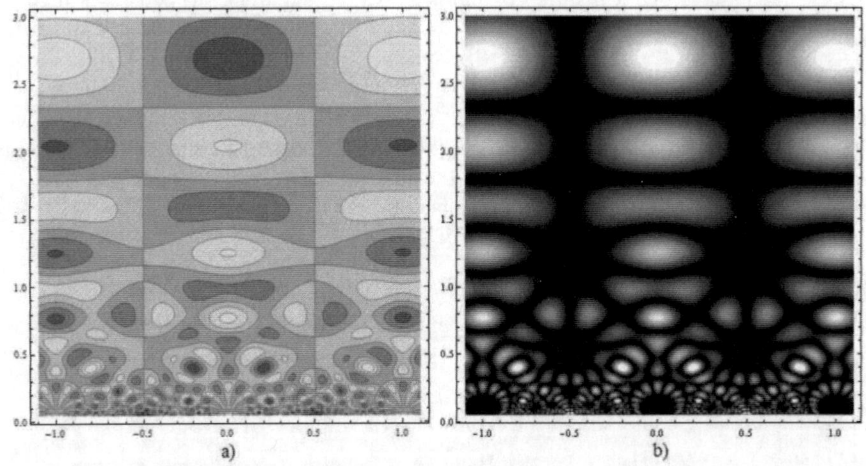

Figure 9. The contour plot (a) and density plot (b) for the eigenfunction corresponding to the eigenvalue $r = 13.7797514$ of triply punctured two-sphere.

5. Conclusion and outlook

We have found bound state solutions for both the singly punctured two-torus and the triply punctured two-sphere. The former has doubly degenerate eigenvalue spectra while the latter has nondegenerate spectra. It is interesting to see whether the degeneracy for the two-torus can be explained physically say through its geometry. This can be further investigated by deforming the torus in a way that the two 1-cycles that define the torus are no longer the same size (see Fig. 10) and studying its spectra.

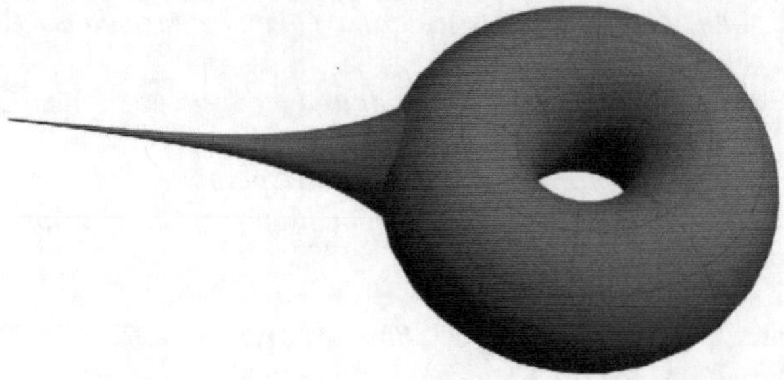

Figure 10. Deforming the torus via changing the size of the cycles.

We have also found the fact that the two-sphere case is more bound than that of the torus from their respective lowest eigenvalues. It is interesting to further examine eigenfunctions with comparable eigenvalues. In Figures 11 and 12, we show respectively the eigenfunctions for the torus for r = 13.420771791 and for the sphere for r = 13.1720749. We can again see the eigenfunction is more delocalised for the torus, while for the sphere, there is some form of directional alignments of the higher probability regions. We hope to study further the influence of the presence of cusps in these surfaces and others.

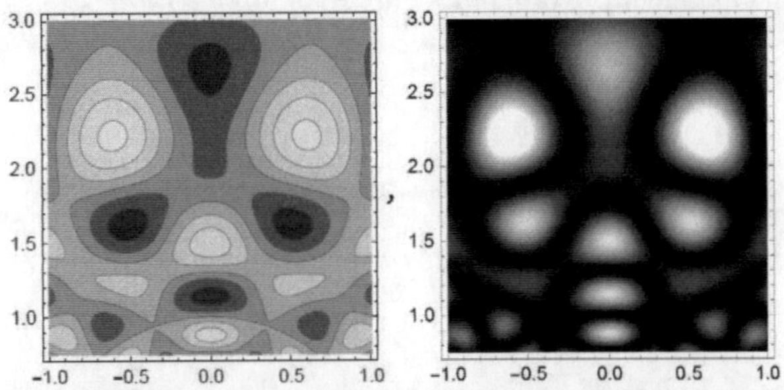

Figure 11. The contour and density plots for the even eigenfunction corresponding to r = 13.420771791 for singly punctured two-torus.

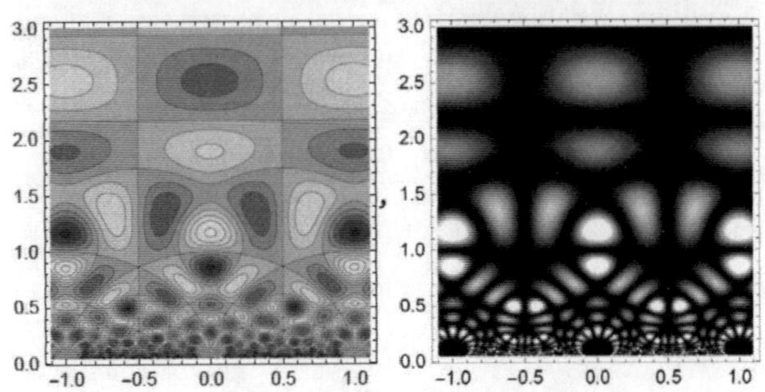

Figure 12. The contour and density plots for the eigenfunction corresponding to r = 13.1720749 for triply punctured two-sphere.

Acknowledgments
This research is partly funded by Malaysian Ministry of Higher Education ERGS grant 1-2013-5527178 and FRGS grant 65877-87504. One of the authors (H.Z.) would like to thank the ICSAS 2016 conference organizers for the invitation and hospitality.

6. References

[1] Gutzwiller M C 1990 *Chaos in Classical and Quantum Mechanics* (New York: Springer-Verlag)
[2] Stöckman H-J 2000 *Quantum Chaos – An Introduction* (Cambridge: Cambridge University Press)
[3] Borthwick D 2007 *Spectral Theory of Infinite-Area Hyperbolic Surfaces* (Boston: Birkhauser)
[4] Venkov A B 1990 *Spectral Theory of Automorphic Functions and its Applications* (Dordrecht: Kluwer Academic Publishers)
[5] Gutzwiller M C 1983 *Physica D* **7** 341–355
[6] Antoine M, Comtet A and Ouvry S 1990 *J. Phys. A: Math. Gen.* **23** 3699–3710
[7] Pnueli A 1994 *Ann. Phys.* **231** 56–83
[8] Levay P 2000 *J. Phys. A: Math. Gen.* **33** 4129–4141
[9] Levay P 2000 *J. Phys. A: Math. Gen.* **33** 4357–4376
[10] Chan K-T, Zainuddin H, Molladavoudi S 2013 *Chin. Phys. Lett.* **30** 010304
[11] Chan K-T, Zainuddin H, Atan K A M, and Siddig A A 2016 *Chin. Phys. Lett.* **33** 090301
[12] McCreary P R, Murphy T J and Carter C 2005 *The Mathematica J.* **9** 564–582
[13] Hilgert J 2005 *Maass Cusp Forms on SL(2,R)* (Lecture notes, Metz University)
[14] Hejhal D A 1999 in *Emerging Applications of Number Theory*, (eds) Hejhall D A, Friedmann J, Gutzwiller M C and Odlyzko M A (New York: Springer-Verlag)
[15] Then H 2005 *Math. Comp.* **74** 363–381
[16] Siddig A A M and Zainuddin H 2009 *Int J Pure Appl Math* **54** 279–295
[17] Chan K-T, Zainuddin H and Kamari H M 2014 *AIP Conference Proceedings* **1588** 230–235

Six to Ten Digits Multiplication Fun Learning Using Puppet Prototype

D'oria Islamiah Rosli[1], Azita Ali[2], Lim Soo Peng[3], Imam Sujardi[4], Budi Usodo[5], Fengky Adie Perdana[6]

[1,2,3] Faculty Of Technical and Vocational Education, Universiti Tun Hussein Onn Malaysia, Parit Raja, 86400 Batu Pahat, Johor Darul Takzim, MALAYSIA.

[4,5,6] Mathematic Department Program of Sebelas Maret University,Jl. Ir. Sutami 36A Kentingan Jebres Surakarta 57126, INDONESIA

E-mail: doria@uthm.edu.my

Abstract. Logic and technical subjects require students to understand basic knowledge in mathematic. For instance, addition, minus, division and multiplication operations need to be mastered by students due to mathematic complexity as the learning mathematic grows higher. Weak foundation in mathematic also contribute to high failure rate in mathematic subjects in schools. In fact, students in primary schools are struggling to learn mathematic because they need to memorize formulas, multiplication or division operations. To date, this study will develop a puppet prototyping for learning mathematic for six to ten digits multiplication. Ten participants involved in the process of developing the prototype in this study. Students involved in the study were those from the intermediate class students whilst teachers were selected based on their vast knowledge and experiences and have more than five years of experience in teaching mathematic. Close participatory analysis will be used in the prototyping process as to fulfil the requirements of the students and teachers whom will use the puppet in learning six to ten digit multiplication in mathematic. Findings showed that, the students had a great time and fun learning experience in learning multiplication and they able to understand the concept of multiplication using puppet. Colour and materials of the puppet also help to attract student attention during learning. Additionally, students able to visualized and able to calculate accurate multiplication value and the puppet help them to recall in multiplying and adding the digits accordingly.

1. Introduction

In Eleventh Malaysia Plan, Ministry of Education Malaysia (MOE) has taken serious actions in regards to increase the quality of teachers and lecturers in several aspects, such as student oriented learning and at the same time enhance education delivery system. Education in Malaysia aims to produce generations that can think globally, committed and retain them to contribute back to the Malaysian society. In fact, MOE continuously review the curriculum in preschool to tertiary levels, as to ensure the school curriculum able to equip the students with knowledge, skills and values in the future.

Additionally, second phase of Malaysia Education Blueprint (2016-2020) also stated that Malaysian government is looking forward in planning and structuring in order to accelerate and improve Malaysia Education System so that the education system will continuously spur with excellence [1]. Mathematic is a discipline that trains human mind to think logically and systematically in solving problems and making decisions. Mathematic in nature inherently promote meaningful learning and challenging thought. In relation to that, mathematic is one of the important efforts to be mastered by an individual to help them to think precisely and therefore perform appropriate actions.

Malaysia National Education Philosophy ensures that the curriculum remains relevant and in line with the current education trend for Malaysian society. For instance, Primary School Standard Curriculum (KSSR) for Mathematic was designed to consider continuous learning overtime. The measures are in line with the need to provide knowledge and skills in mathematic for students who have different background and multiple ranges of abilities in them. Through these knowledge and skills, it is hope that, students able to explore science, makes adaptations, modifications and innovations in 21st century learning.

Additionally, mathematic is one of the ways to develop individual intellectual proficiency in making logical reasoning, spatial visualization, analysis and abstract thinking. To learn mathematic, students able to develop learning skills in numeracy, reasoning, thinking and problem solving through mathematic applications [2].

In relation to that, KSSR goal is also to build students' understanding in numbering concepts, basic skills in calculation, understanding mathematical ideas and competent in applying the knowledge and mathematical skills in everyday life. Furthermore, mathematic provides opportunities for students to perform creative tasks and fun experiences and excited way in learning. Such experiences raise interest and become the driving force for students to learn mathematic in a fun way in the classroom.

Mathematic plays an important role in the development of science. Malaysian government has identified basic skills of numeracy as part of a major initiative to improve student achievement in Malaysia. In fact, LINUS (Literacy and Numeracy Screening) Program is a recovery program for students [3]. As such the programs is driven through the Recovery Program-3M (PROTIM) which was initiated in 2008.

From the program, MOE have found that around 24% of the students do not reach standard 4-targeted numeracy skills. Additionally, the study also claimed that failure to master the literacy and numeracy skills at early stage showed that, students have difficulties in learning. In fact, there were about 32,000 students dropped out from schools at various levels in 2008 [4].

Basic knowledge in mathematic operations such as addition, negation, division and multiplication is crucial in order to master the numeracy skills in problem solving. Students in primary schools needs the basic knowledge skills as to ensure that they are able to apply the skills, in dealing with time, money, measurement, distance, time, scale or any other areas. To date, this study will focus on developing a puppet prototype for six to ten multiplication operation as an initiative to help students to master multplication operation in a fun learning environment.

2. Fun Learning

Edutainment means educating while entertaining. Edutainment is an alternative to the current methods of teaching and learning. In primary schools, the government intends to introduce edutainment approach in order to support the 10th Malaysia Plan in making a transformation to the education system to become a reality. Indeed the focus of the edutainment is to embed interactive teaching methods, so that the teaching and learning will be more interactive and understood by the students.

In addition, edutainment can also reduce the level of anxiety and fear among the students to interact or communicate with the teachers and the students. In fact, edutainment also provide relaxing and entertaining environment. Thus students are always in a state of positive learning and eagerly wait for the learning session the next day.

Edutainment sessions provide the opportunity for the students to speak freely, express opinions and ideas logically but controlled to provide an attractive presentation. In addition, students are free to move, to act, sing and so on but manageable.

Edutainment is considered effective if the learning activities are structured and well planned [5]. However there are several features that should be considered by teachers when planning edutainment activities in the classroom. Activities should be tailored to the level of students to be taught such as the usage of puppet, to master students' multiplication skills and to foster students' interest in learning mathematics. Absorption edutainment requires the implementation and provision of appropriate learning time or flexible hours. In addition, edutainment activities can be implemented resolved with the disciplines without any problems or obstacles. Clear rules, fun layout, easily understood instructions and need to be prepared. Additionally, planned edutainment activities will help students to actively participate in the learning environment and the students felt satisfied in learning [6]. Additionally, teaching method that is likely to maintain good motivation and retain students to be online is a method that involves them actively [7].

Jackmen thought puppet is one way to entertain and attract students towards creative thinking and imagination. Jackmen stated that, the puppet can help the development of affective, cognitive and physical skills [8]. In fact, the use of puppet also give an opportunity for the students to improve their communication skills, produce creative ideas, stimulate the imagination because of the intellectual development of students depends on the natural environment that stimulate [9].

3. Participatory Analysis

In this study, ten participants will closely involve in the process of developing the puppet prototype that help students to learn multiplication. Participants will take note on questions that given to them and feedback from the participants will be recorded in a form of checklist. The data then will be analyzed as an input to improve the puppet prototype. Inputs from the participants were crucial as it will be used by the researcher to design and progressively improved the prototype for the next phase of this study.

Furthermore, before the prototype evaluation proceeds the participants also will be briefed by the researcher on how they use the puppet for six to 10 digits multiplication to learn mathematics. It is hope that the participants will had a great and fun time in using puppet as a medium to help them in learning mathematic especially in learning basic multiplication operations. Figure 2 illustrates numbers that represents each finger. The process of using the puppet to learn six to ten digit multiplications will be elaborated in the next sub sections.

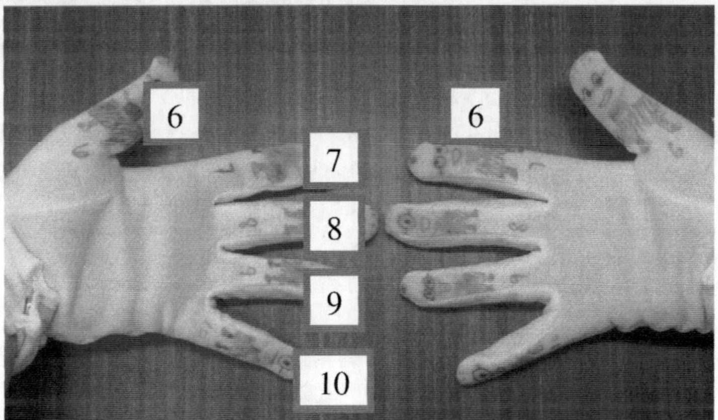

Figure 1: Representation of digits and fingers

i) Representation of digits and finger (Figure 1)
- The thumb represents number 6
- The index finger represents number 7
- Finger ghost representing number 8
- The ring finger represents number 9
- The little finger represents number 10

ii) Value for open and folded fingers
- Each finger that opens equals to 10
- Each finger that folded equals to 1

iii) Mastering the multiplication tables
- Each finger that opens, equals to 10
- Fingers that is folded on the right and left hand need to be multiplied
- Multiplication tables value = total of open fingers value + total of multiplied folded fingers value

Figure 2 illustrates basic multiplication digits for 6x6 values. Two open fingers represents total of 20 values. 4 fingers were folded on the left and right hand side, which equals to 4x4 equivalents to 16 values.

Therefore the 6x6 multiplication tables value = 20 + 16
= 36.

Example 1:
6 x 6

Open Finger	: 2	= 20
Folded Finger	: 4 x 4	= 16
	6 x 6	**= 36**

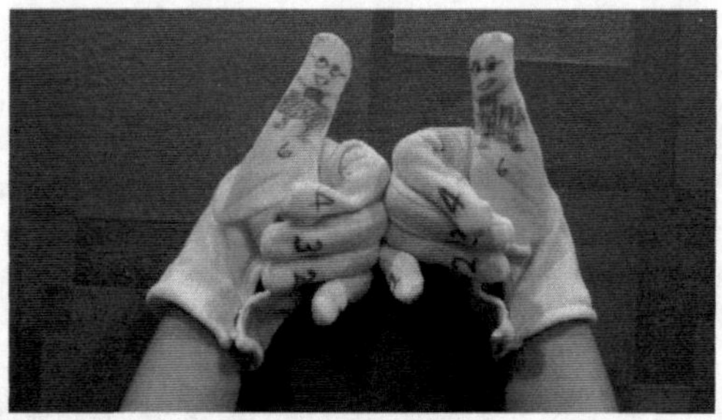

Figure 2: 6x6 values

Figure 3 shows the basic multiplication digits for 8x7 values. Five open fingers represents total of 50 values. 2 fingers on the left and 3 fingers on the right hand side were folded, which equals to 2x3 and equivalents to 6 values.

Therefore the 8x7 6 multiplication tables value = 50 + 6
= 56.

Example 2:
8 x 7

Open Finger:	5	= 50
Folded Finger:	2 x 3	= 6
	8 x 7	**= 56**

Figure 3: 8 x 7 values

4. Results and Discussion

At the earliest stages of designing and developing the puppet prototyping, there were several trials in selecting the best materials for the puppet. Figure 4 illustrates the selection begins from using the existing small puppet inserted on the finger, the hand woven gloves, the hand plastic gloves and finally the smooth white hand gloves. Due to the inconsistencies in using types of finger puppets, most of the participants felt confused because they may think that different finger puppet will represent different values. Indeed, each open finger actually represents the same value of 10 and folded finger represents a value of 1. Consistency in design is crucial as it help users to have a clear overview and understanding in using the system or applications. Furthermore, consistency helps users to understand the basic knowledge and retain the information in their Long Term Memory [10].

Additionally, the participants also reported that the drawing on the hand woven and plastic hand gloves was not clear. This is due to the rough materials and it was difficult to draw an image on the cloth. Whilst for the plastic hand glove, the drawing was easily removed and the participants having difficulties to recall the information. Information requires users to think and recall, and give mental workload for the users. Too much of information will give burden to users' mind and give stress to the students [11]. As for the smooth white hand gloves, the participants reported that they can see a clear drawing on the puppet and they can easily fold their finger as compared to the hand woven and plastic gloves. However, participants claimed that the plastic gloves will sticks together, if the gloves kept in a hot temperature and can get easily torn.

In relation to the concept of using puppet for fun learning, the participants felt excited because by using the hand gloves puppet, the participants able to experience by manipulating their own fingers and do the calculation on their own. In addition these allow the participants to think and at the same time help the participants to memorize the multiplication operations. In fact thinking through actions will help users to master the skills effectively [12]. Cartoon characters and numbers on the puppet help the participants to recall the information such as representations in numbers and calculate the addition and multiplication easily.

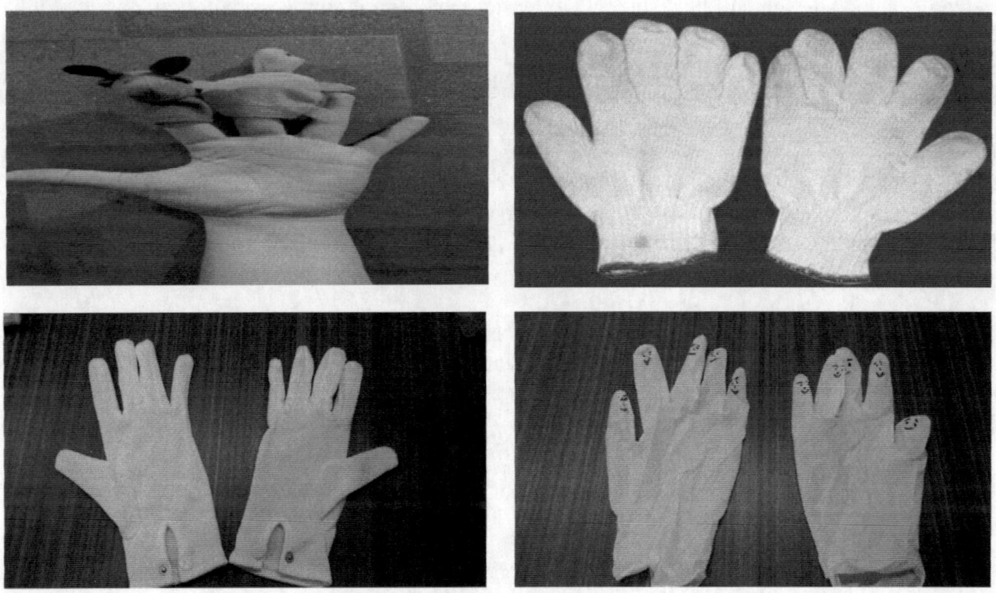

Figure 4: Puppet materials selection

Furthermore, the participants also suggested that, the cartoon character and displayed numbers on the puppet can be improved if the design uses printed images. The participants also informed that they would prefer brighter colour for the cartoon characters because the colour may attract students' attention as to ensure that the students will experience fun learning in the class. In fact, colour is a language capable of expressing emotion, and the drawing helps to convey the information in an artistic creation [13]. Most of the participants claimed that the puppet help them to do the calculation fast and they are having a great time in using the puppet for multiplication learning purposes.

Moreover, the participants also claimed that is easy for them to learn the concept of puppet multiplication operations since it only add the total of the open fingers values with the multiplication of the left and right folded finger values. Since the puppet multiplication steps is simple and easy for the users to understand, the participants does not need a longer time to learn the puppet multiplication operations. In fact students also felt that it is easy for them to memorize the multiplication table if they used the puppet to learn multiplication operations. Moreover user friendly elements may be helpful for the students to learn the system easily.

5. Conclusion

To conclude, the participants had a great time in using the puppet to learn six to ten digit multiplications. Learning in a fun way using puppet give a new dimension of fun learning because it allow direct manipulation with human physical fingers. Direct manipulation help the users for self experiencing the learning process and also help them in memorizing the multiplication table. Indirectly, teachers and students can interact with each other and the positive interactions during teaching and learning will contribute to a peaceful learning environment.

In fact, to come out with a user friendly design, the selection of materials, the image or drawings that used in the development of the product also help the students to have a fun learning in the environment. Information that is not appropriately displayed to the users will restrict them, to get a full access of the information that they need. Moreover, to have fun learning environment, materials or and the usage of colour also plays an important role as it help users to have a better understanding about the topic, a good design will help students to master their learning process. Additionally, thinking that leads to actions allow users to activate human sensory memories effectively and it is hope that, manipulation of puppet in learning multiplication will help students to learn multiplication easily and able to help them to excel in mathematic.

References

[1] Eleventh Malaysia Plan. Strategy Paper 10th: Transforming Education System. Retrieved from: http://rmk11.epu.gov.my/pdf/11MP-Brochure-BI.pdf

[2] Jonsson, B., Norqvist, M., Liljekvist, Y., Lithner, J., 2014. Learning mathematics through algorithmic and creative reasoning. *The Journal of Mathematical Behavior*. Elsevier. 36 20-32.

[3] Luyee, E. O., Roselan, F. I., Anwardeen, N., Mohd Mustapha, F. H., 2015. Suitability of the Literacy and Numeracy Screening (Linus) 2.0 Programme in Assessing Children's early Literacy. *The Malaysian Online Journal of Educational Science*. Vol 3 (2), 36-44.

[4] Ministry of Education Malaysia. (2012). The Malaysia education blueprint 2013-2025: Preliminary report. Putrajaya: Kementerian Pelajaran Malaysia.

[5] Sauer, S., Osswald, K., Wielemans, X., Stifter, M. 2014. U-Create: Creative Authoring Tools for Edutainment Applications. *Technologies for Interactive Digital Storytelling and Entertaintment*. Vol 4326 163-168.

[6] Robertson, D. 2007. Visual scenario representation in the context of a tool for interactive storutelling. *Visual Stroytelling Using Virtual Reality Technology for Storytelling*. Vol 4871. 3-12.

[7] Mohd Hatta., N. 2012. *Keberkesanan Penggunaan Kaedah Bermain Dalam Pengajaran dan Pembelajaran Asuhan Tilawah Al-Quran Tahun 4*. Seminar Penyelidikan Tindakan IPG

KBL, 34-44.

[8] Ab Yazid, N. 2012. *Using Puppets in Classroom Improve Preschool Children Communication Skills*. Seminar Pendidikan Johor 2015

[9] Piaget, J. 1999. Play, *Dreams and Imitation in Childhood*. New York: Taylor & Francis.

[10] James, R., Stephen., T & Lynn., E. 1997. Designing and validating a measure of moral judgement: Stage preference and stage consistency approaches. *Journal of Education Psychology*. Vol 89. 5-28.

[11] Chaouachi, M., & Frasson, C., 2012. Mental Workload, Engagement and Emotions: An Exploratory Study for Intelligent Tutoring Systems. *Intelligent Tutoring Systems*. 65-71.

[12] Tall, D. 2004. Thinking Through Three Worlds of Mathematics. *Proceedings of the 28th Conference of the International Group for the Psychology of Mathematics Education*. Vol 4 281-288.

[13] Zagalo, N., Gobel., Torres., Malkewitz, R., Branco, V. 2006. INSCAPE: Emotion Expression and Experience in an Authoring Environment. *Technologies for Interactive Digital Storytelling and Entertaintment*. Vol 4326 219-230.

Mastering the soft skills in the implementation of work based learning among community college students

[1]Azita Binti Ali, [2]Doria Islamiah Rosli, [3]Imam Sujadi, [4]Budi Usodo, [5]Fengky Adie Perdana

[1,2]Faculty of Technical and Vocational Education, Universiti Tun Hussein Onn Malaysia, *86400 Batu Pahat, Johor, Malaysia*
[3,4]Mathematics Education Department, Sebelas Maret University
[5]Master of Physics Department, Sebelas Maret University

Email: azita@uthm.edu.my

Abstract. Emphasizing the aspects of soft skills among students is an important element to produce graduates who are competitive when facing any situations in the workplace. Various efforts have been taken by the Ministry of Education (MOE) and the Ministry of Higher Education Malaysia (MOHE) to improve the education system in Malaysia. Learning methods were introduced to ensure the education systems achieve the educational goals and to produce individuals who are well-balanced with spiritually, emotionally and physically. However, the issue of unemployment among graduates often being spoken in the community and it was regarded as a failure of educational institutions to produce quality graduates. Thus, the method of Work-Based Learning (WBL) was seen as a way to improve the soft skills among the graduates. The study was conducted using quantitative research survey as the design of the study used a questionnaire that was adapted as an instrument. Data were analysed using Statistical Package of Social Science (SPSS) version 20.0. The respondents were consisted of 97 students who attended WBL programs at the community college. Data were obtained from questionnaires using descriptive statistics for the calculation of the mean and one-way ANOVA test. The findings of the level of soft skills among community colleges were high where the communication skills obtained (mean = 4.1218), critical and problem solving skills (mean = 4.0946), teamwork skills (mean = 4.2297), learning and information management (mean = 4.1219), entrepreneurial skills (mean = 4.0240), professional ethics and moral (mean = 3.9410) and leadership skills (mean = 4.2104). The findings also showed the differences in term of communication skills among the community colleges. This study was significant to the community colleges to identify the level of soft skills among students who performed WBL methods in order to reduce the number of unemployment.

1. Introduction

Education is an important area that contributes to the development of the country. Learning methods are introduced to ensure that the education system reaches its goals as contain in the National Education Philosophy (NEP). According to Community College Management Sector GCMS (2010), Work Based Learning (WBL) is an approach to learning and innovative teaching that involves cooperation between Community College Management Sector and the companies involved. WBL is a learning approach that involves working experience which students have to go through the learning process in their institutions and to increase experience and expertise in industrial applications (Becker,

2013). Issues which are related to unemployment among graduates often being spoken in the community and it is regarded as a failure to educational institutions in order to produce quality graduates (Ismail, 2012). Students with soft skills could compete with the challenges of globalization today, (Buntat, 2008) and it is seen that WBL as one of the ways to improve soft skills in Community College.

The collaboration between industry and educational institutions is a major milestone towards the consolidation of higher education. According to Ahmad (2006), the cooperation between educational institutions and industry in terms of providing training to their students is poorly implemented and it is focused on industrial training which are compulsory to students. This statement is also supported by Abdul Razzaq et al. (2010), and they argue that if an institution wants to produce graduates who can work and succeed in their career, the curriculum must be consistent with the needs of the industry. The cooperation between the educational institutions and industry is the best way to shape the curriculum to determine the skills required.

The phenomenon of unemployment in Malaysia has become a worrying phenomenon in recent years. This problem occurs not only in Malaysia but also in many developing countries like Philippines, Indonesia, India and developed countries such as Britain and the United States; they experienced the same thing (International Labor Organization, 2009). Employers want employees who have technical or hard skills as well as the soft skills. Meanwhile, Rusmin (2010) mentions that a degree is no longer a main attraction to the employers of any company and job market in choosing graduates but the combination of soft skills is indeed required. Tahir (2005) also pointed out that graduates in Malaysia have sufficient technical skills but employers are less satisfied with their soft skills.

Employers and industry in Malaysia have fewer graduates who have knowledge and technical skills and most of them are not able to communicate properly (Mustafa Mohd Nor, Suradin & Ahad, 2010). According to Sulaiman (2010), thinking skills have been introduced in teaching and learning and it is known as Critical and Creative Thinking Skills (CCTS). The Critical and Creative Thinking Skills was emphasized in 1988. However, the problem nowadays is to what extent the level of critical skills and problem-solving skills of students performed at each institution. Hassan (2007) stated that although the university has provided various measures but the students still do not have teamwork skills and experience required by employers. Nordin (2010) stated that entrepreneurship is considered as the transformation to the unemployment problem nowadays.

Professional ethics and moral aspects are also increasingly important due to the occurrence of these crimes very seriously and it involves the students (Rasul, 2008). Leadership skills are important elements but the formation of the nation's cultural identity is given less attention by the students in developing their skills (Esa, 2006). Thus, this study aims to see the mastery of soft skills in the implementation of Work Based Learning (WBL) among community college students; namely communication skills, critical and problem solving skills, teamwork skills, learning and information management, entrepreneurial skills, ethics and moral professional and leadership skills.

2. Metodology

The aim of this study was to ensure the process of data acquisition can be carried out more easily and can produce quality results. In this study, the researcher used a quantitative research to assess the level of soft skills in the implementation of the Work Based Learning (WBL) among community college students.

2.1. Population and sample

This study used a random sampling technique and the respondents were the students of community colleges in Malaysia. The population in this study were 104 people and the selected sample was 97. This generalization was based on sampling schedule by Krejie and Morgan (1970).

2.2. Instrument

The instrument used in this study was the distribution of questionnaires. According to Konting (2009), a questionnaire was used to obtain information about the facts, beliefs, feelings, wishes and many more.

This questionnaire was adapted from the previous studies by study Mohamad Firdaus Abdullah (2011) and Ileyati Mohd Yusoff (2012). The questionnaire consisted of two parts. Part A was the questions related to the background of the respondents. This section contained four items; namely gender, race, education institutions and fields of study.

Part B was aimed to identify the level of soft skills of Community College when implementing WBL. It consisted seven sections which represented each item of the soft skills. The scoring was done on 5-ponit Likert scale for each item of the soft skills. **Table 2.1** below showed the number of questionnaire items.

Table 2.1: Number of Questionnaire Items

Bil.	Part	Factor	Bilangan Item
1	Part A	Demography	4 (item 1-4)
		Gender	
		Race	
		Education Institutions	
		Fields of Study	
2	Part B	Communication Skills	11 (item 1-11)
		Critical skills and Problem Solving Skills	11 (item 12-22)
		Teamwork Skills	12 (item 23-34)
		Continuous Learning and Information Management	6 (item 35-40)
		Entrepreneurship skills	9 (item 41-49)
		Ethics and Professional Moral	7 (item 50-56)
		Leadership skills	12 (item 57-68)

To test the reliability of the instrument, this study was used Statistical for Social Science (SPSS) for Windows 20 to obtain Alpha Cronbach reliability coefficient which indicated the connection of items to each other (Najib 2003). Based on the results of a pilot study conducted, the researchers obtained a Cronbach alpha value at a good level of communication skills; namely 0.737 and 0.724 for the ethics and professional moral.

2.3. Analysis of Data

This section described the results of the analyzed data through questionnaires that were returned to the researcher. The determination of each completed survey data were collected and analyzed prior to the next level of making the right decision based on the data that was analyzed as shown in **Table 2.2** below.

Table 2.2:Method of Data Analysis

Research Questions	Analysis Method
Communication Skills Level	Mean Score
Critical skills and Problem Solving Skills Level	Mean Score
Teamwork Skills Level	Mean Score
Continuous Learning and Information Management Level	Mean Score
Entrepreneurship Skills Level	Mean Score
Ethics and Professional Moral Level	Mean Score
Leadership Skills Level	Mean Score
The Difference between Communication Skills Level among Community Colleges	One-way Anova

The Difference between Level Skills and Problem Solving Skills Critical among Community College	One-way Anova
The Difference between Team Work Skills Community College	One-way Anova
The Difference between Continuous Learning and Information Management among Community College	One-way Anova
The Difference of Entrepreneurship Skills among Community College	One-way Anova
The Difference between Ethics and Professional Moral among Community College	One-way Anova
The Difference of Leadership Skills among Community College	One-way Anova

3. Result

The findings were divided into two parts: part A consisted of four items on the background of the respondents. Table 3.1 showed the background of the respondents based on the number and percentage of respondents.

Table 3.1: Respondents' background

Gender, Race, Education Institutions	Number of people	Percentage (%)
Man	36	37.1
Woman	61	62.9
Malay	95	97.9
Indian	2	2.1
KolejKomunitiSelayang	14	14.4
KolejKomunitiAmpang	10	10.3
KolejKomuniti Kota Melaka	30	30.9
KolejKomunitiSg.Petani	20	20.6
KolejKomunitiBesut	23	23.7
Total	97	100

This section described the descriptive analysis for each item in the construction of the level of soft skills among respondents towards WBL. The overall average mean score for the level of communication skills is (4.1218) which indicated high level. While, the mean score for critical skills and problem solving skills was (4.0946) and teamwork skills was (4.2297). The mean score for continuous learning and information management was (4.1219) and also entrepreneurship skills was (4.0240). The mean score for ethics and professional moral among Community Colleges was (3.9410) and leadership skills was (4.2104). Overall, the level of soft skills among Community College in implementing WBL was high. **Table 3.2** below showed the mean scores for each soft skills among Community College.

Table 3.2 : Mean Score for Soft Skills

Overall Average of Mean Score of Communication Skills	4.1218	High
Overall Average of Mean Score of Critical skills and Problem Solving Skills	4.0946	High
Overall Average of Mean Score of Teamwork Skills	4.2297	High
Overall Average of Mean Score of Continuous Learning and Information Management	4.1219	High
Overall Average of Mean Score of Entrepreneurship Skills	4.0240	High
Overall Average of Mean Score of Ethics and Professional Moral	3.9410	High
Overall Average of Mean Score of Leadership Skills	4.2104	High

Based on the analysis of one-way ANOVA in **Table 3.3** below, there was a significant difference between the level of communication skills among the community colleges. This study showed a significant difference less than 0.05 ($p < 0.05$). Therefore, the null hypothesis (Ho) was rejected. As a result, there was a significant difference in the communication skills among Community College when implementing WBL. As for the other soft skills, there was not a significant difference among Community College in **Table 3.3** below. It showed that the p-value was greater than 0.05 ($p > 0.05$). Thus, the alternative hypothesis (H1) was accepted. Therefore, there was no significant difference between the critical skills and problem solving skills, teamwork skills, learning and information management, entrepreneurial skills, professional ethics and moral and leadership skills among community college students after implementing WBL.

Table 3.3: The Difference between Soft Skills Level among Community College

KI 1	Sum of Squares	Df	Mean Square	F	Sig.
Between Groups	1.803	4	.451	3.433	.012
Within Groups	12.079	92	.131		
Total	13.882	96			
KI 2	Sum of Squares	Df	Mean Square	F	Sig.
Between Groups	.257	4	.064	.648	.630
Within Groups	9.114	92	.099		
Total	9.371	96			
KI 3	Sum of Squares	Df	Mean Square	F	Sig.
Between Groups	.281	4	.070	.691	.600
Within Groups	9.358	92	.102		
Total	9.639	96			
KI 4	Sum of Squares	Df	Mean Square	F	Sig.
Between Groups	1.330	4	.333	2.216	.073
Within Groups	13.809	92	.150		
Total	15.140	96			
KI 5	Sum of Squares	Df	Mean Square	F	Sig.
Between Groups	.981	4	.245	2.241	.071
Within Groups	10.062	92	.109		
Total	11.043	96			
KI 6	Sum of Squares	Df	Mean Square	F	Sig.
Between Groups	.289	4	.072	.551	.699
Within Groups	12.068	92	.131		
Total	12.357	96			
KI 7	Sum of Squares	Df	Mean Square	F	Sig.
Between Groups	.661	4	.165	1.190	.321
Within Groups	12.784	92	.139		
Total	13.446	96			

Findings and Discussion

The Level of Soft Skills

Throughout the analysis of data, the overall findings showed that the level of communication skills among community college students who took WBL program obtained high mean score. The communication skills level obtained highest mean score in English as well. This is because most of the students who attend WBL program in community college were Malay students. Therefore, Community College students are more comfortable communicating with their friends using Malay language. The finding is consistent with the study of Youth Development Research Institute Malaysia (2012) which states that Malay language is a language that is easily learned by students. This is consistent with the study conducted by Hurley (2008), which noted the importance of communication skills in their career by students as 80% of their working time, they will communicate with each other, alone or in their group. However, the analysis that was carried out on students who followed WBL programme at the community colleges shows that the students are less proficient to communicate in English. The finding is also supported by the study by Kamaruddin (2013) which states that the weakness of Malay students lies in the use and mastery of English for speech or communication.

Based on the mean scores obtained, it indicated the critical skills and problem solving skills for each item was high. This shows a similarity with the study by Mulnix (2010) which states that critical thinking is the ability to understand, control, form, identify and practice the ability to find the relationship between the various references in the environment. This statement shows that community college students are able to understand the problems that occur based on their experiences during their learning in the Community College and practical training. These skills are helpful for students when they are working in organizations that real soon. This finding is contradicted to Walker (2003) which states that all people are using critical thinking while solving problems because they lack of self-awareness and other features that enable them to interpret and evaluate the problem. Most previous studies showed the same trend insights that students are not using the critical thinking to analyze the problem properly. Thus, the researcher found that most community college students have good critical skills and problem solving skills. The community college students can also improve their skills in order to be more competitive with other students in order to get a place in the eyes of employers.

The analysis of data showed the level of teamwork skills obtained was high. This is because when a community college students have industrial training, they need to have the teamwork skills in order to cooperate with each other in the companies. This is proven that WBL is able to improve the students' skills with good teamwork. In fact, students can also cooperate with members of the group in sharing ideas, knowledge and improvement. A study conducted by Yusof (2014) which states that teamwork is considered when one can create interdependence situations between each other. The statement showed that the Community College students who perform WBL was considered excellent in terms of teamwork skills in an organization. These findings are also consistent with a study by Shah and Nor (2011), among the features that should be included in teamwork skills are able to work well in a team and various socio-cultural background to achieve the same goal. Therefore, the community college students have high teamwork skills.

The results also showed that the level of continuous learning and information management among community college students was high. The findings obtained in conjunction with a study conducted by Najib (2011) which states that education at the tertiary level is to provide students who receive and seek knowledge. This is because community college students are open-minded to accept anything that brings positive results. As an example of WBL is a new approach in this century and community college students are always ready to undergo the WBL approach. Therefore, it appears that community college students who perform WBL can improve learning and information management with the best. Encouragement and support from the students is necessary for the implementation of WBL, so it can be improved.

Other than that, the results showed that the level of entrepreneurial skills was also high. The results showed that community college students can identify business opportunities properly. The findings are in line with the studies carried out by the Prophet and Holden (2008), graduates can identify business opportunities that endeavor to solve the problem of unemployment. This shows that community college students who apply WBL approach has a good entrepreneurial skills to tackle the problem of unemployment today. This study is supported by Ismail (2009) which states that the graduates need to think as an inventor rather than finding a job. This is because community college students have a family background who have experience in the business. This statement differs from the study done by Chee Hoe Hee (2006), which states that entrepreneurship has no place as a career choice among graduates. There is no doubt that the entrepreneurship skills of community college students could assist in the implementation of good WBL but they should give encouragement in order to ensure the implementation goes strong.

To answer the research questions of this study, seven questions were asked to the respondents. The results showed that the level of ethics and professional moral was high. As we know, there is a lot of immoral activities happen among students. This problem will cause the process of building a first-class human capital will be affected. However, the community college student proved that the level of their ethics and professional moral was high. This shows that Community College students who have a good attitude and work ethic can practice ethical behavior in the global community. The findings stated that the Community College students can practice ethical behavior is consistent with a study conducted by Kamal and Ahmad Kasim (2015) which states that quality employees are not only experts in their work, but also has an admirable character and behavior. These skills are very focused because these skills can show the attitude and behavior of students as well as to expose students to the importance of this skill when they face the real job. This proves that the graduates of community colleges can help to improve the performance of the company in competing with other companies.

To answer a research question in this study, twelve questions were presented to the respondents. Based on analysis of data, the results showed that the level of leadership skills was high. These findings are supported by the study of Lee Barron (2012), which states that when a person has good leadership skills, they can influence and convince people inside and outside organizations to work and help to achieve organizational goals. This shows that the question of leadership is a big thing in determining the success and excellence in the next generation. Based on the findings, the Community College students have excellent leadership skills and they can judge people well. The implementation of WBL at community colleges can produce students to have good leadership skills. To achieve this, leaders should influence followers to carry out his ideas with high ethical to assess, criticize and understand the needs of followers (Greenleaf, 2010). Thus, Community College students can be seen practicing good leadership skills as well as to evaluate other people well. These skills need to be maintained so that the effectiveness of WBL get the attention of many parties.

The Differences of Soft Skills among Community Colleges

Based on the findings, there is a difference between communication skills with community colleges. One-way ANOVA tests were conducted to determine whether there is a significant difference between communication skills with community colleges. Results showed significant differences which the communication skills between two college students are different. The result is consistent with the studies of Hoy &Miskel (2005), which asserting a cause of communication skills gap between education institutions due to the quality of lecturers in delivering information. The quality of lecturers is one of the important factors that determine the quality of students who graduate from an educational institution. They are the executors of the policies and goals that have been set by the government through the Ministry of Education. This may occur due to differences of delivery in the community colleges. When the lecturer does not put emphasis on the confident then it will affect the students' actions. Therefore, the responsibility of a lecturer in integrating various essential skills across the curriculum that has been developed by the curriculum. The community college faculty should take appropriate steps in order to improve the quality of their presentation to the students. This can prevent the differences between Community Colleges.

The results of this study indicated that there is no difference between the critical skills and problem solving skills after the implementation of Work Based Learning (WBL) among the community colleges. One-way ANOVA test was carried out to find out whether there is a significant difference between critical skills and problem solving skills. This illustrates the critical skills and problem solving skills among college has no different. These findings are contradicted with the study done by Ismail (2011) which stated that graduates of universities showed significant gaps in critical skills and problem solving skills. This is true because they have ability in simple non-technical aspects of problem solving skills. Similarly, the findings Abdullah et al (2012) found that graduates of public universities in Malaysia is not well controlled with Critical Thinking. This study is in line with the studies in Australia by Hambur, Rowe and Luc (2002) where all universities in Australia have chosen critical thinking skills, problem solving and communication (interpersonal and writing) as the basis of the skills of others and are transferable skills (transferable) and easily measured. Overall, the study found that community college does not have a difference in critical skills and problem solving skills among college. This proved that the successful implementation of Work Based Learning (WBL) can improve this skill well. This will help students when looking for a job after they graduate. Thus, the implementation of WBL have a positive impact on many people.

The results of this study indicated that there is no difference between the level of teamwork skills after the implementation of Work Based Learning (WBL) between community colleges. This indicates Community College students are proficient in teamwork skills. These findings together with a study conducted by Mahamod et al., (2012) which states that there is no significant difference between the students of the Institute of Higher Learning (IPTA) and the Teacher Training Institute of Malaysia (RRIM). This is due to the students of both institutions have a high level of teamwork. They are more open and willing to accept the diverse view. Students who have a high level of teamwork are able to face any challenges and obstacles. This statement shows that among students who attend WBL in the community college have almost the same and as a result they are able to face any problems among their peers. Yulpisman (2006) stated that teamwork skills are necessary skills for the graduates to help them compete for jobs either in the local or international market. Thus, community college students have good team

work skills as a result of the implementation of Work Based Learning (WBL) that was carried out.

The results of this study showed that there was no significant difference in continuous learning and information management after the implementation of Work Based Learning (WBL) among Community College. It shows each community college has similar findings with other community colleges. The findings also showed Community College students were proficient in continuous learning and information management. This is consistent with the study by Mohamed (2014), stating that there is no significant difference between the level of knowledge of community college students after the course of lifelong learning and information management in accordance with Community College Community College. Global competitiveness and diversity requirements have urged people to constantly update their knowledge and skills. This opinion is supported by the study by Imran (2009) where the challenge of the community college is to ensure the programme of continuous education and information management can be handled easily by many members in the community. However, Community College has demonstrated that they can rise to the challenge to ensure that the program is reached to produce students who have the skills required by the industry. However, these findings differ from the study Rasul et al. (2012) which states that there is a significant difference between continuous learning and information management by institutions. Each institution has a level of continuous learning and information management that varies with each institution and ultimately produce graduates who are able to have these skills. This opinion is supported by Saadan (2011), the culture of lifelong learning is an important element for balancing the needs of the current economy based on the mastery of knowledge and excellence translate talent in the country. Thus, continuous learning and information management enables Community College student to improve their knowledge from time to time with the mastery of a variety of methods and techniques of communication tools and are able to analyze all information.

The results of this study showed no significant differences in the level of entrepreneurship skills in the implementation of Work Based Learning (WBL) among community colleges. This indicates Community College students have more or less in entrepreneurship skills among Community College. Generally, these findings suggest that the location factor does not play a significant role in the level of entrepreneurship skills of students. These findings are supported by Nasharuddin and Aaron (2010) which states that there is no significant difference in perceived entrepreneurship among graduates if disclosure of the subject of entrepreneurship, peers at universities, entrepreneurship courses offered at each institution are the same. Based on this study show that community college students are not significant differences in entrepreneurial skills due to the exposure provided by Community Colleges is the same in every College Komuniti. This study is contradicted to the study done by Rasheed (2001) and Sabarre (2013) where the difference mastery entrepreneurial skills of students rely on student involvement in entrepreneurial activities in institutions that can develop students' competence or entrepreneurship. This indicates that the level of entrepreneurial skills is very dependent on the involvement of students to attend courses or activities at each institution. This is supported by studies

Mitchelmore and Rowley (2010), a person can become entrepreneurs when they are through their various processes in business development. Thus, the approach taken by the institution may indirectly foster a culture of entrepreneurship among students and change their mentality towards work alone. Some of the study results obtained, Community College is seen as a catalyst to produce graduates with entrepreneurial skills who quality.

The results of this study showed no significant difference between the level of professional ethics and moral community college after the implementation of the Work Based Learning (WBL). This shows that all community colleges have similar findings with each other. The findings also showed Community College students have a level of professional ethics and high morals. These findings are not consistent with a study conducted by Esa et al., (2013). The results showed that there were significant differences between the engineering program at the Polytechnic for the application of professional ethics and morality among students. The difference in this skill may exist because the engineering program at the Polytechnic students have a strong peer influence. For example, when we have a friend who drives towards goodness, indirectly we also can adopt an ethical stance. But, if we have that peer influence is not good, then it will affect our moral and behavior. Thus, the Polytechnic should resolve this issue so that there are no gaps students' moral and professional ethics. Overall, the Community College managed to establish ethical and moral professional skills of students through the implementation of the WBL. All community colleges that implement WBL found to have no significant difference. Therefore, based on the findings Community College students have fantastic opportunities in forming themselves to step into real working environment. This is consistent with findings in line with the statement by Shahrulanuar Mohamed (2011) which states that the moral and ethical aspects have a strong link with the development of a country.

The results of this study showed no significant difference between the level of leadership skills with community colleges after the implementation of Work Based Learning (WBL). This shows that all community colleges have similar findings with each other. These findings are similar to a study made by Adnan (2005), students at one university have found that there is no significant difference in students' leadership skills. This is because university students through the application of good communication skills will contribute to the ability to lead, motivate, negotiate, resolve the conflict, solve problems and create a healthy relationship. The study found that college students who follow the implementation of Community Work Based Learning (WBL) have good communication skills and indirectly they can be a good leader. The findings of the study one way ANOVA test is different from the study Anuar&Esa (2010), which states that there is a significant difference between the courses at PusatGiat MARA (MajlisAmanat Rakyat) for elements of leadership skills. The ability of students to apply leadership skills will vary according to their ability level, as described by Gardner (1983); Abraham (2002); and Abdullah (2009) that each individual has their own fitness and fitness level is dependent on the method of application and the level of understanding of each individual. Therefore, the leadership skills to be mastered by all students in preparation before stepping into the real working world.Therefore, the leadership skills to be mastered by all students in preparation before stepping into the real working world. This is consistent with a study by Lussier and Chua (2007) which shows that the successful country is dependent on leaders who have strong leadership skills.

Conclusions

Based on the discussion of the findings, it can be concluded that the level of soft skills is a high Community College after implementing WBL program. The findings showed that the soft skills of students who attend community college WBL is at a high level. The first question of the study, the level of communication skills Community College students are at a high level of communication skills, particularly in languages other than English. This is because most of the students who attend WBL is Malay students. However, the findings also show that community college students still have a poor command in English. For the second research question, the critical skills and problem solving skills are also at a high level. This is because students who attend WBL are exposed to higher order thinking skills in order to be as required by the employer. This shows the effectiveness of the implementation of the WBL at community colleges. In addition, teamwork skills are required in the implementation of the WBL at community colleges. Students can organize group activities to acquire and exchange views and ideas together. Community College students are also very open-minded and receptive to the idea that you can do well to them. As an example of WBL lived by themselves. Overall, the soft skills of Community College students are at a high level. This is in line with the requirements of the Higher Education Ministry to produce students who have strong soft skills.

Overall, the differences in the level of soft skills in Community College students who follow WBL has no significant difference among the Community Colleges. The communication skills showed a significant difference among community colleges. This may be caused by different beliefs or confidence level when communicating with each other, the way the lecturers teach and so on. However, in this study, it showed that the different levels of soft skills have no significant differences. This indicates Community Colleges are successfully implemented WBL to produce students who are needed by the industry. Hopefully, the implementation of WBL will establish flexible students as needed by employers today. Thus, based on the findings, recommendations for Community Colleges provide more programs of Work Based Learning (WBL) at community colleges especially for the certificate. This is because the certificate can help students to experience the real work. Community College will emphasize the mastery of English to each student. This is to prevent them to have a poor command of English. Therefore, these recommendations could help to prevent unemployment problems.

References

[1] Abdullah, S. (2009).Kecerdasan Pelbagai: Aplikasi dalam Pengajaran dan Pembelajaran. Selangor: PTS Publication.

[2] Abraham, S.J. (2002). "Pengajaran Pelbagai Kecerdasan dalam Pendidikan Sains Sosial" dalam A.R. Abd Rashid [ed]. Pengajaran dan Pembelajaran Sains Sosial: Teoridan Amalan. Kuala Lumpur: Penerbit UM [Universiti Malaya], ms.117-125.

[3] Abu Bakar, T. S., Md Ali, A., Omar, R., MdSom, H. &Muktar, S. N., (2010).Penglibatan Kaum Wanita Dalam Aktiviti Keusahawanan Di Negeri Johor: Kajian Terhadap Faktor-Faktor Kritikal Kejayaan Dan Kegagalan Pengendalian Perniagaan.

[4] Anuar, A, &Esa, A (2010). "Penerapan Kemahiran Insaniah di Pusat Giat MARA (PGM): Satu Analisis ATIKAN: Jurnal Kajian Pendidikan, 3(2) Desember 2013 149 Perbandingan" dalam Esa, A & Mustafa, M.Z. Kemahiran Insaniah: Kajian di Institusi-institusi Pengajian. Batu Pahat, Johor: Penerbit UTHM [Universiti Tun Hussein Onn Malaysia], ms.1-19.

[5] Aziz, Y. (2003). Komunikasi Untuk Pengurus. Utusan Publication & Distributor Sdn. Bhd. Kuala Lumpur.

[6] Becker, G. (2013). Development Education and Other Academic Programs. Human Capital, 1-10.

[7] Crockett, C. (2005). The Cultural Paradigm of Virtue. Journal of Business Ethics, 62, 2, 191-208.

[8] Davis DC, Woodward B (2006) An analysis of skills required of graduates of an information systems program. Information Technology, Learning, and Performance Journal 24 (2), 11-21.

[9] Esa, A. (2013). Kemahiran insaniah dalam Proses pembelajaran dan Pengajaran pada Program Kejuruteraan di Politeknik Malaysia .Atikan Jurnal Kajian pendidikan, 139-150.

[10] Gaedeke RM, Tootelian DH (1989) Employers rate enthusiasm and communication as top job skills. Marketing News 23 (7), 14.

[11] Gardner, H. (1983). Frames of Mind: The Theory of Multiple Intelligences. New York: Basic Books.

[12] Greenleaf, W. D. (2010).Moral, social, and technical dimensions of the principalship. Peabody Journal of Education, 63, 138-149.

[13] Hambur, S., Rowe, K., & Luc, L. T. (2002). Graduate skills assessment. Australian Council For

Educational Research. Commonwealth Department of Education Science & Training.
[14] Harun, N. N. (2010). Aspirasi Kerjaya Keusahawanan Dalam Kalangan institusi Pengajian Awam. Jurnal Pendidikan Malaysia, 11-17.
[15] Hassan, A. (2007). Isudan Strategi ke Arah Pembangunan Modal Insan Sektor Pengajian Tinggi Bertaraf Dunia. Pengurusan Awam, 23-44.
[16] Hurley (2008).The Importance of Strong Business Communication Skills. McGraw Hill Inc.
[17] Hussain Othman, B. M. (2009.). Pelaksanaan Pendekatan Pembelajaran Berasaskan Pengalaman (PBL dan POBBL) bagi Meningkatkan Kemahiran Insaniah Pelajar. Pemikiran Kritis dan Kemahiran Menyelesaikan Masalah.
[18] Kaprawi, Noraini. (2010). Kajian Tinjauan Secara Kuantitatif. dlm. Noraini Idris (Pnyt). Penyelidikan dalam Pendidikan, 195-214. Kuala Lumpur. Mc. Graw-Hill (M) Sdn. Bhd.
[19] Kementerian Pengajian Tinggi Malaysia (2006).Modul pembangunan kemahiran insaniah (soft skills) untuk Instutusi Pengajian Tinggi Malaysia.Serdang: Penerbit Universiti Putra Malaysia.
[20] Kinichi, A.J., McKee Ryan, F. M., Schriesheism, C.A & Carson, K.P (2002) Assessing The Construct Validity of the Jobs Descriptive Index: A Review and Meta-Analysis. Journal of Applied Technology, 87(1), 14-32. American Psychological Association.Inc.
[21] Konting, M. M. (2009). Kaedah Penyelidikan Pendidikan. Selangor Darul Ehsan :Dewan Bahasa Dan Pustaka.
[22] Krejcie, R.V, Morgan, D.W. (1970) Education and Psychological Measurement: Determining Sample Size for Research Activities.
[23] Kuratko&Hodgetts (2007). Entrepreneurship: Theory, Process, Practice. USA: South Western Cengage Learning.
[24] Lee Barron, J. R. (2012). Martial Arts Training as a Method of Modifying Attitudes and Behaviors in TheClasroom. International Journal of Combat Martial Arts and Sciences ICMAUA.11 (I), ms. 17-22.
[25] Maswan, S. (2007).The Effects of a Web-Based Guided Inquiry Approach on Students Achievement. Journal of Computers, 38-43.
[26] McQuick, R.W., & Lindsay, C. (2005); Mohd Noor, N., &Sahimin, S. (2010); &Pineteh, E.A (2012), "The Concept of Employability" dalam Journal of Urban Study, 42:2, ms.197-219
[27] Michael, D. T & Diane R. L. (2008). "Training in Australia". Australia. Pearson – Education Australia
[28] Mitchelmore, S., & Rowley, J. (2010). Entrepreneurial competencies: a literature review and development agenda. International Journal of Entrepreneurial Behavior & Research, 16(2), 92 -111.
[29] Shah, I. M. &Nor, Z. M. (2011). Faktor-Faktor Kerja Berpasukan Dan Pengaruhnya Kepada Prestasi. Jurnal Kemanusiaan, 17(2), 42-44.
[30] Shahrulanuar Mohamed, A. A. (2011). Pembangunan Etikadan Moral Dalam Kursus kursus yang Ditawarkan di Universiti Teknikal Malaysia Melaka (UTeM). Journal of Human Capital Development, 141-155.
[31] Simoon, J. M. (2010). Understanding Work Based Learning. Gower.
[32] Sulaiman, N. R. (2010) PengajaranSains KBSM Selangor: Dewan Bahasa dan Pustaka.
[33] Stevens, B. (2014). Career and Technical Education Work- Based Learning Guide.Richmon, Virgnia: Education, Virginia of Department.
[34] Syed, Z.A. (2013). The need for inclusion of entrepreneurship education in Malaysia lower and higher learning institutions.Education + Training. Vol. 55 No. 2, pp. 191-203.
[35] Tahir, K. M. (2005). Tahap Keyakinan Kemahiran Generik di Kalangan Pelajar Kolej Komuniti. Thesis Ijazah Sarjana Pendidikan .

IOP Conf. Series: Journal of Physics: Conf. Series **795** (2017) 012005 doi:10.1088/1742-6596/795/1/012005

High temperature corrosion of cold worked YUS409D bellows of bellow-sealed valve in LBE

A P A Mustari[1*], D Irwanto[1] and M Takahashi[2]

[1]Nuclear Physics and Biophysics, Faculty of Mathematics and Natural Sciences, Institut Teknologi Bandung, Ganesha Street No 10, Bandung 40132, Indonesia

[2]Research Laboratory for Nuclear Reactors, Tokyo Institute of Technology, N1-18, 2-12-1 O-okayama, Meguro-ku, Tokyo, 152-8550, Japan

*E-mail : pramutadi@fi.itb.ac.id

Abstract. Lead-bismuth eutectic (LBE) loop test is highly contributes to the lead-alloy-cooled fast breeder reactor (LFR) and accelerator driven system (ADS) research and development by providing comprehensive results of both corrosion and erosion phenomenon. Bellows-sealed valve is a crucial part in the LBE loop test apparatus, due to its capability of preventing corrosion on valve spring, thus improves the operation time of the system. LBE is very corrosive to stainless steel by formation of oxide layer or elemental dissolution, e.g. Ni. Thus, new type of bellows for bellows-sealed valve made of nickel free material, i.e. YUS409D, is proposed to be used in the LBE. Bellows material undergo heat treatments for mechanical improvement including cold working and annealing. The thickness reduction by the heat treatments is about 90% of initial condition. Corrosion behavior of the bellows has been studied in stagnant LBE at 500 and 600 °C for 500 hours. The oxygen concentration was controlled at about 10^{-7} wt%. Typical oxide layers were developed on the surface. Oxidation rate was sharply increased at 600°C.

1. Introduction

Lead-bismuth eutectic (LBE) is a candidate material for coolant of the lead-alloy-cooled fast breeder reactor (LFR) and, for target and coolant of accelerator driven system (ADS) due to its advantageous. However, LBE is very corrosive to the structural material at high temperature. It is a better condition for ADS, i.e. MYRRHA project, in comparison with LFR since the proposal temperature of the cooling will be set below 450C. For the fabrication the utilization of material treatments, e.g. annealing, cold working, and welding, are inevitable. Therefore, investigation on the material compatibility of the treated material is necessary.

Structural materials are usually subjected to different levels of cold working during final manufacturing stages to increase hardenability. Cold deformation affects the corrosion resistance of stainless steels because planar dislocation arrays [1] and deformation twins are introduced. Other reports related to cold work were mostly done on low cold work degree [2]–[4]. Thus, research on high cold work in liquid metal environment may give comprehensive understanding on its effect, especially, in a liquid metal corrosion. Furthermore, it may be a valuable input for the end user, such as Power Plant Company.

The aim of the present study is to investigate newly proposed ferritic/martensitic type, i.e. YUS409D, of bellows under high-temperature LBE. The static tests were carried out at 500 and 600°C with 2.6×10^{-7} and 4.7×10^{-6} wt%, respectively, for 500 hours. The bellows of a bellow-sealed valve are high cold work component, which plays an important role as to protect a spring from corrosion, to eliminate the effect of back pressure during pressure relief, etc.

2. Experiment

The YUS409D is the new development ferritic/martensitic bellows to be used in LBE environment. Ni content was not used as alloying element to prevent from preferential dissolution. SS316L is popular material to be used as bellows material. In detail of the chemical composition is shown in Table 1. For fabrication of the bellows, the material is undergoing repeated cold working following by annealing. Bellows is a double layer cold worked material with 80% of total cold work degree. However, the cold work treatment was done step-by-step for reducing the thickness and increasing the hardness, and then annealing was done in between of cold work for the purpose of softening the material. The cold work process was carefully done to maintain the ductility of the bellows. The image of the bellows is shown in Fig. 1. In detail process of making the bellows is not revealed due to some restrictions.

Table 1. chemical content of bellows of bellow-sealed valve

Material	Fe	C	Cr	Ni	Mo	Mn	Si	P	S	Ti
YUS409D	Balance	0.005	11	-	-	0.34	0.44	0.024	0.002	0.19

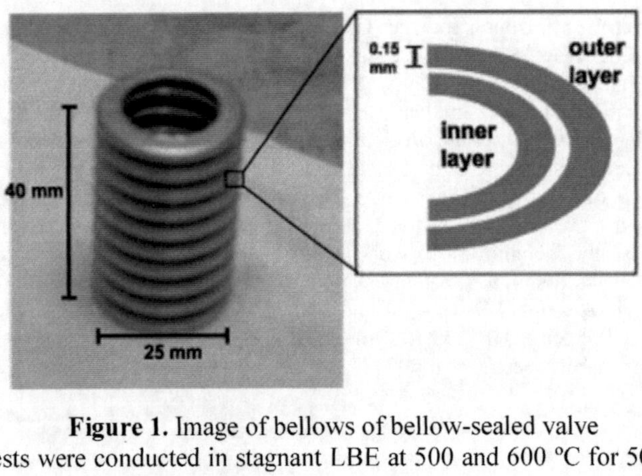

Figure 1. Image of bellows of bellow-sealed valve

The corrosion tests were conducted in stagnant LBE at 500 and 600 °C for 500 hours. The oxygen concentration (OC) was controlled by injection of $4\%H_2$+Ar/Ar/Steam to achieve 2.6×10^{-7} and 4.7×10^{-6} of oxygen partial pressure at 500 and 600 °C, respectively. Figure 2 shows oxygen concentration condition on oxygen potential diagram of oxides. It shows that the test conditions were in oxidation zone, i.e. between Fe_3O_4/FeO line and PbO line. The OC was measured by using oxygen sensor. In detail description of the apparatus is shown in ref. [5].

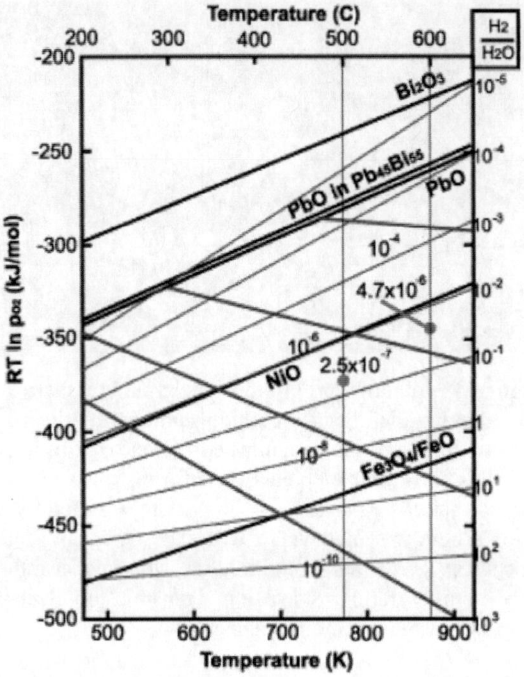

Figure 2. Image of bellows of bellow-sealed valve

After corrosion test, the bellows was rinsed by glycerol at 130°C and then continue with water cleaning at 80°C. The bellows were cut in the middle and mounted in the resin. For observation purpose, the cross-section was polished by abrasive paper up to #2400. For grain boundary observation, the cross section was etched by glyceregia etching solution. The planimetric method of ASTM E 112-10 standard test method was used for determination of average grain size. Optical microscope and SEM/EDX were used for the investigation of the corrosion behavior.

3. Results and discussion

3.1. Microstructure profiles

The grain boundary profiles were measured using 3-5 figures from the different location. In accordance to Fe-Cr system phase diagram, the corrosion test temperature, i.e. 500 and 600 °C, will not change the martensitic phase of the YUS409D (11wt% of Cr).

Figure 3 shows optical micrographs of grain boundary profile of bellows from the post-test specimen. The DIN1.4313, DIN1.4922, and Optifer-IX, that are ferritic/martensitic steel [6], were chosen to be compared with YUS409D result due to comparable chemical content. The DIN1.4313 (FJZ) has 13wt%Cr and 4.0wt%Ni. The DIN1.4922 (FJZ) has 11.5wt%Cr and 0.5wt%Ni, while Optifer-IX has 8.9wt%Cr with no Ni content. It is reported that average grain size for DIN1.4313, DIN1.4922 and Optifer-IX are 76±7, 81±5 and 55±10 μm, respectively. Average grain sizes of the materials are higher than an average grain size of YUS409D, i.e. 39±9 μm. The low percent of cold working induces material by increasing dislocation density. On the other hand, high cold work induces material by recrystallization, which reduces the dislocation density and forms small new grains [7].

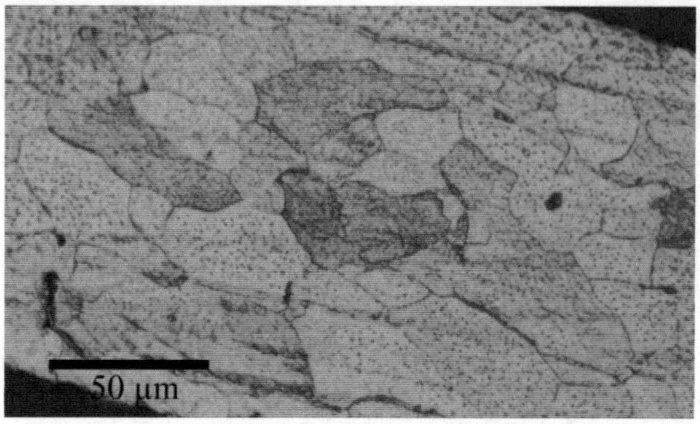

Figure 3. Optical micrograph of the grain boundaries

Titanium is present as an alloying element for carbide stabilization during a welding process. It is reported that adding titanium will prevent intergranular corrosion by forming carbide in preference to chromium-rich precipitates. Thus chromium-depleted zone formation is prevented [8]. This capability of preventing intergranular corrosion is particularly important for bellow of bellow-sealed valve. During operation, bellows of a bellow-sealed valve will experience low-fatigue cycle, in which an intergranular crack may occur [9]. At this condition, i.e. low-cycle fatigue, intergranular crack formation may be accelerated by intergranular corrosion. Transmission electron microscopy (TEM) is needed for observing the titanium carbide in more detail.

3.2. Oxide layer formation

Fig. 4 shows SEM micrographs of cross section of bellows after corrosion in liquid LBE at 500°C for 500 h. The SEM image shows a two-layer on the surface. The outer layer had a porous appearance and penetrated with LBE. The inner layer seems to be dense and no LBE penetration. The average oxide layer thickness was about 13 μm. It also shows the EDX result with confirmation of duplex oxide layers. The outer layer is enriched by Fe and O, while the inner layer is enriched by Fe, Cr and O. According to the previous studies, magnetite and Fe-Cr spinel are most likely to form on the martensitic steels [10]–[13]. LBE penetrated through magnetite layer until the interface of magnetite/spinel. However, LBE was prevented to penetrate into spinel layer to have direct contact with the bulk material. The magnetite is porous and prone to LBE penetration, while Fe-Cr spinel is denser compared to magnetite [14]. Fe-Cr spinel layer had some cracks probably due to mechanical treatment after corrosion test.

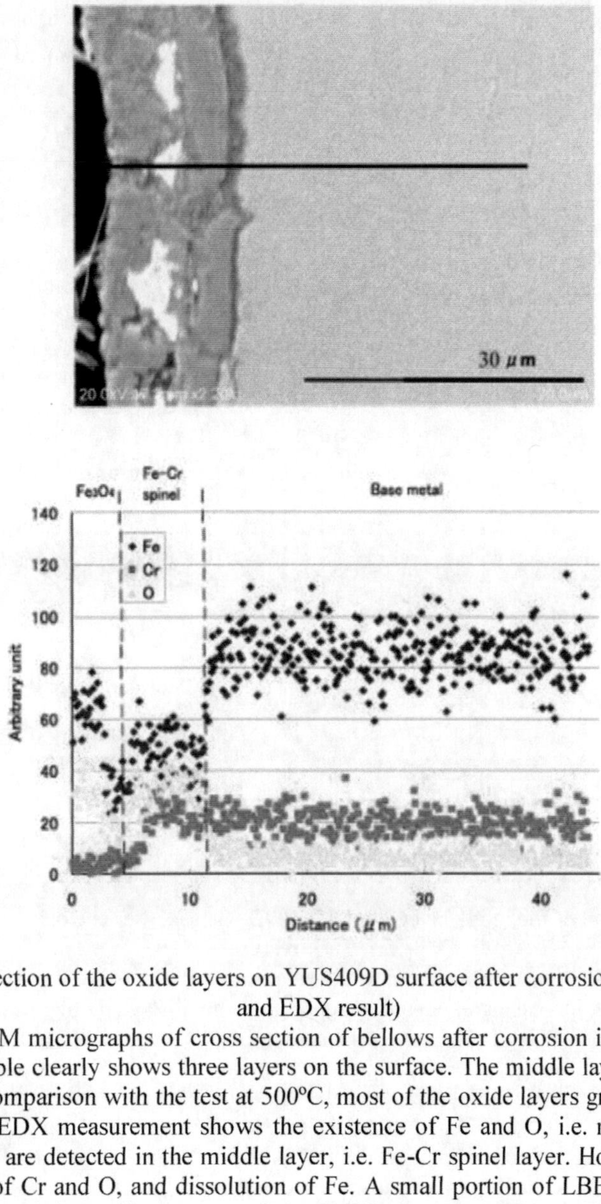

Figure 4. Cross section of the oxide layers on YUS409D surface after corrosion test at 500°C (SEM and EDX result)

Fig. 5 shows SEM micrographs of cross section of bellows after corrosion in liquid LBE at 600°C for 500 h. The sample clearly shows three layers on the surface. The middle layer seems to be denser than the outer. In comparison with the test at 500°C, most of the oxide layers grew of about twice, i.e. about 32 μm. The EDX measurement shows the existence of Fe and O, i.e. magnetite, in the outer layer. Fe, Cr and O are detected in the middle layer, i.e. Fe-Cr spinel layer. However, the inner layer shows enrichment of Cr and O, and dissolution of Fe. A small portion of LBE is penetrated through magnetite. It is reported that uniform penetration of LBE occurs in LBE corrosion test at temperatures 400-600 C [15].

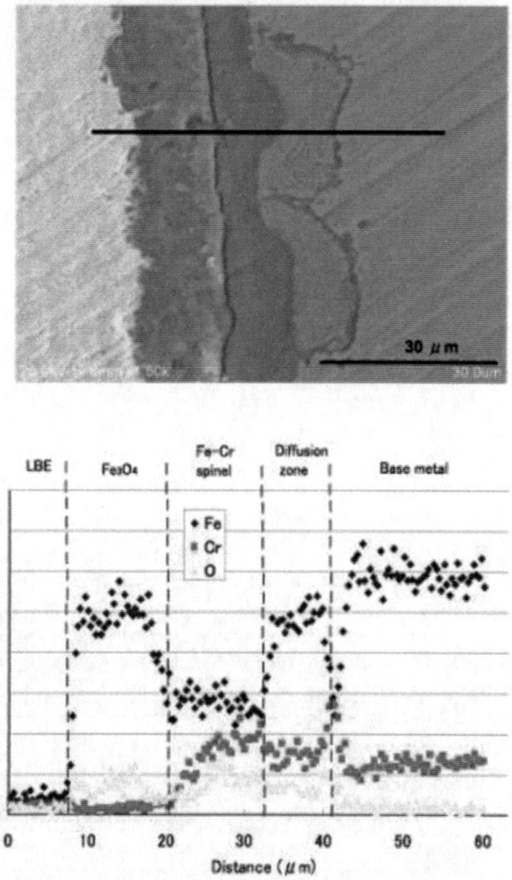

Figure 5. Cross section of the oxide layers on YUS409D surface after corrosion test at 600°C (SEM and EDX result)

The temperature has a great influence on the oxidation rate of YUS409D bellows. The oxidation thickness increases with increasing exposure temperature. The YUS409D bellows show thicker oxide formation at 600°C than at 500°C. The oxide thickness at 600°C is 2-3 times that at 500°C. Diffusivity of alloying elements, e.g. Cr, Fe, increases at higher temperature, thus accelerates the layer formation.

3.3 Oxidation rate

For determination of the corrosion layer thickness as a function of temperature and time of the experiments, backscattered electron images from SEM and X-ray line from EDX were used. The estimated corrosion rate OR then calculated using the determined layer:

OR=h/t μm/year

H is the layer thickness at temperature T and immersion time t. The corrosion layer was summation of combination of the selective metal oxide, i.e. Fe and Cr, and diffusion zone.

The estimation of corrosion rate in comparison to other ferritic/martensitic steel, i.e. HT9 and T91, and to austenitic stainless steel, i.e. AISI 316L, is shown in Table 2. The results of HT9, T91, and AISI 316L were calculated using correlation in reference [16]. The YUS409D bellows show much higher oxidation rate in comparison with other selected results. Also, a hundred centigrade increased of temperature results in double of oxide layer formation on cold worked YUS409D. It is predictable since YUS409D is highly cold-working materials, while the other results are without cold work. The YUS409D, i.e. bellows of bellow-sealed valve material, is presently proposed to be used in Tokyo Institute of Technology's LBE loop apparatus [17]. Time operation of the LBE loop apparatus were about 1000 hours. Thus, utilization of the YUS409D is still applicable for this condition. However, for long time operation, i.e. more than 3600 hours, utilization of the material under oxidation condition should be avoided. The merit of the specimen is on non-utilization of nickel as an alloying element, where previous bellows material was austenitic stainless steel, i.e. nickel base material. Nickel as an alloying element will induce preferential dissolution under LBE environment [15]. Further investigation at reduction or dissolution is needed. Also, investigation on austenitic steel with the same heat treatment is important for comparison.

Table 2. Corrosion rates of YUS409D and other materials

Material	Temperature	Oxidation rate (um/year)	Reference
YUS409D	500	220	this study
	600	570	this study
HT9 and T91	500	49.7	ref.[16] correlation
	600	166.5	ref.[16] correlation
AISI316L	500	80.4	ref.[16] correlation
	600	269.4	ref.[16] correlation

The cold working seems to give effects on oxide layer formation. Previous studies on pure iron show that oxidation rate is increasing due to cold work in several environments, such as O2, water vapor and CO2. Caplan et. al reported that rapid oxidation rate of cold worked pure Fe is due to extra dislocation [2]. Extra dislocation can act as vacancy sink that inhibits the formation of pore belt at the metal/oxide interface. Since, the pore belt can restrain the diffusion of iron, which means a decrease in the oxidation rate. In this study due to high cold work degree, i.e. 80%, most likely phenomena were not only dislocation and small defect but also re-crystallization. Indeed, annealing is included in the process for relieving stress, but due to heavy cold work re-crystallization may form near the surface or in the whole bulk. Grain density per unit area, which was increased due to re-crystallization, leads to increase of oxygen ion transport through grain boundary. Small grain size, i.e. high number of grain boundaries, leads to increasing of oxidation rate due to high diffusion rate [18]. Vicente et. al reported that when the grain size became small due to heavy cold work, i.e. cold work degree more than 42%, high oxygen transport along grain boundaries occurs as well as rapid outward Cr movement along the grain boundaries [7]. Thus, oxidation attack increased on the cold-worked on low alloy ferritic steel 2.25Cr1Mo in air at 550°C.

The condition of the corrosion test of this study is in the static condition. However, in the real situation, the bellows will experience semi-static LBE, low-fatigue cycle, etc. Thus, comprehensive study is needed to ensure the compatibility. The semi-static LBE environment may give similar phenomena to the stagnant condition. However, the low-fatigue condition may affect the corrosion resistance of the steel. Spall-off of the oxide layer may occured due to low-cycle fatigue. This condition will enhance the probability of LME by LBE penetration. Therefore, it is highly recommended to use this type of bellow-sealed valve under a corrosive temperature of LBE (< 450°C).

In addition, longer corrosion test with various temperature profiles is necessary to be conducted for a better understanding of high cold work treatment.

4. Conclusion

Static corrosion test of bellows, i.e. cold-worked steel, YUS409D in LBE were studied at 500 and 600 °C. The grain size of the YUS409D bellows was found smaller than other ferritic/martensitic with similar components due to high cold work treatment. High density of grain boundaries was contributing to rapid inward oxygen and outward Cr along grain boundaries. Enhancement of temperature gave an effect on the formation of oxide layer. Temperature increases from 500 to 600 °C results in acceleration of layer formation by two times. The oxidation rate of YUS409D bellows was found higher than both without cold worked ferritic/martensitic steel and austenitic stainless steel. For short-term operation, the utilization of the cold worked YUS409D in high-temperature LBE was applicable. For long-term utilization, next experiment should be conducted taking into account observation of corrosion and mechanical properties.

Acknowledgment
This research was partly supported by the Asahi Glass Foundation.

References

[1] Y. J. Oh and J. H. Hong, "Nitrogen effect on precipitation and sensitization in cold-worked Type 316L(N) stainless steels," *J. Nucl. Mater.*, vol. **278**, no. 2, pp. 242–250, 2000.
[2] D. Caplan and G. M. Cohen, "Effect of cold work on the oxidation of iron from 400-650 °C*," *Corros. Sci.*, vol. **6**, no. February, 1966.
[3] D. Caplan and M. Cohen, "Effect of cold work on the oxidation of iron in water vapour at 550 °C," *Corros. Sci.*, vol. **7**, no. 10, pp. 725–727, 1967.
[4] W. R. Price, "The effect of cold work on the oxidation of iron from 400 ° to 650 ° C," *Corros. Sci.*, vol. **7**, pp. 473–474, 1967.
[5] A. P. A. Mustari and M. Takahashi, "Study on Corrosion Properties of Welded Ferritic-Martensitic Steels in Liquid Lead-Bismuth at 600°C," *J. Power Energy Syst.*, vol. **5**, no. 1, pp. 69–76, 2011.
[6] H. Zhang, B. Long, and Y. Dai, "Metallography studies and hardness measurements on ferritic/martensitic steels irradiated in STIP," *J. Nucl. Mater.*, vol. **377**, no. 1, pp. 122–131, 2008.
[7] V. B. Trindade, U. Krupp, B. Z. Hanjari, S. Yang, and H.-J. Christ, "Effect of alloy grain size on the high-temperature oxidation behavior of the austenitic steel TP 347," *Mater. Res.*, vol. **8**, no. 4, pp. 371–375, 2005.
[8] A. P. Bond and E. A. Lizlovs, "Intergranular Corrosion of Ferritic Stainless Steels," *J. Electrochem. Soc.*, vol. **116**, no. 9, p. 1305, 1969.
[9] A. P. A. Mustari and M. Takahashi, "Metallurgical analysis of corroded bellows of bellow-sealed valve in lithium flow system," *Fusion Eng. Des.*, vol. **88**, no. 4, pp. 202–208, 2013.
[10] D. Gómez Briceño, L. Soler Crespo, F. J. Martín Muñoz, and F. Hernández Arroyo, "Influence of temperature on the oxidation/corrosion process of F82Hmod. martensitic steel in lead-bismuth," *J. Nucl. Mater.*, vol. **303**, no. 2–3, pp. 137–146, 2002.
[11] C. Fazio, G. Benamati, C. Martini, and G. Palombarini, "Compatibility tests on steels in molten lead and lead-bismuth," *J. Nucl. Mater.*, vol. **296**, no. 1–3, pp. 243–248, 2001.
[12] T. Furukawa, G. Muller, G. Schumacher, A. Weisenburger, A. Heinzel, F. Zimmermann, and K. Aoto, "Corrosion Behavior of FBR Candidate Materials in Stagnant Pb-Bi at Elevated Temperature," *J. Nucl. Sci. Technol.*, vol. **41**, no. 3, pp. 265–270, Mar. 2004.
[13] T. Furukawa, G. Muller, G. Schumacher, A. Weisenburger, A. Heinzel, and K. Aoto, "Effect of oxygen concentration and temperature on compatibility of ODS steel with liquid, stagnant Pb45Bi55," *J. Nucl. Mater.*, vol. **335**, no. 2 SPEC. ISS., pp. 189–193, 2004.
[14] A. Weisenburger, C. Schroer, A. Jianu, A. Heinzel, J. Konys, H. Steiner, G. Müller, C. Fazio, A. Gessi, S. Babayan, A. Kobzova, L. Martinelli, K. Ginestar, F. Balbaud-Célerier, F. J. Martín-Muñoz, and L. Soler Crespo, "Long term corrosion on T91 and AISI1 316L steel in flowing lead alloy and corrosion protection barrier development: Experiments and models," *J. Nucl. Mater.*, vol. **415**, no. 3, pp. 260–269, 2011.
[15] D. Sapundjiev, S. Van Dyck, and W. Bogaerts, "Liquid metal corrosion of T91 and A316L materials in Pb-Bi eutectic at temperatures 400-600 C," *Corros. Sci.*, vol. **48**, no. 3, pp. 577–594, 2006.
[16] I. Soon Hwang, J. Lim, and S. gi lee, "Development of alumina forming ferritic steels for lead-bismuth eutectic-cooled applications," in *Structural Materials for Innovative Nuclear Systems (SMINS-2)*, 2012, pp. 165–174.
[17] M. Kondo, M. Takahashi, S. Yoshida, N. Sawada, A. Yamada, and K. Hata, "Experimental

Studies on Pb-Bi Flow Technology and Steel Compatibility With Pb-Bi," in *12th International Conference on Nuclear Engineering, Volume 1*, 2004, pp. 723–731.

[18] A. Heinzel, M. Kondo, and M. Takahashi, "Corrosion of steels with surface treatment and Al-alloying by GESA exposed in lead-bismuth," *J. Nucl. Mater.*, vol. **350**, no. 3, pp. 264–270, 2006.

The gravitational field equations of a twisted skyrmion string: numerical solution

Miftachul Hadi[1, 2, 4, 5, a], **Malcolm Anderson**[1, b], **Andri Husein**[3]
[1] Department of Mathematics, Universiti Brunei Darussalam, NEGARA BRUNEI DARUSSALAM
[2] Physics Research Centre, Indonesian Institute of Sciences (LIPI), INDONESIA
[3] Physics Department of Post Graduate Program Sebelas Maret University, INDONESIA
[4] School of Physics, Ulsan National Institute of Science and Technology (UNIST), SOUTH KOREA
[5] Institute of Modern Physics, Chinese Academy of Sciences, Lanzhou, CHINA

E-mail: itpm.id@gmail.com

Abstract. We study nonlinear sigma model, especially Skyrme model with twist: twisted Skyrmion string where twist term, mkz, is indicated in vortex solution. To add gravity, we replace $g^{\mu\nu}$ in Lagrangian system with a space-time metric tensor, $g^{\mu\nu}$ which in view of the time-independence and cylindrical symmetry of the assumed vortex solution is taken to be a function of r alone. We use ode45 for numerical calculation, i.e. a tool box in Matlab to solve coupled Einstein field equations which have ordinary differential equations (ODE) form.

1. Introduction to Nonlinear Sigma Model

The Lagrangian density of a free (without potential) nonlinear sigma model on a Minkowski background space-time is defined to be [1]

$$L = \frac{1}{2\lambda^2}\gamma_{AB}(\phi)\eta^{\mu\nu}\partial_\mu\phi^A \cdot \partial_\nu\phi^B \tag{1}$$

where $\gamma_{AB}(\phi)$ is the field metric, $\eta^{\mu\nu}$ is the Minkowski metric tensor, λ is a scaling constant with dimensions of (length/energy)$^{1/2}$ and $\phi = \phi^A$ is the collection of fields.

One of the most interesting examples of a $O(N)$ nonlinear sigma model, due to its topological properties, is the $O(3)$ nonlinear sigma model in (1+1)D, with the Lagrangian density

$$L = \frac{1}{2\lambda^2}\eta^{\mu\nu}\partial_\mu\phi \cdot \partial_\nu\phi \tag{2}$$

[1,a] Miftachul Hadi (itpm.id@gmail.com)
[1,b] Malcolm Anderson (malcolm.anderson@ubd.edu.bn)
[3] Andri Husein (andrihusein08@gmail.com)

where μ and ν range over $\{0,1\}$, and $\phi=(\phi^1,\phi^2,\phi^3)$, subject to the constraint $\phi\cdot\phi=1$, where the dot $(\cdot)$ denotes the dot product on real coordinate space of three dimensions, R^3.

A simple representation of field, Ø, (in the general time-dependent case) is [2]

$$\phi=\begin{pmatrix}\sin f(t,\vec{r})\sin g(t,\vec{r})\\ \sin f(t,\vec{r})\cos g(t,\vec{r})\\ \cos f(t,\vec{r})\end{pmatrix} \tag{3}$$

where f and g are scalar functions on the background space-time, with Minkowski coordinates $x^\mu=(t,\vec{r})$.

If a Skyrme term is added to $O(3)$ nonlinear sigma model, the result is a modified Lagrangian density, i.e. Skyrme model [2]

$$L=\frac{1}{2\lambda^2}\eta^{\mu\nu}\partial_\mu\phi\cdot\partial_\nu\phi-K_s\eta^{\kappa\lambda}\eta^{\mu\nu}(\partial_\kappa\phi\times\partial_\mu\phi)\cdot(\partial_\lambda\phi\times\partial_\nu\phi) \tag{4}$$

where the Skyrme term is the second term on the right hand side of (4).

2. The Gravitational Field of a Twisted Skyrmion String

We are interested in constructing space-time generated by a twisted Skyrmion string. Without gravity, the Lagrangian density of the system is L_1, as given in equation below [3]

$$\begin{aligned}L_1&=\frac{1}{2\lambda^2}\left(\eta^{\mu\nu}\partial_\mu f\partial_\nu f+\sin^2 f\eta^{\mu\nu}\partial_\mu g\partial_\nu g\right)\\&-K_s\left[2\sin^2 f\left(\eta^{\mu\nu}\partial_\mu f\partial_\nu f\right)\left(\eta^{\kappa\lambda}\partial_\kappa g\partial_\lambda g\right)-2\sin^2 f\left(\eta^{\mu\nu}\partial_\mu f\partial_\nu g\right)^2\right]\end{aligned} \tag{5}$$

To add gravity, we replace $\eta^{\mu\nu}$ in L_1 (5) with a space-time metric tensor, $g^{\mu\nu}$, which in view of the time-independence and cylindrical symmetry of the assumed vortex solution is taken to be a function of r alone.

Metric tensor, $g^{\mu\nu}$, is of course the inverse of the covariant metric tensor, $g_{\mu\nu}$, of the space-time where $g^{\mu\nu}=\left(g_{\mu\nu}\right)^{-1}$. We use a cylindrical coordinate system (t,r,θ,z), where t and z have unbounded range, $r\in[0,\infty)$ and $\theta\in[0,2\pi)$. The metric tensor with its components can be written as a matrix below

$$g_{\mu\nu}=\begin{pmatrix}g_{tt}&0&0&0\\0&g_{rr}&0&0\\0&0&g_{\theta\theta}&g_{\theta z}\\0&0&g_{z\theta}&g_{zz}\end{pmatrix} \tag{6}$$

where the components of metric tensor are all functions of r, and the presence of the off-diagonal components $g_{\theta z}=g_{z\theta}$ reflects the twist in the space-time.

The Lagrangian we will be using is [3]

$$\begin{aligned}L_2&=\frac{1}{2\lambda^2}\left(g^{\mu\nu}\partial_\mu f\partial_\nu f+\sin^2 fg^{\mu\nu}\partial_\mu g\partial_\nu g\right)\\&-K_s\left[2\sin^2 f\left(g^{\mu\nu}\partial_\mu f\partial_\nu f\right)\left(g^{\kappa\lambda}\partial_\kappa g\partial_\lambda g\right)-2\sin^2 f\left(g^{\mu\nu}\partial_\mu f\partial_\nu g\right)^2\right]\end{aligned} \tag{7}$$

where

$$f=f(r),\qquad g=n\theta+mkz \tag{8}$$

We need to solve:

i. the Einstein field equations

$$G_{\mu\nu} = -\frac{8\pi G}{c^4} T_{\mu\nu} \tag{9}$$

where the stress-energy tensor of the vortex, $T_{\mu\nu}$, is defined by

$$T_{\mu\nu} \equiv 2\frac{\partial L_2}{\partial g^{\mu\nu}} - g_{\mu\nu} L_2 \tag{10}$$

and

$$G_{\mu\nu} = R_{\mu\nu} - \frac{1}{2} g_{\mu\nu} R \tag{11}$$

with $R_{\mu\nu}$ the Ricci tensor and

$$R = g_{\mu\nu} R^{\mu\nu} = g^{\mu\nu} R_{\mu\nu} \tag{12}$$

the Ricci scalar;.

ii. the field equations for f and g

$$\nabla^{\mu} \frac{\partial L_2}{\partial(\partial f / \partial x^{\mu})} = \frac{\partial L_2}{\partial f}, \qquad \nabla^{\mu} \frac{\partial L_2}{\partial(\partial g / \partial x^{\mu})} = \frac{\partial L_2}{\partial g} \tag{13}$$

However, the field equations for f and g are in fact redundant, as they are satisfied identically whenever the Einstein field equations are satisfied, by virtue of the Bianchi identities (i.e. permuting of the covariant derivative of the Riemann tensor) $\nabla_{\mu} G^{\mu}_{\nu} = 0$. So, only the Einstein field equations will be considered [3].

To simplify the Einstein field equations, we first choose a gauge condition that narrows down the form of the metric tensor. The gauge condition preferred here is that [3]

$$g_{\theta\theta} g_{zz} - (g_{\theta z})^2 = r^2 \tag{14}$$

The geometric significance of this choice is that the determinant of the 2-metric tensor projected onto the surfaces of constant t and z is r^2, and so the area element on these surfaces is just $r dr d\theta$.

As a further simplification, we write $g_{tt} = A^2$, $g_{rr} = -B^2$, $g_{\theta\theta} = -C^2$, $g_{\theta z} = \omega$, where A, B, C, ω are functions of r and so $g_{zz} = -\left(\frac{r^2 + \omega^2}{C^2}\right)$. The metric tensor, $g_{\mu\nu}$, therefore has the form

$$g_{\mu\nu} = \begin{pmatrix} A^2 & 0 & 0 & 0 \\ 0 & -B^2 & 0 & 0 \\ 0 & 0 & -C^2 & \omega \\ 0 & 0 & \omega & -\left(\frac{r^2 + \omega^2}{C^2}\right) \end{pmatrix} \tag{15}$$

Substitute (6), (8) into (7), we obtain L_2 as below [4]

$$\begin{aligned} L_2 = &\frac{1}{2g_{rr}} \left[\frac{1}{\lambda^2} + \frac{4K_s \sin^2 f}{r^2} \left(n^2 g_{zz} - 2nmk g_{\theta z} + m^2 k^2 g_{\theta\theta} \right) \right] \left(\frac{\partial f}{\partial r} \right)^2 \\ &+ \frac{\sin^2 f}{2\lambda^2 r^2} \left(n^2 g_{zz} - 2nmk g_{\theta z} + m^2 k^2 g_{\theta\theta} \right) \end{aligned} \tag{16}$$

From (10), (15) and (16), we derive the non-zero components of the stress-energy tensor, $T_{\mu\nu}$, as below

$$T_{tt} = \frac{A^2}{2B^2}\left(\frac{1}{\lambda^2} + \frac{4NK_s \sin^2 f}{r^2}\right)\left(\frac{\partial f}{\partial r}\right)^2 - \frac{A^2 N \sin^2 f}{2\lambda^2 r^2} \tag{17}$$

$$T_{rr} = \left(\frac{1}{2\lambda^2} + \frac{2NK_s \sin^2 f}{r^2}\right)\left(\frac{\partial f}{\partial r}\right)^2 + \frac{B^2 N \sin^2 f}{2\lambda^2 r^2} \tag{18}$$

$$T_{\theta\theta} = -\frac{C^2}{2B^2}\left(\frac{1}{\lambda^2} + \frac{4NK_s \sin^2 f}{r^2}\right)\left(\frac{\partial f}{\partial r}\right)^2 + \left(n^2 + \frac{C^2 N}{2r^2}\right)\frac{\sin^2 f}{\lambda^2} \tag{19}$$

$$T_{\theta z} = \frac{\omega}{2B^2}\left(\frac{1}{\lambda^2} + \frac{4NK_s \sin^2 f}{r^2}\right)\left(\frac{\partial f}{\partial r}\right)^2 + \left(nmk - \frac{\omega N}{2r^2}\right)\frac{\sin^2 f}{\lambda^2} \tag{20}$$

$$T_{zz} = -\frac{1}{2B^2C^2}\left(1 + \frac{\omega^2}{r^2}\right)\left(\frac{r^2}{\lambda^2} + 4NK_s \sin^2 f\right)\left(\frac{\partial f}{\partial r}\right)^2 + \left(m^2k^2 + \frac{N}{2C^2}\left(1 + \frac{\omega^2}{r^2}\right)\right)\frac{\sin^2 f}{\lambda^2} \tag{21}$$

where $N = n^2 g_{zz} - 2nmkg_{\theta z} + m^2k^2 g_{\theta\theta}$. Note that $T_{\theta z}$ is in general non-zero, provided that either mk or ω is non-zero. In fact, mk acts as a source term for ω. The twist in the vortex is therefore solely responsible for a non-zero circular stress, $T_{\theta z}$.

In components form, the Einstein field equations (11), corresponding to the metric tensor are

$$G_{tt} = R_{tt} - \frac{A^2 R}{2},\ G_{rr} = R_{rr} + \frac{B^2 R}{2},\ G_{\theta\theta} = R_{\theta\theta} + \frac{C^2 R}{2},\ G_{\theta z} = R_{\theta z} - \frac{\omega R}{2},\ G_{zz} = R_{zz} + \frac{\left(r^2 + \omega^2\right)}{2C^2} R \tag{22}$$

3. Solution of the Einstein Field Equations

We use ode45 for numerical calculation, i.e. a tool box in Matlab to solve ordinary differential equations (ODE). Such tool box works based on explicit formula of 4th, 5th order Runge-Kutta. Solution scheme with ode45 (or in general, Runge-Kutta) is transforming one or more than one differential equations of *n* order into *n* differential equations of first order. In order to obtain numerical solution, we need *n* suitable initial values for *n* differential equations of first order. Perhaps, this is the reason why this method is called an initial value method, and system of equations which are solved called an initial value problem (IVP) [4].

The equations (22) can be rearranged as source equations for the first derivatives of B and f and the second derivatives of A, C, ω as follows [4].

$$B' = \frac{BA'}{A} + \frac{r\varepsilon B^3}{\lambda^2}\left[\frac{n^2}{C^2}\left(1 + \frac{\omega^2}{r^2}\right) + \frac{2nmk\omega}{r^2} + \frac{C^2m^2k^2}{r^2}\right]\sin^2 f \tag{23}$$

$$f' = \pm\left(\frac{\varepsilon}{2\lambda^2} - 2\varepsilon K_s\left[\frac{n^2}{C^2}\left(1 + \frac{\omega^2}{r^2}\right) + \frac{2nmk\omega}{r^2} + \frac{C^2m^2k^2}{r^2}\right]\sin^2 f\right)^{-1/2} \times \left(\begin{array}{l} \frac{A'}{rA} + \frac{\omega\omega' C'}{r^2 C} + \frac{C'}{rC} - \frac{C'^2}{C^2}\left(1 + \frac{\omega^2}{r^2}\right) - \frac{\omega'^2}{4r^2} \\ + \frac{\varepsilon B^2}{2\lambda^2}\left[\frac{n^2}{C^2}\left(1 + \frac{\omega^2}{r^2}\right) + \frac{2nmk\omega}{r^2} + \frac{C^2m^2k^2}{r^2}\right]\sin^2 f \end{array}\right)^{1/2} \tag{24}$$

$$A''=-\frac{A'}{r}+\frac{A'^2}{A}+\varepsilon\sin^2 f\left[\frac{n^2}{C^2}\left(1+\frac{\omega^2}{r^2}\right)+\frac{2nmk\omega}{r^2}+\frac{C^2m^2k^2}{r^2}\right]$$
$$\times\left\{\begin{array}{l}\dfrac{rA'B^2}{\lambda^2}-2K_s\left(\begin{array}{l}\dfrac{A'}{r}+\dfrac{\omega A\omega'C'}{r^2C}+\dfrac{AC'}{rC}-\dfrac{AC'^2}{C^2}\left(1+\dfrac{\omega^2}{r^2}\right)-\dfrac{A\omega'^2}{4r^2}\\+\dfrac{\varepsilon AB^2}{2\lambda^2}\left[\dfrac{n^2}{C^2}\left(1+\dfrac{\omega^2}{r^2}\right)+\dfrac{2nmk\omega}{r^2}+\dfrac{C^2m^2k^2}{r^2}\right]\sin^2 f\end{array}\right)\\ \times\left(\dfrac{\varepsilon}{2\lambda^2}-2\varepsilon K_s\left[\dfrac{n^2}{C^2}\left(1+\dfrac{\omega^2}{r^2}\right)+\dfrac{2nmk\omega}{r^2}+\dfrac{C^2m^2k^2}{r^2}\right]\sin^2 f\right)^{-1}\end{array}\right\} \tag{25}$$

$$\begin{aligned}C''=&\frac{C'^2}{C}\left(1+\frac{2\omega^2}{r^2}\right)-\frac{2\omega\omega'C'}{r^2}-\frac{C'}{r}+\frac{C\omega'^2}{2r^2}\\&-\frac{\varepsilon B^2\sin^2 f}{\lambda^2}\left[\frac{n^2}{C^2}\left(1+\frac{\omega^2}{r^2}\right)+\frac{2nmk\omega}{r^2}+\frac{C^2m^2k^2}{r^2}\right]\left(\frac{C}{2}-rC'\right)\\&+\frac{\varepsilon\sin^2 f}{2}\left[\frac{n^2}{C^2}\left(1+\frac{\omega^2}{r^2}\right)+\frac{2nmk\omega}{r^2}+\frac{C^2m^2k^2}{r^2}\right]\\&\times\left\{\begin{array}{l}\dfrac{CB^2}{\lambda^2}-4K_s\left(\begin{array}{l}\dfrac{A'C}{rA}+\dfrac{\omega\omega'C'}{r^2}+\dfrac{C'}{r}-\dfrac{C'^2}{C}\left(1+\dfrac{\omega^2}{r^2}\right)-\dfrac{\omega'^2C}{4r^2}\\+\dfrac{\varepsilon B^2C\sin^2 f}{2\lambda^2}\left[\dfrac{n^2}{C^2}\left(1+\dfrac{\omega^2}{r^2}\right)+\dfrac{2nmk\omega}{r^2}+\dfrac{C^2m^2k^2}{r^2}\right]\end{array}\right)\\ \times\left(\dfrac{\varepsilon}{2\lambda^2}-2\varepsilon K_s\sin^2 f\left[\dfrac{n^2}{C^2}\left(1+\dfrac{\omega^2}{r^2}\right)+\dfrac{2nmk\omega}{r^2}+\dfrac{C^2m^2k^2}{r^2}\right]\right)^{-1}\end{array}\right\}\end{aligned} \tag{26}$$

$$\begin{aligned}\omega''=&\frac{\omega'}{r}+\frac{4\omega C'^2}{C^2}\left(1+\frac{\omega^2}{r^2}\right)-\frac{4\omega^2\omega'C'}{r^2C}-\frac{4\omega C'}{rC}+\frac{\omega\omega'^2}{r^2}\\&-\frac{\varepsilon B^2(\omega-r\omega')\sin^2 f}{\lambda^2}\left[\frac{n^2}{C^2}\left(1+\frac{\omega^2}{r^2}\right)+\frac{2nmk\omega}{r^2}+\frac{C^2m^2k^2}{r^2}\right]\\&+\varepsilon\sin^2 f\left[\frac{n^2}{C^2}\left(1+\frac{\omega^2}{r^2}\right)+\frac{2nmk\omega}{r^2}+\frac{C^2m^2k^2}{r^2}\right]\\&\times\left\{\begin{array}{l}\dfrac{\omega B^2}{\lambda^2}-4K_s\left(\begin{array}{l}\dfrac{\omega A'}{rA}+\dfrac{\omega^2\omega'C'}{r^2C}+\dfrac{\omega C'}{rC}-\dfrac{\omega C'^2}{C^2}\left(1+\dfrac{\omega^2}{r^2}\right)-\dfrac{\omega\omega'^2}{4r^2}\\+\dfrac{\varepsilon\omega B^2\sin^2 f}{2\lambda^2}\left[\dfrac{n^2}{C^2}\left(1+\dfrac{\omega^2}{r^2}\right)+\dfrac{2nmk\omega}{r^2}+\dfrac{C^2m^2k^2}{r^2}\right]\end{array}\right)\\ \times\left(\dfrac{\varepsilon}{2\lambda^2}-2\varepsilon K_s\sin^2 f\left[\dfrac{n^2}{C^2}\left(1+\dfrac{\omega^2}{r^2}\right)+\dfrac{2nmk\omega}{r^2}+\dfrac{C^2m^2k^2}{r^2}\right]\right)^{-1}\end{array}\right\}\end{aligned} \tag{27}$$

The ($\pm$) sign at (24) suggest us to choose it, so that the values of $f'<0$ for most or all values of r. Here we choose the sign (-). Equation (23) to (27) need to be solved simultaneously. To do so, we convert second order ODE from (25) to (27) to be first order ODE. At this stage, one finally get eight coupled first order ODEs which are ready to solve.

We obtain equations to generate initial value for $r = r_{\min}$, i.e. [4]

$$B(r_{\min}) = B_0 + \frac{\varepsilon a^2 B_0}{2}\left(\frac{1}{\lambda^2} - \frac{K_s a^2}{B_0^2}\right) r_{\min}^2 \tag{28}$$

$$f(r_{\min}) = \pi + a r_{\min} \tag{29}$$

$$A(r_{\min}) = A_0 - \frac{\varepsilon K_s a^4 A_0}{2C_0^2} r_{\min}^2 \tag{30}$$

$$C(r_{\min}) = B_0 r_{\min} + \frac{\varepsilon K_s a^4}{2B_0} r_{\min}^3 \tag{31}$$

$$\omega(r_{\min}) = \omega_1 r_{\min}^2 \tag{32}$$

where B_0, a, A_0, ω_1, K_s and λ are unknown constants. For $r \to \infty$, no matter of our choices for the values of such unknown constants, we want to graph $B(r)$, $f(r)$, $A(r)$, $C(r)/r$ and $\left(r^2+\omega^2\right)/C^2$ solutions respectively as graphics which are asymptotic to $B \to 1$, $f \to 0$, $A \to 1$, $C/r \to (const. \neq 0)$, and $\left(r^2+\omega^2\right)/C^2 \to 1$.

The values assumed for B_0 and A_0 affect the solution only by rescaling the metric functions, the radial coordinate, r, and the twist term, mk. This follows from the fact that:

i. All the components of the Einstein tensor, except G_{tt} in (22), depend on A only through A''/A or A'/A, while G_{tt} is proportional to A^2. Similarly, all the components of the stress-energy tensor, except T_{tt} in (17) to (21), do not depend on A, while T_{tt} is proportional to A^2. So, we can multiply A by any non-zero constant without affecting any of the other metric functions or f.

ii. A transformation of $r \to cr$, $\frac{d}{dr} \to \left(\frac{1}{c}\right)\frac{d}{dr}$, $\frac{d^2}{dr^2} \to \left(\frac{1}{c^2}\right)\frac{d^2}{dr^2}$, $B \to \left(\frac{1}{c}\right)B$, $\omega \to c\omega$, $mk \to cmk$, where c is any positive constant, leave the field equations (23) to (27) unchanged.

So, it is always possible to divide A by A_0 and B by B_0 so that the initial values of A and B (at $r = 0$) are both 1. And in doing this, C_0 is also rescale to 1, as $C = C_0 r + \ldots$ is unchanged, and so if $r \to B_0 r$ we must have $C_0 \to C_0/B_0 = 1$. Making this rescaling simplifies equations (28) to (32), i.e

$$B(r_{\min}) = 1 + \frac{\varepsilon a^2}{2}\left(\frac{1}{\lambda^2} - K_s a^2\right) r_{\min}^2 \tag{33}$$

$$f(r_{\min}) = \pi + a r_{\min} \tag{34}$$

$$A(r_{\min}) = 1 - \frac{\varepsilon K_s a^4}{2} r_{\min}^2 \tag{35}$$

$$C(r_{\min}) = r_{\min} + \frac{\varepsilon K_s a^4}{2} r_{\min}^3 \tag{36}$$

$$\omega(r_{\min}) = \omega_1 r_{\min}^2 \tag{37}$$

If we define $r_{\min} = 0.0001$, $r_{\max} = 4\times10^4$, $\varepsilon = 8\pi$, $a = -0.186$, $\omega_1 = 10^{-15}$, $K_s = 4\times10^{-14}$, $\lambda = 2100$, $mk = 10^{-6}$, we obtained graphs of typical solution as follows

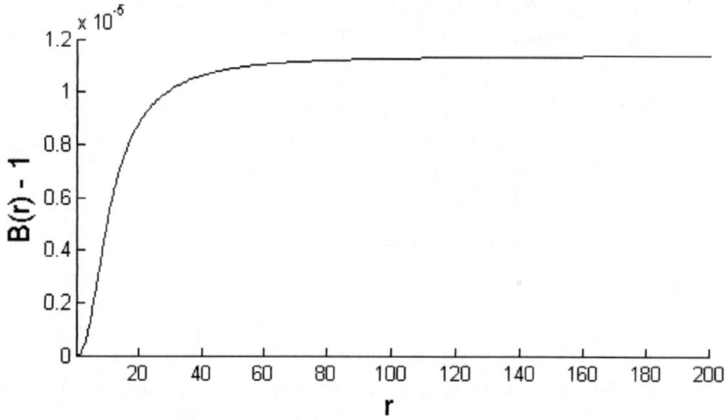

Figure 1 Typical form of the $B(r)-1$ function

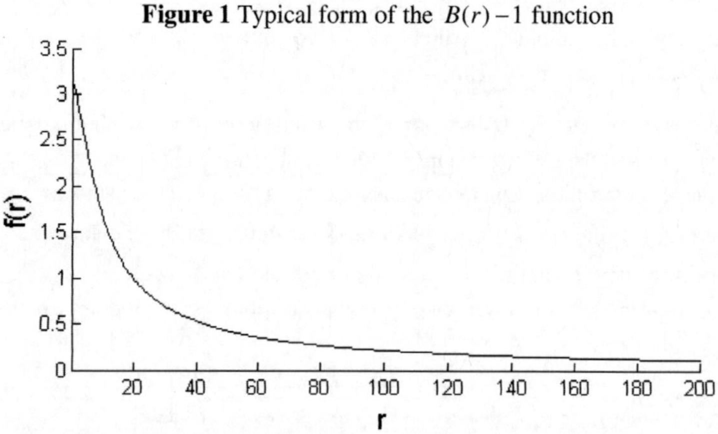

Figure 2 Typical form of the $f(r)$ function

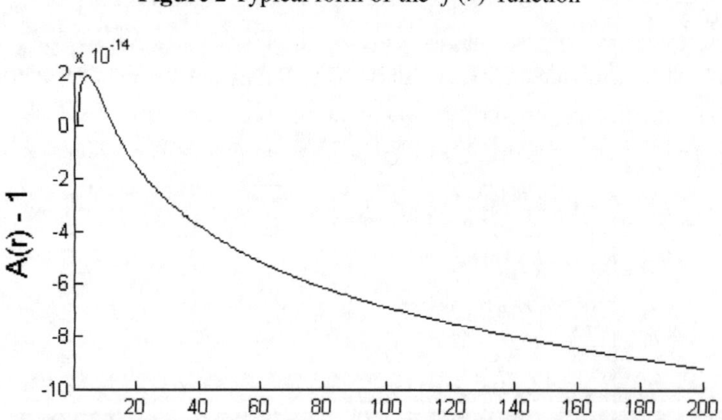

Figure 3 Typical form of the $A(r)-1$ function

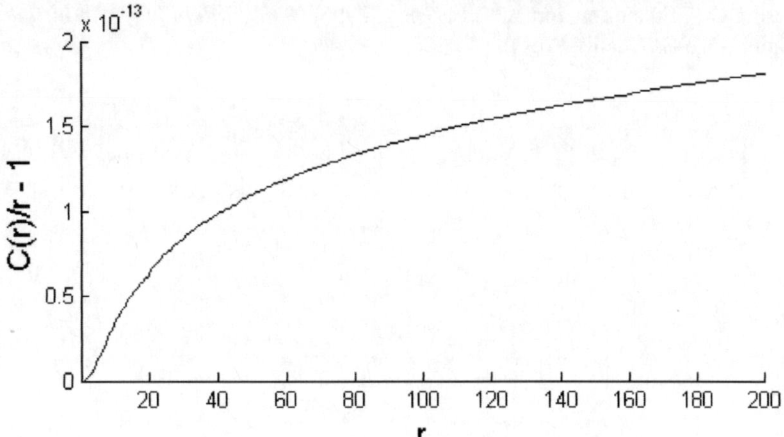

Figure 4 Typical form of the $\frac{C(r)}{r} - 1$ function

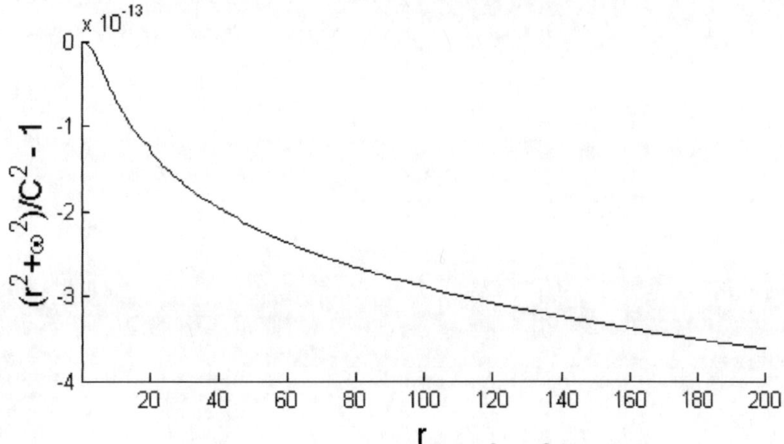

Figure 5 Typical form of the $\frac{r^2 + \omega^2}{C^2} - 1$ function

Acknowledgment
This research is supported fully by Graduate Research Scholarship Universiti Brunei Darussalam (GRS UBD). This support is highly appreciated.

References

[1] C. C. Chen, T. Earnest, 1995, *Nonlinear Sigma Model* (University of Illinois).
[2] M. Hadi, M. Anderson and A. Husein, 2014, *Journal of Physics: Conference Series* **539**, 012008.
[3] Miftachul Hadi, Malcolm Anderson and Andri Husein, 2015, arXiv:1505.05930v1 [gr-qc].
[4] Miftachul Hadi, Malcolm Anderson, Andri Husein, *The gravitational field of a twisted Skyrmion string*, unpublished.
[5] H. J. Wospakrik, 2002, *Ph.D. thesis* (University of Durham).
[6] E. Simanek, 2010, arXiv: 1001.5061v1.
[7] B. M. A. G. Piette, B. J. Schroers and W. J. Zakrzewski, 1995, *Nuclear Physics B* **439**, 205-235.

IOP Conf. Series: Journal of Physics: Conf. Series **795** (2017) 012006 doi:10.1088/1742-6596/795/1/012006

[8] T. Gisiger, M. B. Paranjape, *1996, Physics Letters B* **384**, 207-212.
[9] Y. M. Cho, H. Khim and P. M. Zhang, *2005, Physical Review A* **72**, 063603.
[10] M.P. Hobson, G.P. Efstathiou and A.N. Lasenby, 2006, *General Relativity: An Introduction for Physics* (Cambridge University Press).

Synchrotron WAXS and XANES studies of silica (SiO_2) powders synthesized from Indonesian natural sands

Khairanissa Muchlis[1], Nur Aini Fauziyah[1], Siriwat Soontaranon[2], Wanwisa Limpirat[2] and Suminar Pratapa[1]

[1]Department of Physics, Faculty of Mathematics and Natural Sciences, Institute of Technology Sepuluh Nopember (ITS), Surabaya 60111, INDONESIA

[2]Synchroton Light Research Institute (Public Organization), Nakhon Ratchasima, THAILAND

E-mail: suminar_pratapa@physics.its.ac.id

Abstract. In this study, we have investigated polymorphic silica (SiO_2) powders using, Wide Angle X-ray Scattering (WAXS) and X-Ray Absorption Near Edge Spectroscopy (XANES), laboratory X-Ray Diffraction (XRD) instruments. The WAXS and XANES spectra were collected using synchrotron radiation at Synchrotron Light Research Institute (SLRI), Nakhon Ratchasima, Thailand. The silica powders were obtained by processing silica sand from Tanah Laut, South Kalimantan, Indonesia. Purification process of silica sand was done by magnetic separation and immersion with HCl. The purification step was needed to reduce impurity or undesirable non Si elements. Three polymorphs of silica were produced, i.e. amorphous phase (A), quartz (B), and cristobalite (C). WAXS profile for each phase was presented in terms of intensity vs. 2θ prior to analyses. Both XRD ($\lambda_{CuK\alpha}$=1.54056 Å) and WAXS (λ=1.09 Å) patttern show that (1) A sample contains no crystallites, (2) B sample is monophasic, contains only quartz, and (3) C sample contains cristobalite and trydimite. XRD quantitative analysis using *Rietica* gave 98,8 wt% cristobalite, while the associated WAXS data provided 98.7 wt% cristobalite. Si K-edge XANES spectra were measured at energy range 1840 to 1920 eV. Qualitatively, the pre-edge and edge features for all phases are similar, but their main peaks in the post-edge region are different.

1. Introduction

Silica is a fundamental building block of many rock-forming minerals and naturally occurs substances that comprise the vast majority of the earth's crust. Its structure which is obtained through patterns of Si-O linking and physicochemical properties have been the subject of numerous investigations for decades. Because of their prevalence and commercial applications, SiO_2 are widely used in the ceramic and glass industries and has an enormous application potential in optical fibres, in microelectronics [1-5]. Even if it is a very simple system, its properties and structures can be very complex as a function of temperature and pressure. Silica exists in many different phases at ambient pressure and temperature, e.g. α-quartz, tridymite, a-cristobalite, glass. Most of these structures are composed of corner-sharing tetrahedral SiO_4 units with a fourfold-coordinated silicon at the centre and twofold-coordinated oxygen atoms at the corners [1-4].

Synchrotron facility such as that at Synchrotron Light Research Institute (SLRI), Nakhon Ratchasima, Thailand provides Small/Wide Angle X-Ray Scattering (SAXS/WAXS) and X-Ray

Absorption Near Edge Spectroscopy (XANES) at region between 0-50 eV above the edge of X-Ray Absorption Spectroscopy (XAS) instruments. These instruments can be utilized for phase analysis of crystalline samples and "local structure" studies of samples, respectively.

Phase analyses of silica has been extensively done particularly on their structures, including "local" structure [1]. They used XANES data in their study. Meanwhile, phase composition examination is usually done using XRD data, not WAXS data. The latter data was used mainly for determining degree of crystallinity. Combination of XRD, WAXS and XANES data analysis, therefore, is a subject of phase analysis which will enrich one's knowledge on phase structure and composition as well as local structure.

In this paper, we present polymorphic silica, SiO_2 (amorphous, quartz, and cristobalite) powders using CuKα radiation XRD and synchrotron radiation WAXS and XANES. Phase identification and composition determination of the polymorphs from XRD and XANES data are compared, while the local structure of each phase in the samples is reported

2. Experimental Method

In the experimental procedures of this research. SiO_2 polymorph samples are natural samples provided by processing natural silica sands. The silica sand was taken from Tanah Laut, Pelaihari, Kalimantan Selatan. The purification step is needed to reduce impurity or undesirable non Si element. The steps were cleaning and sieving, extracting by using permanent magnet repeatedly, milling and soaking with hydrochloric acid. Meanwhile, solvent extraction results Na_2SiO_3 titrated with HCL 2 M. The process produces SiO_2 gel. SiO_2 gel was neutralized to pH 7, dried to obtain an amorphous silica powder. Furthermore the amorphous SiO_2 sample was sampled and calcined at 950°C to produce cristobalite powder.

The XRD pattern were recorded by X-Ray Diffractometer Philips X'Pert MPD (Multi Purpose Diffractometer) ($\lambda_{CuK\alpha}$=1.54056 Å) for all samples. The XRD patterns were collected with step size of 0.0170°. The XRD patterns were analysed using *Match!2* for phase identification and using *Rietica* based on Rietveld method to provide accurate phase compositions of the samples. The WAXS profile is measured using WAXS instrument at Beam Line 2.2 (BL 2.2) SLRI. A Double Multilayer Monochromator was adopted to monochromatize synchrotron x-rays within the energy 9 keV. The WAXS measurement was conducted using a CCD detector, 89 cm distance of sample to detector (SDD), and the standard material of 4-Bromobenzoic Acid. The WAXS profile was presented with scattering intensity (I) curves over a range of scattering vector magnitude (q) (I vs q) or two dimensional (2D) scattering intensity maps. Conversion of curves (I vs q) to (I vs 2θ) is needed before analysis.

Si K-edge XANES Spectra were recorded with 0.2 eV step size in fluorescence mode with fixed-exit double crystals X-ray monochromator and Silicon drift detector using X-Ray Absorption Spectroscopy (XAS) instrument at Beam Line 8 (BL8) SLRI. The beam size was 14 mm (width) x 1 mm (height) [12]. XANES spectra were normalized using the Athena software [13]

3. Results and Discussion

Figure 1 shows the XRD and WAXS patterns of all samples, respectively. Phase identification reveals that sample A contains no crystalline phase, sample B comprises only quartz (NO PDF 16-2490) and sample C consist of cristobalite (NO.PDF 76-1390) and tridymite (NO. PDF 21-0442). The difference in Bragg peak positions of each phase in the XRD and WAXS patterns is due to the difference in radiation wavelength. The overall results demonstrate that phase identification using WAXS confirms that of XRD.

Further analyses using *Rietica* [9] gives goodness-of-fit 3.12% and 1.06% with the typical difference plots as given in Figure 2. These parameters indicate that the Rietveld refinements were succesfull and acceptable. Following the succesfull refinement, the phase composition in each sample is presented in Table 1. Phase composition results for A sample is not given since it is amorphous. Difference plot for sample B on XRD and WAXS is similar. This proves that the XRD and WAXS difference plots (shown by green lines) for sample B are similar in terms of number of Bragg peaks for the associated wavelengths, i.e. 8 peaks up to 2θ = 55 for XRD and up to 38 for WAXS. It means only quartz presents in sample B. [Since quartz is the sole phase in sample B, its content is 100%]. Sample C, the Rietveld refinements of XRD and WAXS give goodness-of-fit using *Rietica* are 3.35% and 3.92%. The XRD quantitative analysis gave 98,8 ± 1.9 wt% cristobalite, while the associated WAXS data provided 98.7 ± 2.8 wt% cristobalite. These results confirm that XRD and WAXS are similar.

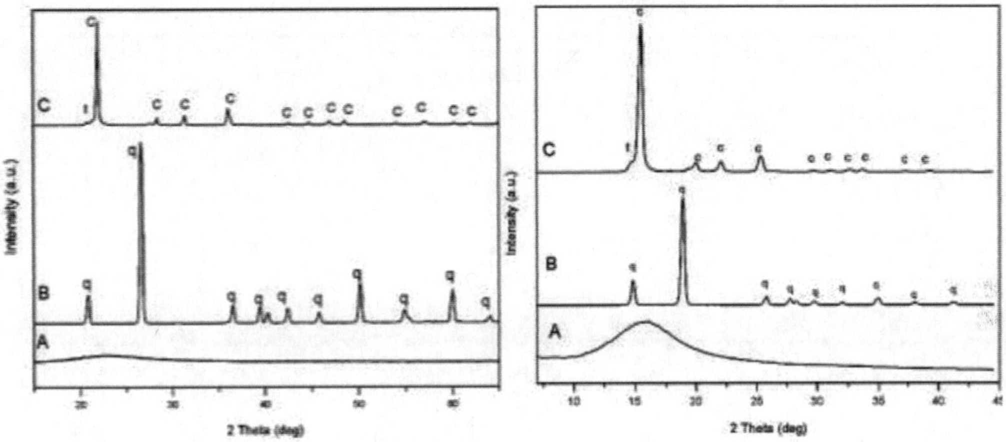

Figure 1 (XRD) Pattern of (A) amorphous, (B) quartz, (C) Cristobalite. XRD & WAXS. Note the difference in Bragg peak positions.

Table 1. Quantitative phase analyses of samples using *Rietica* by Rietveld method

	% weight	
	XRD	WAXS
Sample A	Amorphous	Amorphous
Sample B	100 Quartz	100 Quartz
Sample C	98±1.9 Cristobalite	98.7±2.8 Cristobalite
	1.2±0.6 Tridymite	1.3±0.5 Tridymite

Figure 3 shows the Si K-edge XANES spectra of amorphous (A), quartz (B) dan cristoballite (C) at energy range 1840 to 1920 eV.

IOP Conf. Series: Journal of Physics: Conf. Series **795** (2017) 012007 doi:10.1088/1742-6596/795/1/012007

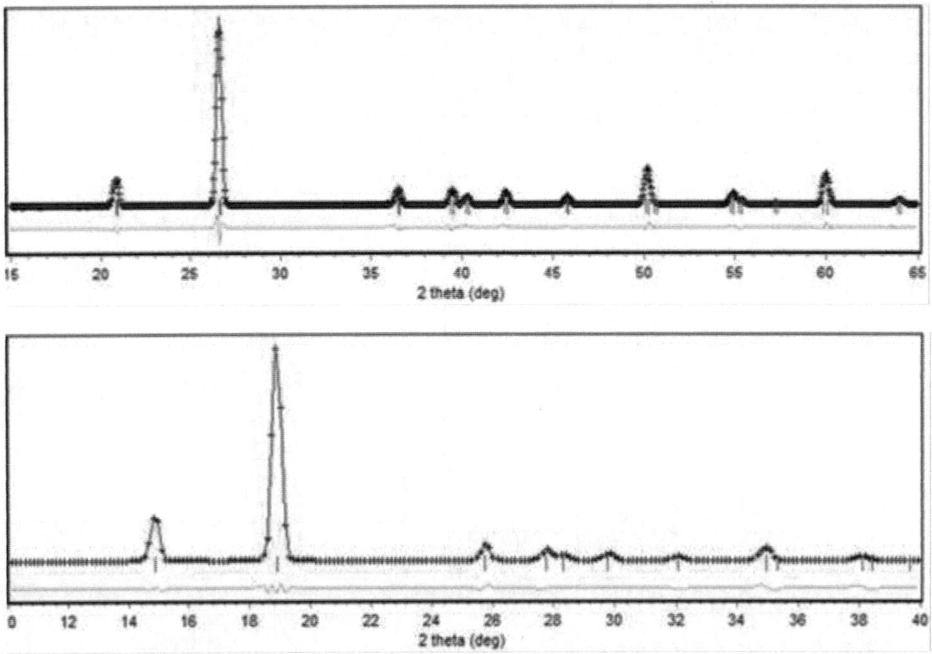

Figure 2. Typical (a) XRD and (b) WAXS Rietveld refinement plots using *Rietica* for quartz sample. The GoF values are 3.12% (XRD) and 1.06% (WAXS). The raw data, calculated pattern and difference between them were marked by black cross, red line and green line, respectively. The blue tick marks underneath the pattern indicates the calculated peak position

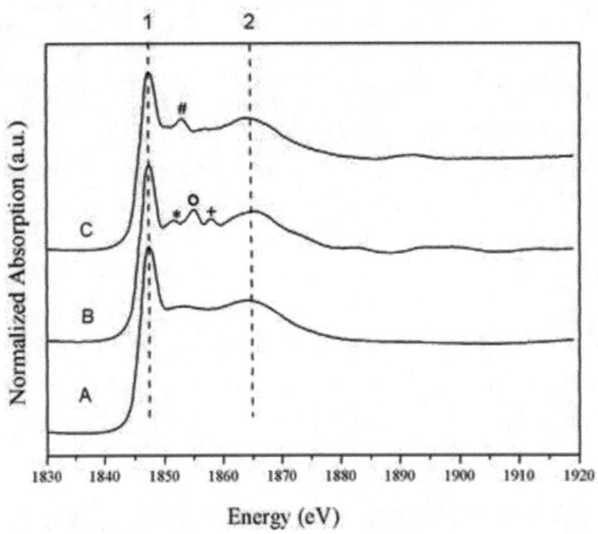

Figure 3 Si K-edge Spectra (A) amorphous, (B) quartz, (C) cristobalite
Noted peak position #=1853,2 eV, *=1853.2, O=1855.2 eV and + =1853.2 eV

Many change are served between energy about 1847 and 1865 eV (it is noticed between 1 and 2 region). The strongest peak, line 1, on the absorption edge about 1847.4 eV are similar for all samples. Meanwhile, Ligny et all in their study obtain the absorption edge at 1847.1 eV. At the region, the peak position for amorphous (A) is totally different, because there is no absorber crystalline atom arrangement. For the sample the peak of quartz (B) exist at 1853.2, 1855.2 eV, and 1853.2 eV and Cristobalite (C) at 1853.2 eV. Line 2 are present for all samples, they can be assigned to Si electronic structure on which is superimposed some multiple scattering [1].

4. Conclusion

The phase identification and phasecomposition was successfully investigated by Rietveld analysis of X-Ray Diffraction (XRD) data using Match2! and *Rietica* software respectively. i.e. amorphous phase (A), quartz (B), and cristobalite (C). Both XRD ($\lambda_{CuK□}$=1.54056 Å) and WAXS (λ=1.09 Å) patttern show that (1) A sample contains no crystallites, (2) B sample is monophasic, contains only quartz (100% quartz), and (3) C sample contains cristobalite and trydimite. XRD quantitative analysis using *Rietica* gave 98,8 wt% cristobalite, while the associated WAXS data provided 98.7 wt% cristobalite. Si K-edge XANES spectra were measured at energy range 1840 to 1920 eV. Qualitatively, the pre-edge and edge features for all phases are similar, but their main peaks in the post-edge region are different.

Acknowledgement
The authors would like to thank Indonesian Endowment Fund for Education (LPDP) scholarship, and Ministry of Research Technology and Higher Education of the Republic of Indonesia and LPPM ITS to support finance through program PUPT-2016 No. 01751/IT2.11/PN.08/2016 given to SP

References

[1] Ligny D D, Neuville D R, Cormier L, Roux J, Henderson G S, Panczer G, Shoval S, Flank A. M, Lagarde P 2009 *Journal of Non-Crystalline solids* 355, 1099-1102
[2] Li D, Bancroft G M, Kasrai M, Fleet M E, Secco R A, Feng X H, Tan K H, and Yang B X 1994 *American Mineralogist* 79, 622-623
[3] Gilbert B, Frazer B H, Naab F, Fournelle J, Valley J W and De Stasio G 2003 *American Mineralogist 88,763–769*
[4] Kuzmin A, and Chaboy J 2014 *IUCrJ* 1, 571-589
[5] Mossman, Brooke T and Glenn, Robert E 2013 *Crit Rev Toxicol*, 632-660
[6] Badro J, Teter D M, Downs R T, Gillet P, Hemley R J, Barrat J L 1997 *Physical Review B* 56, 10
[7] Zhang L, Zhou J, Zhang J, Su J, Zhang S, Chen N, Jia Y, Li J, Wang Y and Wang J Q 2016 *Journal of Synchrotron Radiation* 23
[8] Demuth Th, Jeanvoine Y, Hafner J et al 1999 *J. Phys. Condens. Matter II*, 3833-3874
[9] Hunter B 1998 *International Union of Crystallography Commission on Powder Diffraction Newsletter* 20
[10] Musyarofah, U. Nurbaiti, E. Dewa, Triwikantoro, Mashuri, S. Firdaus, and S. Pratapa, 2015 *Adv. Mater. Res* 1123,383–386
[11] Fauziyah N A, Musyarofah, Triwikantoro, Mashuri, Firdaus S and Pratapa S, *Advanced Materials Research Vol. 1112 (2015) pp 385-388*
[12] Klysubun W, Sombunchoo P, Deenan W and Kongmark C 2012 *J. Synchrotron Rad.* 19, 930–936
[13] Ravel B, Newville M 2005 *J. Synchrotron Rad* 12, 537

Optical properties of double layer thin films zinc oxide doping aluminum (ZnO/Al) were deposited on glass substrates by sol gel method spray coating technique

Anes Permatasari[1], Heri Sutanto[1,*] and Sinta Marito Siagian[1]

[1]Physics Department of Post Graduate Program Diponegoro University,
Jl. Prof. Soedarto, Tembalang, Kota Semarang, Jawa Tengah 50275, INDONESIA

E-mail: anespermatasari@st.fisika.undip.ac.id

Abstract. Thin films of double layer of ZnO/Al has succeeded in deposition on a glass substrate using sol-gel method and spray coating techniques. Variations of doping Al as much as 2%, 4%, 6% and 8%. ZnO precursor synthesized using zinc acetate dehydrate ($Zn(COOCH_3)_2.2H_2O$), isopropanol ($(CH_3)_2CHOH$) and monoethanolamine (MEA) were stirred using a magnetic stirrer for 45 minutes. ZnO precursor get homogeneous and then added of aluminum nitrate nonahydrate predetermined doping concentration and stirred again for 15 minutes. Deposition solution is done by the spray on a glass substrate and then heated at a temperature of 450°C. A layer of ZnO/Al deposited over the ZnO to produce a thin layer of a double layer. Optical properties layer of ZnO/Al characterized using UV-Vis spectrophotometer. Based on data from UV-Vis absorbance was determined the value of the energy band gap. Pure and dopped layers has different energy due the Al dopping. For pure ZnO layer has energy band gap of 3.347 eV and decreased to 3.09 eV for ZnO layer with Al dopant.

1. Introduction

Semiconductor materials as good photocatalysts can be used to degrade pollutants or toxic compounds namely Zinc Oxide (ZnO), Titania (TiO_2), Tungsten Oxide (WO_3) and Zinc stannate (Zn_2SnO_4) [1]. One of the semiconductor material that is readily available and nontoxic nature is ZnO. Some forms of ZnO are in the form of nanoparticles, powder, and a thin film. Several active research on thin layers, one of which is a thin layer of ZnO, largely because of its unique properties and advantages for its application [2]. The form of a thin layer is preferred to avoid the separation of the catalyst after the detoxification process.

Material Zinc Oxide (ZnO) has the energy band gap of 3.37 eV which is able to absorb active in UV light. However, for a more efficient photocatalyst, the material must be able to absorb visible light in the spectrum of sunlight. Thus, for ZnO material to be able to absorb UV light and visible light, the energy band gap of ZnO should be minimized. The energy band gap is getting smaller will increase the transfer of electrons from the valence band to the conduction band so that the hole is very important for the formation of hydroxyl radicals that are used for the degradation of pollutants formed in photocatalytic applications [3]. Methods used to expand the range of absorption of the light spectrum and minimize the energy band gap is by doping the semiconductor oxide.

The optical properties and structure of ZnO are affected by deposition parameters as well as the provision of doping. One of them is causing doping Aluminium seed morphology to increase the surface area coating and efficiency of photocatalytic activity [4]. The number of doping Al provided can affect the characteristics of the thin layers desirable [9,10]. Characteristics of thin layers is expected better with some of the controls provided in addition to the doping concentration, also add layers that form the double layer deposited [6]. A thin film using a double layer of ZnO:Al has a higher absorbance values up toward the infrared waves [5]. ZnO dopped Al thin film of double layer structure is expected to have better properties than single layer so that the photocatalyst can be more efficient. This thin film can be prepared by a method commonly referred as deposition of growing a thin layer.

Many techniques of deposition on thin film are by pulsed laser deposition, RF magnetron sputtering, chemical vapor deposition, spray pyrolysis and sol gel method. Among some of these techniques, sol-gel method is considered better, because the deposition of a thin film of either small or wide area and a complicated shape on a variety of substrates at room temperature so that it does not require air vacuum device [2]. Deposition on a substrate is also be done with the casting method, such as dipcoating, spin coating, slot-die coating and spray coating. One method that is more suitable for the manufacture of thin films is spray coating, because it can be done using equipment that is reasonably under room temperature conditions [8].

2. Experimental

2.1. Preparations of ZnO and ZnO:Al precursor

The process of making a thin film and double layer of ZnO ZnO/ZnO:Al done by sol-gel method of spray coating techniques. Molarity solution of 0.5 M with Al doping variation in molarity percentage by 2, 4, 6 and 8%. The preparation process and the manufacture of precursors to be done with a thin film of several stages which will be explained in the next section.

ZnO precursor namely Zinc acetate dihydrate ($Zn(COOCH_3)_2H_2O$) as much as 2,853 grams were dissolved into 26 ml of Isopropanol, then stirred for 15 minutes. After stirring for 15 minutes, the solution was then added 0.778 ml Monoethanolamine (MEA), and then stirred back on the magnetic stirrer at room temperature (25°C) for 15 minutes until the solution (precursor) homogeneous. If it is homogeneous, the precursor solution is ready for the deposition process.

Furthermore, the formation of precursors ZnO:Al namely Zinc acetate dehydrate ($Zn(COOH_3)_2.2H_2O$) as much as 2,853 grams dissolved in 26 ml Isopropanol was stirred for 15 minutes on magnetic stirrer. Then the solution was added MEA as much as 0.778 ml and stirred again for 15 minutes at room temperature (25°C) until a homogeneous solution. Furthermore, homogeneous solution is added aluminum nitrate nonahydrate ($Al(NO_3)_3.9H_2O$) in accordance with the concentration ratio of the molarity percentage that is consecutive, 2, 4, 6 and 8% (0.099; 0.203; 0.311; and 0.424 gram) and stirred for 15 minutes on magnetic stirrer. ZnO precursor solution and the double layer of ZnO/ZnO:Al that has been created is stored in the bottle and ready for the deposition process.

2.2. Preparations of ZnO:Al thin films

ZnO deposition process and double layer ZnO/ZnO:Al on a glass substrate using spray coating techniques. The glass substrate which has been cleaned is placed on a hot plate at a temperature of 450°C for 10 minutes. Precursor solution put into the spray gun and a precursor sprayed evenly onto a glass substrate. Spraying is done in sequence according to the concentration ratio of doping. Especially for double layer ZnO/ZnO:Al, deposition is done by spraying a solution of ZnO first and then followed by spraying a solution of ZnO:Al with each concentration doping.

After the deposition process is completed, a thin film annealing for 1 hour at a temperature of 450°C. Annealing process is done so that the growth and nucleation in the precursor formed evenly on the surface of the glass substrate. A thin film formed then is cooled to room temperature (25°C) and characterized.

3. Result and Discussion

The energy gap of a thin film is calculated based on the value of absorbance and transmittance of visible uv spectrometer with a wavelength range of 300-800 nm. Absorption peak of ZnO:Al 8% at 364 nm is shown in Figure 1. This indicates that a given doping can shift the absorption wavelength toward larger wavelength of visible light. The highest absorption peak lies at a wavelength of 368 nm ZnO:Al samples with doping 6%. The percentage transmittance of pure ZnO by 79.4% experienced an increase after the increase in doping and the highest in doping Al 4% that is equal to 91.8% and then decreased again on doping Al 8%, ie 81.8% as at show in Figure 2. Improved spectral transmittance of a sample occurs in the visible region as much as to 79-91%.

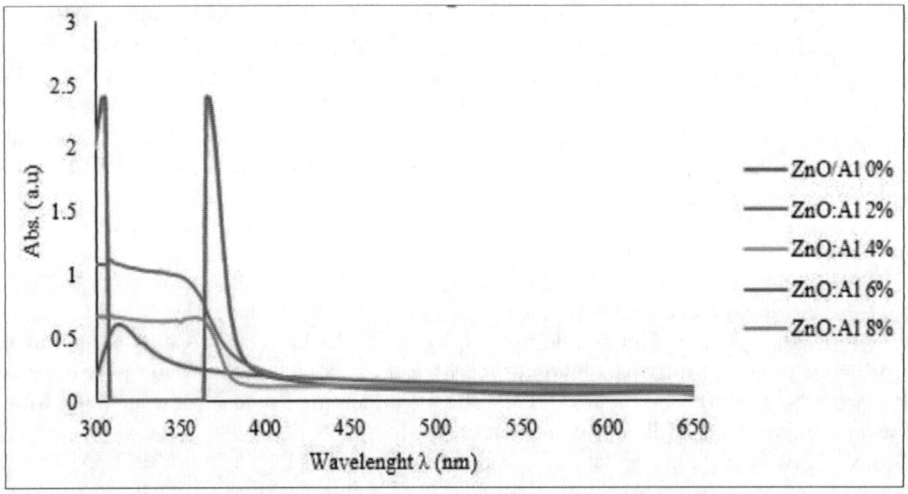

Figure 1. UV Visible absorbance spectra of undoped ZnO and ZnO:Al.

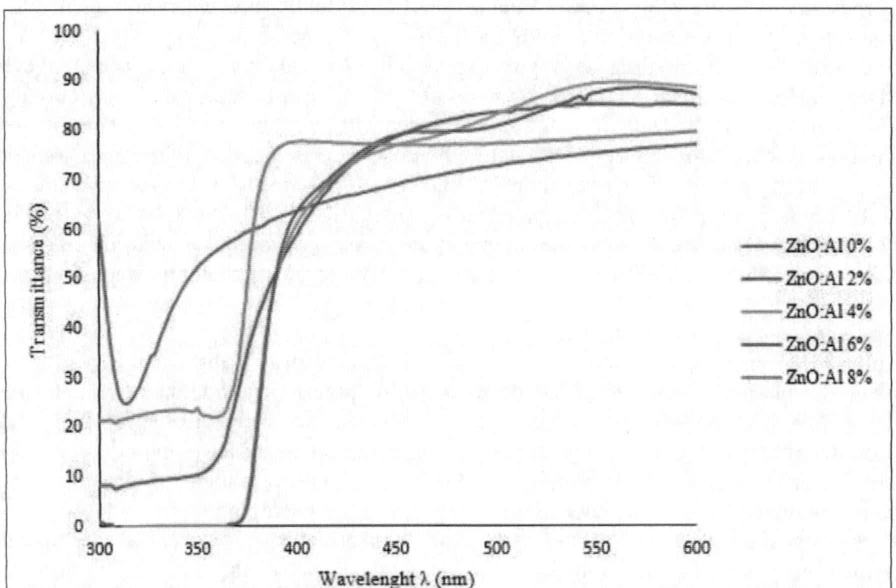

Figure 2. UV Visible transmittance spectra of undoped ZnO and ZnO:Al.

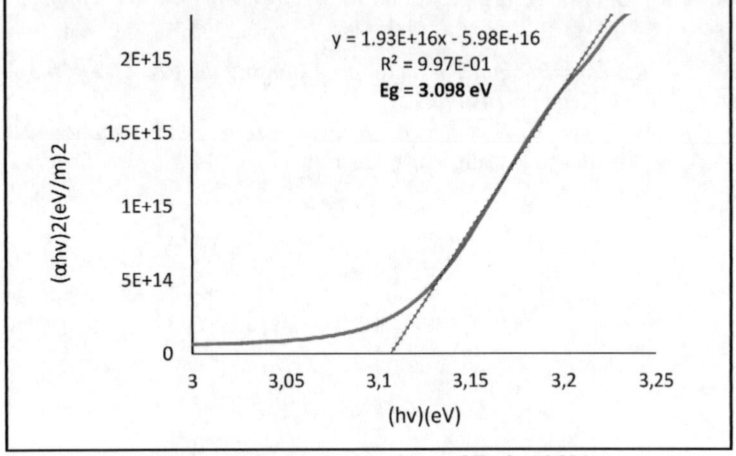

Figure 3. Tauc plot bandgap of ZnO:Al 8%

The addition of Al doping also affects the energy gap of thin film where it is influential to the electron transition. The value of energy band gaps of ZnO thin film with variation of dopped Al are listed in Table 1. The energy gap decreased greatly at 3.098 eV on sample with Al doping of 8% as shown in Figure 2.

Table 1. Optical bandgap values of ZnO with variation of dopped Al

Thin films	Band gap (eV)
ZnO:Al 0%	3.347
ZnO:Al 2%	3.099
ZnO:Al 4%	3.123
ZnO:Al 6%	3.109
ZnO:Al 8%	3.098

4. Conclusions

A thin film of a double layer of ZnO:Al successfully formed and the energy gap was calculated based on the data shown in figure 1, 2 and table 1. After the doping, the value of the energy gap decreased from 3,347 eV into 3,098 eV. The most significant drop occurred in the sample with 8%. Smaller energy gap is needed to accelerate the process of electron transitions that is good for photocatalyst applications. The highest peak of the absorption spectrum lies at a wavelength of 368 nm is the sample with 6% Al doping. It shows the concentration of the sample absorption in the UV region. Transmittance value for samples with doping is 79-91% in range.

References

[1] Baruah S, Pal S K, and Dutta, J 2012 *Nanoscience and Nanotechnology-Asia* 90-102.
[2] Vishwas M, Rao N K, Phani A R, Gowda K V A, Chakradhar R P S 2011 *Solid State Communications* **152** 324-327.
[3] Chen L, Ta T, Thuy T, Huanga C, Li J, Yuana L, and Caia Q 2013 *Appl. Surf. Sci.* **273** pp 82-88.
[4] Mahadik M A, Shinde S S, Hunge Y M, Mohite V S, Khumbar S S, Moholkar A V, and Rajpure K Y 2014 *Journal of Alloys and Compounds* 446-451.
[5] Kang L, Gao Y, Luo H, Wang J, Zhu B, Zhang Z, Du J, Kanehira M, Zhang Y *Solar Energy Material & Solar Cells* **95** 3189-3194.
[6] Zheng L, Dong Y, Bian H, Lee C, Lu J, Li Y 2016 *Electrochimica ActaS00134686* 30848-9.

[7] Zi-qiang X, Hong D, Yan L, Hang C 2006 *Material Science in Semiconductor Processing* **9** 132-135.
[8] Prawara B 2006. *Rancang Bangun Thermal Spray Coating dengan Menggunakan Sistem Hight Velocity Oxygen Fuel.* Kegiatan: **4977.0127**.
[9] Liu X, Pan K, Li W, Hu D, Liu S, and Yude W 2014 *Ceramics International* **40** 9931–9939.
[10] Mohaghegs A, Boudjardar S, Harmit S, and Guerbous L 2014 *Materials Letters* 248-251.

Effect of co doping to the optical properties of ZnO:Co Thin films deposited on glass substrate by sol-gel spray coating technique

Sinta Marito Siagian, Heri Sutanto, Anes Permatasari

Physics Department of Post Graduate Program Diponegoro University,
Jalan Prof. Soedarto, Tembalang, Kota Semarang, Jawa Tengah 50275, INDONESIA

E-mail: sintasiagian@st.fisika.undip.ac.id

Abstract. This study aims are to analyze the effect of cobalt doping concentration to the optical properties ZnO:Co thin films deposited by using sol-gel spray coating technique. Deposition of ZnO and ZnO:Co thin films has been successfully formed on a glass substrate using sol-gel method of spray coating technique with a variation of doping Co as much as 3%, 5%, 7%, 9%, and 11%. Solution of ZnO:Co synthesized by mixing zinc acetate dehydrate (Zn $(COOCH_3)_2.2H_2O$ into a isopropanol $((CH_3)_2CHOH)$ then monoethanolamine (MEA) and cobalt nitrate are added as a dopant at room temperature, then ZnO:Co sprayed on a glass substrate that has been heated at a temperature of 450°C. A thin film further characterized using spectrophotometer Uv-Vis to determine the value of absorbance and transmittance. Energy band gap was determined by using tauc plot that uses the absorbance value. The results show that there has been a difference in the value of the Energy band gap of ZnO with and without of Co doping. The addition of Co doping causes the decrease of small energy band gap. The magnitude of the energy band gap of thin films of ZnO is 3.337 eV and ZnO:Co 3% is 3.129 eV. Thin film that has narrower band gap was applied for material photocatalyst

1. Introduction

Semiconductors are widely used to study the photocatalysts are metal oxides such as TiO2, WO3, ZnO, CDO, In2O3, CdS, CdSe, MoS2 and WS2 [1]. Among the wide range of semiconductors, ZnO has attracted the attention of researchers because of the good chemical stability, non-toxic and low cost. Moreover, ZnO can absorb most of the solar spectrum, making it perfect as a photocatalyst [2]. ZnO is an n-type semiconductor material with a band gap energy of 3.37 eV and a binding energy of 60 MeV. Narrow band gap is expected to be used as an important application is a photocatalyst [3]. Therefore, changes because of the Co doping played an important role in the photocatalysts. Cobalt (Co) is a transition metal that can be used as a dopant in ZnO having a value Eg 2.4 eV smaller than ZnO.

The Eg value corresponding to the energy at a wavelength of visible light, so it is expected the presence of Co doping on ZnO Eg value will be smaller. Various methods have been used by researchers in the deposition of a thin layer of ZnO: Co is spray pyrolysis [4], sol-gel [6] and co-precipitation [5]. Among these techniques, sol-gel method has advantages such as low cost deposition technique is simple [3], is excellent in controlling the composition, having homogeneity which is high on the molecular level, lowering the temperature of crystallization, and are capable of producing a thin layer ZnO surface area is very spacious compared to other methods [7]. One method that is more suitable for the manufacture of thin layers is spray coating. Mechanical spray coating is a process in which the coating material (feedstock) as individual particles driven by the flow of pressurized gas to a surface (substrate).

The particles hit the substrate, paste, and forming a thin surface appropriate [8]. Advantages of spray coating technique is the deposition rate is high, it can be done on the atmospheric conditions, as well as environmentally friendly. This study will be conducted by making a ZnO:Co thin films with sol-gel method that is deposited on a glass substrate by thermal spray coating. Further research also studied the effect of the concentration of CO on ZnO to its optical properties such as transmittance, absorbance and band gap.

2. Method

Deposition of ZnO and ZnO:Co thin films has been successfully formed on a glass substrate using sol-gel method of spray coating technique with a variation of Co doping as much as 3%, 5%, 7%, 9%, and 11%. Solution ZnO:Co synthesized by mixing zinc acetate dehydrate (Zn $(COOCH_3)_2.2H_2O$ into a isopropanol ($(CH_3)_2CHOH$) then monoethanolamine (MEA) and are added cobalt nitrate as a dopant at room temperature synthesized. The solution was stirred using a magnetic stirrer hot plate at room temperature for 45 minutes to obtain a solution that is clear and homogeneous, then ZnO:Co sprayed on a glass substrate that has been heated at a temperature of 450°C.

3. Results and Discussion

Testing characteristics of the optical properties of ZnO:Co thin films were performed using an UV-Vis spectrophotometer in the wavelength range between 300 nm to 800 nm. The data obtained from this test are processed to obtain the value of the energy band gap of each sample, which can be seen the effect of adding a dopant Co based on its concentration, respectively. To determine band gap can use tauc plot. The energy band gap of ZnO:Co plot the data obtained through direct transition absorption equation. Plot $(\alpha h\upsilon)^2$ vs $h\upsilon$ by extrapolating the linear part of the curve to zero absorption lines give the energy band gap for direct transitions shown in Figure 1.

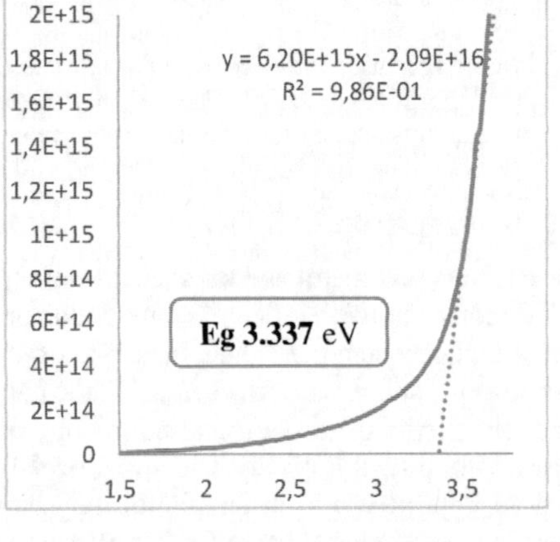

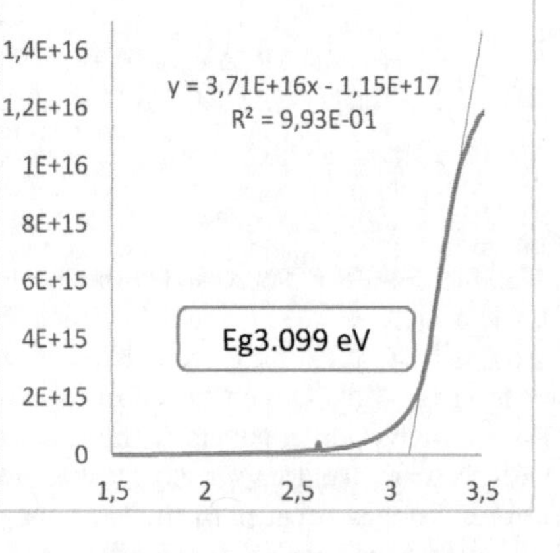

Figure 1. Plot $(\alpha h\upsilon)^2$ vs $h\upsilon$ for undoped **Figure 2.** Plot $(\alpha h\upsilon)^2$ vs $h\upsilon$ for ZnO:Co 11%

The excitation of electrons from the valence band to the conduction band need an energy, energy for excitation through the band gap is the one who called the band gap energy. Cobalt has Eg 2.4 eV smaller than ZnO and these values correspond to the wavelength of visible light energy, so that with the addition of Co dopants Eg value will be small.

Table 1. Values energy band gap undoped and ZnO:Co

Thin films	Band gap (eV)
ZnO:Co (0%)	3.337
ZnO:Co (3%)	3.129
ZnO:Co (5%)	3.125
ZnO:Co (7%)	3.117
ZnO:Co (9%)	3.108
ZnO:Co (10%)	3.099

Based on the data above was obtained that the higher the concentration of the dopant cobalt energy band gap is getting narrower, with a maximum reduction currently on ZnO: Co (11%) which is 3.099 eV. Narrow band gap is expected to be used for photocatalytic applications [3]. The decline in energy band gap because of their long shifts hypercromic wave in an optical absorption of ZnO [9]. The energy band gap shrinks the less photon energy required to excite the electron. The electrons are generated that has the important role in the application of photocatalysts.

Absorbsivitas molar is characteristic of a substance that informs how much light is absorbed by the dye molecules at specific wavelengths. The greater the value of molar absorptivity of a substance, the more light is absorbed by it, or in other words, the greater the absorbance value.

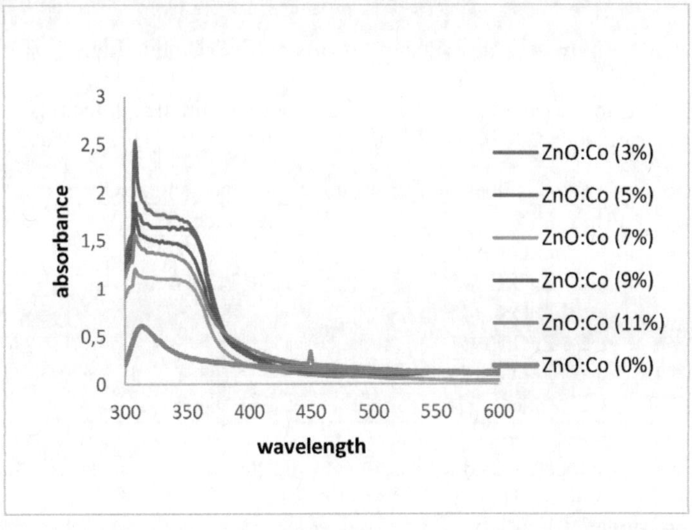

Figure 3. absorption spectra of ZnO and ZnO:Co thin films with different concentrations

The figure shows the absorbance spectrum of ZnO:Co thin films with different variation of dopan concentration. value is measured using a spectrophotometer obtained without doping 0.2090 at a wavelength of 350 nm, for the increase in doping 3% obtained value of its absorbance was 1.2757 at a wavelength of 362 nm, to 5% obtained value 1.1938 in the wavelength range of 360 nm, for the addition of dopants 7% obtained value 0.9065 in the wavelength range 352 nm, while the concentration of 9% while the observed value is approximately 1.1307 nm, and for the highest concentration of 11% was obtained abosrbansi its value is 1.1135 at a wavelength of 350. The different wavelength of absorption caused by increasing dense and regular arrangement of atoms which resulted in the higher value. The highest absorbance peaks were observed in the spectrum at maximum wavelength of 362 nm in exist the blue-shift area, these materials have characteristics could absorb at the wavelengths in this area. Absorbance values were high due to the number of atoms involved in the process of absorption of the light beam with a maximum absorbance value of 1.2757 at a wavelength of 362 nm to a concentration of 3%. This shows that the concentration of the sample can absorb these higher UV rays. The higher value of the material absorption and more light is absorbed. Transmittance of ZnO:Co thin films can be shown in the figure below

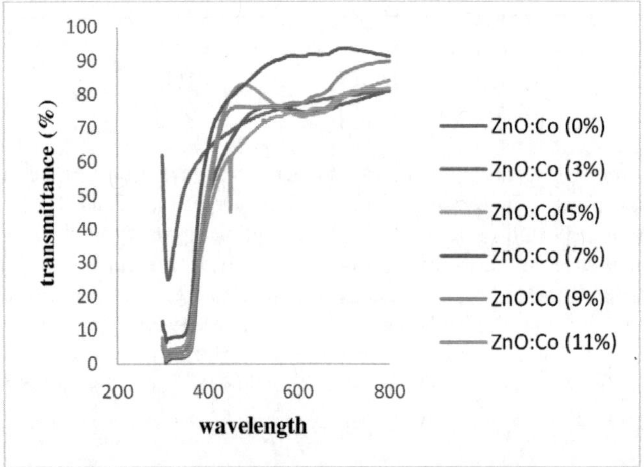

Figure 4. Transmittance Uv-Vis of ZnO:Co thin films

Based on the picture above can be seen that the highest transmittance in the region of a wavelength of 466 nm with a transmittance value is 83.5 % at a concentration 7% is higher than 70% ZnO value.

Table 2. Values of Transmittance undoped and ZnO:Co

Thin Film	Transmittance (%)
ZnO:Co (0%)	68.5 %
ZnO:Co (3%)	74.5 %
ZnO:Co (5%)	83 %
ZnO:Co (7%)	83.5 %
ZnO:Co (9%)	76.3 %
ZnO:Co (10%)	70.6 %

The transmitttance is expected to decrease due to the three factors : oxygen deficiency, surface roughness and impurity centers. Thus, these transitions suggest that Co^{2+} ions are substituted at the Zn^{2+} ions in the zno lattice [10].

4. Conclusions

Based on the data was obtained that the higher the concentration of the dopant cobalt energy band gap is getting narrower, with a maximum reduction currently on ZnO:Co (11%) which is 3.099 eV. The energy band gap shrinks the less photon energy required to excite the electron. The electrons are generated that has the important role in the application of photocatalysts. Absorbance values were high due to the number of atoms involved in the process of absorption of the light beam with a maximum absorbance value of 1.2757 at a wavelength of 362 nm to a concentration of 3%, This shows that the concentration of the sample can absorb these higher UV rays.

References

[1] Djarwanti., Cholid, S., Yuniati, A., 2009, journal of research industry 3 (2), 109-117.
[2] Kaur J, Singhal S, 2013, *Ceramics International* **40**, 7417–7424.
[3] Poongodi G, Anandan P, Kumar R M, Jayavel R, 2015, *Molecular and Biomolecular*

Spectroscopy **148**, 237–243.

[4] Benramache S, Temam H B, Arif A, Guettaf A, Belahssen O, 2014, *Optik* **125**, 1816-1820

[5] Kumar S, Song T K, Gautam S, Chae, K H, Kim S S, Jang K W, 2015, Material research **66**, 76-82.

[6] Ashraf R, Riaz S, Bashir M, Naseem S, 2013, *Advance in nano biomechanics robotics and energy research (ANBRE 13).*

[7] Efendi N H, Sutanto H, 2014, *Youngster Physic Journal* 3 (**2**), 135-142.

[8] Surono, A T, Sutanto, 2014, *Youngster Physics Journal* 2 (**1**), 7- 14.

[9] Ullah R, Dutta J, 2008, *Journal of Hazardous Materials* **156**, 194–200.

[10] Rajendar V, Dayakar T, Shobhan K, Srikanth I, Venkateswara R, 2014, *Superlattices and Microstructures* **75,** 551-563

Characteristic and quality control test in sector collimator gamma knife perfexion at Siloam hospital

Jhon Wesly Manik, Eko Hidayanto and Heri Sutanto

Physics Department of Post Graduate Program Diponegoro University,
Jl. Prof. Soedarto, Tembalang Kota Semarang Jawa Tengah 50275, INDONESIA

E-mail: jhon.wesly@st.fisika.undip.ac.id

Abstract. In this study conducted to evaluate the collimator 4 mm, 8 mm, 16 mm to determine the level of precision Gamma Knife Perfexion there three years and eight months has not carried out measurements after the first year and verify the alignment of sector 1-8. Measurement with three axes (x-axis, y-axis, z-axis) using the film Gafcromic EBT-3, which will be signaled to a central point in the film to be measured and given a dose of 5 Gy for 10 minutes and then scanned and analyzed using ImageJ softwere 1.50 and dose rate in film comparison with measurent dose rate using electrometer. Alignment of Patient-Positioning System (PPS) with Radiation Focal Point (RFP) determination of quality control testing each collimator 4 mm in the standard test with < 0.1 and > 0.4 mm passive voice AAPM 54. The results of the standard are still in conformity item 0.21 mm for 4 mm and corelasi dose rate film and measurent using electrometer collimator 4 mm = 0.965, 8 mm = 0.964, 16 mm = 0.959.

1. Introduction

Stereotactic Radiosurgery is the best option in brain disorder treatment by one-day session dosage addressing. The archieve focus passive voice into something of the brain to cure disorder such as the tumor or functional disorder .One of the Stereotactic Radiosurgery device to cure tumor or functional disorder is Gamma Knife with its fifth generation called Gamma Knife Perfection. This Gamma Knife Perfection is better for it can focus the source of radiation to a tumor or functional disorder with 0,15 mm accuracy. Gamma Knife Perfection treatment is available in Indonesia since 2012. However, the cost is very expensive (Petti L Paula,2006).

Leksell Gamma Knife Perfection is produced on 2006, which is redesigned based on previous model (B and C with 201 Collimator). Gamma Knife Perfection has 192 collimator and Cobalt-60 source which is aligned with the collimation system. Both levels of dosage addressed through every collimator and biological effect passed by is small however, 192 collimator will lead them to focus point and will have high dosage. This radiation is called Gamma Knife Perfection radiation technique which is beneficial for normal tissue.

To point out to focus point, Gamma Knife Perfection has the stereotactic frame to achieve high precision. This stereotactic frame is put on patient head, and then the referential Cartesian coordinate known as Leksell coordinate system is set up. Patient has the an image of skeleton which attached by one of the equipment according to clinical MRI ½ Tesla, CT-Scan, DSA. These images are used by treatment planning system to find target in the coordinate chamber. Once the target's coordinate has been found, the patient can somehow position into a unit to make the target right into radiation focus point.

Each cobalt-60 source is attached on each sector and moved by a motor. Each sector has collimator sized 4 mm, 8 mm, and 16 mm and each sector can be blocked or sealed (Maitz AH, et). With the advancement of collimator size precision, PPS (Positioning Patient System) and geometrical, on Gamma Knife Perfection (where the Cobalt=60 source collimator system is undetermined) and Patient Positioning System (where the frame attached to the treatment table), and the bed can move to one position to another, the head and neck will not change during treatment.

During treatment, eight sector can be moved independently on collimator tungsten ring for various exposure position in order to provide 4, 8, and 16 mm archive size or the combination of the mentioned archive size. Additionally, some sector can be blocked which mean the sources of sector do not align with one of the collimator holes. Further detail of the design of LGK PFX, Dosimetric Characteristic, radiation safety aspect, and clinical use are for evaluating regularity of individual sector output.

This research is conducted in order to evaluate Gamma Knife Perfexion which has not been done for 3,5 years and to verify beam alignment for all collimator (4mm, 8mm, 16mm) along with X-axis, Y-Axis, and Z-axis. It is done by phantom which use Gafronic film that radiated (x Gy), scanned, and analyzed using *software image j 1.50* and quality control test determination for each colimator 4mm, 8mm, 16mm which is tested by standart 0,1 and < 1,0% (Novotny, 2008). The determination of quality control accuration of output factor is recomended by AAPM (American Association of Physicist in Medicine) 54.

2. Method

A special film holder , made with very narrow geometric tolerance, is aligned between with its center in the PPS calibration in the PPS calibration center poin. When the tool is aligned in the PPS, the tip of a sharp needle, located in the tool, exactly points towords the PPS calibration center poin. Just prior to exposure a small piece of radiological film, also located in the tool, is pierced by the tip of the needly. During exposure the film plane coincides with the RFP. Two film are consecutively exposed, one with it's surface oriented perpendicularyrelative to the symmetry axis of the source distribution. The second film is rotated through 900 with respect to the first film.image intensity profiles are scaned by means of an automatic densitometer in three mutually perpendicular direction. the geometrical resolution of the flatbed scanner has been chooseen close to 0.05 mm and the dynamic range was set to 16 bits. the obtained intensity profiles include the shift in intensity caused by the small hole pierced into the film. by measuring the asymmetry of the position of the hole in relation to the density distribution at approximately FWHM, the accuracy of the leksel gamma knife unit is determined.

3. Results and Discussion

3.1. Measurement accuracy stabilitas

The film image measurement gafcromic Collimator 4 mm of XZ axis high and FWHM values are generated through graphs, Figures below.

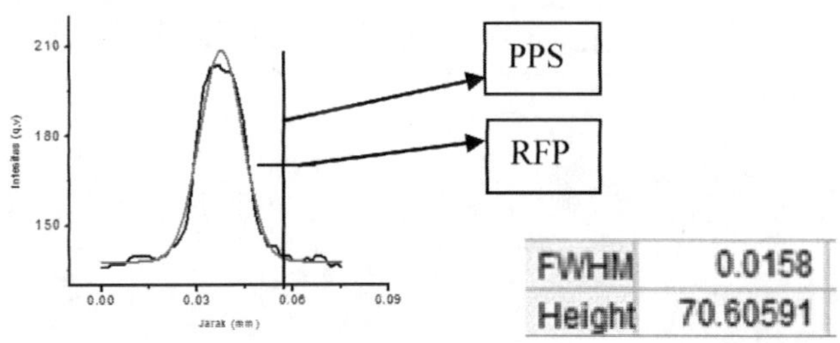

Figure 1. Result collimator 4 mm XZ

The film image measurement gafcromic Collimator 4 mm of YZ axis high and FWHM values are generated through graphs, Figures below.

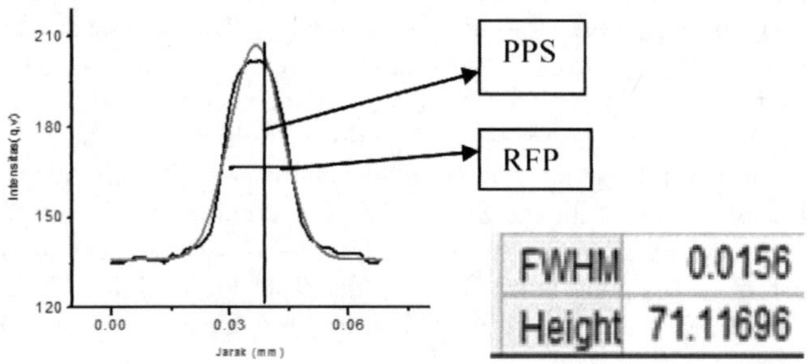

Figure 2. Result collimator 4 mm YZ

Table 1. Result measured image of film gafcromic collimator 4 mm sumbu X, Y, Z

No	ΔX	ΔY	ΔZ
1	-0.050	-0.036	-0.242
2	-0.021	0.032	-0.193
3	-0.079	0.026	-0.103
4	0	0	-0.22
5	0	0	-0.192
6	0	0	-0.255
Mean Deviation	-0.051	0.004	-0.201

$\Delta r = \sqrt{\Delta X^2 + \Delta Y^2 + \Delta Z^2}$

$\Delta r = \sqrt{(-0.051)^2} + (-0.004)^2 + - (0.201)^2$

$\Delta r = 0.21$ mm

The measured distance,i, e. the accuracy PPS and RFP of the Gamma knife, is well within the specifications (0.4 mm). We hereby cerctify that the accuracy of the above Leksel Gamma Knife is in accordance with the Gamma Knife.

3.2. Correlation dose rate film and dose rate electrometer

The dose rate calculation equation

$$D_{DW} = KTP. N_{D_1W} .M$$

Table 2. Comparison dose rate film Gafcromic with calculasion Dose rate Electrometer collimator 4 mm.

Dose rate collimator 4 mm (Gray/ menit)	
Film Gafcromic	Electrometer
0.851	0.849
0.848	0.846
0.844	0.841
0.859	0.854
0.865	0.862
0.85	0.848
0.846	0.846

0.855	0.853

The result measurement film Gafcromic and electrometer will be correlated to view the linearity of the dose rate.

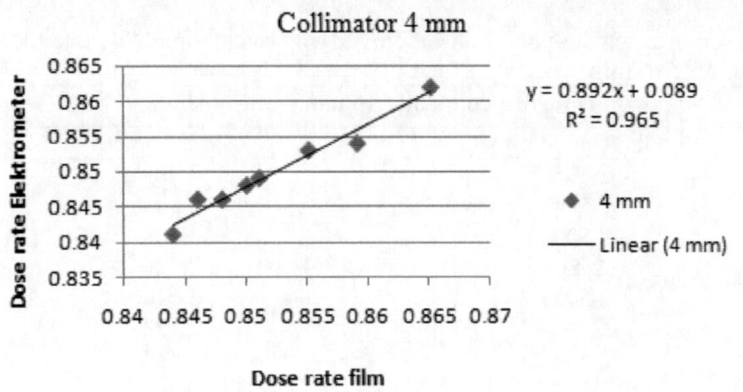

Figure 3. Corelasion dose rate film with Electrometer collimator 4 mm

The correlation results gafcromic the film dose rate and dose rate electrometer shows the correlation values $R^2 = 0.965$ for dose rate collimator 4 mm sector 1 - 8.

Table 3. Comparison dose rate film Gafcromic with calculasion Dose rate Electrometer collimator 8 mm.

Dose rate collimator 8 mm (Gray/ menit)	
Film Gafcromic	Electrometer
1.983	1.984
1.95	1.934
1.89	1.887
1.915	1.897
1.911	1.909
1.898	1.888
1.906	1.903
1.97	1.964

The result measurement film Gafcromic and electrometer will be correlated to view the linearity of the dose rate.

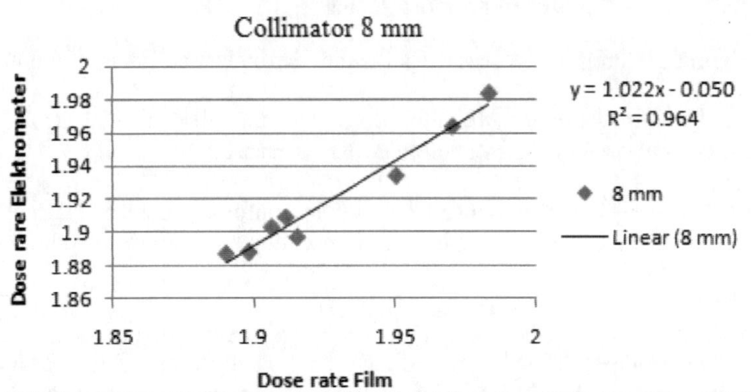

Figure 3. Corelasion dose rate film with Electrometer collimator 8 mm

The correlation results gafcromic the film dose rate and dose rate electrometer shows the correlation values R2 = 0.964 for dose rate collimator 8 mm sector 1 - 8.

Table 4. Comparison dose rate film Gafcromic with calculasion Dose rate Electrometer collimator 16 mm.

Dose rate collimator 16 mm (Gray/ menit)	
Film Gafcromic	Electrometer
2.495	2.483
2.504	2.505
2.503	2.503
2.53	2.526
2.545	2.546
2.53	2.528
2.515	2.514
2.512	2.51

The result measurement film Gafcromic and electrometer will be correlated to view the linearity of the dose rate.

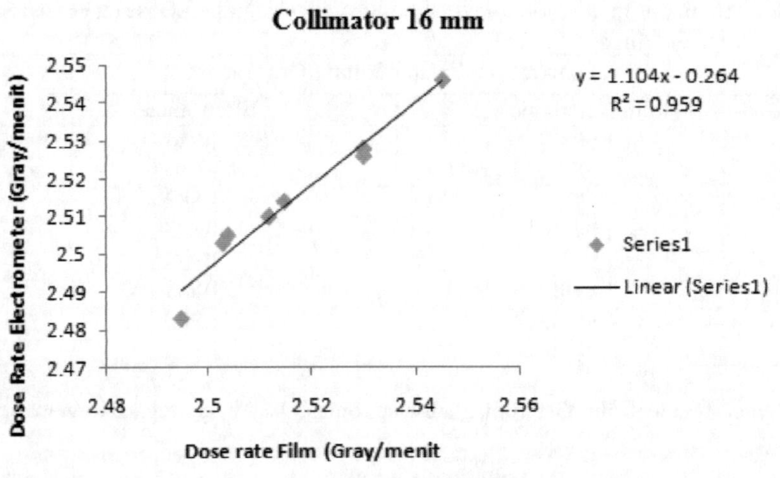

Figure 3. Corelasion dose rate film with Electrometer collimator 16 mm

The correlation results gafcromic the film dose rate and dose rate electrometer shows the correlation values R2 = 0.964 for dose rate collimator 16 mm sector 1 - 8.

Conclusion : The results of the standard are still in conformity item 0.21 mm for 4 mm and corelasi dose rate film and measurent using electrometer collimator 4 mm = 0.965, 8 mm = 0.964, 16 mm = 0.959.

References

[1] American Association of Phisicists in Medicine. 1995. Comprehensive QA for Radiation Oncology; Report of AAPM Radiation Therapy Committee Task Group-54, Medical Physics. 21
[2] Denina, anna, 2010. cobalt-ku.17-7-2016. annadenina.wordpress.com/2010/08/14/cobalt-ku/
[3] Devic S. 2011. Radiochromic film dosimetry: past, present, and future. Phys Med;27:122–34.

[4] Flickinger, John C. 1990. Treatment Planning For Gamma Knife Radiosurgery With Multiple Isocenter. University of Pittsburgh School of Medicine,
[5] Kodrat, henry - Rima Noviriyanthy. 2015. Stereotactic Radiosurgery pada Benign Skul., banda aceh.Departement Radiologi Fakultas Kedokteran Universitas Syiah Kuala
[6] Leksell gamma knife perfexion.2006, system description. Art no 1002703. Stockholm: Elekta, AB;
[7] Low DA, Harms WB, Mutic S, Purdy JA. 1998. A technique for the quantitative evaluation of dose distributions. Med Phys ;25:656–61.
[8] Ma L, Verhey L, Chuang C, et al. 2008 Effect of composite sector collimation on average dose fall-off for gamma knife perfexion. J Neurosurg;109(Suppl.):15-20.
[9] Novotny Jr, Josef dkk, 2008. Dosimetric comparison of the Leksell Gamma Knife Perfexion and 4C. Department of Radiation Oncology, University of Pittsburgh Cancer Institute,17- 7-2016
[10] Petti L Paula,2006,leksel gamma knife perfexion QA Consideration.Taylor McAdam Bell Neuroscience Institute Washinton Hospital Healtcare System Fremon,CA.
[11] Pil Chung,jae, dkk. 2016. An effective calibration technique for radiochromic films using a single-shot dose distribution in Gamma Knife. Department of Medical Physics, Korea University of Science and Technology
[12] Quality Assurace, Radiophysical Part Leksell Gamma Knife 6121,2013. Siloam Hospital Lippo Village
[13] Wu A, Lindner G, Maitz AH, et al. 1990. Physics of gamma knife approach on convergent beams in stereotactic radiosurgery. Int J Radiat Oncol Biol Phys ;18(4):941-9.

IOP Conf. Series: Journal of Physics: Conf. Series **795** (2017) 012011 doi:10.1088/1742-6596/795/1/012011

1-Dimension magnetotelluric data inversion using MOEA/D algorithm

Pramudiana[1], Sungkono[2]

[1,2]Physics Department, Faculty of Mathematic and Natural Sciences, Institut Teknologi Sepuluh Nopember
Jl. Arief Rahman Hakim, Surabaya 60111 Indonesia;

E-mail: [1]pramudyana28@gmail.com ; [2]hening_1@physics.its.ac.id

Abstract. Magnetotelluric (MT) data is used to derive resistivity imaging of subsurface. The subsurface resistivity is obtained by inversion of MT data. Generally, MT data contains two parts, namely: apparent resistivity and phase or real and imaginary parts. Inversion of MT data for reconstructing resistivity value of each layer is to minimize single objective (combination two parameters MT data) which used global or local optimization method. Nerveless, single objective optimization method has several disadvantages, such as; (1) weight value to combine two parameters of MT data is needed, where this weigh value depend on the amplitude of both MT data; (2) there is no validation of the inversion results. In this research, Inversion MT data to estimate 1D resistivity of subsurface uses multi-objective evolutionary algorithm based on decomposition (MOEA/D)to minimize root mean square error (RMSE) of calculated and observed data for apparent resistivity and phase data simultaneously. The algorithm has applied to synthetic and field data. This result shows that MOEA/D algorithm is robust and accurate to determine subsurface resistivity and lithology.

1. Introduction

Magnetotelluric is one of passive geophysical methods which utilizes natural electromagnetic field to understand physical parameters distribution within a subsurface. The magnetotelluric method uses Maxwell law as the basic concept as defined in **[1]**, **[2]**. The observation of this technique is magnetic and electric fields, further analyzed to estimate apparent resistivity and phase or real and imaginary parts.

Forward modeling of one dimension magnetotelluric (MT 1D) can be solved by the analytic method or using recursion equation **[1]**. Forward modeling of MT 1D is quite simply, a correlation between data and parameter model is not linear. However, non-linier inversion using linearization is needed initial value which is close to the solution value. Besides, the solution of linearization method is often trapped in local minimum, where the optimum solution of non-linear inversion always associates with global minimum.

On the other hand, a global minimum can be found by stochastic method. Meanwhile, there are two kinds of the stochastic methods, namely: finding an optimum solution using single objective and multi-objective approaches. Considering that MT 1D has two sorts data, i.e. apparent resistivity and phase or complex number. Thus, optimum solution of MT 1D can be found by minimizing both of them. Minimizing two variable using single objective is probably trapped in local minimum

[3].Whereas, the weight factor is needed. Furthermore, weight factor in every data is different each other. To get the best of weight factor for each data should be done trial and error. In addition, using single objective causes difficult to validate the model with the true geological condition **[4], [5]**. Consequently, Dal Moro and Colleagues suggest that inversion process using more than two variables should use a multi-objective method. One of the best multi-objective methods is multi-objective evolution algorithm based on decomposition (MOEA/D) **[6]**.

In the last decades, MOEA/D is more efficiency and effective than other multi-objective algorithms which base on non-dominating sorting. In this research, MT 1D is inverted using MOEA/D to estimate subsurface resistivity in a way minimizing both of objective functions, namely; the error of observed and calculated data for phase and apparent resistivity. This algorithm has been tested on synthetic and observed data.

2. Methods

MOEA/D purposes to re-decompose multi-objective problem becoming several sub-problem. Generally, this method is started by input parameter and defining search space. Furthermore, the number of parameters has the closest relation. Wider search is needed more individual to evaluate and also time-consuming. The next step is generating initial population using one of the methods which has been chosen, like Weighed sum or Tchebycheff approaches. Then, evaluating objective function, defining the non-dominant individual, reproduction, and cross over process are calculated [6]. Reproduction and cross over aim to avoid solution trapped in the local minimum (exploration process). In this paper, to invert synthetic data 250 populations, 250 maximum non-dominant archives, and 500 times iterations are needed. Meanwhile, 150 populations, 150 maximum non-dominant archives, and 500 iterations are needed for field data inversion.

2.1. Inversion of data synthetic

Data synthetic inversion proposes to know how performance (robust) MOEA/D algorithm for simultaneously MT 1D data inversion. This process is also to determine the reliability of MOEA/D. Conductive layer between the resistive layers of structure is applied as synthetic model in this research. Furthermore, two sorts of approaching layer which are implemented in this study are inversion using same and difference layers.

Forward modeling in this inversion process aims to obtain apparent resistivity and phase from model parameter. As a result, objective function can be calculated. Meanwhile, objective functions conducted in this inversion are root mean square error (RMSE) between true data and calculation data as described below

$$E = \sqrt{\frac{1}{N}\sum_{f}\left(\log 10(d_{obs}^{\rho_a} / d_{pred}^{\rho_a})\right)^2} \quad (1)$$

$$E = \sqrt{\frac{1}{N}\sum_{f}\left(d_{obs}^{\Phi} - d_{pred}^{\Phi}\right)^2} \quad (2)$$

Where $d_{obs}^{\rho_a}$ and $d_{pred}^{\rho_a}$ denote observed and calculated apparent resistivity data, respectively, d_{obs}^{Φ} and d_{pred}^{Φ} are observed and predicted phase data, respectively, N is number of frequency or period.

On the other hand, inversion result on the basis of the multi-objective algorithm is Pareto front model which shows both RMSE of MT data. It can be called Pareto optimum. All of individual non-dominant in Pareto front model (Pareto optimum) probably becomes a solution in this inversion. Thus selection optimum solution is needed. A method to determine a optimum solution is relative distance

IOP Conf. Series: Journal of Physics: Conf. Series **795** (2017) 012011 doi:10.1088/1742-6596/795/1/012011

considering both minimum and maximum values in for each objective function through normalization approach [3]. Mathematically, it is given by eqs. (3) and (4):

$$f_{xi}(j) = \frac{f_i(j)-\min(f_i)}{\max(f_i)-\min(f_i)} \quad i=1,2; j=1,1\ldots Npop \quad (3)$$

$$D = \sqrt{fx_1^2 - fx_2^2} \quad (4)$$

Where f_{xi} and D denote normalization of each objective function and distances between utopia point and Pareto front. Optimum model is identified by the lowest of *D*.

2.2. *Field data inversion*

Field data inversion is done to test the reliability of MOEA/D in field MT 1D data. Sasaki field data [13;14] at first station is used. Inversion process which is tackled in this field data is similar to invert process in the synthetic data described before.

3. Results and Discussion

3.1. *Inversion of Synthetic Data*

First of all, the first step is inversion using the same of true layer. Previously, forward modeling is done to obtain synthetic. Skin depth is calculated to know search space which is applied in the inversion process. In this synthetic data, inversion uses the search space: 1000 and 5000 meter for maximum thickness, 100 and 500 meter for minimum thickness; 500, 1500, and 100 Ωm for maximum resistivity; and also 10, 500, and 1 Ωm for minimum resistivity. Second inversion, inversion synthetic data using difference layer approaching (5 layers) is also done. It proposes to know the influence of defining layer in the inversion results. Search spaces is used in this inversion are 1000, 500, 5000, and 100 meter for maximum thicknesses; 10, 100, 500, 50 meter for minimum thicknesses; 500, 1000, 2000, 1000, and 100 Ωm for maximum resistivity; and also 5, 50, 100, 10, and 1 Ωm for minimum resistivity.

Fig. 1 is the result of synthetic data inversion using three layers which is done by MOEA/D algorithm simultaneously. Fig 1a and 1b are comparing between calculated data (blue) and observed data (red) for apparent resistivity and phase, respectively. Therefore, Fig 1c demonstrates Pareto front model with all of the probable solution of individual non-dominant. Based on this Pareto front model, it can be seen that resistivity and phase error about 0 – $3x10^{-3}$ and 0-$2.5x10^{-3}$ particularly. Fig 1d is comparing between subsurface modeling for the true and inverted model. The figure shows that both of subsurface models are close.

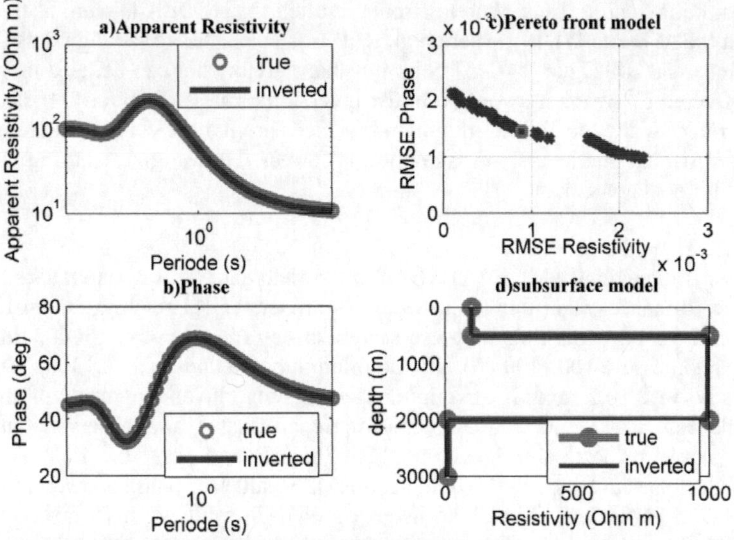

Figure 1 Inversion result of synthetic data using three layers.

Inversion result using five layers approaching is shown by Fig 2. Figs. 2a and 2b illustrate the comparison between observed or field and calculated for apparent resistivity and phase, respectively. The figures show that both the data are very close. Fig.2c illustrates Pareto front model (RMSE for both data) which shows that apparent resistivity has RMSE about 0–0.15x10^{-3}, while phase has RMSE around 0-0.08.Fig 2d is subsurface modeling from true model and inverted model using 5 layers. This figure shows both of data looked similar or close. Even though the second layer shows slightly difference and it does not give significantly influence. Consequently, it can be seen that MT data inversion which is conducted by the higher number of layer than the true layer can obtain subsurface modeling which is quite similar to the true model.

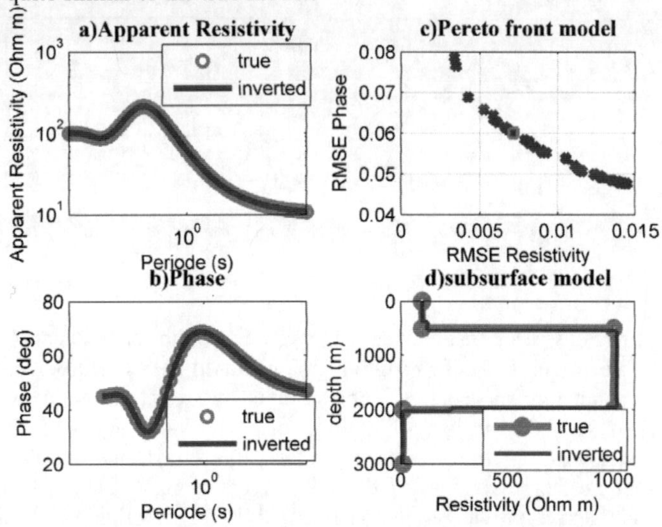

Figure 2 Inversion result of synthetic data using five layers.

Pareto front models for inversion synthetic data MT as described above demonstrate the low number (Figs. 1c and 2c). The low of RMSE for both data is owing to the fact that MT data inverting is free noise data (synthetic data). Thus, Pareto front resulted by MOEA/D can also estimate noise contaminated in the MT data [7], [8], where high RMSE is correlated to the high noise and vice versa. Further, Dal Moro and colleagues [4], [5] point out that Pareto front can be used to validity resulted model which affected by wrong assumption in the inversion process. Generally, the correct inversion process is illustrated by the location of minimum distances around the center of Pareto front. Figs. 1c and 2c can be known that the inversion assumption is correct. Thus, Fig. 1d and Fig. 2d show that the true and inverted model are similar.

3.2. Inversion Field Data

This inversion aims to recognize MOEA/D Algorithm in field data. This research uses field data which applied to test performances algorithm by previous researchers [9], [10] in the station 1. This inversion utilizes four rock layers approaching, and also search spaces are given by: 15000, 10000, and 10000 for maximum thicknesses; 100, 1000, 1000 for minimum thicknesses; 50, 300, 100, and 100 for maximum resistivity; 2,10,2, and 1 for minimum resistivity. Inversion result of the field data is presented in the Fig. 3. Fig. 3a and 3b demonstrate calculated and observed data for apparent resistivity and phase, respectively. Therefore, Fig. 3c is Pareto front the field data which gives information about noise contained in the field data. RMSE value for phase and resistivity are about 2-2.08 and 0.05-0.065 particularly. Similar to invert problem in synthetic data, error value in phase is higher than resistivity error value owing to the fact that phase data is worse than resistivity data [9].

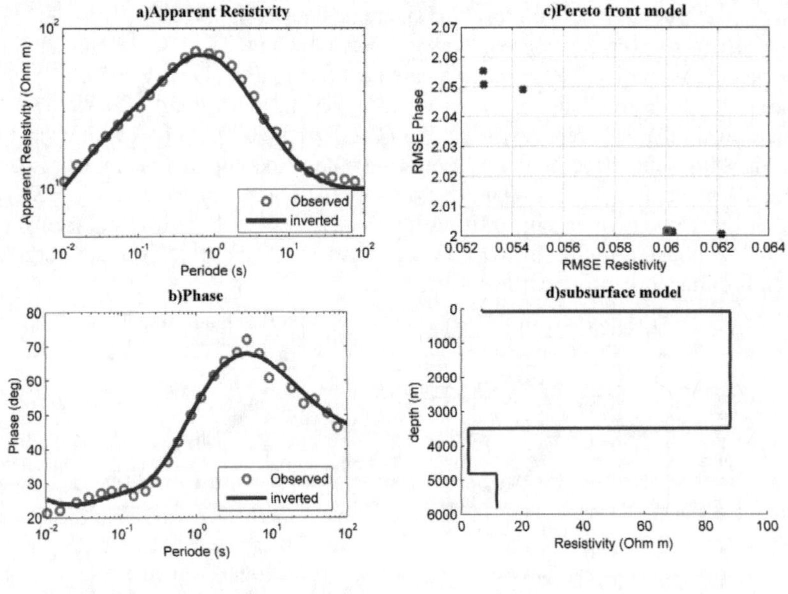

Figure 3 Inversion result using MOEA/D algorithm for field data [7].

Fig 3d shows subsurface modeling for the result inversion. Based on the figure, it can be seen that data in station 1 [10] has about four rock layers. The top layer has resistivity value about 10 Ωm in the shallowest depth. Then, second layer has the highest resistivity value about 95 Ωm. Meanwhile, the following layer has the lowest resistivity value. Therefore, fourth layer has higher value than previous layer just about 10 Ωm. Further, inversion MT data also invers using MT2DinvMatlab software [10] to determine 2D resistivity and to provide comparison resistivity model. Fig. 5 is resistivity 2D resulted by inversion process using MT2DinvMatlab. The figure shows that the first station is located in zero point. The figure demonstrates that the site contains three layers materials within subsurface. The first layer has the highest resistivity value. Meanwhile, the second layer has lower value than the last layer.

This result is similar trend to the inversion using MOEAD result which shows four layers with very thick layer in the first layer, which can be ignored.

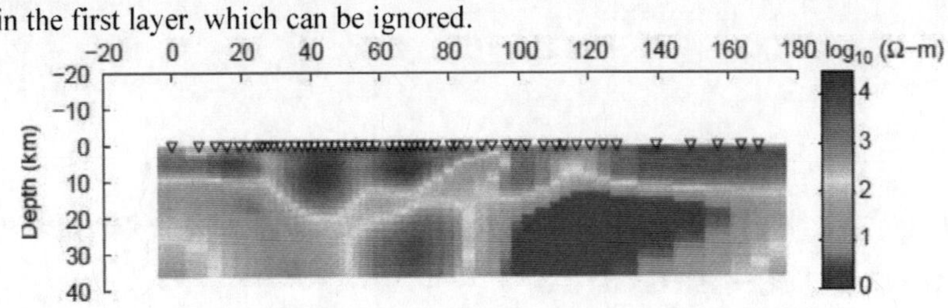

Figure 4. Joint inversion results of field data by MT2DinvMatlab software [10]

Conclusion

MT 1D data inversion to minimize RMSE between measured and calculated data for apparent resistivity and phase simultaneously can be solved by MOEA/D. Using the algorithm, MT 1D inversion do not need weight value in the objective function. Consequently, MOEA/D can be used to invert MT 1D data accurately, easily, and also fast for estimating subsurface resistivity which has been tested on synthetic and field data. Besides, inversion results also provide information about validity model of inversion result which can be described by Pareto front properties and also give information about uncertainty of model (owing to MOEA/D solution as much as populations which are used).

Acknowledgements

The authors would like to give thanks to Yarpiz (www.yarpiz.com) which has shared MOEA/D source code that was used in this research.

References

[1] H. Grandis, *Pengantar pemodelan inversi dalam geofisika*. Himpunan Ahli Geofisika Indonesia, 2009.

[2] F. Simpson and K. Bahr, *Practical magnetotellurics*. Cambridge, UK; New York: Cambridge University Press, 2005.

[3] Sungkono, "Inversi terpisah dan simultan dispersi gelombang Rayleigh dan horizontal-to-vertical spectral ratio menggunakan algoritma genetik," Institut Teknologi Sepuluh Nopember, Surabaya, 2011.

[4] G. Dal Moro, "VS and VP vertical profiling via joint inversion of Rayleigh waves and refraction travel times by means of bi-objective evolutionary algorithm," *J. Appl. Geophys.*, vol. 66, no. 1–2, pp. 15–24, Oktober 2008.

[5] G. Dal Moro and F. Ferigo, "Joint analysis of Rayleigh- and Love-wave dispersion: Issues, criteria and improvements," *J. Appl. Geophys.*, vol. 75, no. 3, pp. 573–589, Nov. 2011.

[6] Q. Zhang and H. Li, "MOEA/D: A Multiobjective Evolutionary Algorithm Based on Decomposition," *IEEE Trans. Evol. Comput.*, vol. 11, no. 6, pp. 712–731, Dec. 2007.

[7] M. Moorkamp, A. G. Jones, and S. Fishwick, "Joint inversion of receiver functions, surface wave dispersion, and magnetotelluric data," *J. Geophys. Res. Solid Earth*, vol. 115, no. B4, p. B04318, Apr. 2010.

[8] M. Moorkamp, A. G. Jones, and D. W. Eaton, "Joint inversion of teleseismic receiver functions and magnetotelluric data using a genetic algorithm: Are seismic velocities and electrical conductivities compatible?," *Geophys. Res. Lett.*, vol. 34, no. 16, p. L16311, Aug. 2007.

[9] Y. Sasaki, "Two- dimensional joint inversion of magnetotelluric and dipole- dipole resistivity data," *GEOPHYSICS*, vol. 54, no. 2, pp. 254–262, Feb. 1989.

[10] S. K. Lee, H. J. Kim, Y. Song, and C.-K. Lee, "MT2DInvMatlab—A program in MATLAB and FORTRAN for two-dimensional magnetotelluric inversion," *Comput. Geosci.*, vol. 35, no. 8, pp. 1722–1734, Aug. 2009.

Influence of water saturation on ultrasonic P-wave velocity in weakly compacted sandstone

Thaqibul Fikri Niyartama[1], Umar Fauzi[2] and Fatkhan[3]

[1] Physics Study Program, Universitas Islam Negeri Sunan Kalijaga Yogyakarta, Jl. MarsdaAdiSucipto No. 1 Yogyakarta 55281, INDONESIA

[2]Physics of Earth and Complex Systems, ITB, Jl. Ganesha 10. Bandung 40132, INDONESIA

[3]Geophysics Engineering, ITB Jl. Ganesha 10. Bandung 40132, INDONESIA

E-mail: thaqibul.niyartama@uin-suka.ac.id

Abstract. Laboratory measurements of Ultrasonic P-wave velocities were conducted in weakly compacted sandstone with varying degree of water saturations. We used P wave transducer at frequency 63 kHz and imbibition technique in order to study the influence of water saturation on the P-wave velocity. Our experiment showed that the P-wave velocity (Vp) was reduced significantly at the beginning of the imbibition process. The variations on travel times and the amplitude changes were detected at any degree of saturation. The first and second amplitude of P wave decreased as water saturation (S_w) increased in the range of 0.1 to 0.6 in B5 sample, the amplitude increased again afterward. The shifting peaks of the signal that indicated attenuation were also observed in the experimental.

1. Introduction

Accurate estimation of fluid saturation in oil reservoir requires a good understanding of the relationship between seismic signal and fluid saturation in porous medium. There are various types of rocks that make up oil reservoirs, one of which is weakly compacted sandstone. In this paper, the influence of fluid saturation of seismic signal has been investigated using water imbibition experiments in the Ngrayong sandstone. Ngrayong sandstone, which is clean quartzose sandstone, is taken from the outcrop of Ngrayong formation in Blora, Central Java [1]. Previous studies have reported the relation of velocity-saturation in the consolidated medium[2-5], and unconsolidated sand[6]. Changes of velocity due to saturation are complicated and depend on many parameters, such as poroelastic parameters and position of the patch in the rock [7][8].There are two theoretical bounds to explain the relation of velocity-saturation [9][10]. In the low frequency limit, rock bulk modulus and fluid bulk modulus are defined by the Gassmann-Wood equation and for the high frequency limit is defined by the Gassmann-Hill equation.

In order to get a deeper understanding of the influence of water saturation to the wave signal in rock reservoir, we use weakly compacted sandstone.Water saturation can be obtained by using the imbibition technique in the laboratory. Imbibition is the process of replacing non-wetting fluid with wetting fluid, followed by depressurization of a saturated sample in a porous medium [2][7].

2. Laboratory Experiments

The Experiments were conducted on four cylindrical samples of Ngrayong sandstones. Rock samples were dried in oven at 120 °C for 48 hours. Then, porosity values were measured from all samples using porosimeter. After the Rock properties of each sample were known (Table 1), imbibition process was performed on the variations of saturation degree. Imbibition was conducted by dropping water on the surface of rock samples with a volume of water according to the degree of saturation and rotated to make water distributed homogeneously in rock samples.

Table 1. Rock properties

Parameter	B2	B3	B5	B7
Porosity (%)	37.45	37.55	38	37.5
Bulk Density (g/cm^3)	2.06	2.44	2.57	2.1
Diameter (cm)	3.82	3.86	3.92	3.9
Length (cm)	4.29	4.2	2.95	3.7
Bulk Volum (cc)	39	32	22	39
Pore Volum (cc)	14.61	12.16	8.56	14.65
Mass (gr)	82.14	78.04	56.3	71.53

P-wave velocity measurements were accomplished using OYO sonic viewer model 5210 at frequency of 63 kHz in a direction parallel to the core axis. The output of ultrasonic waves was displayed on a digital oscilloscope (Tektronix TDS 2022C). Then the data and image could be stored in particular format. When mechanical impulses generated by the pulse of generator propagates along rock samples with length L (meter), and the travel time of pulse T (second), P wave velocity (V) of the propagating wave can be calculated by:

$$V = \frac{L}{t} \tag{1}$$

The errors obtained for travel time and sample height determination were < 3% and <1%, respectively, while for relative velocity value, <9%. Table 2 shows the XRF and SEM investigations of rock samples that mainly composed of silicon (Si), iron (Fe), and potassium (K). Silicon is present in the form of SiO_2, indicated quartz as the major constituent of sandstone. Ferrous iron is the constituent of smectite (nontronite) group, while the main constituents of clay minerals group are chlorite, biotite, carbonate (siderite, ankerite) and sulfide. Smectite clays may inflate when they contact with liquid [11].

Table 2. The chemical composition of Ngrayong sandstone

XRF		SEM and EDS
mineral	Percentage (%)	(mineral and compound)
Si	56.1	Quartz
K	6.52	Fe
Ca	3.54	Ti
Ti	2.77	Ca
V	0.09	K
Cr	0.11	Si
Mn	0,29	Al
Fe	30.3	Mg
Ni	0.12	Na
Cu	0.23	Smectite (Nontronite)

Physically, clay minerals can be distinguished from SEM results using backscattered SEM image CamScan CS-44 LB. Figure 1 (a) shows the presence of quartz mineral covered by nontronite.

Figure 1. Weekly compacted sandstone from the backscattered SEM (CamScan CS-44 LB). (a) quartz mineral that mostly covered by clay nontronite (b) Smectite Group (nontronite)

3. Results and Discussion

P-wave velocity (Vp) as a function of water saturation for all Ngrayong sample is shown in Figure 1. We could not achieve maximum water saturation from imbibition process, revealed that not all pores were filled by water. The measurements of Vp just before imbibition process (dry conditions) were relatively low (880 ms^{-1} to 1023 ms^{-1}), indicated the characteristic of weakly compacted sandstone.

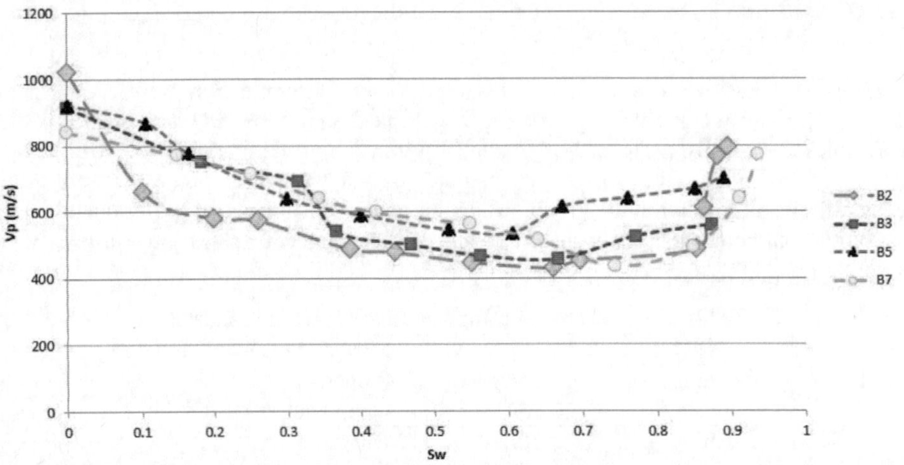

Figure 2. Velocity measurement of rock samples from imbibition experiment. P-wave velocity changes due to the influence of water imbibition on the weekly compacted rocks of Ngrayong sandstone.

When compared to the overall results, the decrease of Vp and the first peak amplitude (Figure 3) occured after the first stage of water imbibition was given to the rock sample. The increase of Vp when water saturation (Sw) decreasing until Sw < 0.7 is associated with the increase of sample density. At this stage, the results of water imbibition appeared in the pore, and gradually with the increase of imbibition, more pores were filled with water (70% pores) though there was still enough space in porous rock between the fluids. The remaining low frequency causes lack of pressure differences. So, there is enough time for the fluid pressure back to normal at one wave cycle. Vp increased gradually from Sw = 0.67 (sample B5); Sw = 0.74 (sample B2); Sw = 0.77 (sample B3), and 0.85 (sample B7). When water imbibition reached an average of Sw = 0.76 and provided uniform water distribution to almost all pores in the rock samples, the Vp increased. The volume of water in all pores contributes to reduce time for fluid pressure to return normally. P-wave frequency is quite high compared to the size of the patch so that the pressure between fluids in the pores does not reach equilibrium, resulting in the rock stiffness that lead to the increase of Vp. The increase of Vp is followed by the recovery of the first peak amplitude (Figure 3). Shifting peaks in Figure 3 indicate of signal attenuation from the seismic signal at an intermediate frequency.The first and second amplitude of P wave decreased as water saturation (S_w) increased at the range of 0.1 to 0.6 in B5 sample. The patches have irregular shape and randomly distributed. In addition, for the case of Ngrayong sample, the presence of smectite clay group can be one factor that lead to the trend of Vp lower than the limit of Gassmann Wood and Gassman Hill.

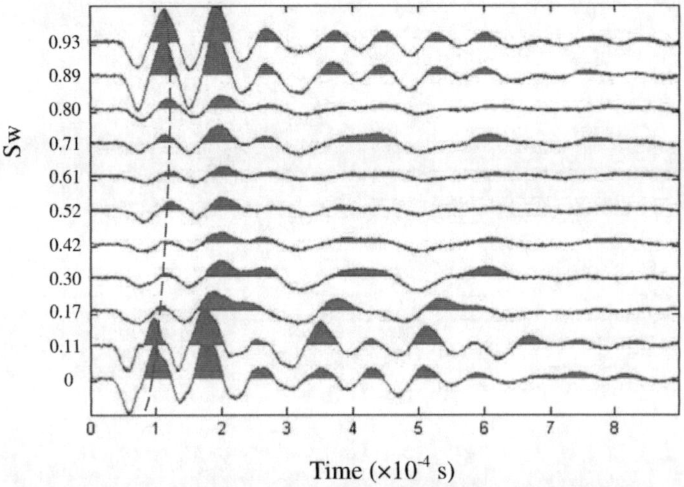

Figure 3. Waveform (B5 sample) recorded from different Sw with a dashed line indicates shifting in the first peak of wave.

4. Conclusion

The experimental results show that the saturation of weekly compacted rock can be detected and observed by the wave travel time and the amplitude of seismic signal. Nontronite, a member of smectite group, has been identified as a part of Ngrayong sandstone minerals. It seems that nontronite contribute to the decrease of Vp because water cause nontronite to swell. The swelling nontronite, then, push the mineral grain of quartz. When the contact between the grains of quartz mineral is weaker, the bulk modulus does not become larger.

Acknowledgement
We thank Tobias Muller and Claudio DellePiane for their contribution and several stimulating discussions in the interpretation section.

References

[1] Ardhana WD 1993 Indonesian Petroleum Association 22nd Annual Convention Proceedings (Jakarta) vol 1 (Tulsa: AAPG) 395-443
[2] Lopes S and Lebedev M 2012 Geophys. Prospect.**60** 572-80
[3] Lopes S, Lebedev M, Mueller T M, Clennell M B and Gurevich B 2014 Geophys. Prospect. **62** 1126-42

[4] Mavko G and Nolen-Hoeksema R 1994 Geophysics **59** 252-8
[5] Knight R and Nolen-Hoeksema R 1990 Geophys. Res. Lett. **17** 1529-32
[6] Barriere J, Bordes C, Brito D, Senechal P and Perroud H 2012 Geophys. J. Int. **191** 1152-70
[7] Muller T M, Gurevich B and Lebedev M 2010 Geophysics **75** 75A147-64
[8] Knight R, Dvorkin J and Nur A 1998 Geophysics **63** 132-8
[9] Mavko G and Mukerji T 1998 Geophysics **63** 918-23
[10] Mavko G, Mukerji T and Dvorkin J 2009 The rock physics handbook: Tools for seismic analysis of porous media (Cambridge: Cambridge University Press)
[11] Bailey L, Lekkerkerker H N W and Maitland G 2015 Soft Matter **11** 222-36

NMR Metabolic profiling of green tea (*Camellia sinensis* L.) leaves grown at Kemuning, Indonesia

D. S. C. Wahyuni[1,2*], M. W. Kristanti[1,2], R. K. Putri[1,2] and Y. Rinanto[3]

[1] Pharmacy Department, Faculty of Mathematics and Sciences, Sebelas Maret University

[2] Natural Medicine Research Group, Pharmacy Department, Faculty of Mathematics and Sciences, Sebelas Maret University

[3] Biology Education Program, Faculty of Teacher Training and Education, Sebelas Maret University

Email: dinarsari_cw@staff.uns.ac.id

Abstract. Green tea (Camellia sinensis L.) has been famous as a beverage and natural medicine. It contains a broad range of primary and secondary metabolites i.e. polyphenols. Nuclear Magnetic Resonance (NMR) has been widely used for metabolic profiling in medicinal plants. It provides a very fast and detailed analysis of the biomolecular composition of crude extracts. Moreover, an NMR spectrum is a physical characteristic of a compound and thus highly reproducible. Therefore, this study aims to profile metabolites of three different varieties of green tea C. Sinensis grown in Kemuning, Middle Java. Three varieties of green tea collected on Kemuning (TR1 2025, Gambung 4/5, and Chiaruan 143) were used in this study. 1H-NMR spectra were recorded at 230C on a 400 MHz Agilent WB (Widebore). The analysis was performed on dried green tea leaves and analyzed by ^{1}H-NMR, 2D-J-resolved and 1H-1H correlated spectroscopy (COSY). MestRenova version 11.0.0 applied to identify metabolites in samples. A 1H-NMR spectrum of tea showed amino acids and organic acids signal at the area δ 0.8-4.0. These were theanine, alanine, threonine, succinic acid, aspartic acid, lactic acid. Anomeric protons of carbohydrate were shown by the region of β-glucose, α-glucose, fructose and sucrose. The phenolic region was depicted at area δ 5.5-8.5. Epigallocatechin derivates and caffeine were detected in the tea leaves. The detail compound identification was observed and discussed in the text.

1. Introduction

Camellia sinensis L. or known as green tea is the popular beverage all over the worlds. It comprises hundreds of both primary and secondary metabolites at different concentrations. The former occurs in tea leaves including sugars, organic acids, amino acids and the latter comprising alkaloids and polyphenols. Those metabolites determine the quality of tea leaves. It is influenced by variety, geography, and climates [1]. Lee [2] reported chemical constituents of tea differed among three varieties naming white tea, green tea and oolong teas. Moreover, the geography of tea leaves grown influences its chemical composition. It revealed that theanine and catechin derivatives were correlated to geographical status, either inter-country or inter-city.

Intermediate and final products of cell metabolism process are noticed as metabolites. Generally, those metabolites are influenced by the current circumstances [2]. Metabolomics as a holistic approach is needed to understand better about primary and secondary metabolites [3]. An advance and powerful analytical tools such as nuclear magnetic resonance (NMR) has been introduced. Nowadays, NMR is a popular method as a metabolome analysis [4]. Despite its low sensitivity, NMR is a promising feature as its non-destructive method and sample preparation [5].

NMR spectroscopy is favored due to its simple pretreatment. Moreover, it is a non-selective method which covers a wide range of metabolites particularly in the field of agriculture [6]. The use of NMR for metabolites profiling has been reported at different food stuff like coffee [7], wine [8] and tea [9].

The aim of the study was to identify the primary and secondary metabolites of green tea leaves grow at Kemuning, Indonesia using ^{1}H-NMR with some two-dimensional NMR techniques, J-resolved and 1H-1H COSY. It used three varieties available at the tea gardens, TRI 2025, Gambung and Cinruan.

2. Materials and Methods

2.1. *Plant Material and Extraction*

Green tea (*Camellia sinensis* L.) were freshly collected from Kemuning green tea garden situated at Kemuning, Karanganyar (070 36' 23.6" S, 1110 07' 02.9" E) Middle Java, Indonesia, on May 2016. Kemuning garden is commercially produced tea product for export and local market. There were three varieties such as TRI 2025, Gambung and Cinruan. Three replicates of each sample were used for NMR metabolomics. Fresh green tea samples were plucked and saved in nitrogen liquid. The standard metabolomics protocol of sample preparation and ^{1H}-NMR profiling described by Kim et al. [10]. Samples were mashed under liquid nitrogen and dried in the freeze-drying for 48 hours. Samples of 30 mg plant material were weighed and extracted with 1.5 ml of a mixture of phosphate buffer (pH 6.0) in deuterium oxide containing 0.05% trimethylsilylproprionic acid sodium salt-d4 (TMSP) and methanol-d4 (1:1). Samples were vortexed at room temperature for 1 min, ultrasonicated for 20 min and centrifuged at 13000 rpm for 10 min. An aliquot of 0.6 ml of the supernatant was transferred to 5 mm NMR tubes for ^{1}H-NMR measurement.

2.2. *^{1}H-NMR Measurements*

^{1}H-NMR spectra were recorded at 21^{0}C on 400 MHz Agilent spectrometer. Deuterated water was used as the internal lock. Each ^{1}H-NMR spectrum consisted of 128 scans requiring 18 min and 26 s acquisition time with the following parameters: 0.16 Hz/point, pulse width (PW) of 60° (11.3 μs), and relaxation delay (RD) of 2s. Two-dimensional J-resolved NMR spectra were acquired using eight scans per 128 increments for F1 (chemical shift axis) and eight k for F2 (spin-spin coupling constant axis) using spectral widths of 66 Hz and 5000 Hz respectively. Both dimensions were multiplied by sine-bell functions (SSB = 0) before double complex Fourier transformation. J-resolved spectra were tilted by 450, symmetrized about F1, and then calibrated to TMSP. 1H-1H correlated COSY spectra were acquired with a 1.0 s relaxation delay and 6361 Hz spectral width in both dimensions. The window function for the COSY spectra was Qsine (SSB = 0). MestRenova version 11.0.0 applied to identify metabolites in samples.

3. Result and Discussion

3.1. *Qualitative analysis of ^{1}H-NMR Spectra*

^{1}H-NMR spectra of three varieties of green teas leaves were analyzed. 2D NMR technique such as J-resolved and COSY (Correlated Spectroscopy) spectra were applied to further metabolites identification. The metabolites identified cover amino acids, organic acids, carbohydrates, alkaloids, flavonoids. All assignments were done by comparing the spectra with other tea leaves reports [2] [1].

A ^{1}H-NMR spectrum is classified into three parts. First, amino acids with few organic peaks placed at the area between δ 0.8 – 4.0 ppm. Secondly, anomeric protons of carbohydrate are found in the region of δ 4.0 – 5.5 ppm and phenolic region placed at the area of δ 5.5 – 8.0 ppm. A comparison of

^{1}H-NMR spectra of three different varieties of green tea, TRI 2025, Gambung and Cinruan depicted at figure 1, figure 2 and figure 3.

3.2. Compound identification

The signals in the amino acids region (δ 0.8 – 4.0 ppm) were useful to identify a number of amino and organic acid signals (Fig. 1). Amino acids like alanine *δ* 1.48 (d, *J* = 7.2), threonine (δ 1.32 (d, *J* = 7.0 Hz); 4.23 (m)) and theanine (δ 1.10 (t, *J* = 7.4 Hz); 3.2 (q, *J* = 6.2 Hz). Carbohydrate region exhibited anomeric proton of α-glucose *δ* 5.17 (d, *J* = 3.8 Hz) and β-glucose *δ* 4.59 (d, *J* = 7.9 Hz). Sucrose was shown by signal at *δ* 3.74 (t, *J* = 9.5 Hz); 3.43 (t, *J* = 9.5 Hz). Signals at *δ* 3.70 (d, *J* = 11.6 Hz); 3.52 (d, *J* = 11.5 Hz) were assigned for fructose (Fig. 2). Phenolic region showed that coffein *δ* 7.76 (s), and epigallocatechin (EGC) and epigallocatechin-3-gallate (EGCG) (*δ* 6.7 (d, *J* = 8.4); 6.9 (dd, *J* = 8.4, 2.7); 7.0 (d, *J* = 2.4)) present in green tea leaves (Fig. 3).

In the current study, we successfully identified the particular amino acids in green tea, theanine. The theanine signals at δ 1.10 (t, J = 7.4 Hz) of H-2 and 3.2 (q, J = 6.2 Hz) of H-1 (Fig. 4) which is correlated in the 1H-1H COSY spectra (Fig. 5). The threonine signals at δ 1.32 (d, J = 7.0 Hz) of H-3 and δ 4.23 (m) of H-4 (Fig. 4) were correlated each other in the 1H-1H COSY spectra (Fig. 5). The epigallocatechin signals at δ 6.7 (d, J = 8.4) of H-5 and δ 6.9 (dd, J = 8.4, 2.7) of H-6 (Fig. 4) were correlated each other in the ^{1}H-^{1}H COSY spectra (Fig. 5). Caffeine, the characteristic alkaloid in green tea, was shown by the singlet at δ 7.8 (s), δ 3.9 (s) and δ 3.5 (s).

Theanine is a characteristic amino acid present in green tea. It is reported to have a beneficial effect in cancer [11], memory ability [12] and blood pressure [13]. Total amino acids such as theanine, alanine, threonine, etc were approximately 1-4% dry weight [14].

Chacko [15] reported catechin derivatives particularly EGCG are important to metabolic syndrome. It occurred in higher amount of EGCG particularly in green tea than other varieties [15]. Catechins and caffeine in green tea act synergistically to increase energy expenditure and fat oxidation [14]. Polyphenols and caffeine content in green tea leaves were reported approximately as 45-90% and 0.4-10%, respectively [14]. It varies due to varieties, plant origin and growing condition [16].

4. Conclusions

The ^{1}H-NMR-based metabolomics approach allowed for simultaneous qualitative analysis in green tea leaves. Amino-organic acids and phenolic metabolites could be detected in the spectra. Green tea (*Camellia sinensis* L.) leaves located in Kemuning, Karanganyar, Indonesia contained characteristic amino acid in green tea, theanine, and characteristic phenolic, epicatechine derivates. The quantitative measurement for each metabolite is further analysis to be done to compare the effect of varieties on green tea.

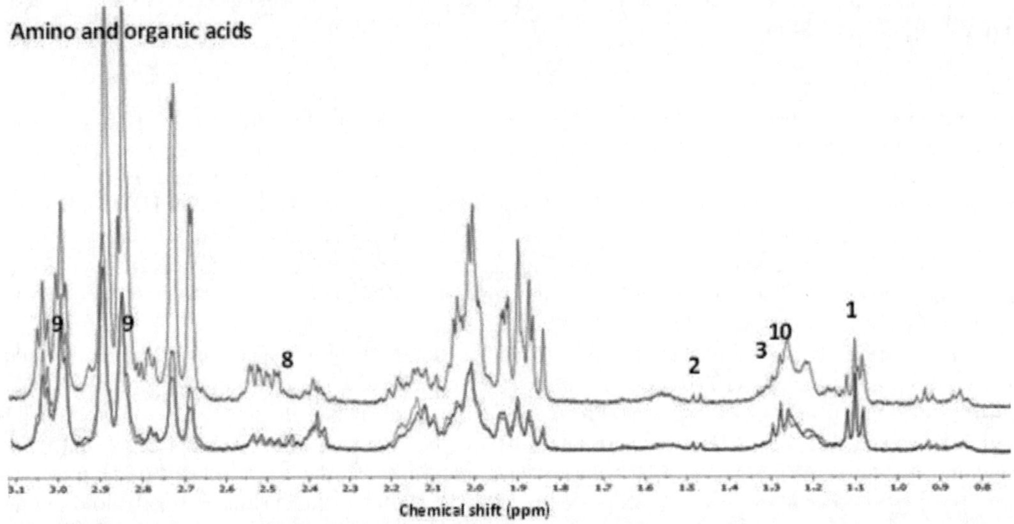

Figure 1. ^{1}H-NMR spectra of Gambung (red), TRI 2025 (blue) and Chinruan (green) at the area between δ 0.8 – 3.0 ppm collected on May 2016. 1: theanine, 2: alanine, 3: threonine, 4: α-glucose, 5: β-glucose, 6: sucrose, 7: fructose, 8: succinic acid, 9: aspartic acid, 10: lactic acid, 11: coffein, 12: EGC/EGCG, 13: gallic acid

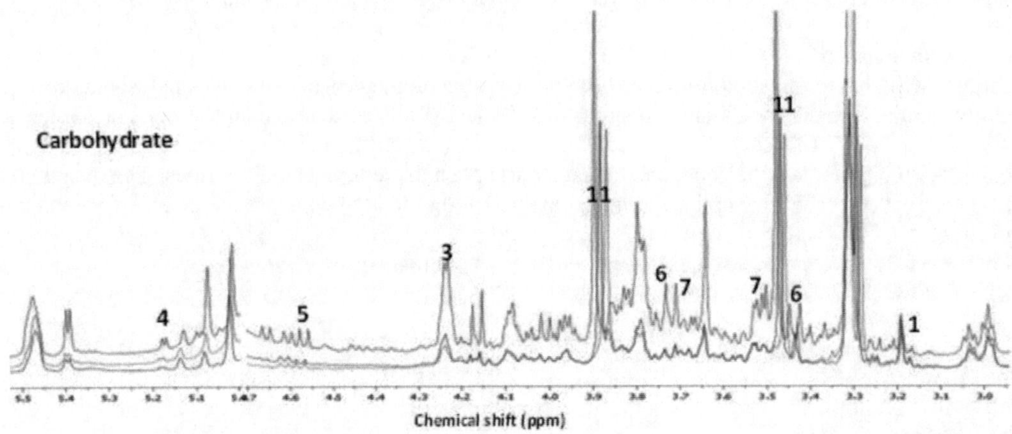

Figure 2. ^{1}H-NMR spectra of Gambung (red), TRI 2025 (blue) and Chinruan (green) at the area between δ 3.0 – 5.5 ppm collected on May 2016. 1: theanine, 2: alanine, 3: threonine, 4: α-glucose, 5: β-glucose, 6: sucrose, 7: fructose, 8: succinic acid, 9: aspartic acid, 10: lactic acid, 11: caffeine, 12: EGC/EGCG, 13: gallic acid

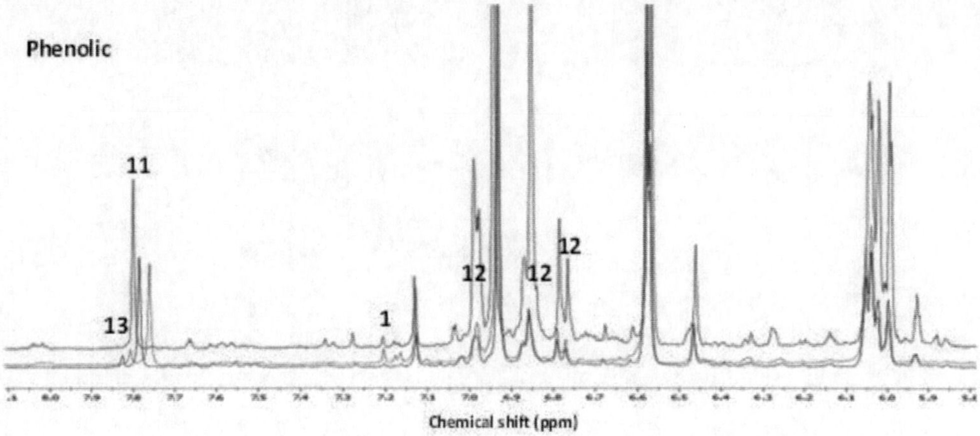

Figure 3. ^{1}H-NMR spectra of Gambung (red), TRI 2025 (blue) and Chinruan (green) at the area between δ 5.5 – 8.0 ppm collected on May 2016. 1: theanine, 2: alanine, 3: threonine, 4: α-glucose, 5: β-glucose, 6: sucrose, 7: fructose, 8: succinic acid, 9: aspartic acid, 10: lactic acid, 11: caffeine, 12: EGC/EGCG, 13: gallic acid

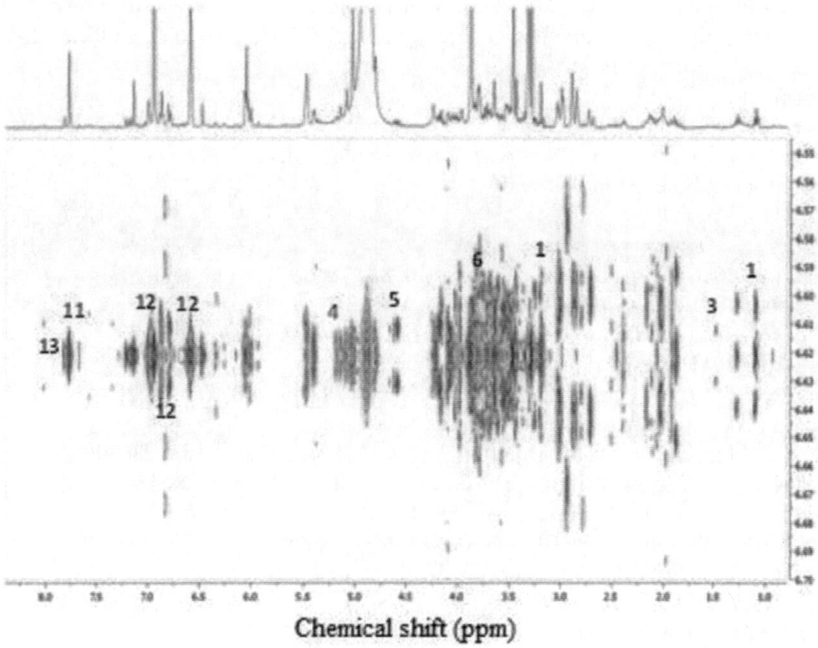

Figure 4. Two dimensional ^{1}H-^{1}H *J*-resolved spectra of Chinruan shows 1: H-2 of theanine, 3: H-4 of threonine, 4: H anomeric of α-glucose, 5: H anomeric of β –glucose.

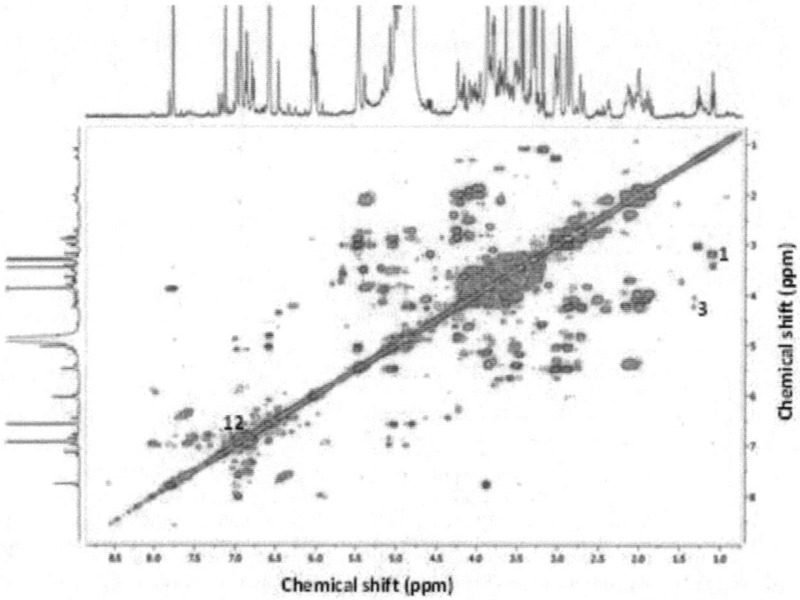

Figure 5. Two dimensional ^{1}H-^{1}H COSY spectra of Chinruan shows correlation between 1: H-1 and H-2 of theanine, 3: H-3 and H-4 of threonine, 12: H-5 and H-6 of epigallocatechin

Acknowledgements

The authors are grateful to PT. Rumpun Sari Kemuning I for sample collection. The work is fully funded by Penelitian Keunggulan (PK-UNS), 2016, Sebelas Maret University. The authors declared no potential conflicts of interest in this study.

References

[1] Lee J E *et al* 2010 *J. Agric. Food Chem.* **58** 10582
[2] Lee J E *et al* 2015 *Food Chem.* **174** 452
[3] Wolfender J Rudaz S Choi Y H Kim H K.2013 *Curr. Med. Chem.* **20** 1056
[4] Lee J E *et al* 2011 *Food Res. Int.* **44** 597
[5] Verpoorte R Choi Y H Mustafa N R Kim H K 2008 *Phytochem. Rev.* **7** 525
[6] Dixon R A *et al* 2006 *J. Agric. Food Chem.* **54** 8984
[7] Charlton A J Farrington W H H Brereton P 2002 *J. Agric. Food Chem.* **50** 3098
[8] Ali K Maltese F Toepfer R Choi Y H Verpoorte R 2011 *J.* Biomol. NMR **49** 255
[9] Lee J E *et al* 2011 *J. Agric. Food Chem.* **59** 10579
[10] Kim H K Choi Y H Verpoorte R 2010 *Nat. Protoc.* **5** 536
[11] Cooper R 2012 *Int. J. Food Sci. Nutr.* **63** 90
[12] Park S K *et al* 2011 *J. Med. Food.* **14** 334
[13] Suzuki Y Miyoshi N Isemura M 2012 *Proc. Japan Acad.* **88** 88
[14] Rains T M Agarwal S Maki K C 2011 *J. Nutr. Biochem.* **22** 1
[15] Chacko S M Thambi P T Kuttan R Nishigaki I 2010 *Chin. Med. BioMed. Central* **5** 1
[16] Khokhar S Magnusdottir S G M 2002 *J. Agric. Food Chem.* **50** 565

IOP Conf. Series: Journal of Physics: Conf. Series **795** (2017) 012014 doi:10.1088/1742-6596/795/1/012014

God particles in the perspective of The AlQuran Surah Yunus: 61 and modern science

Sri Jumini

Physics Department Program of Sains AlQuran University,
Jl. Raya Kalibeber, Km. 03 Mojotengah Wonosobo Central Java, INDONESIA

E-mail: umy fadhil@yahoo.com

Abstract. The Qur'an is the book of Allah revealed to guide human beings, settting the rules of life to enable them to achieve happiness in this world and hereafter. The Qur'an has mentioned various scientific nature detailly and accurately so we are able to find new knowledge which is previously unknown by human being. One was about the God particle (Higgs Boson). This article aims to provide a deeper understanding of the concept of the Higgs Boson, the Higgs Boson explained this concept in detail relatated to 1) Perspective of science 2) Perspective of Al-Qur'an 3) Development of technology or science and technology. This study is a qualitative research using library research (library research) that examines and analyzes the books relating directly or indirectly. The results of the analysis states that 1) The concept of the Higgs Boson particle in terms of basic science is also the reason why almost all elementary particles have a greater mass, 2) The concept of the Higgs Boson in the Qur'an is implied from the results of the comparison interpretation of the commentators in Surah Yunus paragraph 61 related to Atom concepts and smaller particles theory of (Higgs Boson), interpretation of Al-Maraghi, and Al-Misbah. 3) The concept of the Higgs Boson in science and technology provide the most advance technology and it is the greatest achievement in the world of science and technology.

1. Introduction

The progress of science in modern times accelerates the technology and vice versa [1]. As a result of both the developing and growing faster than in other social institutions, frequently cultural gaps are attended by some social tensions and psychology. Gaps and tensions on the one hand due to the delay of man in anticipation of the development of science and on the other because the human delay in meeting the challenges and demands as a result of technological advances.

Science news in the book of the Qur'an proved aligned with the findings of contemporary science and modern science. Although the Qur'an is not an encyclopedia of science, but it gives the general principles of scientific studies and motivation for Muslims to understand the universe and its contents to get closer to God [2], and every word in the sentence which tells the phenomena of science comes from the Possessor of Science, so it has a very precise meaning.

The scope of physics in the teachings of the Qur'an is very confront thinking that underlies his efforts, as well as various conceptions, just as we find some verses that are relevant to the development of physics itself. As in Q.S. Yunus verse 61 that uses the word small ant to refer something smallest, and the prevalent word of these words is small ants or tiny dust particles, since this word is associated with the smallest item refering to an atom, then the small ant in this surah is often interpreted as an atom [3], when refering to the implied meaning in surah Yunus verse 61, the Qur'an since 15 centuries ago

has hinted that the atom as small ant or small. This statement is in accordance with the word of God in the Al-Qamar verse 49
إِنَّا كُلَّ شَيْءٍ خَلَقْنَٰهُ بِقَدَرٍ ٤٩
Meaning: We have created everything according to size
The atom that although there is a small size. God created everything, including quark which is smaller than protons and neutrons which have the size and mass vary according to place and they work. Thus some object has the size and weight (mass), although the object was a small thing though.
It is known that the mass of an object is related to the amount of material it contains, and which forms the material, namely the collection and combination of atoms [4]. And in translation of QS. Yunus verse 61 uses the word does not exist (also) smaller than that and bigger, can mean that there is something smaller than the atoms called protons, neutrons, and electrons. But now the composition is increased because of god particle findings (Higgs Boson) which is now the concept is still developed. It discusses the process of the mass forming in which the beginning of a process that happens in an empty room into a Higgs field which later this Higgs particle boson will attract some of the particles will form larger particles and massless. This quadruple mass of particles fused with other particles includes mass to be a matter that will become an object which is commonly observed.
One element in the development of physics is an analysis of the data collected from the various measurements of the magnitudes of physical looks done through a process of critical thinking, which then proceeded with the evaluation of the results with the reasoning sound to reach the conclusion that a rational [5]. In the theory, if people with enough energy are capable of separating the Higgs Boson particle of an object, then the object cannot be seen or loss of form, but it is not meant missing. It is necessary for an in-depth study of the development of the concept of the God particle Higgs Boson in view point of modern science and Quran Surah Yunus verse 61.

2. Numerical Methods

The approach used in this study is qualitative research library technique, namely the research literature by examining and analyzing the books relating directly or indirectly to the title which the author discussed [6]. Based on the source, the data can be divided into two: 1) the primary data, the source that contains information or data about the God particle theory in the concept of hig Bosson either of the previous research as well as articles in journals and books. The materials relating to the concept of the God particle higgs Bosson in the Qur'an used the method of interpretation of the various sources of commentators, especially for the interpretation of the Qur'an surah Yunus verse 61 [7]; 2) secondary data are data obtained from sources obtained from ungenuine sources, the information or data [8]. Data analysis techniques performed are: 1) Mechanical Inductive Thinking, it is the process of organizing facts or observations are separated into a series of relationships or generalization (Saifudin Azwar, 1998: 40). 2) deductive method, it is a process approach that departs from the general truth about a phenomenon (theory) and generalizes the truth at an event or specific data which typically similar to the phenomenon in question; 3) Mechanical comparative Tafsir, exegesis is a method using a comparison among one another [10]. This method is used to identify similarities and differences of objects, people, working procedures, ideas, and criticism of a case.

3. Results and Discussion

The development of science is driven by the discovery of a phenomenon that cannot be explained by the old theories. The physicists have been researching the origin of the material. Wood fibers, for example, are composed of cells. Most of the cells are made of molecules. The larges composition of molecules is composed of atoms, and lastly, the atoms are made of subatomic particles, and subatomic particles are composed of elementary particles. In other words, the "material" is actually a set of patterns that are not obvious. Searches the base material of the universe ends with the conclusion there was not any.

3.1. Analysis of the concept of God Particles Higgs Boson Science Perspective

IOP Conf. Series: Journal of Physics: Conf. Series **795** (2017) 012014 doi:10.1088/1742-6596/795/1/012014

All matter in the universe is composed of elementary particles, and the behavior of these particles can be understood through four fundamental forces [11]. Matter and the fundamental forces cannot be treated as two different things. Three of the four fundamental forces affect all particles. Except for gravity, the force of gravity was a universal force, because the force of gravity is very weak, the effect is negligible for elementary particles. It is known that there are two classes of elementary particles, namely the fermions and bosons. Fermions are composed of quarks and leptons. While the boson is composed of elementary particles that have some styles, such as the electromagnetic force (photon) and the nuclear force (gluon). This means that the gluons are responsible for holding quarks together, and the weak nuclear forces (W boson and Z) are intermediary by gauge boson. These particles were discovered in the 1980s. Gluons and quarks cannot be observed but there is no evidence for their existence. Photons, W and Z bosons, and particles are observed so that they both theoretically and experimentally proved [12]. Strength carrier particles to gravity are not observed eventhough scientists claim that there must be particles called gravitons which are responsible for gravity. Although the standard model describes the strong nuclear force, the electromagnetic force and the weak nuclear force, the standard model cannot explain gravity. Also, the quantum and relativity cannot be fitted into the standard model together with being mathematically compatible [13]. Therefore, the standard model still needs to be improved. Bosons are elementary particles (or a collection of elementary particles), which can occupy the same quantum state, unlike fermions adhere to the principle of the prohibition of Wolfgang Pauli.

3.1.1. The Fermi-Dirac statistics

This statistics in quantum mechanics is one of two possible ways in which the system of indistinguishable particles and can be distributed among a set of energy states. Fermion, named after Enrico Fermi as they obey Fermi-Dirac statistics, is proposed by Fermi and Paul Dirac independent. Unlike the Bose-Einstein statistics, Fermi-Dirac statistics only apply to the types of particles that adhere to the restriction known as the Pauli principle of prohibition. Such particles have a value of hemispheric spin and named fermions after the correct statistics describe their behavior. Fermi-Dirac statistics apply, for example, for electrons, protons, and neutrons.

3.1.2. The Bose-Einstein statistic

This statistic was first proposed by Satyendra Bose in a letter to Albert Einstein (which he translated into German) in 1925. A real case is "Bose-Einstein condensation." The phenomenon of Bose-Einstein condensation is a condition in which a substance has a new nature, where all the particles that are in the energy minimum [14]. Statistics of Bose-Einstein specify statistical distributions for bosons at various energy levels in iquilibrium thermal. Unlike fermions, bosons are particles berspin round so it does not comply with the principle of the prohibition of Pauli; a large number of particles bosons can occupy the same state at the same time. Statistics of Bose-Einstein was introduced by Bose (for photons). So the electrons are fermions and photons are bosons. Fermi-Dirac and Bose-Einstein statistics are important in quantum mechanics in particular, because it is not possible in principle to say exactly, where very small particles are, and how fast the particles are: we can only talk about a common position and momentum in probabilistic terms (Statistics). It is an improper statement of the Heisenberg uncertainty principle (Werner Heisenberg, 1927).

Boson relationship with the standard model, the three basic styles naturally besides gravity, resulted from the exchange of energy between the particles in a group called Boson, i.e. Gauge boson and the Higgs Boson. Gauge bosons act as an intermediary for the natural three basic styles while Higgs Boson is elementary particles that are the main reasons why almost all elementary particles have a greater mass.

3.1.3. Higgs boson in the mass formation
Higgs boson mass fill and interact with the Higgs field, the massless particles can interact strongly with the Higgs field, which are then observed as a mass. Particles that do not interact with the Higgs field are massless and will pitch kale granted.

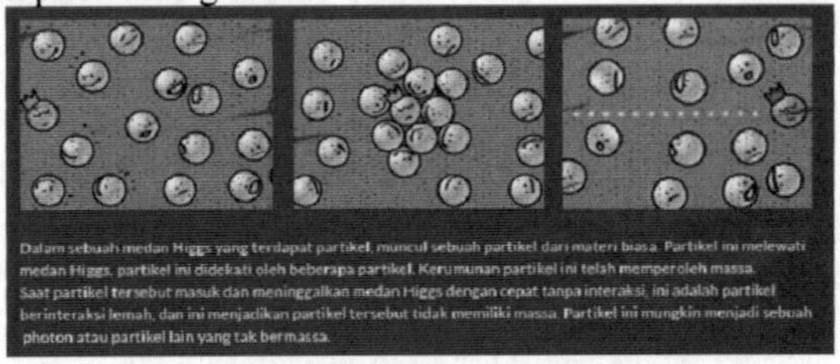

Figure 1. particle interaction to the Higgs field [12]

Illustration Higgs Boson in mass formation in the eyes of physicists.

(a) Professor David Miller of University College London provides a simple analogy. In a room, there is a collection of people who are cool crowd talking. In this analogy, the room was a Higgs field that contained a set of particles. Then a celebrity came. Her presence instantly influenced on the interaction in the room. Animals were then further packed crowd. In this analogy, each in the crowd behaved like the Higgs boson, which gave mass to the particle/celebrity. The stronger the interaction runs, the more severe the masses will be. The phenomena as mantioned means that boson interaction gets stronger, and this also then leads to formulation of the material.

(b) Based on the theory that is stated by Joao Guimaraes da Costsa physicist from Harvard University, particles passing through the Higgs field would acquire mass, like a swimmer moving through the pool which were wet. If there is no such mechanism then arbitrarily will be massless. So he found the Higgs boson discovery further confirms that the Higgs mechanism for particles to acquire mass is correct [15].

The easiest way to prove the existence of the Higgs field is to find the Higgs boson. How to find the Higgs boson is by driving a large number of protons. When two protons collide, they will break into constituent particles, namely quarks and gluons. Two gluons then are fused to form the Higgs boson. However, the age of the Higgs boson is very short: In the blink of Higgs boson decays into other particles directly. There are various ways Higgs boson decays, one of which is by emitting a photon.

In addition to being a pair of photons, Higgs boson can decay into two pairs, among other lepton or quark pairs [16]. Because these reactions are not as what scientists are looking for, then these are collectively called as a response to the background (background) while the feedback of the Higgs boson decays into a pair of photons in receiving to the signal. What can happen toward the two-photon collision reactions known as subatomic, and what percentage chances of each reaction will occur. Higgs boson can decay into other particles. In addition to being a pair of photons, Higgs boson can decompose into two pairs, among other leptons or quark pairs. Other ways of decay of the Higgs boson is also examined, and the results are then combined. Researchers observed that this excess is much different from what is expected from background. From the combined analysis of various ways of decay is concluded that the probability of excesses discerned is due to the background only a billionth, or just 0.0000001% only.

Search Higgs boson thus do not by finding the Higgs boson itself, but by looking for particles decay products of the Higgs boson. Gluon energy in a bind quark into protons is very large. Therefore, the researchers needed a machine that can accelerate particles up to speed greatly, at least it is not up to 99.99998% the speed of light. In this condition, a proton has enormous-momentum and if it collides with another proton, the energy of the collision will be so great, and it can rival the binding energy of the gluons. Higgs Boson is very rarely. Of the approximately 10 billion collisions, it is estimated that there will be only one Higgs boson which appears. Therefore, the number of experiments carried out should be very much and it was done over the years. Within one second about 40 million collisions in the LHC and the experiments took place at any time: 24 hours a day, seven days a week. It is calculated that every day happens > 3.4 trillion collisions.

Some researchers doubted the existence of this particle, but other researchers mainly follow the development of the research, and the researchers who participated directly in the research convince that these particles are real. That the inventor of the Higgs boson particle, the European Nuclear Research Organization or the European Organization for Nuclear Research (CERN), has managed to discover the findings of the so-called 'Particle Higgs boson.' The conviction was upheld because CERN has two independent teams for comparing these discoveries, namely: 1) ATLAS (A Toroidal LHC Apparatus); 2) CMS (Compact Muon Solenoid). They did the same experiment, so the results of the research and the data collected can be mutually tested and research-diversified. The result ranking is from zero to five-sigma (five sigmas). In December of 2012 l, the two teams also have stated that their data indicate two similar levels (two sigmas), which prove that the Higgs-Boson particle was indeed there.

Higgs field unlike another particular branch of study in physics. Although the space is completely emptied, then the space will still contain a terrain that refuses to leave the room. The field is the Higgs field. Higgs field is like air for fish. But without this sphere, humans could not exist, because the particles can only acquire mass through interactions with the Higgs field.

All sorts of elementary particle electrons, quarks, photons, etc. before interacting with the Higgs field does not have a mass (symmetry). Higgs field interaction is different from these elementary particles that damage symmetry and give different masses. Massless particles like photons do not interact with the Higgs field, but other such particles; electrons and quarks interact with the terrain and mass produce by the nature of their interactions. The larger the particle interacts with the Higgs field the greater will be its mass. In elementary particles, photon includes in the boson while the electrons and quarks are the fermions.

The interaction of particles with the Higgs field is analogized with the pieces of papers sucking or absorbing the ink. In comparation of pieces of the paper and the ink, an individual particle is the mass. Just like a piece of paper of different sizes and thickness to absorb varying amounts of ink, different particles "absorb" the various mass number [17]. In quantum mechanics, a boson is a particle that follows Bose-Einstein statistics. An important characteristic of the boson is that statistically, they do not restrict the number of those who occupy the same quantum state. On the other hand, there are fermions obey Fermi-Dirac statistics and the Pauli principle of prohibition. : "Should not there are two electrons in an atom with the same four quantum number." More generally, there are no two identical fermions (particles with fractional spin) may occupy the same quantum state simultaneously. Fermions are sometimes said to be constituents of matter, while bosons said particles that transmit interactions (force carriers), or parts of radiation.

Fermi-Dirac and Bose-Einstein can only speak about the position and momentum in probabilistic terms. This statement is not correct from the Heisenberg uncertainty principle. Although both contain elements of "uncertainty", the basic thinking between the two is very different. In Heisenberg's concept, "uncertainty" is in the context of measurement [18]. It is almost impossible to measure the two quantities simultaneously, such as position and momentum of a particle. However a rigorous measurement is made, there must be certain inaccuracies in scale.

The simple correlation between the particle and its spin as bosons or fermions is that each boson has spin integer: 0, 1, 2, etc while each fermion has a plus half-integer spin: ½, 3/2, 5/2. Etc. The relationship is so close that people often define boson as "particles with a round integer" and fermions as "particles with half-integer spin." Because it is not very precise, we define that the Higgs can be piled on top of each other while fermions take space, is a true difference between the two classes of particles.

IOP Conf. Series: Journal of Physics: Conf. Series **795** (2017) 012014 doi:10.1088/1742-6596/795/1/012014

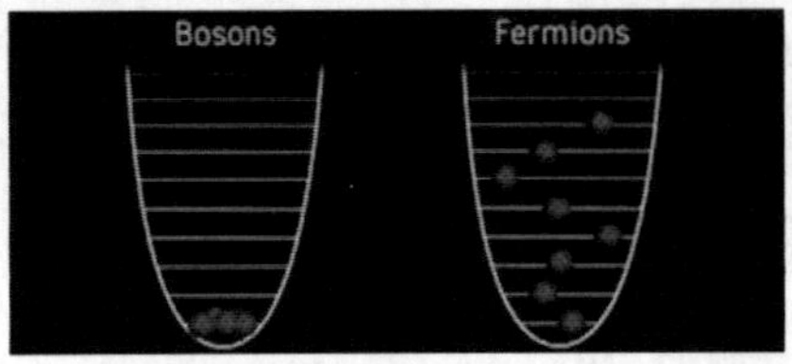

Figure 2. Differentiation fermions from bosons [12]

The distinction of the two is that fermions take up a space, while bosons can be piled on top of each other. Bosons do not take up space at all. Two bosons, or two trillion bosons, can easily sit in the same location, right on top of each other. That's why boson is a particle that carries force because those fermions obey the Pauli Exclusion Principle.

The particles of the Standard Model have a very specific spin. All known fermions charged quark-lepton and neutrino is a spin-1/2. Gauge bosons more photons, gluons, and the W and Z bosons all the spin-1. Higgs boson, stood apart from the rest, is a spin-0. Particles with zero spin are called "scalar," and fields from which they arise are called "scalar field."

Before Higgs came together to break the symmetry, the gauge boson mass of weak interaction partner for left-handed fermions (clockwise) and right-handed. That's the way nature works. All couples of the nuclear force, gravity, and electromagnetic are equally well for left-handed and right-handed particles (counter-clockwise); one of the power couples will be weak but not with others. It also explains why the weak interaction violates the rules. This occurrence can only happen if they move at the speed of light. It means that they should have zero mass as the result. It indicates that the particles which have zero mass are those which move at the speed of light. The phenomena will make the couple power become weak. As the result, there will be a weak interaction toward the Higgs field.

3.2. Analysis of the concept of God Particle Higgs Boson: Al-Qur'an Perspective

The Qur'an has mentioned various nature sciences detailly and accurately. These evedents enable us to find some new knowledge which is previously unknown by human. Although science has grown so fast, none of the scientific theories is contrary to the Qur'an. The theories that have not been scientifically proven yet do not have to be compared with Kalam of God. Perhaps the progress of science and technology in the future will reveal the truth and mistakes. One example is in the word of surah Yunus verse 61 as follows:

وَمَا تَكُونُ فِي شَأۡنٖ وَمَا تَتۡلُواْ مِنۡهُ مِن قُرۡءَانٖ وَلَا تَعۡمَلُونَ مِنۡ عَمَلٍ إِلَّا كُنَّا عَلَيۡكُمۡ شُهُودًا إِذۡ تُفِيضُونَ فِيهِۚ وَمَا يَعۡزُبُ عَن رَّبِّكَ مِن مِّثۡقَالِ ذَرَّةٖ فِي ٱلۡأَرۡضِ
وَلَا فِي ٱلسَّمَآءِ وَلَآ أَصۡغَرَ مِن ذَٰلِكَ وَلَآ أَكۡبَرَ إِلَّا فِي كِتَٰبٖ مُّبِينٍ ٦١

Meaning: You are not in any matter or recite any of the Qur'an or do any action without Our witnessing you while you do it. Not hidden from thy Lord even of zarrah (atoms) in the earth or in the heavens. There is nothing more small and not (also) greater than that, but (all recorded) in a Clear Book (Lawh Mahfuz)

The implied meaning of Qs. Yunus verse 61 the word "not hidden from your Lord even of zarrah (atom." The word (ذرة) "small ant" in this surah is understood by scholars in a variety of meanings. Some of them are the animal which is very small even its head, or flying debris that are only visible in the gap nuclei did not stand alone. They are composed of particles called protons, electrons, and neutrons. But now the composition is increased by the discovery of the Higgs Boson or often called the God Particle. An object formed from some of the particles, without exception, be it a piece of wood, the car to the human body is made up of particles. God created everything, including Quark which is smaller than protons and neutrons. It has the various size and mass according to its place and work.

Table 1. Thought science scientists and commentators of the Qur'an about the Higgs Boson.

Scientists science	Interpreter [Mufassir] (Qs. Yunus verse 61)
1. Peter Higgs	1. Quraish Shihab
Higgs Boson is a particle that makes up an	Kata (ذرة) is understood by scholars in a variety

object. We know that electrons, protons, and neutrons are particles found in atoms. Higgs Boson is the physical mass of the atom, a physics theory which cannot explain from where the mass of an object is finally revealed. Higgs Boson is a particle-forming mass. In its development, with sufficient energy when people can separate the Higgs Boson particles of matter, then in the theory they may not look or loss of form, but the object is not meant to disappear.

2. Leon Lederman

According to Leon Higgs field works like black magic through the particles. These particles go by the name Higgs boson.

And physics scientist named the particles as "God particles" (God Particle). The reason is the publisher would not let us call it Goddam Particle (particle damned) although it may be a title more appropriate, given the nature of evil and the cost of it that causes these particles like damned.

3. Stephen Hawking

According to his opinion "Higgs Boson could destroy the universe, Professor of Mathematics at Cambridge University explained that in the highest energy level, God particles can become unstable and result in catastrophe. There will be no more space and time." Higgs is potential to change the future of the world after becoming unstable at energies above 100 billion giga-electron-volts (GeV), "said Hawking. In these conditions, there will be quantum fluctuations that creates the vacuum bubbles in the universe. Because of the low energy, the gas will inflate with the speed of light and sweep everything. "It can happen anytime. We will never know when disaster comes, "said theoretical physicist.

of meanings, including very small ants, its head, and particles of dust which is only visible at the crack of sunlight. While at the present, [*darroh*] is interpreted in the sense of the atom. And indeed the word that is now used to designate atoms, even during the descent of the Qur'an atom yet known. Previous use of the word was to indicate the smallest things. Therefore, there are various meanings as pointed out above. And on the basis of that we cannot thanks once found particles smaller than atoms, and we are familiar with protons, neutrons, electrons, leptons, quarks and the Higgs Boson, we cannot say that this verse has hinted at the existence of something smaller from the atom by His word: "there is no smaller or larger than the small ant."

2. Al-Maragi

There is no anything that is smaller than the atom among the secrets of nature that you cannot see, nor greater than that, even the magnitude of such 'Throne of Allah Ta'ala, but it is known by Him and counted on His side in the great notebook containing records of all the existing, in order to enhance the natural system and match the entire act.

To convey the same as this paragraph, is the word of God:

ف لآ أق سم ب ما ت بصرون 83 وما لا ت بصرون 93

Meaning: I swear by what you see. And what you do not see (Q.S al-Haqqah: 38-39)

The paragraph indicates a sign thatthere are manythings in the universe that are not visible. Physicists have proven the particles making up the object. The latest discovary is a particle of God (Higgs Boson) which never comes to people mind in the time of Al-Qur'an revelation. However, now it is evident to anyone.

Humans do not realize or may not know much important the Higgs boson is, especially in the formation of the masses. People cannot imagine how the shape of the Higgs boson and what its structure is until it can form a mass or object that shapes the universe. It is all God's great power away from human reasons. Replies of God to man it is by the recognition of deeds, according to the result of achievement scales deeds he has done while living in the world. Word of Allah:

١٠٢ ٱلْمُفْلِحُونَ هُمُ فَأُوْلَٰٓئِكَ مَوَٰزِينُهُۥ ثَقُلَتْ فَمَن ١٠٣ خَٰلِدُونَ جَهَنَّمَ فِي أَنفُسَهُمْ خَسِرُوٓاْ ٱلَّذِينَ فَأُوْلَٰٓئِكَ مَوَٰزِينُهُۥ خَفَّتْ وَمَنْ

Meaning: He who weigh scales [goodness] it then they are the ones that can be luck. And whoever lightweight scales, then they are the people who are harming themselves, they are eternally in Hell. (Q.S. Al-Mukminun: 102-103)

This verse describes the people interacting with his Lord. If they interact hard, then will be higher faith so then more good deeds but if they interact weakly, usually ted not many good deeds. In other words, the Qur'an has explained the interaction of particles toward the Higgs field. If the particles interect strongly, they will have masses and if these particles interact weakly, they have no mass. The Higgs particle is the basic element which creates a mass to form stars and planets, including Earth. Higgs particles interact with the Higgs field to get the mass on other particles. In this case, there are two interactions, namely the weak interaction and the strong interaction. The first mentioned does not get the mass as a discussion of the weak interaction to God which will not produce ligh scales and of weak interactions which will get heavy scales. God will know every behavior even though it is amounted to *dzarah*. This is the purpose of God for man to exploit his life as well as possible, always work, seek knowledge, and grateful all of God's creation. All of this was shown to increase one's faith to the creator of this universe.

References

[1] Armahedi Mahzar, *Islam Masa Depan,* (Bandung: Pustaka Setia, 1993), page. 3
[2] Hasan Baharun, et. all, *Metodologi Studi Islam: Percikan Pemikiran Tokoh Dalam Membumikan Agama*, (Yogyakarta: Ar-Ruzz media, 2011), page. 50
[3] Agus Purwanto, *Ayat-Ayat Semesta.* (Bandung: Mizan Media Utama, 2013), page. 320
[4] Arthur W. Wiggins and Charles M. Wynn, *Five Biggest Unsolved Problems In Science*, (Us: John Wiley &Sons Inc, 2003) page. 14
[5] Samsul Munir Amin, *Mukjizat Al-Qur'an Tentang Arkeologi*, (Wonosobo: LP3M UNSIQ, 2011), page. 11
[6] Mestika Zed. 2008. *Metode Penelitian Kepustakaan*. Jakarta : Yayasan Obor Indonesia, hal. 2
[7] Tatang M. Arifin. 1995. *Menyusun Rencana Penelitian*, Jakarta : PT. Raja Grafindo Persada, hal. 133
[8] Tatang M. Arifin. 1995. *Menyusun Rencana Penelitian*, Jakarta : PT. Raja Grafindo Persada, page. 134
[9] SaifudinAzwar, *MetodePenelitian*. (Yogyakarta: PustakaBelajar, 2011), page. 41
[10] MuchotobHamzah, *Study Al-Qur'anKomprehensif, (*Yogyakarta:Gama Media, 2003), page.268
[11] Seyda Demirok, *The Higgs Boson From an Undergratuate's Viewpoint*, (Bogazici: Universitas Bogazici, 2015). page.2
[12] Sean Carrol, *The Particle at The End of The Universe: How The Hunt For the Higgs Boson Leads Us to The Edge of A New World*, (USA: A Plume Book, 2012). page 1215
[13] M. Cerena, *Status Of Higgs Boson Physics*, (Chicago:Fermi National Accelerator Laboratory, 2014) page. 7
[14] Elsa Kasmi, Syahyona, "*makalah fisika statistic kondensasi bose-einstein*" (Padang: Jurusan Fisika Fakultas Matematika dan Ilmu Pengetahuan Alam di Universitas Negri Padang, 2012) page.2
[15] Kurnia Mukaromah, "*Penemuan Partikel Tuhan (Higgs Boson)*", blogspot.com, 2012 accessed at 01/04/2016.
[16] Tri L astramadja, "*mengenal boson higgs*", Langitselatan.com, 2013 accessed at 16/03/2016.
[17] Martinusj. g. veltman." The higgs boson". *Scientific American*, 1986.
[18] http://ignasiusbagus.blogspot.co.id. *Teori Probabilitas Teori Normalisasi.* 2011 accessed at 1 juni 2016

IOP Conf. Series: Journal of Physics: Conf. Series **795** (2017) 012015 doi:10.1088/1742-6596/795/1/012015

Atmospheric conditions analysis of the heavy rain phenomenon in Biak (case study 4-5 December 2014)

Prayoga Ismail* **and Ejha Larasati Siadari**

Indonesian State College of Meteorology Climatology and Geophysics
Perhubungan Street I No. 5 Meteorology Complex, Pondok Betung, Bintaro 15221, Indonesia

Email: yogamailforalvin@gmail.com

Abstract Heavy rain in Biak on 4-5 December 2014 caused a flood. The accumulation of rainfall until 5 December was 102 mm, it exceeded BMKG's threshold about the intensity of extreme condition. This research aimed to find out the atmospheric conditions related to super heavy rain phenomenon. The atmospheric dynamical analysis used ERA-Interim Reanalysis data 0.125x0.125 degrees resolution. Divergence, relative vorticity, and streamline are the parameters investigated. Their dynamic fluctuations in significant pressure levels proved there were strong convergence and negative vorticity. Streamline analysis showed that there was a convergence wind pattern. While, the physical analysis used NOAA/NESDIS SST data to find sea surface temperature anomaly, RMM index data to find MJO phase and ONI index data to find ENSO phase. At the time, SST around Biak was in warmer condition, and there was a weak El Nino influenced by moderate MJO phase 5. Upper air sounding analysis is used to study the predictability of extreme weather. The indices indicated moderately convective activity and very unstable atmospheric layer before the event, while marginally convective activity and unstable boundary layer with possible inversion aloft during the event. All the conditions supported to produce convective cloud (Cumulonimbus) so that cloud had a long lifetime and produced thunderstorm and heavy rainfall. Biak meteorological station observation and MTSAT imagery used to verify the atmospheric conditions.

1. Introduction

On 4-5 December 2014, super heavy rain caused flooding in the city of Biak and became one of the phenomena disrupted the flight activities at the Frans Kaisiepo Biak airport for several hours, and causing waterlogging even flooding in some areas in Biak. The accumulated rainfall measured in Biak Meteorological Station was 102 mm per day. It is included one extreme weather phenomenon according to the Indonesian Agency for Meteorological, Climatological, and Geophysics (BMKG) is super heavy rain with an intensity of more than 20 mm per hour or more than 100 mm/day [4]. From the rainfall data, the super heavy rain case in the city of Biak with intensity exceeded extreme values threshold of BMKG.

Biak is one of the districts in the province of Papua, Indonesia. The seas and oceans around Biak made variability of weather and climate in the region influenced by the interaction of ocean and

*Corresponding author. Tel.: +62-8224-2285-955.
Email address: yogamailforalvin@gmail.com.

atmosphere. The global phenomena such as El Nino, La Nina, and MJO (Madden-Julian Oscillation) also gave impact to the weather and climate in Biak, as well as regional scale like monsoon circulation, the ITCZ (Inter-Tropical Convergence Zone), and the local influences.

Extreme weather is a state or physical phenomena in the atmosphere of a place, at a particular time-scale and short-term which is BMKG categorized if surface air temperature ≥ 35° C, wind speed ≥ 25 knots, and the rainfall in a day ≥ 50 mm [14]. Extreme weather conditions are output from significant conditions happened to the weather parameters. This paper examined the atmospheric conditions during the super heavy rain phenomenon in Biak on the dates.

2. Methodology

To find out the atmospheric conditions during heavy rain in Biak on 4-5 December 2014, this paper investigated the dynamics and physical aspects of weather parameters. The atmospheric dynamical analysis used ERA-Interim Reanalysis data 0.125x0.125 degrees spatial resolution. Divergence, relative vorticity, and streamline were the parameters investigated. Pressure levels used were significant layers of 1000, 925, 850, 700, 500, 400, 300, 200, 150, and 100 hpa. The 850 hpa level, there was the dynamics of the atmosphere that representative for tropospheric weather variability under highly influenced by the state of the Earth's surface. The 500 hpa level of variability that described the dynamics of the middle troposphere. And 300 hpa level represented the dynamics of the atmosphere at the upper-level troposphere.

While, the physical analysis investigated SST anomaly, MJO phase, and ENSO phase. The SST anomaly data were NOAA/NESDIS SST100 1x1 degrees spatial resolution to found the sea surface temperature anomalies. The data were composite gridded-image derived from 8-km resolution global SST observations and generated daily on a global scale.

OLR anomalies and RMM index data were used to found MJO phase. Outgoing longwave radiation (OLR) is used as a way to identify tall, thick, convective rain clouds. The NOAA OLR anomalies maps showed the difference from expected cloudiness based on the position of the MJO. The violet and blue shading indicated higher than normal, active or enhanced tropical weather, while orange shading indicated lower than normal cloud or suppressed conditions. The MJO diagram illustrated the progression of the MJO through different phases, which coincided with locations along the equator around the globe. RMM1 and RMM2 are mathematical methods that combine cloud amount and winds at upper and lower levels of the atmosphere to provide a measure of the strength and location of the MJO [3]. When the index is within the center circle, the MJO is considered weak, meaning it is difficult to discern using the RMM methods. Outside of this circle the index is stronger and will usually move in an anti-clockwise direction as the MJO moves from west to east. There were eight different MJO phases in the diagram.

While ONI index data used to found ENSO phase. Oceanic Nino Index (ONI) was three months running mean of ERSST.v4 SST anomalies in the Niño 3.4 region (5°N - 5°S and 120° - 170°W), based on centered 30-year base periods updated every five years. In this case, ONI used 1981-2010 base periods. Warm (El Nino) periods based on a threshold of + 0.5°C and cold (La Nina) based on a threshold of - 0.5°C.

Upper air sounding analysis was used to study the predictability of extreme weather. The upper air data used rawindsonde of Biak meteorological station observation provided by the University of Wyoming. The parameters investigated were CAPE, K Index, Lifted Index, Sweat Index, and Showalter Index by analyzing based on the threshold values. MTSAT imagery used to verify the atmospheric conditions.

3. Results and Discussion

3.1. The Atmospheric Dynamical Analysis

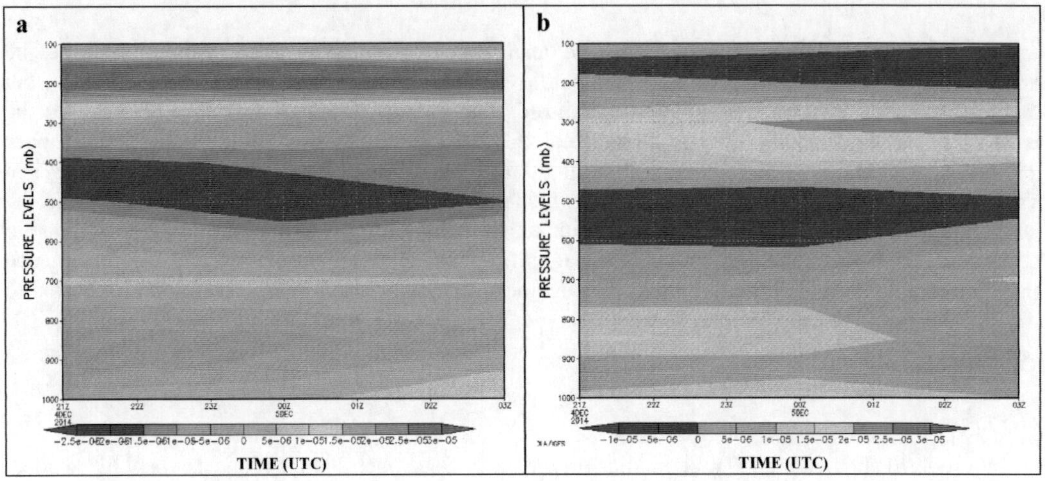

Figure 1. (a) Divergence in Biak; (b) relative vorticity in Biak on 4-5 December 2014

Figure 1 (a) showed that in 850mb level, negative divergence occurred in the range $-0.5\ e^{-5}\ s^{-1}$ to $-1\ e^{-5}\ s^{-1}$. The divergence in each layer varied to heights. In the higher level of 200 hpa, there was strong positive divergence. Divergence is a vector field that showed the flow in the atmosphere. A positive divergence in the upper-level and a negative in the lower level of troposphere indicated rising air movement. The more positive value of divergence, the air mass diverged and sank. The more negative value, the stronger the convergence indicated the growth of clouds. In the area of convergence activities occurred convection that formed many convective clouds.

Relative vorticity indicated the direction of rotation of air mass in the atmosphere on the micro scale. Due to Biak location in the southern hemisphere, the (+) values indicated the anticyclonic movement, while the (-) values indicated cyclonic rotation. The greater the magnitude of vortices, the stronger anticyclonic movement of air mass so that the chances of the formation of clouds getting smaller because of the air mass spread. Figure 1 (right) showed the relative vorticity at 500 hpa level reached $-9\ e^{-6}\ s^{-1}$, indicated a cyclonic movement (clockwise rotation). In the southern hemisphere, the more negative vorticity values, the greater the convection currents that occurred, leading the formation of clouds.

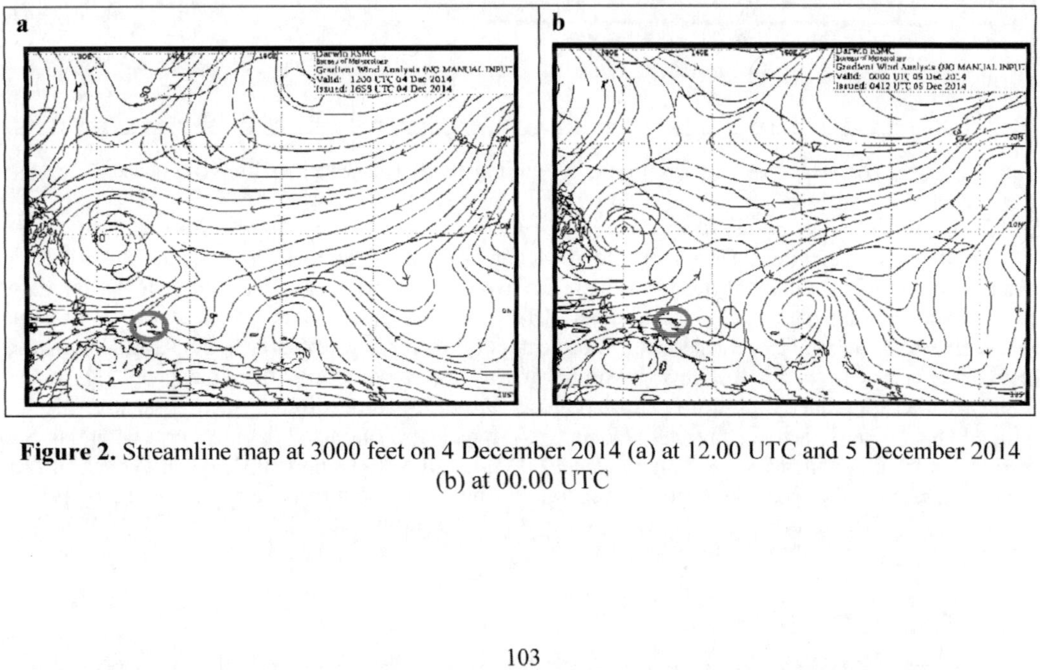

Figure 2. Streamline map at 3000 feet on 4 December 2014 (a) at 12.00 UTC and 5 December 2014 (b) at 00.00 UTC

An extreme weather phenomenon can be analyzed by using streamline map to describe the atmospheric condition. Streamline map included as synoptic scale analysis where the wind data from many observation stations were spatially analyzed. The synoptic analysis is a way of studying the weather system at a specific time on a large scale [15]. Based on the streamline map on 4 December 2014 at 12.00 UTC and 5 December 2014 at 00.00 UTC from BOM (Bureau of Meteorology) in Figure 2, wind patterns at an altitude of 3000 feet showed that there was a low-pressure area in the north of Papua Island and in the eastern of Maluku. In the western of Philippines occurred a synoptic scale weather phenomenon called Typhoon Hagupit. For the Biak region was an area of convergence that caused air masses to converge, or come together to Biak. Some of these conditions supported to the occurrence of the heavy rain in Biak.

3.2. The Atmospheric Physical Analysis

3.2.1. MJO phase analysis

Figure 3 (a) showed the OLR anomalies on the early December 2014. Indonesia was blue shaded, categorized as low OLR level. It indicated that there were many cloud cluster mainly convective clouds covered Indonesian region including Biak. The figure 3 (b) showed RMM index was in the fifth phase, indicated by convective activities in the maritime continent (Indonesia). By early December 2014, the MJO (Madden-Julian Oscillation) phase shifted eastward so that enhanced convection centered over the maritime continent (Indonesia). While, anomalous convection was weak across the central Pacific. So that, the MJO supported convection activities that produced convective clouds in Biak.

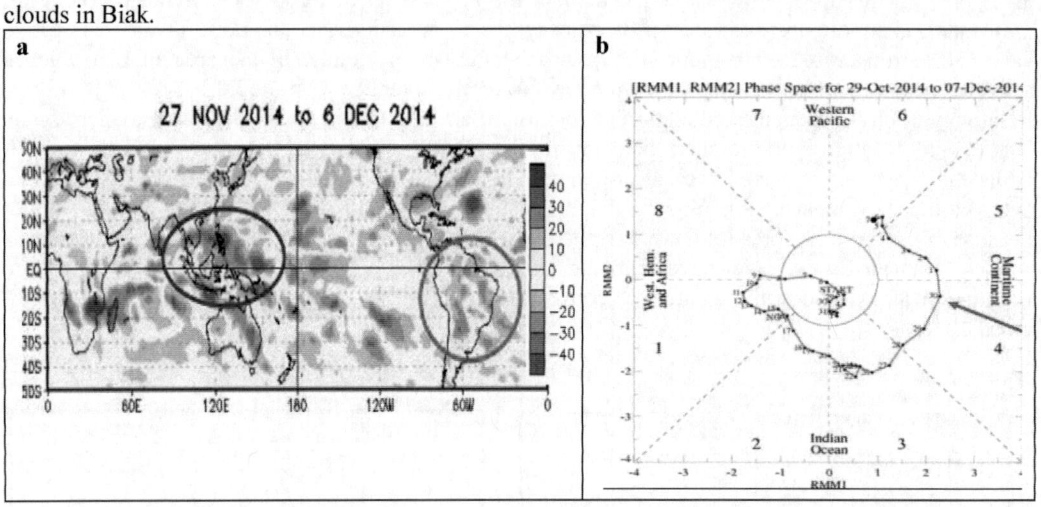

Figure 3. (a) NOAA OLR anomalies composite; (b) RMM index for MJO phase

3.2.2. Sea surface temperature analysis

Based on the satellite measurement conducted by NOAA (National Oceanic and Atmospheric Administration) of the daily average of sea surface temperature and sea surface temperature anomaly, the sea surface temperature on 3-5 December 2014 in the sea around Biak was in the range of value between 29.5°C – 31.5°C. It considered a warm category. Increasing sea surface temperature indicated increasing energy in the sea that gave possibility increasing evaporation in the atmosphere [1]. It also indicated that the water supply was still available in the formation of convective clouds in the sea around Biak. A fairly active convective clouds growth had heavy rain and thunderstorm potential. Sea surface temperature anomaly was in the range +1.0°C - +1.50°C to its normal. The warmer sea surface temperature anomalies generated convective cloud forming because it had enough latent energy [9].

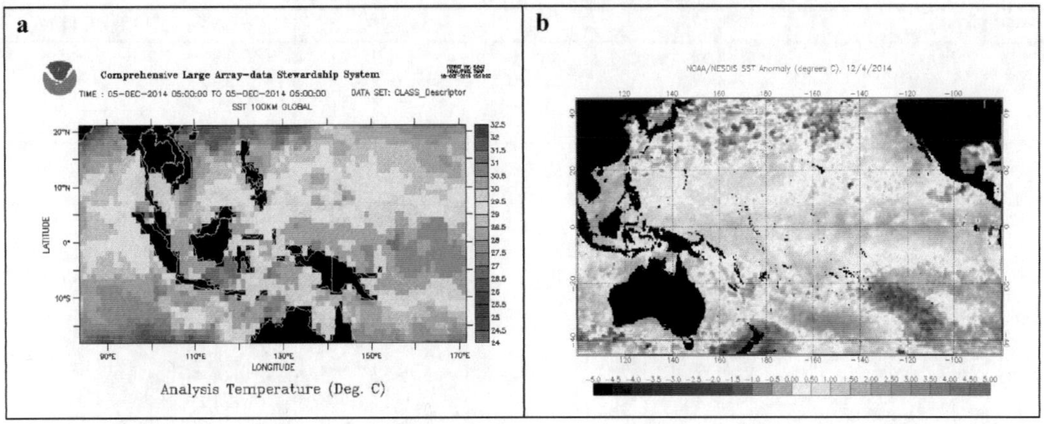

Figure 4. (a) Sea surface temperature on 5 December 2014; (b) sea surface temperature anomaly for three days (4–6 December 2014)

3.2.3. ENSO phase analysis

The SST anomalies based on Oceanic Nino Index, on December 2014 reached + 0.6 ^{o} and considered as weak El Nino. In this condition, the Walker circulation shifted to the eastward in Pacific. There was convergence movement in the central and the eastern Pacific due to increasing rising air in the region. While, in Indonesia, the air mass diverged due to sinking air. The influence of El Nino on eastern maritime continent was decreasing rainfall [8]. So that, in weak El Nino condition, Biak monthly rainfall decreased but not significant.

3.3. Upper Air Sounding Analysis

Local scale is used to determine the convection factor of an area, which is used to determine the vertical cloud growth area [17]. Meteorologists are concerned with static stability parameters to understand convective weather patterns [16]. If the atmosphere is unstable with abundant low-level moisture and a mechanism exists to lift the air (thereby releasing the potential instability), convective weather and rainfall (showers) can develop.

Based on the analysis of stability indices in Table 1 on the day of the super heavy rainfall weather phenomenon occurred and the previous two days showed an increased stability indices values such as LI (Lifted Index) and CAPE (Convective Available Potential Energy). Lifted Index analysis showed that the stability value during heavy rain occurred was quite high than two days before in the amount of -3.76, while two days before only reached -2.49 and -3.09. The negative LI value that reached -3.76 indicated a strong thunderstorm potential. Although the KI index showed a lower value than two days before, KI analysis was used to identify potential convective still showed a high stability value on the day of the super heavy rain occurred in the amount of 33.70. KI value was in a moderate condition which indicated a thunderstorm potential 60-80%. The analysis of SI index showed that the SI Index value was in the amount of -0.05 indicated a weak potential of the thunderstorm. Based on potential energy analysis in Biak, CAPE index value (Convective Available Potential Energy) of 1720 J/kg was in moderate convection category showed the enough energy potential in convective potential. While SWEAT index that commonly used for tornado and thunderstorm system prediction showed a lower value on the day heavy rainfall occurred than the previous two days.

Table 1. Atmospheric stability indices value on 2-4 December 2014

Date/ Index	LI	KI	SI	SWEAT	CAPE [J/kg]
2 Dec	-2.49	36.10	-1.07	248.8	523.7
3 Dec	-3.09	35.20	0.00	252.4	1425.0
4 Dec	-3.76	33.70	-0.05	216.0	1720.0

3.4. Satellite Imagery

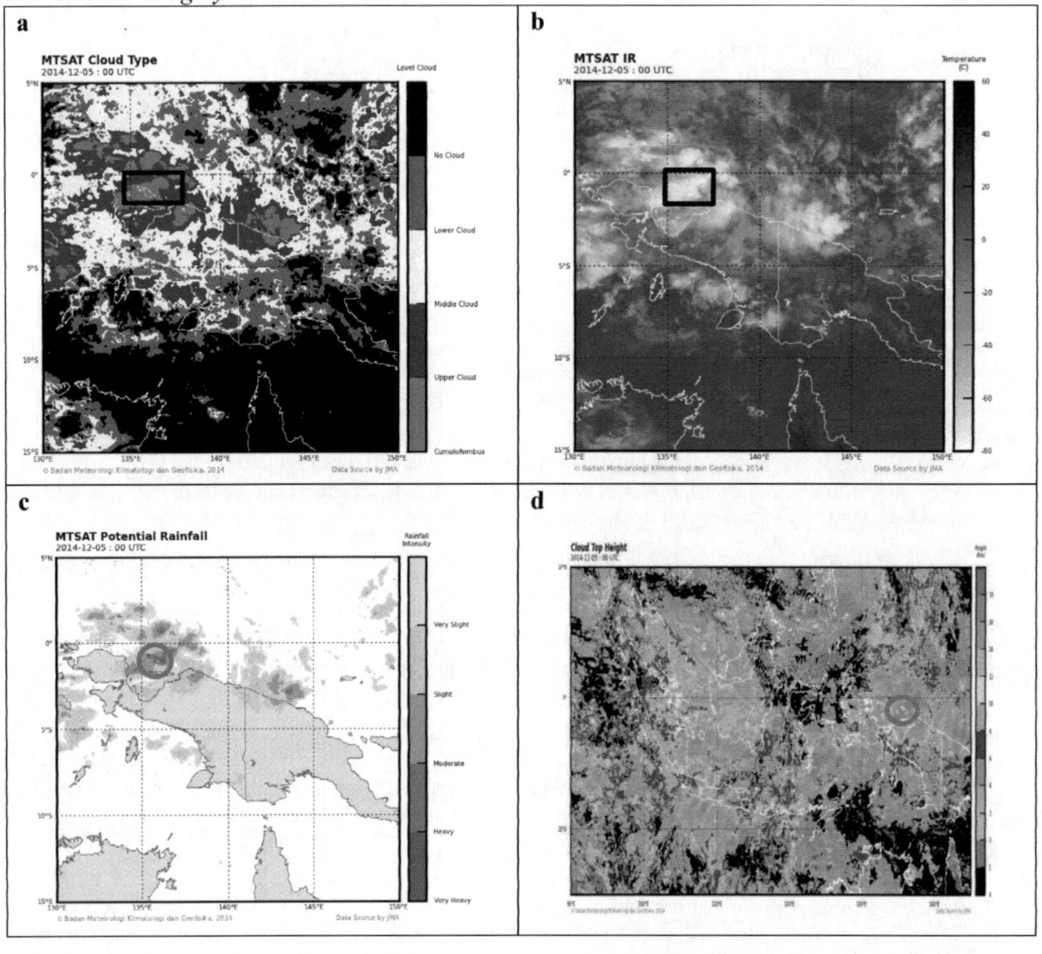

Figure 5. (a) MTSAT Cloud type imagery at 00 UTC, 5 December 2014; (b) MTSAT channel IR at 00 UTC, December 2014; (c) MTSAT Cloud Type Height at 00 UTC, 5 December 2014; (d) MTSAT Potential Rainfall at 00 UTC, 5 December 2014

According to the analysis of satellite imagery MTSAT Cloud Type BMKG on 4 December 2014 at 18:00 UTC until 5 December 2014 at 04:00 UTC showed convective cloud cover, cumulonimbus cloud (red area), above Biak region. This convective cloud cover condition was also supported by the analysis of satellite imagery MTSAT channel IR on 4 December 2014 at 18 UTC until 5 December 2014 at 04.00 UTC. From the cloud temperature that showed in Figure 2 showed that the cloud top temperature was in the range of -40°C to -80°C. The negative cloud temperature associated with cumulonimbus cloud and cirrus cloud. But in this case, the type of cloud above Biak region was cumulonimbus cloud. The negative cloud top temperature values can be used to estimate the height of cumulonimbus cloud in the troposphere. Analysis of satellite imagery MTSAT Cloud Top Height on 4 December 2014 at 18:00 UTC until 5 December 2014 at 04:00 UTC indicated that the cloud top height was in the range of 12-18 km. The rainfall potential that showed from the analysis of satellite imagery MTSAT Potential Rainfall in Biak region indicated the potential was in the category of slight to very heavy with the dominance potential was heavy rainfall.

4. Conclusion

Research on the atmospheric conditions analysis of the super heavy rain phenomenon in Biak on 4-5 December 2014 gave the conclusion. The atmospheric conditions analyzed in dynamical and physical aspects. From the atmospheric dynamical analysis, the negative divergence value in the lower troposphere and negative relative vorticity value in middle troposphere indicated there was air mass convergence movement in Biak. The positive divergence value in higher troposphere indicated a divergence area which identical to rising air movement where the boundary layer was in unstable condition and supported for convection activities. Streamline analysis at 3000 feet showed low-pressure areas in the north of Papua and in the eastern of Maluku. In the western of Philippines occurred Typhoon Hagupit. Besides that, Biak was in an ITCZ area. All these conditions could be a trigger for convective cloud cluster forming.

Atmospheric physical analysis showed a low OLR anomalies value that indicated a convective cloud cluster covered maritime continent region including Biak. MJO was in phase 5, indicated by an enhancement in convection activities that centered over the maritime continent. Sea surface temperature condition and its anomalies analysis showed a warm sea surface temperature and a warmer anomaly. These conditions supported for convective cloud forming in Biak region. While ENSO phase analysis indicated a weak El Nino where Biak Monthly rainfall decreased but not significant. Upper air sounding analysis showed an increased stability indices values such as LI and CAPE. Lifted Index value of -3.76 during heavy rain occurred was quite high than two days before, while two days before only reached -2.49 and -3.09. The negative LI value indicated a strong thunderstorm potential. Moreover, CAPE index value of 1720 J/kg had a huge potential energy which can be a trigger for convective cloud development. On the other hand, other stability indices were not describing upper air condition appropriately for the super heavy rain phenomenon.

Based on analysis of MTSAT satellite imagery showed a convective cloud cover, cumulonimbus type above Biak region where the cloud top temperature was is in the range of -40°C to -80°C and the cloud top height was in the range of 12-18 km. These indicated heavy rainfall potential in Biak region. Most of the dynamics and physical analysis of the atmospheric conditions in Biak on the dates were proved and supported to super heavy rain phenomenon occurred.

References

[1] Aldrian E 2008 *Meteorologi Laut Indonesia* (Jakarta: Badan Meteorologi Klimatologi dan Geofisika)
[2] Aldrian E and Susanto R D 2003 *International Journal of Climatology* **23** 1435-1452
[3] CPC NCEP 2014 *Madden-Julian Oscillation: Recent Evolution, Current Status and Predictions* (United States: National Oceanic and Atmospheric Administration)
[4] Fadholi A, Sari F P, Aji P and Ristiana Dewi 2014 *Jurnal Fisika dan Aplikasinya* **10** 24-30
[5] Hess G D, Spillane K T, and Lourensz R S 1988 *Journal of Applied Meteorology* **27** 305-317
[6] Jacovides C P and Yonetani T 1990 *Weather and Forecasting* **5** 559-569
[7] Kirtman Ben and Vecchi G A 2011 *Why Climate Modelers Should Worry About Atmospheric and Oceanic Weather* (Singapore: World Scientific Publishing) p 511-523
[8] Kubota H, Syamsudin F, Shirooka R and Hamada J I 2011 *Journal of the Meteorological Society of Japan* **89A** 111-122
[9] Nuryanto E D and Imelda U B 2014 *Jurnal Meteorologi dan Geofisika* **15** 147-155
[10] Peatman S C, Matthews A J, and Stevens D P 2014 *Quarterly Journal of the Royal Meteorological Society* **140** 814-825
[11] Qu Tangdong, Du Y, Strachan J, Meyers G, and Slingo J 2015 *Oceanography* **18** 50-61
[12] Schmetz J, Borde R, Holmlund K, and Ko¨nig M 2005 *Geophysical Research Letters* **32** 1-4
[13] Showalter A K 1953 *Bulletin American Meteorological Society* **34** 250-252
[14] Tentang Meteorologi http://meteojuanda.info/index.php?option=com_content&view=article&id=36&Itemid=34. (last accessed 4 November 2016) (Surabaya: Juanda Meteorological Station)
[15] Wirjohamidjojo S and Subagyo Y 2013 *Meteorologi Sinoptik: Analisis dan Penaksiran Hasil Analisis Cuaca Sinoptik* (Jakarta: Pusat Penelitian dan Pengembangan Badan Meteorologi Klimatologi Dan Geofisika) p 9-143
[16] Wilson G S and Scoggins J R 1976 *NASA Contractor Report CR-2678* (Washington: National Aeronautics and Space Administration) p 129
[17] Zakir A, Sulistya W, and Khotimah M K 2009 *Perspektif Operasional Cuaca Tropis* (Jakarta: Pusat Penelitian dan Pengembangan Badan Meteorologi Klimatologi Dan Geofisika) p 67

Design of heart rate monitor based on piezoelectric sensor using an Arduino

Veni Setyowati[1*], Jodelin Muninggar[1], Made R. S. Shanti.N.A[1,2]**

[1]Department of Physics
[2]Department of Physics Education,
Faculty of Science and Mathematics, Universitas Kristen Satya Wacana
Jl. Diponegoro No.52-60, Salatiga, Sidorejo, Kota Salatiga, Jawa Tengah 50711, Indonesia

*) E-mail: 642012004@student.uksw.edu

Abstract. Reading of result heart rate using an acoustic stethoscope needs a particular skill, quiet environment, and hearing sensitivity. This project had the purpose design of a user-friendly automatic heart rate monitor and especially in a noisy area which to eliminate problems and incorrect reading of result. The liquid crystal display shows a heart rate as a result of measurements. The design of the heart rate monitor has two main parts; the signal recorder that a piezoelectric sensor, a filter, and an amplifier as recorder. The second parts was Arduino microcontroller with reinforced. Besides, three supporting buttons provided as the manual switches, the 'on', the 'start', and 'reset' buttons. The values acquired from the heart rate monitor indicate that those were on the Vernier BPS-BTA value range. The measurement error factor of the heart rate monitor then compared to the Vernier BPS-BTA test device was 3.15%. Besides, the value of statistical independent-test indicates that there is no significant difference (P = 0.971) between the heart rate monitor device and the Vernier BPS-BTA. In conclusion, this device was ready to be used because it has almost the same accuracy with the standart device.

** Corresponding author: made.ray@staff.uksw.edu

1. Introduction

The heart is an organ which pumps blood throughout the body and return to the heart as the effect of my card contraction. The tricuspid valves and bicuspid valves can open and to close automatically [7, 15]. This condition is useful to ensure that the blood flows to aortic and periphery area with the pulse characteristics [8, 9]. The hearts distinctive rhythm is an indicator to diagnose of heart diseases[1, 16, 12] which have a distinctive rhythm at frequency 20-800 Hz which can be listened by ears [5]. On a medical check-up, one of the diagnosis devices which have high accuracy existence is acoustic stethoscopes. There are several factors that influence the diagnosis accuracy on the utility of acoustic stethoscope. There is the sound (pulse) on the chest wall or blood vessels which recorded, the heartbeat that can be perceived by human ears and environmental condition [2]. Those factors make the stethoscope utility on diagnostic by medical experts [5]. Based on those factors, the acoustic stethoscope is applied for calculating of heart rate. Therefore, heart rate monitor device to applied and operated in a noisy environment. This study to design a device for calculate heart rate automatically using a piezoelectric sensor which controlled by an Arduino. Hopefully, it used as a learning media for healthcare students or medical physics.

2. Methode

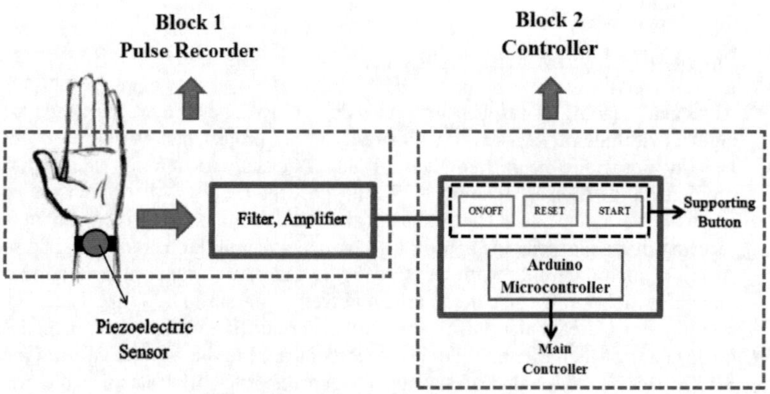

Figure 1. The design of heart rates monitor

Figure 1 show an automatic design of heart rate. The device divided into two main blocks. The first was pulse recorder block, which consists of a piezoelectric sensor, a filter and an amplifier (Op-Amp). The piezoelectric sensor was medically used to convert sound energy in the body into electrical pulse [10-13]. The piezoelectric sensor was put on the heart pulse on the wrist using a mild pressure, the sensor placed as close as possible to the heart rate to record it [3, 14]. The sign of detected heart rate to show by the flashing lights between the filter and amplifier circuits, as shown in Figure 2. The pulse recorded and filtered by the piezoelectric sensor. The function of a filter was to minimize the error during the pulse recording [17]. The filtered pulse was not added a new frequency from the recorded input because the filter converts the size of amplitude small relatively only. Therefore, an amplifier is needed to amplify the output signal produced by the piezoelectric sensor.

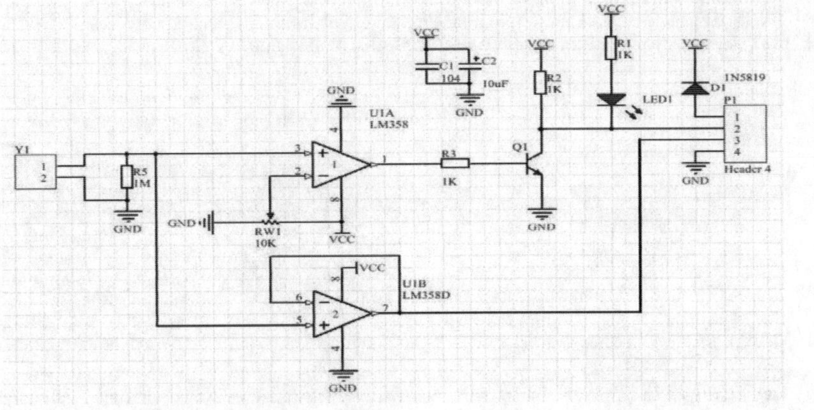

Figure 2. Filter, Amplifier, and Piezoelectric sensor
Source: www. Indo-ware.com [4]

The second was an amplifier controller block using an Arduino microcontroller. The Microcontroller Arduino Hardware was designed for a simple open hardware in an Arduino board using Atmel processor support and onboard I/O [6, 8]. The Arduino microcontroller functions to convert the recording from piezoelectric sensor and to count the heart rate for 60 seconds which displayed on an LCD in BPM (Beat per Minute). Besides, three supporting buttons are available in this device as manual switches: the "on", the "start", and the "reset" buttons. Those three buttons have an important role in calculating the heart rate. When the "on" button, pressed to connect the power supply, the device will prepare every component to be ready. The "start" button pressed, the heart rate calculating runs automatically and stops after 60 seconds. Restarting the calculating will be done by pressed the "reset" button.

3. Results and Discussion

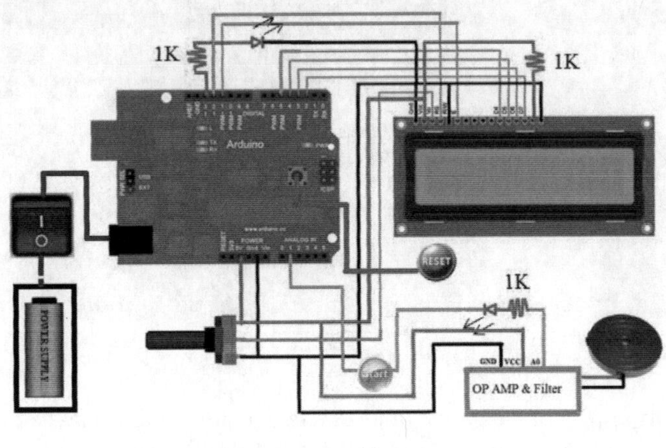

Figure 3. The Device Circuit

Figure 2 show the circuit of the heart rate monitor device design. The device packed in a casing. Then, the heart rate monitor based on piezoelectric sensor using an Arduino, showed on Figure 4.

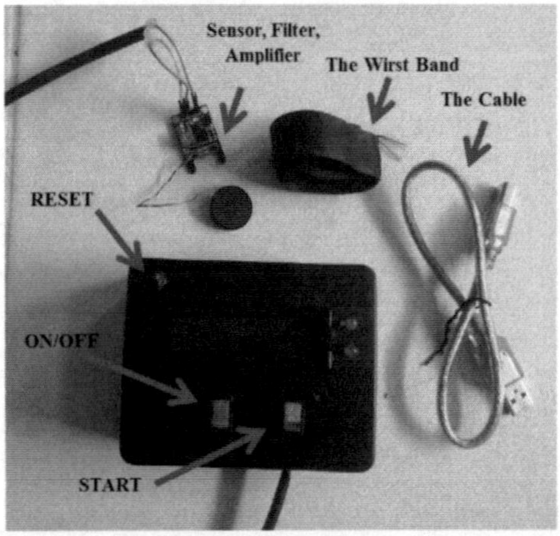

Figure 4.a. The heart rate monitor display

Figure 4.b. Appear "waiting. . ." on the display when the device is going to start the data calculating

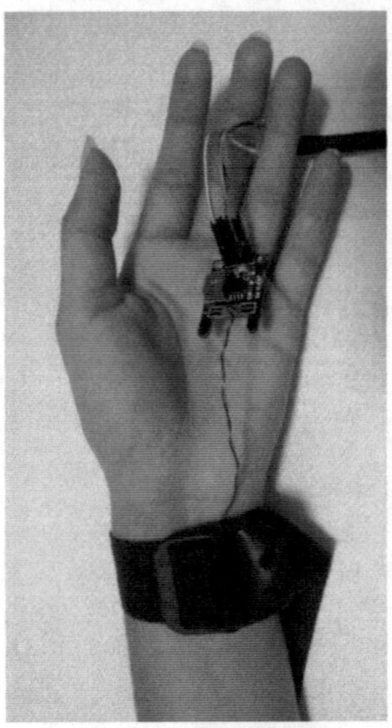

Figure 4.c. The position sensor

Figure 4.d. The counting stops automatically

Figure 4a show the display of the design of heart rate monitor. At the time the device turned on by pressing the "start" button, it says "waiting…" on the monitor before starting the calculating as shown in Figure 4b, then "BPM" will appear on the screen but the calculating has not started. At the moment after "waiting …" appears. The piezoelectric sensor has put on the wirst band within a mild pressure as seen on figure 4.c. Then, the calculating start and stop automatically after 60 seconds, showed on Figure 4.d.

The heart rate monitor device is tested by compared to another device, the Vernier BTS-BTA, to get count of the heart rate in BPM (Beat per Minute). The comparisons of heart beat five samples measuring the automatic heart rate monitor designed with the Vernier BPS-BTA are showed in Table 1. The data from five samples using the heart rate monitor within the standard deviation were: sample 1 (75 ± 5) BPM, sample 2 (86 ± 2) BPM, sample 3 (74 ± 1) BPM, sample 4 (93 ± 3) BPM, and sample 5 (93 ± 3) BPM. The result compared to Vernier BPS-BTA, and the results were: sample 1(75±4) BPM, sample 2 (85 ± 4) BPM, sample 3 (76 ± 5) BPM, sample 4 (91 ± 5) BPM, and sample 5 (91 ± 2) BPM. The values acquired from the heart rate monitor indicate that those count on the Vernier BPS-BTA value range.

Table 1. Comparison result of the heart rate monitor (HRM) device to Vernier BPS-BTA

	The Vernier BPS-BTA	Vernier Standard Deviation	Vernier Error (%)	Heart Rate Monitor	Device Standard Deviation	Heart Rate Monitor Device Error (%)
Sample 1	75	4	5.1	75	5	6.8
Sample 2	85	4	4.8	86	2	1.8
Sample 3	76	5	6.2	74	1	0.8
Sample 4	91	5	5.4	93	3	3.1
Sample 5	91	2	2.5	91	3	3.3
	Mean		4.8		Mean	3.16

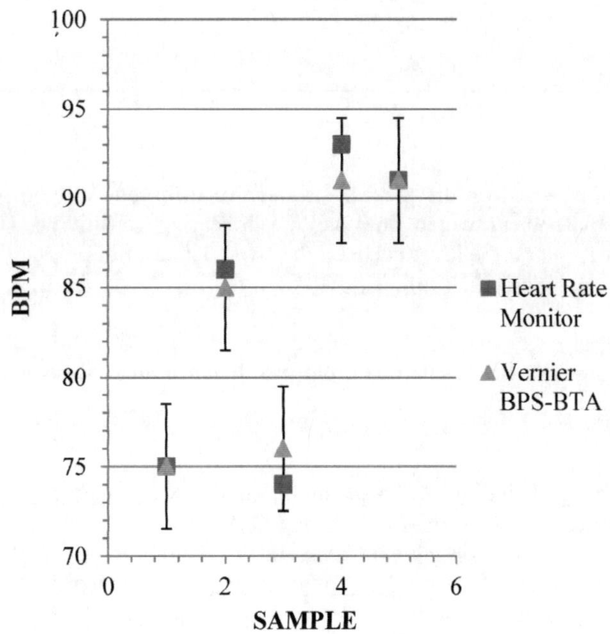

Figure 5. Distribution of the five samples recorded by the heart rate monitor and the comparator device, the Vernier BPS-BTA

The proportion of calculating error factor of the heart rate monitor device was 3.15%, as shown in Table 1. While the distribution of the five samples testing results between the heart rate monitor and the comparator device the Vernier BPS-BTA, showed in Figure 5. The result calculation of the Vernier BPS-BTS has the standard deviation in the range of the data from the calculation result use the automatic heart rate monitor. Therefore, the heart rate monitor was possible as a learning media for healthcare students or medical physics in calculating the heart rate. Besides, independent t-test performed, and the result shows that there was no significance differences (P=0.971) between the heart rate monitor and the Vernier BPS-BTA (Table 2).

Table. 2. The comparison result of Heart Rate Monitor (HRM) and Vernier BPS BTA with Statistical Test used Independent T-Test

		t-test for Equality of Means						
		t	df	Sig. (2-tailed)	Mean Difference	Std. Error Difference	95% Confidence Interval of the Difference	
							Lower	Upper
HRM and Vernier BPS-BTA	Equal variances assumed	-.038	8	.971	-.20000	5.28205	-12.38042	11.98042
	Equal variances not assumed	-.038	7.870	.971	-.20000	5.28205	-12.41541	12.01541

4. Conclusion

The heart rate monitor based on the piezoelectric sensor using an Arduino possible used everyone even in a noisy environment which does not need hearing sensitivity. To use this device, a piezoelectric sensor is put on the heart pulse on the wrist with a mild pressure so that the sensor is as close as possible to the sound of a heartbeat.

5. Acknowledgments

This research was supported by Department physics, Science and Mathematics faculty, Universitas Kristen Satya Wacana.

6. Reference

[1] Shu Yi, Li Cheng, Wang Zhe, Mi Wentian, Li Yuxing, Ren Tian-Ling 2015 *MDPI AG* **15** 3224
[2] Mangion Kenneth. 2007 *Malta Medical Journal* **19** 41.
[3] Sujatha E, Y Radha 2013 *International Journal of Scientific Research* **2** 206
[4] Indoware *Piezoelectric Sensor* https://www.indo-ware.com/produk-2856-piezoelektrik-piezo-elektrik-sensor-getar-indoware.html, accessed on 15 August 2016.
[5] Nolmanee Suranan, Tunkasiri Tawee, Siriwitayakorn Kingkeo, Tantrakoon Jerapong 2007 *MDPI AG* **7** 3192
[6] Sethuramalingam, T.K, Karthighairasan, M 2012 *Bonfring International Journal of Power Sistem Integrated Circuit* **2** 18
[7] Didier C. Combatalade, D.C. 2010. *Heart Rate Variability Applied to Psychophysiology*. Thought Technology Ltd. MAR 953-00
[8] Arduino Store *Arduino Uno*, http://datasheet.octopart.com/A000049-Arduino-datasheet-10591224.pdf, Accessed : 20 September 2016

[9] Anonim. 2011. Anatomy and Physiology of the Cardiovascular System. Jones and Bartlett Publishers. Capter 5; page : 41-43

[10] Luna Michael. E.S, Fullan Scott 2014 *Patent Application Publication* US 20140128753A1

[11] Curran Tony, Sheppard Gill 2011 Cardiologi, *Modul 1: Anatomy and Physiology of The Heart* Canterbury District Health Board Te Te Poari Hauora o Waitaha. 6

[12] Kavitha, K.C, Banu, A. Bazila 2014 *International Journal Of Innovative Research In Science, Engineering And Technology (IJIRSET)* **3** 2417

[13] Park Jang-Ho , Jang Dae-Geun, Park Jung. Wook, Youm Se-Kyoung 2015 *MDPI AG* **15** 23402

[14] Datta Roop. Pratim, Gupta Debopriya 2015 *International Journal of Engineering Research & Technology (IJERT)* **4** 1199

[15] Van Putte Cinnamon. L, Regan Jennifer . L, Russo Andrew .F 2007 *Essentials of Anatomy and Physiology* New York, United State of America McGraw-Hill Global Education, 0072943696 323

[16] Ufoaroh S U, Oranugo C O, Uchechukwu M E 2015 *International Journal of Engineering Research and General Science* **3** 26

[17] Lacanette Kerry 2010 *A Basic Introduction to Filter –Active , Passive , and Switched Capasitor National Semiconductor Corporation* Texas Instrumen

IOP Conf. Series: Journal of Physics: Conf. Series **795** (2017) 012017 doi:10.1088/1742-6596/795/1/012017

Visualization of heat transfer in material for varians of boundary value with Relaxation Iteration Gauss-Seidel method

[1]Imam Basuki, Cari, Suparmi,

Physics Department of Post Graduate Program Sebelas Maret University, Jl. Ir. SUtami 36A KentinganJebres Surakarta 57126, INDONESIA

E-mail: hitakayana2yb@gmail.com

Abstrack— The research was aimed to know the effect of initial boundary value to the heat propagation rate pattern using iterations over Gauss-Seidel relaxation method and to analyze the exact value of each node descritization profile of test material. This study was an analytical study to fine analytical an numerical solution. Result from this study is that the pattern of variation of the boundary or initial conditions of a material with regard conductivity value remains at steady state the exact value of the smallest are in the same iteration value. The indicates that the value of the thermal equilibrium tend to be at the same iteration. Result from study showed that the pattern of initial boundary values that causes steady state of heat propagation of test material that has smallest exact similar to the iteration value.

1. Introduction

Physics concepts are produced through experiments and theoretical works which are in the form of mathematical equation. Physics is said as the father of technology since the advanced development of physics is in the form of technology, and in turn, physics developed rapidly due to the application of technology in physics, theoretically, physics concepts are described using mathematical modeling in the form of mathematical equation tha can be solved analytically or numerically. Physics is closely related to mathematics for theoretical physics often expressed in mathematical notation and mathematical logic can provide a framework in which the laws of physics can be formulated appropriately [1]. One development of the mathematical methods used in the case of the physical and engineering fields is a numerical method [2]. Numerical methods are techniques to solve problems that are formulated mathematically by means of operating a matter of analytic [3]

Heat transfer is one of the physical phenomena that is described partial differential equations Laplace or Poison [4]. In accordance with the usefulness of mathematics in the fields of physics, the solution partial differential equations can be obtained by several methods. The selection of methods and approaches based on the purpose and complexity of the problems.

Heat transfer is a concept to predict the energy transfer occurs due to a temperature difference between objects or material [5].

The heat energy can not be observed directly but the direction of movement and its effects can be observed and measured, as in the event of conduction heat transfer. Conduction is a process that if two objects or two-part of object is contacted with the other temperature it will pass heat transfer [6].

Heat flows from a higher-temperature object to a lower body temperature. Heat transfer rate that passes through solid objects is proportional to the temperature gradient or temperature difference of unity long [7].

As mentioned before that the rate of heat propagation in a variety of geometric shapes material following the equation of state is presented in the form of mathematical equations in the form of differential equations. Solution method or the propagation model of the flow of heat at a material solved analytically and numerically. The analytically solution by using a systematic calculation and the obtained solution in the form of the exact value [8].

However, some form of differential equations, there is difficulty the applying analytical resolution methods so numerical methods is chosen exam alternative to overcome. Numerical methods are used to solve partial differential equations such as Method Crank- Nicholson, methods Milne, Hamming method and Gauss-Seidel method [9].

In this studies obeys the rate of heat flow that follows the model of partial differential equations are solved using numerical methods of Gauss-Seidel and then subsequently visualized using Mathlab with some variation of the initial boundary value.

Shape of specimens analyzed in this study are presented in Figure 1 with an initial boundary values are follow :

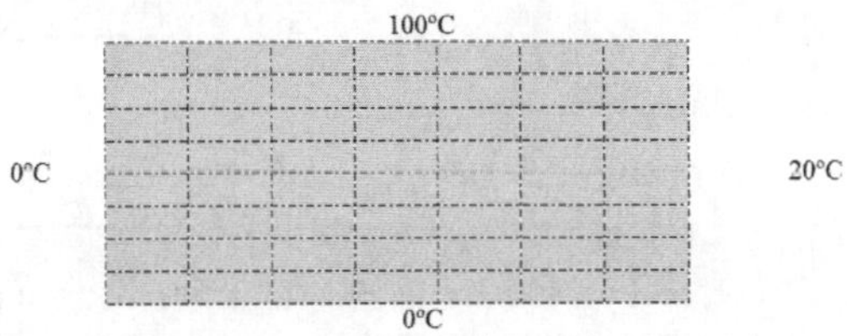

Figure 1. Model workpiece as two-dimensional metal plate with a specific conductivity value in steady state.

Based on the figures, by using Gauss-Seidel methods can obtain the solution of partial differential equations in the hot stream on a metal plate two dimensions in a steady state or heat flow rate system does not change with time (constant), then the temperature at any point does not change.

To apply Gauss-Seidel relaxation method, we set variation value of the initial boundary on the completion of iterations over Gauss Seidel relaxation method to observe how the heat propagation rate pattern seen from the distribution of its exact value on each node discretization test material and visualize the propagation of heat profile.

The mathematical equation that govern the heat propagation rate is given as :

$$\frac{\partial^2 T}{\partial x^2} + \frac{\partial^2 T}{\partial y^2} + g = 0$$

With :
T : temperature
x : absis
y : coordinat
g : $\frac{q(\Delta x^2)}{k}$

To solve the above equation, it is used a Taylor series with two independent variables $T\ (x, y)$ is a way to add additional variables so that the Taylor series with two independent variables $T\ (x, y)$ be [10] :

$$\frac{\partial^2 T}{\partial x^2} + \frac{\partial^2 T}{\partial y^2} + \frac{q(\Delta x^2)}{k} = 0$$

To get the second derivative can be done in the following way:

1) If the first derivative of the advanced differential form, then the second derivative settled in the form of differential retreat.

2) If the first derivative of the differential form of retreat, then the second derivative settled in the form of forward differential.

Now we know the second derivative function of T with respect to x and y, then substituted in equation Poisson equation or temperature distribution, in order to obtain:

$$\frac{\left(\frac{T_{i+1,j}-T_{i,j}}{\Delta x}\right)-\left(\frac{T_{i,j}-T_{i-1,j}}{\Delta x}\right)}{\Delta x}+\frac{\left(\frac{T_{i,j+1}-T_{i,j}}{\Delta y}\right)-\left(\frac{T_{i,j}-T_{i,j-1}}{\Delta y}\right)}{\Delta y}=0$$

$$\frac{T_{i+1,j}-2T_{i,j}+T_{i-1,j}}{\Delta x^2}+\frac{T_{i,j+1}-2T_{i,j}+T_{i,j-1}}{\Delta x^2}+\frac{q}{k}=0$$

For the size of Δx and Δy are the same, then the above equation simplifies to:

$$T_{i+1,j}-2T_{i,j}+T_{i-1,j}+T_{i,j+1}-2T_{i,j}+T_{i,j-1}+\frac{q(\Delta x^2)}{k}=0$$

or

$$T_{i+1,j}+T_{i-1,j}+T_{i,j+1}+T_{i,j-1}-4T_{i,j}+\frac{q(\Delta x^2)}{k}=0$$

Gauss Seidel equation constructed from the above equation becomes:

$$T_{i,j}=\frac{T_{i+1,j}+T_{i-1,j}+T_{i,j+1}+T_{i,j-1}+\frac{q(\Delta x^2)}{k}}{4}$$

and solving iteratively for i = 1 to n and j = 1 to m. In general, from the heat propagation equation with q is the rate heat transfer, T is the temperature distribution at a distance of x and y, which has a high length L and K. Therefore the value of T at the edge of the plate known temperature (boundary conditions) and at the time before propagation, the value at points it is zero (baseline) settlement is counting equation x and y T in particular.

For partial differential equations

$$\frac{\partial^2 T}{\partial x^2}+\frac{\partial^2 T}{\partial y^2}+\frac{q(\Delta x^2)}{k}=0 \qquad 0\le x\le L \ dan \ 0\le y\le K$$

$$T(0,y)=0$$
$$T(L,y)=0$$
$$T(x,0)=100$$
$$T(x,K)=20$$

known:

Δx = 0,1
Δy = 0,1
q = 10000 Btu/hr ft
k = 40 Btu/hr ft

To find out the solution heat propagation at each point of the Gauss Seidel method, can be done by solving iteratively for i = 0 to n, and j = 1 to m.

.

2. Results and Discussion

Completion of the heat propagation equation solved iteratively with over-relaxation equation. Relaxation parameters can be searched using the equation:

$$\omega=\frac{1}{1+\left(\frac{0,1}{0,1}\right)^2}\left[\cos\frac{\pi}{m}+\left(\frac{\Delta x}{\Delta y}\right)^2\cos\frac{\pi}{n}\right]=\frac{1}{1+\left(\frac{0,1}{0,1}\right)^2}\left[\cos\frac{3,14}{50}+\left(\frac{0,1}{0,1}\right)^2\cos\frac{3,14}{50}\right]=0,93$$

Thus, λ can be searched using the following equation:

$$\lambda=\frac{2}{1+\sqrt{1-\omega^2}}=\frac{2}{1+\sqrt{1-0,93^2}}=1,47=15$$

For partial differential equations:

$$\frac{\partial^2 T}{\partial x^2}+\frac{\partial^2 T}{\partial y^2}+\frac{q(\Delta x^2)}{k}=0 \qquad 0\le x\le L\ dan\ 0\le y\le K$$

using boundary conditions and the same coefficient of relaxation. Furthermore solved iteratively for $i = 1$ to n and $j = 1$ to m with over-relaxation following equation:

$$T_{i,j}^{new}=\lambda T_{i,j}^{new}+(1-\lambda)T_{i,j}^{old}$$

Iteration can be stopped if the relative error has reached the limit. The magnitude of the relative error is defined as:

$$|(\varepsilon_a)_{i,j}|=\left|\frac{T_{i,j}^{new}+T_{i,j}^{old}}{T_{i,j}^{old}}\right|\times 100\%$$

Analytical solutions obtained are as follows: In the initial condition or Iteration 0, the solution in the first iteration by means of Gauss Seidel, then at the point $T_{i,j}$, :

$$T_{1,1}=\frac{T_{i+1,j}+T_{i-1,j}+T_{i,j+1}+T_{i,j-1}+\frac{q(\Delta x^2)}{k}}{4}=\frac{0+100+0+50+\frac{10000\times(0.1)^2}{40}}{4}=37{,}75$$

of the equation over-relaxation ($\lambda = 1.5$) was obtained

$$T_{1,1}=1.5(37.75)+(1-1.5)0=56{,}625$$

for $T_{2,1}$:

$$T_{2,1}=\frac{0+56.625+0+50+2.5}{4}=26.906$$

From over-relaxation of the equation is obtained:

$$T_{2,1}=1.5(26.906)+(1-1.5)0=40.359$$

iteration followed by Mathlab until iteration on $T_{7,7}$, point, in order to obtain a value as shown below. From the results of this iteration is then performed visualization of heat propagation using Mathlab [11] with some variation of the initial boundary conditions. And the results are as follows :

a) $T(0,y)=0; T(L,y)=20; T(x,0)=100; T(x,K)=0$

Profile 1
Iteration Matrix
Iteration = 24
Computing times = 0.0042768

100	100	100	100	100	100	100	100
0	481.716	658.982	726.063	738.412	700.802	572.574	20
0	267.883	428.148	506.856	526.784	492.221	389.495	20
0	161.668	278.869	346.431	369.647	351.804	293.184	20
0	99.920	179.230	230.353	253.569	252.165	231.435	20
0	58.781	107.779	142.181	162.109	171.852	180.393	20
0	27.426	50.922	68.484	80.834	92.743	118.284	20
0	0	0	0	0	0	0	0

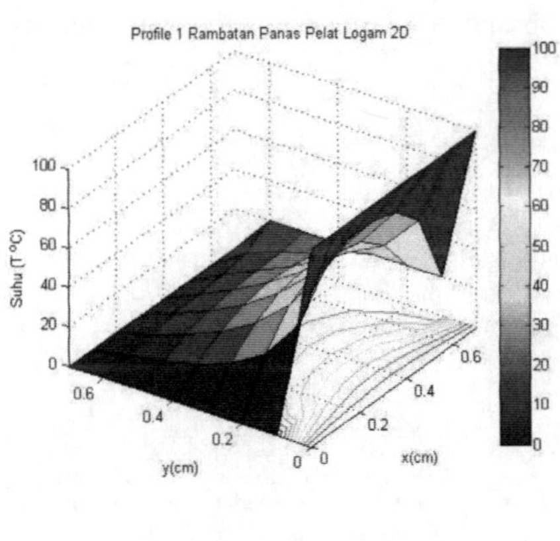

Profile 1A
Iteration matrix with Q = 10000; k = 40
Iteration = 24
Computing times = 0.0076659

100	100	100	100	100	100	100	100
0	481.979	659.382	726.524	738.873	701.203	572.837	20
0	268.284	428.777	507.589	527.517	492.851	389.896	20
0	162.129	279.602	347.289	370.505	352.537	293.645	20
0	100.381	179.963	231.211	254.426	252.898	231.897	20
0	59.182	108.408	142.914	162.842	172.482	180.794	20
0	27.689	51.323	68.946	81.295	93.143	118.547	20
0	0	0	0	0	0	0	0

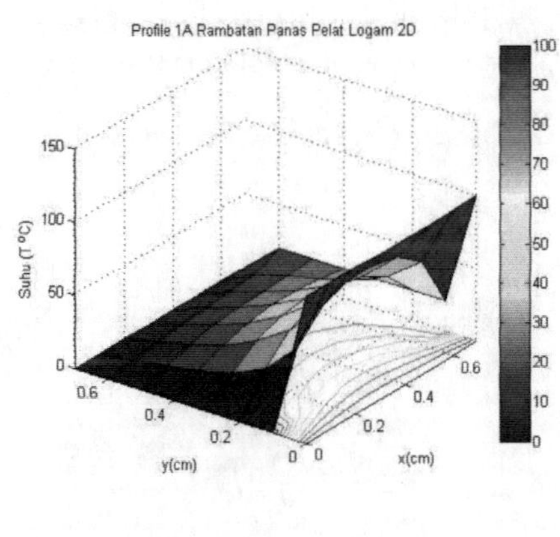

The number of iterations resulting from this limit value is 24 times. Profile distribution of heat to show the temperature decreases in the same area or exactly on point $T_{7,1}$.

b) $T(0,y) = 100; T(L,y) = 20; T(x,0) = 0; T(x,K) = 0$

Profile 2
Iteration Matrix
Iteration = 24
Computing times = 0.0099215

0	0	0	0	0	0	0	0
100	481.716	269.848	169.527	120.128	102.566	118.284	20
100	657.017	428.148	288.131	208.420	171.852	170.569	20
100	718.204	497.594	346.431	253.569	205.854	192.141	20
100	718.204	497.594	346.431	253.569	205.854	192.141	20
100	657.017	428.148	288.132	208.420	171.852	170.569	20
100	481.716	269.848	169.527	120.128	102.566	118.284	20
0	0	0	0	0	0	0	0

Profile 2A
Iteration Matrix with Q = 10000; k = 40
Iteration = 24
Computing times = 0.012379

0	0	0	0	0	0	0	0
100	481.979	270.249	169.988	120.589	102.967	118.547	20
100	657.418	428.777	288.864	209.153	172.482	170.970	20
100	718.665	498.327	347.289	254.426	206.587	192.602	20
100	718.665	498.327	347.289	254.426	206.587	192.602	20
100	657.418	428.777	288.864	209.153	172.482	170.970	20
100	481.979	270.248	169.988	120.590	102.967	118.547	20
0	0	0	0	0	0	0	0

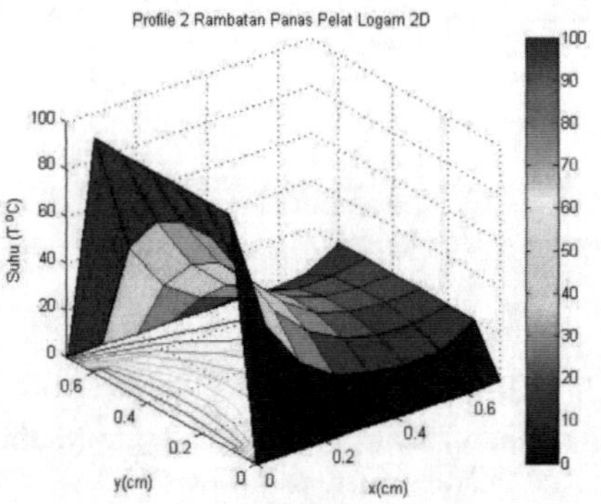

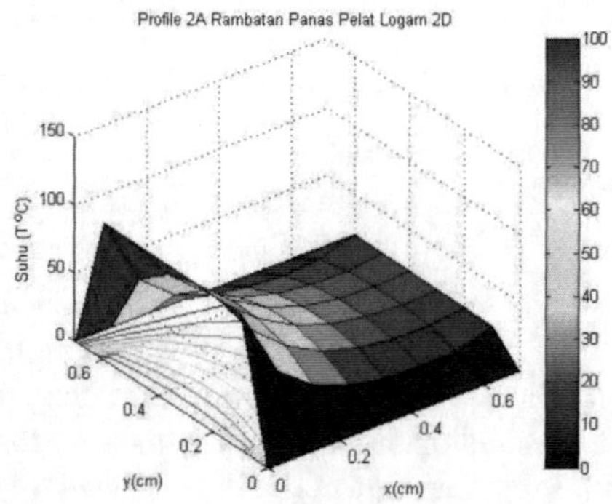

The resulting number of iterations 24 times. Second heat distribution profile to show the temperature decreases in the same area or exactly at a node $T_{5,1}$ dan $T_{5,6}$.
When observed more than a second variation of the above limits, namely: 0; 20; 100; 0 and 10; 20; 0; 0 produce the same number of iterations equilibrium which is 24 times. And when the second iteration matrix deductible would generate diagonal matrix with value 0.

0	389134	556536	618284	598236	454290
-389134	0	218725	318364	320369	218926
-556536	-218725	0	116078	145950	101043
-618284	-318364	-116078	0	46311	39294
-598236	-320369	-145951	-46311	0	9824
-454290	-218926	-101043	-39294	-9823	0

0	389.133	556.536	618.284	598.236	454.290
-389.134	0	218.725	318.364	320.369	218.926
-556.536	-218.725	0	116.079	145.950	101.043
-618.284	-318.364	-116.078	0	46.311	39.295
-598.236	-320.369	-145.950	-46.311	0	9.824
-454.290	-218.925	-101.042	-39.295	-9.824	0

Both matrices show has a positive value on the triangle top and negative on the lower triangle. And both showed the smallest iteration value in the same area.

c) $T(0,y) = 100; T(L,y) = 0; T(x,0) = 20; T(x,K) = 0$

Profile 3
Iteration matrix
Iteration = 28
Computing times = = 0.14655

0	0	0	0	0	0	0	0
0	27.426	50.922	68.484	80.834	92.743	118.284	20
0	58.781	107.779	142.181	162.109	171.852	180.393	20
0	99.920	179.230	230.353	253.569	252.165	231.435	20
0	161.668	278.869	346.431	369.647	351.804	293.184	20
0	267.883	428.148	506.856	526.784	492.221	389.495	20
0	481.716	658.982	726.063	738.412	700.802	572.574	20
100	100	100	100	100	100	100	100

Profile 3A
Iteration Matrix with Q = 10000; k = 40
Iteration = 28
Computing times = 0.010536

0	0	0	0	0	0	0	0
0	27.689	51.323	68.946	81.295	93.143	118.547	20
0	59.182	108.408	142.914	162.842	172.482	180.794	20
0	100.381	179.963	231.211	254.426	252.898	231.897	20
0	162.129	279.602	347.289	370.505	352.537	293.645	20
0	268.284	428.777	507.589	527.517	492.851	389.896	20
0	481.979	659.382	726.524	738.873	701.203	572.837	20
100	100	100	100	100	100	100	100

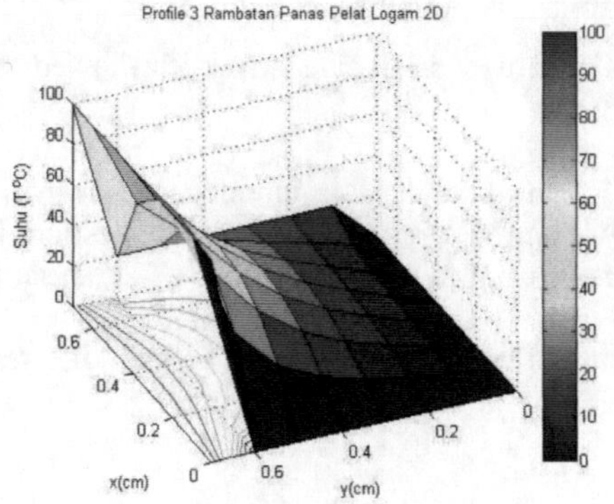

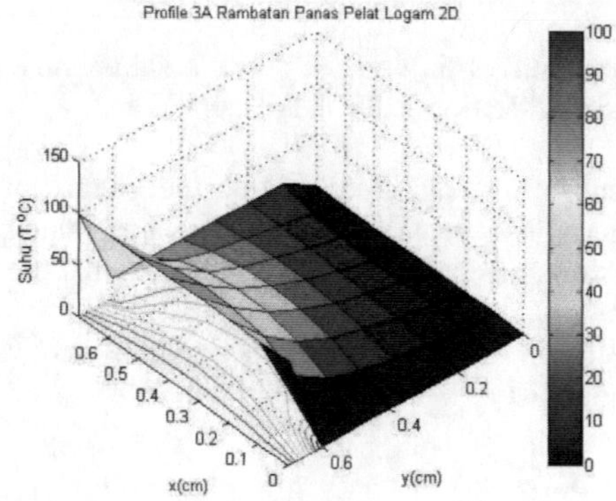

The number of iterations resulting from this limit value is 28 times. Profile distribution of heat to show the temperature decreases in the same area or exactly on point $T_{1,1}$

d) $T(0, y) = 100; T(L, y) = 20; T(x, 0) = 0; T(x, K) = 0$

Profile 4
Iteration Matrix
Iteration = 28
Computing times = 0.0096928

0	0	0	0	0	0	0	0
20	118.284	92.743	80.834	68.484	50.922	27.426	0
20	180.393	171.852	162.109	142.181	107.779	58.781	0
20	231.435	252.165	253.569	230.353	179.230	99.920	0
20	293.184	351.804	369.647	346.431	278.869	161.668	0
20	389.495	492.221	526.784	506.856	428.148	267.883	0
20	572.574	700.802	738.412	726.063	658.982	481.716	0
100	100	100	100	100	100	100	100

Profile 4A
Matrix iteration with Q = 10000; k = 40
Iteration = 28
Computing times = 0. 010513

0	0	0	0	0	0	0	0
20	118.547	93.143	81.295	68.946	51.323	27.689	0
20	180.794	172.482	162.842	142.914	108.408	59.182	0
20	231.897	252.898	254.426	231.211	179.963	100.381	0
20	293.645	352.537	370.505	347.289	279.602	162.129	0
20	389.896	492.851	527.517	507.589	428.777	268.284	0
20	572.837	701.203	738.873	726.524	659.382	481.979	0
100	100	100	100	100	100	100	100

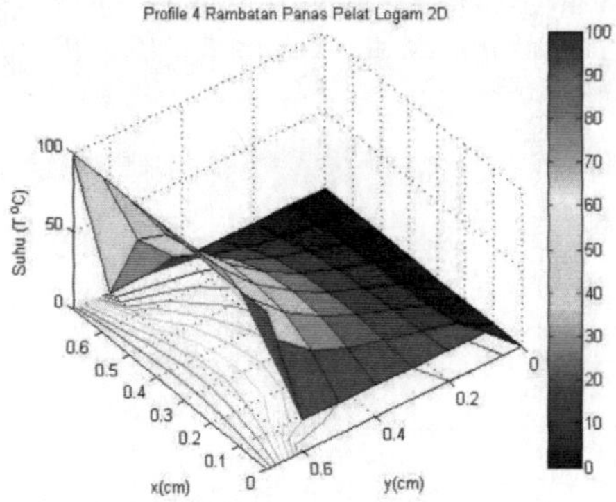

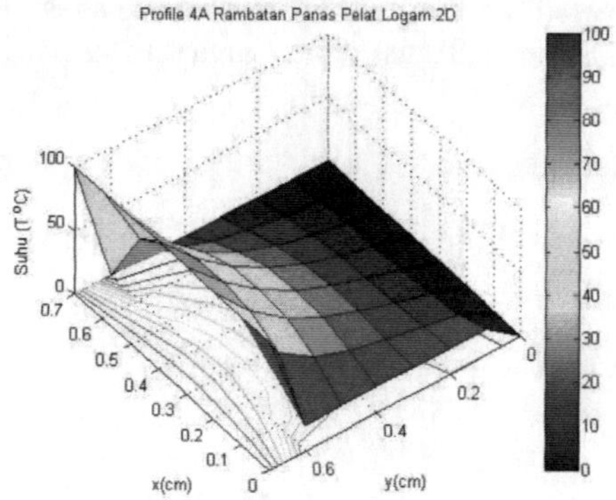

The resulting number of iterations 28 times. Second heat distribution profile to show the temperature decreases in the same area or exactly at a node $T_{6,1}$.
When observed more than a second variation of the above limits, namely: 0; 20; 100; 0 and 20; 0; 0; 100 produce the same number of iterations equilibrium which is 28 times. And when the second iteration matrix deductible would generate diagonal matrix with value 0.

-90.858	-41.821	-12.350	12.350	41.821	90.858
-121.612	-64.073	-19.928	19.928	64.073	121.612
-131.515	-72.935	-23.216	23.216	72.935	131.515
-131.516	-72.935	-23.216	23.216	72.935	131.516
-121.612	-64.073	-19.928	19.928	64.073	121.612
-90.858	-41.820	-12.349	12.349	41.820	90.858

-90.858	-41.820	-12.349	12.349	41.820	90.858
-121.612	-64.074	-19.928	19.928	64.074	121.612
-131.516	-72.935	-23.215	23.215	72.935	131.516
-131.516	-72.935	-23.216	23.216	72.935	131.516
-121.612	-64.074	-19.928	19.928	64.074	121.612
-90.858	-41.821	-12.349	12.349	41.821	90.858

Both matrices show has a positive value on right side and negative on left side. And both showed the smallest iteration value in the same area.

Results from this study is that the pattern of variation of the boundary or initial conditions of a material with regard conductivity value remains at steady state the exact value of the smallest are in the same iteration value. This indicates that the value of the thermal equilibrium tend to be at the same iteration.
A recommendation for further research is by varying the initial conditions more and include different conductivity.

3. Refference

[1] Alatas, Husen. *Buku Pelengkap Fisika Matematika 1*. Jakarta.
[2] Hasimi Pane, Ali., 2011, *Penyelesaian Numerik Perpindahan Panas Konduksi 2-D Pada Bidang Datar Menggunakan Program MS.Excell dan Engineering Equation Solver*, Universitas Sumatra Utara, Medan.
[3] Djojodiharjo, Harijono. 2000. *Metode Numerik.* Jakarta: PT. Gramedia Pustaka Utama.
[4] Kreyszig, E., 1988, *Advance Engineering Mathematics*, John Wiley & Sons, Inc.
[5] Holman. 1997. *Perpindahan Kalor.* Edisi Keenam. Jakarta Erlangga.
[6] Captra, S.C., Canale R.P., 1990, *Numerical Methods for Engineering,* second edition, McGraw-Hill, New York.
[7] Incropera, F.P., et. al., 1981, *Fundamental of Heat Transfer*, John Wiley & Sons, Inc.
[8] Kreith, Frank & Arko Prijono. 1986. *Prinsip-prinsip Perpindahan Panas.* Edisi Ketiga. Jakarta: Erlangga.
[9] James, M.L., et.al., 1993, *Applied Numerical Methods for Digital Computation*, HarperCollins College Publishers.
[10] Smith, G.D., 1985, *Numerical Solution of Partial Defferential Equations: Finite Difference Methods*, third edition, Oxford University Press.
[11] Setiawan, Agus. 2006. *Pengantar Metode Numerik.* Yogyakarta: Andi.

Low-frequency magneto-impedance effect in electrodeposited multilayers [$Ni_{80}Fe_{20}$/Cu]$_N$ on the Cu-PCB

Frendi Maulana[1], Nuryani[1] , B Purnama[1,a)]

Physics Department, Post Graduate program Universitas Sebelas Maret, Jl. Ir. Sutami 36A Kentingan Surakarta INDONESIA 57126

E-mail: frendimaulana270@gmail.com , [1,a)]bpurnama@mipa.uns.ac.id

Abstract. Abstract. It has been observed magneto-impedance effects of electrodeposited [Ni80Fe20/Cu]3 multilayers on substrate Cu PCB. The ratio MI (Magneto-impedance) evaluate the dependence for both of PCB Cu substrates-geometry and the frequency modification. The characterization results showed the geometry dependence of the ratio of the MI tends to increase with the increase of angle. In addition, the MI ratio magnitude increases in increasing the frequency measurements. Finally, the highest MI ratio of 8.23% occurs for the sample C.

1. Introduction

Magneto-impedance effect (MI) is a changed of impedance in a magnetic material under induced magnetic field. The phenomenon correlate to the changes of the skin depth in wire material magnetic as well as in both thin films and multilayers [1-4]. Recently, the MI effect attractively realizes on a flexible substrate that modifies of the crystalline structure with in result the change of magnetic characteristic [5]. The application of the MI sensors such as industry, navigation, vehicle detection, military, geomagnetic, and bio-medical measurements such as heart, neutron systems in the human body [6-8].

In this study, the geometry dependence of the MI effects in multilayered [$Ni_{80}Fe_{20}$/Cu)]$_3$ are evaluated in the low-frequency region (f < 100 kHz) [7]. The samples preparation are carried out by electrodeposition method for both the magnetic layer $Ni_{80}Fe_{20}$ and the spacer layer Cu. A Substrate of a patterned Cu PCB is used for the whole experiments. The MI effect evaluates by measurement of the total impedance Z (Z2 = R2 + X2) under the modified magnetic field. Then, the MI ratio calculates by difference total impedance when the zero and under magnetic field divided by impedance when under magnetic field.

2. Numerical Methods

The samples films used in this experimental are prepared electrodeposition method with Pt electrode. The sample characterization in the multilayered structure of [$Ni_{80}Fe_{20}$(800 nm)/Cu(200 nm)]$_3$ carried out by XRF and XRD characterization (not presented here). Procedure for the samples deposition is done similar with previous method [9]. Modification of patterned Cu PCB (substrate A, B, and C) are used for the whole experimental. Here, the angle of $\alpha = 60°$ is called substrate A. The angle of α for both B and C substrates are 105° and 180° respectively. Substrates are cleaned with ultrasonic cleaner in ethanol and acetone for several time. The sample modification is show in Figure 1.

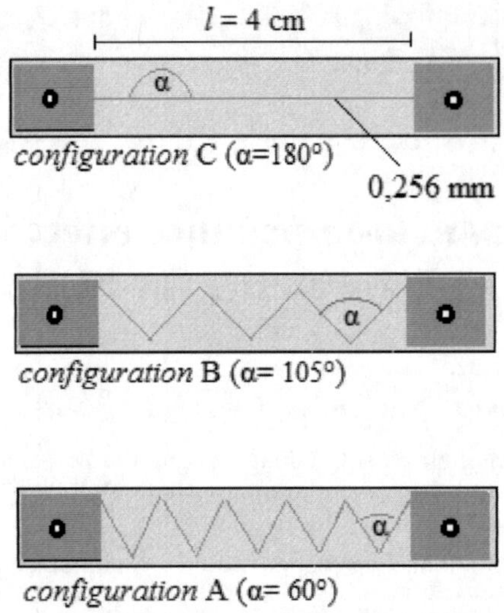

Figure 1. the modified substrates of Cu PCB use in the experiment.

An electrolyte bath for $Ni_{80}Fe_{20}$ layers and Cu spacer layer are displayed in Table 1. During experimental, the electrolytes bath keep for constant pH of 2.6 [10-12]. Diagram schematic of the electrodeposited procedure and the MI measurement set-up present at Figure 2 and Figure 3. The deposition rate for both the $Ni_{80}Fe_{20}$ layer and the Cu layer are 15.5 mA/cm2 and 8 mA/cm2 respectively. Thus, separately a fresh-electrolytes bath are promptly prepared before fabrication process. Electrodeposited are performed at room temperature and atmospheric preassure condition. The obtained multilayered $[Ni_{80}Fe_{20}/Cu)]_3$ are measured the MI characteristic for different geometry substrates and frequency measurement.

Table 1. The composition of the electrolytes bath solution for $[Ni_{80}Fe_{20}/Cu)]_3$.

Desired layer	Composition	Concentration
NiFe	$NiSO_4.6H_2O$	0,009 M
	$FeSO_4.7H_2O$	0,012 M
	H_3BO_3	0,149 M
	$C_6H_8O_3$	0,002 M
	H_2SO_4	1 drop
	Aquades	120 ml
Cu	$CuSO_4.5H_2O$	0,065 M
	$C_6H_{12}O_6$	0,002 M
	H_2SO_4	1 drop
	Aquades	120 ml

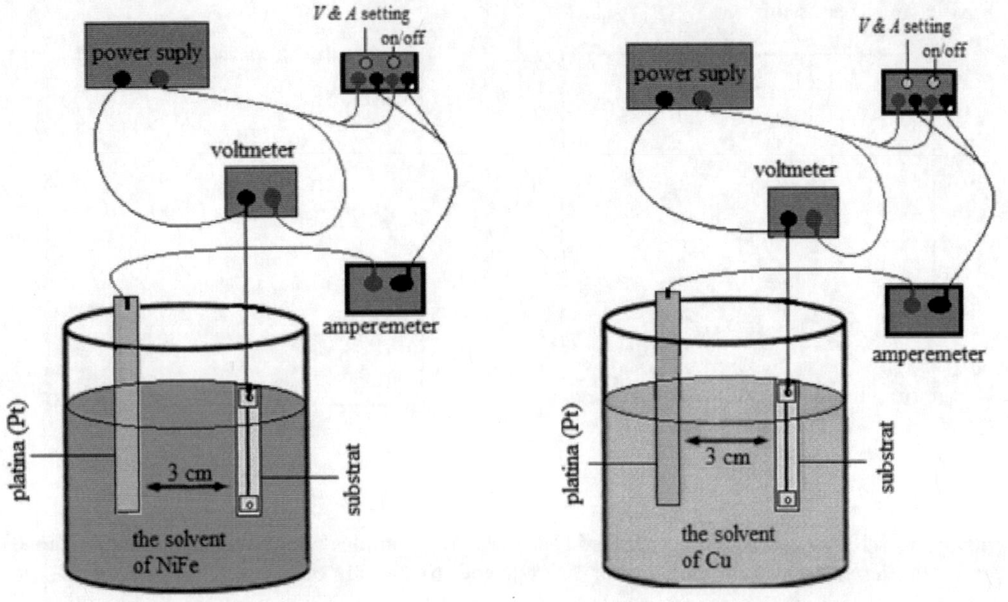

Figure 3. experimental set up for electrodeposition of the multilayers $[Ni_{80}Fe_{20}/Cu)]_3$.

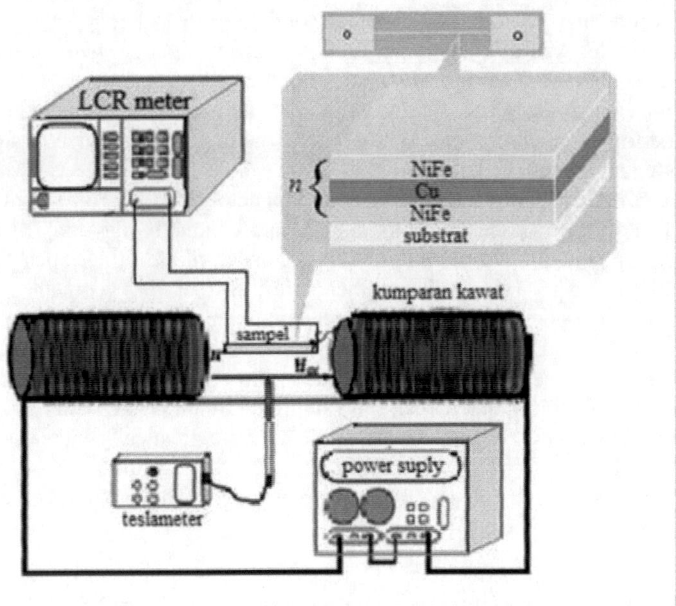

Figure 4. The schematic of magneto-impedance measurement for multilayer samples of $[Ni_{80}Fe_{20}/Cu)]_3$.

3. Results and Discussion

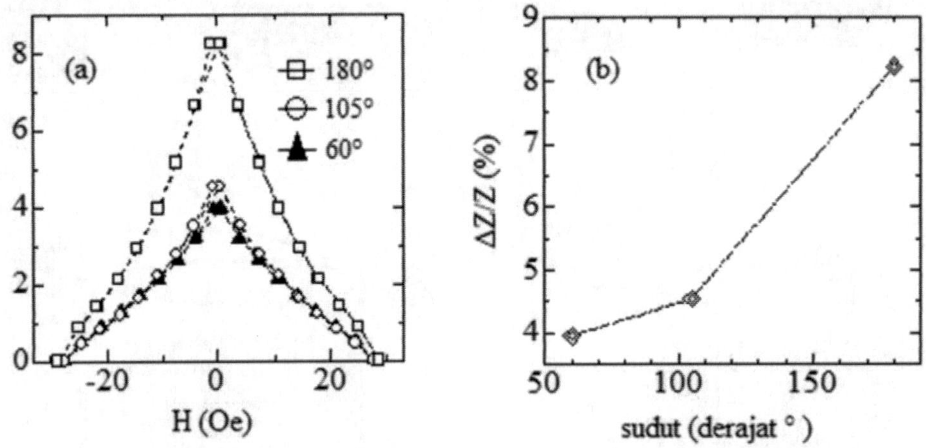

Figure 5. (a) The MI curve for multilayered $[Ni_{80}Fe_{20}/Cu)]_3$ samples with variation angle of geometry modification measure at a frequency of 100 kHz and (b) the ratio of MI as a function of angle.

Figure 5 shows the expression MI value/ratio as a function of the changes of H. For the whole samples, the MI ratio maximum realize when the magnetic field of H = 0. Then the MI ratio gradually decreases with the increase of H. The MI ratio maximum of 8.19% observe for sample C (with the angle α of 180^0). Whereas the maximum ratio of samples A and B are 3.28% and 4.36% respectively. This interesting results may attribute the different configuration of a magnetic domain for individual samples. In case samples A and B, the magnetization realizes in pattern geometry owing compensated region of magnetization for each and other. Contrast, for sample C, the magnetic domain configuration realize in single plane direction. So that, the resultant of magnetic configuration will uniformly respond to the magnetic field. Within result, the high MI ratio due to significantly different configuration under zero magnetic field.

In general, the MI effect dominantly supports by magneto-inductive [6]. For multilayered structure films, the MI effect support by the ratio of magnetic and conductive layers [13] and also skin effect [14]. For the recent report, the MI of multileyer structure attribute by an open flux of the magnetic domain configuration [15]. For our experiments results that the geometry modification of the multilayered samples will potentially form the open-flux magnetic domain configuration within result high MI ratio.

Another expression of the frequency dependence of the MI curve present is depicted Figure 6. Clearly observed from the figure that the MI ratio significantly increases with frequency. The MI ratio of 0.059% occurs at frequency f of 20 kHz. Then the MI ratio drastically increases of 8.23% when the measurements carried out at frequency f of 100 kHz. The MI ratio become 140 times (=8.23/0.059) much larger in increasing 5 times (=100/5) of frequency. This is indicated that the MI effect dominantly realizes by skin effect [14].

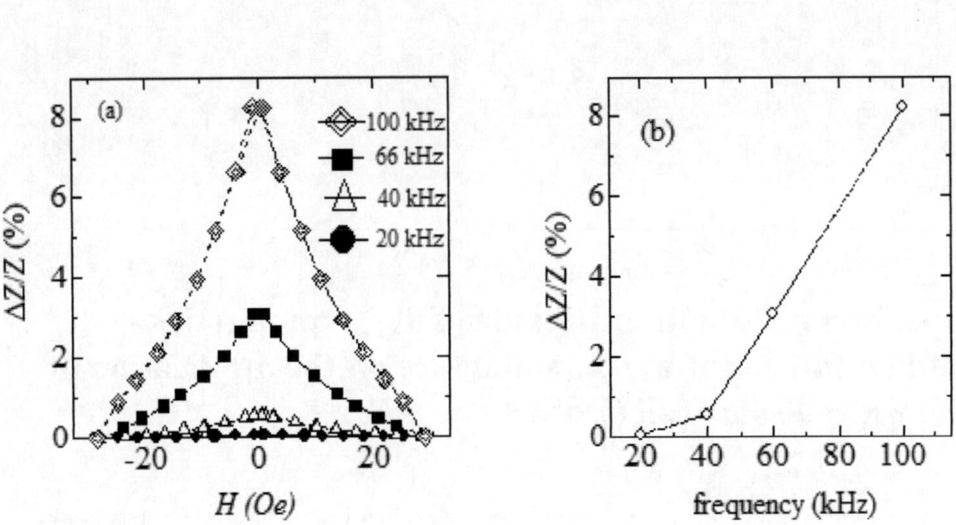

Figure 6. (a) curve MI multilayers $[Ni_{80}Fe_{20}/Cu)]_3$ in the measurement frequency of 20 kHz to 100 kHz sample C, (b) graphs the ratio of MI function of frequency.

4. Conclussion

In conclusion, the magneto-impedance effects in samples of $[Ni_{80}Fe_{20}/Cu)]_3$ multilayers have discussed. The dependence for both of PCB Cu substrates-geometry and the frequency measurement of the ratio MI have evaluated. The experimental results showed the geometry dependence of the ratio of the MI tends to increase with the increase of angle. In addition, the MI ratio magnitude increases in increasing the frequency measurements. Finally, the highest MI ratio of 8.23% occurs for the sample C.

References

[1] Cortes M, Peng T, Woytasik M, and Moulin J 2015 *Journal of The Electrochemical Society* 162 no.7, 129-132
[2] Chaturvedia A, Ruiza A, Leb A T, Tungc M T, Mukherjeea P, Srikantha H, and Phana M H 2014 *Science Jet* 3:48 1-4
[3] Ismail, Nuryani, and Purnama B 2016 *AIP conference Proceedings* 1710 .030002
[4] Wicakson B A, Nuryani, and Purnama B 2015 *Jurnal fisika dan aplikasinya* 11, no.3
[5] Talaat A, Ipatov M,and Zhukova V 2014 *Proceedings of the 8th International Conference on Sensing Technology*, Sep. 2-4, 2014, Liverpool, UK
[6] Phan M H and Peng H X 2008 *Progress in Materials Science* 53 323-420
[7] Kurlyandskaya G V, deCos D, and Volchkov S O, Defektoskopia 2009 45 (6), 13–42 [Russ. J. NonDestructive Testing 45 (6), 377–398
[8] Tuan L H, Huy N T, and Huy P T 2009 *journal of physics*, 187(1), 1-5
[9] Amiruddin M, Utari, and Purnama B 2014 *Jurnal Fisika dan Aplikasinya*, 10 No. 2, 95–98
[10] Bedir M, Bakkaloglu O F, and Oztas M, 2006 *Pramana Journal of Physic*, 66 No. 6,. 1098-1104
[11] Aydogmus E, Kaya H, Atalay F E, Atalay s, and Avcu D, 2014 *Acta physica Polanica A* 125, 227-229
[12] Hanafi E P, Utari, and Purnama B, 2013 *Jurnal Fisika & Aplikasinya*, 9 No.2 63-65
[13] Panina L V and Mohri K 2000 *Sensors and Actuators* 81 71-77
[14] Marques M S, Mori T J A, Schelp L F, Chesman C, Bohn F, and Corrêa M A 2012 *Thin solid films* 520 2173-2177
[15] Garcia-Arribas A, Fernandez E, Svalov A, Kurlyandskaya G V, and Barandiaran J M 2016 *Journal of Magnetism and Magnetic Materials* 321-326

Comparison of Titanium Dioxide (TiO_2) nanoparticle-nanofiber and nanofiber-nanoparticle on the application of dye-sensitized solar cell (DSSC)

Linda Kusumaning Tyas[1], Risa Suryana[1], Fahru Nurosyid[1], Novita Asma Ilahi[1]
Physics Department of Post Graduate Program Sebelas Maret University,
Jl. Ir. Sutami 36A Kentingan Jebres Surakarta 57126, INDONESIA

E-mail: lindakusumant@gmail.com

Abstrak. Dye-sensitized Solar Cell (DSSC) is a solar cell that has great potential in the future because of the cheaper cost of fabrication and environmentally friendly basic ingredients. This study aims to determine the effect of type of screen on the TiO_2 layer as the active electrode DSSC. The active electrode TiO_2 based DSSC fabricated by the method of double layer. Efficiency Dye-sensitized Solar Cell (DSSC) can be obtained from the current-voltage curve I-V meter. Nanofiber on the nanoparticles can reach a highest efficiency of DSSC about 0,015%. The second variation of between nanofiber-nanoparticle layering, and nanoparticle-nanofiber, it appears that the nanofiber layer of nanoparticles above, no significant changes, namely in 10 minutes η = 0.014965; 15 minutes η = 0.011021 and 20 minutes η = 0.013332. This is demonstrated by the nature of the dominant nanofiber as a photon trap covered by the nature of the dominant nanoparticles absorb the dye, so that overtime does not affect the incoming electron. While the results of nanofiber layer on the nanoparticles changed significantly in the variation of 20 minutes, ie η = 0.00283. You can also see the most optimum time was 15 minutes, which is η = 0.01559. This may be due in this 15 minute nanofiber coating has a thickness that is optimum so that electrons can reach the electrode diffuse due to the interaction between photons and the dye more.

1. Introduction

Solar energy is one alternative energy that can be used by humans today. Geographically, Indonesia is located on the equator causes Indonesia can receive solar heat more than other countries, namely 4800 watt /m^2/day. Based on this phenomenon, the solar energy to electrical energy is converted into solar cells [1]. Dye-sensitized solar cells (DSSCs) have been extensively researched over the past Decades due to Reviews their high energy conversion efficiency and especially low production costs as cheaper alternatives to silicon solar cells. DSSC typically consists of a nanocrystalline TiO_2 films covered by a monolayer of dye molecules, redox, electrolyte, and counterelectrode. For the formation of anefficient DSSC, photoanode materials with various morphologies such as nanoparticles [2].

Advantages of using TiO_2 that is environmentally friendly, has inert properties and photoactive, easily manufactured and inexpensive cost. TiO_2 has an energy band gap width is 3.2 eV and is only able to absorb ultraviolet light from the solar spectrum [3]. TiO_2 anatase phase that is used has to have the

[1] To whom any correspondence should be addressed.

ability photoactive high so it is often used as a photocatalyst. Anatase is thermodynamically less stable than rutile, but its formation occurs at lower temperatures [4]. In its application as a photocatalyst DSSC, commonly used in the type of anatase TiO_2 because it has the high photocatalytic ability. In addition, the nanocrystal structure and high surface area of TiO_2 can be used to improve system performance DSSC, especially in increasing the density and electron transfer [5].

Modification of TiO_2 layers needed to make the residence time of photons in layers longer than the planar shaped layer. This is because the photons coming will be trapped between the nanorod, nanowire or nanotube. Photon interaction with electron excitation dye produces further injected into the conduction band of TiO_2. Electrons diffuse towards the electrode on the surface of TiO_2. In addition, there is a modified form of TiO_2 nanofiber. By modifying the shape of TiO_2 layer in the form of nanofiber, the residence time of photons in a nanofiber layer will be longer than the shaped layer is because the photons coming planar. This is because the photons coming will be trapped between the nanorod, nanowire or nanotube. Photon interaction with electron excitation dye produce further injected into the conduction band of TiO_2. Electrons diffuse towards the electrode on the surface of TiO_2. The number of electrons in the electrodes will determine the efficiency of DSSC. The photoelectrode of the DSSC fabricate from TiO_2 Nanostructures commonly. There are many approaches for the DSSC's photoelectrode such as doctor blade method, screen printing, drop-casting, spin-coating [6], spray coating [7], chemical vapour deposition [8], chemical bath deposition [9], hydrothermal method, electrophoretic deposition (EPD) and electrospun [10]. DSSC consists of four main components, namely an n-type semiconductor such as nanoparticles of TiO_2, a dye that is sensitive (Dye sensitizer) on the absorption of visible light, an electrolyte made in one of the semiconductor surface and a counter electrode such as platinum (Pt) , carbon (C), which helps transfer electrons to the electrolyte [11].

Dye-sensitized Solar Cells (DSSC) is a photoelectrochemical solar cell that uses an electrolyte as charge transport medium. DSSC is divided into several parts consisting of nanopores TiO_2, dye molecules adsorbed on the surface TiO_2 dan catalysts which are all deposited between two glass conductive DSSC can not be separated from the dye, by far the dye that is used as a sensitizer can be either synthetic dye and natural dye. Synthetic dye generally using ruthenium-based organic metal complex. In studies that have been done, the dye of the ruthenium complex compounds can achieve efficiencies of 11-12% [12]. However, the amount of ruthenium complex dye is limited and the price is quite expensive. For these reasons, developing research towards finding natural dyes extracted from flowers, leaves, and fruits [13]. Natural dye that is used as a sensitizer in DSSC classified as environmentally friendly, making it too easy and cheap although lifetime is low [14].

Application to the DSSC, TiO_2 used generally phase photoactive anatase because it has high ability. TiO_2 with nanopores structure that the pore size in the nanoscale will increase the performance of the system because nanopores structure has the characteristics of high surface area so that it will increase the amount of dye that has absorbing the implications of which will increase the amount of light that has absorbing. From various studies have been conducted several deposition techniques TiO_2 in DSSC, one of the DSSC fabricated using a double layer coating method electrophoretic deposition (synthesis of TiO_2) and Dr.Blade (TiO_2 nanoparticles), obtained a considerable efficiency is 5.96% [10].

2. Eksperimental

2.1 Materials

TiO_2 used in this study using two types namely TiO_2 nanofiber and TiO_2 nanoparticle synthesis with 21 nm Titanium Dioxide Anatase TiO_2 powder, 99.8% trace metals basis, Titanium (IV) Isopropoxide (TTIP) 97% purity, Acetic Acid ($CH3CO_2H_2$, 99.7% Aldrich, USA), Ethanol (C_2H_5OH, Merck, Germany), and Polyvinyl Pyrrolidone (PVP, Mw ~ 29000 Aldrich). As the dye, dried moss, N-hexane, silica gel.

2.2 Preparation of TiO_2
There are two layers of TiO_2, the TiO_2 nanoparticles, and TiO_2 Nanofiber.
2.2.1 TiO_2 Nanoparticles
Titanium (IV) Oxide nanoparticles, anatase titanium dioxide powder, 99.8% trace metals basis. TiO_2 of 0.5 g dissolved in 2 ml of ethanol, stirred for 30 minutes at 300 rpm using a vortex stirrer [15].
2.2.2 Synthesis of TiO_2 nanofiber
Titanium (IV) Oxide synthesis of TiO_2 Nanofiber material made of titanium (IV) Isopropoxide (TTIP), Acetic Acid, Ethanol, and Polyvinyl Pyrrolidone (PVP). 1 ml acetic acid mixing with ethanol, stirred for 10 minutes. Once it is added Polyvinyl Pyrrolidone (PVP) of 2.5 g, stirred for 1 hour. Then, after 1 hour, is added titanium (IV) Isopropoxide (TTIP) 0.5 ml, then stirred until homogeneous.

2.3 Deposition of TiO_2
2.3.1 TiO_2 nanofiber on TiO_2 nanoparticle
In the treatment of the first, the coating of TiO_2 nanofiber on top of TiO_2 nanoparticles, TiO_2 superimposed on the glass conductive Fluorine Tin Oxide (FTO) with an area of deposition of 1 cm x 2 cm using spin coating, the rotating speed is the same, 1000 rpm in 60 seconds , TiO_2 layer that has been deposited on a hot plate heated to a temperature of 30 °C/10 min. In the deposition of TiO_2 nanoparticles done 3 times coating. [3]. Furthermore, TiO_2 nanofiber layer deposited on surfaces that have been coated with TiO_2 nanoparticles using methods electrospinning. Nanofiber layer made with a variety of plating time of 10 minutes, 15 minutes and 20 minutes at a distance of 25 cm spinnerets with collectors. Voltage electricity used between 5 kV 6 kV using the same annealing temperature in each variation is 450 °C for 3 hours to evaporate the organic material [16].
2.3.2 TiO_2 Nanoparticles on Nanofiber
TiO_2 layer of nanoparticles on TiO_2 nanofiber, done in a way that is similar to the previous one, TiO_2 nanofiber deposited on glass conductive Fluorine Tin Oxide (FTO) with an area of deposition of 1 cm x 2 cm using methods electrospinning made with the time variations layering are 10 minutes, 15 minutes and 20 minutes at a distance of 25 cm spinnerets with collectors. Voltage electricity used between 5 kV-6 kV. After the deposition of TiO_2 nanofiber finish, then TiO_2 nanoparticles deposited on the fiber nanofiber TiO_2 using spin coating, the rotating speed is the same, namely 1000 rpm in 60 seconds. TiO_2 layer that has been deposited on a hot plate heated to a temperature of 30 °C/10 min. In the deposition of TiO_2 nanoparticles done 3 times coating. The second layer of TiO_2 nanofibers and nanoparticles was annealed at a temperature of 450 ° C for 3 hours.

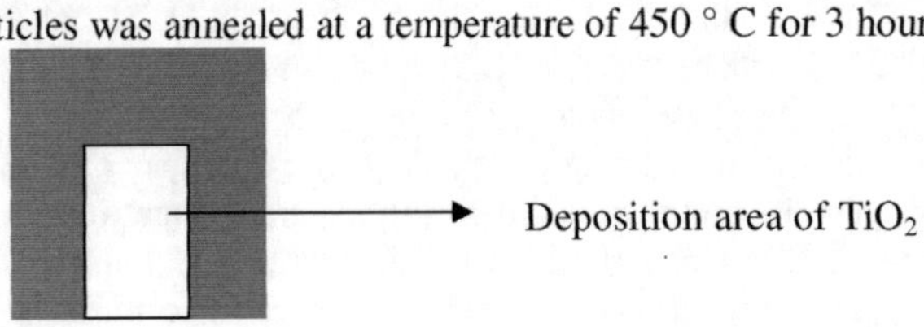

figure 1. Active cell of Fluorine Tin Oxide (FTO: 1cm x 2cm)

2.4 Extraction of Chlorophyll Dye Bryophyta
Wet moss, dried by the sun first. Once dry, moss weighed as much as 20 g and pulverized using a mortar. Moss smooth and then dissolved in 100 ml acetone (ratio 1: 5), stirred using a vortex stirrer for 1 hour. Once homogeneous, the solution was filtered using Whatman filter paper number 42. The solution was filtered, and then do the chromatographic process. This process uses a silica gel 30 g mixed with 100 ml of n-hexane, and then inserted into the chromatography column. Afterwards the solution moss, put in a column chromatography, wait to a solution of moss down until the end of the column and drips[17].

2.5 Characterization absorbance of TiO_2 double layer nanoparticles and nanofiber
TiO_2 layering with a two layering technique, was soaked in moss extract chlorophyll dye, for 24 hours at room temperature so that the dye can absorb perfectly. After the immersion, then tested absorbances using UV-Visible Spectrometer Lambda 25 [10]. The wavelength range of 300-800 nm is used. The characterization results in the form of UV-Vis absorbance wavelength curve [15].

2.6 Preparation of Electrolyte and Counter Electrodes
In making this electrolyte solution using a mixture Polyethylene Glycol (PEG) 400, potassium iodide (KI), and iodine (I2). First, a homogeneous solution prepared from 10 ml of PEG and 0.8 g of KI. The second solution is prepared from 0.127 g I2dan 10 ml PEG. After that, each solution was mixed and stirred using a vortex stirrer for 30 minutes. Value Molecular weight (Mw) of the PEG, KI, and I2 respectively amounted to 380-420 g /mol, 166 g /mol and 254 g /mol [18].
Making the counter electrode using a carbon comprising 2B pencil that has been shaded in active cells evenly and layers have been shaded 2B pencil back layered with soot burning candle wax on the pitch-black color.

2.7 Fabrication of DSSC
Photoanode fabrication of DSSC is made of a two-layer structure of TiO_2. First TiO_2 nanofiber over TiO_2 nanoparticles and TiO_2 nanoparticles over TiO_2 nanofiber. After each had been coated with two layers of TiO_2, is the process of soaking the moss chlorophyll dye. This soaking process is carried out for 24 hours. Then the counter electrode spilled electrolyte solution 2 drops on the active cell, the active electrode positioned above the counter electrode with the active cell position facing each other, are arranged offset in order to facilitate the testing. Sandwich DSSC which has arranged by clamped using binder clips.

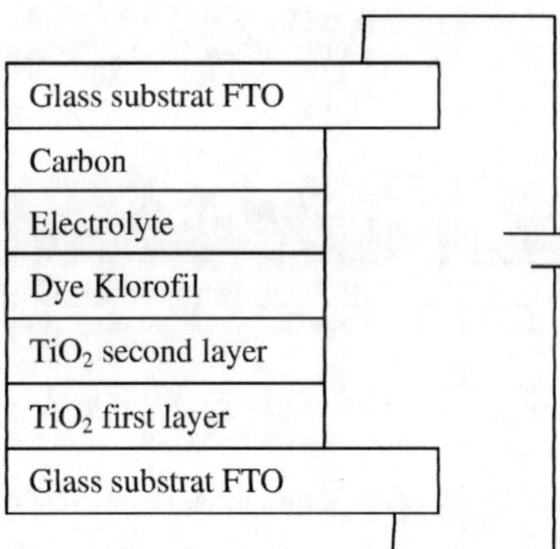

Figure 2. Ilustration of the result assembling DSSC

2.8 Characterization of I-V on DSSC
I-V measurements using the tool Keithley meter 2602A System Source. Probe the positive (red) is connected to the counter electrode substrate (carbon layer) and the negative probe (black) is connected to the active electrode substrate (a layer of TiO_2 and dye). Testing I-V performed in two conditions, namely conditions of dark and bright conditions. In dark conditions, DSSC closed using aluminum foil and in bright conditions, DSSC irradiated using a xenon lamp of 1000 W / m2. Characterization test is indicated by the chart-circuit (I) and voltage (V) obtained from the active cell, which will show the efficiency of DSSC.

3. Result and Discussion

TiO_2 layering capability using the techniques of 2 layers of TiO_2 nanoparticles and TiO_2 nanofiber with moss extraction chlorophyll dye can be proved by the results of characterization by UV-Vis Spectrometer shown in Figure 3. In the red line (B) is TiO_2 nanofiber on top of TiO_2 nanoparticles and the blue line (D) is a layer of TiO_2 nanoparticles on the nanofiber. On the charts very visible absorbance at highest peak 6.25715 a.u with a wavelength of 348 nm. Meanwhile, back when observed high absorbance peak average currently on TiO_2 nanoparticles on the TiO_2 nanofiber coating that is a 15 minutes peak at 2.5481 a.u, a wavelength of 400 nm and 20 minutes 1.91375 a.u peak at a wavelength of 450 nm. This high absorbance can be affected because the TiO_2 layer is thick enough so that the dye absorbed is also higher. From the graph it can be seen, the layering of TiO_2 nanoparticles TiO_2 nanofiber on may affect the absorbance is higher because the dye will be absorbed more perfect in the pores of the TiO_2 nanoparticles.

.

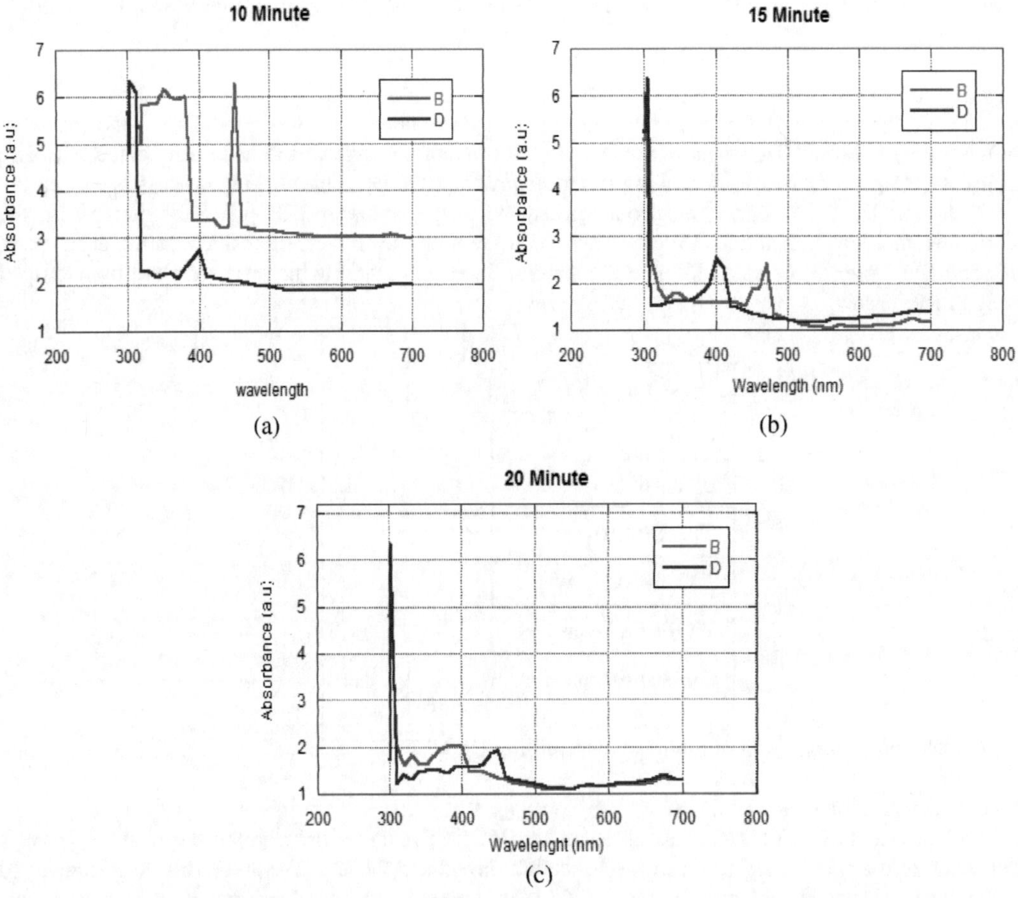

Figure 3. (a) Absorbance TiO_2 double layer, 10 minutes. (b) Absorbance TiO_2 double layer, 15 minutes. (c) Absorbance TiO_2 double layer, 20 minutes.

The next characterization is the current-voltage (I-V) which is a method to determine the ability of DSSC to be able to convert light into electrical energy that is expected to have a high percentage of capabilities. Measurements were made using Keithley 2602A conducted in the dark and the light. A state of light, the light coming from a xenon lamp with an intensity of 1000 W/m2. Results from testing the current-voltage variation TiO_2 double layer with a time of 10 minutes, 15 minutes and 20 minutes. In the picture, there are four colours (yellow (B), green (D), red (F) and blue (H)) The colour chart shows the I-V on each variation. To yellow (B) shows the dark state of TiO_2 nanoparticle on the TiO_2 nanofiber and green (D) indicates the dark state of TiO_2 nanoparticle on the TiO_2 nanofiber. Furthermore, there are two colours that indicate the state of the light that is, red (F) shows the light state of TiO_2 nanoparticle on the TiO_2 nanofiber and blue (H) shows the light state of TiO_2 nanoparticle on the TiO2 nanofiber.

From the pictures I-V curves 10 minutes, 15 minutes and 20 minutes when the light state located in quadrant IV, located under the I-V curve when the dark state. From the I-V characteristic curve shown in Figure 4:

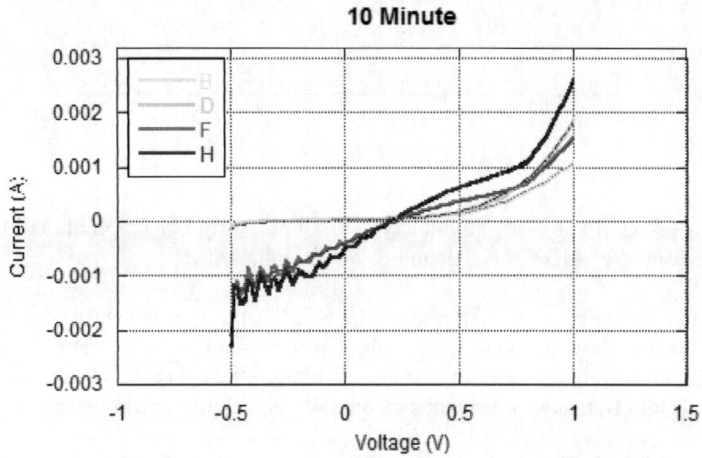

(a)

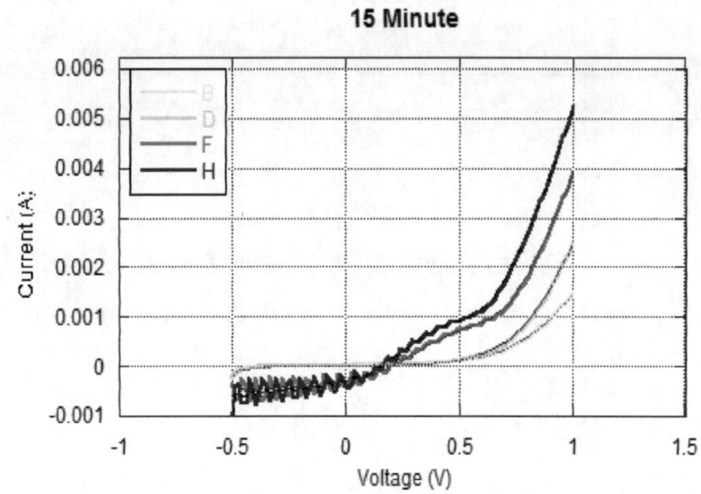

(b)

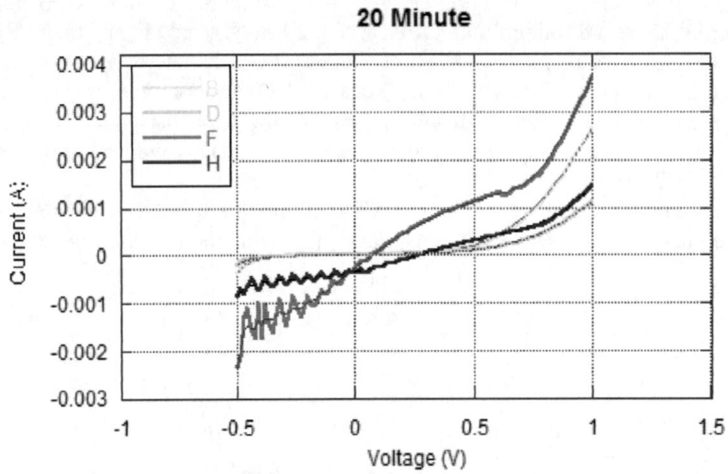

(c)

Figure 4. (a)I-V graph of TiO_2 double layer, 10 minutes. (b) I-V graph of TiO_2 double layer, 15 minutes. (c) I-V graph of TiO_2 double layer, 20 minutes.

From Figure 4 (a), (b) and (c) it can be determined the value of the current (Isc) and voltage (Voc). So based on the calculation can also know the value of maximum power (Pmax) which can also determine the maximum current (Imax) and maximum voltage (Vmax). From these calculations, it can be seen the percentage of efficiency. The results of the I-V characteristics can be seen in table 1:

Table 1. The results of the efficiency of the both TiO2 double layer

Nano Particle	Nano fiber	Voc (Volt)	Isc (Ampere)	Pmax (Watt)	Vmax (Volt)	Imax (Ampere)	Fill factor	Efficiency (%)
TiO_2 Nanofiber on The TiO_2 Nanoparticle								
3 Layer (21 nm)	10 minutes	0.220029	0.000379	2.10E-05	0.084975	0.000247	0.251147	0.010477
	15 minutes	0.190025	0.000386	3.12E-05	0.115106	0.000271	0.425063	0.01559
	20 minutes	0.054994	0.000219	5.66E-06	0.039975	0.000142	0.470532	0.00283
TiO_2 Nanoparticle on The TiO_2 Nanofiber								
3 Layer (21 nm)	10 minutes	0.000488	0.190083	2.99E-05	0.100018	0.000299	0.322958	0.014965
	15 minutes	0.114995	0.000441	2.20E-05	0.070008	0.000315	0.434712	0.011021
	20 minutes	0.250036	0.000352	2.67E-05	0.114996	0.000232	0.302748	0.013332

Based on Table 1 explains that the changes efficiency on both the structure of TiO_2 layer is not too significant, for example, the results of the efficiency of a layer of TiO_2 nanoparticles on the TiO_2 nanofiber, when 10 minutes has an efficiency $\eta = 0.014\%$, 15 minutes $\eta = 0.011\%$ and 20 minutes $\eta = 0.013\%$. It can becaused by the properties of nanofiber that serves to trap photons obscured by the overlaying nanoparticles. As for the TiO_2 nanofiber on the TiO_2 nanoparticles at the time 10 minutes had an efficiency $\eta = 0.010\%$, and a 15 minutes $\eta = 0.015\%$, but seen a significant difference at 20 minutes which has an efficiency $\eta = 0.002\%$, the difference is due to the working electrode not achieve optimum thickness so that electrons can not diffuse reach the electrodes. Based on these results the highest efficiency can be seen in the TiO_2 nanofiber on the TiO_2 nanoparticles in 15 minutes. This can happen because the plating time 15 minutes, nanofiber formed and spread evenly on the above FTO already coated with TiO_2 nanoparticles. So that photons inside nanofiber more and longer stuck. This can increase the absorption of photons, which then interacts with the dye and can generate a lot of excited electrons. The role of the TiO_2 nanoparticles on the TiO_2 nanofiber also has a very important influence. Where TiO_2 nanoparticle acts as an absorber dye while TiO_2 nanofiber role in improving the absorption of light absorbed by the dye so that photons can be trapped inside for longer[19].

In conclusion, comparison of a double layer of TiO_2 nanoparticles is a layer on the TiO_2 nanofiber had less good results because at anytime change does not significantly increase efficiency with this case due to the pure chareacteristic of nanofiber lost by characteristic of the nanoparticles. As for TiO_2 nanofiber layer on the TiO_2 nanoparticles better efficiency results in 15 minutes. Consider that in the 15 minutes have an optimum thickness so that electrons can reach the electrode diffuse.

References

[1] Rakhman, D.F., Pramono, S.H., Maulana, E. Skripsi, Universitas Brawijaya, Malang, 2014.
[2] Rapsomanikis A., Papoulis D., Panagiotaras D., Kaplani E., Stathatos E. 2014. Global NEST Journal, Vol 16 No 3, pp.485-489.
[3] Grätzel, M. 2005.Inorg. Chem.,44(20). pp. 6841-6851. DOI:10 1021/ic0508371.
[4] Michael Hoffmann. 1995. Journal of American Chemical Society, 95. pp.69-96
[5] Zhang, H., Banfield, J.F. 2000. Journal of Physics Chemistry B. vol.104. pp.3481
[6] E. Edri, S. Kirmayer, M. Kulbak, G. Hodes, D. Cahen. The Journal of Physical Chemistry Letters.5(2014) 429-433.
[7] S.Y. Kim, Y. Kim, K.M. Lee, W.S. Yoon, H.S. Lee, J.T.Materials and Interfaces 6 (2014) 13430-13437.
[8] A.K. Chandiran, A. Yella, M.T. Mayer, P. Gao, M.K. Nazeeruddin, M. Grätzel, Advanced Materials.26 (2014) 4309-4312.
[9] E.Edri, N. Kedem, H. Cohen, P. Barnes, G. Hodes, The Journal of Physical Chemistry.118 (2014). 16884-16891.
[10] Pirhadi, M. Feshki, S., Marandi, M., Davarnejad, R. Journal Nanostruct.,6(1):32-37, Winter 2016.DOI:10.7508/jns.2016.01.005.
[11] Stathatos, E. 2012. Journal of Engineering Science and Technology, Review 5, pp.9-13.
[12] Chiba Y, Islam A, Watanabe Y, Komiya R, Koide N, Han L.Y. Japanese Journal of Applied Physics. 45: L640 (2006). DOI:10.1143/JJAP.45.L638.
[13] Zhou, H, et al. 2011. Journal of Photochemistry and Photobiology A Chemistry, pp.188-194.
[14] Lee, J.K. and Mengjin, Y., 2011. Material Science and Engineering B 176, pp.1142-1160
[15] Deni Yulika, Skripsi, Universitas Sebelas Maret, Surakarta, 2014
[16] Diah A.E, Skripsi, Universitas Sebelas Maret, Surakarta, 2016.
[17] Dinasti D.P, Skripsi, Universitas Sebelas Maret, Surakarta, 2016.
[18] Nafi,ah Ardhani, Skripsi, Universitas Sebelas Maret, Surakarta, 2013.
[19] S.W. Lee, D. Zhang. I.P. Herman, Applied Physics Letters 104 (2014) 053113.
[20] Grätzel, M. 2003. Journal of Photochemistry and Photobiology. Vol. 4, pp. 145-153.
[21] Z. Huang. 2003. Composites Science and Technology, vol. 63, no.15, pp.2223-2253.

IOP Conf. Series: Journal of Physics: Conf. Series **795** (2017) 012020 doi:10.1088/1742-6596/795/1/012020

Challenges for nanofluid applications in heat transfer technology

Diah Hidayanti Sukarno

Assessment Center for Regulatory System and Technology of Nuclear Installation and Material (BAPETEN), Jl. Gadjah Mada 8 Jakarta Pusat 10120

d.hidayanti@bapeten.go.id

Abstract. Nanofluid has a potential to become a promising coolant in many diverse industrial processes. However, that opportunity faces several challenges that need to be solved through a long road of nanofluid research programs. Three kinds of the challenges that will be studied in this paper are: 1) determination of nanofluid thermophysical properties, 2) heat transfer characteristics of nanofluid, and 3) the stability factor of nanofluid. This paper also assesses the issue that must be addressed when nanofluid is utilized in nuclear technology applications. The radiation safety aspect of nanofluid utilization in nuclear reactor technology must be taken into account. The comprehensive and multidisciplinary research and assessment are crucial to be carried out in order to ensure the practical applications of nanofluid as new and potential heat transfer fluid.

1. Introduction

The term of 'nanofluid' is firstly proposed by Choi and Eastman in 1995 while presenting the new way to enhance the thermal conductivity of heat transfer fluid (HTF) [1]. In the past, the effort to increase the thermal conductivity of fluid has been done by dispersing the solid particles in millimeter or micrometer-sizes into conventional HTF (water, ethylene glycol, oil, etc.) as base fluids. However, the inclusion of solid particles in millimeter or micrometer sizes brings several problems, such as sedimentation, channel clogging, and abrasion. Then, those problems are solved by reducing the size of suspended solid particles into nanometer sizes, about 1-100 nm, producing nanofluid. The kinds of nanoparticles used are usually metals (Cu, Au), oxide metals (Al_2O_3, TiO_2, CuO, Fe_3O_4), and non-metallic element (carbon).

There are two methods of nanofluid preparation namely single-step method and two-step method [2]. The single-step method is also known as bottom-up process since the nanoparticles are created from the atomic or molecular components that grow up in size becoming nanometer-sized particles. In this process, base fluids are prepared simultaneously with the nanoparticles preparation. This kind of method is typically applied in laboratory scale. In the two-step method, nanoparticles are made from a bulk material that is disintegrated into smaller parts (top-down process). At the second step, the resulted nanoparticles are suspended in their base fluid. The typical method used in practical applications is the two-step method due to the easiness factor.

The inclusion of nanoparticles aims to increase the thermal conductivity of nanofluid since the added nanoparticles have higher thermal conductivity than base fluids. Figure 1 shows us that water and other conventional HTF have much lower thermal conductivity than common metallic and non-metallic solid.

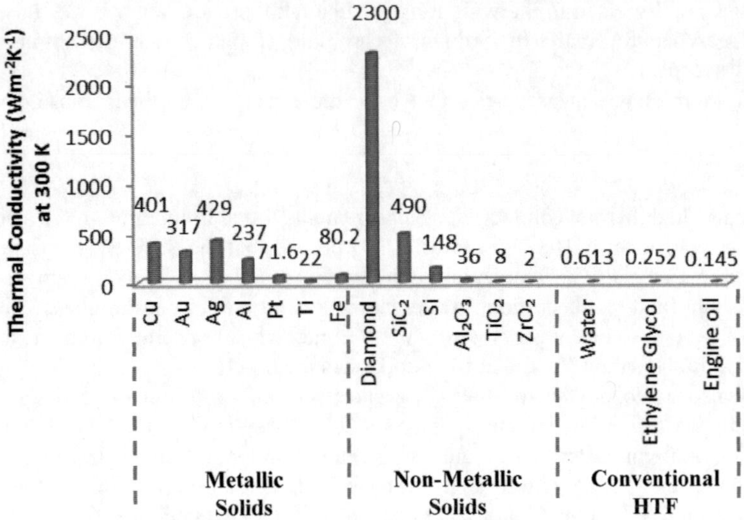

Figure 1. Thermal conductivity values of materials [3]

The higher the thermal conductivity of nanoparticles, the higher the expected heat transfer performance of nanofluid. According to the purpose of nanofluid creation, the major topic that is firstly explored is the thermal conductivity of nanofluid. The existing researches have reached the unified conclusion that thermal conductivity of nanofluid is higher than that of base fluid. The increase of thermal conductivity varies depending on many parameters, such as nanofluid concentration, nanoparticle type, temperature, and nanoparticle shape. The prediction of nanofluid thermal conductivity is performed through the experimental research and theoretical modelling. Due to the increase of thermal conductivity and promising heat transfer performance, nanofluid is potential to be applied in many various fields, e.g.:

- As the coolant in microchips cooling and other electronics
- As the mixture of diesel fuel to increase the total combustion heat and to decrease the pollution of exhaust emission
- As the coolant in the various industrial cooling process
- As the coolant in nuclear reactor technology, as both primary coolant and emergency coolant at the reactor safety features
- As the drug delivery for medical application. The size of nanoparticle makes the delivery to be more effective.

However, the research on nanofluid has given the insight that the application of nanofluid as heat transfer fluid still meets a lot of challenges. The purpose of this paper is to assess several challenges on the application of nanofluid as the alternative coolant, i.e.:

- determination of nanofluid thermophysical properties
- analysis of heat transfer characteristics of nanofluid
- stability factor of nanofluid.

As complement, this paper will also assess the challenge of nanofluid application in nuclear technology.

2. Determination of Nanofluid Thermophysical Properties

Thermophysical properties data are crucial in analyzing the heat transfer performance of nanofluid. Determination of nanofluid properties becomes an important issue because the measurement results of several properties are dispersed in the wide range. Although there are a number of theoretical models proposed by researchers to predict the values of nanofluid properties, there is no an accurate model that is generally accepted.

Some of the thermophysical properties that have the effects on the heat transfer performance of nanofluid are:

2.1. Thermal conductivity

The results of nanofluid thermal conductivity measurements found in literature have a certain range. A benchmark study was done by Buongiorno et al. [4] proved that the different measurement methods tend to produce the systematic difference in the measurement results. The measurement of the thermal conductivity of nanofluid has been done by researchers using various techniques. The most popular technique used is the transient hot wire (THW) method. Other measurement techniques are temperature oscillation method [5] and cut-bar apparatus method [6].

Researchers also try to predict the thermal conductivity of nanofluid theoretically. The classical models [7], such as Maxwell model, Bruggeman model, Jeffrey model, and Rayleigh model, used for predicting the thermal conductivity of liquid suspension containing mm or μm sized solid particles tend to underpredict the thermal conductivity of nanofluid. Therefore, a lot of new models appear by taking into account several factors that are not accounted in the classical model. Various possible heat transport mechanisms responsible for the thermal conductivity of nanofluid are considered in the new models, such as conduction through nanolayer, nanoparticle Brownian motion, nanoparticle vibration, and thermophoresis mechanism. The summary of classical and theoretical models of nanofluid thermal conductivity can be found in Zhang's book [2]. The proposed theoretical models combine the static models and the dynamic models. The static model likes classical model, assumes that nanoparticles and base fluid molecules are in static positions. Oppositely, the dynamic models consider the effects of nanoparticles movement due to Brownian motion to the thermal conductivity of nanofluid. However, a general model for predicting nanofluid thermal conductivity has not been available yet. It means that the real heat transport mechanisms prevail in the nanofluid system that contribute to the thermal conductivity of nanofluid are still not known clearly.

2.2. Viscosity

The number of theoretical models for predicting nanofluid viscosity is not as much as that for predicting nanofluid thermal conductivity. The conventional models are usually used to predict the dynamic viscosity of nanofluid, such as Einstein equation [8], Brinkman equation [9], and Batchelor equation [9]. The empirical correlations come from the measurement activities are also not found much.

Viscosity is the important factor in considering the reliability of nanofluid in practical applications. The inclusion of nanoparticles must increase the viscosity of nanofluid compared to that of base fluid. Hence, it will bring the impact to the pumping power increment. The benefits of nanofluid as heat transfer fluid are determined based on the consideration between the increase of heat transfer performance and the increase of pumping power. The properties comparative method [10] showed that viscosity has the important role in natural convective heat transfer. It becomes a factor that is responsible for the decrease of natural convective heat transfer of nanofluids.

2.3. Density

Density (ρ) of nanofluid is typically determined using the mixture rule as below:

$$\rho_{eff} = (1-\phi)\rho_f + \phi\rho_p \tag{1}$$

The subscripts of p and f represent nanoparticle and base fluid, respectively.

2.4. Specific heat

Specific heat (c_p) of nanofluid could be determined using mixing rule (Eq. (2)) and thermal equilibrium principle (Eq. (3)).

$$c_{p,eff} = (1-\phi)c_{p,f} + \phi c_{p,p} \quad (2)$$

$$\rho_{eff} c_{p,eff} = (1-\phi)\rho_f c_{p,f} + \phi \rho_p c_{p,p} \quad (3)$$

2.5. Expansion thermal coefficient

Similar to specific heat, expansion thermal coefficient (β) of nanofluid is determined either using Eq. (4) and Eq. (5).

$$\beta_{eff} = (1-\phi)\beta_f + \phi\beta_p \quad (4)$$

$$\rho_{eff}\beta_{eff} = (1-\phi)\rho_f\beta_f + \phi\rho_p\beta_p \quad (5)$$

The determination of nanofluid properties needs to be explored further, especially through experimental measurements. The measurement data obtained are important to build and validate the accurate theoretical models.

3. Heat Transfer Characteristics of Nanofluid

Research on heat transfer using nanofluid aims to compare the heat transfer performance of nanofluid and that of base fluid. Unfortunately, researchers gave the different results, primarily in natural convective heat transfer. Putra et al. [11] conducted an experiment of natural convective heat transfer using Al_2O_3-water nanofluid and CuO-water nanofluid in a horizontal tube. The result showed that the heat transfer performances of both nanofluids are lower than those of water. The numerical researches performed by Khanafer et al. [12] and Mahmoudi et al. [13] give the contradictive results. Several researches performed numerically [14], theoretically [15], and experimentally [16] showed the existence of nanoparticles optimum concentration. At that optimum concentration, the heat transfer performance of nanofluid changes from higher than the heat transfer performance of base fluid to lower than that of base fluid. The viscosity is assumed to be the most important factor that is responsible for the degradation of the nanofluid heat transfer.

In forced convective heat transfer using nanofluids, generally the results showed the increase of heat transfer performance. The increase of heat transfer performance varies depending on many factors. One of the contributing factors for the increase of nanofluid heat transfer is the thermal conductivity of nanofluid. Further research is needed to assess other influential factors of heat transfer performance of nanofluid. The characteristics of mixed convective heat transfer of nanofluid are also need to be studied further due to its complexity. Both the experimental and numerical studies of mixed convective heat transfer using nanofluid are still rarely found.

The heat transfer performance of nanofluid compared to that of base fluid could be qualitatively predicted using the properties comparative method. In the previous paper [10], we have shown that the natural convective heat transfer of Al_2O_3-water nanofluid and ZrO_2-water nanofluid tend to be lower than that of water. In this paper, we will use the properties comparative method to analyze the forced convective heat transfer of nanofluid. The ratio between the heat transfer coefficient (h) of nanofluid to that of water in forced convection mode based on their thermophysical properties, is given by Eq. (6) and (7).

$$Nu = \frac{hd}{k} = \mathrm{Re}^m \mathrm{Pr}^n \quad (6)$$

$$\frac{h_{eff}}{h_f} \approx \left(\frac{k_{eff}}{k_f}\right)^{1-n} \left(\frac{\rho_{eff}}{\rho_f}\right)^{m} \left(\frac{c_{p,eff}}{c_{p,f}}\right)^{n} \left(\frac{\mu_{eff}}{\mu_f}\right)^{n-m} \tag{7}$$

where *Nu* is Nusselt number, *Re* is Reynolds number, *Pr* is Prandtl number, k is thermal conductivity, d is diameter or characteristic length, and μ is dynamic viscosity. The typical values of m (0.5 or 0.8) are higher than those of n (0.3). Hence, the contributions of properties effects within Eq. (7) in consecutively decreasing order are k, ρ, c_p, and μ. The enhancement of the forced convective heat transfer coefficient of nanofluid is proportional to the increase of thermal conductivity, density, and specific heat. The forced convective heat transfer coefficient of nanofluid is deteriorated by the enhancement of dynamic viscosity. It is clear that for types of nanofluids which have low increase of viscosity compared to water, their forced convective heat transfer performances should be higher than those of water.

Here, we can see that the valid values of nanofluid thermal properties are important in analyzing the heat transfer performance of nanofluids accurately. In addition, more experimental researches are needed to reach the conclusive results.

4. Stability Factor

The ability of nanoparticles to agglomerate is a critical problem faced in the practical application of nanofluid. The agglomeration brings the effects to the properties of nanofluid and hence impacts on the heat transfer performance of nanofluid. The agglomerates formed could have various sizes and configurations depending on many factors, primarily elapsed time. Keblinski [17] showed that the nanoparticles configuration in an agglomerate affects the thermal conductivity of nanofluid. The different time duration between nanofluid preparation and the experimental study as well as the different time duration of experimental study performed also determine the agglomerates characteristics. The number, size, and configuration of agglomerates can change every time. That is why it becomes to be difficult to control the agglomeration effects to the thermal conductivity of nanofluid.

Different nanofluid preparations, such as dispersion method, the time duration for ultrasonic vibration process, etc. affect the stability level of nanofluid. Fedele et al. [18] showed that the selection of dispersant type and dispersion technique could improve the nanofluid stability. In their study, the high-pressure homogenization was found to be the best method of dispersion technique. They also found that the ***n***-dodecyl sulphate and polyethylene glycol are the best dispersants to improve the stability of SWCNHs-water and TiO_2-water nanofluids. Other ways to increase the stability of nanofluid are the using of surfactant and the arrangement of nanofluid acidity. However, the stability that is reached in low pH condition makes nanofluid to be difficult to be applied in many application systems.

Finally, it can be analyzed that the dispersed measurement data of nanofluid properties, especially thermal conductivity and viscosity, may be caused by the unknown number, sizes, and configurations of formed nanoparticles agglomerates.

5. Nuclear Application Aspect

The application of nanofluid in nuclear reactor system must be assessed from the safety aspect, especially radiation safety aspect. Nanoparticles could be activated by neutrons through neutron capture reaction producing the nucleus at the excited states. The unstable activation products reach their stable states by emitting gamma rays, beta particles, alpha particles, fission products or neutrons (in nuclear fission). The higher the energy of emission, the more hazardous the impact received. Another considerable aspect is the amount of neutron population in the reactor core. Every type of nanoparticle has certain neutron absorption cross sections. The type of nanoparticle used should have low neutron absorption cross section to maintain the population number of neutron in the core. However, the existence of nanoparticles in the reactor core is prohibited if it increases the criticality value (neutron multiplication number) deviating from the limiting operational condition. The increase of criticality can cause the uncontrolled thermal power increase. Several kinds of metallic substances that have low neutron absorption cross section are Al, Mg, Zr, Si, and Bi [19]. Those substances can be considered to be used as nanoparticles.

Finally, it is clearly seen that the development of nanofluid application in many fields needs a comprehensive research program that covers the basic research and the applied research. The nanofluid research is a multidisciplinary research that collaborates various fields of sciences, such as chemistry, physics, engineering, and mathematics. One of the main topics of basic research needed is research on molecular physics to investigate the real mechanisms of momentum and energy transport within the nanofluid system from a molecular point of view. Based on that knowledge, we can know all factors which affect the nanofluid thermophysical properties and hence we can develop ways to improve the heat transfer performance of nanofluid. There are also a lot of applied research topics needed to be explored, for example, the heat transfer characteristics of nanofluid, the increase of pressure drop, the feasibility of nanofluid applications in various technical equipments, etc.

6. Conclusion

The limitation of knowledge and scientific literature on nanofluid hydrodynamics and thermal characteristics brings several challenges in nanofluid applications. More comprehensive studies and multidisciplinary research are absolutely needed to perform for solving those challenges. A lot of research topics related to nanofluid have existed to be implemented in the future. The research should combine and collaborate the experimental research, the numerical research, and the theoretical research.

References

[1] Choi SUS and Eastman JA 1995 Int. Mechanical Engineering Congress and Exhibition San Francisco, USA.
[2] Zhang Y 2013 Nanotechnology Science and Technology Series Nova Science Publishers, Inc. New York.
[3] Shanthi R, Anandan SS and Ramalingam V 2012 Thermal Science **16** 423-444.
[4] Buongiorno J et al. 2009 J. Appl. Phys. **106** 094312.
[5] Das SK, Putra N, Thiesen P and Roetzel W 2003 Int. J. Heat Transfer **125** 567-574.
[6] Li CH and Peterson GP 2006 J. Appl. Phys. **99** 1-8.
[7] Abareshi, M., Goharshadi, E.K., Zebarjad, S.M., Fadafan, H.K. and Youssefi, A. 2010 J. Magnetism and Magnetic Materials **322** 3895–3901.
[8] Kakac S and Pramuanjaroenkij A 2009 Int. J. Heat and Mass Transfer **52** 3187-3196.
[9] Yang L, Du K, Ding YH, Cheng B and Li YJ 2012 Powder Technology 210–218.
[10] Hidayanti D 2016 Prosiding Seminar Keselamatan Nuklir 2016.
[11] Putra N, Roetzel W and Das SK 2003 Int. J. Heat and Mass Transfer **39** 775-784.
[12] Khanafer K, Vafai K and Lightstone M 2003 Int. J. Heat and Mass Transfer **46** 3639-3653.
[13] Mahmoudi AH, Shahi M, Raouf AH and Ghasemian A 2010 Int. Communications in Heat and Mass Transfer **37** 1135-1141.
[14] Mahrood MRK, Etemad SG and Bagheri R 2011 Int. Communications in Heat and Mass Transfer **38** 1449-1454.
[15] Corcione M 2010 Int. J. Thermal Sciences **49** 1536-1546.
[16] Ho CJ, Liu WK, Chang YS and Lin CC 2010 Int. J. Thermal Sciences **49** 1345-1353.
[17] Keblinski P, Phillpot SR, Choi SUS and Eastman JA 2002 Int. J. Heat and Mass Transfer **45** 855-863.
[18] Lamarsh JR 1983 Introduction to Nuclear Engineering 2nd Edition, Addison-Wesley Publishing Company Inc., Massachusetts.
[19] Fedele L, Colla L, Bobbo S, Barison S and Agresti F 2011 Nanoscale Research Letters.

Solvent and extraction methods effects on the quality of eel *(Anguilla bicolor)* oil

H Sasongko[1,3*], N R Efendi[1], A Budihardjo[2,3], Y Farida[1], T Amartiwi[1], A A Rahmawati[1], A Wicaksono[1], and Sugiyarto[2,3]

[1] Study program of Pharmacy, Faculty of Mathematics and Natural Sciences, Universitas Sebelas Maret. Jl. Ir. Sutami 36A Surakarta 57126, Central Java, INDONESIA

[2] Study program of Biology, Faculty of Mathematics and Natural Sciences, Universitas Sebelas Maret. Jl. Ir. Sutami 36A Surakarta 57126, Central Java, INDONESIA

[3] Biodiversity Research Group, Study program of Bioscience of Post Graduate Program, Universitas Sebelas Maret. Jl. Ir. Sutami 36A Surakarta 57126, Central Java, INDONESIA

E-mail: heru_sasongko@staff.uns.ac.id

Abstract. Eel (*Anguilla* bicolor*)* is a general fish consumption in many countries, especially Japan, China, Germany, and France. Besides its vitamin rich, eel oil is also known to contain fatty acids that are necessary for pharmaceutical purposes and as food a supplement. This research was aimed to evaluate the quality of eel oil by different solvent and extraction methods. In this study, fresh eels were extracted using maceration and reflux methods.Chloroform was used as the solvent in the maceration while water used in the reflux method. The oil quality was examined based on the Official Methods of Analysis of the Association of Official Analytical Chemist (AOAC).The result showed that the yield of eel oil using maceration method was 5.44% ± 0.64with a specific gravity of 0.915 g/mL, while reflux method obtained the yield of 5.33 % ± 0.84 and specific gravity of 0.8575 g/mL. The physicochemical parameters of oil quality used in this study were acid, peroxide, saponification, and iodine value. The maceration method obtained the acid value of 17.389 mgKOH/g, the peroxide value of 7.021meqO_2/kg, the saponification value of 111.16mgKOH/g, and the iodine value of 65.14 WIJS. While the reflux method produced the acid value of 9.116 mgKOH/g, the peroxide value of 6.088 meqO_2/kg, the saponification value of 70 mgKOH/g, and the iodine value of 87.74 WIJS.

1. Introduction

Eel *(Anguilla bicolor)* is a fish that inhabits Indonesian waters with a regional distribution comprise Sumatra, Java, Borneo, Sulawesi, Bali, Nusa Tenggara, Maluku, and Papua[1–3]. Eel is a fish consumption that is popular in many countries, especially Japan, China, Germany, and France. Besides its unique taste, eel also rich in vitamin A, B1, B2, B6, C, D, protein albumin, Docosahexaenoic Acid and Eicosapentaenoic acid as well as several other minerals[1,4]. Therefore, this fish has a potential export commodity in Indonesia. In addition to vitamins, minerals, and fatty acids, eels also known to

contain albumin that can heal an open wound[5]. Eel *(Anguilla bicolor)* oil contains palmitic acid (Hexadecanoic acid) 48.55% and oleic acid (9-octadecenoic acid) 51.44% [6,7].
To obtain high quality and quantity of eel oil, there must be an appropriate method to avoid minor or major alteration of the oil contents. The solvent extraction technique can perform the extractions on a liquid or a solid sample with the minimal effort. An appropriate extraction method is important to extract compounds of interest in high amounts of quantity[8]. Chloroform is generally considered an extraction the best extraction solvent for most of the edible oils in laboratory-scale[9]. While water is the most suitable solution that can be reached by small industries with the fewer side effects. The methods used to obtain high amounts of compounds from natural products are pressing matters for product quality[10]. The purpose of this research was to determine the quality of eel oils extracted from different solvent and methods.

2. Materials and methods

2.1. Materials
The materials of this study were eels *(Anguilla bicolor)* that were collected from UPT UNS, distilled water, NaCl, KOH (Merck®), phenolphthalein indicator, glacial acetic acid, KI, chloroform, $Na_2S_2O_3$, iodine bromide reagent and ethanol 96% p.a. The equipment used in this study were analytical scales, reflux, oven, heating mantle, knife, blender, water bath, centrifugation and glasswares (Pyrex Iwaki®).

2.2. Oil extractions
The extraction of eel oil was carried out using two different methods, maceration, and reflux. The maceration was done by macerating 1 kg of fresh eels that has been cut into small pieces for 24 hours using chloroform as the solvent. While reflux extraction was performed by using distilled water as the solvent. To achieve this, 1 kg of fresh eels were placed inside a 500-ml round bottom flask. The sample was extracted for 6 hours at the solvent's boiling point. The oils were separated from the solvents using centrifugation. All methods were carried out triplicately.

2.3. Characterization

2.3.1. Acid value. The acid was calculated using the formula below[11]:

$$\text{Acid value} = \frac{56{,}1 \times \text{KOH (M)} \times \text{titration (mL)}}{\text{The sample (g)}}$$

2.3.2. Peroxide value. The peroxide value was calculated using the formula below[11]:

$$\text{Peroxide value} = \frac{\text{Na2S2O3 (mL)} \times \text{Na2S2O3 (N)} \times 1000}{\text{The sample (g)}}$$

2.3.3. Saponification value. Saponification value was calculated using the formula below[11]:

$$\text{Saponification value} = \frac{28{,}05 \times (\text{blank titran (mL)} - \text{sample titran (mL)})}{\text{The sample (g)}}$$

2.3.4. Iodine value. The iodine value was calculated by the formula below[11]:

$$\text{Iodine value} = \frac{\text{Blank titran (mL)} - \text{sample titran (mL)}}{\text{The samples (g)}} x\text{N Na2S2O3} \times 12{,}691$$

3. Result and discussion

Table 1. Parameters of eel oils quality

Parameter	Maceration (chloroform)	Reflux (water)	Unit
Specific gravity	0.915	0.8575	g/mL
Acid value	17.389	9.116	mg KOH/g
Peroxide value	7.021	6.088	meq O_2/ kg
Saponification value	111.160	70.000	mg KOH/g
Iodine value	65.140	87.740	WIJS

In this study, the yield of eel oil from maceration doesn't significantly different from reflux method. Reflux is an extraction method using thermal condition. The temperature and solvent are affecting the result[12].The higher temperature in reflux method produces much yield of oils. Maceration method usually taking more time to gain many oils because it's temperature is lower than reflux method. Also, the similar polarity of solvent with target compounds will affect the oil quantity. In this study, oils extracted by both maceration and reflux method obtained unsignificantly different yield of oils. Maceration method has a darker color than reflux method. But the consistency, odor, and flavor are relatively same. From the results of the organoleptic, it is known that oil color is yellowish, it is like the color of DHA that is yellowish[13].

Tabel 2. The yield of eel oils

Extraction method	Result ± SD
Maceration	5.44 % ± 0.64
Reflux	5.33 % ± 0.84

The acid value is the number of total acidity lipids, which includes contributions from all the constituent fatty acid glycerides which form a molecular compound. As oil-fats become rancid, triglycerides are converted into fatty acids and glycerol and causing an increase in acid values. Besides, better information on the acidity of glycerides could be obtained from the acid value, which takes into account the contribution of all the constituent fatty acids in the oil or fat[14]. The acid value of oil takes effect on the quantity of free fatty acids present or the degree of hydrolysis of the oil. The acid value of oil suitable for edible purposes should not exceed 4 mg/g [15]. In this study, the acid value of maceration was higher than reflux method. This indicates the free fatty acids in maceration method was higher than reflux method.

Table 3. Organolepticsof eel oils

Parameters	Reflux	Maceration
Consistency	Liquid	Liquid
Color	Yellow	Brown yellowish
Odor	Fishy	Fishy
Flavor	Tasteless	Tasteless

Peroxide value is a measure of the reaction rate of lipid oxidation, which causes defective [16]. Normally, oils become rancid when the peroxide value 30.0 mg/g or ranges from 20.0 mg/g oil to 40.0 mg/g oil [17]. From this study, it is known that the peroxide value of both maceration and reflux method are relatively good. But due to the regulation of the The National Food and Drug Agency, this number hasn't yet fulfilled the criteria of good peroxide value in fish oil (max. 2 meq O_2/ kg).

Saponification numbers measure the amount of fatty acids in fish oil and its size. Saponification numbers will be higher if the oil contains a lot of saturated fatty acids for determining the carbon chain lengths present in fish oil [14].It is known that saponification value on both methods didn't fulfill the criteria from The National Food and Drug Agency (min. 170 mgKOH/g).

The iodine value is a measure of the unsaturation of fats and oils. The higher iodine value indicates, the higher unsaturation of fats and oils[18]. This value can be used to infer the extent of double bonds present in the oil i.e. unsaturation, which reflects the susceptibility of oil to oxidation. From this study, it is known that iodine value from reflux method is higher than maceration method. This means that unsaturated fatty acid in eel oil produced by reflux was higher than maceration method.

4. Conclusion

Eel oils extracted from reflux method using water solvent obtained a better quality than maceration method using chloroform solvent.

Conflict of interest

The authors declare no conflict of interests

References

[1] Hakim A A, Kamal M M, Butet N A and Affandi R 2016 *J. Ilmu Dan Teknol. Kelaut. Trop.* **7**
[2] Fahmi M R 2015 *Pros. Sem. Nas. Masy. Biodiv. Indon.* Vol 1 (Surakarta: Biology Department of Sebelas Maret University) p 38-43
[3] Ohee H L 2016 *Biodiversitas* **17** 665–72
[4] Roos N, Chamnan C, Loeung D, Jakobsen J and Thilsted S H 2007 *Food Chem.* **103** 1104–11
[5] Putri A A B, Yuliet Y and Jamaluddin J 2016 *J. Farm. Galen. Galen. J. Pharm.* **2** 35–41
[6] Sandita A, Maulana I T and Syafnir L 2015 *Pros. Farm.* **0** 388–96
[7] Clara M. Kusharto, R. A. Hangesti Emi Widyasari, Budywiryawan, Eko Sri Wiyono and Sugengherisuseno 2014 *J. Sains Kesihat. Malays.* **12** 41–6
[8] Turkmen N, Sari F and Velioglu Y S 2006 *Food Chem.* **99** 835–41
[9] Koski A, Psomiadou E, Tsimidou M, Hopia A, Kefalas P, Wähälä K and Heinonen M 2002 *Eur. Food Res. Technol.* **214** 294–8
[10] Mezzomo N, Mileo B R, Friedrich M T, Martínez J and Ferreira S R S 2010 *Bioresour. Technol.* **101** 5622–32
[11] Cunniff P and Association of Official Analytical Chemists 1995 *Official methods of analysis of AOAC International* (Washington, DC: Association of Official Analytical Chemists)
[12] Lin J-T, Liu S-C, Hu C-C, Shyu Y-S, Hsu C-Y and Yang D-J 2016 *Food Chem.* **190** 520–8
[13] Yi T, Li S-M, Fan J-Y, Fan L-L, Zhang Z-F, Luo P, Zhang X-J, Wang J-G, Zhu L, Zhao Z-Z and Chen H-B 2014 *Lipids Health Dis.* **13** 190
[14] Jayeoye T J 2015 *Pak. J. Sci. Ind. Res. Ser. Phys. Sci. Print* **58** 122–9
[15] Esuoso K O and Odetokun S M (Akure U (Nigeria) D of C 1995 *Riv. Ital. Delle Sostanze Grasse Italy*
[16] Ozogul İ, Polat A, Özogul Y, Boga E K, Ozogul F and Ayas D 2014 *Int. J. Food Sci. Technol.* **49** 847–53
[17] Babalola T and Apata D 2011 *Agric. Biol. J. N. Am.* **2** 935–43
[18] Knothe G 2002 *J. Am. Oil Chem. Soc.* **79** 847–54

The Solutions of the D-dimensional Schrodinger Equation for the Potential $V(r) = ar^{-6}+br^{-5}+cr^{-4}+dr^{-3}+er^{-2}+fr^{-1}$

Wahyulianti[1], **Suparmi, Cari, Fuad Anwar**

Physics Department of Post Graduate Program Sebelas Maret University,
Jl. Ir. Sutami 36A Kentingan Jebres Surakarta 57126, INDONESIA

E-mail: wahyuliantiyuli@yahoo.com

Abstract. The solutions of the D-dimensional Schrodinger equation was solved by using Wavefunction Ansatz method. We used the form of potential as $V(r) = ar^{-6}+br^{-5}+cr^{-4}+dr^{-3}+er^{-2}+fr^{-1}$. The D-dimensional Schrodinger equation was separated to radial part and angular part. The radial part of D-dimensional Schrodinger equation solved using Wavefunction Ansatz method. We obtained the energy equation and the radial wavefunctions.

1. Introduction

Schrodinger equation was the fundamental equation to solve the wave equation of physic. In quantum mechanics, Schrodinger equation described the behavior of electrons or particles. The solution of the Schrodinger equation is the equation of the energy and wave functions [1]. Schrodinger equation developed became D-dimensional forms for complex systems in quantum mechanics [2]. The wave equation in quantum mechanics for higher dimensions consisted of Klein-Gordon equation, Dirac equation, and Schrodinger equation [3]. Generally, the solution of the wave equation could be obtained by several approach methods, i.e. Romanovski polynomial method [4], Asymptotic Iteration Method (AIM) [5], Supersymmetry (SUSY) Quantum Mechanics [6], factorization method [2], Nikiforov-Uvarov method [8], and Wavefunction Ansatz method [3,9-11]. In this paper, we would solve the D-dimensional Schrodinger equation using Wavefunctions Ansatz method. Wavefunction Ansatz method was a method that can be used to analyze the part radial of D-dimensional Schrodinger equation. In Section 2 we would discuss the wavefunction Ansatz method and section 3 discuss about the several results.

2. Wavefunction Ansatz Method

In this section, we review the solution of general Schrodinger equation in D-dimensional with Wavefunction Ansatz method. The General Schrodinger equation in D-dimensional shown as followed equation [3,12]:

$$-\frac{\hbar^2}{2\mu}\nabla_D^2\,\psi(r,\Omega)+V(r,\Omega)\psi(r,\Omega)=E\psi(r,\Omega) \qquad (1)$$

The radial part of Schrodinger equation in D-dimensional obtained by the transformation Eq. (1):

$$\psi(r,\Omega)=\frac{1}{r^{\frac{D-1}{2}}}R_m(r)Y^{l}_{l_{D-1}\cdots l_1}(\hat{x}) \qquad ;\hat{x}=\theta_1,\theta_2,\cdots\theta_{D-1}$$

We have the form of radial Schrodinger equation with $(\hbar = \mu = 1)$ as:

$$\left[\frac{d^2}{dr^2} + 2[E - V(r)] - \frac{\kappa^2 - 1/4}{r^2}\right] R_m(r) = 0 \tag{2}$$

In this paper we used the form of potential with general equation as:

$$V(r) = \frac{a}{r^6} + \frac{b}{r^5} + \frac{c}{r^4} + \frac{d}{r^3} + \frac{e}{r^2} + \frac{f}{r}\,. \tag{3}$$

The ansatz solution of radial function R_m can be expressed as:

$$R_m(r) = f_m(r) \exp g(r) \tag{4}$$

where

$$f_m(r) = \begin{cases} 1 & \text{jika } m = 0, \\ \prod_{j=1}^{m} (r - \sigma_j^{(m)}) & \text{jika } m = 1,2,3,\ldots \end{cases} \tag{5}$$

and

$$g(r) = \frac{1}{2}\frac{A}{r^2} + \frac{B}{r} + Cr + D\ln r \tag{6}$$

By substituting Eq.(4) to Eq.(2), we have:

$$-2E + 2V(r) + \frac{\kappa^2 - 1/4}{r^2} = g''(r) + (g'(r))^2 + \frac{f_m''(r) + 2f_m'(r)g'(r)}{f_m(r)} \tag{7}$$

3. Results and Discussion

Firstly, for the ground state $m = 0$, we use $f_0(r) = 1$. Substituting Eq. (3), (5), and (6) to Eq. (7) we obtain:

$$\begin{aligned} -2E + 2\left(\frac{a}{r^6} + \frac{b}{r^5} + \frac{c}{r^4} + \frac{d}{r^3} + \frac{e}{r^2} + \frac{f}{r}\right) &= \frac{A^2}{r^6} + \frac{2AB}{r^5} + \frac{B^2 - 2AD + 3A}{r^4} + \\ \frac{2B(1-D) - 2AC}{r^3} &+ \frac{D(D-1) - 2BC - \left(\kappa^2 - 1/4\right)}{r^2} + \frac{2CD}{r} + C^2 \end{aligned} \tag{8}$$

by comparing right the side and left the side of Eq.(8), we obtained:

$$\begin{aligned} &2a = A^2, \quad 2b = 2AB, \quad 2c = B^2 - 2AD + 3A, \quad 2d = 2B(1-D) - 2AC, \\ &2e = D(D-1) - 2BC - \left(\kappa^2 - 1/4\right), \quad 2f = 2CD, \quad -2E = C^2 \end{aligned} \tag{9}$$

and

$$\lambda = 2e + \kappa^2 - 1/4 \tag{10}$$

where

$$\lambda = D(D-1) - 2BC \tag{11}$$

from Eq. (9) We obtain $A = \pm\sqrt{2a}$. We choose $A = -\sqrt{2a}$ and Eq. (11) becomes

$$\lambda = \left(\frac{3}{2} + \tau\right)\left(\frac{1}{2} + \tau\right) + \frac{2b}{\sqrt{2a}} \frac{f}{\left(\frac{3}{2} + \tau\right)} \tag{12}$$

With $D = \frac{3}{2} + \tau$.

By substituting Eq. (9) to (11) we obtained equation:

$$\frac{2b}{\sqrt{2a}}C^2 - \lambda C + \left(\frac{1}{2} - \frac{b^2}{4a\sqrt{2a}} + \frac{c}{\sqrt{2a}}\right)f = 0 \tag{13}$$

This equation could be solved with algebraic equation as:

$$C = \frac{\lambda \pm \sqrt{\lambda^2 - \left(\frac{4c - 2\sqrt{2a}}{a} - \frac{b^2}{a^2}\right)bf}}{\frac{4b}{\sqrt{2a}}} \tag{14}$$

Substituting Eq. (9) into Eq. (14) and we find the equation of energy for the ground state:

$$E_0^{\pm} = -\frac{1}{16}\frac{a}{b^2}\left[\lambda \pm \sqrt{\lambda^2 - \left(\frac{4c - 2\sqrt{2a}}{a} - \frac{b^2}{a^2}\right)bf}\right]^2 \tag{15}$$

and from Eq.(4), the corresponding eigenfunction can be written as

$$R_0(r) = N_0 r^D \exp\left[\frac{1}{2}\frac{A}{r^2} + \frac{B}{r} + Cr\right] \tag{16}$$

Where N_0 is normalization constant.

Secondly, for $m = 1$, we used $f_1(r) = r - \sigma_1^{(1)}$. By substituting Eq. (3), (5), and (6) to Eq. (7) we obtain

$$\left[-2E + 2V(r)\right]\left(r - \sigma_1^{(1)}\right) = \left(\begin{array}{c}\frac{A^2}{r^6} + \frac{2AB}{r^5} + \frac{B^2 - 2AD + 3A}{r^4} + \frac{2B(1-D) - 2AC}{r^3} + \\ \frac{D(D-1) - 2BC - \left(\kappa^2 - \frac{1}{4}\right)}{r^2} + \frac{2CD}{r} + C^2\end{array}\right)\left(r - \sigma_1^{(1)}\right)$$
$$-\frac{2A}{r^3} - \frac{2B}{r^2} + 2C + \frac{2D}{r} \quad . \tag{17}$$

A solution of Eq. (17) is

$$\begin{gathered} A = \pm\sqrt{2a}, \quad B = -\frac{b}{\sqrt{2a}}, \quad 2c = B^2 - 2AD + 3A, \\ \lambda = D(D+1) - 2C(B - \sigma_1^{(1)}), \quad f = 2C(D+1), \quad -2E = C^2 \end{gathered} \tag{18}$$

and from Eq. (18) we find

$$\lambda = \left(\frac{3}{2} + \tau\right)\left(\frac{5}{2} + \tau\right) + 2\eta\left(\frac{b}{\sqrt{2a}} + \sigma_1^{(1)}\right). \tag{19}$$

with $D = \frac{3}{2} + \tau$ and $\eta = \frac{f}{D+1}$.

We obtained algebraic equation and its solution from Eq. (18) and Eq. (19), i.e.

$$2\left(\sigma_1^{(1)}+\frac{b}{\sqrt{2a}}\right)C^2-\lambda C+\left(\frac{3}{2}+\frac{c}{\sqrt{2a}}-\frac{b^2}{4a\sqrt{2a}}\right)f=0 \tag{20}$$

and

$$C=\frac{\lambda\pm\sqrt{\lambda^2-8\left(\sigma_1^{(1)}+\frac{b}{\sqrt{2a}}\right)\left(\frac{3}{2}+\frac{c}{\sqrt{2a}}-\frac{b^2}{4a\sqrt{2a}}\right)f}}{4\left(\sigma_1^{(1)}+\frac{b}{\sqrt{2a}}\right)} \tag{21}$$

Substituting Eq. (21) into Eq. (18) and we find the energy

$$E_1^{\pm}=-\frac{1}{32}\left(\frac{\lambda\pm\sqrt{\lambda^2-8\left(\sigma_1^{(1)}+\frac{b}{\sqrt{2a}}\right)\left(\frac{3}{2}+\frac{c}{\sqrt{2a}}-\frac{b^2}{4a\sqrt{2a}}\right)f}}{\left(\sigma_1^{(1)}+\frac{b}{\sqrt{2a}}\right)}\right)^2 \tag{22}$$

and from Eq.(4), the corresponding eigenfunction can be written as

$$R_1(r)=N_1(r-\sigma_1^{(1)})r^D\exp\left[\frac{1}{2}\frac{A}{r^2}+\frac{B}{r}+Cr\right]. \tag{23}$$

With N_1 is normalization constant.
The energy vs. λ for *m = 0* and *m =* 1 is the plotted in Fig. 1 and Fig. 2.

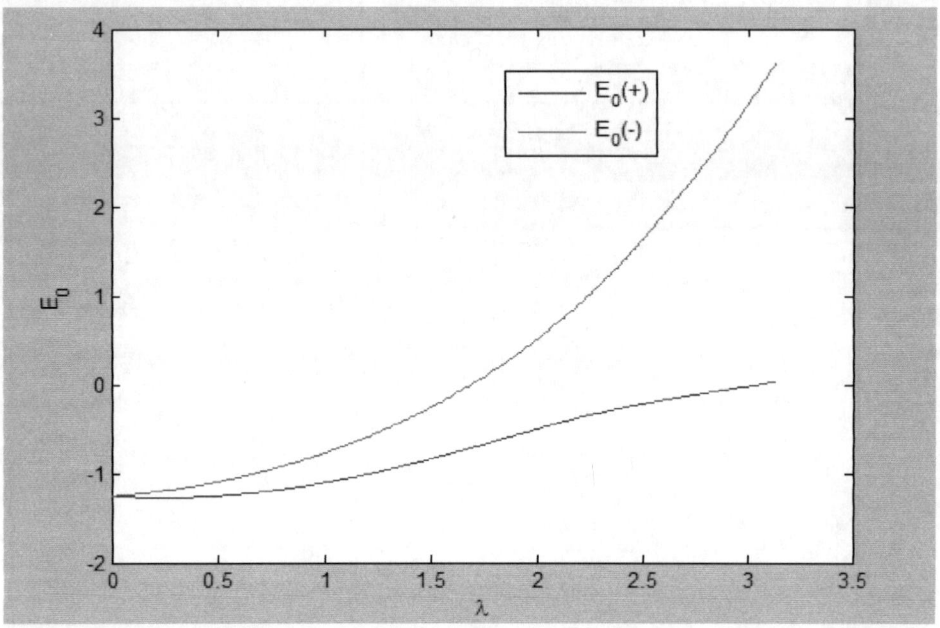

Figure 1. Energy vs. λ for *m = 0* with *a = b = c = -1* dan *d = e = f = 2.*

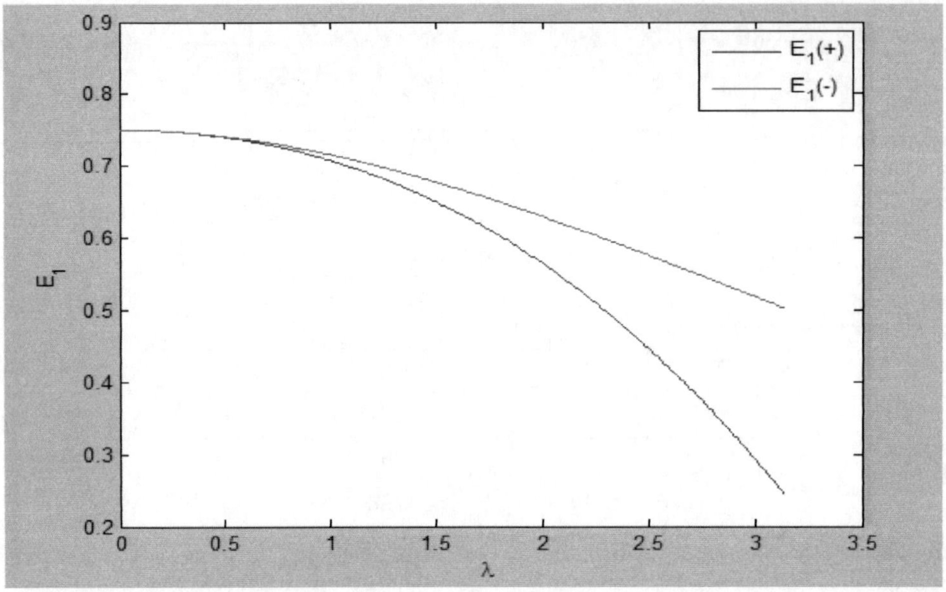

Figure 2. Energy vs. λ for m = 1 with $a = b = c = -1$ dan $d = e = f = 2$.

and because the normalization constant in equation (16) is difficult to determine, so we can rewrite Eq. (16) as $R_0(r) \approx r^D \exp\left[\frac{1}{2}\frac{A}{r^2} + \frac{B}{r} + Cr\right]$. Wavefunction vs. r for $m=0$ and $m=1$ is

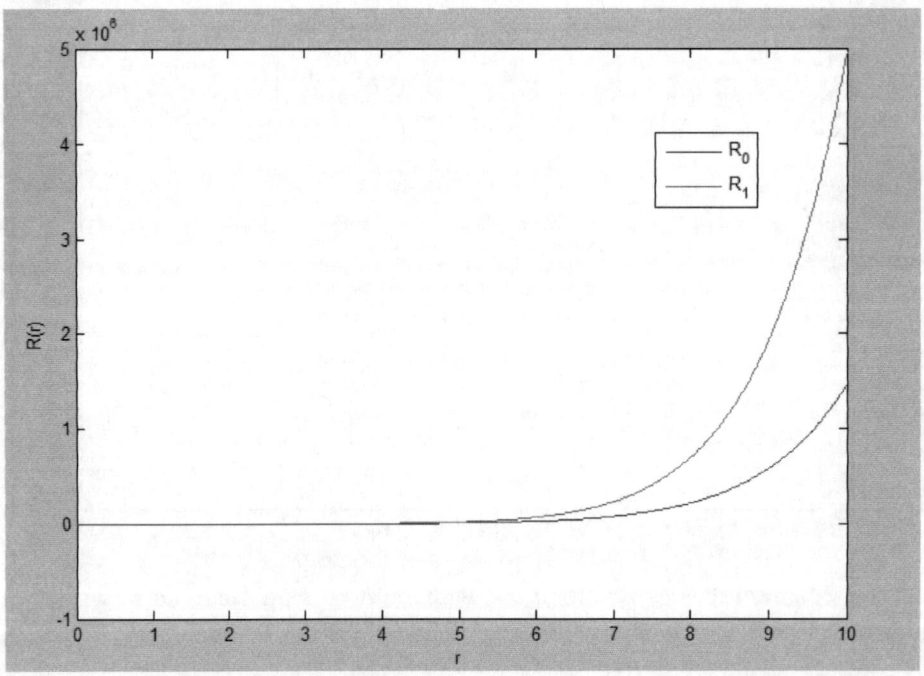

Figure 3. Wavefunction vs. r for $a = b = c = 1$ and $d = e = f = 2$.

4. Conclusion

In conclusion, we get solutions using Wavefunction Ansatz method for the potential of Eq. (3). The solutions of the equation of energy and wave function shown by Eq. (15), (16), (22), and (23). When the parameter f vanishes ($f = 0$), then $\lambda = \left(\frac{3}{2}+\tau\right)\left(\frac{1}{2}+\tau\right)$ for $m = 0$, and $\lambda = \left(\frac{3}{2}+\tau\right)\left(\frac{5}{2}+\tau\right)$ for $m = 1$. The energy increases with the increasing the value of λ for $m = 0$, while the energy decreases with increasing value of λ for $m = 1$. Wavefunction (eigenfunction) was increases from $m = 0$ to $m = 1$.

Acknowledgement

This research was partly supported by Higher Education Project Grant with contract no.632/UN27.21/LT/2016

References

[1] U A Deta, A Suparmi, and C Cari 2014 *Journal of Physics: Conference Series, CTPTN 2014* **539** 012018
[2] Ituen B Okon, Eno E Ituen, Oyebola Popoola, and Akaninyene D Antia 2013 *International Journal of Recent Advances in Physics (IJRAP)* **2** 1
[3] Shi-Hai Dong 2011 *Wave Equations in Higher Dimensions* (New York: Springer) Chapter 8 pp 97-108
[4] A Suparmi, C Cari, and U A Deta 2014 *Chin. Phys. B* **23** 090304
[5] R A Sari, A Suparmi, and C Cari 2016 *Chin. Phys. B* **25** 010301
[6] Suparmi dan Cari 2014 *J. Math. Fund. Sci.* **46** 205
[7] Suparmi, Cari, B N Pratiwi, and U A Deta 2016 *AIP Conference Proceeding* **1710** 030010
[8] B H Yazarloo, L Lu, G Liu, S Zarrinkamar, and H Hassanabadi 2013 *Advances in High Energy Physics* **2013** 5
[9] Shi-Hai Dong 2001 *Physica Scripta* **64** 273
[10] Suleyman O and Mehmet S 1991 *Physics* **152** 3
[11] H Hassanabadi, B H Yazarloo, S Zarrinkamar and H Rahimov 2012 *Commun. Theor. Phys.* **57** 339
[12] Akpan N Ikot, O A Awoga, and A D Antia 2013 *Chin. Phys. B* **22** 020304

The Solution of Schrodinger Equation with the Anharmonic Potential $V(r) = ar^4 + br^3 + cr^2 + dr + er^{-1} + fr^{-2}$ Using Ansatz Method

Kristiana Nathalia Wea[1], **Suparmi, Cari and Fuad Anwar**
Physics Department of Post Graduate Program Sebelas Maret University,
Jl. Ir. Sutami 36A Kentingan Jebres Surakarta 57126, INDONESIA

E-mail: nataliawea@gmail.com

Abstract. Using an ansatz for eigenvalue, we obtain the solution of Schrodinger equation D-dimensional with the potential $V(r) = ar^4 + br^3 + cr^2 + dr + er^{-1} + fr^{-2}$. We can determine the eigenfunction and the eigenvalue from that Schrodinger equation.

1. Introduction

Exact solution of Schrodinger equation with the physical potential is one of important part in quantum mechanics. Some methods have been developed to obtain the solution of Schrodinger equation. For examples Super Symmetry (SUSY) method [1], Nikiforov-Uvarov method [2], asymptotic iteration method (AIM) [3], factorization method [4,5], ansatz method [6,7] and other [8,9]. In recent time the ansatz method becomes a popular method to solve the Schrodinger Equation. Asantz method used to solve the radial Schrodinger wave equation with anharmonic potential. Many studies about Schrodinger equation with anharmonic potential has been done. For examples $V(r) = Ar^{-4} + Br^{-3} + Cr^{-2} + Dr^{-1}$ [10], $V(r) = ar^2 + br^{-4} + cr^{-6}$ [11] and $V(r) = ar^2 + br^4 + cr^6$ [12]. In this study, we present an approximate analytic solution of the radial Schrodinger equation D-dimensional with mixture potential $V(r) = ar^4 + br^3 + cr^2 + dr + er^{-1} + fr^{-2}$ by applying a factorization ansatz wave equation to the eigenfunctions to obtain the exact solution. The wavefunction Ansatz method is used to solve the radial Schrodinger equation with anharmonic potential.

2. The wavefunction ansatz

The Schrodinger equation in D-dimensional space with spherically symmetric central fields [6]

$$\left\{-\frac{\hbar^2}{2m}\left[\frac{1}{r^{D-1}}\frac{\partial}{\partial r}\left(r^{D-1}\frac{\partial}{\partial r}\right)-\frac{L_{D-1}^2}{r^2}\right]+V(r)\right\}\Psi(r)=E\Psi(r) \quad (1)$$

For simplicity, $\hbar = m = 1$, so Equation (1) can we write as

$$\left\{-\frac{1}{2}\left[\frac{1}{r^{D-1}}\frac{\partial}{\partial r}\left(r^{D-1}\frac{\partial}{\partial r}\right)-\frac{L_{D-1}^2}{r^2}\right]+V(r)\right\}\Psi(r)=E\Psi(r) \quad (2)$$

with

$$\psi(r) = r^{-(D-1)/2} U_m(r) Y^l_{l_{D-1}\ldots l_1}(\hat{X}) \quad (3)$$

where $Y^l_{l_{D-1}\ldots l_1}(\hat{X})$ are the generalized spherical function. Substitution this to Schrodinger equation, we obtain the radial Schrodinger equation as

$$\left\{\frac{d^2}{dr^2} - \frac{l(l+D-2)+(D-1)(D-3)/4}{r^2}\right\} U(r) = -2[E - V(r)]U(r) \quad (4)$$

with

$$\kappa = |l - 1 + D/2| \quad (5)$$

equation (4) can we write as

$$\left\{\frac{d^2}{dr^2} - \frac{\kappa^2 - 1/4}{r^2}\right\} U_m(r) = -2[E - V(r)]U_m(r) \quad (6)$$

This is the general radial Schrodinger equation that we use in this study.
Here the potential is taken as

$$V(r) = ar^4 + br^3 + cr^2 + dr + \frac{e}{r} + \frac{f}{r^2} \quad (7)$$

where κ and E are the angular momentum and energy. From the work [3], we make an ansatz for

$$U_m(r) = h_m(r)\exp[P(r)] \quad (8)$$

where

$$h_m(r) = \begin{cases} 1 \\ \prod_{j=1}^{m}\left(r - \sigma_j^m\right) \end{cases} \quad (9)$$

$h_m(r) = 1$ when $m = 0$ and $h_m(r) = \prod_{j-1}^{m}\left(r - \sigma_j^m\right)$ when $m = 1,2,3,\cdots$

and by using SUSY approximation we have

$$P(r) = \frac{1}{3}Ar^3 + \frac{1}{2}Br^2 + Cr + D\ln r \quad (10)$$

3. Results and Discussion

Substituting Equation (8) to Equation (6), we can obtain

$$\left[P''(r) + (P'(r))^2 + \frac{h''_m(r) + 2h'_m(r)P'(r)}{h_m(r)}\right] = -2E + 2V(r) + \frac{\kappa^2 - 1/4}{r^2} \quad (11)$$

Equation (11) is the most fundamental formula for the following analysis.
First, we study for $m = 0$ so that $h_m(r) = 1$. By substituting Eq. (7), (8) and (10) to Eq. (6) we have

$$2Ar + B - \frac{D}{r^2} + \left(Ar^2 + Br + C + \frac{D}{r}\right)^2 = -2E + 2\begin{pmatrix} ar^4 + br^3 + cr^2 + \\ dr + \frac{e}{r} + \frac{f}{r^2} \end{pmatrix} + \frac{\kappa^2 - 1/4}{r^2} \quad (12)$$

Using variables separation method we obtain

$$A^2 = 2a \quad (13a)$$

$$2AB = 2b \quad (13b)$$

$$B^2 + 2AC = 2c \quad (13c)$$

$$A + AD + BC = d \quad (13d)$$

$$CD = e \quad (13e)$$

$$D^2 - D = 2f + \kappa^2 - 1/4 \quad (13f)$$

$$B + C^2 + 2BD = -2E \tag{13g}$$

The parameters *A, B, C, D* and κ^2 can be determined by using Eq. (13).
From Eq. (13*a*), we have

$$A = \pm\sqrt{2a} \tag{14}$$

The physically acceptable solution are chosen as $A = -\sqrt{2a}$.
By subtituting Eq. (14) in to Eq. (13*b*), we get

$$B = -\frac{b}{\sqrt{2a}} \tag{15}$$

By subtituting Eq. (14) and (15) in to Eq. (13*c*), we obtain

$$C = \frac{1}{\sqrt{2a}}\left(\frac{b^2}{4a} - c\right) \tag{16}$$

From Eq. (13*f*) we obtain relation between D and κ^2

$$D = \frac{1 \pm \sqrt{8f + \kappa^2}}{2} \tag{17}$$

By subtituding Eq (14), (15) and (16) in Eq. (13*d*), we have

$$D = \left(-\frac{1}{2\sqrt{2a}}\right)\left(d + \sqrt{2a} + \sqrt{\left(d + \sqrt{2a}\right)^2 - 4be}\right) \tag{18}$$

and so, we can find κ^2 by the use of Eq. (13*f*) and Eq. (18) given as

$$\kappa^2 = \left[\left(-\frac{1}{2\sqrt{2a}}\right)\left(d + \sqrt{2a} + \sqrt{\left(d + \sqrt{2a}\right)^2 - 4be}\right)\right]^2 + \left(\frac{1}{2\sqrt{2a}}\right)\left(d + \sqrt{2a} + \sqrt{\left(d + \sqrt{2a}\right)^2 - 4be}\right) - 2f + 1/4 \tag{19}$$

Putting Eq. (15), (16) and (17) in to Eq. (13*g*) and let the parameter κ^2 was given by Eq. (19), so the ground state energy given as

$$E_0 = \frac{b}{\sqrt{2a}}\left(1 \pm \sqrt{\begin{array}{l}\left[\left(-\frac{1}{2\sqrt{2a}}\right)\left(d + \sqrt{2a} + \sqrt{\left(d + \sqrt{2a}\right)^2 - 4be}\right)\right]^2 \\ + \left(\frac{1}{2\sqrt{2a}}\right)\left(d + \sqrt{2a} + \sqrt{\left(d + \sqrt{2a}\right)^2 - 4be}\right) + 1/4\end{array}}\right) - \frac{1}{64a^3}\left(b^2 - 4ac\right)^2 \tag{20}$$

The eigenfunction for $m = 0$ is can write as

$$U_0(r) = N_o r^D \exp\left[\frac{1}{3}Ar^3 + \frac{1}{2}Br^2 + Cr\right] \tag{21}$$

with N_0 is a normalization constant.

Second, for $m = 1$, by following Eq. (9) in wich we take $h_1 = \left(r - \sigma_1^{(1)}\right)$ and corresponding to the first calculation where $m = 0$ leads to

$$2Ar+B-\frac{D}{r^2}+\left(Ar^2+Br+C+\frac{D}{r}\right)^2+\frac{2\left(Ar^2+Br+C+D/r\right)}{r-\sigma_1^{(1)}}=$$
$$-2E+2\left(ar^4+br^3+cr^2+dr+\frac{e}{r}+\frac{f}{r^2}\right)+\frac{\kappa^2-1/4}{r^2} \quad (22)$$

Both sides of Eq. (18), multiplied by $r-\sigma_1^{(1)}$ and by using variables separation method, we have

$$A^2=2a \quad (23a)$$
$$2AB+\left(2a-A^2\right)\sigma_1^{(1)}=2b \quad (23b)$$
$$B^2+2AC+\left(2b-2AB\right)\sigma_1^{(1)}=2c \quad (23c)$$
$$4A+2AD+2BC=2d \quad (23d)$$
$$3B+C^2+2BD+2A\sigma_1^{(1)}=-2E \quad (23e)$$
$$2CD+2C+\left(-B-C^2-2BD\right)\sigma_1^{(1)}=2e+2E\sigma_1^{(1)} \quad (23f)$$
$$D^2-D+\left(2e-2CD\right)\sigma_1^{(1)}=2f \quad (23g)$$
$$-D^2+D=-2f-(\kappa^2-1/4) \quad (23h)$$

The parameters A, B, C, D and κ^2 can be determined by using Eq. (23).
From Eq. (23*a*), we get

$$A=\pm\sqrt{2a} \quad (24)$$

The physically acceptable solution are chosen as $A=-\sqrt{2a}$.
Putting Eq. (24) in to Eq. (23*b*), we have

$$B=-\frac{b}{\sqrt{2a}} \quad (25)$$

By subtituting Eq. (24) and (25) in to Eq. (23*c*), we obtain

$$C=\frac{1}{\sqrt{2a}}\left(\frac{b^2}{4a}-c\right) \quad (26)$$

From Eq. (23*h*) we obtain the relation between D and κ^2

$$D=\frac{1\pm\sqrt{8f+\kappa^2}}{2} \quad (27)$$

By using Eq. (23*d*), (23*e*) ang (23*f*) we have

$$D=\left(-\frac{1}{2\sqrt{2a}}\right)\left[d+3\sqrt{2a}+\sqrt{\left(d+3\sqrt{2a}\right)^2+4\sqrt{2a}\left(\begin{matrix}-d-2\sqrt{2a}-\frac{eb}{\sqrt{2a}}-\\ \frac{b}{\sqrt{2a}}\left(\frac{b}{\sqrt{2a}}\sqrt{2a}\sigma_1^{(1)}\right)\sigma_1^{(1)}\end{matrix}\right)}\right] \quad (28)$$

and so, we can find parameter κ^2 by subtituting Eq. (28) in to Eq. (23*h*) given as

$$\kappa^2 = \left[\left(-\frac{1}{2\sqrt{2a}}\right)\left[d+3\sqrt{2a}+\sqrt{\left(d+3\sqrt{2a}\right)^2+4\sqrt{2a}\left(\begin{matrix}-d-2\sqrt{2a}-\frac{eb}{\sqrt{2a}}-\\ \frac{b}{\sqrt{2a}}\left(\frac{b}{\sqrt{2a}}\sqrt{2a}\sigma_1^{(1)}\right)\sigma_1^{(1)}\end{matrix}\right)}\right]\right]^2 -$$
$$\left(-\frac{1}{2\sqrt{2a}}\right)\left[d+3\sqrt{2a}+\sqrt{\left(d+3\sqrt{2a}\right)^2+4\sqrt{2a}\left(\begin{matrix}-d-2\sqrt{2a}-\frac{eb}{\sqrt{2a}}-\\ \frac{b}{\sqrt{2a}}\left(\frac{b}{\sqrt{2a}}\sqrt{2a}\sigma_1^{(1)}\right)\sigma_1^{(1)}\end{matrix}\right)}\right]-2f+1/4 \quad (29)$$

Putting Eq. (24), (25), (26) and (28) in to Eq. (23*g*), so the eigenvalue for *m=1* given as

$$E_1^{\kappa} = \frac{b}{2\sqrt{2a}}\left(4\pm\sqrt{8f+\kappa^2}\right)-\frac{1}{64a^3}\left(b^2-4ac\right)^2+\sqrt{2a}\sigma_1^{(1)} \quad (30)$$

with parameter κ^2 was given by Eq. (29).
By using Eq. (8), eigenfunction can be written as

$$U_1(r) = N_1\left(r-\sigma_1^1\right)r^D \exp\left[\frac{1}{3}Ar^3+\frac{1}{2}Br^2+Cr\right] \quad (31)$$

with N_1 is a normalization constant.
In this study the eigenfunction is not normalized because it difficult to determined.
The visualisation of the eigenfunction in Eq. (21) and (31) were shown as

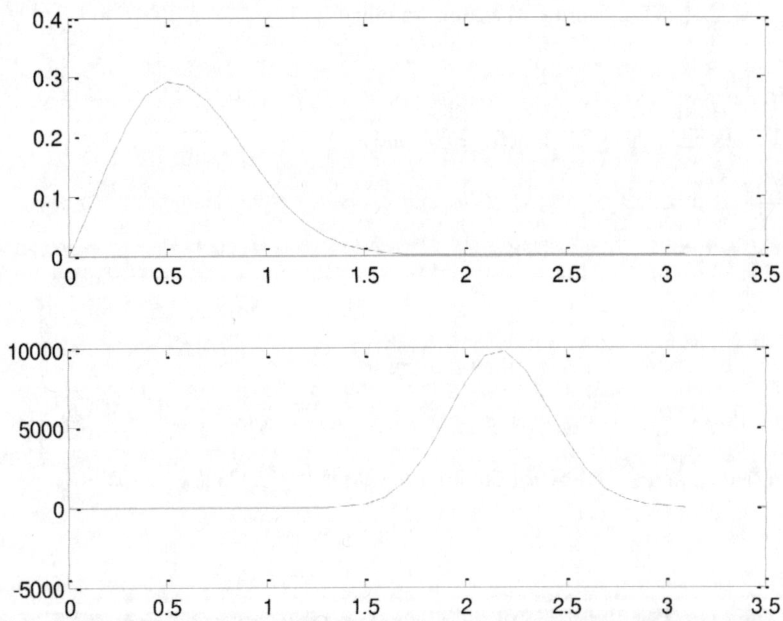

Figure 1. Wave function *U* vs *r*, for *m = 0* and *m = 1*, with *a = b = 1*, *c = d =* 2, and *e = f = 3*

The eigenvalue for $m = 0$ and $m = 1$ were shown in Table1.

Table 1. The eigenvalue with units Joule for $m = 0$ and $m = 1$

Value of constants	E_0	E_1
a=1		
b=1		
c=2	1.5133	10.8435
d=2	-1.6303	-6.7179
e=3		
f=3		

4. Concluding Remarks

We have solved the radial Schrodinger equation using ansatz wave function method in this paper. The eigenfunction for this potential we obtained in Equation (21) and (31). The eigenvalue for this potential we obtained in Equation (20) and (30). The visualisation of the eigenfunction as shown in Figure (1) and the eigenvalue as shown in Table (1). This study is in progress.

Acknowledgement

This work is partly supported by the Higher Education Project Grant with the number of contract 632/UN27.21/LT/2016.

References

[1] Feizi H, Rajabi AA and Shojaei MR 2011 *Physica Polonica*. 42 2147
[2] Hassabadi H, Maghsoodi E, Ikot AN and Zarrinkamar S 2013 *Apllied Mathematics and Computation.* 219 9388
[3] Bayrak O, Bostoosum I and Ciftci H 2006 *International Journal of Quantum Chemistry.* 107 540
[4] Antia AD, Ituen EE, Obong HP and Isonguyo CN 2015 *International Journal Recent advances in Physics.* 4 55
[5] Ikhdair SM and Sever R 2008 *International Journal of Modern Physics.* 19 221
[6] Dong SH 2011 *Wave Equation in Higher Dimensions* (New York: Spinger). P 97
[7] Ikot AN, Akpabio LE and Umoren EB 2011 *Journal of Scientific Research.* 3 23
[8] Ikot AN, 2012 *Chinese Physics Letter.* 29 060307
[9] Hassanabadi H , Yazarlo BH, Zirrinkamar S and Rajabi AA 2011 *Physiscs Revision C.* 84 064013
[10] Shi HD 2001 *Pyisica Scipta*. 64 273
[11] Znojil M 1990 *Journal of Mathematical Physics.* 23 108
[12] Child MS, Dong SH and Wang XG 2000 *Journal og Pyisics A: Math. Gen.* 33 5653

IOP Conf. Series: Journal of Physics: Conf. Series **795** (2017) 012024 doi:10.1088/1742-6596/795/1/012024

The iterative solution of wave propagation in transverse magnetic mode for graded positive-negative

Beta Nur Pratiwi, A Suparmi, C Cari and Dewanta Arya Nugraha

Physics Department, Graduate Program, Sebelas Maret University, Indonesia

Email: pratiwi.bn@gmail.com

Abstract. The iterative solution was used to obtain the electromagnetic wave propagation in transverse magnetic (TM) mode for a graded positive-negative refractive index. The graded graphs of negative permittivity and negative permeability were obtained in hyperbolic functions. By using hyperbolic function for permittivity and permeability in Maxwell equation and by separation variable, we obtained the electromagnetic differential equation. From the differential equation, we used the approachment using MacLaurin series to obtain the wave vector and magnetic fields equation. The distribution of the magnetic fields were given in graph visualization using Matlab software.

1. Introduction

A new class of artificial materials, namely a negative refractive index material (NIM) or so-called left-handed materials (LHM) has attracted the interest of many scientists. NIM or LHM is better known as a metamaterial. [1-2] The theoretical concept of metamaterial was brought in the first time, in 1968 by Vaselago. [3] And then, Pendry, et al brought the concept of metamaterial into the experiment in 1996 and in 1999 by research which regarding to negative permeability in material.[4-5] A year later, Pendry investigated about the perfect lens with negative refraction index material .[6] In the same year, Smith et al.[7] studied about the simultaneously of negative permeability and negative permittivity.

Metamaterial has unique properties with negative refractive index value in a certain wavelength range. The refraction index of a material depends on the permittivity and permeability. LHM has the negative value of permittivity and negative value of permeability $(\varepsilon < 0 \& \mu < 0)$, while for the common material in nature with the positive value of permittivity and positive value of permeability $(\varepsilon > 0 \& \mu > 0)$ is called right-handed material (RHM) [1]. Permittivity and permeability are the basic properties to know the material's response due to the presence of electromagnetic waves. [3] The electromagnetics wave propagation was mostly done through modeling by software. [8-9]

The electromagnetic waves equation in various permeability and permittivity profile can be reduced into the second orde differential equation, like the solution of Schrodinger equation or Dirac equation. The second orde differential equation of electromagnetic wave can be solved to obtain the fields distribution and wave vector using the various methods. Some methods to solve the second order differential equation, such us Nikiforov-Uvarov method [14-15], Supersymmetry Quantum Mechanics (SUSY QM) [16], the polynomial Romanovski method [17] and Asymptotic Iteration Methods (AIM) [18-21]. The solution for the case where the refraction index profile is a hyperbolic tangent function has been investigated by Dalarsson and Tassin [10]. The gradation profile are mostly used in hyperbolic tangent [11-12]. In recently, Husein et al. investigated the solution of wave propagation in a graded interface positive-negative for the hyperbolic cosecant and cotangent functions of permittivity and permeability using AIM. [2-13]

In this research, we considered the wave propagation in transverse magnetic mode for a graded negative-positive refraction index. By using iterative solution which is called AIM, we obtained the fields equation and wave vector.

2. The fields equation

Maxwell equation for dielectric, are given as

$$-\nabla\times\vec{E}=i\omega\mu(\omega,y)\vec{H},\quad \nabla\times\vec{H}=i\omega\varepsilon(\omega,y)\vec{E} \tag{1}$$

where $\varepsilon(\omega,y)$ and $\mu(\omega,y)$ are frequency and spatial (as functions of material thickness, y) dependent dielectric permittivity and magnetic permeability, respectively. [2,13] We let the *yz* plane as the plane of ray incident, so $(d/dx)=0$. By considering the geometry of the material, we obtain a second order differential equation of electromagnetic waves in inhomogeneous materials on mode transverse magnetic (TM). Actually, in two-dimensional study, there are two independent modes i.e. TM mode, the group of fields: $\{E_x,E_y,H_z\}$, $\{E_y,E_z,H_x\}$,$\{E_z,E_x,H_y\}$ and transverse electric (TE) mode, the group of fields: $\{H_x,H_y,E_z\}$, $\{H_y,H_z,E_x\}$,$\{H_z,H_x,E_y\}$.In principle, these two modes (TM and TE) are similar to each other, so we only need to do one calculation, e.g. the TM mode. By using the plane wave solution which is harmonic in time, is given

$$H_x=H(y,z)e^{i\omega t} \tag{2}$$

For the group fields in TM mode which we use i.e.$\left[E_y,E_z,H_x\right]$.From equation (1) and equation (2) we obtain

$$\frac{\partial^2H_x}{\partial y^2}-\frac{\varepsilon'}{\varepsilon}\frac{\partial H_x}{\partial y}+\omega^2\mu\varepsilon H_x+\frac{\partial^2H_x}{\partial z^2}=0 \tag{3}$$

and by using $H_x=H(y,z)=H(y)H(z)$ in equation (3), we obtained

$$\frac{\partial^2H(z)}{\partial z^2}+k_z^{\,2}H(z)=0 \tag{4}$$

$$\frac{\partial^2H(y)}{\partial y^2}-\frac{\varepsilon'}{\varepsilon}\frac{\partial H(y)}{\partial y}+\left(\omega^2\mu\varepsilon-k_z^{\,2}\right)H(y)=0 \tag{5}$$

where k_z is the wave vector in *z* direction. Equation (4) is the simple differential equation which it has the solution,

$$H(z)=e^{\pm ik_z z} \tag{6}$$

and, by using AIM we solved equation (5) with the graded interface permittivity and permeability.

3. Iterative method using Asymptotic Iteration Method

AIM is used to solve second order differential equation in terms: [18-21]

$$f_n''(o)-\lambda_0(o)f_n'(o)-s_o(o)f_n(o)=0 \tag{7}$$

where $\lambda_0(o)\neq0$ and $s_o(o)$ are coefficients of the differential equation and are well-defined functions as well as sufficiently differentiable.The solution of equation (7) can be obtained by using iteration of λ_j and s_j:

$$\lambda_j(o)=\lambda_{j-1}'+\lambda_{j-1}\lambda_0+s_{j-1},\ s_j(o)=s_{j-1}'+s_o\lambda_{j-1},\ j=1,2,3,\ldots \tag{8}$$

Eigenvalues can be obtained using equation (9) below,

$$\lambda_j(o)s_{j-1}(o) - \lambda_{j-1}(o)s_j(o) = 0 = \Delta_j, j = 1,2,3 \dots \quad (9)$$

For the eigenfunctions, following in [22] we use the Hermite differential equation. Hermite differential equation is given as,

$$f''(o) - 2of'(o) + 2jf(o) = 0 \quad (10)$$

Equation (10) has the general solution in polynomial Hermite:

$$f_j(o) = (-1)^j e^{o^2} \frac{d^j}{ds^j}\left(e^{-o^2}\right); j = 0,1,2,3\dots \quad (11)$$

From equation (11) we can get some the first polynomial Hermite functions are given as

(12)

Equation (12) are polynomial Hermite for *j=0,1,2,3,4,5,6,7* and for *j=8*.

4. Solution of the field equations

The permittivity and permeability equation are expressed as

$$\mu(\omega, y) = -\mu_0\mu_r\left(\sinh(\rho y) + \tanh(\rho y)\right) \quad (13)$$

$$\varepsilon(\omega, y) = -\varepsilon_0\varepsilon_r \tanh(\rho y) \quad (14)$$

with ρ is abruptness's parameter describing the transition of wave propagation from RHM to LHM.. The graphs of permeability (Figure 1.a) and permittivity (Figure 1.b) vary according to the material thickness *y* where the interface between RHM and LHM is located at *y=0*.

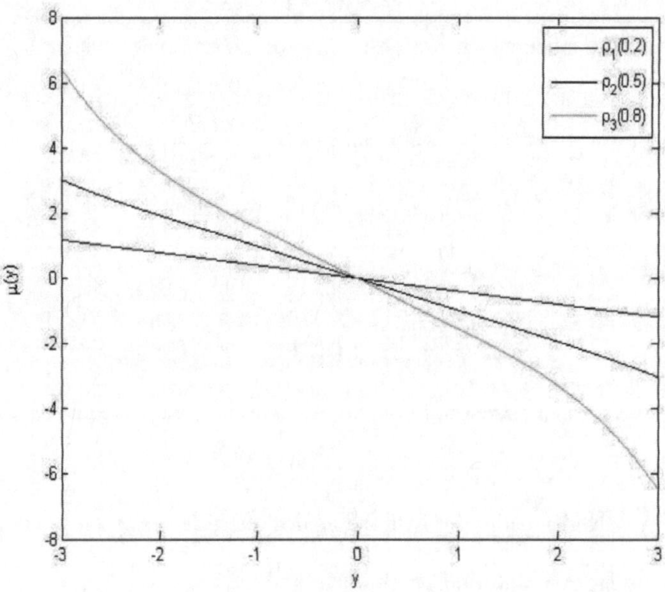

(a)

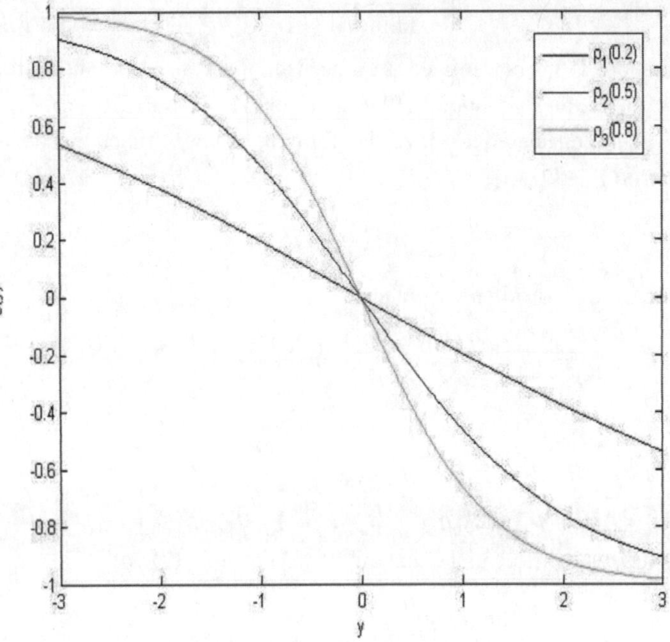

(b)

Figure 1. the graded graphs of (a) permeability equation (b) permittivity equation

By inserting equations (13-14) into equation (5) we obtained the electromagnetic wave equation as follows

$$\frac{\partial^2 H(y)}{\partial y^2} - \frac{\rho\,\mathrm{sech}^2(\rho y)}{\tanh(\rho y)}\frac{\partial H(y)}{\partial y} + \left(\omega^2 \mu_0 \mu_r \varepsilon_0 \varepsilon_r \tanh(\rho y)\left(\sinh(\rho y) + \tanh(\rho y)\right) - k_z^{\ 2}\right) H(y) = 0 \tag{15}$$

by using $c = \left(1/\sqrt{\varepsilon_0 \mu_0}\right)$; $\mu_r \varepsilon_r = n_\omega^2$; $k_0 = (\omega/c)$ where c is light velocity in vacuum, n_ω is the refraction index and k_0 is wave vector in vacuum in equation (15), we get

$$\frac{\partial^2 H(y)}{\partial y^2} - \frac{\rho\,\mathrm{sech}^2(\rho y)}{\tanh(\rho y)}\frac{\partial H(y)}{\partial y} + \left(k_0^{\ 2} n_\omega^2 \tanh(\rho y)\left(\sinh(\rho y) + \tanh(\rho y)\right) - k_z^{\ 2}\right) H(y) = 0 \tag{16}$$

If we set $q = \rho y$ in equation (16),

$$\frac{d^2 H(y)}{dq^2} - \frac{\mathrm{sech}^2(q)}{\tanh(q)}\frac{dH(y)}{dq} + \frac{1}{\rho^2}\left(k_0^{\ 2} n_\omega^2 \tanh(q)\left(\sinh(q) + \tanh(q)\right) - k_z^{\ 2}\right) H(y) = 0 \tag{17}$$

and in the solution domain close to zero, equation (17) can be approached using the MacLaurin series,as follows

$$\tanh(q) \approx q;\quad \mathrm{sech}(q) \approx 1;\quad \sinh(q) + \tanh(q) \approx 2q \tag{18}$$

so we have

$$\frac{\operatorname{sech}^2(q)}{\tanh(q)} \approx \frac{1}{q} \tag{19}$$

But, at $q = 0$ equation (19) is not defined, so it needs to make an assumption. If β is the value of q that close to zero then using equation (19) can be obtained intervals $-\beta \le q \le \beta$. Using this interval, equation (19) is considered to have the form of a linear function connecting the points $(-\beta, -1/\beta)$ and $(\beta, 1/\beta)$, ie [2, 13]

$$b(q) = \frac{1}{\beta^2} q \tag{20}$$

Using that assumption in equation (17), we obtained

$$\frac{d^2 H(y)}{dq^2} - \frac{1}{q}\frac{dH(y)}{dq} + \frac{1}{\rho^2}\left(2k_0{}^2 n_\omega^2 q^2 - k_z{}^2\right) H(y) = 0 \tag{21}$$

By setting $s = \left(2k_0{}^2 n_\omega^2 / \rho^2\right)^{1/4} q$, into equation (21), we got

$$\frac{d^2 H(y)}{ds^2} - \frac{s}{\beta^2}\frac{\rho}{\sqrt{2}k_0 n_\omega}\frac{dH(y)}{ds} + \left(s^2 - \frac{k_z{}^2}{\sqrt{2}k_0 n_\omega \rho}\right) H(y) = 0 \tag{22}$$

Equation (22) is AIM-type differential equation, like in equation (7) with

$$\lambda_0 = \frac{s}{\beta^2}\frac{\rho}{\sqrt{2}k_0 n_\omega}; s_0 = \frac{k_z{}^2}{\sqrt{2}k_0 n_\omega \rho} - s^2 \tag{23}$$

for *s* is close to zero, we approached equation (23) using Maclaurin series, given as

$$\lambda_0(s) \approx \frac{s}{\beta^2}\frac{\rho}{2k_0 n_\omega}; s_0(s) \approx \frac{k_z{}^2}{\sqrt{2}k_0 n_\omega \rho} \tag{24}$$

in equation (22), so we obtained

$$\frac{d^2 H(y)}{ds^2} - \frac{s}{\beta^2}\frac{\rho}{\sqrt{2}k_0 n_\omega}\frac{dH(y)}{ds} - \frac{k_z{}^2}{\sqrt{2}k_0 n_\omega \rho} H(y) = 0 \tag{25}$$

By setting $r = s\sqrt{\frac{\rho}{2\sqrt{2}k_0 n_\omega \beta^2}}$ in equation (25), we got

$$\frac{d^2 H(y)}{dr^2} - 2r\frac{dH(y)}{dr} - \frac{2k_z{}^2\beta^2}{\rho^2} H(y) = 0 \tag{26}$$

Equation (26) is AIM-type differential equation, and by using equation (9) we obtained,

$$\Delta_j = \prod_{j=0}^{m}\left[-\frac{2k_z{}^2\beta^2}{\rho^2} - 2j\right]; m = 1, 2, 3... \tag{27}$$

according to equation (27), the eigenvalues can be obtained as follows

$$-\frac{k_z{}^2\beta^2}{\rho^2} = j \to k_z{}^2 = -\frac{\rho^2 j}{\beta^2} \to k_z = i\frac{\rho}{\beta}\sqrt{j}; j = 0, 1, 2, 3, ... \tag{28}$$

so equation (26) become

$$\frac{d^2 H(y)}{dr^2} - 2r\frac{dH(y)}{dr} + 2jH(y) = 0 \tag{29}$$

which known as Hermite differential equation. The approximate solution of equation (29) is

$$H_j(y) = (-1)^j e^{r^2} \frac{d^j}{dr^j}\left(e^{-r^2}\right) \tag{30}$$

where $$r = \frac{\rho}{\beta\sqrt{2}} y, \quad -\frac{1}{\sqrt{2}} \leq r \leq \frac{1}{\sqrt{2}}, \quad \beta \ll 1 \tag{31}$$

Then for the fields distribution, we plotted from the total eigenfunctions using equations (6,12,30-31) for *j=5* and *j=8*, and we take the positive sign for *H(z)*, we obtained the magnetic fields distribution which is shown in Figure 2.

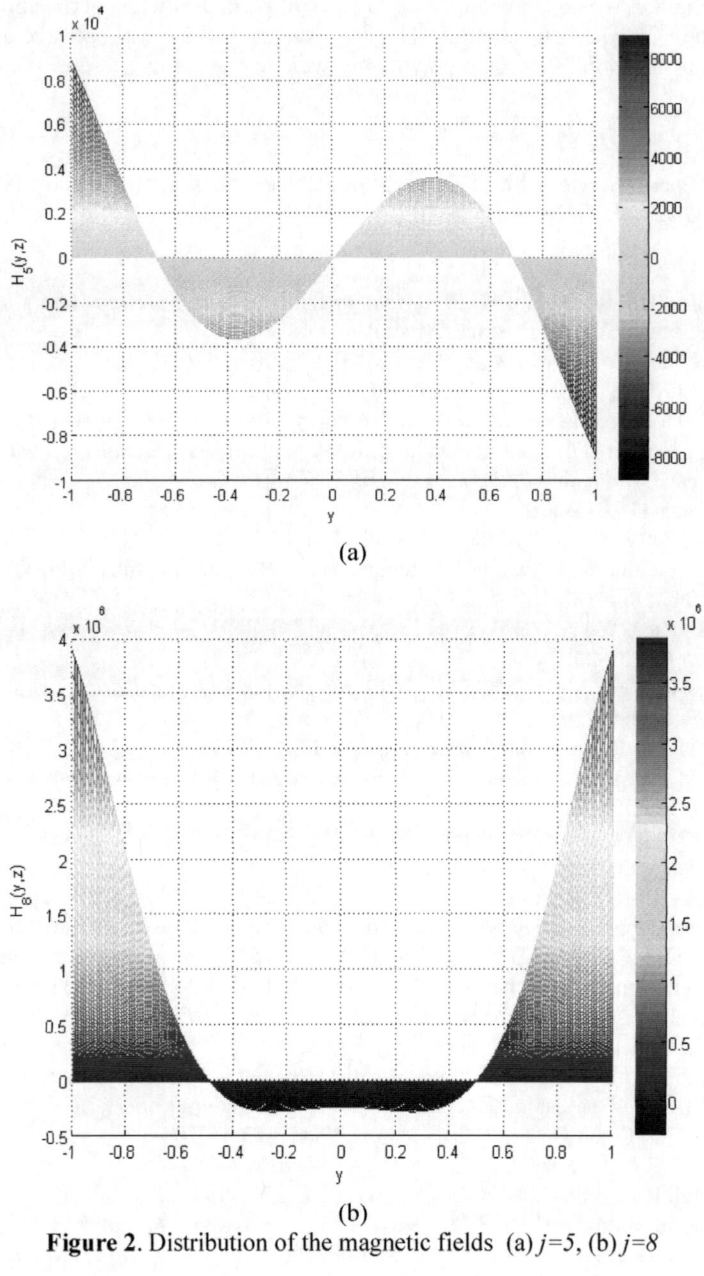

(a)

(b)

Figure 2. Distribution of the magnetic fields (a) *j=5*, (b) *j=8*

Figure 2.a is the five level *j=5* while Figure 2.b for *j=8*. The magnetic field has the different distribution in various the material thickness *y* which in higher level, the amplitude of magnetic fields increase. While for the wave vector k_{z_j} in equation (28) depend on the abruptness's parameter, the material thickness and the level *j*. From equation (28), we can show that the level *j* is proportional to the wave vector, which the value of wave vector increase in the higher level.

5. Conclusion

Iterative method was used to investigated the wave propagation for TM mode in a graded interface negative-positive index. We obtained the graded graphs of permittivity and permeability as a function of material thickness in hyperbolic function. The wave vector equation was obtained using AIM, while the magnetic fields equation in Hermite polynomial were expressed in two-dimensional graph fields versus *y*.

Acknowledgement

This research was partly supported by Sebelas Maret University Higher Education Project Grant with contract no. 632/UN27.21/LT/2016

References

[1] Cui T J, Smith D R and Liu R 2010*Metamaterials: Theory, Design, and Applications* (New York: Springer Science Business Media)
[2] Husein A S,Cari C, Suparmi A and Hadi M 2015 arXiv: 1506.02346v1 [cond-mat.mtrl-sci]
[3] Vaselago V G 1968 *Soviet Physics Uspekhi* **10** 509-514
[4] Pendry J B, Holden A J, Stewart W J and Youngs I 1996 *Physical Review Letters* **76** 4773–4776
[5] Pendry J B, Holden A J, Robbins D J and Stewart W J 1999 *Magnetism from Conductors And Enhanced Nonlinear Phenomena. IEEE Transactions onn Microwave Theory And Techniques* **47** 2075-2084
[6] Pendry J B 2000 *Physical Review Letters* **85** 4184-4187
[7] Smith D R, Padilla W J, Vier D C, Nemat-Nasser S C and Schultz, S 2000 *Physical Review Letters* **84** 4184-4187
[8] Wang M Y, Xu J, Wu J, Wei B, Li H L, Xu T and Ge D B. 2008 *Progress In Electromagnetics Research* **81** 253-265
[9] Mohapatra B N and Mohapatra R K 2013 *International Journal of Scientific & Engineering Research* **4** 2890-2897
[10] Dalarsson M and Tassin P 2009 *Optics Express* **17** 6747-6752
[11] Dalarsson M, Norgren M, Asenov T and Doncov N 2013 *Progress in electromagnetics research* **137** 527-538
[12] Dalarsoon N 2005 Proc.1st *International Workshop on Nanoscience & Nanotechnology IWON*(Belgrade, Serbia and Montenegro) p 194
[13] Husein A S,Cari C, Suparmi A and Hadi M 2016 *AIP Conference Proceedings*vol **1719** (The 4th International Conference on Theoretical and Applied Physics (ICTAP), Bali) p 030048
[14] Suparmi A, Cari C, Pratiwi B N and Deta U A 2016 *AIP Conference Proceeding*vol **1710** (6TH Nanoscience and Nanotechnology Symposium (NNS), Solo Indonesia) p 030010
[15] Cari C, Suparmi A, Yunianto M and Husein A S 2016 *AIP Conference Proceeding*vol **1710** (6TH Nanoscience and Nanotechnology Symposium (NNS), Solo Indonesia) p 030009
[16] Cari C and Suparmi A 2014 *AIP Conference Proceeding*vol **1615** (4th International Conference of Advances in Nuclear Science and Enginerring , Bandung) p 101
[17] Suparmi A, Cari C and Deta U A 2014 *Chin. Phys. B* **23** 090304-1
[18] Ciftci H, Hall R L and Saad N 2003 *J.Phys. A: Math* **36** 11807
[19] Ciftci H, Hall R L and Saad N 2005 *J.Phys. A: Math* **38** 1147
[20] Rostami A and Motovali H 2008 *Progress In Electromagnetics Research B* **4**171 – 182.
[21] Soylu A, Bayrak O and Boztosun I 2008 *Journal of Physics A: Mathematical and Theoretical*

IOP Conf. Series: Journal of Physics: Conf. Series **795** (2017) 012024 doi:10.1088/1742-6596/795/1/012024

41

[22] Spiegel M R 1968 *Schaums Mathematical Handbook of Formulas and Table* (Rensselaer Polytechnic Institute)

IOP Conf. Series: Journal of Physics: Conf. Series **795** (2017) 012025 doi:10.1088/1742-6596/795/1/012025

Asymptotic iteration method for analytical solution of Klein-Gordon equation for trigonometric Pöschl-Teller potential in D-dimensions

Dewanta Arya Nugraha, A Suparmi, C Cari and Beta Nur Pratiwi

Physics Department, Graduate Program, Sebelas Maret University

E-mail: dewanta.an@gmail.com

Abstract. Klein-Gordon equation for Trigonometric Pöschl-Teller Potential in D-dimensions was obtained within framework of a centrifugal term approximation. Asymptotic iteration method was used to obtain the relativistic energy spectrum and wave functions. The value of relativistic energy was calculated numerically and the results have shown that in higher dimension the energy level is increased with positive energy states. The wave functions were expressed in hypergeometric term.

1. Introduction

Relativistic wave equation solution is very important in many aspects of modern physics. Wave equation of Klein-Gordon and Dirac are frequently used for describing the particle dynamics in relativistic quantum mechanics [1]. Klein-Gordon equation allows us to study spin-zero particles [2]. Klein-Gordon Solutions have been done by various techniques such as Standard Method [3], Supersymmetric Quantum Mechanics (SUSY)[4], Nikiforov Uvarov (NU) Method [5], Factorization Method [6], Asymptotic Iteration Method (AIM) [7][8] and others. In higher dimensional, Klein-Gordon solution have done for various potential, they are Poschl-Teller potential including centrifugal term [3], Hylleraas Potential[9], Hulthen potential [10], Manning- Rosen potential [8], Morse potential [11] and other.

In this paper, the radial part solution of D dimension Klein-Gordon equation is studied for Trigonometric Pöschl-Teller Potential in D-dimensional. In some physics area, the higher dimensional spaces extension for some physical problems is very important. The D-dimensional system has been constructed to explain the unification of gravitation and electromagnetic fields [12]. It is suspected that the dimensional system is applicable for gravitation field since it is involved in such huge universe. The relativistic and non-relativistic physical systems in D-dimensional have been investigated by many authors for various methods and potentials [13]–[16]. The Poschl-Teller potential can be applied to study the effect of the complex vibration-rotation energy structure of multi-electron atoms in the relativistic system.[17], [18]. We used the AIM to obtain the energy eigenvalues and radial wave functions for Klein-Gordon equation in D-dimensions.

This paper is organized as follows. In section 2 presented a brief review of D-dimensional Kein-Gordon equation. The methods we used are presented in Section 3 and the results and discussion are presented in Section 4. The conclusion is presented in Section 5.

2. Klein-Gordon Equation in D Dimensions

The D-Dimensional time independent KG equation with potentials of vector and scalar $V(r)$ and $S(r)$, respectively, with $r=|\mathbf{r}|$ describing a spinless particle which in the general write as [19], [20]

$$\left\{\nabla_D^2+\frac{1}{\hbar^2c^2}\left[\left[E-V(r)\right]^2-\left[Mc^2+S(r)\right]^2\right]\right\}\psi_{\ell_1\ldots\ell_{D-2}}^{(\ell_{D-1}=\ell)}(\mathbf{x})=0, \tag{1}$$

With

$$\nabla_D^2=\frac{1}{r^{D-1}}\frac{\partial}{\partial r}\left(r^{D-1}\frac{\partial}{\partial r}\right)+\frac{1}{r^2}\sum_{j=1}^{D-2}\frac{1}{\sin^2\theta_{j+1}\sin^2\theta_{j+2}\ldots\sin^2\theta_{D-1}}\times\left\{\frac{1}{\sin^{j-1}\theta_j}\left(\frac{\partial}{\partial\theta_j}\sin^{j-1}\theta_j\frac{\partial}{\partial\theta_j}\right)\right\}+ \tag{2}$$
$$\frac{1}{r^2}\left\{\frac{1}{\sin^{D-2}\theta_{D-1}}\left(\frac{\partial}{\partial\theta_{D-1}}\sin^{D-2}\theta_{D-1}\frac{\partial}{\partial\theta_{D-1}}\right)\right\},$$

and

$$\psi_{\ell_1\ldots\ell_{D-2}}^{(\ell)}(\mathrm{x})=R_\ell(r)Y_{\ell_1\ldots\ell_{D-2}}^{(\ell)}(\theta_1,\,\theta_2,\,\cdots,\,\theta_{D-1}) \tag{3}$$

Where E, M, and ∇_D^2 denoted KG energy, the mass and the Laplacian for D-dimension, respectively. And $\mathbf{x}$ in equation (1) is a position vector for D-dimension. The wave function for radial part $R_\ell(r)$ is decomposed as

$$R_\ell(r)=r^{-(D-1)/2}F(r), \tag{4}$$

According to Alhaidari et al [21] ,only the choice $S=V$ can produce a non-trivial nonrelativistic limit with a potential function $2V$ and not V. Accordingly, it would be natural to scale the potential terms in Eq. (1) then the potential becomes V in nonrelativistic limit the interaction and not $2V$. Thus equation (1) is reduced to D-Dimensional radial KG equation. (in the relativistic atomic units $\hbar=c=1$)

$$\nabla_D^2\psi_{\ell_1\ldots\ell_{D-2}}^{(\ell_{D-1}=\ell)}(\mathrm{x})+\left[E^2-M^2-(E+M)V(r)\right]\psi_{\ell_1\ldots\ell_{D-2}}^{(\ell_{D-1}=\ell)}(\mathrm{x})=0, \tag{5}$$

3. Asymptotic Iteration Method (AIM)

Asymptotic Iteration Methods (AIM) is an alternative method which has accuracy and high efficiency to determine eigen energies and eigen functions for analytically solvable hyperbolic like potential. Asymptotic Iteration Methods is also giving solution for exactly solvable problem. [22].

AIM is used to solve the second order homogeneous equation of form [7], [8], [23], [24].

$$y_n''(x)=\lambda_0(x)y_n'(x)+s_0(x)y_n(x) \tag{6}$$

with $\lambda_0(x)\neq 0$ and prime symbol denotes the derivative with respect to x. The others parameter n is interpreted as the radial quantum number. The other variables, $s_0(x)$ and $\lambda_0(x)$ are differentiable. To get the solution, we have to differentiate Eq. (6) along x, and find

$$y_n'''(x)=\lambda_1(x)y_n''(x)+s_1(x)y_n(x) \tag{7}$$

Where

$$\lambda_1(x)=\lambda_0'(x)+s_0(x)+\lambda_0^2(x) \tag{8}$$

$$s_1(x)=s_0'(x)+s_0(x)\lambda_0(x) \tag{9}$$

where $\lambda_0(x) \neq 0$ and $s_0(x)$ is a function of $C\infty$ (coefficient of the differential equation) Asymptotic Iteration Method and can be applicated exactly in the different problem if the wave function has been known and fulfill boundary condition zero (0) and infinity (∞).
Equation (6) can be simple itterated until (k+1) and (k+2), k = 1, 2, 3, ... and then we get

$$y_n^{k+1}(x) = \lambda_{k-1}(x) y_n'(x) + s_{k-1}(x) y_n(x) \tag{10}$$

$$y_n^{k+2}(x) = \lambda_k(x) y_n'(x) + s_k(x) y_n(x) \tag{11}$$

where

$$\lambda_k(x) = \lambda_{k-1}'(x) + s_{k-1}(x) + \lambda_0(x)\lambda_{k-1}(x) \tag{12}$$

$$s_k(x) = s_{k-1}'(x) + s_0(x)\lambda_{k-1}(x) \tag{13}$$

Which is called recurrence relation. Eigen value can be found using this equation

$$\Delta_k(z) = \lambda_k(z)s_{k-1}(z) - \lambda_{k-1}(z)s_k(z) = 0 \tag{14}$$

where $k = 1, 2, 3, \ldots$ is the iteration number and n is the representation of radial quantum number.
Equation (6) is the second orde of homogen linear equation which is can be solved with comparison it with the second orde linear equation as follow [8]:

$$y''(z) = 2\left(\frac{az^{N+1}}{1-bz^{N+2}} - \frac{t+1}{z}\right)y'(z) - \frac{wz^N}{1-bz^{N+2}}y(z) \tag{15}$$

where

$$\sigma = \frac{2t+N+3}{N+2}, \quad \rho = \frac{(2t+1)b+2a}{(N+2)b} \tag{16}$$

The wave function of equation (15) can be determined by using the equation below :

$$y_n(z) = (-1)^n C_2 (\mathrm{N}+2)^n (\sigma)_n \, {}_2F_1(-n, \rho+n; \sigma; bz^{N+2}) \tag{17}$$

4. Results and Discussion

Pöschl Teller Potential is defined as[18]

$$V(\mathrm{r}) = \left(\frac{\kappa(\kappa-1)}{\sin^2 \alpha r} + \frac{\eta(\eta-1)}{\cos^2 \alpha r}\right) \tag{18}$$

Substituting equations (2 – 4, 18) into equation (5) we obtain

$$\frac{1}{r^{D-1}}\frac{\partial}{\partial r}\left(r^{D-1}\frac{\partial}{\partial r}\frac{F(r)}{r^{(D-1)/2}}\right) - (E+M)\left\{\frac{\kappa(\kappa-1)}{\sin^2 \alpha r} + \frac{\eta(\eta-1)}{\cos^2 \alpha r}\right\}\frac{F(r)}{r^{(D-1)/2}} + \left(E^2 - M^2\right)\frac{F(r)}{r^{(D-1)/2}} = \lambda_{D-1}\frac{F(r)}{r^2 r^{(D-1)/2}} \tag{19}$$

Where λ_{D-1} separation constant. By using centrifugal approximation for $\alpha r << 1$, then $\frac{1}{r^2} \cong \frac{\alpha^2}{\sin^2 \alpha r}$ for equation (19) become

$$\left[\frac{d^2}{dr^2} - \frac{\alpha^2}{\sin^2 \alpha r}\left\{\left(\frac{D-1}{2}\right)\left(\frac{D-3}{2}\right) - \lambda_{D-1} - \frac{\kappa(\kappa-1)(E+M)}{\alpha^2}\right\} - \frac{\eta(\eta-1)}{\cos^2 \alpha r}(E+M) + \left(E^2 - M^2\right)\right]F(r) = 0 \tag{20}$$

By setting,

$$A = \left(\frac{D-1}{2}\right)\left(\frac{D-3}{2}\right) - \lambda_{D-1} - \frac{\kappa(\kappa-1)(E+M)}{\alpha^2} \tag{21}$$

$$B = \frac{\eta(\eta-1)(E+M)}{\alpha^2} \tag{22}$$

$$E_s = \frac{\left(E^2 - M^2\right)}{\alpha^2} \tag{23}$$

Equation (20) become

$$\frac{d^2F(r)}{dr^2} - \frac{A\alpha^2}{\sin^2 \alpha r}F(r) - \frac{B\alpha^2}{\cos^2 \alpha r}F(r) + E_s\alpha^2 F(r) = 0 \tag{24}$$

We can solve equation (24) by letting $\cos^2 \alpha r = z$, we get

$$z(1-z)\frac{\partial^2 F(r)}{\partial z^2} + \left(\frac{1}{2} - z\right)\frac{\partial F(r)}{\partial z} + \frac{A}{4(1-z)}F(r) + \frac{B}{4z}F(r) - \frac{E_s}{4}F(r) = 0 \tag{25}$$

Equation (25) is the hypergeometric differential equation which must be transformed to AIM type second differential equation. Let,

$$F(z) = z^{\delta}(1-z)^{\gamma} f_{n_r}(z) \tag{26}$$

By substituting (26) to (25), we get

$$z(1-z)f''_{n_r}(z) + \left\{2\delta - (2\delta + 2\gamma + 1)z + \frac{1}{2}\right\}f'_{n_r}(z) + \left\{\frac{4\delta(\delta-1) + 2\delta + B}{4z} + \frac{4\gamma(\gamma-1) + 2\gamma + A}{4(1-z)} - \delta^2 - 2\delta\gamma - \gamma^2 - \frac{E_s}{4}\right\}f_{n_r}(z) = 0 \tag{27}$$

By setting,

$$B = -4\delta^2 + 2\delta$$
$$A = -4\gamma^2 + 2\gamma \tag{28}$$

Equation (28) are inserted into equation (27), we obtain

$$f''_{n_r}(z) = \frac{(2\delta + 2\gamma + 1)z - \left(2\delta + \frac{1}{2}\right)}{z(1-z)}f'_{n_r}(z) + \frac{(\delta+\gamma)^2 + \frac{E_s}{4}}{z(1-z)}f_{n_r}(z) = 0 \tag{29}$$

Equation (29) is AIM type differential equation. Equation (29) compared with equation (6), we get

$$\lambda_0 = \frac{(2\delta + 2\gamma + 1)z - \left(2\delta + \frac{1}{2}\right)}{z(1-z)} \tag{30}$$

$$s_0(z) = \frac{(\delta+\gamma)^2 + \frac{E_s}{4}}{z(1-z)} \tag{31}$$

From equations (12 – 14), we found the eigenvalue of equation (29) in general term as follows

$$E_{sn} = -4(\delta + \gamma + n)^2 \tag{32}$$

With

$$\delta = \frac{1}{4} + \frac{1}{2}\sqrt{\frac{1}{4} - B} \tag{33}$$

$$\gamma = \frac{1}{4} + \frac{1}{2}\sqrt{\frac{1}{4} - A} \tag{34}$$

Where E_n is eigenvalue which n is interpreted radial quantum number, with $n=0,1,2,\ldots$. By inserting equation (21 – 23, 33 – 34) to equation (32) we get relativistic energy equation as follows

$$\left(M^2-E_n^2\right)=\alpha^2\left\{2n+1-\sqrt{\frac{1}{4}-\left(\frac{D-1}{2}\right)\left(\frac{D-3}{2}\right)-\lambda_{D-1}-\frac{\kappa(\kappa-1)(E_n+M)}{\alpha^2}}-\sqrt{\frac{1}{4}-\frac{\eta(\eta-1)(E_n+M)}{\alpha^2}}\right\}^2 \quad (35)$$

From equation (35) we can calculate the relativistic energy value numerically by using matlab. The result is shown in Table 1.

Table 1. The relativistic energy E_n

		E_n (eV)			
		$M=\alpha=\lambda=1$			
D	n	$\kappa=\eta=1.5$	$\kappa=\eta=2$	$\kappa=\eta=2.5$	$\kappa=\eta=3$
3	0	4.006385	9.003597	16.002213	25.001479
	1	4.009589	9.008375	16.005805	25.004080
	2	4.016115	9.017870	16.012966	25.009275
	3	4.026149	9.031961	16.023653	25.017047
	4	4.039869	9.050475	16.037801	25.027376
	5	4.057367	9.073195	16.055327	25.040229
4	0	4.016315	9.008588	16.005153	25.003401
	1	4.019481	9.013352	16.008741	25.006001
	2	4.025931	9.022819	16.015894	25.011194
	3	4.035847	9.036869	16.026570	25.018963
	4	4.049406	9.055330	16.040704	25.029287
	5	4.066704	9.077986	16.058213	25.042135
5	0	4.030125	9.015564	16.009265	25.006092
	1	4.033241	9.020308	16.012848	25.008691
	2	4.039585	9.029736	16.019992	25.013880
	3	4.049341	9.043729	16.030652	25.021644
	4	4.062683	9.062117	16.044766	25.031962
	5	4.079709	9.084686	16.062250	25.044802

From Table 1., it shows that the potential constant of trigonometric Pöschl-Teller potential gives the influent to the energy value which is increasing due to the larger potential constant. The value of relativistic energy in higher dimension is increase with positive energy states.
By comparing equation (15) and (29) and using equation (16) and insert it to (17)

$$f_n(z)=(-1)^n C_2 (1)^n \left(2\delta+\tfrac{1}{2}\right)_n \, {}_2F_1\left(-n,2\delta+2\gamma+1+n,2\delta+\tfrac{1}{2},z\right) \quad (36)$$

By inserting equation (36) to $F_n(z)=(z)^\delta(1-z)^\gamma f_n(z)$, we get

$$F_n(z)=(z)^\delta(1-z)^\gamma f_{n_r}(z)(-1)^n C'(1)^n\left(2\delta+\tfrac{1}{2}\right)_n \, {}_2F_1\left(-n,2\delta+2\gamma+1+n,2\delta+\tfrac{1}{2},z\right) \quad (37)$$

where C' is normalization constant and ${}_2F_1(a,b,c,z)$ is hypergeometry function. In Table 2., are expressed unnormalized radial wave functions in any states solved from equation (37) and in Figure 1 are the graphs of radial wave functions in a various value of quantum number n following the Table 2.

IOP Conf. Series: Journal of Physics: Conf. Series **795** (2017) 012025 doi:10.1088/1742-6596/795/1/012025

Table 2. The wave functions ($n = 0,1,2$)

n	$F_n(r)$
0	$F_0(r) = (\cos^2 \alpha r)^{\delta} (\sin^2 \alpha r)^{\gamma} C'$
1	$F_1(r) = -(\cos^2 \alpha r)^{\delta} (\sin^2 \alpha r)^{\gamma} C'\left(2\delta + \frac{1}{2}\right)\left\{1 - \frac{(2\delta + 2\gamma + 2)}{\left(2\delta + \frac{1}{2}\right)}(\cos^2 \alpha r)\right\}$
2	$F_2(r) = (\cos^2 \alpha r)^{\delta} (\sin^2 \alpha r)^{\gamma} C'\left(2\delta + \frac{1}{2}\right)\left(2\delta + \frac{3}{2}\right) \times$ $\left\{1 - \frac{2(2\delta + 2\gamma + 3)}{\left(2\delta + \frac{1}{2}\right)}(\cos^2 \alpha r) + \frac{(2\delta + 2\gamma + 3)(2\delta + 2\gamma + 4)}{\left(2\delta + \frac{1}{2}\right)\left(2\delta + \frac{3}{2}\right)}(\cos^2 \alpha r)^2\right\}$

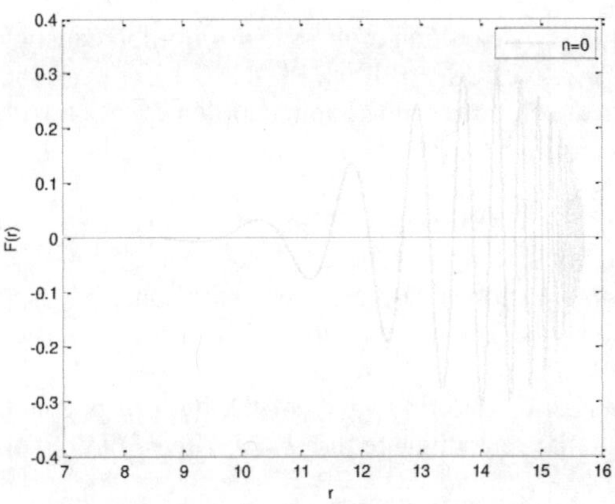

(a)

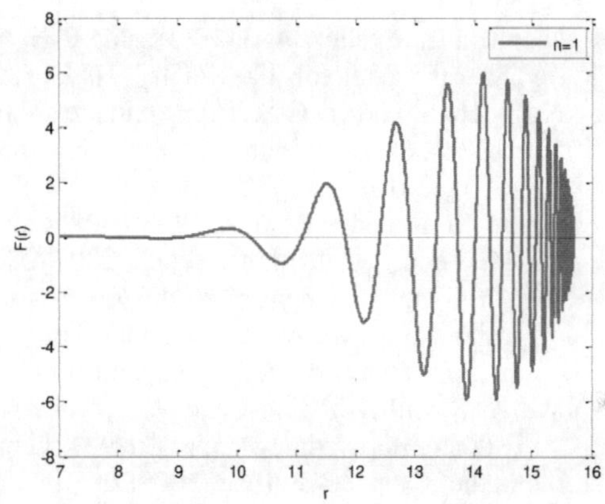

(b)

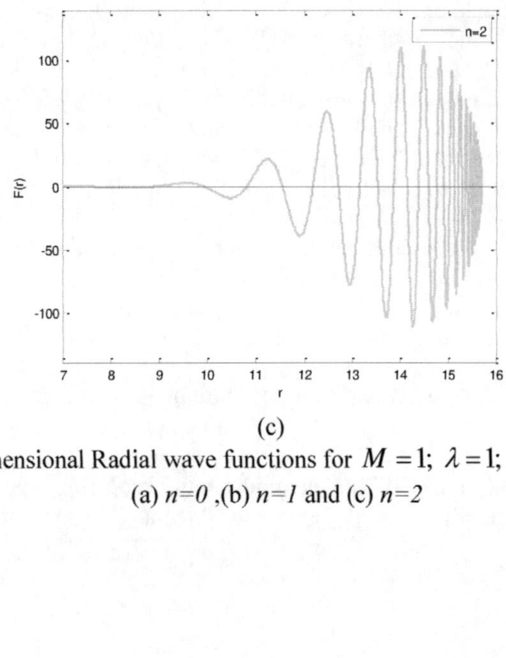

(c)

Figure 1. Four Dimensional Radial wave functions for $M = 1;\ \lambda = 1;\ \alpha = 0.1;\ \kappa = 1.5;\ \eta = 1.5$
(a) *n=0* ,(b) *n=1* and (c) *n=2*

Figure 1 are the graphs for various value of quantum number *n*. It is shown that for the higher value of *n* give the increasing value for amplitude wave functions. The amplitude explains about the deviation of radial wave functions depend on radial dimension.

5. Conclusion

The radial relativistic energy of Klein-Gordon equation for Trigonometric Pöschl-Teller Potential in D-dimensional is obtained by using asymptotic iteration method. The results show that the relativistic energy is increasing with the increase of potential constant value. In higher dimension, the energy is increasing with positive energy.

Acknowledgement

This research was partly supported by SebelasMaret University Higher Education Project Grant with contract No. 632/UN27.21/LT/2016

References

[1] Ikhdair S M and Sever R, "Solution of the D-dimensional Klein-Gordon equation with equal scalar and vector ring-shaped pseudoharmonic potential," *arXiv:0808.1002v1 [quant-ph]*, pp. 1–14, 2008.

[2] Momtazi E, Rajabi A A, and Yazarloo B H, "Analytical solution of the Klein – Gordon equation under the Coulomb-like scalar," *Turkish J. Phys.*, vol. 38, no. 1, pp. 81–85, 2014.

[3] Xu Y, He S, and Jia C-S, "Approximate analytical solutions of the Klein–Gordon equation with the Pöschl–Teller potential including the centrifugal term," *Phys. Scr.*, vol. 81, no. 4, p. 45001, 2010.

[4] Setare M R and Nazari Z, "FIVE-PARAMETER EXPONENT-TYPE POTENTIAL," *Acta Phys. Pol. B*, vol. 40, no. 10, pp. 2809–2824, 2009.

[5] Ikhdair S M, "Bound states of the Klein-Gordon equation in D-dimensions with some physical scalar and vector exponential-type potentials including orbital centrifugal term," *arXiv:1110.0943v1 [quant-ph]*, pp. 1–25, 2011.

[6] Wei G F, Liu X Y, and Chen W L, "The relativistic scattering states of the hulthén potential with an improved new approximate scheme to the centrifugal term," *Int. J. Theor. Phys.*, vol. 48, no. 6, pp. 1649–1658, 2009.

[7] Barakat T, "The asymptotic iteration method for dirac and klein gordon equations with a linear scalar potential," *Int. J. Mod. Phys. A*, vol. 21, no. 19, pp. 4127–4135, 2006.

[8] Das T, "Exact Solutions of the Klein-Gordon Equation for q-Deformed Manning-Rosen Potential via Asymptotic Iteration Method," *arXiv:1409.1457v1 [quant-ph]*, pp. 1–11, 2014.

[9] Ikot A N, Awoga O A, and Ita B I, "Exact Solutions of the Klein-Gordon Equation with Hylleraas Potential," *Few-Body Syst.*, vol. 53, no. 3–4, pp. 539–548, 2012.

[10] Ikot A N, Akpabio L E, and Uwah E J, "Bound State Solutions of the Klein Gordon Equation with the Hulth ′ en Potential," *Electron. J. Theor. Phys.*, vol. 25, no. 25, pp. 225–232, 2011.

[11] Soylu A, Bayrak O, and Boztosun I, "Exact Solutions of Klein – Gordon Equation with Scalar and Vector," *Chinese Phys. Lett.*, vol. 25, no. 8, pp. 2754–2757, 2008.

[12] Dong S H, "The Ansatz Method for Analyzing Schrödinger's Equation with Three Anharmonic Potentials in D Dimensions," *Found. Phys. Lett.*, vol. 15, pp. 385–395, 2002.

[13] Ikot A N, Antia A D, Akpan I O, Awoga O A, and Group T P, "Bound state solutions of schrodinger equation with modified hylleraas plus exponential rosen morse potential," *Rev. Mex. Fis.*, vol. 59, pp. 46–53, 2013.

[14] Panahi H, Zarrinkamar S, and Baradaran M, "Solutions of the *D* -dimensional Schrödinger equation with Killingbeck potential: Lie algebraic approach," *Chinese Phys. B*, vol. 24, no. 6, p. 60301, 2015.

[15] Suparmi A, Cari C, and Pratiwi B N, "Thermodynamics properties study of diatomic molecules with q-deformed modified Poschl-Teller plus Manning Rosen non-central potential in D

dimensions using SUSYQM approach," *J. Phys. Conf. Ser.*, vol. 710, no. 1, 2016.

[16] Suparmi A, Cari C, Pratiwi B N, and Deta U A, "Solution of D dimensional Dirac equation for hyperbolic tangent potential using NU method and its application in material properties," *AIP Conf. Proc.*, vol. 1710, p. 30010, 2016.

[17] Suparmi S and Cari C, "Bound State Solution of Dirac Equation for Generalized Pöschl-Teller plus Trigomometric Pöschl-Teller Non- Central Potential Using SUSY Quantum Mechanics," *J. Math. Fundam. Sci.*, vol. 46, no. 3, pp. 205–223, 2014.

[18] Liu X Y, Wei G F, Cao X W, and Bai H G, "Spin Symmetry for Dirac Equation with the Trigonometric Pöschl-Teller Potential," *Int. J. Theor. Phys.*, vol. 49, no. 2, pp. 343–348, 2010.

[19] Bayrak O and Boztosun I, "Arbitrary ℓ-state solutions of the rotating Morse potential by the asymptotic iteration method," *J. Phys. A. Math. Gen.*, vol. 39, no. 22, p. 6955, 2006.

[20] Qiang W C and Dong S H, "Arbitrary l-state solutions of the rotating Morse potential through the exact quantization rule method," *Phys. Lett. Sect. A Gen. At. Solid State Phys.*, vol. 363, no. 3, pp. 169–176, 2007.

[21] Alhaidari A D, Bahlouli H, and Al-Hasan A, "Dirac and Klein-Gordon equations with equal scalar and vector potentials," *Phys. Lett. Sect. A Gen. At. Solid State Phys.*, vol. 349, no. 1–4, pp. 87–97, 2006.

[22] Suparmi A, Cari C, Deta U A, Husein A S, and Yuliani H, "Exact Solution of Dirac Equation for q-Deformed Trigonometric Scarf potential with q-Deformed Trigonometric Tensor Coupling Potential for Spin and Pseudospin Symmetries Using Romanovski Polynomial," *J. Phys. Conf. Ser.*, vol. 539, p. 12004, 2014.

[23] Sari R A, Suparmi A, and Cari C, "Solution of Dirac equation for Eckart potential and trigonometric Manning Rosen potential using asymptotic iteration method," *Chinese Phys. B*, vol. 25, no. 1, 2015.

[24] Pramono S, Suparmi A, and Cari C, "Relativistic Energy Analysis of Five-Dimensional q - Deformed Radial Rosen-Morse Potential Combined with q -Deformed Trigonometric Scarf Noncentral Potential Using Asymptotic Iteration Method," *Adv. High Energy Phys.*, vol. 2016, 2016.

Characteristics of dye Rhoeo spathacea in dye sensitizer solar cell (DSSC)

Sri Sumardiasih[1], Wilfrida M. Obina[2], Cari[2], Agus Supriyanto[2], Trio Y. Septiawan[2] and Khairuddin[2]

[1,2]Physics Department of Post Graduate Program Sebelas Maret University, Jl. Ir. Sutami 36A Kentingan Jebres Surakarta 57126, INDONESIA

E-mail: usagi20389@gmail.com

Abstract. Dye-sensitized solar cell (DSSC) is a device that converts solar energy into electrical energy. The magnitude of the efficiency of DSSC is mainly based on the amount of dye absorbed by the surface of TiO_2. In this work, used natural dye extracted from leaves Rhoeo spathacea. The dye partially used to immerse of TiO_2 as working electrodes, and the rest are directly mixed TiO_2 paste to obtain dye titanium dioxide.The paste TiO_2 and dye titanium dioxide coated onto the fluorine-doped tin oxide (FTO) glass plate by spin coating method. The absorbance spectra of the dye, dye titanium dioxide and TiO_2 were obtained by UV-Vis spectroscopy. The conductivity of the dye, dye titanium dioxide, and TiO_2 was measured by two point probe El-Kahfi 100. The DSSC based on dye titanium dioxide that stirring for 5 hours the highest efficiency of 0,0520 % whereas those based on TiO_2 immersed for 36 hours showed achieved 0,0501 % obtained from I-V characterization.

1. Introduction

Dye-Sensitized Solar Cell (DSSC) is a device that converts solar radiation into electric energy. DSSC was devided of two section that consisting of nanopori TiO_2 dye molecules are adsorbed on the surface of TiO and catalysts which are all deposited between two conductive glass [1,2]. When the light impinges the surface of the dye, the photon energy is absorbed by the electron dye thus get energy for the excited electron. Furthermore the excited dye injects electrons into the conduction band of the TiO_2 nanoporous films and forwarded to the conductive electrode (anode). Electrons flow through an external load to reach the cathode (counter electrode). Electrons are transferred to the electrolyte in the cathode. Electrolyte containing I^- / I_3^- which acts as an electron mediator so that it can produce in the cell cycle process. Thus, the dye is oxidized redox ions accept electrons from I- to replace the missing electron, and iodide molecules then oxidized to Tri-iodide ion (I_3^-) which compensate electron is missing from the counter electrode [3,6].

Dye-sensitizer is important in DSSC due to can absorbing sunlight and transforming it into electrical energy. One of the types the natural dyes has been used as a sensitizer is anthocyanin because can attach with TiO_2 surface and exhibit broad region of the visible light spectrum [4,8]. Titanium dioxide (TiO_2) is the most common used in DSSC especially the anatase phase because having better photo catalytic activity. TiO_2 have superior properties, like non-toxic, inexpensive, good chemical stability, biocompatibility, etc[5,9].

In this work, we examined the influence of the process for fabrication DSSC using Rhoeo spathacea leaves. Rhoeo spathacea selected because they contain anthocyanins with a range of the visible spectrum is quite wide[10]. The process of synthesis performed in this work is submersion of TiO_2 and the direct mixing to obtain dye titanium dioxide. The characterization of dye titanium dioxide and TiO_2 immersed using UV-Vis spectroscopy system (*Lambda 25)*, two point probe *El-Kahfi 10*, and I-V characterization *(Keithley 2602A meter)*. The purpose is to study the optical, conductivity and conversion efficiency from dye titanium dioxide and TiO_2 immersed using dye Rhoeo spathacea .

2. Experiment

2.1. Dye Extraction

In this work using dyes from anthocyanin extracted from the leaves Rhoeo spathacea. 10 g leaf Rhoeo spathacea is mixed in 50 ml ethanol. The mixture was kept for half an hour at temperature 60°C and kept for 24 hours. The extracted was filtered and used for the synthesis of natural dye mixed TiO_2 and submersion of TiO_2.

2.2. Synthesis of Dye Titanium Dioxide

Dye titanium dioxide obtained from direct mixing 8 ml dye and 1 ml TiO_2. Dye was added dropwise into TiO_2 under vigorous stirring at room temperature. After completing the dropwise addition of dye, the solution was stirred to obtain a homogeneous solution. For the paste of TiO_2 using 0.5 g the anatase phase TiO_2 is mixed 2.5 ml ethanol. The mixture was kept for an hour. The paste of TiO_2 and dye titanium dioxide was coated onto the FTO glass plate using the spin-coating method.

2.3. DSSC Fabrication

The DSSC is composed of two conductive glass, TiO_2 layer, molecules dye, electrolyte and carbon layer[11]. In this work, there are two ways to fabricate o DSSC :

(i) TiO_2 paste with nanoparticle sizes was coated onto the FTO conductive glass with the spin-coating method and sintering at 500°C for 30 minutes. After cooling, the TiO2 electrode was immersed in a solution of dye for 3, 6, 22, 24, and 36 hours.

(ii) The paste of dye titanium dioxide was coated onto the FTO glass with the *spin-coating* method and sintering at 60°C for for 30 minutes.

To prepare the catalytic counter electrode,the surface of FTO glass is rubbed with carbon and then heated over a candle light until blackened. The DSSC are assembled injected with the electrolyte solution and then sealed using the clipboard. To obtain voltage and current of the DSSC used a Keithley 2602A meter (USA) under illumination by a lamp of 1000 W/m2.

3. Results and Discussion

3.1. UV-Vis and Conductivity Analysis

The absorbance spectra were measured by UV-Vis spectroscopy system *(Lambda 25)*. The absorption spectra of dye, dye titanium dioxide, and TiO_2 immersed in dye are given in Figure 1.

IOP Conf. Series: Journal of Physics: Conf. Series **795** (2017) 012026 doi:10.1088/1742-6596/795/1/012026

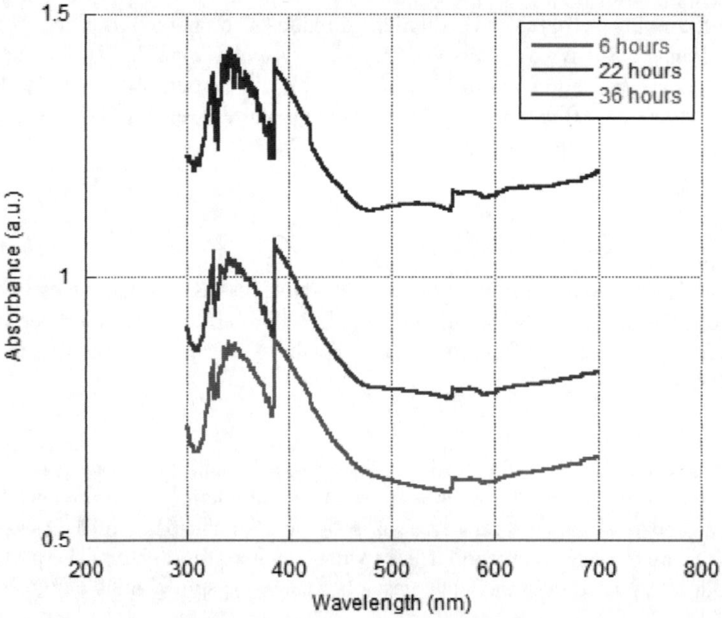

(a)

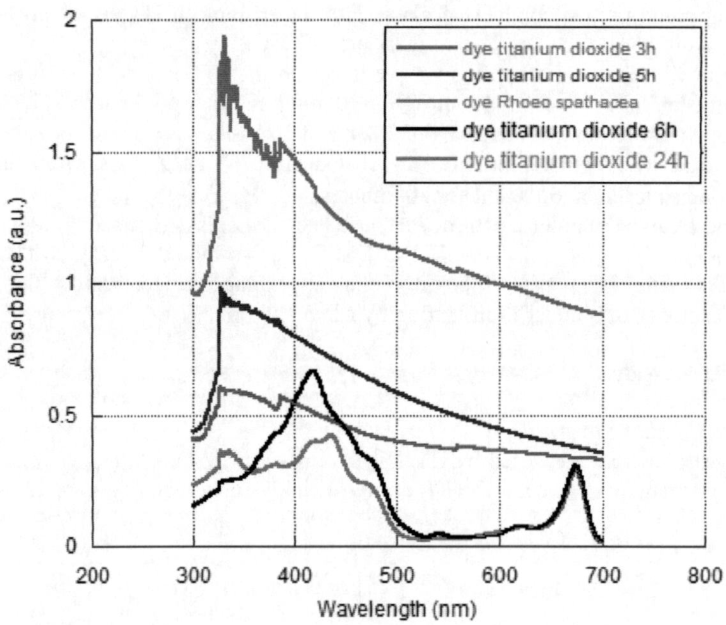

(b)

Figure 1. (a) Absorption spectra of TiO_2 submersion
(b) Absorbtion spectra of dye Rhoeospathacea and dye titanium dioxide

Figure 1 showed that the dye Rhoeo spathacea have a maximum absorption peak at 436 nm which is the visible region. After being mixed directly with TiO_2 and used to immersed, the absorption peak is shifted. It shows that the TiO_2 needs a dye sensitizer to become a good solar cell photoanode material. It is because the pure of TiO_2 does not absorb the solar radiation above 320 nm, so that when the TiO_2 submersion or directly mixed with the dye, the dye binds to the surface of TiO_2. Consequently the number of electrons produced will be increased because the dye is able to absorb light in the visible region. The quality of bonding was influenced by a homogeneous solution. The directly mixed process, the homogeneity of the solution can be arranged through the length of the stirring. But the length of the stirring can make dye titanium dioxide drying due to the process stirring doing with temperature 60° and can lead to break the bond between the dye anthocyanin and TiO_2 compounds. While in the immersion process, the quality of bonding is done by prolonging the time immersed process. Hence, the directly mixed more efficiently than the submersion process and the quality of bonding is better. It is because the dye get suppressed when directly mixed with TiO_2.

The conductivity of the dye, dye titanium dioxide, and TiO_2 was measured by two point probe El-Kahfi 100 and are given in Figure 2.

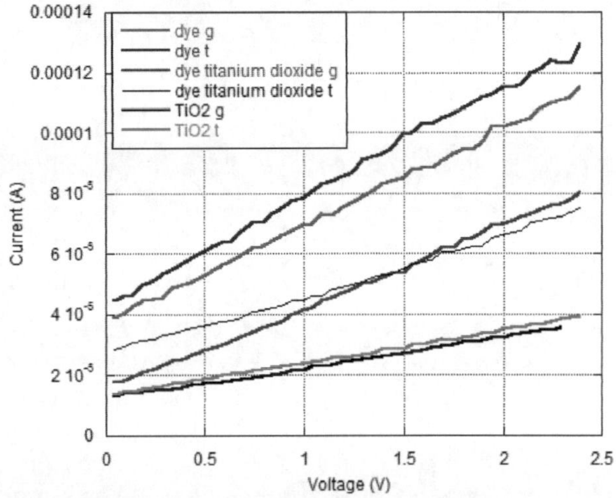

Figure 2. Conductivity of the dye, dye titanium dioxide, and TiO_2

The conductivity is the ability of a material to conduct electricity when there is a difference potential, so the electrons flowing and produce an electric current. Figure 2 shown the conductivity of dye Rhoe spatahcea is 0,0010 $\Omega^{-1}m^{-1}$, dye titanium dioxide 0,003 $\Omega^{-1}m^{-1}$, and the TiO_2 0,0012 $\Omega^{-1}m^{-1}$.

Table 1. The conductivity of dye Rhoeo spathacea, dye tita

Material	Dark of the conductivity ($\Omega^{-1}m^{-1}$)	Light of the conductivity ($\Omega^{-1}m^{-1}$)	$\Delta\sigma$ ($\Omega^{-1}m^{-1}$)
Dye Rhoe spathacea	0.002	0.003	0.001
Dye titanium dioxide	0.002	0.005	0.003
TiO_2	0.001	0.002	0.001

Table 1 shown that the difference value of between light and dark conductivity has the dye titanium dioxide is the highest, which means the dye titanium dioxide molecules has the smallest resistant. It was meant the dye titanium doxide has the highest current and this comparable with the electrons flowed when the absorption of light energy.

IOP Conf. Series: Journal of Physics: Conf. Series **795** (2017) 012026
doi:10.1088/1742-6596/795/1/012026

3.2. Performance of DSSC

Characterization of the current-voltage (I-V) is a method for determining the ability of the DSSC to convert solar radiation into electrical energy. To obtain current-voltage characterization used a Keithley 2602A conducted in the dark and the light is under illumination by a lamp of 1000 W/m2. The current-voltage of the dye titanium dioxide and TiO_2 has been immersed for 24 hours is shown in Figure 3.

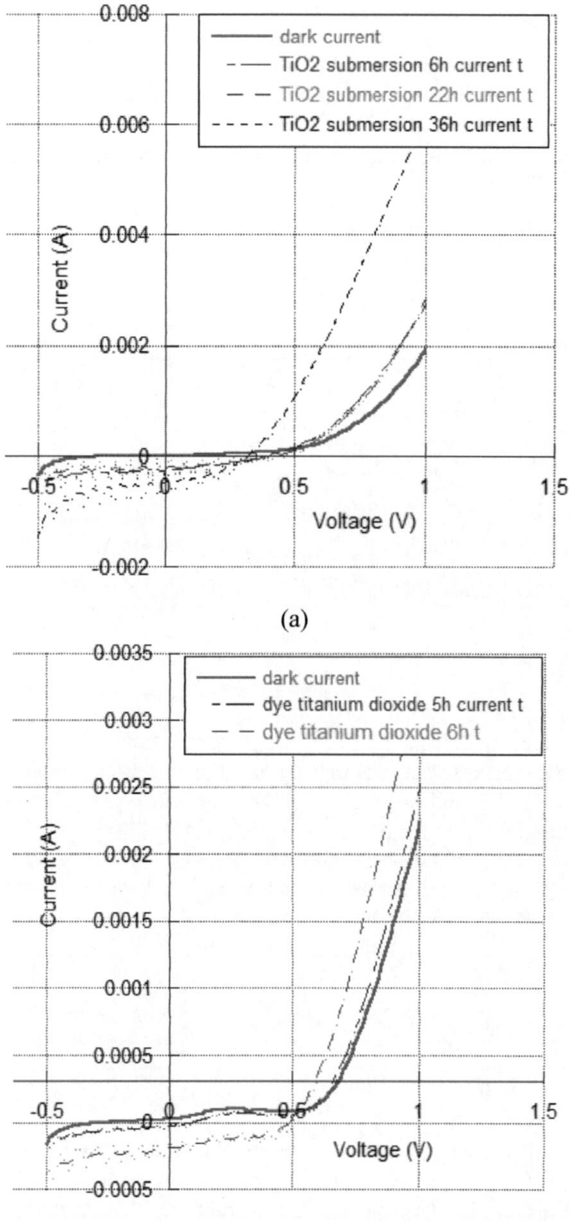

Figure 3. (a) Current-voltage curves of TiO_2 with variation of time immersed
(b) Current-voltage curves of dye titanium dioxide with variation of time stirring

Figure 3 showed that the dye titanium dioxide have a greater curve with the dark current than the TiO_2 has been immersed for 36 h. It means that the efficiency of the dye titanium dioxide is greater than the TiO_2 has been immersed for 36 h. The efficiency of the solar cell is a factor of the size quality of performance of DSSC. The higher of efficiency is mean a better performance. One of the reasons the curve are not neat was electrolytes used fast-drying, so need to inject electrolytes every characterization I-V. The value of efficiency for dye titanium dioxide and TiO_2 shown in Table 2.

Table 2. Efficiency of DSSC with Dye Rhoe spathacea as a Sensitizer

Material	Dye Titanium Dioxide		TiO_2 has been immersed		
	5h	6h	6h	22h	36h
I_{max} (mA)	0.0002	0.0001	0.0002	0.0004	0.0002
V_{max} (mV)	0.1000	0.3700	0.4150	0.4451	0.4751
I_{sc} (mA)	0.0003	0.0002	0.0002	0.0002	0.0001
V_{oc} (mV)	0.0700	0.0505	0.2200	0.2050	0.3400
η (%)	0.0020	0.0520	0.0423	0.0494	0.0501

The conversion of energy is related of the excitation of electrons that produced by the sample during the absoprtion of the light process. The electrons have been produced will be captured by the anode (TiO_2) and the hole in the dye will be compesated by electrons from the electrolyte. Furthermore, electrons flow through an external load (I-V characterization) and obtained characteristics such as Figure 3. The next process is the travel back of electrons to solar cells that accepted by the counter electrode contained carbon layer. The carbon layer as a catalyst, which accelerates the flow of electrons, as well as to be able to accelerate the reaction in the electrolyte. The reaction redox of the electrolyte occurs repeatedly, which generates new electrons to replace the missing electron in the dye (*hole*).

4. Conclusion

The dye Rhoeo spathacea has been used in DSSC as a sensitizer. To fabricate of DSSC using two ways, the TiO2 electrode was immersed in a solution of dye and the directly mixed with TiO_2 nanoporous. From the result UV-Vis spectroscopy, two point probe El-Kahfi 100 and I-V characterization were obtain :

(i) The absorption peak of dye Rhoe spathacea is 436 nm. After being mixed directly with TiO_2 and used to immersed, the absorption peak is shifted.
(ii) The conductivity of dye Rhoe spatahcea is 0,0010 $\Omega^{-1}m^{-1}$, dye titanium dioxide 0,003 $\Omega^{-1}m^{-1}$, and the TiO_2 0,0012 $\Omega^{-1}m^{-1}$
(iii) The efficiency of the dye titanium dioxide that stirring for 5 hours 0,0520% and the TiO_2 was immersed for 36 hours in the dye 0,0501%

In this work shown that the directly mixed was a promising method for uniform dye absorption hence the conversion efficiency of the light to electron can be improved.

Acknowledgement
This Research was partly supported by Higher Education Project Grant with contract no. 343/UN27/HK/2016.

References

[1] Gong J, Liang J, and Sumanthy K 2012 *Renewable and Sustainable Energy Reviews* 16 58485860
[2] Ludin NA, Mahmoud A.M.A, Mohamad AB, Kadhum AAH, Sopian K, and Karim NSA 2014 *Renewable and Sustainable Energy Reviews* 31 386396
[3] Cari, Supriyanto A, Fadli UM, Prasada AB 2016 *Journal of Physics: Conference Series* 710

012027
[4] Syafinar R, Gomesh N, Irwanto M, Fareq M, and Irwan YM 2015 *Energy Procedia* 79 896902
[5] Ananth S, Arumanayagam T, Vivek P, Murugakoothan P 2014 *Optik* 125 495498
[6] Hemmatzadeh R and Mohammadi A 2013 *Journal of Theoretical and Applied Physics* 7 5764
[7] Su'ait MS, Rahman MYA, and Ahmad A 2015 *Solar Energy* 02 3892
[8] Gratzel M 2003 *Journal of Photochemistry and Photobiology C:Photochemistry* 04 145153
[9] Hamdanian M, Safei GJ, Hosseinpur M, Masoomi R, and Jabbari V 2014 *Material Science in Semiconductor Processing* 27733-739
[10] Kartikasari 2014 *Prosiding Ilmiah XXVIII HFI Jateng dan DIY* 261263
[11] Suyitno S, Saputra JT, Supriyanto A, and Arifin, Z 2015 *Spectrochimica Acta Part A* 148 99104

IOP Conf. Series: Journal of Physics: Conf. Series **795** (2017) 012027 doi:10.1088/1742-6596/795/1/012027

The Determination of the Spectrum Energy on the model of DNA-protein interactions using WKB approximation method

Edy Syahroni[1], A Suparmi dan C Cari
Physics Department, Graduate Program, Sebelas Maret University,
Jl. Ir. Sutami 36A Kentingan Jebres Surakarta 57126, INDONESIA

E-mail: edy.syahroni1988@gmail.com

Abstract. The spectrum energy's equation for Killingback potential on the model of DNA and protein interactions was obtained using WKB approximation method. The Killingbeck potential was substituted into the general equation of WKB approximation method to determine the energy. The general equation required the value of critical turning point to complete the form equation. In this research, the general form of Killingbeck potential was causing the equation of critical turning point turn into cube equation. In this case we take the value of critical turning point only with the real value. In mathematical condition, it was satisfied with requirement Discriminant was less than or equal to 0. If D=0, it would give two values of critical turning point and if D<0, it would give three values of critical turning point. In this research we present both of those requirements to complete the general Equation of Energy.

1. Introduction

The molecular Interaction between DNA molecule and protein is phenomenon of utmost important for all of biological regulation. Almost all aspect of cellular function, such as transcriptional regulation chromosome maintenance, replication and DNA repair depend on the interaction of protein with DNA. DNA together with histone and non-histone proteins is the main constituent of the gene which forms the threads of chromatin then forming a chromosome [1]. At some cases, DNA-binding protein has diverse roles and functions such as making up the nucleosome, enzyme modulation chromatin structure to control the gene expression, becoming transcription factors, and as well as being cofactors. Before the DNA either replicates or transcribes, there is a certain protein which triggers the signals so that the DNA can replicate and/or transcribe, this protein is called as transcription factor [2].
All of system proses couldn't be separated from how the DNA and protein interaction. There are many efforts to describe the model of interaction between DNA and protein. Some of them were described in 2016 by Dwiputra D *et al* and Syahroni E *et al.* [3,4]. They were described the simple model of nonlinear interaction such as the model description and stability system. In their model, Hamiltonian was used as a method to describe the total energy of the system [5]. Consider an object an atom, molecule and rock, in mechanically stable state and well isolated from its surroundings. In classical or quantum theory the object has a definite energy, which is conserved. In mechanics, energy can assume any value from some

minimum to the maximum allowed by stability. In quantum theory the possible values of energy are discrete, or quantized [6]. In this study we review the interaction between the proteins to DNA which is influenced by the potential Killingbeck only in the direction of one dimension [4]. We assumed the energy value of system was discrete. Several methods have been proposed to determine the wave function and energy [7]. In this paper we used WKB (Wentzel, Kramer and Brillouin) approximation method to determine the spectrum energy. This approximation could be used for the potential energy V(x) of system changed slowly at a small range [8]. WKB approximation is often used to solve problems in quantum mechanics either to determine the wave function and energy [9]. Several studies using WKB approach have been proposed [10, 11]

2. WKB Apprpximation Method

In this research, Killingbeck potential V(x) is potential that represented the interaction between DNA and the energy of protein represented with E. We used the WKB approximation to determine the spectrum energy from the interaction as equation [8]:

$$\int_a^b k(x)dx = (n+\frac{1}{2})\pi \tag{1}$$

with n=0,1,2,3,..., $k(x)=\frac{1}{\hbar}\sqrt{2m(E-V(x))}$, *a* and *b* are classical turning point, the value of *a* and *b* can be obtained with condition $E-V(x)=0$ (*a* and *b* are the roots value of x).

The general form of Killingbeck potential given by equation [12, 13]:

$$V=-\frac{c}{x}+dx+\beta x^2 \tag{2}$$

If we substituting Killingbeck potential eq. (2) to eq. (1) we have the equation to determine the spectrum energy as:

$$\int_a^b \sqrt{\left(E+\frac{c}{x}-dx-\beta x^2\right)}dx=\frac{\hbar}{\sqrt{2m}}(n+\frac{1}{2})\pi \tag{3}$$

The classical turning point *a* and *b* can be obtained by the following equations:

$$x^3+\frac{d}{\beta}x^2-\frac{E}{\beta}x-\frac{c}{\beta}=0 \tag{4}$$

The eq. (4) is a cubic equation. As we know a cubic equation will give a solution with 3 kind of condition [14]. The solutions of the eq. (4) are:

$$\begin{cases} x_1=S+T-\frac{d}{3\beta} \\ x_2=-\frac{1}{2}(S+T)-\frac{d}{3\beta}+\frac{1}{2}i\sqrt{3}(S-T) \\ x_3=-\frac{1}{2}(S+T)-\frac{d}{3\beta}-\frac{1}{2}i\sqrt{3}(S-T) \end{cases} \tag{5}$$

with:

$$Q=\frac{1}{9\beta^2}(3\beta E-d^2) \qquad S=\sqrt[3]{R+\sqrt{Q^3+R^2}}$$
$$R=\frac{1}{18\beta^3}(3\beta Ed-9c\beta^2-d^3) \qquad T=\sqrt[3]{R-\sqrt{Q^3+R^2}} \tag{6}$$

The solutions of eq. (4) will give the values of classical point. Before we obtained the value of classical point, we need to take a condition to classify how many classical points we need. The eq. (4) can be solved with three kinds of conditions. If c, d and β are real and if discriminant D=Q^3+R^2, than All roots are real and at least two roots are equal if D=0, one root is real and two are complex conjugate if D>0 and all roots are real and unequal if D<0. We take the condition with D=0 and D>0. Our aim is to get the real roots of x. After we set the condition of the classical point, the values can be substituting into the general solution of spectrum energy. The general solution of spectrum energy can be obtained from eq. (3) as followed:

$$\frac{(Eb^2+cb-db^3-\beta b^4)^{\frac{3}{2}}}{(cb+db^3+2\beta b^4)}-\frac{(Ea^2+ca-da^3-\beta a^4)^{\frac{3}{2}}}{(ca+da^3+2\beta a^4)}=-\frac{3}{2}\frac{\hbar}{\sqrt{2m}}(n+\frac{1}{2})\pi \tag{7}$$

2.1. *Solution with condition D=0*

The first condition is D=0. This condition will give three real roots with two roots are equal. The roots are given as:

$$x_1=X=S+T-\frac{d}{3\beta} \quad \text{and} \quad x_{2,3}=-\frac{1}{2}(X+\frac{d}{\beta}) \tag{8}$$

If D=0 than Q^3 will equal with R^2. By using the solutions of quadratic equation, we obtain the following equations of c from eq. (4) as:

$$c_1=\frac{(9Ed\beta+2d^3)+2\sqrt{d^6+3^2Ed^4\beta+3^3E^3\beta^3+3^3E^2d^2\beta^2)}}{27\beta^2}\ ,\ c_2=\frac{(9Ed\beta+2d^3)-2\sqrt{d^6+3^2Ed^4\beta+3^3E^3\beta^3+3^3E^2d^2\beta^2)}}{27\beta^2} \tag{9}$$

The discriminant D<0 will be satisfied with condition of eq. (9). By substituting eq. (6) and (9) to eq. (8), we obtain the equation of classical points expressed as:

$$x_1=-\frac{1}{3\beta}\left(\sqrt[3]{\left(20d^3+8\sqrt{d^6+3^2Ed^4\beta+3^3E^3\beta^3+3^3E^2d^2\beta^2)}\right)}+d\right)$$
$$x_{2,3}=\frac{1}{6\beta}\left(\sqrt[3]{\left(20d^3+8\sqrt{d^6+3^2Ed^4\beta+3^3E^3\beta^3+3^3E^2d^2\beta^2)}\right)}-d\right) \tag{10}$$

or

$$x_1=-\frac{1}{3\beta}\left(\sqrt[3]{20d^3-\left(8\sqrt{d^6+3^2Ed^4\beta+3^3E^3\beta^3+3^3E^2d^2\beta^2)}\right)}+d\right)$$
$$x_{2,3}=\frac{1}{6\beta}\left(\sqrt[3]{20d^3-8\sqrt{d^6+3^2Ed^4\beta+3^3E^3\beta^3+3^3E^2d^2\beta^2)}}-2d\right) \tag{11}$$

These results give two variation of classical turning point. The eq. (10) is conditions obtained with c_1 and eq. (11) is conditions obtained with c_2.

2.2. *Solution with condition D<0*

The second condition is D<0. This condition will give three real roots and all roots are unequal. If D<0, computation is simplified by use of trigonometry. The roots are given as:

$$\begin{cases} x_1 = 2\sqrt{-Q}\cos(\frac{1}{3}\theta) - \frac{1}{3}A \\ x_2 = 2\sqrt{-Q}\cos(\frac{1}{3}\theta + 120°) - \frac{1}{3}A \\ x_3 = 2\sqrt{-Q}\cos(\frac{1}{3}\theta + 240°) - \frac{1}{3}A \end{cases} \quad (12)$$

With $\cos\theta = \frac{R}{\sqrt{-(Q^3)}}$ or $\theta = \arccos\left(\frac{R}{\sqrt{-(Q^3)}}\right)$. If D<0 than $Q^3 + R^2 < 0$, by using the solutions of quadratic equation, we obtain the following equations of c as:

$$c_2 > \frac{(9Ed\beta + 2d^3) + 2\sqrt{d^6 + 3^2 Ed^4\beta + 3^3 E^3\beta^3 + 3^3 E^2 d^2 \beta^2)}}{27\beta^2}$$
$$c_2 > \frac{(9Ed\beta + 2d^3) - 2\sqrt{d^6 + 3^2 Ed^4\beta + 3^3 E^3\beta^3 + 3^3 E^2 d^2 \beta^2)}}{27\beta^2} \quad (13)$$

The discriminant D<0 will be satisfied with condition of eq. (13). By substituting eq. (6) to eq. (12), we obtain the equation of classical points expressed as:

$$x_1 = \frac{2}{3\beta}\left[\sqrt{(d^2 - 3\beta E)}\cos(\frac{1}{3}\theta) - 1\right], x_2 = \frac{2}{3\beta}\left[\sqrt{(d^2 - 3\beta E)}\cos(\frac{1}{3}\theta + 120°) - 1\right], x_3 = \frac{2}{3\beta}\left[\sqrt{(d^2 - 3\beta E)}\cos(\frac{1}{3}\theta + 240°) - 1\right] \quad (14)$$

3. Result and Discussion

In this section, we discuss several obtained in the previous section. The result differentiated based on the value of discriminant. Each condition of discriminant requires a condition that must be satisfied.

3.1. *Condition D=0*

By substituting the classical turning point for each condition c_1 and c_2, from eq. (10) and (11) to eq. (7), we obtain the equation to determine the energy. We use the value of each parameter as $\hbar$ =1.0546, *d*=3, β=-2 for condition c_1 and β=2 for condition c_2. The value of energy listed in Table1. By inspecting Table 1, show that increase of value *n* give different change of energies value. Both condition gives regular value of energy. In this case we take the condition with c_2 that corresponding to condition of a particle-bound, as electrons bound to nucleus. The energy turned toward positive corresponding with quantum state *n*.

Table 1. Energy values corresponding to several state *n* of molecule DNA under the influence of Killingbeck potential with condition c_1 and c_2.

n	Energy	
	$c_1(\beta=-2)$	**$c_2(\beta=2)$**
0	-0.0640	0.1053
1	-0.0962	0.1898
2	-0.1150	0.2093
3	-0.1259	0.2161
4	-0.1328	0.2191
5	-0.1375	0.2206
6	-0.1410	0.2216
7	-0.1436	0.2222

In this section, we also investigate the energy value for variation of linear potential's parameter and coupling potential's parameter. In Table 2 shown the energy values corresponding to several variations of linear potential's parameter d and the energy values corresponding to several variation of coupling potential's parameter β.

Table 2. Energy values corresponding to several state *n* of molecule DNA under the influence of Killingbeck potential with variation of linear parameter d and coupling parameter β

n	Energi			Energi		
	d=3	*d=4*	*d=5*	$\beta=2$	$\beta=3$	$\beta=3.5$
0	0.1053	-4.2571	-6.6787	-4.2571	0.0206	0.0111
1	0.1899	0.2684	0.3118	0.2683	0.0372	0.0174
2	0.2094	0.3314	0.4338	0.3314	0.0407	0.0185
3	0.2161	0.3589	0.4999	0.3589	0.0418	0.0189
4	0.2191	0.3728	0.5381	0.3728	0.0424	0.0190
5	0.2207	0.3805	0.5615	0.3805	0.0426	0.0191
6	0.2216	0.3853	0.5766	0.3853	0.0428	0.0191
7	0.2222	0.3884	0.5868	0.3883	0.0429	0.0192

By inspecting Table 2, show that correlation between increasing the value of each parameter. by increasing the linear potential's parameter increase the range interval of energy value, whereas increasing the coupling potential's parameter causing the range interval of energy become smaller.

3.2. Condition D<0

For condition discriminant D<0, by substituting the classical turning point for each condition c_1 and c_2, from eq. (14) to eq. (7), we obtain the equation to determine the energy. With similar value of each parameter we obtained the value of energy as Table 4. Both of each condition gives a regulars value of energy. In this case we take the condition with c_1 that corresponding to condition of a particle-bound, as electrons bound to nucleus. The energy turned toward positive corresponding with quantum state *n*.

Table 3. Energy values corresponding to several state n of molecule DNA under the influence of Killingbeck potential with condition c_1 and c_2.

n	Energy	
	c_1(β =-2)	**c_2 (β =2)**
0	-5513.0877	5439.1151
1	-5512.8399	5438.8722
2	-5512.5920	5438.6294
3	-5512.3442	5438.3866
4	-5512.0963	5438.1438
5	-5511.8485	5437.9009
6	-5511.6006	5437.6581
7	-5511.3527	5437.4153

We also investigate the value of energy corresponding with variation of linear potential's parameter and coupling potential's parameter shown in Table 4.

Table 4. Energy values corresponding to several state n of molecule DNA under the influence of Killingbeck potential with variation of linear parameter d and coupling parameter β

n	Energi			Energi		
	d=2	*d=3*	*d=4*	β =-2	β =-3	β =-4
0	-5487.2644	-5513.0877	-5690.5127	-5487.2644	-2579.7821	-1564.6055
1	-5487.0134	-5512.8399	-5690.1979	-5487.0134	-2579.6888	-1564.5564
2	-5486.7625	-5512.5920	-5689.9201	-5486.7625	-2579.5956	-1564.5073
3	-5486.5115	-5512.3442	-5689.6442	-5486.5115	-2579.5023	-1564.4582
4	-5486.2605	-5512.0963	-5688.3067	-5486.2605	-2579.4090	-1564.4092
5	-5486.0095	-5511.8485	-5688.0162	-5486.0095	-2579.3158	-1564.3601
6	-5485.7585	-5511.6006	-5687.7035	-5485.7585	-2579.2225	-1564.3110
7	-5485.5075	-5511.3527	-5687.4713	-5485.5075	-2579.1292	-1564.2619

4. Concluding remarks

In this paper, we have determined the spectrum energy for a model of a nonlinear DNA-protein interaction system with Killingbeck potential. We used the WKB approximation to determine the spectrum energy of interaction. The form of Killingbeck potential allows the energy values divided into two specific conditions. Such as with the value of discriminant D=0 and D<0. Each condition required a specific condition to satisfy. The result has shown for each condition, the value of coupling potential's parameter β was decisive to satisfying the condition. For D=0 required the value of β greater than 0, whereas for D<0 required the value of β lower than 0.

Acknowledgement

This Research was partly supported by Higher Education Project Grant with contract no. 632/UN27.21/LT/2016

IOP Conf. Series: Journal of Physics: Conf. Series **795** (2017) 012027 doi:10.1088/1742-6596/795/1/012027

References

[1] Bulyk M L, Gentalen E, Lockhart D J and Church G M, 1999 *Nat Biotechnology* **17** 573–577
[2] Brivanlou A H and Darnell Jr J E 2002 *Signal Transduction and Control of Gene Expression science's Compass Review* **295** 813-818
[3] Dwiputra D, Hidayat W, Khairani R and Zen F P, 2016 *J. phyc* **694** 012076
[4] Syahroni E, Suparmi A, Cari C and Fuad A, 2016 A model of a nonlinear DNA-protein interaction system with Killingbeck potential and its stability (unpublished) *Submitted at the 8th International Conference on Physics and Its Applications*
[5] Greiner W, 2000 *Relativistic Quantum Mechanics (*New York Verlag Berlin Heidelberg: Springer).
[6] Peebles P J E,1992 *Quantum Mechanic* (United Kingdom: Princeton University Press) p 17
[7] Suparm A, Cari C and Deta U A, 2014 *Chin. Phys. B* **23** 090304
[8] Suparmi, 2011*mekanika Kuantum II* (Surakarta : Sebelas Maret University) p 67-86
[9] Cari, Suparmi and Marini H, 2012 *Indonesian journal of Applied Physics* **2** 112
[10] Morsbøl J O, Sorokin S V and Peake N, 2016 *J. Sound and Vibration* **375** 162-186
[11] Cheng K M, Leung P T and Pang C S, 2003 *J. phyc.* A **36**. 5044-5060
[12] Killingbeck J, 1978 *Phys. Lett.* A **65** 87
[13] Plante G and Antippa A F, 2005 *J. Math. Phys.* **46** 062108
[14] Spiegel R M, Lipschutz and Lui J, 2009 *mathematical Handbook of Formulas and Tables Third edition* (USA:McGraw-Hill) p 13
Note: dirac equation jurnal scarf and new coupling tensor

Biochemical fiber sensor based on evanescent field for detection persistent organic pollutants (POPs)

Edi Prasetyo, Martia Putri Gitrin, Ahmad Marzuki*, Venty Suryanti

*Optic and Photonic Laboratory Physics Department
Faculty of Mathematics and Natural Science Sebelas Maret University
Jl. Ir. Sutami No 36A Surakarta

*amarzuki@mipa.uns.ac.id

Abstract. Fiber optic is a light waveguides media that are cylindrical. Optical fiber has certain properties when it transmits light so it can be developed to be a sensing device or sensor. Evanescent wave phenomena appear when there are total internal reflections from many modes in an optical fiber. In this research, the Biochemical Fiber Sensor (BFS) using polishing cladding and some of the core fiber will be fabricated. BFS is used to interact with a biochemical compound.The principle of BFS is based on evanescent absorption which absorbs the typical spectrum of a biochemical compound. By measuring the spectrum from the light output in the BFS, evanescent absorption spectra can be analyzed an optical fiber. In this study, the biochemical compounds that used are lindane that is one of the Persistent Organic Pollutants (POPs). The result showed that there is a change from BFS spectra when it was exposed by POPs compound with various concentration. That change showed that there is evanescent absorption in BFS. Concentration of POPs compound is proportional with evanescent absorption of the POPs compound.

1. Introduction

The environment is a place to grow and reproduce for all organism. Environment conditions will determine the quality of life for living things, especially humans. Recently the environmental pollution is a problem occurred in almost all places in the world. The development of science and technology have had an impact on the appearance of harmful substances to the environment. Human behavior is irresponsible make harmful substances released into the environment that causes environmental pollution. One of the harmful substances are often found is Persistent Organic Pollutants (POPs) from the disposal of industry [1]. POPs are organic compounds that are produced by human activity where the compound cannot be decomposed by the environment [2]. POPs compounds indicated the risk that can affect human health and the environment. The affect in humans include leukemia, aplastic anemia, allergies and asthma. Also, contamination of POPs in the environment cause the extinction of certain species, the decline of fish populations, the emergence of resistance or pest resurgence, and the killing of birds or insects that are useful [3].

Relating to the problem of optical fiber-based sensor technology may seem to be a solution. Fiber optic is a light waveguides media that are cylindrical. Optical fiber has certain properties when it transmits light so it can be developed to be a sensing device or sensor. Various methods based on optical sensing techniques have been developed to detection biochemical compounds, such as fiber bragg gratings [4], grating coupled waveguide[5], surface plasmon resonance (SPR) [6].

In this research, we created biochemical fiber sensor (BFS) with optical spectroscopy method based on evanescent field that occurs in optical fiber. Evanescent phenomena appears when there are total internal reflection from many modes light in optical fiber. Evanescent phenomena occur in boundaries between core and cladding [7]. POPs compounds were placed in the sensing area interact with evanescent field which resulted from the absorption at spectrum light that propagates in the fiber optic. Every molecule on earth has the emission spectrum and absorption spectrum that has a unique spectrum. The Emission spectrum is spectrum radiated by an element when the excitation of atoms. Emission spectrum radiated having the spectrum that contains only certain wavelengths. The Absorption spectrum is the spectrum of the radiation absorption results of the molecule. When polychromatic light is passed through a molecule, it turns out the molecule absorb light of a certain wavelength of wavelengths contained in the emission spectrum [8]. Therefore with knowing typical absorption we can detect POPs compounds. This research used a lindane as sample POPs compounds.

2. Method

2.1. Fabrication Biochemical Fiber Sensor

Fiber sensor is optical fiber used for sensing purposes. BFS based on the evanescent field created by removing the cladding and reducing some of the core diameters in some long fiber optic. Cladding and some of the core diameter core removed by using a polishing machine. After that, the fiber sensor is cleaned by water. The length of fiber sensor sensing area is 3 cm and depth of polish is 1,5 mm shown in Figure 1.

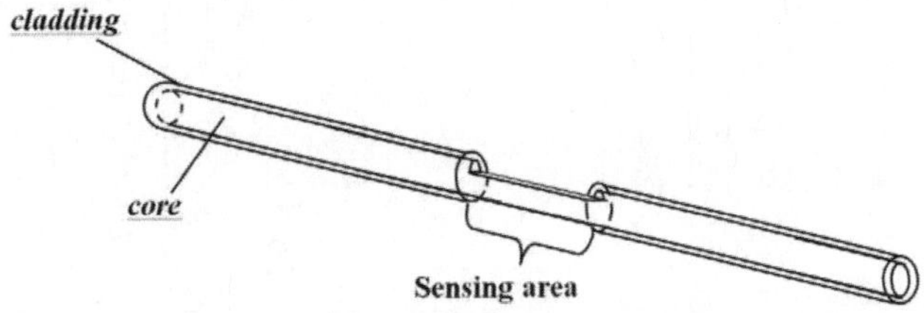

Figure 1. fiber sensor scheme

2.2. Experiment

Experimental set up is shown in Figure 2. The devices used in the experimental setup consists of a light source form halogen lamp, fiber sensors, spectrophotometer and software acquisition. Evanescent phenomena occur in boundaries core and cladding when incoming light and propagates through fiber sensor. In the sensing area, the light interact with lindane because evanescent field phenomena and absorbsition evanescent was occurred. Spectrophotometer is used for measurement spectrum wavelength light that propagates through fiber sensor. The Measurement result is displayed a personal computer with software acquisition.

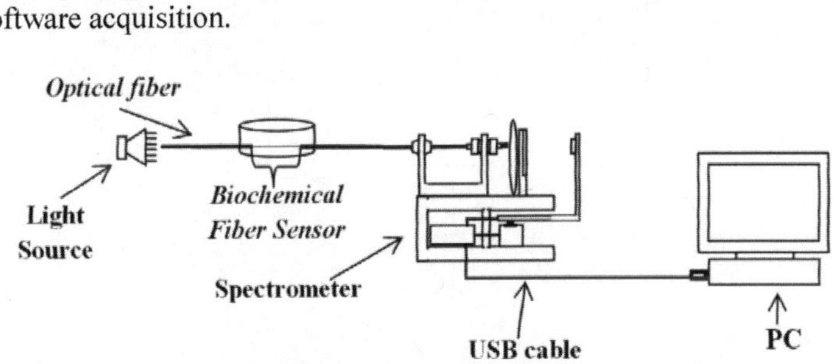

Figure 2. Experiment set up schema.

This research used a lindane as sample POPs compounds. In this experiment, a pure lindane compounds was dissolved with methanol 70% with variations concentration 2,06x10^{-3} ;1,06x10^{-3} ; 5,15x10^{-4}; 3,43x10^{-4} ; 2,58x10^{-4}. For calculating absorption evanescent value, use formula 5.

$$A = -\log T = -log\frac{I_t}{I_{ref}} \quad (1)$$

Where I_{ref} is methanol 70% intensity and I_t is lindane compounds [9,10].

3. Result

The light that propagate through optical fiber cause total internal reflection. Total internal reflection occurs when light propagates in the optical fiber toward the boundary between the core and the cladding have the angle of incidence exceeds the critical angle, the light will be reflected back into the core.When light propagates in the optical fiber and reaches the boundary between the core and the cladding, light penetrate the cladding before returning to the core. The light that penetrate the cladding is called evanescent field. The graph in figure 3. shows the spectrum of light in fiber sensor when given a lindane. Lindane solution fills cladding and some of core diameter fiber optic that removed so that lindane function as cladding. Phenomena evanescent cause light propagates in fiber optic interact with lindane and occurs absorption of light propagates in fiber sensor by lindane compounds or can be called by the absorption evanescent.

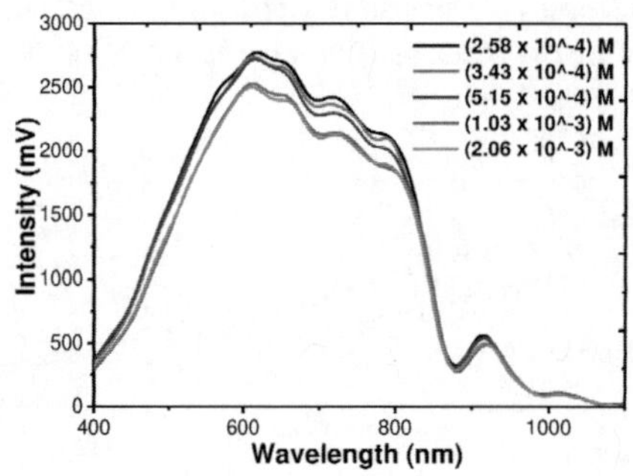

Figure 3. Graph spectrum light fiber sensor.

Absorption evanescent is calculated with use formula 1 and be obtained graph absorption as wavalenght function. Graph absorption evanescent is shown in Figure 4. Absorbance evanescent according to the characteristics of lindane solution absorbs certain wavelengths that are characteristic compound lindane. The results obtained showed absorbance evanescent lindane absorption peak at 913 nm and 1003 nm.

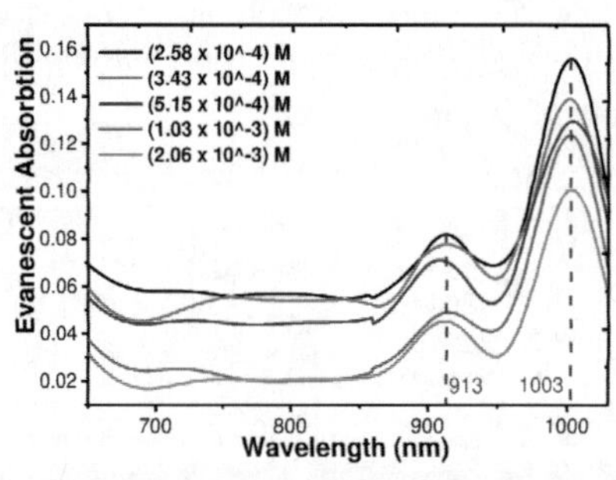

Figure 4. Graph absorption evanescent from lindane.

IOP Conf. Series: Journal of Physics: Conf. Series **795** (2017) 012028 doi:10.1088/1742-6596/795/1/012028

Coefficient absorption evanescent given by

$$A = \gamma LC \tag{2}$$

Where A is the absorbance, C is the concentration of the solution, L is the length of the sensing area and is the coefficient absorption evanescent[11]. Based on the equation 2 can be obtained the relationship between the absorption evanescent with the concentration of lindane as shown in Figure 5.

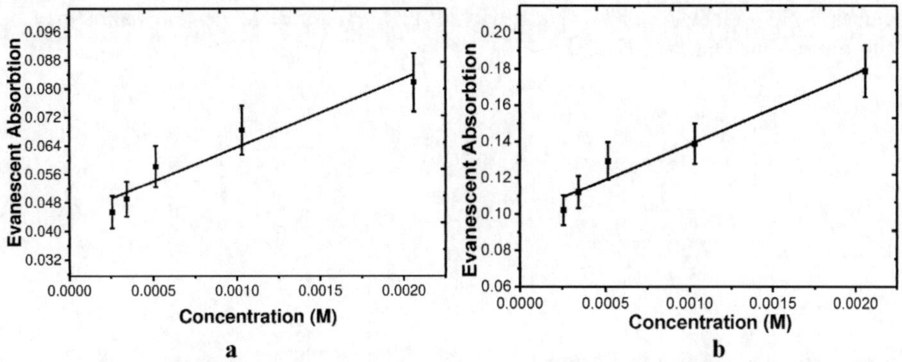

Figure 5. Graph absorbance *evanescent* as concentration function *lindane* at (a. 913nm b.1003 nm).

The results obtained graph shows that the greater the concentration of lindane that dissolved the greater the absorbance which occurs. The coefficient of absorption evanescent calculated based on the equation 2 is shown in Table 1.

Tabel 1. Wavelength and Coefficient absorption*Evanescent*

Wavelength (λ)	Coefficient absorption Evanescent (γ)
(*nm*)	($M\ cm^{-1}$)
913	5,97
1003	14,22

4. Conclusions

Biochemical fiber sensor works based on evanescent field has been demonstrated and used for detection evanescent absorption spectra from POPs compounds. The evanescent absorbance observed corresponds with the characteristics of POPs compounds.

References

[1] Undeman.E, Brown. T.N, Wania. F, & McLachlan. M.S, *Environ. Sci. Technol*. 44. 6249-6255.
[2] Ma. J, Hung. H, Tian. C, & Kallenborn. R, *Nat. Clim. Chang*.255-260.
[3] Jones. K.C., & Voogt. P.de,Environ. Pollut., 209-221.
[4] M.P.DeLisa, Z.Zhang, M.Shiloach, S.Pilevar, C.C.Davis, J.S.Sirkis, W.E.Bentley, *Anal.Chem*. 72(2000)2895–2900.
[5] R. Horvath, H.C.Pedersen, N.Skivesen, D.Selmeczi, N.B.Larsen, *Opt.Lett*. 28 (2003)12331235.
[6] A.D.Taylor, J.Ladd, Q.M.Yu, S.F.Chen, J.Homola, S.Y.Jiang, *Biosens.Bioelec-tron*. 22(2006)752–758.
[7] Yin, S., Ruffin, P.B., & Yu, F.T. 2008. *Fiber optic sensor*. USA: CRC Press.

[8] Beiser. A, 1981,*Concepts of modern physics third editio,* New York: McGraw-Hill International Book.

[9] Saleh. B. E. A, and Teich, M C, 2007, F*undamentals of Photonics Second Edition*, A John Wiley and Sons, Inc. Hoboken, New Jersey.

[10] Senior. J.M, 2009, Optical fiber communication: Principles and practice third editio, England: Pearson Prentice Hall.

[11] Kumar. P.S, Vallabhan. C.P.G, Nampoori. V.P.N, Pillai. V.N.S, & Radhakrishnan. P, J.Opt A:pure Appl. Opt., 4, 247-250.

IOP Conf. Series: Journal of Physics: Conf. Series **795** (2017) 012029 doi:10.1088/1742-6596/795/1/012029

Fabrication and variation layers of Cu/TiO_2 nanocomposite and its applications in Dye-Sensitized Solar Cell (DSSC)

Wilfrida Mayasti Obina[1,2], Cari[1], Agus Supriyanto[1], Sri Sumardiasih[1], Trio Yuda Septiawan[1], Khairuddin[1]

[1]Physics Departement of Post Graduate Program Sebelas Maret University,
[1]Jl. Ir. Sutami 36A Kentingan, Surakarta 57126, INDONESIA

[2]Email: wilfridakiki@gmail.com

Abstract. Fabrication of Cu/TiO_2 nanocomposite had been successfully performed. A nanocomposite was prepared by the insertion of metal (copper) in the TiO_2 nanoparticles. Cu/TiO_2 nanocomposite was deposited on the FTO glass used the spin coating method. Layers of Cu/TiO_2 nanocomposite were prepared in variations at 1, 2 and 3 layers. *Spektrofotometer UV-VIS Lamda 25* was used to determine the optical characteristics and the energy gap of DSSC materials. *El kahfi 100/I-V Meter* and *Keithley 2602A* was used to determine the electrical properties of Cu/TiO_2 nanocomposite. UV-VIS result was showed that there increased absorption wavelength of Cu/TiO_2 nanocomposite and were able to decrease the energy gap. El kahfi was showed that the conductivity of the Cu/TiO_2 nanocomposite was $3.71x10^{-2}$ ($\Omega^{-1}m^{-1}$). Keithley was showed that the highest efficiency in the 3 layers of Cu/TiO_2 nanocomposite of 0.011%.

1. Introduction

Dye-Sensitized Solar Cell (DSSC) was one of the solar cells devices which used natural dyes to convert solar energy into electrical energy. DSSC was consists of a dye, counter electrode, the active layer, electrolyte, and FTO glass as a substrate [1]. The performance of DSSC was increased if the performance of its constituent materials increased. For example, the active layer of TiO_2 was modified. TiO_2 was a semiconductor that stable, transparent, environmentally friendly and large surface [2]. The results showed that TiO_2 had an energy gap of 3.2 *eV* [3]. TiO_2 was selected for the insertion of metal because it had the large surface. Large Surface would cause the hole or gap between the particles of TiO_2. Metal was inserted between the particles to overcome these gaps. Metals had good conductivity properties. The purpose was to facilitate the transfer of electrons when passed a layer of TiO_2. Research had been done about metal doped on TiO_2 through several methods such as hydrothermal, sputtering, electroplating and nanocomposite [4,5]. Fabrication of Au/TiO_2 nanocomposite could improve the efficiency of up to 6% [6]. Gold (Au) and silver (Ag) had electrical conductivity properties were better, but they were expensive and limited supply in nature.

In this study was used metal of copper (Cu). Copper doped into natural dye could increase the efficiency from DSSC of 71% [7]. Copper was good thermal conductivity and there was approximately 68 *ppm* in the earth's crust [8]. Copper had a size of 128 *pm* or 0.128 *m* and TiO_2 nanoparticles had a size of 21 *nm*. So that copper would be inserted on TiO_2 through the fabrication of Cu/TiO_2 nanocomposite. The nanocomposite was a merger of nano-sized material and had a different phase. Research had been made a Cu/TiO_2 nanocomposite and could lower the value of electric resistance. The resistance value of TiO_2 layer was 7.714-kilo ohm and Cu/TiO_2 nanocomposite layer was 6.624-kilo ohm [9].

Variations at 1 layer, 2 layers and 3 layers of Cu/TiO_2 nanocomposite was made in this study. Spin coating method was used in the deposition of Cu/TiO_2 nanocomposite. The spin coating method was a method shed a solution of the film on the substrate and rotated at a certain speed. In this research would be tested the absorbance, the energy gap, conductivity and efficiency of Cu/TiO_2 nanocomposite. Fabrication of Cu/TiO_2 nanocomposite was expected to increase the performance of the DSSC.

2. Experimental Details

2.1. Preparation of Cu/TiO_2 nanocomposite

Cu/TiO_2 nanocomposite was prepared by dissolved 0.1 *g* of Polyvinyl Alcohol (PVA) in 10 *ml* of distilled water while stirred with heated at 80°*C* for 2 *h*. TiO_2 as much as 1 *g* and 0.1 *g* of copper ($CuSO_4.5H_2O$) was added to the mixture while stirred for 2 *h*.

2.2. Extraction of Natural Dyes

Moringa oleifera leaves of 10 *g* were smoothed by porcelain mortar. Ethanol of 50 *ml* was added in the leaves. Dye was extracted by a magnetic stirrer for 1 *h* and then allowed to stand for 24 *h* at room temperature. The result of the soaking was filtered to obtain the pure extract dye.

2.3. Fabrication of DSSC

The conductive glass was Fluorine Tin Oxide (FTO) by the active cell area of 1 *cm* x 1 *cm*. FTO glass was cleaned with distilled water for 15 minutes and alcohol 70% for 15 minutes by the ultrasonic bath. Variation of layers was made by varying the scotch tape to make the active areas. A number of variation layers were equal to the variation of scotch tape. The active layer was the layer of Cu/TiO_2 nanocomposite. Deposition of Cu/TiO_2 nanocomposite was done by varying 1, 2, and 3 layers. Layers of Cu/TiO_2 nanocomposite on FTO glass was heated at 100 °*C* for 30 minutes to evaporate the distilled water and 450 °*C* for 30 minutes. Next, the active layer was soaked in the dye for 24 *h*. The counter electrode was used 2.5 *ml* solution of platinum. The platinum solution was obtained by mixing 1 ml Hexachloroplatinic (IV) acid 10% (Platinum) with 207 *ml* of isopropanol stirred for 30 minutes. The electrolyte solution has used a mixture of 2 *g* of Natrium Iodid (NaI) with 3.68 *ml* of acetonitrile, 1 *ml* Propylene Carbonate, 14.56 *ml* Polyethylene glycol and 0.2 *g* of Iodine (I_2) which stirred for 1 *h*. Last was assembled DSSC with the sandwich method.

2.4. Characterization

The absorbance of materials was tested by a *UV-Visible Spectrophotometer Shimadzu 1601 PC*. The result of absorbance was used to determine the energy gap. The calculation could follow the energy gap in equation (1) [10]. Where h was Planck's constant ($h = 6.63 \times 10^{-34}$ *Js*), c was the speed of light ($c = 3 \times 10^8$ *m/s*), λ was the wavelength, and E was the energy gap ($1 eV = 1.6 \times 10^{-19} J$).

$$E = \frac{hc}{\lambda} \quad (1)$$

The electrical characteristics (conductivity) was tested by *Elkahfi 100 / I-V Meter* and *I-V* characteristics of DSSC by a *Keithley 2602A*. The intensity of the light (Keithley) was used of 1000 W/m^2 from halogen lamps. The test results of Keithley was calculated by Equation (2-3). Where V_{max}= maximum voltage (*V*); I_{max} = maximum current density (mA/cm^2); V_{oc} = Open circuit voltage (*V*) and P_{in} = power of the incoming light.

$$FF = \frac{P_{max}}{P_{in}} = \frac{I_{max}\, x\, V_{max}}{I_{sc}\, x\, V_{oc}} \quad (2)$$

$$\eta = \frac{FF\, x\, I_{sc}\, x\, V_{oc}}{P_{in}}\, x\, 100\% \quad (3)$$

3. Results and Discussion

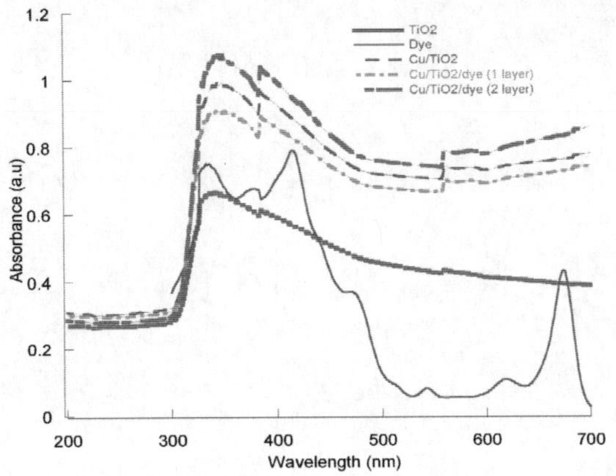

Figure 1. Optical characteristics of Cu/TiO_2 nanocomposite

Figure 1 showed the optical characteristics of dye, TiO_2 paste, Cu/ TiO_2 nanocomposite, Cu/TiO_2/dye (1 layer) and Cu/TiO_2/dye (2 layers). There were three absorbance peaks by the dye which located in the wavelength range 333 *nm* - 674 *nm*. The highest absorbance peak of the *Moringa oleifera* was a wavelength of 415 *nm* and the highest absorbance peak of pure TiO_2 was 341 *nm*. This means that TiO_2 was also able to absorb visible light very well. Absorbance peak of Cu/TiO_2 nanocomposite was the wavelength of 344 *nm*. It was seen that the light absorption increased from the range of 341 *nm* to 344 *nm*. This showed the highest absorption by Cu/TiO_2 nanocomposite. The highest absorbance peak of Cu/TiO_2 nanocomposite for 1 layer was 343 *nm* and 2 layers were 345 *nm*. From the results of the absorbance and the layers could be seen that the increased of the member of layers causes increasing the absorbance. If more light was absorbed, then more electrons could be transferred dan the performance of the DSSC increased.

Table 1. Energy gap of Cu/TiO_2 nanocomposite

Materials	λ (*nm*)	*Eg* (*eV*)
Dye	415	2.99
TiO_2	341	3.64
Cu/TiO_2	344	3.61
Cu/TiO_2/dye (1 layer)	343	3.61
Cu/TiO_2/dye (2 layers)	345	3.60

Table 1 showed that the doped copper on TiO_2 nanoparticles through fabrication nanocomposite could lower energy gap. The energy gap was the energy required by an electron to jump to the conduction band [11]. The presence of copper doped would help facilitate the transferred of electrons from the dye (*LUMO*) to the conduction band (*CB*). From the results could be seen that the energy gap of TiO_2 decreased after was made Cu/TiO_2 nanocomposite. It was from 3.64 *eV* to 3.61 *eV*. Energy gap decreased of 0.03 *eV* would show that copper had been successfully inserted between the TiO_2 nanoparticles. Could also be said that the fabrication of Cu/TiO_2 nanocomposite serves to trap electrons from the dye to be forwarded to the FTO glass.

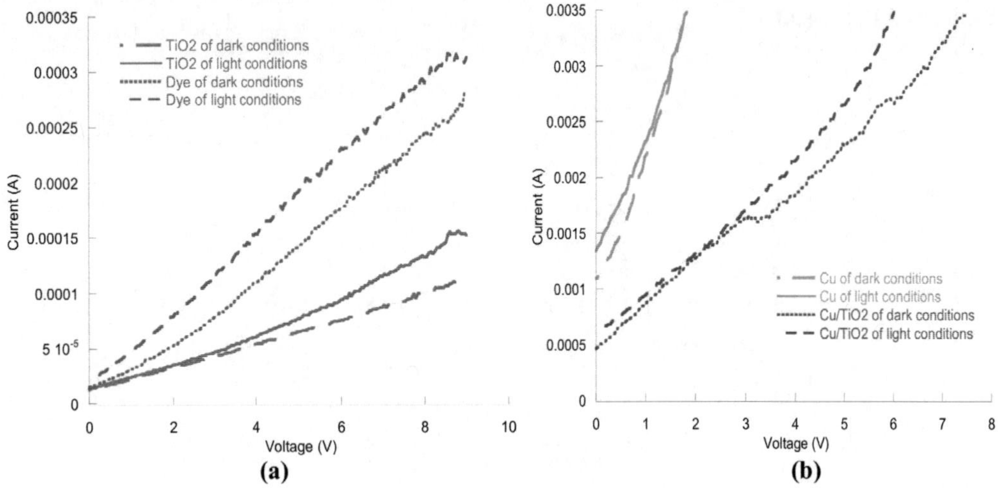

Figure 2. Electrical Characteristics for; (a) TiO_2 and Dye, (b) Cu and Cu/TiO_2 Nanocomposite

Table 2. Electrical conductivity of Cu/TiO_2 nanocomposite

Material	Conductivity of Dark Condition ($\Omega^{-1}m^{-1}$)	Conductivity of light Condition ($\Omega^{-1}m^{-1}$)	Δ Conductivity ($\Omega^{-1}m^{-1}$)
Dye	0.13 x 10^{-2}	0.18 x 10^{-2}	0.05 x 10^{-2}
TiO_2	0.07 x 10^{-2}	0.08 x 10^{-2}	0.01 x 10^{-2}
Cu	16.21 x 10^{-2}	18.99 x 10^{-2}	2.78 x 10^{-2}
Cu/TiO_2	2.93 x 10^{-2}	3.71 x 10^{-2}	0.78 x 10^{-2}

Figure 2 showed the electrical characteristics of the materials in the dark condition or light condition. From the graph showed that all the materials had the ability to conduct electricity in the dark condition or light condition. Figure 2 and Table 2 showed that there were an increased in conductivity from the dark condition to light condition for all materials. The highest conductivity in the condition of light was copper of 18.99 x 10^{-2} $\Omega^{-1}m^{-1}$. This showed that the copper had good conductivity properties, so the right to insert in the TiO_2 nanoparticles. The conductivity of the Cu/TiO_2 nanocomposite increased by 3.63 x 10^{-2} $\Omega^{-1}m^{-1}$ from the conductivity of the TiO_2 was 0.08 x 10^{-2} $\Omega^{-1}m^{-1}$. This showed that the Cu/TiO_2 nanocomposite had a good ability to conduct electricity. So it could help increased the performance of DSSC.

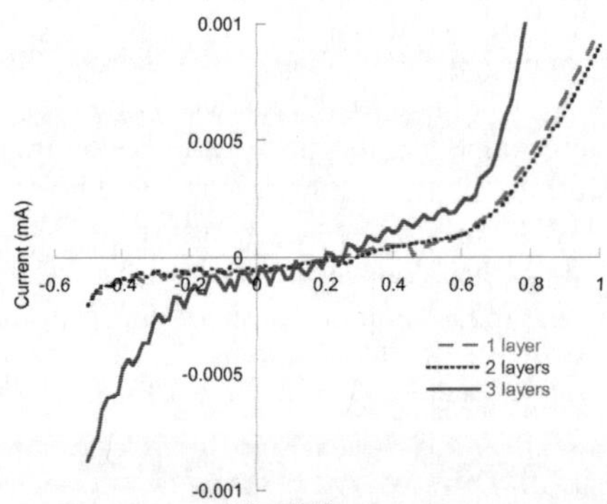

Figure 3. *I-V* characteristics from DSSC for 1 layer, 2 layers and 3 layers of Cu/TiO_2 nanocomposite

Tabel 3. *I-V* characteristics from DSSC

Various layers of Cu/TiO_2 nanocomposite	*Voc* *(mV)*	*Isc* *(mA)*	*FF*	η *(%)*
1 layer	2.35×10^{-1}	4.77×10^{-5}	4.58×10^{-1}	0.005
2 layers	2.95×10^{-1}	5.83×10^{-5}	4.99×10^{-1}	0.008
3 layers	2.04×10^{-1}	4.23×10^{-5}	12.93×10^{-1}	0.011

Characteristics of *I-V* could be seen in Figure 3 and Table 3. From the graph showed that there was an increased the area formed from 1 layer to 3 layers. Efficiency from 1 layer of Cu/TiO_2 nanocomposite was 0.005% , 2 layers of Cu/TiO_2 nanocomposite was 0.008% and 3 layers of Cu/TiO_2 nanocomposite was 0.011%. Copper doped TiO_2 nanoparticles had been increased the performance of DSSC [5,9,11]. Efficiency increased from 1 layer until 3 layers of Cu/TiO_2 nanocomposite. The highest efficiency was the 3 layers of Cu/TiO_2 nanocomposite of 0.011%. Results efficiency explained that there was good contact between the copper and TiO_2 nanoparticles in the fabrication of nanocomposite.

4. Conclusion

Fabrication of nanocomposite with TiO_2 nanoparticles was inserted by copper had been successfully performed. Copper could occupy the space between the TiO_2 particles very well. It was seen from the results of the optical characterization of Cu/TiO_2 nanocomposite had the absorbance peak at 345 nm and decreased energy gap until 3.60 *eV*. The results electrical characteristics of Cu/TiO_2 nanocomposite had a conductivity value of 3.71×10^{-2} ($\Omega^{-1}m^{-1}$) and the highest efficiency in the 3 layers of Cu/TiO_2 nanocomposite of 0.011%. Efficiency from 1 layer of Cu/TiO_2 nanocomposite was 0.005%, 2 layers of Cu/TiO_2 nanocomposite was 0.008% and 3 layers of Cu/TiO_2 nanocomposite was 0.011%.

References

[1] Yasuo C, Ashraful I, Yuki W, Ryoichi K, Naoki K and Liyuan H *2006 JJAP* 45
[2] Brian O R and Michael G 1991 *Nature* 353
[3] Mori S and Yanagida S 2006 *Nanostructured Materials for Solar Energy Conversion* 565-871
[4] Ansari Z A, Umar A, fouad H and Ansari S G 2015 *Nanoelectronics and Optoelectronics* 10
[5] Shrul S, Rita P, Marina I H, Pepen A, Khairurrijal and Mikrajuddin A 2011 *IJBAS-IJENS* 11
[6] Subas M, Onkar G, Vivek D, K Vijayamohanan, K A Bogle, N Valanoor and Satishchandra B O 2012 *Solar Energy* 86
[7] Fadli U M, Prasa A B, Cari and Supriyanto A 2016 *International Conference on Physics and Its Applications (ICOPIA)* Agustus
[8] Kristian HS and Retno DS 2010 *Kimia Anorganik Logam* Yogyakarta: Graha Ilmu
[9] Rita P, Hadriansyah A and Marfuatun 2013 *Jurnal Fisika* 3
[10] R Syafinar, N Gomesh, M Irwanto, M Fareq and YM Irwan 2015 *Energy Procedia* 79
[11] C C Vidyasagar, Y A Naik, T G Venkatesha and R Viswanatha 2012 *International Journal of Nanomaterials and Biostructures* 2(3) 22773851

Shannon Information Entropy for Modified Anisotropic Non-quadratic Potential in Axial Part for Cylindrical Coordinates System

Miftahul Ma'arif*, Suparmi, Cari
Physics Department, Faculty of Mathematics and Natural Sciences,
Sebelas Maret University
Jl. Ir. Sutami 36A Kentingan Jebres Surakarta 57126, Indonesia

E-mail: miftahulmaarif15@student.uns.ac.id

Abstract. Entropy information in quantum mechanics is important to get the quantify of quantum entanglement and correlations. Modified anisotropic non-quadratic potential in Schrodinger equation which has been solved before, and now the solution of wave function in axial part will be used to get its Shannon information entropy. Numerical results are investigated to get the position S_z and the momentum S_p Shannon information entropy.

1. Introduction

Developing correlation measures is very important things to increase our understanding of correlation effects. Recently, information measures has been considerable interest with the applications in atomic and molecular structures. Entropic measures are used to find quantum correlations and entanglement. The development in quantum information, quantum computation, quantum cryptography and quantum teleportation are interest thins which quantum entanglement plays a central roles in it [1-4]. In 1948, Claude E. Shannon had introduced the entropy to measures the uncertainty and formalized the idea of assigning probabilities to the outcome of uncertain events [5]. In 1975, Beckner, Bialynicki-Birula and Mycieslki presented an entropic uncertainty relation as given by following equation [6],

$$S_x + S_p \geq D(1+\ln\pi) \tag{1}$$

where D represents the spatial dimension. In one dimensional coordinate system, the position-space (S_x) and momentum-space (S_p) information entropies are defined by

$$S_x = -\int_{-\infty}^{\infty} |\psi(x)|^2 \ln|\psi(x)|^2 \, dx \tag{2}$$

$$S_p = -\int_{-\infty}^{\infty} |\phi(p)|^2 \ln|\phi(p)|^2 \, dp \tag{3}$$

where $\psi(x)$ is normalized eigenfunction and $\phi(p)$ is normalized Fourier transform. Interval of integration for the variable x depend on the concrete quantum system [7].

The Shannon entropy information for a few molecular potential has been obtained, i.e., harmonic oscillator [8], Pöschl-Teller [9, 10], Morse [9, 11], Coulomb [12], isospectral to the Pöschl-Teller [13], classical orthogonal polynomials [14], infinite circular well [7], and the other potentials and studies[15-17]. In this paper, the modified anisotropic non-quadratic potential which has been got the analytical solution by us before [18]. Now, we will review the solution of this potential before especially in the axial part that has been modified in exponential form and tried to get the Shannon entropy information in position-space, momentum-space, total entropy information and its correlation with uncertainty that presented by BBM.

2. Review to The Solution of Axial Parts

Modified anisotropic non-quadratic potential is expressed as

$$V(\bar{r}) = V(r,\theta,z) = -\frac{V_0}{\sqrt{x^2+y^2}} + \frac{V_1}{x^2} + \frac{V_2 y}{x(x^2+y^2)} + \frac{V_3 e^{-\alpha z}}{(1-e^{-\alpha z})^2} + V_4\frac{1+e^{-\alpha z}}{1-e^{-\alpha z}} \tag{4}$$

We are only interested in the axial part because it is the modified object. By using cylindrical coordinates system so the axial part in Schrodinger equation can be separated directly, and the exponential form must be simplified by change it in to hyperbolic form and each potential constants has been changed , so the Schrodinger equation for axial part can be written as follows

$$-\frac{\hbar^2}{2m}\frac{\partial^2 Z}{\partial z^2} + \frac{\hbar^2}{2m}\left(\gamma^2\frac{\nu(\nu-1)}{\sinh^2\gamma z} - 2q\coth\gamma z\right)\Phi = E'Z \tag{5}$$

The Schrodinger equation that expressed in Equation (5) is solvable by using SUSY QM method and shape invariant approach [19-21], we had got the solutions for Equation (5). The un-normalized ground state wave function for axial part is given by

$$Z_0 = C'\sinh^{\gamma(\nu-1)}\gamma z\, e^{-\frac{q}{\gamma(\nu-1)}z} \tag{6}$$

The un-normalized ground state wave function have to normalized using the following equation

$$\int_{-\infty}^{\infty} Z_0^* Z_0 dz = 1 \tag{7}$$

By substitute Equation (6) in to Equation (7) and solve the integration by using integral tables [22] so we get the constant value is given by

$$C' = 2\gamma\sqrt{\nu-1} \tag{8}$$

The study case in this paper is the system with the quantum number $n = 0$ and a few quantum number ν which would be the independent variable to get its correlation between information entropy.

3. Information Entropy

The normalized ground state wave function can be written as follows

$$Z_0 = 2\gamma\sqrt{\nu-1}\sinh^{\gamma(\nu-1)}\gamma z\, e^{-\frac{q}{\gamma(\nu-1)}z} \tag{9}$$

Axial ground state wave function in Equation (9) for $\nu = 1-3$ are displayed in Figure (1). If the quantum number $\nu = 1$ the eigenstate would be zero it can be proven directly using Equation (9) or we can look at the graph in Figure (1). But it would be different if the quantum number $\nu = 2$ it looks like logarithmic graphic but for $\nu = 3$ becomes like exponential graphic.

The position-space and momentum-space information entropies which has been defined by Equation (2) and (3) can be calculated. However, in arbitrary case we should use these following equations

$$S_z = -\int |\psi(z)|^2 \ln|\psi(z)|^2 d^D z \tag{10}$$

$$S_p = -\int |\phi(p)|^2 \ln|\phi(p)|^2 d^D p \tag{11}$$

IOP Conf. Series: Journal of Physics: Conf. Series **795** (2017) 012030 doi:10.1088/1742-6596/795/1/012030

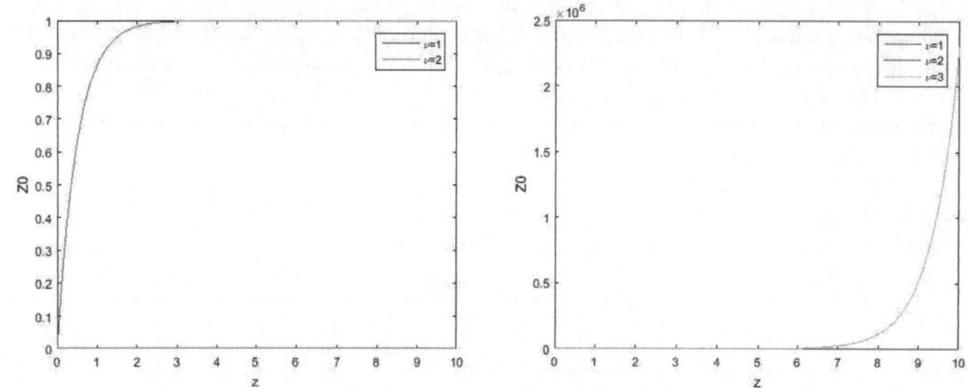

Figure 1. Axial ground state wave function for quantum numbers $\nu = 1-3$.

Before we calculate the momentum-space information entropy, firstly we have to change the the wave function obtained in position-space into momentum-space by using Fourier transform.

$$\phi_0(p) = \frac{1}{2\pi}\int Z_0(z) e^{-ip.z} dz \tag{12}$$

By substituting Equation (9) into Equation (12) so it becomes

$$\begin{aligned}
\phi_0(p) &= \frac{\gamma\sqrt{\nu-1}}{\pi}\int \sinh^{\gamma(\nu-1)} \gamma z\, e^{-\left(\frac{q}{\gamma(\nu-1)}+ip\right)z} dz \\
&= \frac{\gamma\sqrt{\nu-1}}{\pi 2^{\gamma(\nu-1)}}\int \left(e^{\gamma z} - e^{-\gamma z}\right)^{\gamma(\nu-1)} e^{-\left(\frac{q}{\gamma(\nu-1)}+ip\right)z} dz \\
&\simeq \frac{\gamma\sqrt{\nu-1}}{\pi 2^{\gamma(\nu-1)}}\int \left(1-\gamma(\nu-1)e^{-2\gamma z}\right) e^{\left(\gamma^2(\nu-1)-\frac{q}{\gamma(\nu-1)}-ip\right)z} dz \\
&\simeq \frac{\gamma\sqrt{\nu-1}}{\pi 2^{\gamma(\nu-1)}}\left[\frac{\left(e^{\left(\gamma^2(\nu-1)-\frac{q}{\gamma(\nu-1)}-ip\right)z}-1\right)}{\gamma^2(\nu-1)-\frac{q}{\gamma(\nu-1)}-ip} - \frac{\gamma(\nu-1)\left(e^{\left(\gamma^2(\nu-1)-\frac{q}{\gamma(\nu-1)}-ip-2\gamma\right)z}-1\right)}{\gamma^2(\nu-1)-\frac{q}{\gamma(\nu-1)}-ip-2\gamma}\right]
\end{aligned} \tag{13}$$

The third process to get Equation (13) can be obtained by using binomial rules approximation. The momentum-space is displayed in Figure (2).

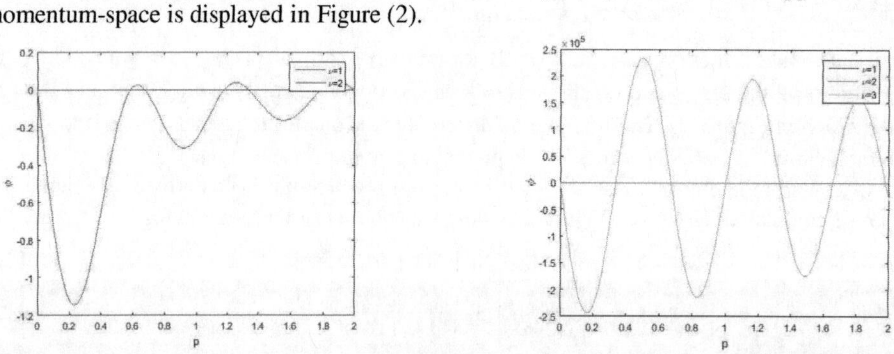

Figure 2. Ground state wave functions in momentum-space for different quantum numbers $\nu = 1-3$.

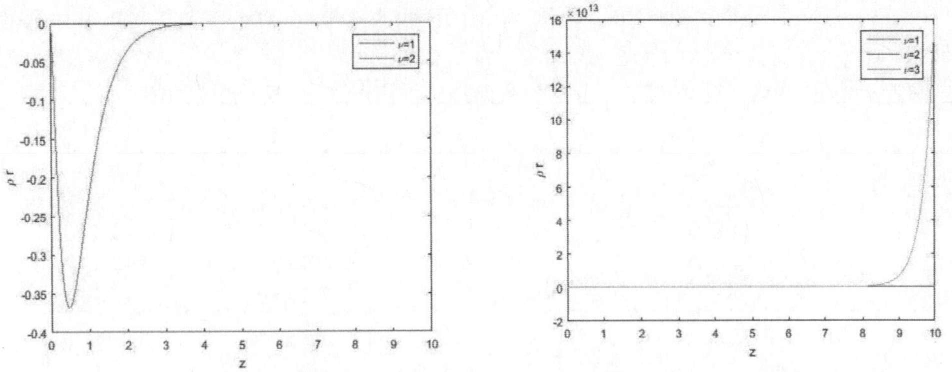

Figure 3. Position entropy densities as a function of the position z .

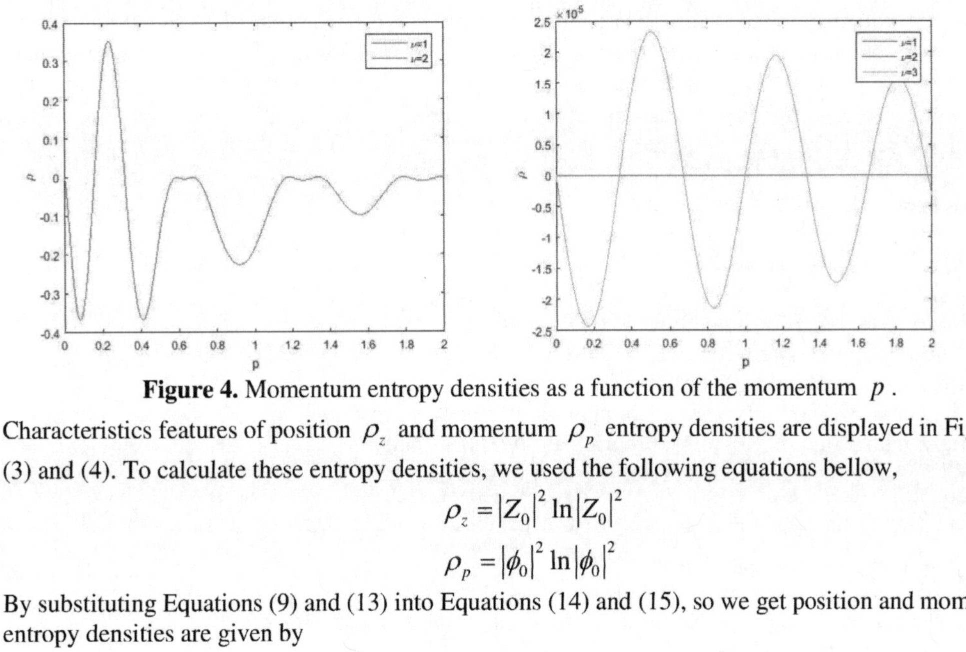

Figure 4. Momentum entropy densities as a function of the momentum p .

Characteristics features of position ρ_z and momentum ρ_p entropy densities are displayed in Figure (3) and (4). To calculate these entropy densities, we used the following equations bellow,

$$\rho_z = |Z_0|^2 \ln |Z_0|^2 \tag{14}$$

$$\rho_p = |\phi_0|^2 \ln |\phi_0|^2 \tag{15}$$

By substituting Equations (9) and (13) into Equations (14) and (15), so we get position and momentum entropy densities are given by

$$\rho_z = 4\gamma^2(\nu-1)\sinh^{2\gamma(\nu-1)}\gamma z e^{-\frac{2q}{\gamma(\nu-1)}z} \ln 4\gamma^2(\nu-1)\sinh^{2\gamma(\nu-1)}\gamma z e^{-\frac{2q}{\gamma(\nu-1)}z} \tag{16}$$

$$\rho_p = \frac{\gamma^2(\nu-1)}{\pi^2 2^{2\gamma(\nu-1)}}\left[\frac{\left(e^{\left(\gamma^2(\nu-1)-\frac{q}{\gamma(\nu-1)}-ip\right)z}-1\right)}{\gamma^2(\nu-1)-\frac{q}{\gamma(\nu-1)}-ip}-\frac{\gamma(\nu-1)\left(e^{\left(\gamma^2(\nu-1)-\frac{q}{\gamma(\nu-1)}-ip-2\gamma\right)z}-1\right)}{\gamma^2(\nu-1)-\frac{q}{\gamma(\nu-1)}-ip-2\gamma}\right]^2 \ln\frac{\gamma^2(\nu-1)}{\pi^2 2^{2\gamma(\nu-1)}}\left[\frac{\left(e^{\left(\gamma^2(\nu-1)-\frac{q}{\gamma(\nu-1)}-ip\right)z}-1\right)}{\gamma^2(\nu-1)-\frac{q}{\gamma(\nu-1)}-ip}-\frac{\gamma(\nu-1)\left(e^{\left(\gamma^2(\nu-1)-\frac{q}{\gamma(\nu-1)}-ip-2\gamma\right)z}-1\right)}{\gamma^2(\nu-1)-\frac{q}{\gamma(\nu-1)}-ip-2\gamma}\right]^2 \tag{17}$$

In Table (1) we present position-space, momentum-space and total Shannon information entropies, it is obtained by using numerical analysis using trapezoidal integration method with 100 times iterations. We just obtained the Shannon information entropies for quantum number $\nu = 2-6$. The higher quantum number ν for S_p couldn't be obtained because the output value is too high and time-consuming in the calculations.

Table 1. Information entropies for $n = 0$ states with respect to different quantum number ν.

ν	S_z	S_p	$S_z + S_p$	$D(1+\ln\pi)$
2	-0.1038	0.0300	-0.0738	4.28946
3	241.9233	1.6344e+10	1.63E+10	4.28946
4	1.00E+04	3.3904e+19	3.39E+19	4.28946
5	2.48E+05	7.4903e+27	7.49E+27	4.28946
6	4.96E+06	2.9664e+35	2.97E+39	4.28946

4. Conclusions

The position S_z and the momentum S_p Shannon information entropies for modified anisotropic nonquadratic potential in axial parts has been investigated. We have taken the ground state wave function with some different value of quantum number ν. The interesting features like position and momentum entropy densities for different quantum numbers $\nu = 1-3$ has been displayed and it has different properties. Next study of Shannon information entropy for this case is needed with the higher quantum number n.

5. References

[1] Y.-C. Lin, C.-Y. Lin, Y.K. Ho, Phys, Rev.A 87 (2013) 022316.
[2] Y.-C. Lin, Y.K Ho, Can. J. Phys. 93 (2015) 646.
[3] C.H. Lin, Y.-C. Lin, Y.K. Ho, Few-Body Syst. 54 (2013) 2147.
[4] C.H. Lin, Y.K. Ho, Few-Body Syst. 55 (2014) 1141.
[5] C.E. Shannon, Bell Syst. Tech. J. 27 (1948) 623.
[6] W. Beckner, Ann. Math. 102 (1975) 159;
I. Bialynicki-Birula, J. Mycielski, Commun. Math. Phys. 44 (1975) 129;
I. Bialynicki-Birula, L. Kudnicki, preprint, arXiv:1001.4668.
[7] Xu-Dong Song, Gua-Ho Sun, Shi-Hai Dong, J. Phys. A. (2015).
[8] V. Marjenik, T. Opartny, J. Phys. A. 29 (1996) 2187.
[9] J.S. Dehesa, A. Martinez-Finkelshtein, V.N. Sorokin, Mol. Phys. 104 (2006) 613.
[10] M.W. Coffey, Can. J. Phys. 85 (2007) 733.
[11] E. Aydiner, C. Orta, R. Sever, Int. J. Mod. Phys. B 22 (2008) 231.
[12] J.S. Dehesa, W. Van Assche, R.J. Yanez, Methods Appl. Math. 4 (1997) 91.
[13] A. Kumar, Indian J. Pure Appl. Phys. 43 (2005) 958.
[14] J.S. Dehesa, R.J. Yanez, A.I. Aptekarev, V. Buyarov, J. Math. Phys. 39 (1998) 3050;
V.S. Buyarov, J.S. Dehesa, A. Martinez-Finkelshtein, J. Approx. Theory 99 (1999) 153;
V.S. Buyarov, P. Lopez-Artes, A. Martinez-Finkelshtein, W. Van Assche, J. Phys. A 33 (2000) 6549.
[15] J. Katriel, K.D. Sen, J. Comput. Appl. Math. 233 (2010) 1399.
[16] S.H. Patil, K.D. Sen, Int. J. Quant. Chem. 107 (2007) 1864.
[17] K.D. Sen, J. Chem. Phys. 123 (2005) 074110.
[18] Miftahul Ma'arif, Suparmi, Cari, IOP J. Phy. Conf. Series. (2016).
[19] C. Cari, A. Suparmi, AIP Conf. Proc. 101 (2004) 1615.
[20] A Inomata, J Junker, A Suparmi, J. Phys. A : Math. Gen. 26 2261.
[21] C Cari et al 2016 J. Phys.: Conf. Ser. 739 012023.
[22] I.S. Gradshteyn, I. M. Ryzhik, Ac. Press Elsevier (2007) ISBN-13: 978-0-12-373637-6.

Solution of generalized Dirac equation for Eckart plus Hulthen central potential combined with modified Hylleraas like tensor potential using Romanovski polynomial method

Dyah Ayu Dianawati, A.Suparmi, and C.Cari

Physics Departement Sebelas Maret University,
Jl. Ir. Sutami 36A Kentingan Jebres Surakarta 57126, INDONESIA

E-mail: dyahayu@student.uns.ac.id

Abstract. Romanovski Polynomial method was used to analyze of Dirac equation with Eckart plus Hulthen central potential combined with modified Hylleraas like tensor potential on spin symmetry case. By using parameter and variable substitution, the Dirac equation was reduced into one dimensional Schrodinger like equation with centrifugal approximation that can be solved using Romanovski polynomial. The relativistic energy spectra is obtained from the relativistic energy equation by using Matlab R2008b software. The relativistic energy with quantum number and spin orbit quantum number variations were obtained. By using matlab R2008b software, visualisasion of the relativistic energy spectra with quantum number variasion were obtained.

1. INTRODUCTION

Classical physics is based on Newton's law. [1] A series of experiment had been done to prove that the porpose of Newton's physics which can not explain some microscopics behaviors and only explain macroscopic materials behavior. So it need to bring new theory in physics, that was the theory of quantum. Quantum mechanics is branch of physics which explains behavior of microscopics materials and its intraction with the energy in atomic scale and subatomics particles. [2]

In quantum mechanics, the Dirac equation is used to explain dynamic particle with spin $^1/_2$. [3] In Dirac equation consists of spin and pseudospin symmetris cases. Spin symmetry case happens when a scalar potential $S(r)$ is equal to vector potential $V(r)$ and for pseudospin symmetry case (p-spin) happens when $S(r) = -V(r)$. [4] The concept of spin symmetry is used to study the change of core in nuclear physics, that is applicable at meson and antinucleon spectrum. [5]

Some methods used to solve the Dirac equation, one of which is Romanovski polynomial method. The Romanovski polynomial method suggested by S.J.Routh in 1884, and after 45 years, 1929, developed by V.I.Romanovski. So that, this polynomial equation is familiar with "*Romanovski*". [6] Romanovski polynomial method is traditional method consisting differentials Schrodinger equation by variable change from general equation hypergeometric. [7,8]

The aim of the research is to calculate relativistic energies spectra with Matlab R2008b software from relativistic energy equation.

In this paper, there several sections the introduction with the basic theory of Dirac equation is presented in section 1, method of research that review the Romanovski polynomial in section 2, result and discussion in section 3, and conclusion in section 4.

2. BASIC THEORY OF DIRAC EQUATION

Particle in Dirac equation has spin $^1/_2$ which moved to scalar potential $S(r)$, vector potential $V(r)$, and tensor potential $U(r)$ on relativity unit $\hbar = c = 1$ is [9] :

$$[\overline{\boldsymbol{\alpha}}.\overrightarrow{\boldsymbol{p}} + \beta(M + S(r)) - i\beta\overline{\boldsymbol{\alpha}}.\hat{\boldsymbol{r}}U(r)]\psi(\vec{r}) = [E - V(r)]\psi(\vec{r}) \quad (1)$$

where E is the relativity energy system, $\overrightarrow{\boldsymbol{p}} = -i\boldsymbol{\nabla}$ is the three dimensional momentum operator, and M is mass of particle. Futher, $\overline{\boldsymbol{\alpha}}\beta$ is matrices element Dirac given by :

$$\overline{\boldsymbol{\alpha}} = \begin{pmatrix} 0 & \boldsymbol{\sigma} \\ \boldsymbol{\sigma} & 0 \end{pmatrix} \quad \text{and} \quad \beta = \begin{pmatrix} I & 0 \\ 0 & -I \end{pmatrix} \quad (2)$$

where σ is matrices vector Pauli and I is 2×2 identity matrices. [10]

The total angular momentum of operator $\vec{j}$ and spin orbit $K = (\vec{\sigma}.\vec{L} + 1)$, where $\vec{L}$ is orbital angular momentum. The eigenvalues of spin orbit coupling operator are $\kappa = \left(j + \frac{1}{2}\right) > 0$ and $\kappa = -\left(j + \frac{1}{2}\right) < 0$. The unaligned spin $j = l - \frac{1}{2}$ and the aligned spin $j = l + \frac{1}{2}$. [11]

The spinnor Dirac equation for radial quantum number and spin orbit number is [9] :

$$\psi_{n\kappa}(r) = \begin{pmatrix} f_{n\kappa}(\boldsymbol{r}) \\ g_{n\kappa}(\boldsymbol{r}) \end{pmatrix} = \frac{1}{r}\begin{pmatrix} F_{n\kappa}(r)Y_{jm}^{l}(\theta,\varphi) \\ iG_{n\kappa}(r)Y_{jm}^{\tilde{l}}(\theta,\varphi) \end{pmatrix} \quad (3)$$

where the component of Dirac equation for spin symmetries are $F_{n\kappa}(r)$ for upper component and $G_{n\kappa}(r)$ for lower component. $Y_{jm}^{l}(\theta,\varphi)$ and $Y_{jm}^{\tilde{l}}(\theta,\varphi)$ are spin and pseudospin spherical harmonics, m is the projection of the angular momentum on z axis, and $l(l+1) = \kappa(\kappa+1)$ for spin symmetry, $\tilde{l}(\tilde{l}+1) = \kappa(\kappa-1)$ for pseudospin symmetry.

The Dirac equation consists of spin and pseudospin symmetries cases. For spin symmetry case happens when $\Sigma(r) = V(r) + S(r)$, while for pseudospin symmetry case happens when $\Delta(r) = V(r) - S(r)$. [3,11,12]

From equation (1) until equation (3), we get solution of the wave function radial part upper $(F_{n\kappa}(r))$ and lower $(G_{n\kappa}(r))$ components are :

$$\left(\frac{d}{dr} + \frac{\kappa}{r} - U(r)\right)F_{n\kappa}(r) = (M + E_{n\kappa} - \Delta(r))G_{n\kappa}(r) \quad (4)$$

$$\left(\frac{d}{dr} - \frac{\kappa}{r} + U(r)\right)G_{n\kappa}(r) = (M - E_{n\kappa} + \Sigma(r))F_{n\kappa}(r) \quad (5)$$

Substitution equations (4) and (5), we get two dimensional Schrodinger like equation. [3,10,12]

$$\left\{\frac{d^2}{dr^2} - \frac{\kappa(\kappa+1)}{r^2} + \frac{2\kappa U(r)}{r} - \frac{dU(r)}{dr} - U^2(r) - [M + E_{n\kappa} - \Delta(r)][M - E_{n\kappa} + \Sigma(r)] + \frac{\frac{d\Delta(r)}{dr}\left(\frac{d}{dr} + \frac{\kappa}{r} - U(r)\right)}{M + E_{n\kappa} - \Delta(r)}\right\}F_{n\kappa}(r) = 0 \quad (6)$$

Equation (6) is the Dirac equation for spin symmetry case or upper commponent of wave function, with $\kappa(\kappa+1) = l(l+1)$.

Like in equation (3), the wave function of radial part of Dirac equation $F_{n\kappa}(r)$ for spin up is [13] :

$$\frac{\left(\frac{d}{dr} - \frac{\kappa}{r} + U(r)\right)}{(M - E_{n\kappa} + \Sigma(r))}G_{n\kappa}(r) = F_{n\kappa}(r) \quad (7)$$

The dirac equation for spin symetry case happens when $\Sigma(r) = V(r) + S(r) = C_s$ =constant with $\frac{d\Delta}{dr} = 0$. So that, equation (6) become [9] :

$$\left\{\frac{d^2}{dr^2} - \frac{\kappa(\kappa+1)}{r^2} + \frac{2\kappa U(r)}{r} - \frac{dU(r)}{dr} - U^2(r) - [M + E_{n\kappa} - \Delta(r)][M - E_{n\kappa} + \Sigma(r)]\right\}F_{n\kappa}(r) = 0 \quad (8)$$

IOP Conf. Series: Journal of Physics: Conf. Series **795** (2017) 012031 doi:10.1088/1742-6596/795/1/012031

3. METHOD

In this paper, Romanovski polynomial method which used to solved the Dirac equation in the spin symmetry case. The Dirac equation was reduced into one dimensional schrodinger like equation. One dimensional Schrodinger equation for the potential was reduced into differential equation of Romanovski polynomial by approximate variable and wave function substitution.

Generalized hypergeometric equation [7] :

$$\sigma(s)\frac{d^2y_n}{ds^2}+\tau(s)\frac{dy_n(s)}{ds}+\lambda_n y_n(s)=0 \quad (9)$$

with :

$$\sigma(s)=as^2+bs+c,\ \tau(s)=ds+e,\text{ and }\lambda_n=-\{n(n-1)+2n(1-p)\} \quad (10)$$

The Pearson differential equation :

$$\frac{d(\sigma(s)w(s))}{ds}=\sigma(s)w(s) \quad (11)$$

Weight function ($w(s)$) obtained from the solution of the Pearson differential equation :

$$w(s)=\exp(\int\frac{(d-2a)s+(e-b)}{as^2+bs+c}ds\ ) \quad (12)$$

Wave function in equation (12) from Rodrigues representation is :

$$y_n=\frac{1}{w(s)}\frac{d^n}{ds^n}\left((as^2+bs+c)w(s)\right) \quad (13)$$

The value of parameter in equation (13) is :

$$a=1,b=0,\ c=1,d=2(1-p),\text{ and }\ e=q\text{ with }\ p>0 \quad (14)$$

Substituion equation (14) to equation (12), we get weight function :

$$w(s)=(1+s^2)^{-p}e^{q\tan^{-1}(s)} \quad (15)$$

The equation can be written :

$$\int_{-\infty}^{\infty}w^{(p,q)}R_m{}^{(p,q)}(s)R_{m`}{}^{(p,q)}(s)ds \quad (16)$$

can be convergen, when :

$$m`+m<2p-1 \quad (17)$$

The differential equation of Romanovski polynomial obtained with substitution equations (15) and (17) to equation (16) :

$$(1+s^2)\frac{\partial^2 R_n{}^{(p,q)}(s)}{\partial s^2}+\{2s(-p+1)+q\}\frac{\partial R_n{}^{(p,q)}(s)}{\partial s}-\{n(n-1)+2n(1-p)\}R_n{}^{(p,q)}(s) \quad (18)$$

with $y_n=R_n{}^{(p,q)}(s)$.

When $r=f(s)$, eigen function in Romanovski polynomial is :

$$\Psi(r)=g_n(s)=(1+s^2)^{\frac{\beta}{2}}e^{\frac{-\alpha}{2}\tan^{-1}(s)}D_n{}^{(\beta,\alpha)}(s) \quad (19)$$

with :

$$D_n{}^{(\beta,\alpha)}(s)=R_n{}^{(p,q)}(s) \quad (20)$$

Romanovski polynomial from Rodrigues formula with weight function in equation (15) is :

$$D_n{}^{(\beta,\alpha)}(s)=R_n{}^{(p,q)}(s)=\frac{1}{(1+s^2)^{-p}e^{q\tan^{-1}(s)}}\frac{d^n}{ds^n}(1+s^2)^n\,(1+s^2)^{-p}e^{q\tan^{-1}(s)} \quad (21)$$

The wave function in equation (19) become :

$$\Psi_n(r)=\frac{1}{\sqrt{\frac{df(s)}{ds}}}(1+s^2)^{\frac{-p}{2}}e^{\frac{q}{2}\tan^{-1}(s)}R_n{}^{(p,q)}(s) \quad (22)$$

The orthogonal integral of Romanovski polynomial is :

$$\int_0^{\infty}\Psi_n(r)\Psi_{n`}(r)dr=\int_{-\infty}^{\infty}w^{(p,q)}R_n{}^{(p,q)}(s)R_{n`}{}^{(p,q)}(s)ds \quad (23)$$

In this case, the value of p and q do not depend on polynomial degree n.

4. RESULT AND DISSCUSION

The equation of Eckart plus Hulthen central potential combined with modified Hylleraas like tensor potential, can be written [14,15]:

$$V(r) = -V_0\frac{e^{-2\alpha r}}{1-e^{-2\alpha r}} + 4V_1\frac{e^{-2\alpha r}}{(1-e^{-2\alpha r})^2} - V_2\frac{1+e^{-2\alpha r}}{1-e^{-2\alpha r}} + V_3\frac{(a+e^{(-2\alpha r)})}{(e^{(-2\alpha r)}-1)} \quad (24)$$

And the equation of Eckart plus Hulthen central potential combined with modified Hylleraas like tensor potential in hyperbolic function is :

$$V(r) = -V_0\frac{\cosh(\alpha r)-\sinh(\alpha r)}{2\sinh(\alpha r)} + 4V_1\,\mathrm{cosech}^2(\alpha r) - V_2\coth(\alpha r) - V_3\frac{a(\cosh(\alpha r)-\sinh(\alpha r))}{2\sinh(\alpha r)} - V_3\frac{(\cosh(\alpha r)-\sinh(\alpha r))}{2\sinh(\alpha r)} \quad (25)$$

Dirac equation for spin symmetry case contained in equation (8), can be written :

$$\left(\frac{d^2}{dr^2} - \frac{\kappa(\kappa+1)}{r^2} + \frac{2\kappa U(r)}{r} - \frac{dU(r)}{dr} - U^2(r) - (M+E_{n\kappa}-C_s)\big(M-E_{n\kappa}+\Sigma(r)\big)\right)F_{n\kappa}(r) = 0 \quad (26)$$

Then, substitution equation (10) to equation (26) :

$$\Bigg(\frac{d^2}{dr^2} - \frac{\kappa(\kappa+1)}{r^2} - \frac{a\kappa}{r}V_3\coth(\alpha r) - \frac{a\kappa}{r}V_3 - \frac{\kappa}{r}V_3\coth(\alpha r) + \frac{\kappa}{r}V_3 - \frac{\alpha V_3}{2\sinh^2(\alpha r)}(1+a) - \frac{{V_3}^2a^2}{4\sinh^2(\alpha r)} - \frac{{V_3}^2a^2}{2}\coth(\alpha r) - \frac{{V_3}^2a^2}{2}\coth^2(\alpha r) + \frac{{V_3}^2a^2}{2} + \frac{{V_3}^2a^2}{4\sinh^2(\alpha r)} + \frac{{V_3}^2a^2}{2}\coth(\alpha r) + {E_{n\kappa}}^2 - M^2 + \frac{V_0}{2}\coth(\alpha r)(M+E_{n\kappa}-C_s) - \frac{V_0}{2}(M+E_{n\kappa}-C_s) - 4V_1\,cosech^2(\alpha r)(M+E_{n\kappa}-C_s) + V_2\coth(\alpha r)(M+E_{n\kappa}-C_s) + C_s(M-E_{n\kappa})\Bigg)F_{n\kappa}(r) = 0 \quad (27)$$

with variable approach was contained in equation (27), where $ra << 1$, so that with centrifugal approximation $\frac{1}{r^2}$, is :

$$\frac{1}{r^2} = \frac{a^2}{sinh^2(\alpha r)} \quad (28)$$

Substitution equation (28) to equation (27), we get :

$$\left(\frac{d^2}{dr^2} - \left(\begin{matrix}\kappa(\kappa+1) + \frac{4V_1}{a^2}(M+E_{n\kappa}-C_s) + \\ \frac{\alpha V_3(1+a)}{2a^2} - \frac{\kappa V_3}{a}\sinh(\alpha r) + \kappa V_3\sinh(\alpha r) + \kappa V_3\cosh(\alpha r) + \frac{\kappa V_3}{a}\cosh(\alpha r)\end{matrix}\right)\frac{a^2}{\sinh^2(\alpha r)} - \left(\frac{{V_3}^2a^2}{2}\coth(\alpha r) - \left(\frac{V_0}{2}+V_2\right)(M+E_{n\kappa}-C_s)\right)\coth(\alpha r)\right)F_{n\kappa}(r) = \left(M^2 - {E_{n\kappa}}^2 + \frac{V_0}{2}(M+E_{n\kappa}-C_s) - C_s(M-E_{n\kappa}) - \frac{{V_3}^2a^2}{2}\right)F_{n\kappa}(r) \quad (29)$$

In equation (29) was reduced into one dimensional Schrodinger equation, is :

$$\left(\frac{d^2}{dr^2} - \frac{a^2A_s}{\sinh^2(\alpha r)} - a^2A_s\coth(\alpha r)\right)F_{n\kappa}(r) = -a^2E'_sF_{n\kappa}(r) \quad (30)$$

where,

$$A_s = \left(\kappa(\kappa+1) + \frac{4V_1}{a^2}(M+E_{n\kappa}-C_s) + \frac{\alpha V_3(1+a)}{2a^2} - \frac{\kappa V_3}{a}\sinh(\alpha r) + \kappa V_3\sinh(\alpha r) + \kappa V_3\cosh(\alpha r) + \frac{\kappa V_3}{a}\cosh(\alpha r)\right) \quad (31)$$

$$B_s = \frac{\left(\frac{{V_3}^2a^2}{2}\coth(\alpha r) - \left(\frac{V_0}{2}+V_2\right)(M+E_{n\kappa}-C_s)\right)}{a^2} \quad (32)$$

$$E'_s = -\frac{\left(M^2 - {E_{n\kappa}}^2 + \frac{V_0}{2}(M+E_{n\kappa}-C_s) - C_s(M-E_{n\kappa}) - \frac{{V_3}^2a^2}{2}\right)}{a^2} \quad (33)$$

Then, in equation (30) with approximation variable $\cot(\alpha r) = ix$, where $-\infty < x < \infty$, we obtained :

$$\left((x^2+1)\frac{d^2}{dx^2}F_{n\kappa}(r) + 2x\frac{d}{dx}F_{n\kappa}(r)\right) - \left(A_s - B_s\frac{\coth(\alpha r)}{(x^2+1)} + \frac{E_s'}{(x^2+1)}\right)F_{n\kappa}(r) = 0 \quad (34)$$

Wave equation in equation (19), when $\Psi(r) = g_n(s) = F_{n\kappa}(r)$, we get :

$$(x^2+1)\frac{d^2}{dx^2}(1+x^2)^{\frac{\beta}{2}}e^{\frac{-\alpha}{2}tan^{-1}(x)}D_n{}^{(\beta,\alpha)}(x) + 2x\frac{d}{dx}(1+x^2)^{\frac{\beta}{2}}e^{\frac{-\alpha}{2}tan^{-1}(x)}D_n{}^{(\beta,\alpha)}(x) -$$

$$\left[A_s - B_s\frac{\coth(\alpha r)}{(x^2+1)} + \frac{E_s'}{(x^2+1)}\right](1+x^2)^{\frac{\beta}{2}}e^{\frac{-\alpha}{2}tan^{-1}(x)}D_n{}^{(\beta,\alpha)}(x) = 0 \quad (35)$$

we obtained :

$$(1+x^2)\frac{d^2D}{dx^2} + [2x(\beta+1)-\alpha]\frac{dD}{dx} - \left[\frac{\alpha\beta x+\beta^2-\frac{\alpha^2}{4}+ixB_s-E_s'}{(1+x^2)} + A_s - \beta^2 - \beta\right]D = 0 \quad (36)$$

In equation (36), if the coeficient $\frac{1}{(1+x^2)} = 0$, we have :

$$-\frac{\alpha^2}{4} + \beta^2 - E_s' = 0 \quad \text{and} \quad \alpha\beta + iB_s = 0 \quad (37)$$

Equation (36) become :

$$(1+x^2)\frac{d^2D}{dx^2} + [2x(\beta+1)-\alpha]\frac{dD}{dx} - [A_s - \beta^2 - \beta]D = 0 \quad (38)$$

With compare paramater between equation (38) and (18), we get relationship equation :

$$A_s - \beta^2 - \beta = n(n-1) + 2n(1-p) \quad (39)$$

$$2x(\beta+1) - \alpha = 2s(-p+1) + q \quad (40)$$

So that, $\alpha = -q$ dan $p = -\beta$, dengan $p > 0$ (41)

Using equation (39) and (41), we get :

$$\beta = \beta_n = \sqrt{A_s + \frac{1}{4}} - n - \frac{1}{2} \quad (42)$$

$$\text{So that, } \beta_n = \sqrt{\begin{pmatrix}\kappa(\kappa+1) + \frac{4V_1}{a^2}(M + E_{n\kappa} - C_s) + \frac{\alpha V_3(1+a)}{2a^2} - \frac{\kappa V_3}{a}\sinh(\alpha r) \\ +\kappa V_3\sinh(\alpha r) + \kappa V_3\cosh(\alpha r) + \frac{\kappa V_3}{a}\cosh(\alpha r) + \frac{1}{4}\end{pmatrix}} - n - \frac{1}{2} \quad (43)$$

Substitution equation (43) to equation (37) :

$$\alpha = \alpha_n = \frac{-iB_s}{\beta} = \frac{iB_s}{\left(-\sqrt{A_s+\frac{1}{4}}+n+\frac{1}{2}\right)} \quad (44)$$

$$\text{So that, } \alpha_n = \frac{iB_s}{-\sqrt{\begin{pmatrix}\kappa(\kappa+1)+\frac{4V_1}{a^2}(M+E_{n\kappa}-C_s)+\frac{\alpha V_3(1+a)}{2a^2}-\frac{\kappa V_3}{a}\sinh(\alpha r) \\ +\kappa V_3\sinh(\alpha r)+\kappa V_3\cosh(\alpha r)+\frac{\kappa V_3}{a}\cosh(\alpha r)+\frac{1}{4}\end{pmatrix}}+n+\frac{1}{2}} \quad (45)$$

Using equations (37) and (44), we get :

$$\beta^4 - E_s'\beta^2 + \frac{B_s{}^2}{4} = 0 \quad (46)$$

The eigen value of energy is : $\sqrt{A_s + \frac{1}{4}} - n - \frac{1}{2} = \frac{E_s' \pm \sqrt{E_s'^2 - B_s{}^2}}{2}$ (47)

Substitution E_s value in equation (33) to equation (47), we get :

$$(M + E_{n\kappa} - C_s) = \frac{a^2\left(4+8n+\frac{2V_3{}^2a^2}{a^2}\right)\left(\sqrt{A_s+\frac{1}{4}}\right)+2nV_3{}^2a^2-4a^2\left(n^2-n-\left(A_s+\frac{1}{4}\right)\right)-B_s{}^2}{\left((4(M-E_{n\kappa})+2V_0)\left(n+\sqrt{A_s+\frac{1}{4}}\right)\right)} \quad (48)$$

Equation (48) is relativistic energy spectrum equation for spin symmetry case from Eckart plus Hulthen central potential combined with modified Hylleraas like tensor potential using Romanovski polynomial method.

In Table 1. Is the relativistic energy spectrum was numerically calculated with the quantum number (n_l) also depends on the relativistic energy value. The greater value of quantum number (n_l) causes the grater value of relativistic energy, and the *state* value based on n, l, and j values, with $j = l + \frac{1}{2}$.

Table 1. Relativistic Energy (fm^{-1}) for Generalized Eckart plus Hulthen Central Potential Combined with Modified Hylleraas Like Tensor Potential on Spin Symmetry Case with $M = 10fm^{-1}, Cs = 5fm^{-1}, a = 0.05\,fm, V_0 = 5.2fm^{-1}, V_1 = 4.6fm^{-1}, V_2 = 4fm^{-1}, V_3 = 3.6fm^{-1}$

l	n	κ	State	$E_{n\kappa}$	l	n	κ	State	$E_{n\kappa}$
4	0	4	$0_{g_{9/2}}$	0.2496920085	4	0	-5	$0_{g_{9/2}}$	0.1629084067
5	0	5	$0_{h_{11/2}}$	0.2795647223	5	0	-6	$0_{h_{11/2}}$	0.1922223671
6	0	6	$0_{i_{13/2}}$	0.3096197128	6	0	-7	$0_{i_{13/2}}$	0.2217360393
7	0	7	$0_{j_{15/2}}$	0.3398579117	7	0	-8	$0_{j_{15/2}}$	0.2514365592
8	0	8	$0_{k_{17/2}}$	0.3702811056	8	0	-9	$0_{k_{17/2}}$	0.2813199462
9	0	9	$0_{l_{19/2}}$	0.4008915843	9	0	-10	$0_{l_{19/2}}$	0.3113856303
10	0	10	$0_{m_{21/2}}$	0.4316919666	10	0	-11	$0_{m_{21/2}}$	0.3416346086
4	1	4	$1_{g_{9/2}}$	0.2759760674	4	1	-5	$1_{g_{9/2}}$	0.1865639858
5	1	5	$1_{h_{11/2}}$	0.3064297333	5	1	-6	$1_{h_{11/2}}$	0.2170002082
6	1	6	$1_{i_{13/2}}$	0.336986563	6	1	-7	$1_{i_{13/2}}$	0.2473694978
7	1	7	$1_{j_{15/2}}$	0.3676700587	7	1	-8	$1_{j_{15/2}}$	0.2777573725
8	1	8	$1_{k_{17/2}}$	0.3984970842	8	1	-9	$1_{k_{17/2}}$	0.3082162636
9	1	9	$1_{l_{19/2}}$	0.4294803804	9	1	-10	$1_{l_{19/2}}$	0.3387799445
10	1	10	$1_{m_{21/2}}$	0.4606300416	10	1	-11	$1_{m_{21/2}}$	0.3694714308
4	2	4	$2_{g_{9/2}}$	0.2912607685	4	2	-5	$2_{g_{9/2}}$	0.1933474663
5	2	5	$2_{h_{11/2}}$	0.3234056545	5	2	-6	$2_{h_{11/2}}$	0.2275888496
6	2	6	$2_{i_{13/2}}$	0.3553604582	6	2	-7	$2_{i_{13/2}}$	0.2606784412
7	2	7	$2_{j_{15/2}}$	0.3872294758	7	2	-8	$2_{j_{15/2}}$	0.2931512027
8	2	8	$2_{k_{17/2}}$	0.4190829154	8	2	-9	$2_{k_{17/2}}$	0.3252811847
9	2	9	$2_{l_{19/2}}$	0.4509703845	9	2	-10	$2_{l_{19/2}}$	0.3572285097
10	2	10	$2_{m_{21/2}}$	0.4829283771	10	2	-11	$2_{m_{21/2}}$	0.3890949277
4	3	4	$3_{g_{9/2}}$	0.2983853231	4	3	-5	$3_{g_{9/2}}$	0.1885580348
5	3	5	$3_{h_{11/2}}$	0.3328501868	5	3	-6	$3_{h_{11/2}}$	0.2282466052
6	3	6	$3_{i_{13/2}}$	0.3667276105	6	3	-7	$3_{i_{13/2}}$	0.2650941292
7	3	7	$3_{j_{15/2}}$	0.4002307369	7	3	-8	$3_{j_{15/2}}$	0.3004253299
8	3	8	$3_{k_{17/2}}$	0.4335004528	8	3	-9	$3_{k_{17/2}}$	0.334847967
9	3	9	$3_{l_{19/2}}$	0.4666349827	9	3	-10	$3_{l_{19/2}}$	0.3686984965
10	3	10	$3_{m_{21/2}}$	0.4997056986	10	3	-11	$3_{m_{21/2}}$	0.4021845862
4	4	4	$4_{g_{9/2}}$	0.2989690072	4	4	-5	$4_{g_{9/2}}$	0.1740761265
5	4	5	$4_{h_{11/2}}$	0.3361901401	5	4	-6	$4_{h_{11/2}}$	0.2209726508
6	4	6	$4_{i_{13/2}}$	0.3723476536	6	4	-7	$4_{i_{13/2}}$	0.2624331502
7	4	7	$4_{j_{15/2}}$	0.4077905075	7	4	-8	$4_{j_{15/2}}$	0.3011871209

8	4	8	$4_{k_{17/2}}$	0.4427444318	8	4	-9	$4_{k_{17/2}}$	0.3383329222
9	4	9	$4_{l_{19/2}}$	0.4773646918	9	4	-10	$4_{l_{19/2}}$	0.3744405252
10	4	10	$4_{m_{21/2}}$	0.5117630624	10	4	-11	$4_{m_{21/2}}$	0.4098493842

The visualisasion of relativistic energy spectra as shown as in Figure 1 :

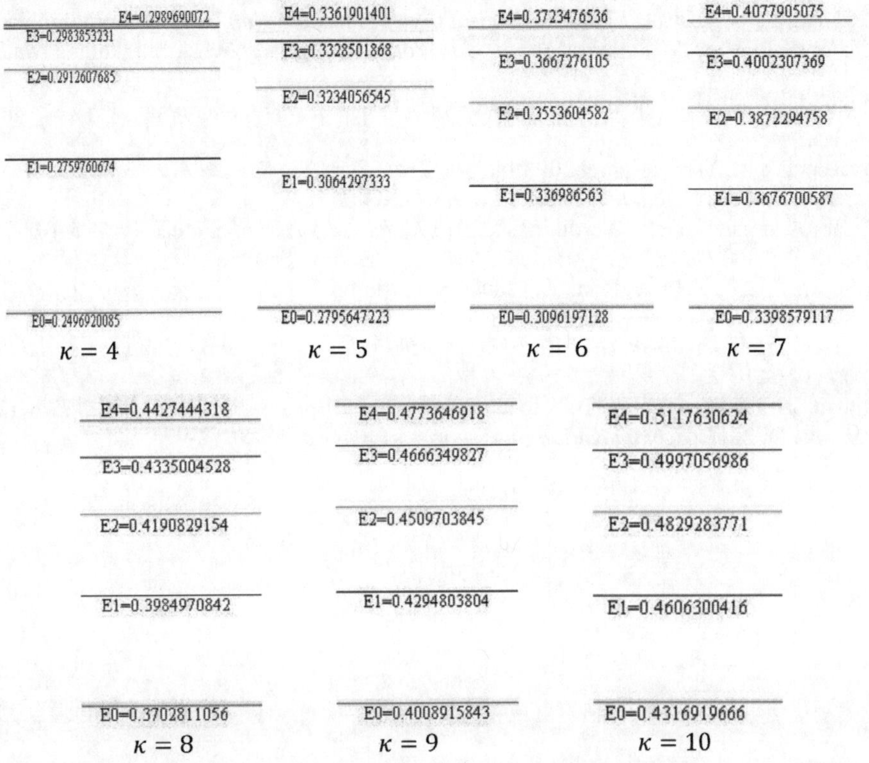

Figure 1. Relativistic Energi Spectra with Variation n, κ

In figure 1. As shown as relativistic energy spectra with variation n value with κ value is constans. For $\kappa = 4$ to $\kappa = 10$ with variation $n = 0$ to $n = 4$

5. CONCLUSION

The Dirac equation for Generalized Eckart plus Hulthen central potential combined with modified Hylleraas like tensor potential on spin symmetry case can be solved using Romanovski polynomial method to calculate and visualisasion relativistic energies spectra with Matlab R2008b software. The relativistic energy analitically value can be solved by centrifugal approximation. The relativistic energy value depend on quantum number (n_l) and the spin orbit quantum number (κ) values, positive. The gap energy for $\kappa = 4$ and $\kappa = 5$ relative small, while the gap energy for $\kappa = 6$ to $\kappa = 10$ relative large.

ACKNOWLEDGEMENT

This research was partly supported by progress Higher Education Project Grant with contract no.632/UN27.21/LT/2016.

REFERENCES

[1] Beiser, A. 1992 *Konsep Fisika Modern*. Jakarta : Erlangga
[2] A.Suparmi 2011 *Mekanika Kuantum I* Surakarta : Jurusan Fisika Fakultas MIPA UNS
[3] A.Suparmi & C.Cari 2013 *Atom Indonesia*. **39** 3 112-123
[4] Ginocchio, J.N. 2005 *Physical Review C*. **414** 4-5 165-261
[5] Ginocchio, J.N., Leviatan, A., Meng, J., & Zhou, S.G. 2004 *Physical Review C* **69** 3
[6] C.Cari 2013 *Mekanika Kuantum Penyelesaian Potensial Non-Sentral dengan Supersimetri, Hypergeometri, Nikivarov- Uvarov, dan Polinomial Romanovski*. Surakarta Jawa Tengah : UPT.
[7] Nikiforov A. F. & Uvarov U.B. 1988 *Special Function in Mathematical Physics*. Birkhausa : Basel
[8] W. Koepf & M. Masjed-jamei 2006 *Integral Transform and Special Functions* **17** 559
[9] Akpan N.Ikot 2014 *Quantum Physics Letters* **1** 7-14
[10] C.Cari, A.Suparmi, Deta, Werdiningsih 2013 *Makara Journal of Science* **17** 3 93-102
[11] Greiner W 2000 *Relativistic Quantum Mechanics*. Berlin : Spinger
[12] A.Suparmi, C.Cari, J.Handhika, C.Yanuarief, H.Marini 2012 *IOSR Journal of Applied Physics* 2 **2** 43-51
[13] Hamzavi H. & Rajabi A.A. 2013 *Hindawi Publishing Corporation Advances in High Energy Physics* **12**
[14] Ikhdair Sameer M. & Falaye Babatunde J. 2014 *The European Physical Journal Plus* **129**.
[15] Agboola, D. 2011 *Communication in Theoritical Physics* **55** 6 972-976

Analysis of wave equation in electromagnetic field by Proca equation

Oky Rio Pamungkas[1], Soeparmi[2], Cari[2]

[1]Physics Departement, Faculty of Mathematics and Natura Science, Sebelas Maret University, Surakarta, Indonesia

[2] Physics Departement, Post Graduate Program, Sebelas Maret University, Surakarta, Indonesia

E-mail: okyriopamungkas@student.uns.ac.id

Abstract. This research is aimed to analyze wave equation for the electric and magnetic field, vector and scalar potential, and continuity equation using Proca equation. Then, also analyze comparison of the solution on Maxwell and Proca equation for scalar potential and electric field, both as a function of distance and constant wave number.

1. Introduction

At the beginning of the 19th century was found a variety of elementary particles such as electrons, protons, neutrons, etc. which had a mass of relatively small[12]. Quantum mechanics describes the statistical behavior of the particles by analyzing the formula of energy and wave function[8]. Energy and wave function of particles can be obtained by solving Schrödinger equation directly[13]. Quantum field theory describes the behavior of particles which have relatively high speed and small mass[11].
Proca equation can also be combined with Maxwell equation to describe behavior of photons[2] and massive photons[21][4]. Bose-Einstein gravity condensate of particles which have spin 1 also can be explained by modifications of Proca equation[3]. Electromagnetic consists of the electric and magnetic fields that can spread in the vacuum in the universe[9].
Proca equation was used in various of teoritical studies, such as; non-Abelian supersymmetric N=1[10], solutions of Dirac-Proca equation[20], local gauge invariance test in electrodynamics[22], to explain the phenomena in the universe, namely the Tachyonic Cherenkov spectrum radiation[19] and Tachyonic synchrotron radiation[18] from relativistic thermal electrons in Jupiter's belt radiations[16]. Radiation was also detected at the Crab Nebula or explosion of supernova[15] and detection of Tachyonic gamma ray at the galactic core[17]. Proca equation is also used for basic research to obtained *GravitoElectroMagnetic* (GEM) equation[6] and London-Proca-Hirsch equation[5] for superconducting case in electrodynamics field[14].

2. Tensor Fields and Quantum Field Theory

Mathematical equation in tensor field using Lorentz transformations explains the case of relativistic particles[7]. According to Griffiths (1999) the relativistic case explain that the electric and magnetic fields are part of the electromagnetic field tensor $F^{\mu\nu}$ and $G^{\mu\nu}$ that is shown in equation (1) and (2) below:

$$F^{\mu\upsilon} = \begin{pmatrix} 0 & E_x/c & E_y/c & E_z/c \\ -E_x/c & 0 & B_z & -B_y \\ -E_y/c & -B_z & 0 & B_x \\ -E_z/c & B_y & -B_x & 0 \end{pmatrix} \tag{1}$$

$$G^{\mu\upsilon} = \begin{pmatrix} 0 & B_x & B_y & B_z \\ -B_x & 0 & -E_z/c & E_y/c \\ -B_y & E_z/c & 0 & -E_x/c \\ -B_z & -E_y/c & E_x/c & 0 \end{pmatrix} \tag{2}$$

Equation (1) and (2) can be written in a general form, namely as a form 4-vectors of Maxwell's equation shown by equation (3) and (4) below:

$$\frac{\partial F^{\mu\upsilon}}{\partial x^{\upsilon}} = \mu_0 J^{\mu} \tag{3}$$

$$\frac{\partial G^{\mu\upsilon}}{\partial x^{\upsilon}} = 0 \tag{4}$$

Equation (3) is called inhomogeneous Maxwell's equation, whereas equation (4) is called homogeneous Maxwell's equation. Inhomogeneous and homogeneous Maxwell's equation can be reduced to Ampere, Gauss, and Faraday equations by means of varying the value of μ. If the value of μ in equation (3) is zero thus obtained form of the Gauss equation that describes the divergence of the electric field caused by the density of charge at a point. Gauss equation can be shown by equation (5) as:

$$\nabla.\boldsymbol{E} = \frac{\rho}{\varepsilon_0} \tag{5}$$

If the value of μ in equation (3) are 1, 2, 3 thus obtained equation (3) can be expressed in 3-vector form as an Ampere equation as: $\nabla \times \boldsymbol{B} - \frac{1}{c^2}\frac{\partial \boldsymbol{E}}{\partial t} = \mu_0 \boldsymbol{J}$. Ampere equation explains that the change of the electric field and current density will produce a magnetic field.

If the value of μ in equation (4) is zero, then we obtained Gauss equation for the magnetic field as:

$$\nabla.\boldsymbol{B} = 0 \tag{6}$$

If the value of μ in equation (4) are 1, 2, 3 thus we obtained the formula which is expressed in 3-vector form of Faraday equation as: $\nabla \times \boldsymbol{E} = -\frac{\partial \boldsymbol{B}}{\partial t}$. Faraday equation explain the relationship between the electric and magnetic fields, the change of the magnetic field will produce an electric field.

3. Proca Equation

Proca equation explain the behavior of particles with spin 1. A photon is one of particles that have spin 1[1]. The photon is also the electromagnetic waves that describes by Maxwell equation[8]. According Poenaru (2012) the electromagnetic field to massive photon is described by Proca equation shown in equation (7), (8), (9), and (10) as:

$$\nabla.\boldsymbol{E} = \frac{\rho}{\varepsilon_0} - \mu_\gamma^2 \phi \tag{7}$$

$$\nabla.\boldsymbol{B} = 0 \tag{8}$$

$$\nabla \times \boldsymbol{E} = -\frac{\partial \boldsymbol{B}}{\partial t} \tag{9}$$

$$\nabla \times \boldsymbol{B} = \mu_0 \boldsymbol{j} + \mu_0 \varepsilon_0 \frac{\partial \boldsymbol{E}}{\partial t} - \mu_\gamma^2 \boldsymbol{A} \tag{10}$$

In a Proca equation, $\mu_\gamma^{-1} = \hbar/(m_\gamma c)$ describes the Compton's wavelength of the photon with mass m_γ [12]. According to Aquino (2011), a photon has imaginary mass which is expressed by the equation (11) below:

$$m_\gamma = \frac{2}{\sqrt{3}}\left(\frac{hf}{c^2}\right) i \tag{11}$$

Equation (11) states that photon has imaginary mass which expressed by m_γ. Quantitation of imaginary mass can be explained by a factor of Planck's constant *h*. Equation (11) is identical to the de-Broglie equation which expressed $\lambda = h/p$ that describes the correlation between mass and wavelength. De-Broglie equation describes the correlation between the mass of material and wavelength. Linear momentum p is the value of massive particles that moving at a certain speed. The linear momentum which is described by $p = mv$ in classical mechanics. Proca equation have additional correction factor magnetic field potential $\boldsymbol{A}$ and mass factor μ_γ^2. The value of $\mu_\gamma = \frac{m_\gamma c}{\hbar}$ describes the imaginary mass of the photon. The real mass of the photon will be explained in equation (12) below:

$$\mu_\gamma{}^2 = \frac{4}{3}\left(\frac{2\pi}{\lambda}\right)^2 \quad (12)$$

Equation (12) is explain is a constant value of wave propagation that described by $\frac{2\pi}{\lambda}$. Wave propagation is defined by $k = \frac{2\pi}{\lambda}$, so that equation (12) becomes:

$$\mu_\gamma{}^2 = \frac{4}{3}k^2 \quad (13)$$

Substituting equation (13) into Proca equation, so that equation (7), (8), (9), and (10) becomes:

$$\nabla.\boldsymbol{E} = \frac{\rho}{\varepsilon_0} - \frac{4}{3}k^2\phi \quad (14)$$

$$\nabla.\boldsymbol{B} = 0 \quad (15)$$

$$\nabla \times \boldsymbol{E} = -\frac{\partial \boldsymbol{B}}{\partial t} \quad (16)$$

$$\nabla \times \boldsymbol{B} = \mu_0 \boldsymbol{J} + \mu_0 \varepsilon_0 \frac{\partial \boldsymbol{E}}{\partial t} - \frac{4}{3}k^2\boldsymbol{A} \quad (17)$$

Proca equation does not contradict with the classical theory of Maxwell in electrodynamics. If the value of imaginary mass was taken zero ($m_\gamma = k = 0$) then Proca equation will return to common form of Maxwell equation. Generally form of 4-dimensional form of Proca equation can be expressed in equation (18) below:

$$\left(\nabla^2 - \frac{1}{c}\frac{\partial^2}{\partial t^2} - \frac{4}{3}k^2\right)\boldsymbol{A}_\mu = -\mu_0 \boldsymbol{J}_\mu \quad (18)$$

Vector $\boldsymbol{A}_\mu$ and $\boldsymbol{J}_\mu$ is 4-vector magnetic potential that expressed by $(\boldsymbol{A}, i\frac{\phi}{c})$ and current density that expressed as $(\boldsymbol{J}, i\frac{\phi}{c})$. If equation (18) applied in vacuum that is assumed there is no charge ($\boldsymbol{J}_\mu = 0$) then equation (18) as:

$$\left(\nabla^2 - \frac{1}{c}\frac{\partial^2}{\partial t^2} - \frac{4}{3}k^2\right)\boldsymbol{A}_\mu = 0 \quad (19)$$

Equation (19) looks like to wave equation in general form. If it is considered existence of photons is in point of charge and with the use of separation variable then equation (19) can be written as:

$$\left(\nabla^2 - \frac{1}{c}\frac{\partial^2}{\partial t^2} - \frac{4}{3}k^2\right)\phi = -\frac{\rho}{\varepsilon_0} \quad (20)$$

From equation (20) we get solution of scalar potential as space function in Proca's case given as:

$$\phi(r) = \frac{1}{4\pi\varepsilon_0}\frac{q}{r}e^{-\frac{2}{\sqrt{3}}kr} \quad (21)$$

Equation (21) looks like scalar potential named Coulomb's potential $\phi(r) = \frac{1}{4\pi\varepsilon_0}\frac{q}{r}$ in Maxwell's case. The different between Proca and Maxwell about scalar potential will be explained using graph shown in Fig. 4.1:

IOP Conf. Series: Journal of Physics: Conf. Series **795** (2017) 012032 doi:10.1088/1742-6596/795/1/012032

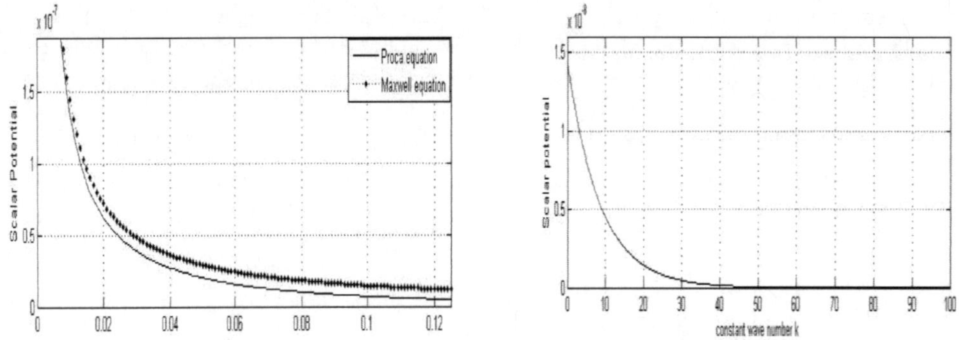

(a) (b)

Figure 4.1 Graph of scalar potential as function: (a) distance, (b) constant wave number in Proca equation

Graph (a) in figure 4.1 showed the different of value of scalar potential between Proca and Maxwell for distance variation. It is caused by factor of mass from the photon which described by $e^{-\frac{2}{\sqrt{3}}kr}$ in equation (21). At the distance that is close to zero then the scalar potential of Proca or Maxwell shown the same value. But, by increasing the distance will cause the difference in value of scalar potential between Maxwell and Proca equation. This indicates a correction value of Proca equation for a relatively great distance. While the graph (b) in figure 4.1 showed the graph of a scalar potential for variation of constant wave number in equation (21). If the value of k=0 then the value of a scalar potential will be shown like Coloumb potential. Reduction of equation (21) against to function of space will produce the value of electric field $\boldsymbol{E}$ which is expressed by equation (22) below:

$$\boldsymbol{E}(r) = \frac{q}{4\pi\varepsilon_0 r^2}\left(1 + \frac{2}{\sqrt{3}}kr\right)e^{-\frac{2}{\sqrt{3}}kr} \tag{22}$$

If the value of k=0 then equation (22) will reduce to Maxwell equation for electric field as $\left(\boldsymbol{E}(r) = \frac{q}{4\pi\varepsilon_0 r^2}\right)$. To describe visually of electric field in Proca and Maxwell equation (22) can be plotted graph as a function of distance and constant wave number shown in Fig. 4.2:

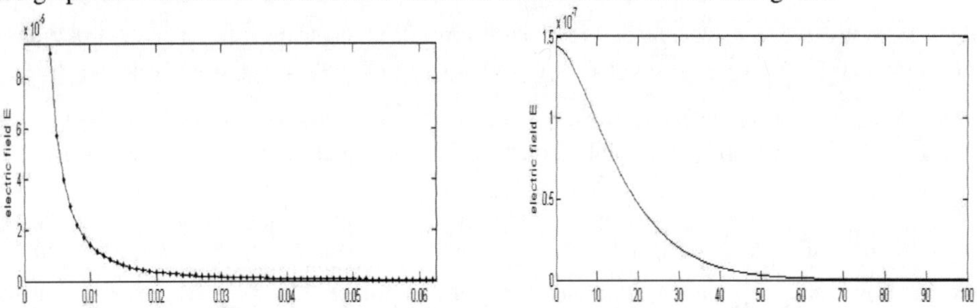

(a) (b)

Figure 4.2 Graph electric field as function of: (a)distance and (b)constant wave number in Proca equation

The graph (a) in figure 4.2 is the graph of a electric field as function of space for Proca and Maxwell equation. The graph (a) explain that Maxwell or Proca equation have the same value of the electric field. This is due to the small effect of the value $\left(1+\frac{2}{\sqrt{3}}kr\right)e^{-\frac{2}{\sqrt{3}}kr}$ in equation (22). The graph (b) in figure 4.2 is the graph of a electric field as function of the constant wave number for Proca equation. If equation (22) have value $k = 0$ then it would be reduced to Coloumb equation for the point charge that shown by ($\boldsymbol{E}(r) = \frac{q}{4\pi\varepsilon_0 r^2}$). This indicates to ignore the presence of a photon. But if a presence of the photon is considered ($k \neq 0$) that describes of a effect of the massive photon on the value of a electric field.

4. Effect of Proca equation in Electromagnetic equation

Proca equation describes the massive photon in electrodynamics [1]. Therefore, Proca equation affects to the equations in electrodynamics cases. Here is an explanation of some of the equation in electrodynamics field that caused of Proca equation:

4.1. Wave equation for electric and magnetic fields

Wave equation for Proca equation can be traced from the electric field $\boldsymbol{E}$ and magnetic field $\boldsymbol{B}$. Wave equation for electric field $\boldsymbol{E}$ is derived from Proca equation, equations (9) and (10). By operating the *curl* operator to equation (9), we get:

$$\nabla^2\boldsymbol{E} - \frac{1}{c^2}\frac{\partial \boldsymbol{E}}{\partial t} + \frac{4}{3}k^2\frac{\partial \boldsymbol{A}}{\partial t} = \mu_0 \boldsymbol{J} \tag{23}$$

The wave equation for the electric field in vacuum $\boldsymbol{J} = 0$ then equation (23) becomes:

$$\nabla^2\boldsymbol{E} - \frac{1}{c^2}\frac{\partial \boldsymbol{E}}{\partial t} + \frac{4}{3}k^2\frac{\partial \boldsymbol{A}}{\partial t} = 0 \tag{24}$$

Equation (24) explain to wave equation with massive photon. Meanwhile, the wave equation for magnetic field $\boldsymbol{B}$ can be derived from Proca equation by using *cross product* operator on equation (10) and substituting equation (9) into equation (10) it is obtained:

$$\nabla^2\boldsymbol{B} - \frac{1}{c^2}\frac{\partial^2 \boldsymbol{B}}{\partial t^2} + \frac{4}{3}\mu_0\boldsymbol{B} = \mu_0(\nabla\times\boldsymbol{J}) - \boldsymbol{A}\times\nabla k^2 \tag{25}$$

Wave equation for vacuum condition $\boldsymbol{J} = 0$ and for k=constant, so equation (25) becomes to:

$$\nabla^2\boldsymbol{B} - \frac{1}{c^2}\frac{\partial^2 \boldsymbol{B}}{\partial t^2} + \frac{4}{3}\mu_0\boldsymbol{B} = 0 \tag{26}$$

If value of constant wave number $k=0$, then equation (24) and (26) will be returned to standard form of the wave equation. The value of $k=0$ indicates the absence of imaginary mass of a photon.

4.2. Scalar and vector potential

Scalar and vector potentials for Proca condition can be derived by substituting equation $\boldsymbol{E} = -\nabla\phi - \frac{\partial \boldsymbol{A}}{\partial t}$ into equation (7), so that obtained *dot* operations are shown in equation (27):

$$\nabla^2\phi - \mu_\gamma^2\phi + \frac{\partial}{\partial t}(\nabla\cdot\boldsymbol{A}) = -\frac{\rho}{\varepsilon_0} \tag{27}$$

Equation (27) describes scalar potential. It can be expressed in the form of wave propagation, thus becoming:

$$\nabla^2\phi - \frac{4}{3}k^2\phi + \frac{\partial}{\partial t}(\nabla\cdot\boldsymbol{A}) = -\frac{\rho}{\varepsilon_0} \tag{28}$$

Equation (28) explain to wave equation with massive photon. Substituting equation $\boldsymbol{B} = \nabla\times\boldsymbol{A}$ and $\boldsymbol{E} = -\nabla\phi - \frac{\partial \boldsymbol{A}}{\partial t}$ into equation (10), it will obtain the following form:

$$\left(\nabla^2\boldsymbol{A} - \mu_0\varepsilon_0\frac{\partial^2 \boldsymbol{A}}{\partial t^2}\right) - \nabla\left(\nabla\cdot\mathbf{A} + \mu_0\varepsilon_0\frac{\partial V}{\partial t}\right) = -\mu_0\boldsymbol{J} + \mu_\gamma^2\boldsymbol{A} \tag{29}$$

Equation (29) describes the relationship between current density with 4-vector potential. Equation (29), has factor correction of mass as expressed by vector potential. If a value of massive photon is ignored $\mu_\gamma^2 = \frac{4}{3}k^2 = 0$, then equation (29) will be reduced according to the theory of electrodynamics for scalar and vector potential.

4.3. Continuity equation

In the same way as Maxwell's equation, it will obtain a new equation that described about continuity equation by Proca. We have derived equation (7) and equation (10) with vector operation, so that results the formula that shown in equation (30):

$$\mu_0\left(\frac{\partial\rho}{\partial t}+\nabla\cdot\boldsymbol{J}\right)=\mu^2\left(\nabla\cdot\boldsymbol{A}+\frac{1}{c^2}\frac{\partial\phi}{\partial t}\right) \tag{30}$$

Equation (30) is relevant to continuity equation in Maxwell's condition ($\frac{\partial\rho}{\partial t}+\nabla.\boldsymbol{j}=\mathbf{0}$) if values of $\mu^2=0$, which means effect imaginary mass of photon ignored.

Equation (30) can be written becomes to:

$$\left(\frac{\partial\rho}{\partial t}+\nabla\cdot\boldsymbol{J}\right)=\frac{\mu^2}{\mu_0}\left(\nabla\cdot\boldsymbol{A}+\frac{1}{c^2}\frac{\partial\phi}{\partial t}\right) \tag{31}$$

If equation (31) compared to continuity equation in Maxwell's condition, then the value of $\frac{\mu^2}{\mu_0}\left(\nabla\cdot\boldsymbol{A}+\frac{1}{c^2}\frac{\partial\phi}{\partial t}\right)=0$, so that continuity equation is fulfilled. It can be explained by equation (32) below:

$$\frac{\mu^2}{\mu_0}\left(\nabla\cdot\boldsymbol{A}+\frac{1}{c^2}\frac{\partial\phi}{\partial t}\right)=0 \tag{32}$$

From equation (32), we have two condition. First, the value of $\frac{\mu^2}{\mu_0}=0$, this condition illustrates the absence of imaginary mass propagation of the photon. Second, the value of $\nabla\cdot\boldsymbol{A}+\frac{1}{c^2}\frac{\partial\phi}{\partial t}=0$, this is a form of the equation describing an eternity. The change from a scalar potential proportional to the divergence of vector potential.

5. Conclusion

Proca equation can be analyzed based on the electric or magnetic field. The result are shown by equation (24) for electric field and equation (26) for a magnetic field. Equation (24) and (26) are wave form equation, with parameters of massive photon. If the mass is ignored then the formula will be returned to being standard wave equation.

From analysis, it has been obtained of vector and scalar potential shown by equation (27) and (29), and also is obtained a new formula from continuity equation shown by equation (31). There equation are obtained as result of an application of Proca equation in electrodynamics. If the mass of photon is ignored, the equation will be reduced back to Maxwell's equation in classical electrodynamics.

References

[1] FD Aquino. (2011). *Proca Equation And The Photon Imaginary Mass*. Brazil : Maranhao State University.
[2] Belin Jr H, ST Costa, MM Ferreira, JA Helayël-Neto. (2005). *The European Physical Journal*, 42, 127-137.
[3] R Brito, V Cardoso, CA Herdeiro, E Radu. (2016). *Physics Letter B*, 752, 291-295.
[4] A Caccavano, PT Leung. (2013). *Physics Letters A*, 377, 2777-2779.
[5] V Christianto. (2015). ISSN: 2349-915X, Vol. 4, pp 1-7.
[6] V Christianto. (2014). URL: http://independent.academia.edu/VChristianto.
[7] R Ferraro. (2007). *An Introduction to Special and General Relativity*. USA: Springer.
[8] DJ Griffiths. (1999). *Introduction to Electrodynamics 3rd edition*. USA: Prentice-Hall, Inc.
[9] MY Han. (2004). *A Story of Light An Introduction to Quantum Field Theory of Quarks and Leptons*. USA: World Scientific Publishing.
[10] H Nishino, S Rajpoot. (2013). *Nuclear Physics B,* 872 213-227.
[11] K Peeters. (2011). *A Guide to Quantum Field Theory*. United Kingdom: University of Durham.
[12] DN Poenaru. (2005). *arXiv:physics/0508195v1[physics.hist-ph]*.
[13] Suparmi. (2011). *Mekanika Kuantum II*. Surakarta: Universitas Sebelas Maret.
[14] M Tajmar. (2008). *Physics Letters A,* 372 3289-3291.

[15] R Tomaschitz. (2015). *Journal of High Energy Astrophysics*, 8, 10-20.
[16] R Tomaschitz. (2013). *Physics Letters A*, 377, 3247-3253.
[17] R Tomaschitz. (2008). *Physics Letters A*, 372 4344-4349.
[18] R Tomaschitz. (2004). *Physics Letters A*, 355 577-610.
[19] R Tomaschitz. (2001). *Physics Letters A*, 293 247-272.
[20] Y Tsutsumi. (2003). *Journal of Mathematical Analysis And Applications,* 278 485-499.
[21] LC Tu, J Luo, GT Gillies. (2005). *Institute of Physics Publishing*. Rep. Prog. Phys. 68 , 77-130.
[22] LC Tu, CG Shao, J Luo, J Luo. (2006). *Physics Letters A,* 352 267-271.

Electrospinning Titanium Dioxide (TiO_2) nanofiber for dye sensitized solar cells based on Bryophyta as a sensitizer

Novita Asma Ilahi, Risa Suryana, Fahru Nurrosyid, Linda Kusuma N.T

Physics Department of Postgraduate Program, Sebelas Maret University
Jl. Ir. Sutami 36A Kentingan Jebres Surakarta 57126, INDONESIA

Email: nasmailahi@gmail.com

Abstract. From an engineering and economic perspective, immobilized TiO_2 nanocatalysts are preferred in a variety of applications. In this study, TiO_2 polymer solution was synthesized using ethanol, acetic acid, polyvinylpyrrolidone (PVP), and titanium tetra isopropoxide (TTIP). TiO_2 solution was deposited on the FTO substrate by electrospinning method to obtain nano-sized layer. Capillary of syringes given a positive DC voltage of 6 kV to produce nanofiber, then annealed at 450 °C for 3 hours. Chlorophyll has obtained from extracted moss through a chromatographic process to used for dye. TiO_2 nanofiber layer manufactured with varied by time and characterized by UV-Vis and IV-meter. The result exhibited a maximum efficiency of 0,0036% and significant absorption at 350 nm-500 nm wavelength.

1. Introduction

Currently, power conversion efficiencies of up to 24% was achieved by silicon-based solar cells, although this high efficiency was offset by the high cost.[1] DSSC is promising alternative to silicon analogues because of inherent advantages such a simple manufacturing processing, low cost, and ease of large-scale of production.[2] DSSC have high conversion efficiencies up to 12% have been reported.[3-6]

Key factor to achieve DSSC high performance is to choose of mesoporous TiO_2 working electrode and light scattering layer. The concept is to incorporated light scattering layer so that the lower conversion of photon due to thin mesoporous TiO_2 layer can be compensated.[7] The charge carrier mobility, electron transfer efficiency and the rate of recombination of charge carriers are performance related size-dependent an improvement of the performance with respect to micrometric-size materials as a result of the increased specific surface area and surface reactivity.[8] The nanofibers showed increase light absorpted in the visible of region. This is partly derived optical path length extending from the wavelength of incident light caused by scattering of light in a layer of TiO_2 nanofibers.[9]

The sol-gel electrospinning derived TiO_2 nanofibers provide the unique advantage of high specific surface area with enchanced catalytic property.[10] A comparative among nanoparticles and sol-gel derived nanofibers suggested that enhanced catalytic property of nanofibers is often attributed to the higher specific surface area, for mesoporosity and particle alignment.[11-12] The method of electrospinning has been drawing attentions in the field of research TiO_2 nanofibers in preparation for the DSSC application for a careful selections of raw materials with different ratios can produce in TiO_2 desired shape with a uniform control diameter of the fiber.[9]

In this research, we success to fabricated and synthesize TiO_2 nanofibers by electrospinning method as the working electrode and DSSC scattered layer. The rate of time of electrospinning to fabricated TiO_2 nanofibers was varied from 10, 15, 30, and 45 min.

2. Experimental details

2.1. Materials

The materials used in manufacturing of polymer solution was made by using absolute ethanol (C_2H_5OH, Merck, Jerman), acetic acid ($CH3CO_2H$, 99.7% Aldrich) was obtained from Sigma-Aldrich, USA. Powder polyvinylpyrrolidone (PVP, Mw ~29000 Aldrich) and titanium tetra isopropoxide (TTIP, 97% purity, Aldrich, USA) were used as a precursor in the electrospinning procedure.

2.2. Preparation of photocatalysts

Flourine doped tin oxide (FTO) glass was prepare as follow: FTO glass was cutted into 2.5 cm × 2 cm pieces, successively rinsed with soap, alcohol 90% in an ultrasonicator for 15 min, and dried at temperature of room.

2.2.1. Preparation of electrospinning solution. First, TiO_2 nanofiber solution was made by mixed 3.5 mL absolute ethanol with 1 mL acetic acid under stirring for 10 minutes. Subsequently, 2.5 g of PVP were successively added to a stirred solution and dissolved under magnetic stirring for 1 hour. Finally, 0.5 mL TTIP was mixed in solution to obtain a homogeneous solution after the PVP was completely dissolve and stirred at room temperature for 24 hours.

2.2.2. Preparation of TiO_2 nanofiber. Electrospinning consist of a high-voltage DC power supply positive, infusion pumps and syringe with a needle tip diameter in 0.45 mm and collector. Figure 1 shows an electrospinning illustration for preparation TiO_2 nanofibers.

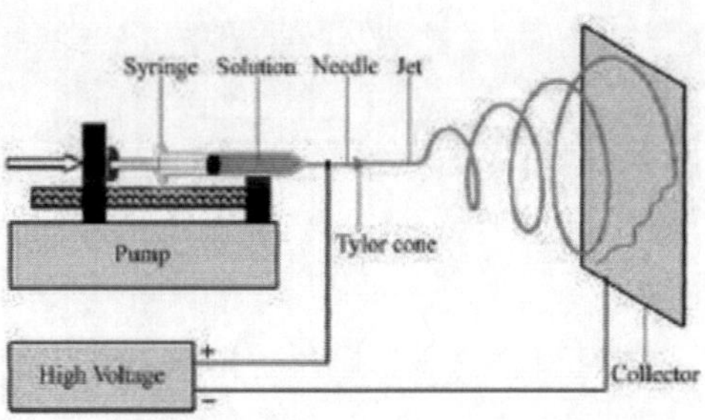

Figure 1. Electrospinning illustration for preparation of TiO_2 nanofiber

The syringe was connected to high-voltage power supply and the solution was loaded into the syringe. Electrospinning was performed at a voltage of 6 kV with a high-voltage supply. The flow rate was controlled at 3.0 mL/h by the syringe pump, and aluminum foil collector was set at a distance 25 cm. In this experiment, the rate of time of the electrospinning solution was varied from 10, 15, 30, and 45 minutes. The nanofibers electrode obtained was consequently subjected to calcinated in a furnace at 450 ^{O}C for 3 hours with healing rate 10 ^{o}C/min for removal of residual solvent and allow further condensation of the structure. Table 1 shows list the conditions of experimental for the TiO_2 nanofiber preparation.

Table 1
The Condition for the TiO_2 nanofiber preparation

Code	Acetic acid (ml)	Ethanol (ml)	PVP (g)	TTIP (ml)	Time (min)
C-1	1.0	3.5	2.5	0.5	10
C-2	1.0	3.5	2.5	0.5	15
C-3	1.0	3.5	2.5	0.5	30
C-4	1.0	3.5	2.5	0.5	45

2.3. Characterization of the photocatalysts

Before characterization of photocatalysts, TiO_2 nanofiber was submerged in dye solution for 24 hours. Dye solution made from the moss (Bryophyta) extracted by chromatography process to obtain purified chlorophyll. The solution of dye was a mixture from bryophyta and acetone with a weight ratio of 1:5, and stirring for 1 h.

For efficiency tests, the finished DSSC was cut into 2 cm × 1 cm pieces, and their *I-V* curves were measure using 1000 w/m^2 xenon lamp with Keithley 2602A *I-V* meter. The light absorption edges of photocatalysts was observed by UV-vis Lambda 25 spectrophotometer.

3. Results and discussion

3.1 Photocatalytic activity

Absorbance spectrum measurement of photocatalysts layer was measured using a UV-Visible Lambda 25 spectrophotometer with a determined wavelength in the range 300-700 nm. Based on the result of these measurements, obtained the following results:

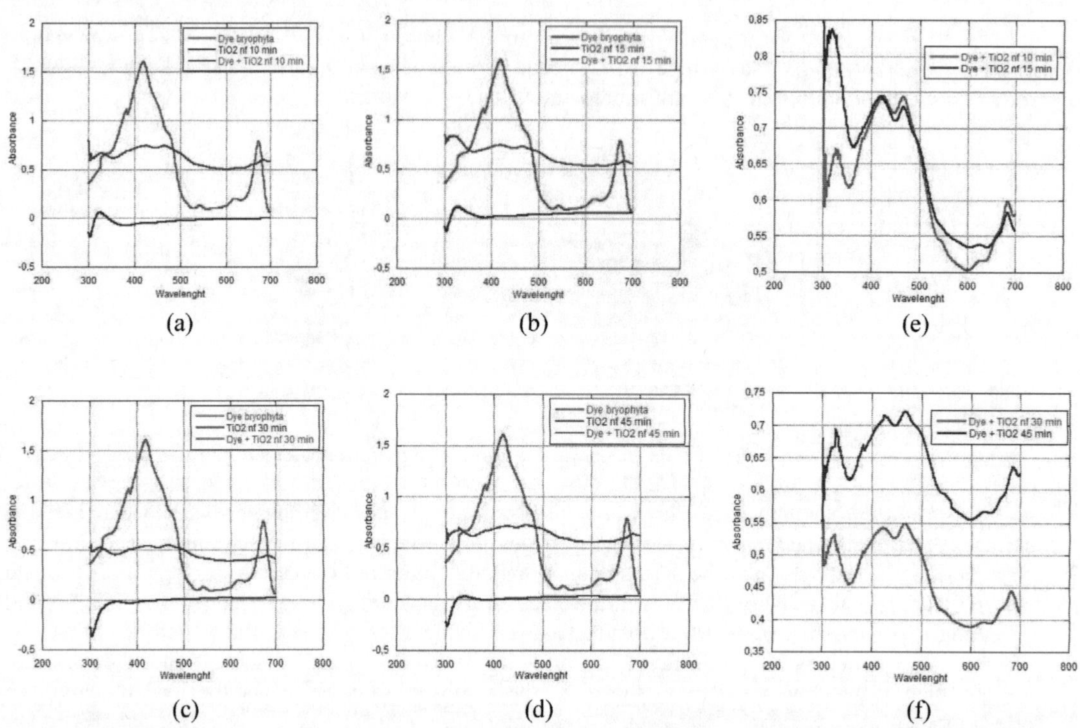

Figure 2. Comparison of absorbance and wavelength of TiO_2 nanofibers, and dye + TiO_2 nanofiber with rate of time electrospinning (a) 10 min, (b) 15 min, (c) 30 min, (d) 45 min; Comparison of

absorbance and wavelength dye + TiO_2 nanofiber for electrospinning (e) 10 and 15 min, (f) 30 and 45 min.

Figure 2 shows that the Bryophyta dye has the highest absorbance in range of the wavelength 350-500 nm and the second absorbance peaks appeared in range of the wavelength 650-700 nm. It shows that the absorbance of Bryophyta dye has sizeable in the wavelength range of visible light. It also shows that there was a displacement in the maximum absorbance of the mixtured dye titanium dioxide nanofiber. Bryophyta dye has significant effect to enlarge absorbance and expand wavelength range of TiO_2 nanofibers.

In this study, the absorbance of TiO_2 nanofibers been obtained are small. However, all the data showed an increased in range of the wavelength 300-350 nm. So it can be seen that the maximum absorbance of TiO_2 nanofiber occurs in range of the wavelength 300-350 nm, with a maximum absorbance of 0.12 Au. The significant result was obtained from displacement wavelength of the TiO_2 nanofiber to dye TiO_2 nanofibers absorbance. The ability to absorb the light of dye TiO_2 nanofibers expand to the wavelength range of 300-500 nm and 650-700 nm. This displacement indicates that TiO_2 requires dye as a photosensitizer to extend the light absorption range, because when dye and TiO_2 mixed the dye molecules will firmly anchor to the surface of TiO_2.[13]

Maximum absorbance at the dye TiO_2 nanofibers layer is obtained at the rate of time TiO_2 nanofiber 15 min, of 0.85 Au. The result of a significant increase in absorbance shown in Table 2.

Table 2
Significant increase of absorbance

Code	Rate of time (min)	Absorbance of TiO_2 nanofibers (Au)	Absorbance of Dye TiO_2 nanofibers (Au)	Beginning wavelength range (nm)	Final wavelength range (nm)
C-1	10	0.06	0.80	300-350	300-500
C-2	15	0.12	0.85	300-350	300-500
C-3	30	0.02	0.56	300-350	300-500
C-4	45	0.03	0.72	300-350	300-500

Although there was a decrease of absorbance at 30 min electrospinning variety, increased return on variations within 45 min. As for yield reduction absorbance of 30 min and 45 min can be caused by many factors, such as the thickness, the evaporation of TiO_2 nanofibers layer, the effects of temperature, and the degree of homogeneity of the solution. Thus, it can be concluded that the higher the TiO_2 nanofibers electrospinning rate of time, the absorbance layer will tend to increase.

3.2 Performance of photocatalytic

The ability of DSSC was determined by converting the light into electrical energy by using characterizing current-voltage (I-V) method. Characterization performed by using a Keithley 2602A in dark and light condition under irradiation intensity of halogen lamps 1000 W/m^2, this intensity is the highest intensity radiation standards in Indonesia. The resulting test of current and voltage on photovoltaics with a rate of time variation electrospinning nanofibers as the working electrode is shown in Figure 3.

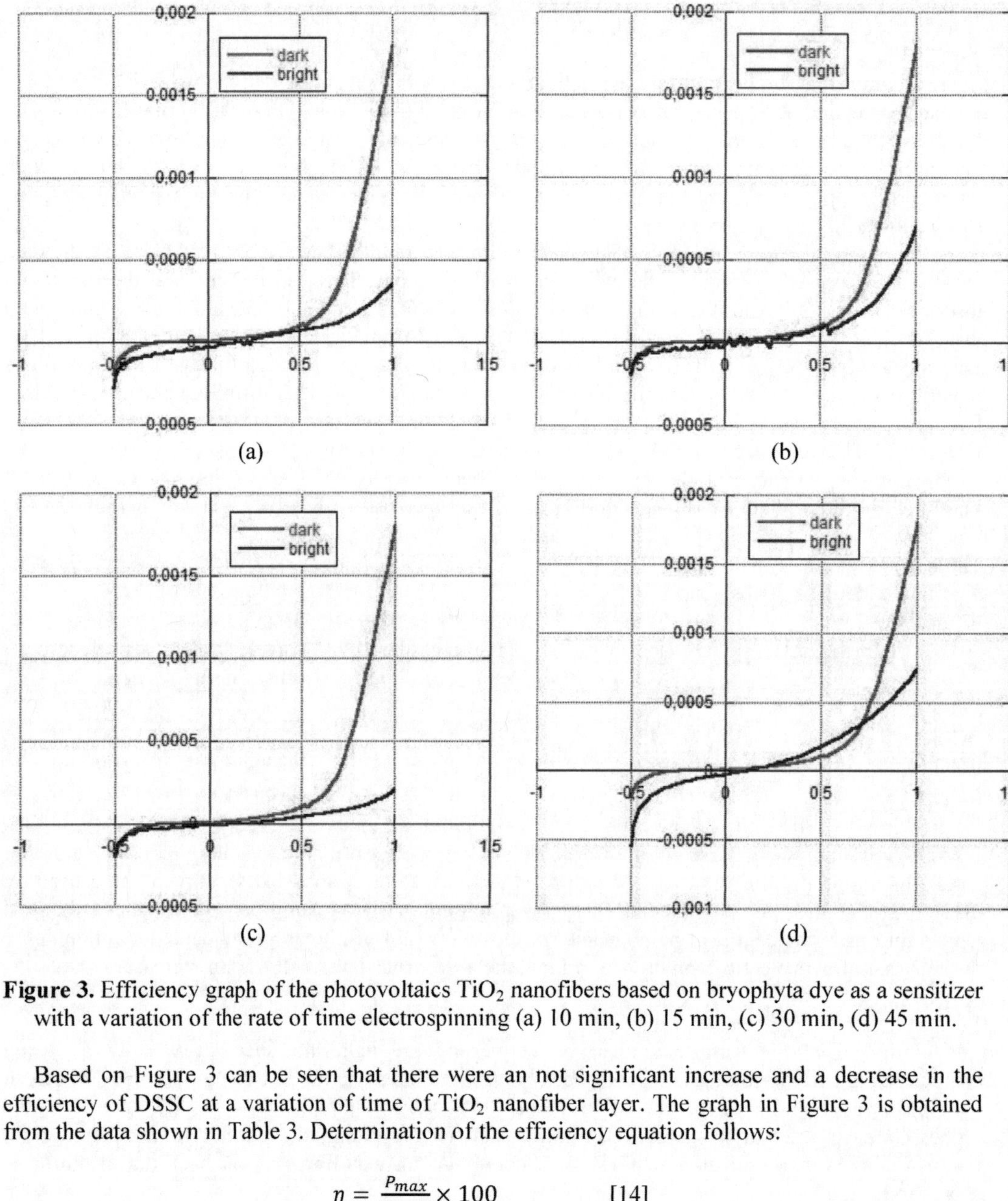

Figure 3. Efficiency graph of the photovoltaics TiO_2 nanofibers based on bryophyta dye as a sensitizer with a variation of the rate of time electrospinning (a) 10 min, (b) 15 min, (c) 30 min, (d) 45 min.

Based on Figure 3 can be seen that there were an not significant increase and a decrease in the efficiency of DSSC at a variation of time of TiO_2 nanofiber layer. The graph in Figure 3 is obtained from the data shown in Table 3. Determination of the efficiency equation follows:

$$\eta = \frac{P_{max}}{P_{light}} \times 100 \qquad [14]$$

Table 3

Efficiency of Photovoltaics

Rate of time	10 min	15 min	30 min	45 min
Isc	2.7 mA	5.2 mA	9.2 mA	2.2 mA
Voc	0.3 mV	0.3 mV	0.3 mV	0.2 mV
P max	5.5 Watt	6.5 Watt	5.3 Watt	4.0 Watt
Efficiency	0.0028 %	0.0036%	0.0027%	0.002%

In this work, can be determined that the efficiencies generated stable and did not increase significantly. This can be caused by many factors, including the influence of immersion layer in dye time, a low level of homogeneity in the mixtured layer, layer thickness, the electrolyte used quick-drying, saturation levels of Bryophyta dye was very quickly, and sandwich system dssc that not working optimally.

Maximum efficiency photovoltaic layer obtained on dye TiO_2 nanofiber 15 min of 0.0036%. While the efficiency for variation electrospinning layer of 10 min, 30 min, and 45min respectively 0.0028%, 0.0027%, and 0.002%.

4. Conclusion

The alteration from the rate of time of electrospinning nanofibers TiO2 layer affects the absorbance value and efficiency DSSC. Bryophyta dye has a maximum absorbance in the wavelength range of 350-500 nm and 650-700 nm. After TiO_2 nanofibers layer dissolve in bryophyta dye, the absorbance of TiO_2 nanofibers layer displaced from range 300-350 nm to 300-500 nm.

The efficiency obtained on the variation rate of time electrospinning 10 min of 0.0028%, the variation rate of time electrospinning 15 min of 0.0036%, the variation rate of time electrospinning 30 min of 0.0027%, and the variation rate of time electrospinning 45 min of 0.002%. It means, the greater rate of time electrospinning TiO_2 nanofibers layer then the absorbance and the efficiency of DSSC will tend to increase.

References

[1] M. Gratzel, Nature (London) 414, 338 (2001)
[2] M. Gratzel, J. Photochem. Photobiol., A 164, 3 (2004)
[3] A. Yella, H.W. Lee, H.N. Tsao, C. Yi, A.K. Chandiran, M.K. Nazeeruddin, E.W.G. Diau, C.Y. Yeh, S.M. Zakeeruddin, M. Gratzel, Science 334(2011) 629-634
[4] J. Kroon, N. Bakker, H. Smit, P. Liska, K. Thampi, P. Wang, S. Zakeeruddin, M. Gratzel, A. Hinsch, S. Hore, Progress in Photovoltaics: Research and Applications 15 (2007) 1-18
[5] Y. Chiba, A. Islam, Y. Watanabe, R. Komiya, N. Koide, L. Han, Japanese Journal of Applied Physics Part 2 Letters 45 (2006), L638
[6] B.E. Hardin, H.J. Snaith, M.D. McGeHee, Nature Photonics 6 (2012) 162-169
[7] S. Hore, C. Vetter, R. Kern, H. Smith A. Hinsch, Influence of Scattering Layers on Efficiency of Dye-Sensitized Solar Cells. Sol. Energy Mater. Sol. Cells 90, (2006) 1178-1188
[8] J.H. He, L. Xu, Y. Wu, Y. Liu, Polym. Int. 56 (2007) 1323-1329
[9] S. Chuangcote, T. Sagawa, S. Yoshikawa, Applied Physics Letters 93 (2008) 033310
[10] B. Caratao, E. Carneiro, P. Sa, B. Almeida, S. Carvalho, J. Nanotecnol. 2014 (2014) 472132 (pp. 5)
[11] S.K. Choi, S.Kim, S.K. Kim, H. Park, J. Phys. Chem. C, 114 (2010) 16475-16480
[12] W. Wang, Q. Yuan, Y. Chi, C.I. Shao, N. Li, X-T.Li, Chem. Res. Chinese Univ. 28 (2012) 727-731
[13] S. Ananth, T. Arumanayagam, P. Vivek, P. Marugakoothan, Direct Synthesis of Natural Dye Mixed Titanium Dioxide Nano Particles by Sol-gel Method for Dye Sensitized Solar Cells Application. Optik, vol 125 (2014) 495-498
[14] J. Gong, J. Liang, K. Sumanthy, Review on Dye Sensitizer Solar Cells (DSSC): Fundamental Concepts and Novel Materials. Renewable and Sustainable Energy Reviews, vol 16 (2012)

Approximate theory the electromagnetic energy of solenoid in special relativity

I Prastyaningrum, S. Kartikaningsih

Teaching and Education Institute PGRI Madiun
Jl Setia Budi 85 Madiun 63118, INDONESIA

Email: ihtiari.prastya8911@gmail.com

Abstract. Solenoid is a device that is often used in electronic devices. A solenoid is electrified will cause a magnetic field. In our analysis, we just focus on the electromagnetic energy for solenoid form. We purpose to analyze by the theoretical approach in special relativity. Our approach is begun on the Biot Savart law and Lorentz force. Special theory relativity can be derived from the Biot Savart law, and for the energy can be derived from Lorentz for, by first determining the momentum equation. We choose the solenoid form with the goal of the future can be used to improve the efficiency of the electrical motor.

1. Introduction

The currents which arise due to the motion of charges are the source of magnetic fields. When the charges move in a wire at a certain velocity can produce a current *I*, the magnetic field at any point *P* due to the current can be calculated by adding a small part of magnetic field $d\mathbf{B}$, from small segments of the wire $d\mathbf{s}$, (Figure 1.1)

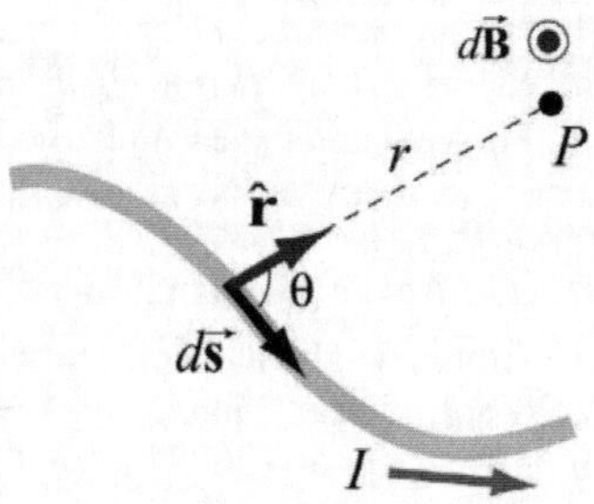

Figure 1. Magnetic field $d\mathbf{B}$ at point *P* because the effect of the current on the wire

These segments can be assummed as a vector quantity. It has magnitude of the length of the segment and pointing in the direction of the current flow in the wire. The source of current is verry small and be written as $\mathbf{I}d\mathbf{s}$ [1].

Let *r* denote as the distance from the current source (in this case is a wire) to the field point *P*, and $\hat{r}$ the corresponding unit vector. The Biot-Savart law gives an expression for calculate the magnetic field contribution, $d\mathbf{B}$, from the current source, $Id\mathbf{s}$, and the equation of Biot-Savart Law is

$$dB = \frac{\mu_o}{4\pi}\frac{Ids\times r}{r^2} \quad (1)$$

Where μ_o is a constant called the permeability of free space

$$\mu_o = 4\pi \times 10^{-7} \text{ T.m/A} \quad (2)$$

Biot-Savart law is used to determine the magnetic field strength either on a straight wire, circular, solenoids or toroid. By an integral part, we can find the magnetic field strength either on a straight wire, circular, solenoids or toroid [2].

Previously some analysis related to electromagnetic theory has been done. As performed by Babson et al. Babson analyze about momentum and the electromagnetic impulses [3]. Electromagnetic Momentum can also be written in the form of relativistic, in conjunction with the speed and energy [4]. Besides momentum Faraday's law of induction can also be analyzed in the form of special relativity [5].

In this case, we just use the magnetic field strength in solenoid form. The magnetic field strength of solenoid is applied to electromagnetic force and is brought to the relativistic theory. In the last season, we must set the relativistic energy by using a momentum equation.

2. Numerical Method of Electromagnetic Force for Relativistic Theory

It was explained earlier that the Biot Savart law can be applied to a straight wire, circular, solenoids, or toroids. In this study only focused on a circular wire on the solenoid, given for most of the electronic device that works by a strong magnetic field using a solenoid wire. For example transformer, electric motor, dynamo and so on.

To analyze the electromagnetic field in relativistic, we can use two assumptions call inertial frame S and S'. Assume both have a frame of reference in which the initial state $(x = y = z = 0)$ at $t = 0$. While conditions for inertial frame S' has a velocity relative to frame S is parallel to the x-axis. Look at the figure bellows,

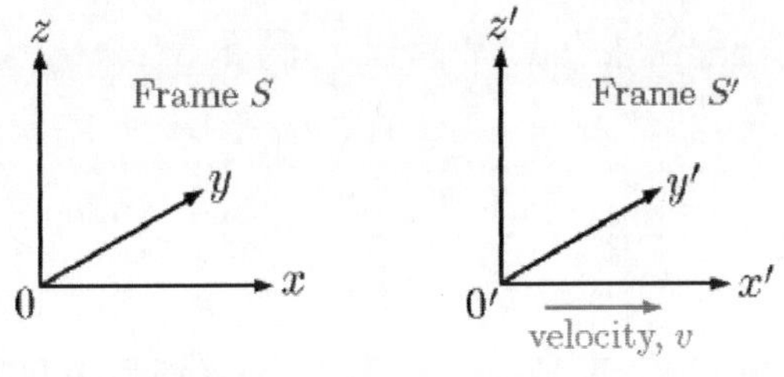

Figure 1. Two inertial frame S and S'

Inertial frame S with an overview of the time t and the coordinate axes (x, y, z), while for the inertial frame S 'has reviews time t' and the coordinates (x ', y', z '). Relations between the two inertial frame both for time and coordinates described using Lorentz transformations. By using the Lorentz transformation we can the relationship between the inertial frame S and S' for each dimension, namely by using the constant γ where,

$$\gamma = \frac{1}{\sqrt{1-\frac{v^2}{c^2}}} \quad (3)$$

Where v is the relative speed of S' with respect to S.

With reviews inertial frame *S* then we can perform an analysis of the movement of cargo with a certain speed if using a review inertial frame *S'*. For the relativistic case analysis that we use is the inertial frame *S'* so that the charge is looked at is the charge contained in the inertial frame *S'* moving with velocity v circular [6].

In connection with the relativistic case, the number of negative charge per unit length is γN while the number of positive charge per unit length is N/γ, so the total number of charge is,

$$\lambda' = q + (-q) \tag{4}$$

$$\begin{aligned} \lambda' &= \frac{N}{\gamma} - \gamma N \\ &= N\left(\frac{1}{\gamma} - \gamma\right) \\ &= -\gamma N\left(1 - \frac{1}{\gamma^2}\right) \\ &= -\gamma N \frac{v^2}{c^2} \end{aligned} \tag{5}$$

If the total charge is used to determine the total current, where the current total is the total amount that flows from the positive charge with a speed v, then I can be written as a form of equation

$$I = Nv \tag{5}$$

So that the charge density is equal,

$$\lambda' = -\gamma \frac{v}{c^2} I \tag{6}$$

The value of the total electric field is

$$E' = \frac{\lambda\prime}{2\pi\epsilon_0 r} = \frac{-\gamma\frac{v}{c^2}I}{2\pi\epsilon_0 r} = -\gamma\frac{1}{2\pi\epsilon_0 r}\frac{v}{c^2}I \tag{7}$$

By setting the equation $1/_{c^2} = \mu_0\epsilon_0$, so the Eq. (7) can be rewritten,

$$E' = -v\gamma\frac{\mu_0 I}{2\pi r}$$

The electric field equations we will use to determine the electromagnetic force equation for the solenoid. Starting from Biot-Savart equation for circular wire, then developed into the form of the equation for the solenoid. If the electromagnetic force equation in the form of relativistic known, then using the momentum equation, can be determined equation to electromagnetic energy in the form of relativistic theory.

Based on the explanation before, we try to find the equation of electromagnetic energy in a relativistic theory. The first we start from Biot-Savart Law in solenoid form. At a radial distance denote by *a* (in this case *a* equals the radius of the wire) the long of solenoid is *l*, and the number of coils denotes by *N*, we see a magnetic field :

$$\boldsymbol{B} = \frac{\mu_0 IN}{L} \tag{3}$$

From the equation of magnetic field strength at the solenoid, we can find the equation for the vector potential that will be used to determine the equation of the electromagnetic momentum.

IOP Conf. Series: Journal of Physics: Conf. Series **795** (2017) 012034 doi:10.1088/1742-6596/795/1/012034

3. Electromagnetic Momentum and Electromagnetic Energy

Electromagnetic Momentum is the result of multiplying the charge with a potential vector [7]. To define the vector potential in the solenoid, let's do an analysis of the solenoid long has the winding number N, where the radius of solenoid denoted by a, and a point along the solenoid with the distance r, to determine potential vector we use the surface integral of **B** which is perpendicular to the solenoid [8]. The potential vector equations obtained is,

$$\boldsymbol{A} = \frac{1}{2}\boldsymbol{\mu N r I} \tag{9}$$

To determine the equation of the electromagnetic momentum, we can multiply the vector potential with electric charge q. So that we will find the equation of electromagnetic momentum in the function of v.

$$\boldsymbol{p} = \frac{1}{2}\boldsymbol{q}^2\boldsymbol{\mu N v} \tag{10}$$

The electromagnetic momentum of the equation we can derive the equations of the electromagnetic energy in a relativistic theory. Electromagnetic energy can be obtained by first determining the relativistic mass (in this case the mass used is an electric charge). Electromagnetic energy in the form of relativistic obtained by integrating the force versus distance, where the distance is the length of the solenoid is denoted by L. So we obtain,

$$\boldsymbol{E} = \int_0^v \boldsymbol{F}.\boldsymbol{dL} \tag{11}$$

Where F is a differential of momentum with respect to time, so the Eq. (11) can be rewritten as,

$$\boldsymbol{E} = \int_0^v \frac{\boldsymbol{dp}}{\boldsymbol{dt}}.\boldsymbol{v}\,\boldsymbol{dt} \tag{12}$$

Eq. (12) can be solved by partial integral models. The charge that is used is relativistic charge, so we get the equation,

$$\boldsymbol{E} = \int_0^v \frac{d\left(\frac{1}{2}q^2\mu N v\right)}{dt}.\boldsymbol{v}\,\boldsymbol{dt}$$

Where $\boldsymbol{\mu}$ and N is constants, so it can get out of the integral form

$$E = \frac{1}{2}\mu N v \int_0^v \frac{d(q^2 v)}{dt} . v\, dt \tag{13}$$

Because we use the relativistic charge, so

$$q = \gamma q_0 \tag{14}$$

Where q_0 is the rest charge, and the value of $\gamma = \frac{1}{\sqrt{1-\frac{v^2}{c^2}}}$. By subtitude Eq. (14) to the Eq. (13) we can obtain,

$$E = \frac{1}{2}\mu N v \int_0^v \frac{d(\gamma^2 q_0^2 v)}{dt} . v\, dt \tag{15}$$

The rest charge can be assumed as constants, so can go out from the part of integral, and we obtain,

$$E = \frac{1}{2}\mu N v q_0^2 \int_0^v v\, d(\gamma^2 v) \tag{16}$$

Eq. (16) can be solved by partial integral,

$$E = \frac{1}{2}\mu N v q_0^2 \left(v^2\gamma^2 - \int_0^v \gamma^2 v\, dv\right) \quad (17)$$

By completing the process of the integral in Eq. (17) by changing the value of $\gamma = \frac{1}{\sqrt{1-\frac{v^2}{c^2}}}$ and sets the $v = c \sin x$ then we will obtain the equation of the electromagnetic energy in relativistic theory

$$E = \frac{1}{2}\mu N v q_0^2 \left(v^2\gamma^2 - c^2 \ln \sqrt{1 - \frac{v^2}{c^2}} \right) \tag{18}$$

Eq. (18) is a model of the theoretical approach to the energy equation for the relativistic electromagnetic solenoid. Further development of this theory research can be done on a toroid or some other device that uses the principle of the magnetic field.

Reference

[1] Herman A. Haus and James L. Melscher. 1998 *Electromagnetic Field and Energy.* Massachusetts Institute of Technology.
[2] Rustam Efendi *et al.* 2007. *Medan Elektromagnetika Terapan*. Jakarta: Erlangga.
[3] David Babson *et al.* 2009. *Hidden Momentum, Field Momentum, and Electromagnetic Impulse*. Am. J. Phys. 77. American Association of Physics Teachers.
[4] Department of Physics. 2013.Relativistic Dynamics: The Relations Among Energy, Momentum, and Velocity of Electrons and the Measurement of *e/m*.

[5] Alexander L. Kholmetskii. 2003. The Faraday Induction Law and Field Transformations in Special Relativity. *Apeiron*, Vol. 10, No. 2, April 2003
[6] Anonymous. *Advanced Electromagnetism*. The University of Liverpool.
[7] David J. Griffiths.2003. *Electromagnetic Momentum*. American Association of Physics Teachers.
[8] Ben Yu-Kuang Hu. 2012. *Introducing Electromagnetic Field Momentum*. European Journal of Physics. 33 (2012) 873-881.

Dye characteristics of *Zingiber officinale var rubrum*, *Cinnamomum zaylanicum*, *Curcuma longa L.*, *Oryza sativa L. Indica* in dye sensitized solar cell (DSSC)

Cari[1], U. Mahfudli Fadli[2], A. Bayu Prasada and A. Supriyanto.

Physics Graduate Program, Faculty of Mathematics and Science, Sebelas Maret University, St. Ir. Sutami 36A Kentingan Surakarta 57126, INDONESIA

E-mail: cari@staff.uns.ac.id[1], ulfa.mfadli@yahoo.com[2]

Abstract. The aims of the research to were know performance of DSSC using the dye of *Zingiber*, *Cinnamomum*, *Curcuma*, and *Oryza* as a photosensitizer with a variation of dye deposition area with spin coating techniques. The structure of the samples as a sandwich consisting of the working electrode (TiO_2), dye, electrodes of platinum (Pt) and the electrolyte sandwiched between two electrodes. Test absorbance dye using UV-Visible Spectrophotometer Lambda 25, using a two-point conductivity test probes El Kahfi 100 and characterization test I-V using a Keithley 2602A. For *Zingiber* results showed that absorbance at 243 nm and 279 nm, photoconductivity of 0.29 $\Omega^{-1}m^{-1}$ and the efficiency is 0.015% on 0.5 cm^2. *Cinnamomum* results showed that absorbance at 253 nm and 403 nm, photoconductivity of 0.11 $\Omega^{-1}m^{-1}$ and the efficiency is 0.002% on 3 cm^2. *Curcuma* results showed that absorbance at 243 nm and 422 nm, photoconductivity of 0.177 $\Omega^{-1}m^{-1}$ and the efficiency is 0.072% on 3 cm^2. *Oryza* results showed that absorbance at 240 nm and 423 nm, photoconductivity of 0.21 $\Omega^{-1}m^{-1}$ and the efficiency is 0.04% on 2.25 cm^2. Best absorbance value was obtained from *Oryza* dye; the highest photoconductivity was obtained from *Zingiber* dye, and the highest efficiency was obtained from *Curcuma* dye.

1. Introduction

The third generation of solar cells which can be an alternative to cell conventional solar silicon is DSSC. Materials of DSSC was used because the cost of purchasing a cheap, easy in production, but the efficiency is still small [1]. There are four main components in DSSC, which is a semiconductor as working electrode (TiO_2), a dye sensitive to the absorption of light, an electrolyte and a counter electrode such as platinum (Pt). Studies with ruthenium complex dyes was achieved an efficiency of 11.1% [2]. Price and availability is limited so switch to the organic, such as *Zingiber*, *Cinnamomum*, *Curcuma*, and *Oryza*.

TiO_2 nanoparticles in FTO serves as the electron catcher for providing the hole. The function of counter electrode is a catalyst while transferring electrons to the electrolyte. Electrolyte generates electron when the redox reaction, which will be forwarded to fill the hole in the dye. The use of glass to work forward light so that the light energy absorbed by electrons in the dye atoms, then towards the hole of TiO_2 layers. Furthermore, electrons move to the FTO layers and external sircuit, then the electrons towards the platinum layer and electrolyte.

To improve the performance of DSSC, using the more dye and more variation to limit the highest efficiency. Energy gap (Eg) of TiO_2 rutile and TiO_2 anatase is 3.0 eV and 3.2 eV. Eg anatase is the highest because the active surface area is larger and more photocatalyst [3]. The counter electrode is a double function, the first; it can stream of electrons from the external circuit (lamp) back to the redox system (electrolyte) and the second as a mediator to catalyze the redox reaction [4]. The counter electrode of platinum (Pt) has been the material of choice because it is an excellent catalyst for reducing triiodide [5]. Platina serves to decrease the resistance of the FTO glass. Electrolytes used in this study is a mixture of iodine (I_2) with Potassium Iodide (KI) or Sodium Iodide (NaI) into the polyethylene glycol (PEG). Electrolytes were used as the electron mediator between the working electrode and the counter electrode. The efficiency value for electrolyte KI is the greatest then NaI, and TEAI each mixed into I_2 [6].

This study aims to determine the absorbance value, conductivity and efficient of DSSC. Determination of absorbance in this study in accordance the law of Lambert-Beer, with The Increase value of absorbance of a material which suggested a linear relationship between concentration and absorbance according to the equation 1.

$$A = \varepsilon . b . c \qquad (1)$$

Where A is the absorbance and ε is the molar absorptivity coefficient ($molar^{-1}.L^{-1}$), then b is the thickness of the medium through which the beam (cm), and c is the concentration of the sample solution (ppm) [7]. Determination of the conductivity of a material can be calculated by equations 2.

$$\sigma = d/R.A \qquad (2)$$

With σ is the conductivity (Ωm^{-1}) and R is the resistance of materials (Ω). While d is the distance between the two electrodes (m) and A is the cross-sectional area (m^2).

Performance solar cells are the ability to convert light into electrical energy. Determination of performance solar cells is measured using Keithley 2602A to find the value of short circuit current (I_{sc}) and open circuit voltage (V_{oc}).The value of I_{sc} is determined based on the data in which the voltage value starts from a negative to a positive. Similarly for the value of V_{oc} is determined by the current value starts from a negative to a positive value. Then the value maximum voltage (V_{in}) and maximum current (I_{in}) are obtained by multiplication of voltage and current. Determine of performance solar cells can by comparing maximum power from multiplication result V_{oc} and I_{sc}.

2. Experiment

2.1. Preparation of TiO_2 Solution

TiO_2 used in this study is Titanium (IV) Oxide nanoparticle (anatase) with 99.8% trace metals basis. TiO_2 solution of 0.5 grams dissolved in 3 ml of ethanol, then stirred at 500 rpm for 30 minutes.

2.2. Preparation of Natural Dyes Sensitizer

This study using a dyes extract from black rice (*Oryza sativa L. indica*), red ginger (*Zingiber officinale var rubrum*), cinnamon (*Cinnamomum zaylanicum*) and turmeric (*Curcuma longa L.*). Then taken as many as 10 grams was crushed to a fine powderusing a blender. Then the materials were dissolved in ethanol, acid acetate and distilled water used to have a ratio of 25:4:21 so that the high value absorbance of the solution because the best anthocyanin concentrate [8]. Then stirred (300 rpm) at 30°C for 30 minutes. Materials was allowed to stand for 24 hours, and it were filtered with a filter paper 103. Then dyes were filtered with a filter paper 103, it were tested of absorbance using UV-Vis *Spectrometer Lambda 25* and conductivity using El Kahfi *I-V meter 2602A*. Dyes said to be able to work if it has absorption in the ultraviolet and visible area, while the conductivity states how large can pass an electric current.

2.3. Preparation of Electrolyte

The mix of Potassium iodide (KI) solids of 0.8 grams and 10 ml of polyethylene glycol, then stirred at 300 rpm for 30 minutes. Then was added iodine (I_2) of 0.127 grams and stirred at 300 rpm for 30 minutes.

2.4. Preparation of Counter Electrode

The counter electrode in the form of platinum (Hexachloroplatinic (IV) acid10%) and it was mixed with isopropanol (207 ml) was stirred at 300 rpm for 30 minutes. Then the solution was dropped on FTO glass with the variation of the deposition area of 0.5 cm^2, 1 cm^2, 1.5 cm^2, 2.25 cm^2 and 3 cm^2 and the side was masking tape affixed, then the FTO glass was heated at 250^0C for 15 minutes, and the solution of 3 ml was dropped on FTO.

2.5. Preparation of TiO_2

The TiO_2 solution was deposited on the FTO conductive with spin coating techniques. At the FTO glass with the variation of the deposition area of 0.5 cm^2, 1 cm^2, 1.5 cm^2, 2.25 cm^2 and 3 cm^2, and FTO side was masking tape affixed. Pasta TiO_2 dripped on FTO glass that has been glued in the spinner, then stirring with a speed of 200-300 rpm. The next is it was heated using a hotplate at 500^0C for 60 minutes, then cooled at room temperature.

2.6. Assembly of Sandwich DSSC

The composition of the DSSC in the form of glass FTO that has been coated with TiO_2 and has been soaked in dye, and then called the working electrode. The working electrode is etched with an electrolyte solution and then it was covered of platinum layer or called the counter electrode. Then the composition of DSSC is clamped on both sides of the right and left. Once formed system Sandwich DSSC then characterized current and voltage (I-V), which indicates how much the DSSC capable of converting light energy.

3. Result and Discussion

3.1. Absorption Spectra of natural dye doped Cu

The data of absorbance using UV-Vis *Spectrometer* lambda 25 with a wavelength range between 200-800 nm. Figure 1 shows that the maximum absorbance of *Zingiber* dye from 243 nm to 279 nm, *Cinnamomum* dye approximately from 253 nm to 403 nm, *Curcuma* dye from 243 nm to 422 nm, while *Oryza* dye from 240 nm to 423 nm. From the figure 1, the widest range of absorption is *Oryza* dye, so that *Oryza* can absorb more energy ultraviolet and visible light.

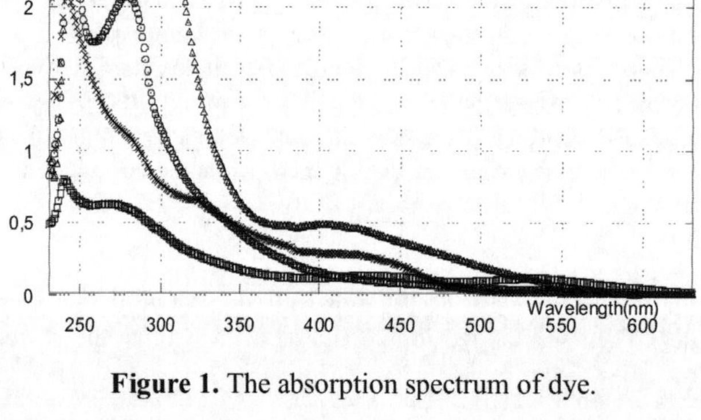

Figure 1. The absorption spectrum of dye.

Absorbance on figure 1 can work in the ultraviolet (UV) and visible (VIS) light, which is the area of energy absorption to excite electrons in the dye molecule. The result will be more electron are released from Higher Occupied Molecular Organic (HOMO) to the bandgap and then headed to Lower Unoccopied Molecular Organic (LUMO) in the dye molecule. The result will be formed holes in the HOMO and free electron in the LUMO, then the free electron flow to conducting TiO_2. The conduction band of TiO_2 can be the trapping of electrons from the dye. Increase the value of absorbance in accordance with the law of Lambert-Beer (equation 1).

3.2. Conductivity of dye+Cu

Characterization is electrical conductivity using El-kahfi 100 / IV-Meter with irradiation from halogen lamp of 680.3 W/m^2. The electrical conductivity is the ability of a material to conduct electricity when a potential difference was placed at the ends of a conductor, then the electron move through the material.

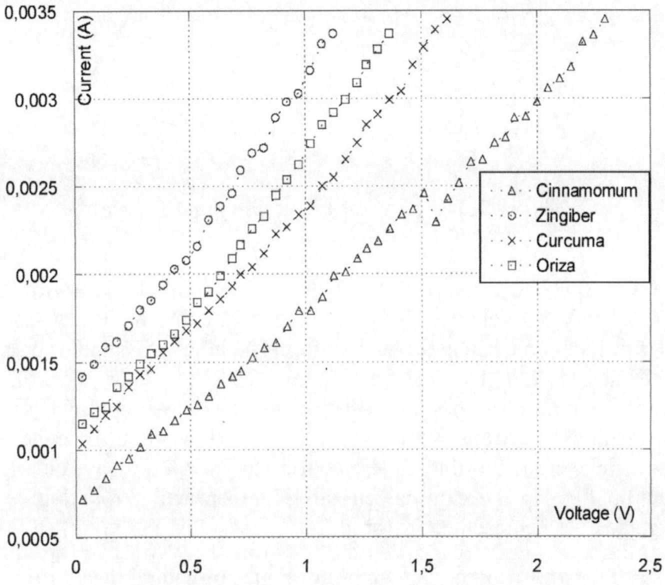

Figure 2. The Conductivity curve of dye.

The TiO_2 is a conductor, which has functions can delivering electron generated by the dyes. Figure 2 shows that the *Zingiber* conductivity value by using the equation two, worth of 0.29 $\Omega^{-1}m^{-1}$, *Cinnamomum* of 0.11 $\Omega^{-1}m^{-1}$, *Curcuma* of 0.177 $\Omega^{-1}m^{-1}$, and *Oryza* dye of 0.21 $\Omega^{-1}m^{-1}$. From the four materials, the ability of *Zingiber* dye to conduct electrons is the best, so it increase of power solar cells. The greater of conductivity value is indicates that the resistance it worth getting smaller, so the produce of electrons can increased.

3.3. Photovoltaic Properties

Characterization of the current-voltage (I-V) is a method to determine how much ability DSSC can convert light into electrical energy. Measurements using Keithley 2602A in under irradiation with a halogen light 1000 W/m^2. The test with immersion 24 hours in FTO area of 0.5 cm^2, 1 cm^2, 1.5 cm^2, 2.25 cm^2 and 3 cm^2.

The highest light energy was absorbed by the dye, it is proportional to the number of excited electros. Excitation electron of dyes molecule occur from HOMO to band gap and then passed to LUMO. After that, the electron will flow to the TiO_2, or forwarded to the working electrode. Working electrode layer is TiO_2, that functioning as an electron acceptor from the dye. Besides being able to forward the generated electrons dye, TiO_2 semiconductor can also produce electron as a result of absorbing light energy, but relatively small so that it was ignored.

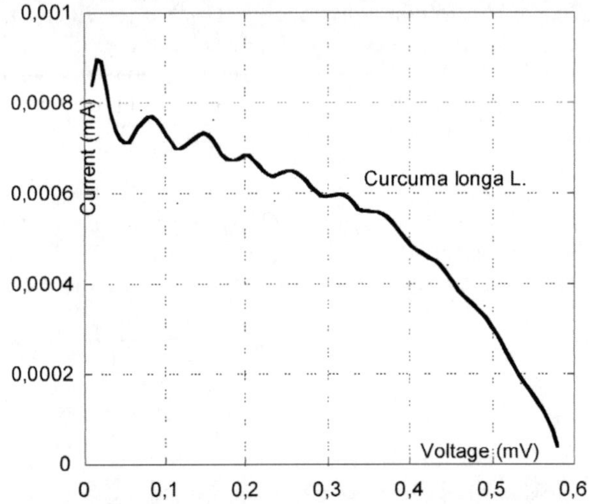

Figure 3. I-V curve of *Curcuma longa L.* Dye.

The thickness of TiO_2 on FTO glass rather FTO will influence of electrons flow. TiO_2 is a kind of a semiconductor with a band gap value of 3.2 eV which is suitable for use in solar cells. TiO_2 will produce a hole and will function as an electron dye trap. The hole will catch or pass on electrons from the dyes. If the thickness is too thin, the amount of TiO_2 little hole, then the electrons will be forwarded to an external circuit is not optimal. With proper TiO_2 coating thickness, hence maximizing the trapped electrons with the electrons flowed, so the current flow will also be the maximum, that will produce the highest efficiency. If too thick, the electrons trapped too many, then flow to the external circuit, so that electron flow to the external circuit is not optimal. From Figure 3, found that the efficiency (η) of organic dyes is presented in Table 1.

Table 1. Photovoltaic performance of natural dyes

Dye	V_{OC} *(mV)*	I_{SC} *(mA)*	V_{MAX} *(mV)*	I_{MAX} *(mA)*	*FF*	η
Zingiber	0,370	$5,9 \times 10^{-5}$	0,175	$4,7 \times 10^{-5}$	0,38	0,006
Cinnamomum	0,230	$7,9 \times 10^{-5}$	0,140	$4,5 \times 10^{-5}$	0,35	0,002
Curcuma	0,580	8×10^{-4}	0,370	$5,8 \times 10^{-4}$	0,46	0,072
Oriza	0,430	$3,3 \times 10^{-4}$	0,310	$2,8 \times 10^{-4}$	0,61	0,039

The result can conclud that the DSSC that produces the best performance in *Curcuma longa L.*dye with efficiency is 0.072%, while *Cinnamomum* dye only 0.002%. The value efficiency of the material of organic compounds is still small (under 1%). For increase the performance DSSC, we can be using different counter electrode with polyaniline (PANI), because PANI was increased efficiency from 6.90% (platinum) to 7.15% using PANI [9], using of electrolyte which has viscosity is smaller than iodine such as PEO polymer gel [10].

4. Conclusion

The results were showed that the *Zingiber* dye has a maximum absorbance from 243 nm to 279 nm, photoconductivity of 0.29 $\Omega^{-1}m^{-1}$ and the efficiency is 0.005% on 0.15 cm^2. While *Cinnamomum* dye were showed absorbance from 253 nm to 403 nm, photoconductivity of 0.11 $\Omega^{-1}m^{-1}$ and the efficiency is 0.002% on 3 cm^2 area. In *Curcuma longa L.*dye were showed absorbance from 243 nm to 422 nm, photoconductivity of 0.177 $\Omega^{-1}m^{-1}$ and the efficiency is 0.072% on 3 cm^2. The *Oryza sativa L. indica* dye were showed absorbance from 240 nm to 423 nm, photoconductivity of 0.21 $\Omega^{-1}m^{-1}$ and the efficiency is 0.039% on 2.25 cm^2. So it can be concluded that the best absorbance value is in the *Oryza sativa L. indica* dye, the highest photoconductivity is in *Zingiber* dye, and the highest efficiency is in *Curcuma longa L.*dye.

Acknowledgement
Thanks to the managers of Physics Materials and Chemistry Laboratory in Faculty of Science UNS Surakarta. This research were supported by LPPM Postgraduate UNS Grant with contract no. 343/UN27/HK/2016, April 07th 2016.

References

[1] Stathatos, E.2012.Jurnal of Engineering Science and Technology Review 5. pp 9-13.
[2] Chiba.Y., Islam. A., Watanabe.Y., Komiya R., Koide N., Han LY.2006. J. Appl. Phys. 45 (2006) L638-L640.
[3] Kuznesof, P.M.2006.Chemical And Technical Assessment.Page 1(8)-8(8).
[4] Ting, C-C, and Chao, W-S. 2010.Applied Energy, Vol. 87 (2010), pp 2500–2505.
[5] Calandra, P., Calogero. G., Sinopoli, A., and Gucciardi, P. G. 2010. International Journal of Photoenergy, Vol. 2010, Article ID 109495, 15 pages, doi:10.1155/2010/109495.
[6] Prasetyo, Y.H, Wahyuningsih, S., Suryana, R.2014.Indonesian Physics Journal No: 53, Vol XVIII, August, ISSN: 1410-2994. Page:47-49.
[7] Dian, A., Yanlinastuti, Noviarty, Masrukan.2012.UraniaVol. 18 No 2. Juni: 59-119.
[8] Misbachudin, M.C. et all.2013.ISBN 978-602-8047-80-7 LPF 1350: page 1-5.
[9] Li, Q., Wu, J., Tang, Q., Lan, Z., Li, P., Lin, J., Fan, L.2008. Elsevier.
[10] Duangkaew, P., Chindaduang, A., Tumcharem.2010.Science Jurnal Ubonratchathani University.Vol 1.No 1 (January-June).page: 9-14.

Study of imbalanced internal resistance on drop voltage of $LiFePO_4$ battery system connected in parallel

Fengky Adie Perdana[1], Agus Supriyanto[2], Agus Purwanto[3], Anif Jamaluddin[4]

[1,2]Physics Department, Sebelas Maret University
Jl.Ir Sutami 36 A Surakarta Indonesia
[3]Chemical Engineering, Sebelas Maret University
Jl.Ir Sutami 36 A Surakarta Indonesia
[4]Physics Education Program, Sebelas Maret University
Jl.Ir Sutami 36 A Surakarta Indonesia

[1]fengkyadieperdana@gmail.com, [2]agusf77@yahoo.com, [3]aguspur@uns.ac.id, [4]elhanif@uns.ac.id

Abstract. The purpose of this research focuses on the effect of imbalanced internal resistance for the drop voltage of $LiFePO_4$ 18650 battery system connected in parallel. The battery pack has been assembled consist of two cell battery $LiFePO_4$ 18650 that has difference combination of internal resistance. Battery pack was tested with 1/C constant current charging, 3,65V per group sel, 3,65V constant voltage charging, 5 minutes of rest time between charge and discharge process, 1/2C Constant current discharge until 2,2V, 26 cycle of measurement test, and 4320 minutes rest time after the last charge cycle. We can conclude that the difference combination of internal resistance on the battery pack seriously influence the drop voltage of a battery. Theoretical and experimental result show that the imbalance of internal resistance during cycling are mainly responsible for the drop voltage of $LiFePO_4$ parallel batteries. It is thus a good way to avoid drop voltage fade of parallel battery system by suppressing variations of internal resistance.

1. Introduction

$LiFePO_4$ battery is the future of energy storage systems and become the most commonly adopted to many systems. To develop li-ion battery cell into battery packs, we must assemble them that connect with multiple series or parallel in order to fulfill the voltage required and power demand of battery packs. [1]. To achieve the required voltage level, the battery cells connected in series. Then to achieve the required energy capacity, the battery cells connected in parallel. [2], [3]. The module battery that assembly in parallel if comparing with any single battery during high power discharge it has higher discharge efficiency and had better discharge performance. Then, a module battery that assembly in a battery pack that connect with series or parallel consists of several battery cells. a battery pack that assembly in series or parallel can be considered as one single battery with large capacity and high voltage, if there is no difference between battery cells. The variations of battery cell parameter cannot be neglected, so the inconsistent manufacturing processes and the in homogeneous operating environments.[4], [5].

The state of health (SOH) and state of charge (SOC) are two main streams that usage to monitoring battery cell and battery pack. [6]. The percentage of the remaining charge to the maximum capacity that show the real information about a battery's remaining energy called State of Charge (SOC). [7], [8]. Then, State of Health (SOH) is used to estimate the remaining performance of a battery and to characterise the current health status. Because it can for describes the physical health condition of a battery compared to a another battery fresh one. To know the State of Health/State of Health requires the methodes for ampere-hour counting and precise capacity evaluation of battery cell that frequently used for this.[9]. The adaptive extended Kalman filter used to measuring voltage and battery current for estimating the battery state-parameter proposed an equivalent of circuit models the battery cell.[10].

Initial SOC, difference branch of capacity and internal resistance are affected the state of health (SOH) battery pack with parallel components. Internal resistance is a number that states the value of resistance that exists within the battery components, so that will determine the speed of ion exchange from the anode to the cathode. The detrimental effect of internal resistance imbalance between parallel-connected cells arises because differences in internal resistance lead to uneven current distribution within the cells, the resulting unexpectedly high currents decrease battery pack life. [11].

Herein, the author studied the detrimental effect of imbalance internal resistance on drop voltage of $LiFePO_4$ battery pack that connected with a double cell in parallel. The capacity fade of battery pack 18650 lithium-ion that connected in parallel using 1/2C discharge rates up to 26 cycles. Parameters obtained through internal resistance tester and battery analyzer 8-channel used to identify the drop voltage of the whole lithium-ion battery pack.

2. Numerical Methods

For identical lithium ferric phosphate ($LiFePO_4$) performance, we used internal resistance (IR) tester EQ-MSK-BK300 for measure the internal resistance (IR) of battery pack $LiFePO_4$, furthermore to know the performance of battery pack we used battery analyser BST8-3, and software TC5.3. There is 5 type of battery pack that connected in parallel which has difference combination of internal resistance that observed. Figure 1. Showed the experimental method in this study.

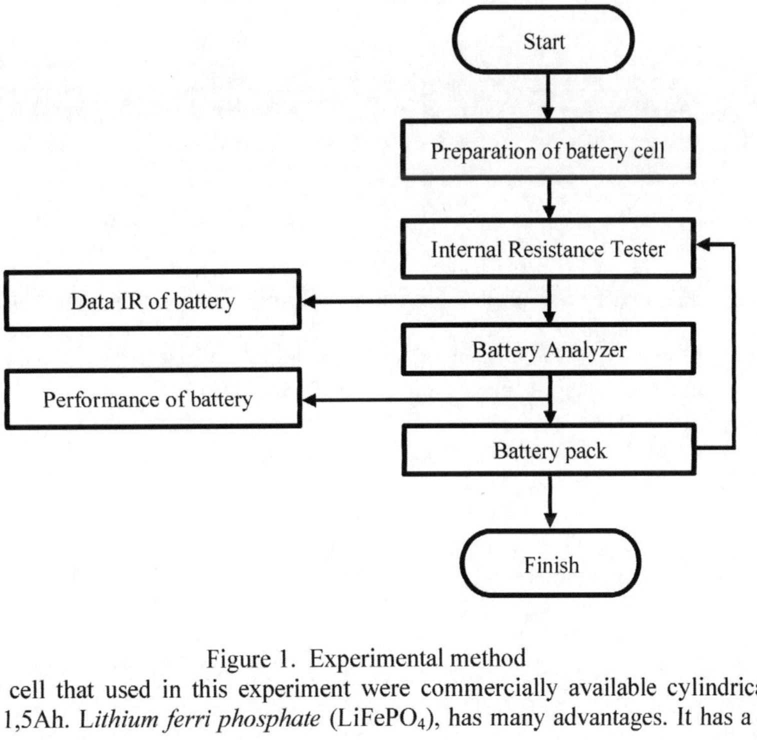

Figure 1. Experimental method

The battery cell that used in this experiment were commercially available cylindrical LiFePO4 3,2Volt 18650 1,5Ah. *Lithium ferri phosphate* ($LiFePO_4$), has many advantages. It has a good cycle, high safety, stable, long cycle life, inexpensive material, relatively low lifetime costs and environment-friendly. [12].

The internal resistance of battery cell was tested with EQ MSK BK300 internal resistance tester and the performance battery was tested with battery analyzer BST8.3. Internal resistance EQ MSK BK300 is a precision internal resistance meter tool for Quality Control (QC) or Research and Development (RnD) process of all rechargeable batteries. It can measure the Open Circuit Voltage (OCV) and Internal Resistance of almost all of the commercial batteries such as Li-ion battery. The accuracy of this internal resistance tester up to 1 mΩ and can measure another battery from 0-19.99V and applicable for measure the internal resistance of conventional capacitance single battery and battery pack.

BST8.3. It is an eight-channel battery analyser that can analyse from 6.0 mA up to 3000 mA, and 5V voltage of battery pack. BST8.3 has independent constant-voltage source and constant-current and each channel of battery analyzer programmed, controlled and calibrated by TC5.3 computer software that includes to setting various working modes: constant voltage charge (CCV), constant current charge (CC), constant current discharge (CD), rest time, and battery cycles. Then, limited threshold conditions include : voltage, capacity, current, time, and the negative voltage slope. The test of battery performance is one of the applications in battery testing fields used battery analyser BST8.3. Moreover, it can use to electrode materials research, capability grading, the formation of a small-scale battery, single battery testing, and etc. The monitoring of data that integrated with graph/data windows are observed with real time, and the testing process can be observing more efficient and more directly.

From the data result of the battery cell, then we assembly a battery cell into the battery pack that consist of two battery cell. We assembly battery pack with different internal resistance for each battery. The specification of internal resistance and the capacity of battery pack that the tested shown in Tabel 1. Battery pack 1 has lowest difference of combination the internal resistance, then battery pack 5 has biggest difference combination of internal resistance.

Table 1. Capacity and Internal resistance battery cell

No	Battery	Internal Resistance (mΩ) (Battery 1-Battery 2)	IR Pack (mΩ)	Capacity (mAh)
1	Pack 1	30-30	16	3000
2	Pack 2	30-40	18	3000
3	Pack 3	30-50	20	3000
4	Pack 4	30-60	21	3000
5	Pack 5	30-70	24	3000

Figure 2. shows a circuit of LiFePO4 battery cell 18650 group tested in this experiment. 2 cell were grouped to build the used battery pack. Considering 3,2-volt for the cell nominal voltages, the pack has 3,2 volt nominal voltage and capacity 3000mAh. Five samples of battery packs that containing two cels connected in parallel were tested using the method described below. The data that recorded with real time is: current, voltage, capacity, energy, DCIR and time.
Parameters of the testing cycle:

- 1/2C (1500mAh) of Constant current charging (CC),
- 3,65 Volt of Constant voltage charging (CV),
- 5 minutes of rest time between charge and discharge process,
- 1/2C Constant current discharge until 2,2V,
- 26 cycle of measurement test,
- 4320 minutes rest time after the last charge cycle.

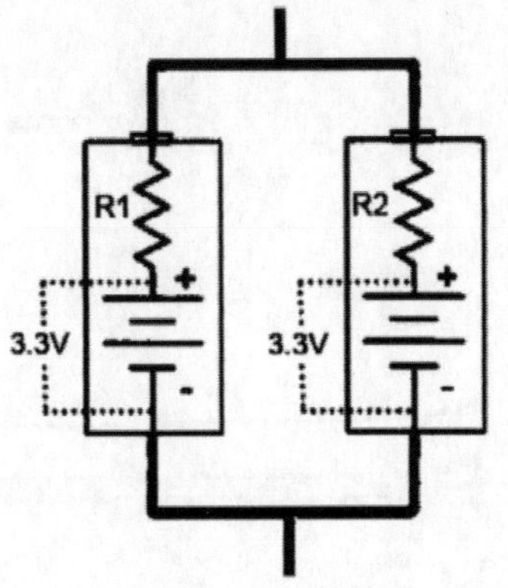

Figure 2. Circuit diagram of battery cell combination

3. Results and Discussion

The performance of battery pack can express by their resistance and capacity variations. [13]. All of the battery pack listed in Table 1. are assembled in parallel to develop 3,2 Volt, 3000mAh. The battery pack are charged and discharged under constant current charge and discharge condition. The capacities and the internal resistance of each battery cell and each battery pack are calibrated before the validation test.
Then, the current of battery pack and the voltage of battery pack are synchronously recorded 26 cycles test during every second. The capacities of five battery pack shown in Table 2. The capacities of the first battery pack is the maximum that same with initial battery pack capacities is 3000mAh. Linear with the difference of internal resistance, the last battery pack has minimum capacities. The initial at the 2nd cycle, and the end of 25th cycle.

Tabel 2. Comparison of capacity&voltage at initial&end cycles

No	Sample	Capacity (mAh)		Voltage Charge (v)		Voltage Discharge (v)	
		Initial	End	Initial	End	Initial	End
1	Sample 1	2922.7	2946.3	2.6680	3.6535	3.3804	2.1754
2	Sample 2	2914.6	2933.0	2.7374	3.6517	3.3556	2.1974
3	Sample 3	2498.9	2472.2	2.7619	3.6514	3.3457	2.1893
4	Sample 4	2463.5	2444.9	2.8537	3.6511	3.2641	2.1896
5	Sample 5	2307.3	2302.4	2.9086	3.6507	3.1959	2.1992

Table 2. shows the comparison inconsistent of start voltage and end voltage battery pack after charge and discharge process. The battery ageing is influenced by the imbalance of internal resistance that consists on the battery pack. This phenomenon can be leads to aged the voltage of the battery from standard one. Battery charge voltage V can be found by Kirchoff's voltage law [14]:

$$V = E_m + V_{Ro} + \sum_{i=1}^{n} V_{Ri}$$

Where V_{Ro} is the voltage of battery across the ohmic internal resistance R_0. V_{Ri} represents the voltage of battery across each polarisation resistance Ri, $i = 1 \dots n$.

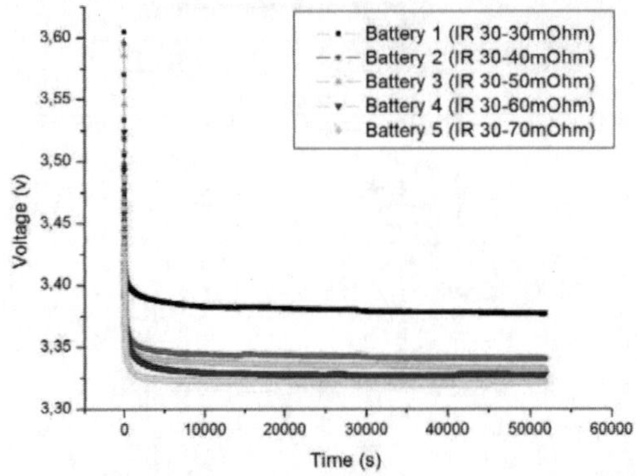

Figure 3. characteristic the rest of battery pack

A comparison of the drop voltage battery and time at rest time cycles is given by Figure.3. It presents the voltage drop at rest time after the last charge for 4320 minutes. The voltage drop changes drastically during rest time for each of battery pack. From the graph, the result drop voltage curves at rest time of all battery pack samples can be detected. In fact, The drop voltage of battery pack 1 that has same internal resistance combination is smallest from each other. Furthermore, the drop voltage of battery pack 5 that has a large combination of internal resistance is biggest from each other.

Table 3. Comparison the average of start voltage and end voltage

No	Sample	Rest after charge		Rest after discharge	
		Start voltage (v)	End voltage (v)	Start voltage (v)	End voltage (v)
1	Sample 1	3,6067	3,4240	2,3077	2,6149
2	Sample 2	3,5935	3,3990	2,3248	2,6453
3	Sample 3	3,5896	3,3946	2,3362	2,7235
4	Sample 4	3,5403	3,3895	2,3741	2,7112
5	Sample 5	3,5009	3,3698	2,4289	2,7750

Table 3 shows the average comparison of start voltage and ends voltage at rest time after charge and after discharge process. the start voltage end the ends voltage for each battery pack are different. The recorded data of drop voltage and the range of voltage can be described as deep of discharge (DOD) battery.

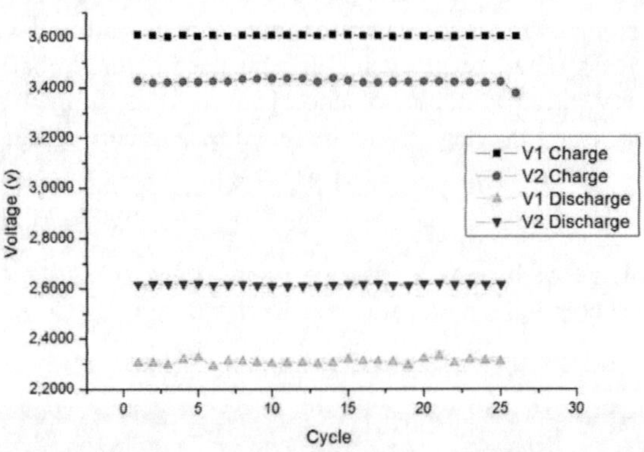

Figure 4. graph of battery voltage on sample 1 at rest cycle after charge and discharge process

In order to show the voltage drops at rest cycle after charge and discharge process, figure 5 give the comparison between voltage battery at each cycle. It shows voltage for 26 cycles of rest time battery. V1 charge that showing by black notation is the voltage battery after finish charge process, and the V2 Charge that showing by red notation is the last voltage from the rest time cycle. Furthermore, V1 discharge that showing by green notation is the voltage battery after finish discharge process, and the V2 discharge that showing by green notation is the last voltage from the rest time cycle.

4. Conclusion

In conclusion, this research study focuses on the effect of imbalanced internal resistance for the drop voltage of $LiFePO_4$ 18650 battery system connected in parallel. The imbalanced can lead to predicting the life cycle and drop voltage of battery pack. This experiment showed the impact of imbalanced internal resistance for drop voltage of battery pack that consist of two battery and assembly in parallel are more important to know than any effect of the resistance of single battery cell.

Binning and shorting the internal resistance of battery cell before assembly the battery pack is very important for a control of battery pack performance including charge and discharge profiles that their battery pack will experience, the average ability of battery pack life can be increasing, and the probability of exposure charge and discharge with very high C rates can be reduced. There are two obvious areas that very important and should be done then would be valuable. First, the imbalanced of internal resistance would affect the capacity of the battery pack. The degradation of capacity would avoid the lifetime battery decreased. Second, the imbalanced of internal resistance would give very interesting fundamental information of drop voltage profiles of parallel connected cells.

This work is potential future experiments, that used to optimise the processes of manufacturing. Combining the knowledge of how the imbalance of internal resistance affects the lifetime, capacity and drop voltage with the software and application of the battery testing that would make it to possible, for measuring the maxsimum tolerance of internal resistance of battery cell that will be an assembly in battery pack. So that, this is can reduced cost of manufacturing battery cell in high-volume, and can be used to increasing the performance ability of battery pack.

Acknowledgment
The work is supported by LPDP affirmation program for National Electric Vehicle (MOLINA) Research and Development. (PRJ-933/LPDP/2014, August 17, 2014).

References

[1] L. Wang, Y. Cheng, and X. Zhao, *Appl. Energy*, vol. 147, pp. 353–360, 2015.
[2] W. Shi, X. Hu, C. Jin, J. Jiang, Y. Zhang, and T. Yip, *J. Power Sources*, vol. 313, pp. 198–204, 2016.
[3] L. Wang, Y. Cheng, and X. Zhao, *Appl. Energy*, vol. 142, pp. 293–302, 2015.
[4] Y. Zheng, L. Lu, X. Han, J. Li, and M. Ouyang, *J. Power Sources*, vol. 226, pp. 33–41, 2013.
[5] S. Thanagasundram, R. Arunachala, K. Makinejad, T. Teutsch, and A. Jossen, no. November, pp. 1–13, 2012.
[6] D. Andre, C. Appel, T. Soczka-guth, and D. Uwe, *J. Power Sources*, vol. 224, pp. 20–27, 2013.
[7] Y. He, X. Liu, C. Zhang, and Z. Chen, *Appl. Energy*, vol. 101, pp. 808–814, 2013.
[8] Y. Wang, C. Zhang, and Z. Chen, *Appl. Energy*, vol. 135, pp. 81–87, 2014.
[9] Y. Hua, A. Cordoba-arenas, N. Warner, and G. Rizzoni, *J. Power Sources*, vol. 280, pp. 293–312, 2015.
[10] H. He, Z. Liu, and Y. Hua, *Energy Procedia*, vol. 75, pp. 1950–1955, 2015.
[11] R. Gogoana, M. B. Pinson, M. Z. Bazant, and S. E. Sarma, *J. Power Sources*, vol. 252, pp. 8–13, 2014.
[12] K. S. Dhindsa, B. P. Mandal, K. Bazzi, M. W. Lin, M. Nazri, G. A. Nazri, V. M. Naik, V. K. Garg, A. C. Oliveira, P. Vaishnava, R. Naik, and Z. X. Zhou," vol. 253, pp. 94–100, 2013.
[13] H. Wang, Y. He, T. Sun, and G. Li, vol. 7, no. 3, pp. 85–92, 2014.

[14] T. Tanaka, S. Ito, M. Muramatsu, T. Yamada, H. Kamiko, N. Kakimoto, and Y. Inui, *Appl. Energy*, vol. 143, pp. 200–210, 2015.

Solution of five dimensional Dirac equation with Asymtotic Iteration Method in the case of pseudo spin symmetry

Y.A Kurniawan, A Suparmi and C Cari

Physics Department , Sebelas Maret University,
Jl. Ir. Sutami 36A Kentingan Jebres Surakarta 57126, Indonesia

E-mail: yosuaardi@student.uns.ac.id

Abstract. The relativistic energies of 5 dimensional Dirac equation in the case of pseudo spin symmetry which governed by a sparable non central using the asymtotic iteration method (AIM).The separable five dimensional shape invariant potentials consisted of Hulthen radial potential and Manning-Rosen angular potentials. The relativistic energies were calculated numerically from the relativistic energy equation.

1. Introduction

Exact solutions for relativistic energies can be obtained from the five dimensional Dirac equation. The Dirac equation use to describe a half spin particle such as electrons and positrons. Unlike Schrodinger [1] and Klein Gordon equation [2], this equation add relativistic field and particle spin characteristic to get positive probability in every condition.

Hulthen Potential [3,4] behaves like Coloumb potential in the short range [5]. Hyperbolic Hulthen potential is written as

$$V(r) = -\frac{V_0}{2}\left(\coth\frac{\alpha r}{2} - 1\right) \tag{1}$$

The Hulthen Potential for the Dirac Equation [5, 6] can be solved exactly using Adlrich approximation. Manning-Rosen potential [7, 8] is used in angular section. Since five dimensional Dirac Equation it has 4 angular parts. Here we use trigonometric Manning-Rosen potentials

$$V(\theta_1) = \frac{V_1}{\sin^2\theta_1} - V_2\cot\theta_1 \tag{2}$$

$$V(\theta_2) = \frac{V_3}{\sin^2\theta_2} \tag{3}$$

$$V(\theta_3) = \frac{V_4}{\sin^2\theta_2} - V_5\cot\theta_3 \tag{4}$$

$$V(\theta_4) = -V_6\cot\theta_4 \tag{5}$$

Where V_1, V_2, V_3, V_4, V_5 and V_6 are constant. Manning-Rosen potential is used as a mathematical model in the description of diatomic molecular vibrations and it constitutes a convenient model for other physical situations. It is known that for this potential the Dirac equation can be solved exactly using suitable substitution.

Dirac Equation in the Case of Pseudo Spin Symmetry [9] with combination of Hulthen Potential and Manning-Rosen Potential can be written

$$\left[\begin{array}{l} -\frac{1}{r^{D-1}}\frac{\partial}{\partial r}\left(r^{D-1}\frac{\partial}{\partial r}\right)-\frac{1}{r^2}\left[\begin{array}{l}\frac{1}{\sin^2\theta_2\sin^2\theta_3\sin^2\theta_4...\sin^2\theta_{D-1}}\left(\frac{\partial^2}{\partial\theta_1{}^2}\right)\\ +\frac{1}{\sin^2\theta_3\sin^2\theta_4...\sin^2\theta_{D-1}}\frac{\partial}{\partial\theta_2}\left\{\frac{1}{\sin\theta_2}\left(\sin\theta_2\frac{\partial}{\partial\theta_2}\right)\right\}\\ +\frac{1}{\sin^2\theta_4...\sin^2\theta_{D-1}}\frac{\partial}{\partial\theta_3}\left\{\frac{1}{\sin^2\theta_3}\left(\sin^2\theta_3\frac{\partial}{\partial\theta_3}\right)\right\}\\ +\frac{1}{\sin^2\theta_{D-1}}\frac{\partial}{\partial\theta_{D-1}}\left\{\frac{1}{\sin^{D-2}\theta_{D-1}}\left(\sin^{D-2}\theta_{D-1}\frac{\partial}{\partial\theta_{D-1}}\right)\right\}\end{array}\right] gnk(r) \\ +\frac{h^2}{2m}\left[V(\vec{r})+\frac{1}{r^2}\left(\begin{array}{l}\frac{V(\theta_1)}{\sin^2\theta_2\sin^2\theta_3\sin^2\theta_4...\sin^2\theta_{D-1}}\\ +\frac{V(\theta_2)}{\sin^2\theta_3\sin^2\theta_4...\sin^2\theta_{D-1}}\\ +\frac{V(\theta_3)}{\sin^2\theta_4...\sin^2\theta_{D-1}}+...+V(\theta_{D-1})\end{array}\right)\right](E-M)\end{array}\right] \tag{6}$$

$$=(E^2-M^2)gnk(r)$$

Where gnk is wave function of Dirac equation in case of pseudo spin symmetry. E represent of relativistic energy and M represent of relativistic mass. Pseudo Spin Symmetry happen when difference between scalar potential and vector potential is constant. If we take constant is zero, the value of scalar potential is negative vector potential.

2. Asymptotic Iteration Method

AIM [10, 11] solve homogeneous linear second order differential equation to get exact solution of the form

$$y_n''(x)=k_0(x)y_n'(x)+s_0(x)y_n(x) \tag{7}$$

Where $k_0 \neq 0$, $s_0(x)$ is coefficient differential equation and n clarifying quantum number. We differentiate Eq (5) with respect to x :

$$y_n'''(x)=k_1(x)y_n'(x)+s_1(x)y_n(x) \tag{8}$$

Where

$$k_1(x)=k_0'(x)+k_0^2(x)+s_0(x)$$
$$s_1(x)=s_0'(x)+s_0k_0 \tag{9}$$

And the second derivative of Eq (5) gives :

$$y''''(x)=k_2(x)y_n'+s_2(x)y_n(x) \tag{10}$$

Where

$$k_2(x)=k_1'(x)+k_1(x)k_0(x)+s_1(x)$$
$$s_1(x)=s_1'(x)+s_0(x)k_1(x) \tag{11}$$

It can iterate up to $(i+1)th$ and $(i+2)th$ derivatives, with $i=1,2,3...$

So we have

$$y_n^{(i+1)}(x)=k_{i-1}(x)y_n'(x)+s_{i-1}(x)y_n(x)$$
$$y_n^{(i+2)}(x)=k_i(x)y_n'(x)+s_i(x)y_n(x) \tag{12}$$

Where

IOP Conf. Series: Journal of Physics: Conf. Series **795** (2017) 012037 doi:10.1088/1742-6596/795/1/012037

$$k_i(x) = k'_{i-1}(x) + k_{i-1}(x)k_0(x) + s_{i-1}(x)$$
$$s_i(x) = s'_{i-1}(x) + s_0(x)k_{i-1}(x) \tag{13}$$

From Eq (13) we got relation :

$$\frac{y_n^{(i+2)}(x)}{y_n^{(i+1)}(x)} = \frac{k_i\left[y'_n(x) + \frac{s_i}{k_i}y_n(x)\right]}{k_{i-1}\left[y'_n(x) + \frac{s_{i-1}}{k_{i-1}}y_n(x)\right]} \tag{14}$$

Asymptotic aspect of the method makes large $i(>0)$, Eq (14) becomes

$$\frac{s_i}{k_i} = \frac{s_{i-1}}{k_{i-1}} \equiv \Delta \tag{15}$$

For given potential, we convert five dimensional Dirac equation to the form of Eq (5). Than determine k_0 and s_0. Using Eq (11) $s_i(x)$ and $k_i(x)$ are obtained. Energy eigenvalue we get from

$$k_i(x)s_{i-1}(x) - k_{i-1}(x)s_i = \Delta_i = 0,\ i = 1,2,3... \tag{16}$$

Where k is the iteration number, and radial quantum number n is equal with iteration number k for this case

3. Bound State Solution

We separated Eq (6) in to one dimensional equation of radial, θ_1, θ_2, θ_3, and θ_4.

3.1 Radial Energy Equation

From the separation of Eq (23) one dimensional radial equation written as

$$\frac{1}{R}\left(-r^2\nabla_r{}^2R\right) + r^2V(r)(E-M) - r^2(E^2-M^2) = \lambda_4 \tag{17}$$

Where λ_4 is constant and r is distance a half particle with source of potential field. ∇_r represent of radial laplacian D dimensional is written as

$$\nabla_r{}^2 = \frac{1}{r^{D-1}}\frac{\partial}{\partial r}\left(r^{D-1}\frac{\partial}{\partial r}\right) \tag{18}$$

By using $R(r) = \frac{\chi(r)}{r^{(D-1)/2}}$, $\coth\left(\frac{\alpha r}{2}\right) = 1-2z$ and $\chi(z) = (z)^{\delta}(1-z)^{\gamma}f_n(z)$ and inserting Eq (1) and Eq (18) to Eq (17) we have

$$z(1-z)f''_n(z) + [1+2\delta-(2\delta+2\gamma+2)z]f'_n(z) + [-(\delta+\gamma)^2 - \delta - \gamma + A_s]f_n(z) = 0 \tag{19}$$

Where

$$\delta = \sqrt{E_s - B_s},\ \gamma = \sqrt{B_s + E_S} \tag{20}$$

$$A_s = 2-\lambda_4,\ B_s = \frac{V_0(E-M)}{2\alpha^2},\ E_s = \frac{V_0}{2\alpha^2}(E+M) + \frac{(E^2-M^2)}{\alpha^2} \tag{21}$$

s_0 and k_0 are obtained from Eq (26), so we can easy to find Δ_n, written as

$$(\delta+\gamma+n)(\delta+\gamma+(n+1)) - A_{sn} = \Delta_0 = 0 \tag{22}$$

By inserting Eq (20) and (21) to equation (22) we have

$$\left[\sqrt{2-\lambda_4+\frac{1}{4}}-\left(n+\frac{1}{2}\right)\right]^2+\frac{4\left[\frac{V_0(E-M)}{2\alpha^2}\right]^2}{\left[\sqrt{2-\lambda_4+\frac{1}{4}}-\left(n+\frac{1}{2}\right)\right]^2}=4\left[\frac{V_0}{2\alpha^2}(E-M)+\frac{\left(E^2-M^2\right)}{\alpha^2}\right] \tag{23}$$

and the wave function for $n=1$ is

$$\chi_{r1}(r)=\left(\frac{1-\coth(\alpha r/2)}{2}\right)^{\delta}\left(1-\frac{1-\coth(\alpha r/2)}{2}\right)^{\gamma}\left[-C_2(2\delta+1)\left(1+\frac{(-2\delta-2\gamma-2)}{2\delta+1}\left(\frac{1-\coth(\alpha r/2)}{2}\right)\right)\right] \tag{24}$$

3.2 θ_1 Equation

The result of separation Eq (6) in variable θ_1 is written as

$$\frac{1}{P_1}\left(\frac{d^2P_1}{d\theta_1^{\,2}}\right)-V(\theta_1)(E-M)=\lambda_1 \tag{25}$$

Where λ_1 is constant and P_1 is an equation respect to θ_1. Eq (25) is reduced to homogeneous linear second order differential equation by inserting Eq (2) and taking $\cot\theta_1=(1-2z_1)i$, $P_1=z_1^{\,\delta}(1-z_1)^{\gamma}f_n(z_1)$, than solve second differential of P_1 and we have

$$z_1(1-z_1)f_n''+[1+2\delta-z_1(2\delta+2\gamma+2)]f_n'-\left[(\delta+\gamma)^2+\delta+\gamma+V_1(E-M)\right]f_n=0 \tag{26}$$

We can determine s_0 and k_0 from Eq (26). From Eq (19) we find Δ_n

$$(\delta+\gamma+n)(\delta+\gamma+n+1)+V_1(E-M)=\Delta_3=0 \tag{27}$$

Can be rewrite as

$$(\delta+\gamma+n+1)^2-(\delta+\gamma+n+1)+V_1(E-M)=\Delta_n=0 \tag{28}$$

Where

$$\delta=\sqrt{\frac{V_2i(E-M)+\lambda_1}{4}},\ \gamma=\sqrt{\frac{-V_2i(E-M)+\lambda_1}{4}} \tag{29}$$

By inserting Eq (29) to Eq (28) we have

$$\lambda_1=\frac{[V_2i(E-M)]^2}{4\left[\sqrt{-V_1(E-M)+\frac{1}{4}}-\left(n+\frac{1}{2}\right)\right]^2}+\left[\sqrt{-V_1(E-M)+\frac{1}{4}}-\left(n+\frac{1}{2}\right)\right]^2 \tag{30}$$

the wave function for θ_1 section for $n=1$ is

$$P_{11}=\left(\frac{1}{2}+\frac{i\cot\theta_1}{2}\right)^{\delta}\left(\frac{1}{2}-\frac{i\cot\theta_1}{2}\right)^{\gamma}\left[-C_2(2\delta+2)\left(1-\frac{2\delta+2\gamma+2}{2\delta+1}z_1\right)\right] \tag{31}$$

3.3 θ_2 Equation

Variable θ_2 from separation of Eq (6) written as

$$\frac{\lambda_1}{\sin^2\theta_2}+\frac{1}{P_2}\left(\frac{1}{\sin\theta_2}\frac{d}{d\theta_2}\left(\sin\theta_2\frac{dP_2}{d\theta_2}\right)\right)-V(\theta_2)(E-M)=\lambda_2 \tag{32}$$

By using $P_2=\frac{Q_2(\theta_2)}{\sin^{1/2}\theta_2}$, $\cot\theta_2=i(1-2z_2)$, and $Q_2(z_2)=z_2^{\,\alpha}(1-z_2)^{\beta}f_n(z_2)$ Eq (32) become

$$z_2(1-z_2)f_s''(z_2)=-[1+2\alpha-z_2(2+2\alpha+2\beta)]f_s'(z_2)+\left\{(\alpha+\beta)^2+\alpha+\beta+\lambda_1-V'(E-M)+\frac{1}{4}\right\}f_s(z_2) \tag{33}$$

Where

$$\alpha=\sqrt{\left(\frac{1}{4}-\lambda_2\right)/4},\ \beta=\sqrt{\left(\frac{1}{4}-\lambda_2\right)/4} \tag{34}$$

We can find Δ_n from Eq (33). We have

$$\Delta_n=(\alpha+\beta+n+1)^2-(\alpha+\beta+n+1)+\lambda_1-V'(E-M)+\frac{1}{4} \tag{35}$$

So we have eigenvalue from θ_2 equation

$$\lambda_2=\frac{1}{4}-\left[\sqrt{-\lambda_1+V'(E-M)}-\left(n+\frac{1}{2}\right)\right]^2 \tag{36}$$

the wave function for θ_2 section for $n=1$ is

$$P_{21}(z_2)=\left(\frac{1}{2}+\frac{i\cot\theta_2}{2}\right)^a\left(\frac{1}{2}-\frac{i\cot\theta_2}{2}\right)^b\left[-C_2(2a+2)\left(1+\frac{-(2a+2b+2)}{2a+1}\left(\frac{1}{2}+\frac{i\cot\theta_2}{2}\right)\right)\right] \tag{35}$$

3.4 θ_3 Equation

Separation variable of Eq (6) for θ_3 is

$$\frac{\lambda_2}{\sin^2\theta_3}+\frac{1}{P_3}\left(\frac{1}{\sin^2\theta_3}\frac{d}{d\theta_3}\left(\sin^2\theta_3\frac{dP_3}{d\theta_3}\right)\right)-V(\theta_3)(E-M)=\lambda_3 \tag{35}$$

Using AIM, consider $P_3=\frac{Q_3(\theta_3)}{\sin^{2/2}\theta_3}$, $\cot\theta_3=i(1-2z_3)$, $Q_3=z_3^{\alpha}(1-z_3)^{\beta}f_n(z_3)$ we reduce Eq (35) to be homogeneous linear second order differential equation

$$z_3(1-z_3)f_n''+[2\alpha+1-z_3(2+2\alpha+2\beta)]f_n'-\{(\alpha+\beta)^2+\alpha+\beta-4V_1''(E-M)+4\lambda_2\}f_n=0 \tag{36}$$

From Eq (36), we have

$$\Delta_n=(\alpha+\beta+n+1)^2-(\alpha+\beta+n+1)-4V_1''(E-M)+4\lambda_2 \tag{37}$$

Where

$$\alpha=\sqrt{V_2'''i(E-M)-(\lambda_3+1)},\ \beta=\sqrt{-V_2'''i(E-M)-(\lambda_3+1)} \tag{38}$$

By inserting Eq (38) to Eq (37) we obtain

$$\lambda_3=-\frac{[V_2'''i(E-M)]^2}{\left[\sqrt{4V_1''(E-M)-4\lambda_2+\frac{1}{4}}-\left(n+\frac{1}{2}\right)\right]^2}-\frac{\left[\sqrt{4V_1''(E-M)-4\lambda_2+\frac{1}{4}}-\left(n+\frac{1}{2}\right)\right]^2}{4}-1 \tag{39}$$

the wave function for θ_3 section for $n=1$ is

$$P_{31}(z_3)=\left(\frac{1}{2}+\frac{i\cot\theta_3}{2}\right)^a\left(\frac{1}{2}-\frac{i\cot\theta_3}{2}\right)^b\left[-C_2(2\alpha+2)\left(1+\frac{-(2\alpha+2\beta+2)}{2\alpha+1}z_3\right)\right] \tag{40}$$

3.5 θ_4 Equation

The last equation from separating Eq (6) is

$$\frac{1}{\sin^3\theta_4}\frac{d}{d\theta_4}\left(\sin^3\theta_4\frac{dP_4}{d\theta_4}\right)-\left(V(\theta_4)(E-M)-\lambda_4+\frac{\lambda_3}{\sin^2\theta_4}\right)P_4=0 \tag{41}$$

With $P_4=\frac{Q_4}{\sin^{3/2}\theta_4}$, $\cot\theta_4=i-2iz_4$, and $Q_4=z_4^{\delta}(1-z_4)^{\gamma}f_n(z_4)$ we have

$$z_4(1-z_4)f_n''(z_4)+[1+2\delta-z_4(2+2\delta+2\gamma)]f_n'(z_4)+\left\{-(\delta+\gamma)^2-\gamma-\delta+\lambda_3+\frac{3}{4}\right\}f_n(z_4)=0 \tag{42}$$

Than we find Δ_n

$$(\delta+\gamma+n+1)^2-(\delta+\gamma+n+1)-\lambda_3-\frac{3}{4}=\Delta_n=0 \tag{43}$$

Where

$$\delta=\sqrt{\frac{V_3^{''}l(E-M)+\left(\frac{9}{4}+\lambda_4\right)}{4}},\ \gamma=\sqrt{\frac{-V_3^{''}l(E-M)+\left(\frac{9}{4}+\lambda_4\right)}{4}} \tag{44}$$

By inserting Eq (44) to Eq (43) the eigenvalue represent as

$$\left[\sqrt{\lambda_3+1}-\left(n+\frac{1}{2}\right)\right]^2+\frac{\frac{1}{4}\left(V_3^{''}l(E-M)\right)^2}{\left[\sqrt{\lambda_3+1}-\left(n+\frac{1}{2}\right)\right]^2}-\frac{9}{4}=\lambda_4 \tag{45}$$

the wave function for θ_2 section for $n=1$ is

$$P_{41}=\left(\frac{1}{2}-\frac{\cot\theta_4}{2i}\right)^p\left(1-\frac{1}{2}+\frac{\cot\theta_4}{2i}\right)^q\left(-C_2(2p+2)\left(1-\frac{2p+2q+2}{2p+1}z_4\right)\right) \tag{46}$$

4. Result and Discussion

Energy bound state obtained by calculate mathematically the Eq (23), and substituting λ_4 in Eq (39)

Table 4.1 Eigenvalue of 5 dimensional Dirac equation with Hulthen potential and Manning-Rosen potential in the case of pseudo spin symmetry, with variation n_r and α

n_r	$E_{n_r}\left(fm^{-1}\right)$		
	$\alpha=0.25\,fm^{-1}$	$\alpha=0.5\,fm^{-1}$	$\alpha=0.75\,fm^{-1}$
1	4.998354	4.993412	4.985164
2	4.993412	4.973592	4.940374
3	4.985164	4.940374	4.864756
4	4.973592	4.893468	4.756757
5	4.958673	4.832453	4.613969

Where n_r is the orbit of electron, and α is screening parameter of Hulthen potential that work on radial section. Table 4.1 shows positive value of energy, that mean particles feel a repulsive force. The increasing of n_r has effect of decreasing of bound state energy. Higher the orbit of electron the bound state become lower. The reason is the higher orbit of electron, the distance electron from nucleus become longer. The increasing of screening parameter doesn't affect with increasing the number of n_r. It shows that decreasing energies analogously with increasing of n_r and wider range of energies.

Eigen function for radial section shown in Eq (24). Eigen function θ_1 shown in Eq (31). Eigen function for section θ_2 shown in Eq (35), for section θ_3 and section θ_4 shown in Eq (40) and Eq (46).

References

[1] Saregar A. 2013. *International Journal of Enginerring And Science. ISSN: 2278-4721, Vol. 2, Issue 3.*
[2] Das T. 2014. *arXiv:1409.1457v1 [quant-ph].*
[3] Agboola D 2009 *arXiv:0811.4441v2 [math-ph].*
[4] Bhagwati N and Singh N N 2014 *Indian J Phys DOI 10.1007/s12648-014-0509-3.*
[5] Falaye B J and Oyewumi K J 2011 *The African Review of Physics 6:0025.*
[6] Hall R L and Zorin P 2016 *CUQM 03.65.Pm, 03.65.Ge, 36.20.Kd.*
[7] Bahar M K and Yasuk F 2012 *DOI 10.1007/s00601-012-1461-8.*
[8] Ikhdiar S M and Sever R 2013 *arXiv:0807.2085v1 [quant-ph].*
[9] Soylu A, Bayrak O, Boztosun I 2007 *Journal of Mathemathical Physics 48, 082302.*
[10] E. Ateser, H. Cifti, and M. Ugurlu, Chinese Journal Of Physics Vol. 45, No. 3.
[11] O. Bayarak, I. Boztosun, H. Ciftci 2006 *Wiley InterScience DOI 10.1002/qua.21141.*

Automatic Detection of Atrial Fibrillation Using Basic Shannon Entropy of RR Interval Feature

Adfal Afdala[1], Nuryani Nuryani[2] and Anto Satriyo Nugroho[3]

[1]Physics Department of Post Graduate Program SebelasMaret University

[2]Informatics Department of Under Graduate Program SebelasMaret University
Jl. Ir. Sutami 36A KentinganJebres Surakarta 57126, INDONESIA.

E-mail: afdala17@gmail.com

Abstract. Atrial Fibrillation is one of heart disease, that common characterized by irregularity heart beat. Atrial fibrillation leads to severe complications such as cardiac failure with the subsequent risk of a stroke. A method to detect atrial fibrillation is needed to prevent a risk of atrial fibrillation. This research uses data from physionet in atrial fibrillation database category. The performance of Shannon entropy has the highest accuracy if a threshold is 0.5 with accuracy 89.79%, sensitivity 91.04% and specificity 89.01%. Based on the result we get a conclusion, the ability of Shannon entropy to detect atrial fibrillation is good.

1. Introduction

Atrial fibrillation (AF) is the most common cardiac arrhythmia, whose prevalence is 0.4%–1% of the general population and increases with age. The prevalence in individual over the age of 80 is about 8% in a developed country [1]. This caused by food and a lifestyle of prevalence. The another factor is atrial fibrillation difficult to detect. Atrial fibrillation can occur in a range a few seconds to hours, that means the short term ECG can't give more information about atrial fibrillation [2], that purpose automatic detection is needed to detect atrial fibrillation.

One way to detect atrial fibrillation is to study the morphology of electrocardiogram. Atrial fibrillation has morphological characteristics in the electrocardiogram that was missed of P wave, and irregularity of heart rate which is represented by R peaks [3][4]. Atrial fibrillation can be detected by observing morphological ECG signal. They are R-R Interval and P waves [5]. R-R interval used as a feature in this research to characterized atrial fibrillation in a time interval.

2. Numerical Methods

Database of atrial fibrillation from physionet was used. Database consist of 23 patient with frequency sampling 250 Hz. A segment consist of 10^4 data, if we convert to the time domain, each segment equals to 40 s ECG measurement. Process on extraction features divided three parts, they are finding R peaks, calculate R-R Interval, and finding maximum entropy occur in each segment. Algorithm of Pan-Tomkins is one of a method to finding R peaks. The ability of this method to detect R peaks has accuracy 99.3% [6].

Pan-Tomkins algorithm consist of some process, they are:

1. Bandpass filter
2. Derivative
3. Squaring Function
4. Integration
5. Adjust threshold
6. Decision

R-R interval is a difference of position R(n) from position R(n-1), it is the main characteristic of atrial fibrillation.

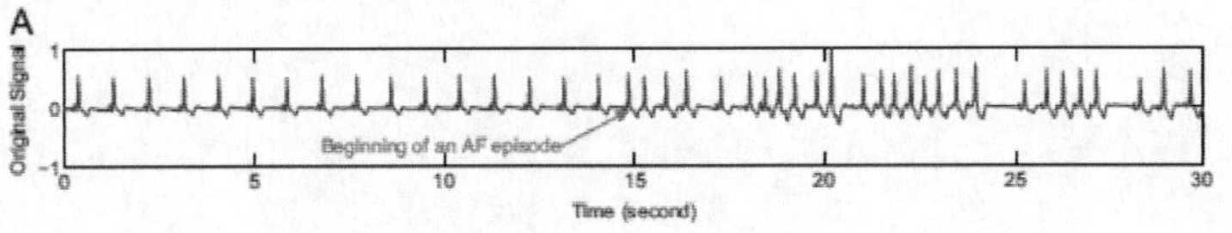

Figure 1. R peaks are characteristic of AF [6]

Figure 1 shows R-R interval as characteristic of atrial fibrillation. RR interval feature has been proposed and investigated by some researcher. Automatic detection of atrial fibrillation using stationary wavelettransform and support vector machine presented in [7]. A simple method for detection atrial fibrillation using RR interval presented in [8]. The next process is finding out the maximum beat in each segment in this process. Shannon entropy applied to get entropy of each segment. Shannon entropy calculates based on equation 1 [9][10].

$$S = -\sum_{i}^{n} P_i log P_i \qquad (1)$$

The next process classifies the feature data using artificial neural network backpropagation. This process divided two parts, they are training process and testing process. Each process using random technique sampling for data training and data testing. The ratio of data training and data testing is 10% and 90%. The flow chart of this article represents in figure 2.

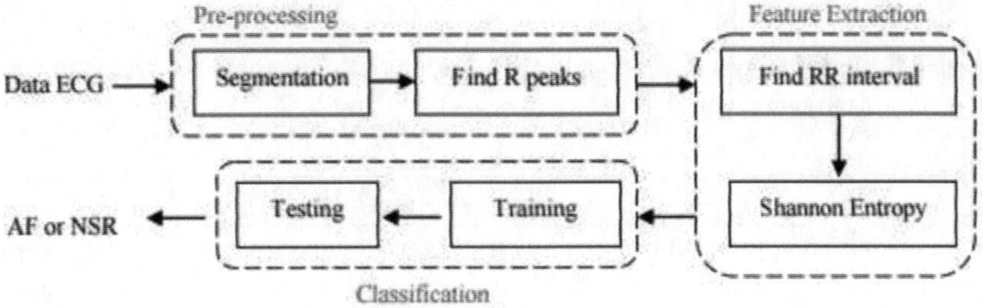

Figure 2. Procedure of the research

3. Results and Discussion

Each patient has different length of a segment where the dominant length is 921 segment. Each segment divided by nine intervals, that means one patient has 921 segments and each segment has ninefeature. To knows the performance of feature, the principal component analysis method applied to this research. The result of this process presented in figure 3.

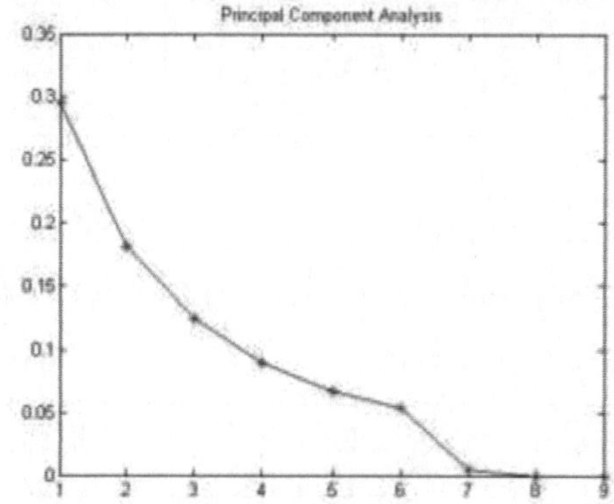

Figure 3. Eigenvector PCA of feature patient 5

Figure 3 shows the eigen vector of each feature, feature 7 to 9 has a lower value than a previous feature. Feature who has lower value reduced from data. That will be minimized computational weight process.

The important process in this research is feature extraction process. Shannon entropy of RR interval is one of feature which will be tested using artificial neural network backpropagation to detect atrial fibrillation. The result of Shannon entropy for each segment presented in table 1.

Table 1. Entropy

Segment	H(1)	H(2)	H(3)	H(4)	H(5)	H(6)	H(7)	H(8)	H(9)	Label
10	0	0	0	0	0	0	0	81.29	143.8	NonAF
15	0	0	0	0	0	0	1.255	316.7	214.1	NonAF
30	0	0	0	0	0	0	0	43.94	222.2	NonAF
400	0	0	0	158.3	28.14	47.74	59.64	9.053	0	AF
450	0	0	2.593	27.91	2.593	8.962	8.962	36.95	0.197	AF
550	0	0	2.78	9.536	106	21.4	49.56	9.536	0.213	AF

Table 1 shows the calculation of Shannon entropy in segment AF and nonAF. Based on table 1, each segment has maximum entropy. If we apply binarization for entropy, we set maximum entropy is 1 and else is zero. Entropy in table 1 becomes presented in table 2.

Table 2. Binarization of entropy

Segment	H(1)	H(2)	H(3)	H(4)	H(5)	H(6)	H(7)	H(8)	H(9)	Label
10	0	0	0	0	0	0	0	0	1	NonAF
15	0	0	0	0	0	0	0	1	0	NonAF
30	0	0	0	0	0	0	0	0	1	NonAF
400	0	0	0	1	0	0	0	0	0	AF
450	0	0	0	0	0	0	0	1	0	AF
550	0	0	0	0	1	0	0	0	0	AF

Table 2 shows segment AF and nonAF have characteristic a part of maximum entropy. Based on table 2 we get a feature to apply in artificial neural network backpropagation. Parameters in the artificial neural network are learning rate, epoch, and momentum. Learning rate is 0.05, epoch 5000, and momentum 0.9.The result of training process presented in figure 4.

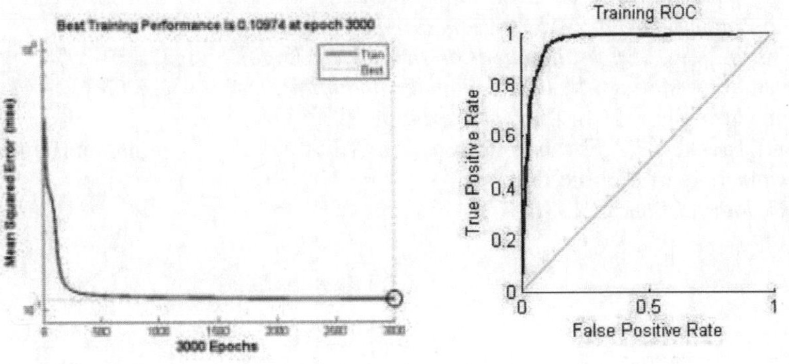

(a) (b)

Figure 4.(a) Performance of MSE training process (b) ROC training process

Figure 4a shows a performance of mean square error in the training process. The smallest value of themean square error is 0.109. The mean square error of training process decreases significantlybetween epoch 1 to 300. Training will be stopped if a mean square error has been convergent.

Figure 4b shows performance training process to knows how well system training predicts AF and NonAF. Based on figure 4b the system has good performance, but the testing process in the difference data set is needed to detect how well the system performance in new data set. This process called testing process. We applied system on training process to detect AF or nonAF in a new data set.This is an output of this research. The result of this process presented in table 3.

Table 3. Performance training system in new data

	Threshold										
Performance	0	0.1	0.2	0.3	0.4	0.5	0.6	0.7	0.8	0.9	1
Accuracy	38.51	77.46	78.21	85.39	85.39	84.83	84.79	84.24	79.59	70.95	61.49
Sensitivity	100	95.86	95.22	85.45	85.31	73.68	72.56	70.79	54.43	28.57	0
Specificity	0	65.45	67.06	84.72	84.82	91.15	91.78	91.98	94.66	96.79	99.27

Table 3 shows the output of this system in some threshold to get the decision AF or NonAF. Accuracy ,sensitivity, and specificity are 85.39%, 85.31%, and 84.82% respectively. The result has a good level to detect atrial fibrillation.

4. Conclusion

This article presents a new feature for detecting atrial fibrillation. This feature is Shannon entropy for RR interval and classifier by artificial neural network. The result of performance is good. Shannon entropy has the ability to detect atrial fibrillation with accuracy, sensitivity, and specificity. The best obtained accuracy, sensitivity, and specificity were 85.39%, 85.31%, and 84.82% respectively.

References

[1] Sarath R .Dessertation.Atrial Fibrillation Signal Analysis. University of Florida,2011.
[2] Dizon L.Thesis.Atrial Fibrillation Algorithm Evaluation and Implementation in Java. Royal Institute of Technology.Swaden.2014.
[3] Camm A J 2012*Europan Society Of Cardiology*.**14**,1385-1413.
[4] Babaeizadeh S *et al* 2009 *JournalElectrocardiology***42** 522.
[5] Padmavathi K and Ramakrishna K 2015*Procedia Computer Science***46**53-59.
[6] Pan J and Tomkins W 1985*IEEE Transactions on Biomedical Engineering***32** 230-235.
[7] Shadnaz A, Alireza M, & Maryam M 2015 *Computers in biology and medicine***60** 132-142.
[8] Lian J, Wang L and Muessig D 2014 *Journal Cardiology***107** 1494-1497.
[9] Bromiley P A and Tachker N.A.Shannon Entropy,Renyi Entropy and Information.Biomedical Engineering.University of Manchester.2010.
[10] Zhou X *et al* 2015 *Journal Pone***10** 1371.

IOP Conf. Series: Journal of Physics: Conf. Series **795** (2017) 012039 doi:10.1088/1742-6596/795/1/012039

Genetics algorithm optimization of DWT-DCT based image Watermarking

Gelar Budiman, Ledya Novamizanti, and Iwan Iwut

Electrical Engineering Faculty, Telkom University
Jl. Telekomunikasi 1 Dayeuhkolot Bandung, Indonesia

gelar.budiman@gmail.com, iwan.tritoasmoro@gmail.com, ledyamizan@gmail.com

Abstract. Data hiding in an image content is mandatory for setting the ownership of the image. Two dimensions discrete wavelet transform (DWT) and discrete cosine transform (DCT) are proposed as transform method in this paper. First, the host image in RGB color space is converted to selected color space. We also can select the layer where the watermark is embedded. Next, 2D-DWT transforms the selected layer obtaining 4 subband. We select only one subband. And then block-based 2D-DCT transforms the selected subband. Binary-based watermark is embedded on the AC coefficients of each block after zigzag movement and range based pixel selection. Delta parameter replacing pixels in each range represents embedded bit. +Delta represents bit "1" and –delta represents bit "0". Several parameters to be optimized by Genetics Algorithm (GA) are selected color space, layer, selected subband of DWT decomposition, block size, embedding range, and delta. The result of simulation performs that GA is able to determine the exact parameters obtaining optimum imperceptibility and robustness, in any watermarked image condition, either it is not attacked or attacked. DWT process in DCT based image watermarking optimized by GA has improved the performance of image watermarking. By five attacks: JPEG 50%, resize 50%, histogram equalization, salt-pepper and additive noise with variance 0.01, robustness in the proposed method has reached perfect watermark quality with BER=0. And the watermarked image quality by PSNR parameter is also increased about 5 dB than the watermarked image quality from previous method.

1. Introduction

Digital watermarking is a method to embed any specific information into a content either sound, image, or video [1]. Any specific information embedded into the content is called watermark. And the content as a place for embedding the data is known as host. The transparency of watermark depends on the watermarking application needs. It consists of 2 kind of transparency, such as : perceptible and imperceptible. Generally, watermark is embedded in the way that the human can not sense or the watermark is imperceptible. The imperceptibility of watermark in the content is important for keeping the content quality into human perception. Human still assume that the content is original and they enjoy the content without unsatisfaction because of the lack of content quality due to watermark existing.

Image watermarking is defined as a watermarking in which the host is a digital image. The watermark could be any digital information, as example text, audio or image having ownership information of the content. There are several parameter that controls the image watermarking performance, such as [2]:

a. Imperceptibility – In this paper we design the image watermarking in which the perceptibility of the watermarks is imperceptible. It means that the watermarks embedded into host image have to be invisible. The invisibility parameter is represented by Peak Signal To Noise Ratio (PSNR).
b. Robustness – The watermarks must have good robustness to face the watermarked image attack. Stirmark benchmark is a tool representing image or audio attack. It is useful to examine the robustness of embedded watermarks. Stirmark benchmark described all image watermarking attacks types which consists of compression, geometric transformation, and enhancement techniques. The performance parameter used for robustness is Bit Error Rate (BER). If the watermark is an image, the performance parameter could be (Structural Similarity) SSIM or PSNR.
c. Capacity – The number of bits which is embedded into the image is also a parameter for watermarking performance. Although the capacity of watermark is not as well as robustness but it is needed for controlling the trade-off between robustness and imperceptibility.

DCT-based image watermarking were already published in several papers. J.R. Hernadez et all [3] Yoonki et all [4] proposed image watermarking utilizing inter-block correlation of DCT coefficients. DCT-based image watermarking method and mixed with time-domain was proposed by Dongyang et all [5]. Dongyang generated watermarked image by using both domain, time domain and frequency domain. Qiusheng Wang et all [6] proposed DCT-based image independent image watermarking. He generated the watermarks from original host image characteristic first, and then embedded it into the host image in frequency domain similar as JPEG compression. Juan et all [3] proposed a spread-spectrum-like DCT watermarking technique for copyright protection of still digital images. He made the watermark in such a way that the watermark can have information to track illegal misuses.

There are several authors who proposed DWT-based image watermarking. Qiudong Sun et all [7] used HVS as selected color scale for hiding the watermark after selecting subband from DWT decomposition. The embedding method in his published paper is just noticeable difference (JND) in detail subband feature adjustment. Chi Ma et all [8] proposed a self-adaptive DWT digital watermarking algorithm. She decomposed three-level wavelet of image before embedding the watermark by adaptive selected coefficients of detail subband. Shih-Hsuan Yang [9] proposed DWT-domain image watermarking by multiresolution analysis. She used seven biorthogonal DWT kernels for embedding the watermark. In another case, Alessandro Piva et all [10] used DWT as transform method before embedding watermark in frame-basis at video watermarking in MPEG-4 standard. Khrisna et all [11] proposed block based robust blind image watermarking using DWT. He used DWT decomposition and SVD before embedding is applied. Yuan Xu et all [12] proposed scale invariant feature transformation (SIFT) after DWT decomposition for embedding watermark. Optimized DWT-based image watermarking in Genetics Algorithm was proposed by Ali Al Haj et all [13]. He used selected subband and specific watermark-amplification parameter value for embedding watermark, and optimized both parameters by GA.

DWT-DCT based image watermarking were also already published by several authors. Afroja Akter et all [14] proposed DWT-DCT based image watermarking by two until four level DWT decomposition. He decomposed host image into 2 until 4 level DWT decomposition, then transformed by DCT before embedding. The watermark first is transformed to frequency domain by DCT before embedded into host audio. Baisa et all [15] proposed multilayer secured DWT-DCT and YIQ color space based image watermarking technique. He embedded watermark at HL region of DWT and used 4x4 DCT.

Image watermarking optimization by GA on DWT-DCT method has been proposed by Abduljabbar et all [16]. But in his paper, the method was not clear described. He only described that the watermark is embedded at DC coefficient after DCT.

In this paper we propose image watermarking by DWT and DCT transform method in certain color space and layer, and optimized by Genetics Algorithm. First, the host RGB image is converted to certain color space. The available and chosen color spaces are RGB, YCbCr or NTSC. The layer in which the watermark is embedded also can be selected. The available choices are 1st layer, 2nd layer, 3rd layer, 1st & 2nd layer, 2nd & 3rd layer, 1st & 3rd layer and all layers. After the selected layer of image in certain color space is transformed by DWT, one subband from subband LL, LH, HL, and HH is selected and then transformed again in block based to frequency domain by DCT. Then one bit watermark is embedded on the AC component of each block such a way that the bit is represented by specific value called delta in a zigzag and vary length of pixel. The vary parameters optimized by Genetics Algorithm are selected color space, selected layer, selected subband, block size, length of pixel to be embedded by one bit watermark, and delta. Bit "1" is represented by +delta, and bit "0" is represented by –delta in vary length of pixel after zigzag.

This paper is one of series paper as the research result funded by high education ministry of Indonesia at 2016. The previous published paper by Iwan Iwut [17] described DCT-based image watermarking designed is optimized by GA without DWT decomposition. In this paper we propose DWT decomposition added in the embedding and extraction before DCT conversion to improve the performance of image watermarking.

This paper is organized as follow : section 1 describes image watermarking introduction, section 2 describes the image watermarking model containing DWT, DCT, and embedding-extraction algorithm, section 3 presents GA optimization model, and section 4 describes evaluation of performance and discussion and the conclusion is presented in section 5.

2. Image Watermarking Model

In this section, image watermarking at embedding and extraction stage with the preprocessing of the image is described. Preprocessing before embedding stage consists of color space and layer choice for embedding watermark, DWT decomposition, DCT transformation and AC coefficient selection. The rest of this section describes embedding and extraction algorithm.

2.1. Color Space, Layer, DWT Subband and DCT

In this proposed method there are 3 color spaces used and selected only one of them. And there are also 7 layers used and selected only one of them. Three color spaces used are RGB, YCbCr, and NTSC color space. The conversion from RGB into YcbCr, NTSC and vice versa are as follow [15] [18] [17]:

$$Y=0.299*R + 0.587*G + 0.114*B \quad (1)$$
$$I=0.596*R - 0.274*G - 0.322*B \quad (2)$$
$$Q=0.211*R - 0.522*G + 0.311*B \quad (3)$$
$$R=Y + 0.956*I + 0.621*Q \quad (4)$$
$$G=Y - 0.272*I - 0.647*Q \quad (5)$$
$$B=Y - 1.106*I + 1.702*Q \quad (6)$$
$$Y=16 + 65.481*R + 128.553*G + 24.966*B \quad (7)$$
$$Cb=128 - 37.797*R - 74.203*G + 112*B \quad (8)$$
$$Cr=128 + 112*R - 93.786*G - 18.214*B \quad (9)$$
$$R=1.1644*Y + 0*Cb + 1.596*Cr - 222.921 \quad (10)$$
$$G=1.1644*Y - 0.3918*Cb - 0.7856*Cr + 135.576 \quad (11)$$
$$B=1.1644*Y + 2.0172*Cb + 0*Cr - 276.836 \quad (12)$$

The choice layer used from 3 layer image are : layer 1, layer 2, layer 3, layer 1-2, layer 1-3, layer 2-3, and layer 1-2-3. Layer 1, 2 and 3 for RGB is R, G, and B. Layer 1, 2 and 3 for NTSC is Y, I, and Q. And layer 1, 2 and 3 for YCbCr is Y, Cb, and Cr. Selected color space and selected layer used are two parameters optimized by GA. Thus, color space parameter value is limited on range 1-3, and layer parameter value is limited on range 1-7.

DWT is time to time-based transformation. For image, DWT is calculated in 2 dimensions and convert the spatial image to spatial image with separated frequency. A digital image converted by DWT will obtain 4 subbands with different frequency, such as : LL, LH, HL, and HH subband. Only one subband will be selected for embedding process. Thus, the parameter value for subband parameter is limited on range 0-3.

DCT is time to frequency-based transformation at real number. Due to the host is a digital image, DCT used is 2 dimension. DCT will be executed after selecting one subband of DWT decomposition result. DCT calculation is block-based processing with size MxM. The DCT formula is as follow [19] [17]:

$$C(u,v) = \alpha(u)\alpha(v)\sum_{x=0}^{M-1}\sum_{y=0}^{M-1} f(x,y)cos\left[\frac{\pi(2x+1)u}{2M}\right]cos\left[\frac{\pi(2y+1)v}{2M}\right] \quad (13)$$

$$f(x,y) = \sum_{x=0}^{M-1}\sum_{y=0}^{M-1} \alpha(u)\alpha(v)C(u,v)cos\left[\frac{\pi(2x+1)u}{2M}\right]cos\left[\frac{\pi(2y+1)v}{2M}\right] \quad (14)$$

Where :

u & v = 0, 1, ... M-1

x and y is respectively vertical and horizontal pixel position of the image.

$$\alpha(u) = \begin{cases} \sqrt{\frac{1}{M}}, & \text{for} \quad u = 0 \\ \sqrt{\frac{2}{M}}, & \text{for} \quad u \neq 0 \end{cases} \quad (15)$$

2.2. Embedding process

Embedding process of image watermarking contains several steps as follow :

a. The host image is read and converted to selected color space from RGB to selected color space.
b. Select only one layer from 7 layer used as described in section 2.1.
c. The selected layer is decomposed into 4 subband by DWT.
d. Select only one subband from subband LL, LH, HL and HH.
e. The selected subband is block based segmented into *MxM.*
f. Each segment from step 5 is transformed by DCT to frequency domain.
g. Each segment in frequency domain is zigzag processed excluding DC component produces the vector which its size is (1 x (M^2-1)).
h. The selected vector from zigzag vector scheme is assumed as $s(n)$ and the watermark bit is $w(n)$. $s(n)$ length for embedded is POS2-POS1+1. Where, POS1 is initial range position and POS2 is last range position of $s(n)$. In embedding stage, the selected vector of zigzag vector scheme or $s(n)$ is replaced by one bit watermarks or $w(n)$ in such way as follow:

   If $w(n)$ is "0" then
   $s(n) = -delta$
   if $w(n)$ is "1" then (16)
   $s(n) = +delta$

   delta is a rasional number parameter which can be changed, and it is also optimized by Genetics Algorithm to find the optimal trade off performance between robustness and imperceptibility. The range of *delta* is from 0.1 to 127. Different bit will be embedded in different block.
i. After all bits are embedded and modify $s(n)$, inverse DCT is applied to each block.
j. Combine all block from IDCT result, and IDWT is applied by combining modified subband and unmodified subband.
k. Combine the embedded layer with unused layer and convert back to the RGB color space from selected color space.

2.3. Extraction process

Extraction process of image watermarking consists of several steps. The detailed steps of extraction stage is as follow :

1. The watermarked image is read and converted to color space same as the color space used in embedding process.
2. Select the layer used for extraction same as the layer used in embedding process.

3. The selected layer is decomposed into 4 subband by DWT.
4. Select only one subband from subband LL, LH, HL and HH.
5. The selected subband is block based segmented into *M*x*M*.
6. Each segment from 5 is transformed by DCT to frequency domain.
7. Each segment in frequency domain is zigzag processed excluding DC component produces the vector *s*(*n*) which its size is (1 x (M^2-1)).
8. Select the *s*(*n*) in the range from POS1 until POS2 and do the extraction stage below:

$$\begin{aligned} &\text{If } \textstyle\sum_{n=POS1}^{POS2} s(n) < 0 \text{ then} \\ &\widehat{w}(n) = 0 \\ &\text{if } \textstyle\sum_{n=POS1}^{POS2} s(n) \geq 0 \text{ then} \\ &\widehat{w}(n) = 1 \end{aligned} \quad (17)$$

9. After all segments or blocks are extracted as point 8, recover the bits into a vector of bit, then we get the watermarks back.

3. GA optimization model

The GA model refers to our previous paper in [17]. The parameters to be optimized by Genetics Algorithm such as: color index (*CI*), layer index (*L*), *M*x*M* block segment (*M*), beginning of selected zigzag vector (POS1), end of selected zigzag vector (POS2), and *delta*. The additional parameter for this proposed method is selected subband (*SS*). The output parameters will be BER and PSNR that is combined to one output parameter called Fitness Function (*FF*). In this paper we ignore capacity due to the previous research result in [17] showed that capacity parameter decreased the impact of robustness and imperceptiblity. In fact, robustness and imperceptibility are two more important parameter than capacity. We propose new *FF* which will focus to robustness and imperceptibility only. Thus, the *FF* formula will be as follow :

$$FF = 1 - BER + PSNR/60 \quad (18)$$

Formulation of *BER* and *PSNR* is as follow :

$$BER = \frac{Number\ of\ Bit\ Error\ from\ extraction}{Watermark\ Bit\ Number} \quad (19)$$

$$PSNR = 10 * log\left(\frac{255^2}{MSE}\right) \quad (20)$$

$$MSE = \frac{1}{PQ}\sum_{x=0}^{P-1}\sum_{y=0}^{Q-1}\left(f(x,y) - \hat{f}(x,y)\right)^2 \quad (21)$$

The GA procedure for finding the optimum performance refers to our previous paper [17] displayed in figure 1 is described as follow :

a. Initialization of all parameter inputs and then execute embedding and extraction process to get *FF*
b. Sort the outset population matrix in descending order at *FF*.
c. Execute the tasks below until *FF* is stable for hundred generations :
    1) Select the top ten rows of updated *FF*. This matrix is also called as elite matrix.
    2) Do crossover 3 times producing 2 children from 2 parents in each iteration. And do mutation with specific mutation probability. Crossover and mutation result combines with elite matrix and obtain updated *FF*.
    3) If there is no change of highest value of updated *FF* for more than 100 generations, the parameters are already optimum. Thus we get the optimum parameters at top of last updated *FF* matrix.

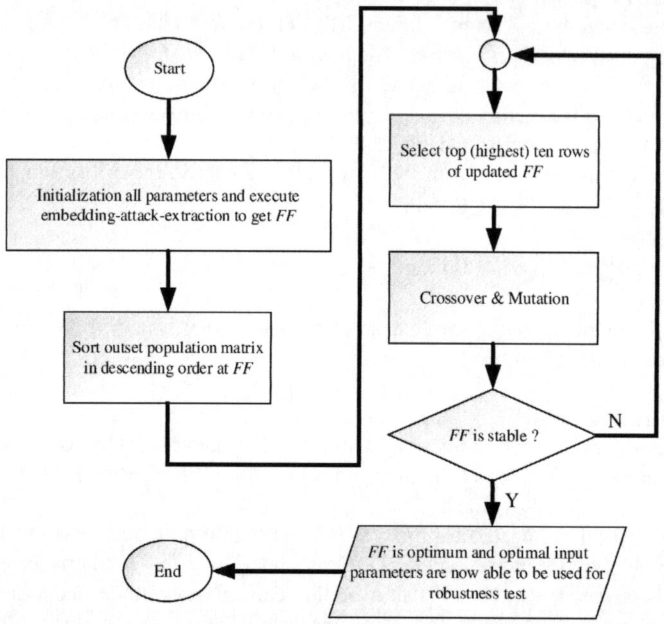

Figure 7. Genetics Algorithm Procedure [17]

4. Performance Evaluation

The host and watermark image to be tested and GA parameters refers to our previous research [17]. There are 5 host images with free copyright and binary watermarks to be tested. "Fruits.png" image will be tested image for optimizing the watermarking parameter at no attack and JPG-rescale attack condition.

Table 1. Parameter Optimization Result for "Fruits.png" from No attack and JPG/Rescale attack

FF	Delta	Blok	Warna	Layer	Pos1	Pos2	Subband	BER	C	PSNR	Attack	Method
1.53	28	5	3	2	11	16	-	0	113	31.54	JPEG 75%	[17]
1.81	2	7	2	3	23	26	-	0	196	48.46	No Attack	[17]
1.60	41	10	3	0	42	53	LL	0	112	36.03	JPEG 50% + Rescale 50%	Proposed
1.90	3	6	3	1	14	18	HH	0	185	53.91	No Attack	Proposed

The simulation result of finding optimum parameter by GA with proposed method comparing with previous method [17] is displayed in table 1. In previous method, GA was applied to no attack and JPEG 75% attack obtaining PSNR 48.46 dB and 31.54 dB with perfect robustness (BER=0). With additional DWT in the embedding-extraction process, proposed method improves the performance. By no attack and stronger double attack (JPEG 50%+Rescale 50%), GA by proposed method obtains PSNR 53.91 dB and 36.03 dB with perfect robustness. The robust subband used for stronger and double attack is LL.

By optimized parameter from table 1, image watermarking with 5 host images and 5 attacks are applied. Five attacks include JPEG compression 75% quality, salt-pepper and additive noise with variance 0.01, resizing 50% and histogram equalization attack. And simulation result of proposed method comparing with the previous simulation research [17] is displayed in table 2. By same attacks as the previous research [17], the proposed method reach perfect result. The proposed method is robust perfectly to all attacks with BER=0. Additional DWT in the embedding-extraction algorithm has significantly improved the robustness of the image watermarking. Table 2 also displays the imperceptibility/fidelity result of watermarked image by optimized parameter from table 1. Comparing with previous research, the imperceptibility or PSNR also has improved to more than 35 dB. In the previous research PSNR only reached in about 30-31 dB.

Table 2. Robustness and imperceptibility test result for "Fruits.png" compared to previous research

Image	PSNR		BER									
			JPEG Compression 75%		Salt & Pepper 0.01		Additive Noise 0.01		Resizing 50%		Histogram Equalization	
	[17]	prop.	[17]	prop.	[17]	prop.	[17]	prop.	[17]	prop.	[17]	prop.
Tulips	30.8934	35.0406	0	0	0.02	0	0.01	0	0.18	0	0	0
Airplane	30.9459	35.1692	0	0	0	0	0.02	0	0.02	0	0	0
Baboon	31.1132	35.3904	0	0	0	0	0.02	0	0.22	0	0	0
Peppers	31.0832	35.5149	0.01	0	0	0	0.01	0	0.22	0	0	0
Fruits	31.5359	36.0329	0	0	0	0	0.02	0	0.125	0	0	0

Table 3. Image watermarking robustness against attack with vary value parameters

Image	BER												
	JPEG Compression				Additive Noise			Salt & Pepper			Resizing		
	Quality Factor				Variance			Variance			Scaling Factor		
	50%	25%	10%	0%	0.01	0.05	0.1	0.01	0.1	0.5	40%	30%	25%
Tulips	0	0.0104	0.0313	0.3646	0	0.0104	0.0417	0	0.0208	0.2604	0	0.1146	0.3125
Airplane	0	0	0.0833	0.3021	0	0	0.0625	0	0.0208	0.2396	0	0.0417	0.1458
Baboon	0	0	0.0313	0.0963	0	0	0.0625	0	0.0208	0.2188	0	0.0521	0.3438
Peppers	0	0	0.0625	0.3542	0	0.0104	0.0625	0	0.0104	0.1875	0	0.0729	0.2292
Fruits	0	0	0.0729	0.3854	0	0.0208	0.0729	0	0.0104	0.2396	0	0.0313	0.1354

Table 3 displays image watermarking robustness against several attack with vary value of its parameter. Image watermarking applies the optimized parameter from table 1. JPEG attack has quality factor (QF) 50%, 25%, 10% and 0%. Additive noise attack has variance 0.01, 0.05, and 0.1. Salt & Pepper noise attack has variance 0.01, 0.1, and 0.5. And resizing attack has scaling factor 40%, 30%, and 25%. The robustness against attack result shows normal tendency. For JPEG compression attack, lower the quality factor will be much more of loss image information, and it causes the robustness will be worse. The quality factor of JPEG that can still be resisted by the image watermarking is 25%. Only one image, "tulips.png" has BER 0.0104 when it's attacked by JPEG with QF 25%. Variance of noise, either salt-pepper and additive noise will have any effect to robustness. The higher of variance will be the lower of robustness. And for resizing attack, tolerable scaling factor for this optimized parameter is at 40% which still has perfect quality watermark, that is BER=0.

5. Conclusion

Additional DWT decomposition in DCT based image watermarking optimized by GA has improved the performance of image watermarking. Either imperceptibility or robustness has been improved to significant performance. By five attacks: JPEG 50%, resize 50%, histogram equalization, salt-pepper and additive noise with variance 0.01, robustness in the proposed method has reached perfect watermark quality with BER=0. And the watermarked image quality by PSNR parameter is also increased about 5 dB than the watermarked image quality from previous method.

Acknowledgement
This research is funded by high education ministry of Indonesia in applied research grant at 2016.

References

[1] I. J. Cox, J. Kilian, F. T. Leighton, and T. Shamoon, *IEEE Transactions on Image Processing*, vol. 6, no. 12, pp. 1673–1687, 1997.

[2] C.-S. Lu, S.-K. Huang, C.-J. Sze, H.-Y. M. Liao, S.-K. Huang, and C.-S. Lu, in *Multimedia Image and Video Processing*, CRC Press, 2000, pp. 1–32.

[3] J. R. Hernández, M. Amado, and F. Pérez-González, *IEEE Transactions on Image Processing*, vol. 9, no. 1, pp. 55–68, 2000.

[4] Y. Choi and K. Aizawa, in *International Conference on Image Processing*, 1999, vol. 2.

[5] D. Teng, R. Shi, and X. Zhao, in *Proceedings 2010 IEEE International Conference on Information Theory and Information Security, ICITIS 2010*, 2010, pp. 826–830.

[6] Q. Wang and S. Sun, in *ICSP 2000*, 2000, pp. 942–945.

[7] Q. Sun, P. Guan, Y. Qiu, and W. Yan, *TELKOMNIKA Indonesian Journal of Electrical Engineering*, vol. 11, no. 7, pp. 4151–4158, 2013.

[8] C. Ma and Y. Zhu, *TELKOMNIKA Indonesian Journal of Electrical Engineering*, vol. 11, no. 11, pp. 6281–6289, 2013.

[9] S.-H. Yang, *Electronic Letter*, vol. 39, no. 24, pp. 2–3, 2003.

[10] A. Piva, R. Caldelli, and A. De Rosa, in *International Conference on Image Processing*, 2000, vol. 3.

[11] K. R. Kakkirala and S. R. Chalamala, in *International Colloquium on Signal Processing & its Applications*, 2014, pp. 58–61.

[12] Y. Xu, Q. Zhang, and C. Zhou, *TELKOMNIKA Indonesian Journal of Electrical Engineering*, vol. 11, no. 1, pp. 191–198, 2013.

[13] A. Al-haj and A. Abu-Errub, *Journal of Computer Science*, pp. 210–212, 2008.

[14] A. Akter and M. A. Ullah, in *International Conference on Informatics, Electronics & Vision*, 2014.

[15] B. L. Gunjal and S. N. Mali, *International Journal of Computer Science, Engineering and Information Technology (IJCSEIT)*, vol. 1, no. 3, pp. 36–44, 2011.

[16] A. Shaamala, S. M. Abdullah, and A. A. Manaf, *International Journal of Computer Science Issues*, vol. 8, no. 5, pp. 220–225, 2011.

[17] I. Iwut, G. Budiman, and L. Novamizanti, *TELKOMNIKA Indonesian Journal of Electrical Engineering*, vol. 4, no. 1, pp. 1–16, 2016.

[18] E. Prathibha, S. Yellampalli, and P. A. Manjunath, *International Journal of Computer Science Issues*, vol. 1, no. 1, pp. 13–18, 2011.

[19] A. Khamrui and J. K. Mandal, *Procedia Technology*, vol. 10, pp. 105–111, 2013.

Runge-Kutta and rational block methods for solving initial value problems

Sudi Mungkasi and Agung Christian

Department of Mathematics, Faculty of Science and Technology, Sanata Dharma University, Mrican, Tromol Pos 29, Yogyakarta 55002, Indonesia

E-mail: sudi@usd.ac.id, christian.0309@yahoo.com

Abstract. Three methods to solve initial value problems are considered. The methods are the first order Euler's, second order Heun's, and rational block methods. The Euler's and Heun's methods are of the Runge-Kutta type. Numerical results show that the rational block method is more robust than Runge-Kutta type methods in solving initial value problems.

1. Introduction

Mathematical models are very useful to solve real problems [1-4]. A class of mathematical models is ordinary differential equation (ODE). To solve ODEs in practice, we must have an initial condition and they form an initial value problem. There are a number of methods to solve initial value problems, such as Runge-Kutta methods and rational block methods.

In this paper, we solve initial value problems using the first order Runge-Kutta (Euler's) method, second order Runge-Kutta (Heun's) method and rational block method. Runge-Kutta methods are standard in the fields of numerical ODEs. The rational block method is a combination of rational methods [5]. In this work, we test the performance of the rational block method in comparison to Runge-Kutta methods. We note that the rational block method calculates approximate values of solution at two points in each iteration [6-9]. Runge-Kutta methods calculate approximate values of solution at one point in each iteration. A robust ODE solver is needed in the process of finding a more difficult problem, such as solving partial differential equations numerically [10-14].

The paper is structured as follows. We describe the problem formulation of the rational block method in Section 2. All three numerical methods to be tested are written in Section 3. Numerical results are presented in Section 4. The paper is concluded with some remarks in Section 5.

2. Problem formulation

We recall the general formulation of initial value problems using the rational block method [6-8].

Given the initial value problem:

$$y'(x) = f(x,y), \;\; y(a) = \eta \tag{1}$$

where $f(x,y)\colon \mathbb{R} \times \mathbb{R}^m \to \mathbb{R}^m$ and the initial value problem (1) has a unique solution. In this paper we want to solve the initial value problem from the starting point to the final point. Therefore, we have to form the interval of integration. Let $x \in [x_0, x_b]$ and the interval $[x_0, x_b]$ is descretised for the numerical integration as $\{x_0, x_1, \ldots x_n, x_{n+1}, \ldots, x_b\} \subset \mathbb{R}$.

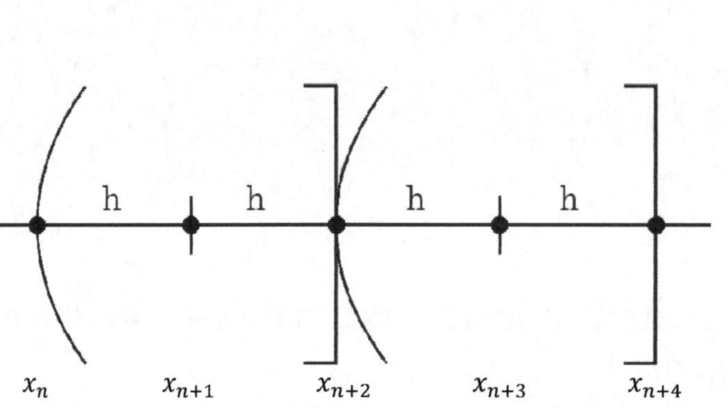

Figure 1. Discretisation of a 2-point explicit rational block method.

The discretisation in the rational block method makes a series of blocks with each point of these blocks is separated by a constant of step-size h, as shown in Figure 1 for a 2-point explicit rational block method. The k-th block contains three points, namely, x_n, x_{n+1} and x_{n+2}. We use y_n at x_n to calculate the approximate value y_{n+1}. Then, we use y_n and y_{n+1} to calculate the approximate value y_{n+2}. On this block, the calculation procces is run in one iteration. That is, the rational block method calculates the approximate values y_{n+1} and y_{n+2} simultaneously in one iteration. Likewise, the $(k+1)$-th block contains three points, namely, x_{n+2}, x_{n+3} and x_{n+4}. We use y_{n+2} at x_{n+2} to calculate the approximate value y_{n+3}. Then, we use y_{n+2} and y_{n+3} to calculate the approximate value y_{n+4}. Like in the k-th and $(k+1)$-th blocks, we calculate the y value on the next blocks until at the point $x = b$ is reached.

The discrete values of x are given by x_n, x_{n+1} and x_{n+2} as

$$x_n = x_0 + nh \tag{2}$$

$$x_{n+1} = x_0 + (n+1)h = x_n + h \tag{3}$$

$$x_{n+2} = x_0 + (n+2)h = x_n + 2\,h \tag{4}$$

where $h = \frac{x_b - x_0}{N}$ is a step-size or the distance from a point to the next one, N is the number of integration steps.

In rational block method, the solution of initial value problem (1) is locally given in the interval $[x_n\,, x_{n+1}]$ by the rational approximant

$$R(x) = \frac{a_0 + a_1 x}{b_0 + x} \tag{5}$$

where a_0, a_1 and b_0 are unknown coefficients. In the calculation process, this rational approximant (5) passes through points (x_n, y_n) and (x_{n+1}, y_{n+1}). To obtain the y_{n+1} value, we assume that the derivative values at x_n is given by $y'_n = f(x_n, y_n)$ and $y''_n = f'(x_n, y_n)$. Therefore, four equations must be satisfied as follows:

$$y_n = \frac{a_0 + a_1 x_n}{b_0 + x_n}\,, \tag{6}$$

$$y_{n+1} = \frac{a_0 + a_1 x_{n+1}}{b_0 + x_{n+1}}, \tag{7}$$

$$f_n = \frac{a_1 b_0 - a_0}{(b_0 + x_n)^2}, \tag{8}$$

$$f_n' = \frac{-2\,(a_1 b_0 - a_0)}{(b_0 + x_n)^3}. \tag{9}$$

Equations (6)-(9) contain three unknown coefficients a_0, a_1 dan b_0. Eliminating these three unknown coefficients, we have

$$y_{n+1} = y_n + \frac{2\,h(f_n)^2}{2\,f_n - hf_n'}. \tag{10}$$

Equation (10) is a one-step second order rational method (see Lambert [5] for details). This method is a formula to calculate the approximate value y_{n+1} by using previous information at point (x_n, y_n).

Furthermore, to calculate the approximate value y_{n+2}, we once again assume that the approximate solution to problem (1) is locally given in the interval $[x_n, x_{n+2}]$ by the rational approximant (5). In the calculation process, this rational approximant in equation (5) passes through points (x_n, y_n), (x_{n+1}, y_{n+1}) and (x_{n+2}, y_{n+2}). The derivative values at x_n and x_{n+1} are given by $y_n' = f(x_n, y_n)$ and $y_{n+1}' = f(x_{n+1}, y_{n+1})$. Therefore, five equations must be satisfied as follows:

$$y_n = \frac{a_0 + a_1 x_n}{b_0 + x_n} \tag{11}$$

$$y_{n+1} = \frac{a_0 + a_1 x_{n+1}}{b_0 + x_{n+1}} \tag{12}$$

$$y_{n+2} = \frac{a_0 + a_1 x_{n+2}}{b_0 + x_{n+2}} \tag{13}$$

$$f_n = \frac{a_1 b_0 - a_0}{(b_0 + x_n)^2} \tag{14}$$

$$f_{n+1} = \frac{a_1 b_0 - a_0}{(b_0 + x_{n+1})^2} \tag{15}$$

We notice that equations (11)-(15) contain four unknown coefficients a_0, a_1, b_0 and f_n. Eliminating these four unknown coefficients, we have

$$y_{n+2} = y_{n+1} + \frac{hf_{n+1}(y_{n+1} - y_n)}{2\,(y_{n+1} - y_n) - hf_{n+1}}. \tag{16}$$

Equation (16) is a two-step third order rational method (see Lambert [5] for details). This method is a formula to calculate the approximate value y_{n+2} by using previous information at points (x_n, y_n) and (x_{n+1}, y_{n+1}).

The calculation of the rational block method is based on the rational approximant (5). By the elimination of unknown coefficients, we have formulas (10) and (16). These formulas are used to find the approximate values y_{n+1} and y_{n+2}. If y_n value is known, the rational block method calculates the approximate value y_{n+1} using formula (10), and then it calculates the approximate value y_{n+2} using formula (16). This means that, in each block, the rational block method obtains the y_{n+1} and y_{n+2} values in one iteration.

3. Numerical method

We write three methods to solve the initial value problem (1). The methods are the first order Runge-Kutta (Euler's method), the second order Runge-Kutta (Heun's method) and the rational block method.

The first order Runge-Kutta (Euler's method) [1] has the iteration formula:

$$y_{n+1} = y_n + h\, f_n\,.$$

Here h is the step size.

The second order Runge-Kutta (Heun's method) [1] has the iteration formula:

$$y_{n+1} = y_n + \frac{h}{2}(k_1 + k_2)\,,$$

where

$$k_1 = f(x_n, y_n)\,,$$

and

$$k_2 = f(x_n + h, y_n + h\, k_1).$$

The rational block method is given by the iteration formulas:

$$y_{n+1} = y_n + \frac{2\,h(f_n)^2}{2\,f_n - h\,f_n'}\,,$$

$$y_{n+2} = y_{n+1} + \frac{h\,f_{n+1}(y_{n+1} - y_n)}{2\,(y_{n+1} - y_n) - h\,f_{n+1}}\,.$$

We note that the rational block method is a two-step method.

4. Numerical results

In this section, we give some examples to assess the performance of the Runge-Kutta (Euler's and Heun's) and rational block methods. We want to see the maximum error of each method with the definition of the error is as follows

$$\text{error} = \max_{0\le n\le N}\{|y_{\text{exact}} - y_{\text{numeric}}|\}$$

where N is the number of integration steps. For computation, we use the Matlab software for numerical programming and ploting the results. As follows, we have three problems considered by Ying *et al.* [7] to solve.

Problem 1

Consider the initial value problem

$$y'(x) = -10\,y(x), \qquad y(0) = 1, \qquad x \in [0,1].$$

The exact solution is given by $y(x) = e^{-10x}$.

Table 1. Maximum errors for Problem 1.

N	Euler	Heun	Rational block
32	0.066654	0.007616	0.003021
64	0.030792	0.001683	0.000749
128	0.014851	0.000397	0.000187
256	0.007304	0.000096	0.000047

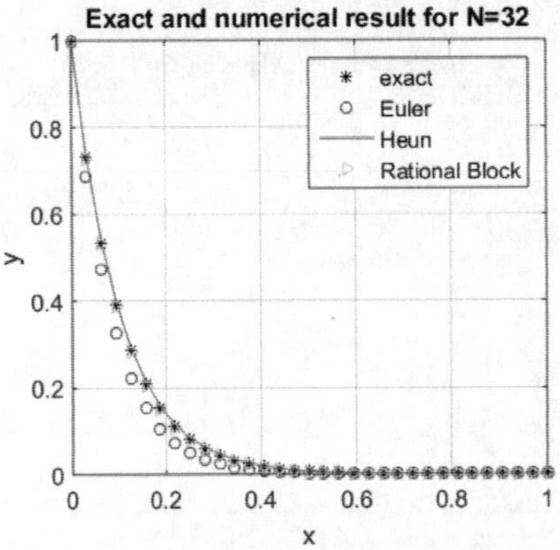

Figure 2. Exact and numerical results for Problem 1.

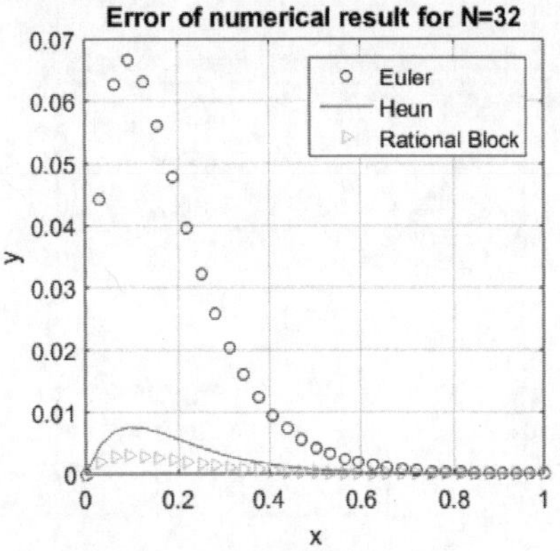

Figure 3. Errors of numerical results for Problem 1.

As given in Table 1 we obtain that the maximum error of the rational block method is smaller than the Euler and Heun methods. All three methods are convergent. That is, if the value of N is greater, the error of the numerical solution gets smaller. From Table 1, Figure 2, and Figure 3 we find that the numerical results are accurate, as numerical solutions are close to the exact solution. The error of each method is less than 0.07. We infer that all three numerical methods solve the problem quite well.

Problem 2

Consider the initial value problem as follows:

$$y''(x) + 101\, y'(x) + 100\, y(x) = 0, \qquad y(0) = 1.01, \qquad y'(0) = -2, \qquad x \in [0,1].$$

This problem can be rewritten as a system of first order ordinary differential equations:

$$y'(x) = z(x), \qquad y(0) = 1.01,$$

$$z'(x) = -100\, y(x) - 101\, z(x), \qquad z(0) = -2,$$

where $x \in [0,1]$. The exact solutions to this system are given by:

$$y(x) = 0.01\, e^{-100\, x} + e^{-x},$$

$$z(x) = y' = -e^{-100x} - e^{-x}.$$

Table 2. Maximum errors for Problem 2.

N	Euler	Heun	Rational block
32	298872461.45	1.253348 x 10^12	0.017842
64	0.007843	0.004487	0.003982
128	0.002421	0.000661	0.000940
256	0.000880	0.000126	0.000233

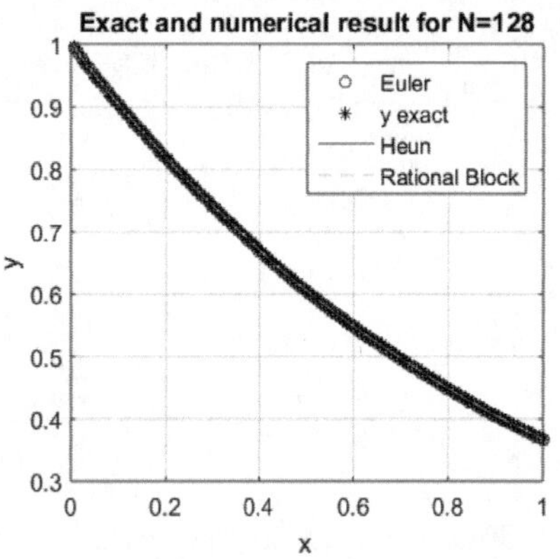

Figure 4. Exact and numerical results for Problem 2.

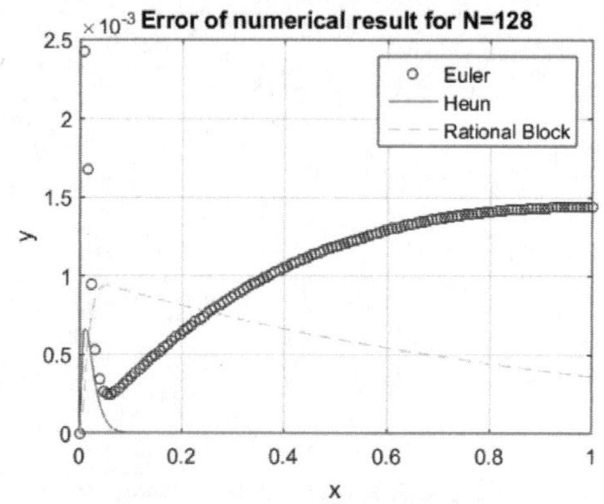

Figure 5. Errors of numerical results for Problem 2.

From Table 2, we observe that if the number of discrete points is too small, such as $N = 32$ (that is, the step size is too large) the Euler's method and Heun's method are not stable, so their numerical errors are very large. However, the Euler's error and Heun's error get smaller for larger number of discrete points. For this problem, the Heun's method performs best giving smallest error for large number of discrete points. All three methods are convergent. For this problem, illustration of exact and numerical results as well as their numerical errors are shown in Figure 4 and Figure 5, respectively.

Problem 3

Consider the following initial value problem:

$$y'(x) = 1 + y(x)^2, \qquad y(0) = 1, \qquad x \in [0,1].$$

The exact solution to this problem is given by

$$y(x) = \tan\left(x + \frac{\pi}{4}\right).$$

This exact solution has a singularity point.

Table 3. Maximum errors for Problem 3. Here inf stands for infinity.

N	Euler	Heun	Rational Block
32	186471279.48	inf	13.92
64	inf	inf	3.64
128	inf	inf	1.20
256	inf	inf	67.13

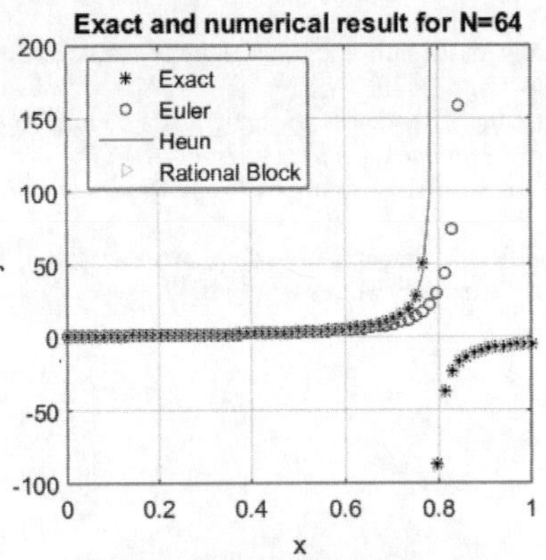

Figure 6. Exact and numerical results for Problem 3.

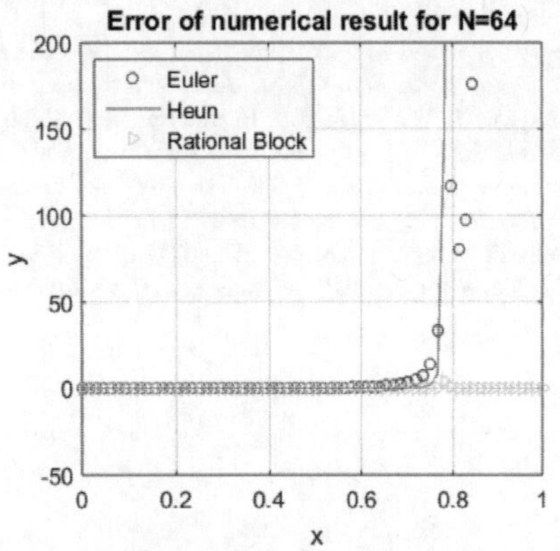

Figure 7. Errors of numerical results for Problem 3.

We have noticed that the solution has a singularity point at $x = \pi/4 \approx 0.7854$. Due to the singularity of the problem, Runge-Kutta methods are divergent, as indicated in Table 3. That is, the Euler's and Heun's methods are not able to solve this problem. In contrast, the rational block method is still able to solve this problem, even though its approximate value is very not accurate. For this problem, illustration of exact and numerical results as well as their numerical errors are shown in Figure 6 and Figure 7, respectively.

5. Conclusion

We have solved initial value problems using the first order Runge-Kutta, second order Runge-Kutta, and rational block methods. The rational block method is fast in computation, because this method can solve two points in one iteration. From numerical results, we find that rational block method is more robust than the other two methods. That is, the rational block method is able to solve a wider range of problems including those with singularity in their solutions.

Acknowledgment

This work was financially supported by Sanata Dharma University. The financial support is gratefully acknowledged by both authors.

References

[1] Burden R L and Faires J D 1993 *Numerical Analysis* 5th Edition (Boston: PWS Publishing)
[2] Nesticò A and Galante M 2015 An estimate model for the equalisation of real estate tax: a case study *International Journal of Business Intelligence and Data Mining* **10** 19
[3] Jovanovič U, Štimec A, Vladušič D, Papa G and Šilc J 2015 Big-data analytics: a critical review and some future directions *International Journal of Business Intelligence and Data Mining* **10** 337
[4] Ansari Z A and Syed A S 2016 Discovery of web usage patterns using fuzzy mountain clustering *International Journal of Business Intelligence and Data Mining* **11** 1
[5] Lambert J D 1974 Two unconventional classes of methods for stiff systems *Stiff Differential Equations* edited by Willoughby R A (New York: Plenum Press) 171
[6] Mungkasi S 2014 Metode rasional eksplisit untuk masalah nilai awal *Prosiding Seminar Nasional Sains dan Pendidikan Sains IX,* 21 Jun 2014, UKSW, Salatiga, Indonesia, 629
[7] Ying T Y, Omar Z and Mansor K H 2014 An A-stable explicit rational block method for the numerical solution of initial value problem *Proceedings of the International Conference on Analysis and Mathematical Applications in Engineering and Science (AMAES),* 19-22 Jan 2014, CSRI, Curtin University, Sarawak, Malaysia, 233
[8] Ying T Y, Omar Z and Mansor K H 2016 Rational block method for the numerical solution of first order initial value problem I: Concepts and ideas *Global Journal of Pure and Applied Mathematics* **12** 3787
[9] Ying T Y, Omar Z and Mansor K H 2016 Rational block method for the numerical solution of first order initial value problem II: A-stability and L-stability *Global Journal of Pure and Applied Mathematics* **12** 3809
[10] Mungkasi S 2016 Adaptive finite volume method for the shallow water equations on triangular grids *Advances in*

Mathematical Physics **2016** 7528625

[11] Supriyadi B and Mungkasi S 2016 Finite volume numerical solvers for non-linear elasticity in heterogeneous media *International Journal for Multiscale Computational Engineering* **14** 479

[12] Mungkasi S, Sambada F A R and Puja I G K 2016 Detecting the smoothness of numerical solutions to the Euler equations of gas dynamics *ARPN Journal of Engineering and Applied Sciences* **11** 5860

[13] Mungkasi S and Roberts S G 2016 A smoothness indicator for numerical solutions to the Ripa model *Journal of Physics: Conference Series* **693** 012011

[14] Budiasih L K, Wiryanto L H and Mungkasi S 2016 A modified Mohapatra-Chaudhry two-four finite difference scheme for the shallow water equations *Journal of Physics: Conference Series* **693** 012012

Jin–Xin relaxation method for solving a traffic flow problem in one dimension

Bernadetta Ambar Sulistiyawati and Sudi Mungkasi

Department of Mathematics, Faculty of Science and Technology,
Sanata Dharma University, Mrican, Tromol Pos 29, Yogyakarta 55002, Indonesia

E-mail: ambarbernadetta@gmail.com, sudi@usd.ac.id

Abstract. We test the performance of the Jin–Xin relaxation and Lax–Friedrichs finite volume numerical methods in solving a traffic flow problem. In particular, we focus on traffic flow at a traffic light turning from red to green. Numerical solutions are compared with the analytical solution to the mathematical model. We find that the Jin–Xin relaxation solution is more accurate than the Lax–Friedrichs finite volume solution.

1. Introduction

Mathematical models are either linear or nonlinear equations [1-4]. Mathematical models may be solved either analytically or numerically [5-7]. Many equations are difficult to solve analytically, so we have to solve it numerically [8-10]. Some examples of models given by partial differential equations are the traffic flow model for busy streets [11], blood flow model for an elastic artery [12], models for gas [13] and hydraulic dynamics [14-16], elasticity in heterogeneous media [17], etc.

Traffic flow and transportation have been studied by a number of authors in the literature. The traffic light is expected to overcome traffic jams on the road and accelerate traffic flow. In this paper we focus on simulations for the traffic flow at a traffic light which turns from red to green.

We consider the Lax–Friedrichs finite volume method and Jin-Xin relaxation method in this paper. We choose those methods because they are simple to implement. We shall compare the errors of the numerical solutions produced by these two methods.

The rest of this paper is organised as follows. We provide the problem to solve in Section 2. Numerical methods are written in Section 3. Numerical results and discussion are given in Section 4. Finally some remarks conclude the paper in Section 5.

2. Problem formulation

We consider the mathematical model for traffic flow:

$$\frac{\partial \rho}{\partial t} + \frac{\partial(\rho u(\rho))}{\partial x} = 0, \tag{1}$$

where x is the free variable of the space, t is the free variable of time, $\rho = \rho(x,t)$ is the density of the traffic on the road, $u = u(x,t)$ is the vehicle velocity. The space domain is a closed interval $[a,b]$. In this paper we want to find the density of vehicles after the traffic light turns from the red to green (see Figure 1 for illustration).

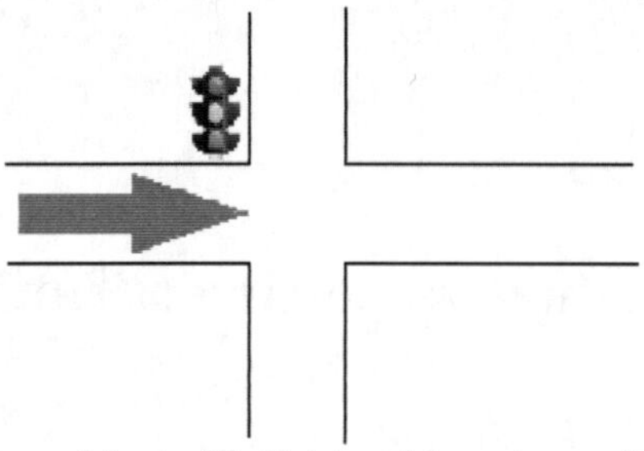

Figure 1. Illustration of the traffic light problem at a road intersection [11].

As used by Mattheij *et al.* [8], the velocity function is given as

$$u(\rho) = u_{\max}\left(1 - \frac{\rho}{\rho_{\max}}\right) \tag{2}$$

where $u_{\max}$ is the maximum vehicle velocity and $\rho_{\max}$ is the maximum density of the traffic on the road. If the velocity approaches zero, then the density tends to the maximum density. Conversely, if the density tends to zero, then the velocity approaches the maximum velocity.

3. Numerical method

In this section, we present the Lax–Friedrichs finite volume and Jin–Xin relaxation methods for solving the traffic flow model (1).

3.1. Lax–Friedrichs finite volume method

The traffic flow model (1) is a conservation law in the form:

$$\frac{\partial \rho}{\partial t} + \frac{\partial f(\rho)}{\partial x} = 0\,. \tag{3}$$

The fully explicit finite volume method for conservation laws is

$$\rho_i^{n+1} = \rho_i^n - \frac{\Delta t}{\Delta x}\left(F_{i+\frac{1}{2}}^n - F_{i-\frac{1}{2}}^n\right) \tag{4}$$

where $\rho_i^n \approx \rho(x_i, t^n)$ is the conserved density and $F_{i+1/2}^n \approx f\left(\rho(x_{i+1/2}, t^n)\right)$ is the flux function computed in the finite volume framework. Here Δt is the time step and Δx is the cell-width. $F_{i+1/2}^n$ is the flux at time t^n at the space point $x_{i+1/2}$.

The Lax–Friedrichs fluxes for equation (1) are

$$\begin{aligned} F_{i+\frac{1}{2}}^n &= \frac{1}{2}\left(f(\rho_{i+1}^n) + f(\rho_i^n)\right) - \frac{\Delta x}{2\Delta t}(\rho_{i+1}^n - \rho_i^n) \\ &= \frac{1}{2}\left(\rho_{i+1}^n u_{\max}\left(1 - \frac{\rho_{i+1}^n}{\rho_{\max}}\right) + \rho_i^n u_{\max}\left(1 - \frac{\rho_i^n}{\rho_{\max}}\right)\right) - \frac{\Delta x}{2\Delta t}(\rho_{i+1}^n - \rho_i^n)\,, \end{aligned} \tag{5}$$

and

$$\begin{aligned} F_{i-\frac{1}{2}}^n &= \frac{1}{2}\left(f(\rho_i^n) + f(\rho_{i-1}^n)\right) - \frac{\Delta x}{2\Delta t}(\rho_i^n - \rho_{i-1}^n) \\ &= \frac{1}{2}\left(\rho_i^n u_{\max}\left(1 - \frac{\rho_i^n}{\rho_{\max}}\right) + \rho_{i-1}^n u_{\max}\left(1 - \frac{\rho_{i-1}^n}{\rho_{\max}}\right)\right) - \frac{\Delta x}{2\Delta t}(\rho_i^n - \rho_{i-1}^n)\,. \end{aligned} \tag{6}$$

3.2. Jin–Xin relaxation method

Equation (3) can be modified to

$$\frac{\partial \rho}{\partial t} + \frac{\partial v}{\partial x} = 0\,, \tag{7}$$

where $v := f(\rho) = \rho\, u(\rho)$ and v is assumed to be a smooth function.

The Jin–Xin relaxation system for equation (7) is

$$\begin{cases} \dfrac{\partial \rho}{\partial t} + \dfrac{\partial v}{\partial x} = 0\,, \\ \dfrac{\partial v}{\partial t} + a\dfrac{\partial \rho}{\partial x} = -\dfrac{1}{\varepsilon}(v - f(\rho))\,, \end{cases} \tag{8}$$

in which we can take $a = f'(\rho)^2$ and ε is a small positive constant. The iteration for ρ is given by

$$\frac{\rho_j^{n+1} - \rho_j^n}{\Delta t} + \frac{1}{\Delta x}\left(v_{j+\frac{1}{2}}^n - v_{j-\frac{1}{2}}^n\right) = 0\,, \tag{9}$$

or

$$\rho_j^{n+1} = \rho_j^n - \frac{\Delta t}{2\Delta x}(v_{j+1}^n - v_{j-1}^n) + \frac{\sqrt{a}\,\Delta t}{2\Delta x}(\rho_{j+1}^n - 2\rho_j^n + \rho_{j-1}^n)\,. \tag{10}$$

The iteration for v is given by

$$\frac{v_j^{n+1} - v_j^n}{\Delta t} + \frac{a}{\Delta x}\left(\rho_{j+\frac{1}{2}}^n - \rho_{j-\frac{1}{2}}^n\right) = -\frac{1}{\varepsilon}(v_j^n - f(\rho_j^n))\,, \tag{11}$$

or

$$v_j^{n+1} = v_j^n - \frac{a\,\Delta t}{2\Delta x}(\rho_{j+1}^n - \rho_{j-1}^n) + \frac{\sqrt{a}\,\Delta t}{2\Delta x}(v_{j+1}^n - 2v_j^n + v_{j-1}^n) - \frac{1}{\varepsilon}(v_j^n - f(\rho_j^n))\,. \tag{12}$$

In other words, if the Jin–Xin relaxation method is written in the fully discrete explicit finite volume method, then the fluxes are given by

$$\rho_{j+\frac{1}{2}}^n = \frac{1}{2}(\rho_{j+1}^n + \rho_j^n) - \frac{1}{2\sqrt{a}}(v_{j+1}^n - v_j^n), \tag{13}$$

$$\rho_{j-\frac{1}{2}}^n = \frac{1}{2}(\rho_j^n + \rho_{j-1}^n) - \frac{1}{2\sqrt{a}}(v_j^n - v_{j-1}^n), \tag{14}$$

and

$$v_{j+\frac{1}{2}}^n = \frac{1}{2}(v_{j+1}^n + v_j^n) - \frac{\sqrt{a}}{2}(\rho_{j+1}^n - \rho_j^n), \tag{15}$$

$$v_{j-\frac{1}{2}}^n = \frac{1}{2}(v_j^n + v_{j-1}^n) - \frac{\sqrt{a}}{2}(\rho_j^n - \rho_{j-1}^n). \tag{16}$$

Once again, here ρ_i^m is an approximation of ρ at time t^m at the space point x_i. The notation Δt gives the time step. The notation Δx is the cell width of the discretised space domain.

4. Numerical results

In this section we present our numerical results.

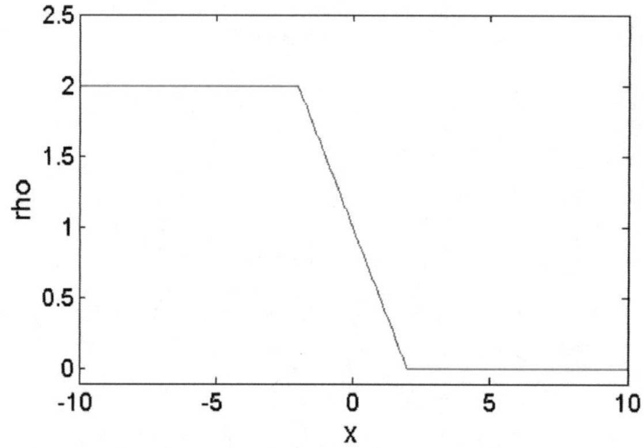

Figure 2. Analytical results for 1 second after a traffic light turns from red to green.

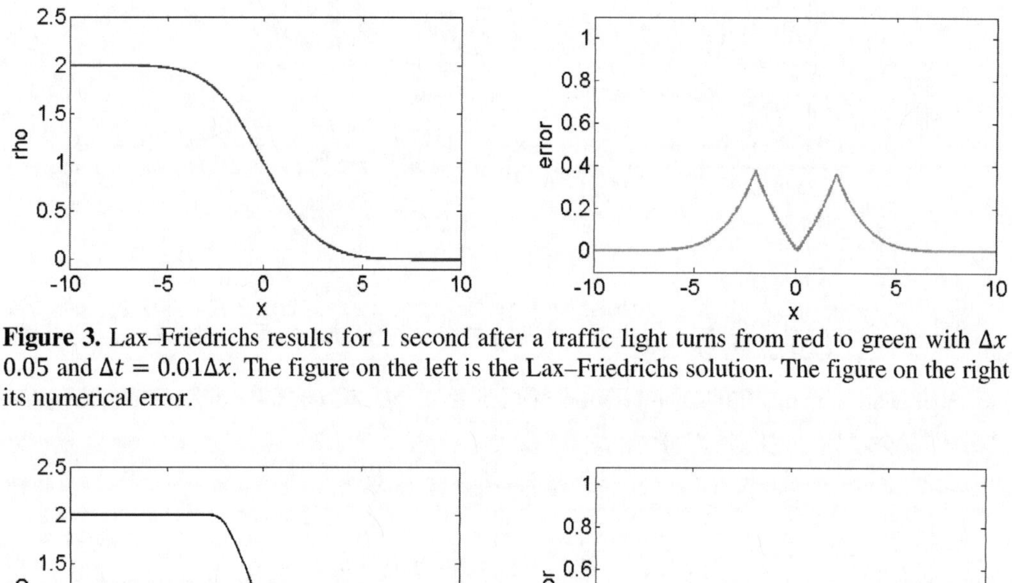

Figure 3. Lax–Friedrichs results for 1 second after a traffic light turns from red to green with $\Delta x = 0.05$ and $\Delta t = 0.01\Delta x$. The figure on the left is the Lax–Friedrichs solution. The figure on the right is its numerical error.

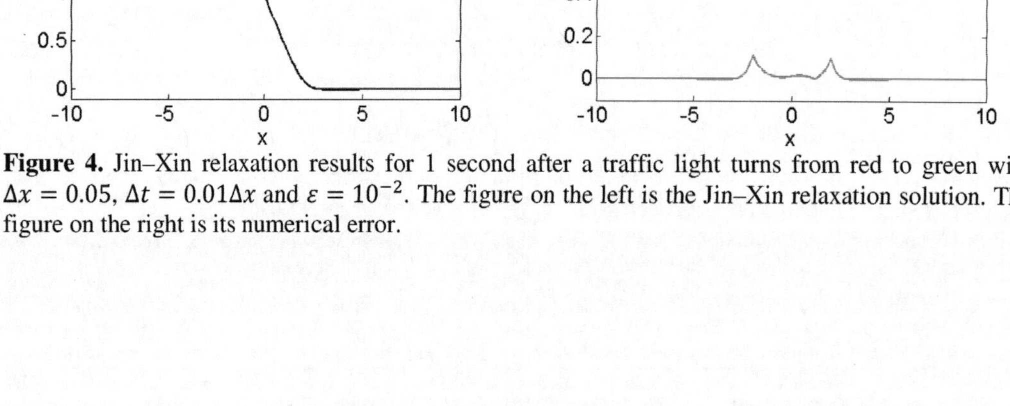

Figure 4. Jin–Xin relaxation results for 1 second after a traffic light turns from red to green with $\Delta x = 0.05$, $\Delta t = 0.01\Delta x$ and $\varepsilon = 10^{-2}$. The figure on the left is the Jin–Xin relaxation solution. The figure on the right is its numerical error.

Following Gunawan [11], we consider the space domain $-10 \leq x \leq 10$ and the time domain $t \geq 0$. The initial condition is

$$\rho(x,0) = \begin{cases} 2 & \text{if } x < 0, \\ 0 & \text{otherwise.} \end{cases} \tag{17}$$

The boundary condition is $\rho(-10,t) = 2$ and $\rho(10,t) = 0$ for all t. We assume that $\rho_{\max} = 2$ and $u_{\max} = 2$. With the initial condition (17), the analytical solution to this problem (see Mattheij *et al.* [8] and Gunawan [11]) is:

$$\rho(x,t) = \begin{cases} 2 & \text{if} \quad \frac{x}{t} < f'(2), \\ \frac{1}{2}\rho_{\max}\left(1 - \frac{x}{u_{\max}t}\right) & \text{if } f'(2) \leq \frac{x}{t} < f'(0), \\ 0 & \text{if} \quad \frac{x}{t} \geq f'(0). \end{cases} \tag{18}$$

The results are shown in Figures 2, 3, and 4. The analytical solution is shown in Figure 2. The numerical solutions are shown in Figures 3 and 4. Figure 3 plots the Lax–Friedrichs solution and its numerical error for 1 second after a traffic light turns from red to green with $\Delta x = 0.05$ and $\Delta t = 0.01\Delta x$. Using the same discretisations, Figure 4 plots the Jin–Xin relaxation solution and its numerical error.

From the numerical results compared to the analytical solution, we obtain that the numerical results have the same behaviour as the physical real problem. The density spreads to the right direction for positive time value. In addition, the Jin–Xin relaxation method performs better than the Lax–Friedrichs method. From Figures 3 and 4, we observe that the numerical error produced by the Jin–Xin relaxation method is much smaller than the error produced by the Jin–Xin relaxation method.

5. Conclusion

Our numerical results show a physically correct behaviour of the traffic flow, after the traffic light turns from red to green. We obtain that the traffic moves from left to right. Furthermore, the density at the left side of the traffic light decreases over time. In addition, the Jin–Xin relaxation solution is more accurate than the Lax–Friedrichs finite volume solution.

Acknowledgment

This work was financially supported by Sanata Dharma University. The financial support is gratefully acknowledged by both authors.

References

[1] Haberman R 1998 *Mathematical Models: Mechanical Vibrations, Population Dynamics, and Traffic Flow* (SIAM, Philadephia)
[2] Ahmad H W, Zilles S, Hamilton H J and Dosselmann R 2016 Prediction of retail prices of products using local competitors *International Journal of Business Intelligence and Data Mining* **11** 19
[3] Simões P, Shrubsall S and Natário I 2015 A spatial econometrics analysis for road accidents in Lisbon *International Journal of Business Intelligence and Data Mining* **10** 152
[4] Truong C D and Anh D T 2015 An efficient method for motif and anomaly detection in time series based on clustering *International Journal of Business Intelligence and Data Mining* **10** 356
[5] Jin S and Xin Z 1995 The relaxation schemes for systems of conservation laws in arbitrary space dimensions *Communications on Pure and Applied Mathematics* **48** 235
[6] Banda M K and Seaid M 2005 Higher-order relaxation schemes for hyperbolic systems of conservation laws *Journal of Numerical Mathematics* **13** 171
[7] LeVeque R J 2002 *Finite Volume Methods for Hyperbolic Problems* (Cambridge University Press, Cambridge)
[8] Mattheij R M M, Rienstra S W and Boonkkamp J H M 2005 *Partial Differential Equation: Modeling, Analysis, Computation* (SIAM, Philadelphia)
[9] Toro E F 1999 *Riemann Solvers and Numerical Methods for Fluid Dynamics* (Springer, Berlin)
[10] Yohana E 2012 Adjoint-based optimization for optimal control problems governed by nonlinear hyperbolic conservation laws *MSc Thesis* (Johannesburg: University of the Witwatersrand)
[11] Gunawan P H 2014 The conservative upwind scheme for simple traffic flow model *Prosiding Seminar Nasional Matematika 2014 of Udayana University Denpasar* 67

[12] Budiawan I W and Mungkasi S 2017 Finite volume numerical solution to a blood flow problem in human artery *Journal of Physics: Conference Series* accepted

[13] Mungkasi S, Sambada F A R and Puja I G K 2016 Detecting the smoothness of numerical solutions to the Euler equations of gas dynamics *ARPN Journal of Engineering and Applied Sciences* **11** 5860

[14] Mungkasi S 2016 Adaptive finite volume method for the shallow water equations on triangular grids *Advances in Mathematical Physics* **2016** 7528625

[15] Mungkasi S and Roberts S G 2016 A smoothness indicator for numerical solutions to the Ripa model *Journal of Physics: Conference Series* **693** 012011

[16] Budiasih L K, Wiryanto L H and Mungkasi S 2016 A modified Mohapatra-Chaudhry two-four finite difference scheme for the shallow water equations *Journal of Physics: Conference Series* **693** 012012

[17] Supriyadi B and Mungkasi S 2016 Finite volume numerical solvers for non-linear elasticity in heterogeneous media *International Journal for Multiscale Computational Engineering* **14** 479

Finite volume numerical solution to a blood flow problem in human artery

Inge Wijayanti Budiawan and Sudi Mungkasi

Department of Mathematics, Faculty of Science and Technology,
Sanata Dharma University, Mrican, Tromol Pos 29, Yogyakarta 55002, Indonesia

E-mail: theresia.inge0303@gmail.com, sudi@usd.ac.id

Abstract. In this paper, we solve a one dimensional blood flow model in human artery. This model is of a non-linear hyperbolic partial differential equation system which can generate either continuous or discontinuous solution. We use the Lax–Friedrichs finite volume method to solve this model. Particularly, we investigate how a pulse propagates in human artery. For this simulation, we give a single sine wave with a small time period as an impluse input on the left boundary. The finite volume method is successful in simulating how the pulse propagates in the artery. It detects the positions of the pulse for the whole time period.

1. Introduction

Blood is one of body components which has important functions. One of blood functions is to distribute the nutritions to all human body tissues. In some cases, blood flow may be hindered because of some problems, such as plugging and artery cavity stiffening [1-2]. It is a dangerous condition that must be overcome. In this case, medical treatment can affect the blood flow pattern. Blood flow in human artery can be represented in a mathematical model. We can investigate blood flow patterns from the solution of its model [3-5]. Numerical methods are considered, as they are powerful [6-8] to solve mathematical models.

In this paper, our focusses are to find and simulate numerical solution of the one dimensional blood flow model in human artery. There are a number of methods which can be used to solve this model, but we use the Lax–Friedrichs finite volume method [9-12] because of its simplicity. Moreover, finite volume methods can be used to find either continuous or discontinuous solution [13-14].

The paper is organised as follows. Section 2 gives the problem that we want to solve. Research method is presented in Section 3. Numerical results are provided in Section 4. Conclusion is drawn in Section 5.

2. Problem formulation

In this section, we describe the problem of blood flow in human artery, which we want to solve.

We consider a straight cylindrical tube with circular cross section and z coordinate is the axis of cylinder (see Figure 1). The one dimensional blood flow model for human artery [2] is

$$\frac{\partial A}{\partial t}+\frac{\partial Q}{\partial z}=0\,, \tag{1}$$

$$\frac{\partial Q}{\partial t}+\frac{\partial}{\partial z}\left(\propto\frac{Q^2}{A}\right)+\frac{A}{\rho}\frac{\partial p}{\partial z}+K_R\frac{Q}{A}=0\,, \tag{2}$$

for $z \in (0, l)$ and $t > 0$, where A, Q, and p are the cross section area of artery, the blood discharge, and the blood pressure, respectively. In addition, $\propto$ is the velocity function in every cross section artery which is assumed as a constant (in this paper we set its value equals to one), ρ is the density of blood, K_R is the coefficient which relates to the blood viscosity, z is the space variable, and t is the time variable.

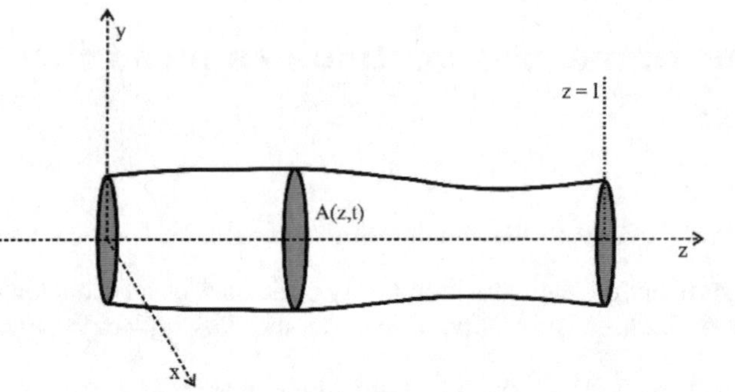

Figure 1. Ilustration of human artery.

In this model, there are three dependent variables (A, Q, dan p) and two equations. In order to have two equations with two unknowns, we define a relation that links the blood pressure with the cross section area of artery

$$p = p_{\text{ext}} + \beta\left(\sqrt{A} - \sqrt{A_0}\right). \tag{3}$$

Here p_{ext} is the external pressure and A_0 is the artery cross section area at initial time $t = 0$. In this paper, p_{ext} is assumed to be zero and A_0 is constant. Furthermore, β is a parameter relating to the artery wall elastic properties:

$$\beta(z) = \frac{4\sqrt{\pi} h_0 E(z)}{3A_0}, \tag{4}$$

where $E(z)$ is the elastic Young's modulus.

3. Numerical method

In this section, we explain the finite volume method for solving blood flow model (1)-(2). We use the flux formulation of Lax–Friedrichs [9].

To find numerical solutions of this blood flow model, let us consider the space domain discretisation (as shown in Figure 2) where $\Delta z = z_{i+1/2} - z_{i-1/2}$ or $\Delta z = z_i - z_{i-1}$, and the time domain discretisation $t^n = n\,\Delta t$ for integers n.

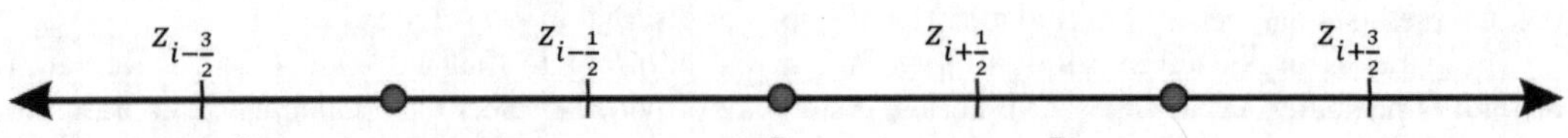

Figure 2. Ilustration of space domain discretisation.

From equation (3), we obtain

$$\begin{aligned} \frac{\partial p}{\partial z} &= \frac{\partial}{\partial z}\left(\beta\left(\sqrt{A} - \sqrt{A_0}\right)\right) \\ &= \frac{d\beta}{dz} A^{\frac{1}{2}} + \frac{\beta}{2} A^{-\frac{1}{2}} \frac{\partial A}{\partial z} - \frac{d\beta}{dz} A_0^{\frac{1}{2}}. \end{aligned}$$

If we multiply both sides of the previous equation with A/ρ , we get

$$\frac{A}{\rho}\frac{\partial p}{\partial z} = \frac{d\beta}{dz}\frac{A^{\frac{3}{2}}}{\rho} + \frac{\beta}{2}\frac{A^{\frac{1}{2}}}{\rho}\frac{\partial A}{\partial z} - \frac{A}{\rho}\frac{d\beta}{dz}A_0^{\frac{1}{2}}. \tag{5}$$

We note that β is a function of z, and A is a function of z and t. Therefore, we have

$$\frac{\partial}{\partial z}\left(\frac{\beta A^{\frac{3}{2}}}{3\rho}\right) + \frac{A}{\rho}\frac{d\beta}{dz}\left(\frac{2}{3}\sqrt{A} - \sqrt{A_0}\right) = \frac{d\beta}{dz}\frac{A^{\frac{3}{2}}}{\rho} + \frac{\beta}{2}\frac{A^{\frac{1}{2}}}{\rho}\frac{\partial A}{\partial z} - \frac{A}{\rho}\frac{d\beta}{dz}A_0^{\frac{1}{2}}. \tag{6}$$

From equations (5) and (6), we get

$$\frac{A}{\rho}\frac{\partial p}{\partial z} = \frac{\partial}{\partial z}\left(\frac{\beta A^{\frac{3}{2}}}{3\rho}\right) + \frac{A}{\rho}\frac{d\beta}{dz}\left(\frac{2}{3}\sqrt{A} - \sqrt{A_0}\right). \tag{7}$$

Based on equation (7), equation (2) can be rewritten as

$$\frac{\partial Q}{\partial t} + \frac{\partial}{\partial z}\left(\frac{Q^2}{A} + \frac{\beta}{3\rho}A^{\frac{3}{2}}\right) = -K_R\frac{Q}{A} + \frac{A}{\rho}\frac{d\beta}{dz}\left(\sqrt{A_0} - \frac{2}{3}\sqrt{A}\right). \tag{8}$$

From the above derivation, blood flow model (1)-(2) are balance laws in the form of

$$\bar{v}_t + \bar{f}\,(\bar{v})_z = \bar{s}(\bar{v}) \tag{9}$$

where the conserved quantities, flux functions, and source terms are

$$\bar{v} = \begin{bmatrix} A \\ Q \end{bmatrix}, \bar{f}(\bar{v}) = \begin{bmatrix} Q \\ \frac{Q^2}{A} + \frac{\beta}{3\rho}A^{\frac{3}{2}} \end{bmatrix}, \text{ and } \bar{s}(\bar{v}) = \begin{bmatrix} 0 \\ -K_R\frac{Q}{A} + \frac{A}{\rho}\frac{d\beta}{dz}\left(\sqrt{A_0} - \frac{2}{3}\sqrt{A}\right) \end{bmatrix}$$

respectively.

Using the finite volume framework, we assume that $\bar{V}_i^n \approx \bar{v}(z_i, t^n)$, $\bar{f}(\bar{V}_i^n) \approx \bar{f}(\bar{v}(z_i, t^n))$, and $\bar{S}_i^n \approx \bar{s}(\bar{v}(z_i, t^n))$. We define the following vectors

$$\bar{V}_i^n = \begin{bmatrix} A_i^n \\ Q_i^n \end{bmatrix}, \bar{f}(\bar{V}_i^n) = \begin{bmatrix} Q_i^n \\ \frac{(Q_i^n)^2}{A_i^n} + \frac{\beta}{3\rho}(A_i^n)^{\frac{3}{2}} \end{bmatrix} \text{ and } \bar{S}_i^n = \begin{bmatrix} 0 \\ -K_R\frac{Q_i^n}{A_i^n} + \frac{A_i^n}{\rho}\frac{d\beta}{dz}\left(\sqrt{A_0} - \frac{2}{3}\sqrt{A_i^n}\right) \end{bmatrix}.$$

The fully discrete finite volume method [9] to get the numerical solution to balance laws (9) is

$$\bar{V}_i^{n+1} = \bar{V}_i^n - \frac{\Delta t}{\Delta z}\left(\bar{F}_{i+\frac{1}{2}}^n - \bar{F}_{i-\frac{1}{2}}^n\right) + \Delta t\, \bar{S}_i^n, \tag{10}$$

with the definition of the Lax–Friedrichs flux

$$\bar{F}_{i+\frac{1}{2}}^n = \frac{\bar{f}(\bar{V}_{i+1}^n) + \bar{f}(\bar{V}_i^n)}{2} - \frac{\Delta z}{2\Delta t}(\bar{V}_{i+1}^n - \bar{V}_i^n), \tag{11}$$

and

$$\bar{F}_{i-\frac{1}{2}}^n = \frac{\bar{f}(\bar{V}_i^n) + \bar{f}(\bar{V}_{i-1}^n)}{2} - \frac{\Delta z}{2\Delta t}(\bar{V}_i^n - \bar{V}_{i-1}^n). \tag{12}$$

From the finite volume scheme (10)-(12), we get the numerical scheme for blood flow model (1)-(2) as follows. The numerical scheme for equation (1) is

$$A_i^{n+1} = A_i^n - \frac{\Delta t}{\Delta z}\left(F_{i+\frac{1}{2}}^n - F_{i-\frac{1}{2}}^n\right), \tag{13}$$

with the definition of Lax–Friedrichs fluxes

$$F_{i+\frac{1}{2}}^n = \frac{1}{2}(Q_{i+1}^n + Q_i^n) - \frac{\Delta z}{2\Delta t}(A_{i+1}^n - A_i^n), \tag{14}$$

and

$$F_{i-\frac{1}{2}}^n = \frac{1}{2}(Q_i^n + Q_{i-1}^n) - \frac{\Delta z}{2\Delta t}(A_i^n - A_{i-1}^n). \tag{15}$$

In addition, the numerical scheme for equation (2) is

$$Q_i^{n+1} = Q_i^n - \frac{\Delta t}{\Delta z}\left(\mathcal{F}_{i+\frac{1}{2}}^n - \mathcal{F}_{i-\frac{1}{2}}^n\right) + \Delta t\left(-K_R\frac{Q_i^n}{A_i^n} + \frac{A_i^n}{\rho}\frac{d\beta}{dz}\left(\sqrt{A_0} - \frac{2}{3}\sqrt{A_i^n}\right)\right), \tag{16}$$

with the definition of Lax–Friedrichs fluxes

$$\mathcal{F}^n_{i+\frac{1}{2}} = \frac{1}{2}\left[\frac{(Q^n_{i+1})^2}{A^n_{i+1}} + \frac{\beta}{3\rho}(A^n_{i+1})^{\frac{3}{2}} + \frac{(Q^n_i)^2}{A^n_i} + \frac{\beta}{3\rho}(A^n_i)^{\frac{3}{2}}\right] - \frac{\Delta z}{2\Delta t}(Q^n_{i+1} - Q^n_i), \tag{17}$$

and

$$\mathcal{F}^n_{i-\frac{1}{2}} = \frac{1}{2}\left[\frac{(Q^n_i)^2}{A^n_i} + \frac{\beta}{3\rho}(A^n_i)^{\frac{3}{2}} + \frac{(Q^n_{i-1})^2}{A^n_{i-1}} + \frac{\beta}{3\rho}(A^n_{i-1})^{\frac{3}{2}}\right] - \frac{\Delta z}{2\Delta t}(Q^n_i - Q^n_{i-1}). \tag{18}$$

We are now ready to present our results of numerical simulations of the blood flow model.

4. Numerical simulation

In this section, we investigate how a pulse propagates in human artery. Numerical settings, initial and boundary conditions are taken as follow.

4.1. Numerical settings

For this numerical simulation, we take $l = 15$ cm and $t \in [0, 0.035]$ second. Figure 3 shows the layout of numerical simulations. We set E as a constant, so it implies β is a constant. It means that the value of E and β do not change for all $z \in (0, l)$. So, the value of $d\beta/dz$ equals to zero. And then, we take $\Delta z = 0.005$, and $\Delta t = 0.002\Delta z$. As long as Δt satisfies the Courant–Friedrichs–Lewy (CFL)'s condition, the method will be stable.

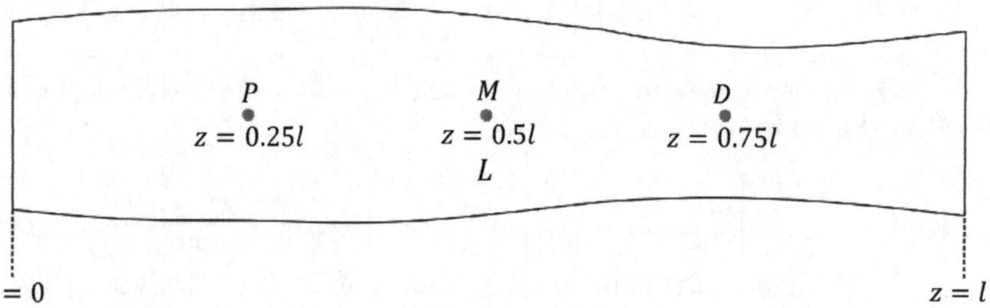

Figure 3. A layout of human artery for numerical simulation.

We investigate at the three monitoring points (*P*, *M*, and *D*) for the pressure variations, where points *P*, *M*, and *D* are proximal, medium, and distal points, respectively. Consider that point *P* is the nearest point from the heart and point *D* is the farthest point from the heart. Table 1 shows the values of coefficients that we use in all simulations.

Table 1. Coefficients value for numerical simulation.

Coefficient	Value
Blood density ρ	1 g/ cm^3
Young's modulus E_0	3x10^6 dyne/ cm^2
Wall artery thickness h	0.05 cm
Initial cross section area A_0	$\pi 0.5^2$ cm^2
Blood viscosity K_R	10^{-8} dyne s/ cm^2

4.2. Initial and boundary conditions

Given the initial values of $A(z,0) = A_0$, $Q(z,0) = 0$, and $p(z,0) = 0$, for every $z \in (0,15)$. The boundary values for this simulation is taken as follows. At the left boundary, we give an impluse input in the form of a single sine wave with a small time period:

$$p(0,t) = 10^3 \cdot \sin\left(\frac{\pi t}{0.0025}\right). \tag{19}$$

For the left boundary value of A and Q, let us consider the characteristic variables of model (1)-(2) which has been explained in [2]: W_1 and W_2 are given by

$$W_2 = \frac{Q}{A} - 2\sqrt{\frac{2\beta}{\rho}}A^{\frac{1}{4}}, \tag{20}$$

$$W_1 = W_2 + 4\sqrt{\frac{2}{\rho}}\left(\sqrt{p + \beta\sqrt{A_0}}\right), \tag{21}$$

$$A = \left(\frac{\rho}{\beta}\right)^2 \frac{(W_1 - W_2)^4}{4^5}, \tag{22}$$

$$Q = A\frac{W_1 + W_2}{2}. \tag{23}$$

We set W_2 as a constant and equals to its initial value. And the last, for each right boundary value of A, p and Q equals to the corresponding value of the nearest neighbour in the domain.

4.3. Numerical results

In this subsection, we summarise our numerical results.

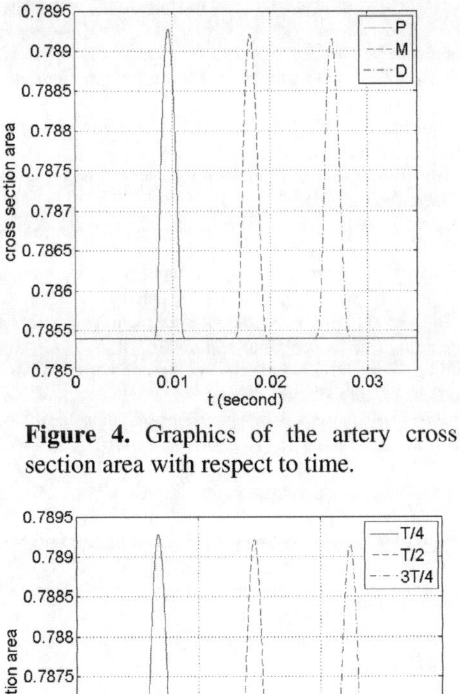

Figure 4. Graphics of the artery cross section area with respect to time.

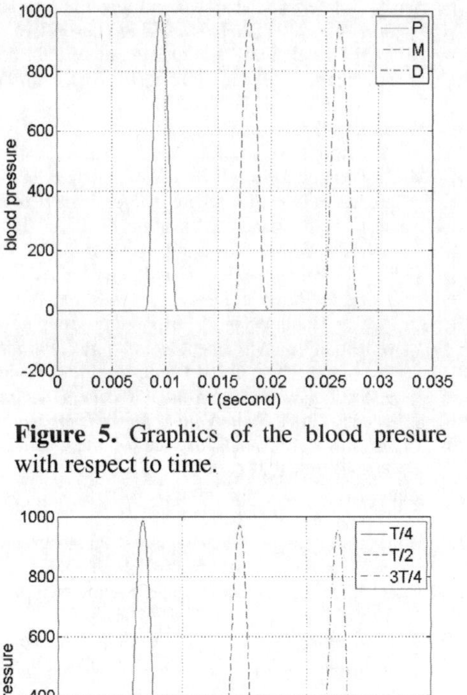

Figure 5. Graphics of the blood presure with respect to time.

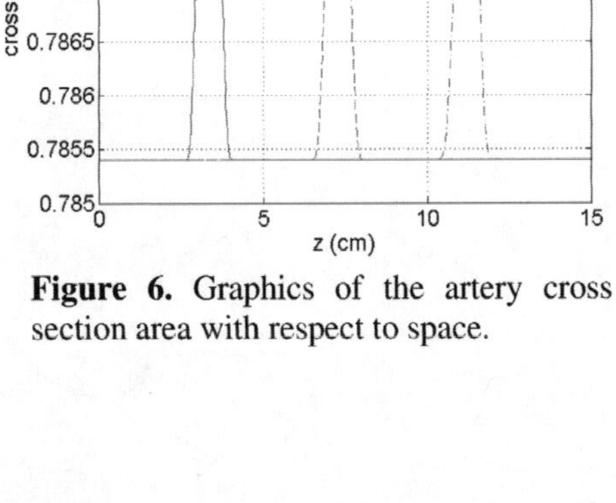

Figure 6. Graphics of the artery cross section area with respect to space.

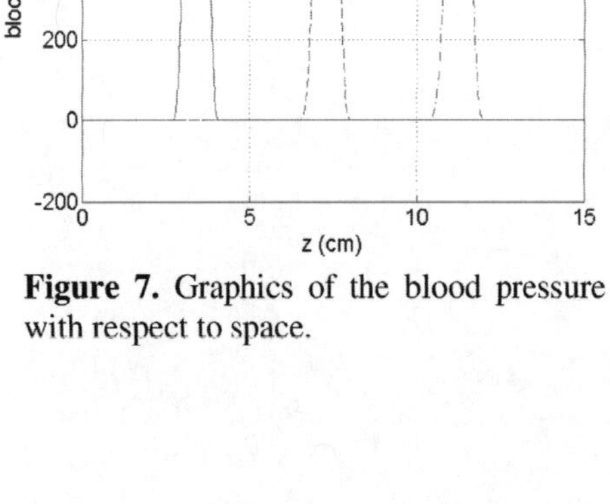

Figure 7. Graphics of the blood pressure with respect to space.

For this numerical simulation, where the elastic Young's modulus is constant, the blood pressure is directly propotional with the artery cross section area (see Figure 4 and Figure 6). Furthermore, the blood pressure amplitude decreases with respect to time and space (see Figure 5 and Figure 7). We infer that the decrease is influenced by the dissipation of the numerical method. We note that as the cell width is taken smaller, the amplitude can be maintained almost the same as the original one, as long as the solution is continuous.

5. Conclusion

Based on the numerical results, the form of pulse does not change but the amplitude decreases with respect to time and space due to numerical dissipation. The dissipation can be small if the numerical cell width is small. This research is limited to problems solved using the first order Lax–Friedrichs finite volume method. Future work could extend the method to higher order ones.

Acknowledgment

This work was financially supported by Sanata Dharma University. The financial support is gratefully acknowledged by both authors.

References

[1] Acosta A, Puelz C, Penny D J and Rusin C G 2015 Numerical method of characteristics for one-dimensional blood flow *Journal of Computational Physics* **294** 96
[2] Formaggia L, Nobile F and Quarteroni A 2002 A one dimensional model for blood flow: application to vascular prosthesis, in Babuška I, et al. (eds.), *Mathematical Modeling and Numerical Simulation in Continuum Mechanics* pp. 137-153 (Springer, Berlin)
[3] Montecinos G I, Müller L O and Toro E F 2014 Hyperbolic reformulation of a 1D viscoelastic blood flow model and ADER finite volume schemes *Journal of Computational Physics* **266** 101
[4] Müller L O and Toro E F 2013 Well-balanced high-order solver for blood flow in networks of vessels with variable properties *International Journal for Numerical Methods in Biomedical Engineering* **29** 1388
[5] Watanabe S M, Blanco P J, and Feijóo R A 2013 Mathematical model of blood flow in an anatomically detailed arterial network of the arm *ESAIM: M2AN* **47** 961
[6] Malik O U, Hilderman R J, Hamilton H J and Dosselmann R 2016 Retail price time series imputation *International Journal of Business Intelligence and Data Mining* **11** 49
[7] Lu Z, Yan J and Wang X 2015 Using grouping strategy and pattern discovery for delta extraction in a limited collaborative environment *International Journal of Business Intelligence and Data Mining* **10** 378
[8] Nesticò A and Pipolo O 2015 A protocol for sustainable building interventions: financial analysis and environmental effects *International Journal of Business Intelligence and Data Mining* **10** 199
[9] LeVeque R J 2002 *Finite Volume Methods for Hyperbolic Problems* (Cambridge University Press, Cambridge)
[10] Mungkasi S 2016 An adaptive mesh finite volume method for the Euler equations of gas dynamics *AIP Conference Proceedings* **1737** 040002
[11] Mungkasi S and Ningrum G I J 2016 Numerical solution to the linear acoustics equations *AIP Conference Proceedings* **1746** 020056
[12] Mungkasi S and Sari I P 2016 Numerical solution to the shallow water equations using explicit and implicit schemes *AIP Conference Proceedings* **1746** 020064
[13] Mungkasi S 2016 Adaptive finite volume method for the shallow water equations on triangular grids *Advances in Mathematical Physics* **2016** 7528625
[14] Supriyadi B and Mungkasi S 2016 Finite volume numerical solvers for non-linear elasticity in heterogeneous media *International Journal for Multiscale Computational Engineering* **14** 479

Jin–Xin relaxation solution to the spatially-varying Burgers equation in one dimension

Sudi Mungkasi and Fioretta Laras

Department of Mathematics, Faculty of Science and Technology,
Sanata Dharma University, Mrican, Tromol Pos 29, Yogyakarta 55002, Indonesia

E-mail: sudi@usd.ac.id, fiorettalaras2@gmail.com

Abstract. We consider the spatially-varying Burgers equation in one dimension. We take the Lax–Friedrichs finite volume method and Jin–Xin relaxation method in solving the equation. According to our research, the Jin–Xin relaxation method produces a more accurate solution, as they produce smaller error than the Lax–Friedrichs finite volume method. However, the Lax–Friedrichs finite volume method is faster in computation than the Jin–Xin method.

1. Introduction

A partial differential equation is an equation which contains its partial derivatives involving two or more independent variables. There are a large number of examples of partial differential equations in mathematical modelling, such as shallow water wave equations [1-3], advection equations [4-5], gas dynamics [6], elasticity equation [7] and Burgers equation [8].

In this paper, we focus on numerical solutions $u(x,t)$ to the spatially-varying Burgers equation in one dimension with the flux function depends on the space variable x and time variable t, where the flux function is of the form $F(x,u) = k(x)f(u)$ for some function f. Details of this problem is found in the work of Prnić [9].

A number of numerical methods are available in the literature, as computer simulations are desired for large scale computations [10-12]. In this paper numerical methods are used to solve the spatially-varying Burgers equation. We focus on the Lax–Friedrichs finite volume method [4-5] and Jin–Xin relaxation method [13-14]. We choose Lax–Friedrichs finite volume method because of its simplicity. The Jin–Xin relaxation method offers linearity in the model, so they are easy to compute. The main task is to compare our results between the Lax–Friedrichs finite volume method and Jin–Xin relaxation method. We research for which method resulting in better performance than the other.

The paper structure is as follows. Section 2 writes the mathematical model that we want to solve. Section 3 presents numerical methods that we use to solve the model. Section 4 provides numerical results. Conclusion is drawn in Section 5.

2. Problem formulation

We consider the spatially-varying Burgers equation in one dimension

$$\frac{\partial u}{\partial t} + \frac{\partial (k(x)f(u))}{\partial x} = 0\,, \tag{1}$$

where

$$f(u) = \frac{u^2}{2}, \tag{2}$$

and

$$k(x) = \frac{1}{1+x^2}. \tag{3}$$

The initial condition for $x > 0$ is

$$u(x,0) = \frac{1}{\sqrt{1+x^2}}. \tag{4}$$

Here x is the space variable, t is the time variable, u is the conserved quantity and $F(x,u) = k(x)f(u)$ is the flux function. The boundary condition is supposed to be known.

3. Numerical methods

Numerical methods that we use in this paper are the Lax–Friedrichs finite volume method and Jin–Xin relaxation method.

3.1. Lax–Friedrichs finite volume method

Burgers equation (1) can be written in a compact form as

$$u_t + [k\,f(u)]_x = 0. \tag{5}$$

Equation (5) is solved using the finite volume method with an explicit numerical scheme

$$U_i^{n+1} = U_i^n - \frac{\Delta t}{\Delta x}\left(F_{i+\frac{1}{2}}^n - F_{i-\frac{1}{2}}^n\right) \tag{6}$$

where $U_i^n \approx u(x_i, t^n)$ is an approximation of the conserved quantity and $F_{i+1/2}^n \approx f\big(u(x_{i+1/2}, t^n)\big)$ is the flux function (see LeVeque [4-5]). Here Δt is the time step, Δx is the cell-width, i is defined as the index of space and n is the index of time. Formulas for calculation of fluxes in equation (6) using the Lax–Friedrichs finite volume method are given by

$$F_{i+\frac{1}{2}}^n = \frac{1}{2}[k_{i+1}\,f(U_{i+1}^n) + k_i\,f(\,U_i^n)] - \frac{\Delta x}{2\Delta t}(U_{i+1}^n - U_i^n), \tag{7}$$

and

$$F_{i-\frac{1}{2}}^n = \frac{1}{2}[k_i\,f(\,U_i^n) + k_{i-1}\,f(\,U_{i-1}^n)] - \frac{\Delta x}{2\Delta t}(U_i^n - U_{i-1}^n). \tag{8}$$

3.2. Jin–Xin relaxation method

The Jin–Xin relaxation system for equation (1) is

$$\frac{\partial u}{\partial t} + \frac{\partial v}{\partial x} = 0, \tag{9}$$

and

$$\frac{\partial v}{\partial t} + a\frac{\partial u}{\partial x} = -\frac{1}{\varepsilon}(v - k\,f(u)), \tag{10}$$

where v is an artificial variable defined by $v = k\,f(u)$, with a is a positive constant of the relaxation system which can be given by $a = k\,(f'(u))^2$ and ε is a small positive constant.

Equations (9) and (10) with the discretised space can be written as

$$\frac{\partial}{\partial t}u_i + \frac{1}{\Delta x}\left(v_{i+\frac{1}{2}} - v_{i-\frac{1}{2}}\right) = 0, \tag{11}$$

and

$$\frac{\partial}{\partial t}v_i + \frac{1}{\Delta x}a\left(u_{i+\frac{1}{2}} - u_{i-\frac{1}{2}}\right) = -\frac{1}{\varepsilon}(v_i - k\,f(u_i)). \tag{12}$$

Furthermore, the first order fully explicit scheme of the Jin–Xin relaxation method (refer to Jin and Xin [13]) to solve the Burgers equation is

$$u_i^{n+1} = u_i^n - \frac{\Delta t}{2\Delta x}\left(v_{i+1}^n - v_i^n - \sqrt{a}\,(u_{i+1}^n - 2u_i^n + u_{i-1}^n)\right), \tag{13}$$

and

$$v_i^{n+1} = v_i^n - \frac{\Delta t}{2\Delta x}\left(a\,(u_{i+1}^n - u_i^n) - \sqrt{a}\,(v_{i+1}^n - 2v_i^n + v_{i-1}^n)\right) - \frac{1}{\varepsilon}\left(v_i^n - k\,f(u_i^n)\right). \tag{14}$$

In this paper, we have used the Euler's time stepping method [15] for the integration with respect to the time variable.

4. Numerical results

In this section we present our research results and give some discussion. All quantities are assumed to have SI units.

4.1. Results of the Lax–Friedrichs finite volume method

For computation in the Lax–Friedrichs finite volume method, we take $x \in [0,2]$ with cell width $\Delta x = 0.02$. We calculate the numerical solution from the initial time $t = 0$ until the final time $t = 1.5$ with time step $\Delta t = 0.5\Delta x$. For the initial condition we assume $u(x,0) = 1/\sqrt{1+x^2}$. The boundary condition at the left-end is $u(0,t) = 1/(1+t)$, and at the right-end $u(2,t)$ is set to be transmissive. The flux function for the Burgers equation here is $F(x,u) = k(x)f(u)$ where $k(x) = 1/(1+x^2)$ and $f(u) = u^2/2$.

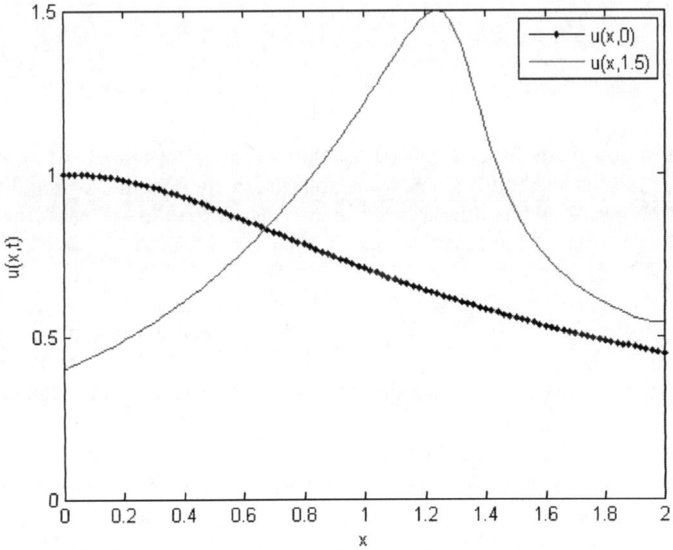

Figure 1. Numerical solutions produced using the Lax–Friedrichs finite volume method at $t = 1.5$.

We obtain that the Lax–Friedrichs finite volume method is fast in computation. The Lax–Friedrichs finite volume method only needs 21.4 seconds to compute numerical solutions, even though this running time can still be optimised. Numerical solution of the Lax–Friedrichs finite volume method is shown in Figure 1.

4.2. Results of the Jin–Xin relaxation method

Using the Jin–Xin relaxation method, we compute from the starting point $x = 0$ to final point $x = 2$ with cell width $\Delta x = 0.05$. We compute the solution from the initial time $t = 0$ until the final time $t = 1.5$ with time step $\Delta t = 0.01\Delta x$. The initial and boundary conditions are the same as the previous simulation. The small positive constant (relaxation rate) that we used here is $\varepsilon = 10^{-3}$.

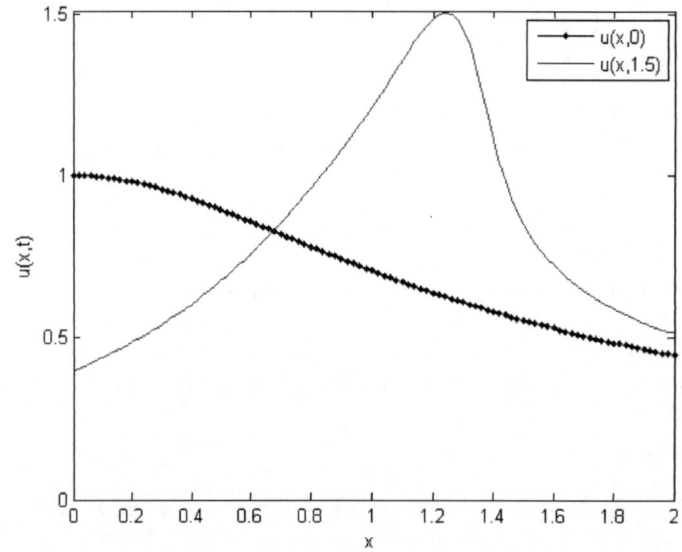

Figure 2. Numerical solutions produced using the Jin–Xin relaxation method at $t = 1.5$.

The Jin–Xin relaxation method needs 300.0 seconds to compute numerical solutions. But if we choose Δx too big, then the method becomes unstable. Numerical results of the Jin–Xin relaxation method are shown in Figure 2 giving similar behaviour to that of the Lax–Friedrichs results.

5. Conclusion

According to our research, the Lax–Friedrichs finite volume method is faster than the Jin–Xin relaxation method in the computation. However, the Jin–Xin relaxation method is easy to compute due to its linear approach. With these results, we suggest that researchers should be aware of these trade-off when they want to take one of these two methods to solve the spatially-varying Burgers equation.

Acknowledgment

This work was financially supported by Sanata Dharma University. The financial support is gratefully acknowledged by both authors.

References

[1] Wiryanto L H and Mungkasi S 2015 Internal monochromatic wave propagating over two bars *Applied Mathematical Sciences* **9** 4479

[2] Mungkasi S 2016 Adaptive finite volume method for the shallow water equations on triangular grids *Advances in Mathematical Physics* **2016** 7528625

[3] Wiryanto L H and Mungkasi S 2016 Numerical solution of waves generated by flow over a bump *Far East Journal of Mathematical Sciences* **100** 1717

[4] LeVeque R J 1992 *Numerical Methods for Conservation Laws* (Basel: Springer)

[5] LeVeque R J 2002 *Finite Volume Methods for Hyperbolic Problems* (Cambridge University Press: Cambridge)

[6] Mungkasi S, Sambada F A R and Puja I G K 2016 Detecting the smoothness of numerical solutions to the Euler equations of gas dynamics *ARPN Journal of Engineering and Applied Sciences* **11** 5860

[7] Supriyadi B and Mungkasi S 2016 Finite volume numerical solvers for non-linear elasticity in heterogeneous media *International Journal for Multiscale Computational Engineering* **14** 479

[8] Mungkasi S 2015 Some advantages of implementing an adaptive moving mesh for the solution to the Burgers equation *IOP Conference Series: Materials Science and Engineering* **78** 012031

[9] Prnić Ž 1995 Godunov scheme for the scalar nonlinear conservation laws with flux depending on the space variable *Mathematical and Computer Modelling* **22** 1

[10] Jakawat W, Favre C and Loudcher S 2016 Graphs enriched by cubes for OLAP on bibliographic networks *International Journal of Business Intelligence and Data Mining* **11** 85

[11] Morano P, Marco Locurcio M, Tajani F and Guarini M R 2015 Fuzzy logic and coherence control in multi-criteria evaluation of urban redevelopment projects *International Journal of Business Intelligence and Data Mining* **10** 73

[12] Tour G and Tangman D Y 2014 Option pricing under a Markov modulated model using a cubic B-spline collocation method *International Journal of Business Intelligence and Data Mining* **9** 356

[13] Jin S and Xin Z 1995 The relaxation schemes for systems of conservation laws in arbitrary space dimensions *Communications on Pure and Applied Mathematics* **48** 235

[14] Banda M K and Seaid M 2005 Higher-order relaxation schemes for hyperbolic systems conservation laws *Journal of Numerical Mathematics* **13** 171

[15] Mungkasi S and Christian A 2017 Runge-Kutta and rational block methods for solving initial value problems *Journal of Physics: Conference Series* accepted

IOP Conf. Series: Journal of Physics: Conf. Series **795** (2017) 012044 doi:10.1088/1742-6596/795/1/012044

Variational iteration method for solving the population dynamics model of two species

Benedictus Dwi Yuliyanto[1] and Sudi Mungkasi[2]

[1]Postgraduate Program in Mathematics Education, Faculty of Teacher Training and Education, Sanata Dharma University, Yogyakarta, Indonesia
[2]Department of Mathematics, Faculty of Science and Technology, Sanata Dharma University, Yogyakarta, Indonesia

E-mail: bene384@gmail.com, sudi@usd.ac.id

Abstract. This paper applies the variational iteration method for solving systems of nonlinear ordinary differential equations. The model under consideration in this work is the population dynamics model of two species. Our results show that the variational iteration method provides formulas to approximate the exact solution at every time value with a very cheap computation.

1. Introduction

Systems of differential equations do not only have important roles in mathematics, but they also play essential roles in other fields of study, such as economics, physics, biology, computer sciences etc. [1-3]. Furthermore, non-linear phenomena often occurs in real problems. A system of non-linear ordinary differential equations are ordinary differential equations which satisfy that the unknown functions only rely on one independent variable and fulfill at least one of the following requirements: consists of dependent variables and/or derivatives to the power of except one, contains multiplication of dependent variable and/or its derivatives.

In biology, differential equations occur in the model of population growth. This paper solves systems of non-linear ordinary differential equations. In particluar, we solve the population dynamics model of two species with the variational iteration method.

Variational iteration method has been a well-known technique to solve mathematical equations [4-9]. It is an analytical approach to solving differential equations. Its greatest advantages are that the method is meshless, the solution is an explicit function, and the iterations are convergent to the exact solution very rapidly. Readers interested in other type of meshless method are referred to the Adomian decomposition method [10-12].

The paper is written in the following structure. In Section 2, we write the population dynamics model, which is the problem that we want to solve. The variational iteration method to solve the model is presented in Section 3. Computational results are provided in Section 4. We conclude the paper with some remarks in Section 5.

2. Population dynamics model

This section provides the general form of the model of the population dynamics of two species:

$$\frac{dx}{dt} = x(a_1 + b_1 x + c_1 y)\,, \tag{1}$$

$$\frac{dy}{dt} = y(a_2 + b_2 y + c_2 x)\,, \tag{2}$$

where:

- x represents the population of the first species,
- y represents the population of the second species,
- a_1 is constant denoting the intrinsic growth rate of species x,
- a_2 is constant denoting the intrinsic growth rate of species y,
- b_1 is constant denoting the rate of the declining in growth of species x due to the increase in the population of species x,
- b_2 is constant denoting the rate of the declining in growth of species y due to the increase in the population of species y,
- c_1 is constant denoting the growth rate of the species x due to interaction with species y,
- c_2 is constant denoting the growth rate of the species y due to interaction with species x.

The free variable is time t. A predator and prey model related to equations (1)-(2) can be found in the work of Sharma and Samanta [13].

3. Variation iteration method

Variational iteration method consists of three basic concepts, that is: the correction functionals, the restricted variations, and the Lagrange multipliers. For the variational iteration method, in this section we follow the work of Batiha *et al.* [4]. Details of the original method can be found in the work of Wazwaz [14].

As an illustration of the basic concept of the variational iteration method, we are given the following non-linear differential equations:

$$Lu + Nu = g(t)\,, \tag{3}$$

with L is linear operator, N is non-linear operator, and $g(t)$ is a function. Variational iteration method can be established and analysed using a correction functional as follows:

$$u_{n+1}(t) = u_n(t) + \int_0^x \lambda(\xi)\,[Lu_n(\xi) + N\tilde{u}_n(\xi) - g(\xi)]d\xi \tag{4}$$

with λ is a Lagrange multiplier, u_n is an approximate solution at the n-th interation, $\tilde{u}_n$ is the restricted variation with $\delta\tilde{u}_n = 0$, and δ is a variational derivative [4].

System (1)-(2) can be rewritten as follows:

$$\frac{dx}{dt} = a_1 x + b_1 x^2 + c_1 xy, \tag{5}$$

$$\frac{dy}{dt} = a_2 y + b_2 y^2 + c_2 xy. \tag{6}$$

The correction functionals of the system (5)-(6) are

$$x_{n+1}(t) = x_n(t) + \int_0^t \lambda_1(s)\left[\frac{dx_n(s)}{ds} - a_1 x_n(s) - b_1\tilde{x}_n^2(s) - c_1\tilde{x}_n(s)\tilde{y}_n(s)\right]ds, \tag{7}$$

$$y_{n+1}(t) = y_n(t) + \int_0^t \lambda_2(s)\left[\frac{dy_n(s)}{ds} - a_2 y_n(s) - b_2\tilde{y}_n^2(s) - c_2\tilde{x}_n(s)\tilde{y}_n(s)\right]ds, \tag{8}$$

where $\tilde{x}_n$ and $\tilde{y}_n$ is the restricted variations with $\delta\tilde{x}_n = 0$ and $\delta\tilde{y}_n = 0$. From equations (7) and (8) we obtain

$$\begin{aligned}\delta x_{n+1}(t) &= \delta x_n(t) + \delta\int_0^t \lambda_1(s)\left[\frac{dx_n(s)}{ds} - a_1 x_n(s) - b_1\tilde{x}_n^2(s) - c_1\tilde{x}_n(s)\tilde{y}_n(s)\right]ds \\ &= \delta x_n(t) + \delta\int_0^t \lambda_1(s)\left[\frac{dx_n(s)}{ds} - a_1 x_n(s)\right]ds\,, \end{aligned}\tag{9}$$

$$\delta y_{n+1}(t) = \delta y_n(t) + \delta \int_0^t \lambda_2(s) \left[\frac{dy_n(s)}{ds} - a_2 y_n(s) - b_2 \tilde{y}_n^2(s) - c_2 \tilde{x}_n(s) \tilde{y}_n(s) \right] ds$$

$$= \delta y_n(t) + \delta \int_0^t \lambda_2(s) \left[\frac{dy_n(s)}{ds} - a_2 y_n(s) \right] ds\,. \tag{10}$$

Using integration by parts, equation (9) becomes

$$\delta x_{n+1}(t) = \delta x_n(t) + \delta \left(\lambda_1(s) x_n(s) - \int_0^t \lambda_1'(s) x_n(s)\, ds - \int_0^t \lambda_1(s) a_1 x_n(s)\, ds \right)$$

$$= (1 + \lambda_1(t)) \delta x_n(t) - \delta \int_0^t [\lambda_1'(s) x_n(s) + a_1 \lambda_1(s) x_n(s)]\, ds\,,$$

$$= (1 + \lambda_1(t)) \delta x_n(t) - \delta \int_0^t [(\lambda_1'(s) + a_1 \lambda_1(s)) x_n(s)]\, ds\,. \tag{11}$$

Using integration by parts, equation (10) becomes

$$\delta y_{n+1}(t) = \delta y_n(t) + \delta \left(\lambda_2(s) y_n(s) - \int_0^t \lambda_2'(s) y_n(s)\, ds - \int_0^t \lambda_2 a_2(s) y_n(s)\, ds \right)$$

$$= (1 + \lambda_2(t)) \delta y_n(t) - \delta \int_0^t [\lambda_2'(s) y_n(s) + a_2 \lambda_2(s) y_n(s)]\, ds$$

$$= (1 + \lambda_2(t)) \delta y_n(t) - \delta \int_0^t [(\lambda_2'(s) + a_2 \lambda_2(s)) y_n(s)]\, ds. \tag{12}$$

The Lagrange multipliers $\lambda_1(t)$ and $\lambda_2(t)$ can be obtained by solving the following system as the stationary conditions:

$$1 + \lambda_1(t) = 0\,, \qquad \lambda_1'(s) + a_1 \lambda_1(s)|_{s=t} = 0, \tag{13}$$

$$1 + \lambda_2(t) = 0\,, \qquad \lambda_2'(s) + a_2 \lambda_2(s)|_{s=t} = 0. \tag{14}$$

Therefore, the Lagrange multipliers are $\lambda_1(t) = -e^{-a_1(s-t)}$ and $\lambda_2(t) = -e^{-a_2(s-t)}$. The solution to system (1)-(2) in linearized forms (taking $b_1 = b_2 = c_1 = c_2 = 0$) is as follows:

$$x(t) = C_1 e^{a_1 t}\,, \tag{15}$$

$$y(t) = C_2 e^{a_2 t}\,. \tag{16}$$

The variational iterations for system (1)-(2) with $\lambda_1(t) = -e^{-a_1(s-t)}$ and $\lambda_2(t) = -e^{-a_2(s-t)}$ is given by:

$$x_{n+1}(t) = x_n(t) + \int_0^t -e^{-a_1(s-t)} \left[\frac{dx_n(s)}{ds} - a_1 x_n(s) - b_1 x_n^2(s) - c_1 x_n(s) y_n(s) \right] ds\,, \tag{17}$$

$$y_{n+1}(t) = y_n(t) + \int_0^t -e^{-a_2(s-t)} \left[\frac{dy_n(s)}{ds} - a_2 y_n(s) - b_2 y_n^2(s) - c_2 x_n(s) y_n(s) \right] ds\,. \tag{18}$$

Equations (17) and (18) compute the series of solutions to the population dynamics model of two species. The series converges to the exact solution.

4. Computational results

From the previous section, values of C_1 and C_2 can be obtained from the initial value. We assume that $x(0) = 4$ and $y(0) = 10$. We obtain $C_1 = 4$ and $C_2 = 10$. Therefore, $x_0(t) = 4e^{a_1 t}$ and $y_0(t) = 10e^{a_2 t}$.

In this section, we provide variational iteration solutions to examples of mutualism, parasitism, and competition of two species.

4.1. Mutualism model

Below are given the solution of the system (1)-(2) the model of mutualism using the variational iteration method. We assume that $a_1 = 0.1$; $a_2 = 0.08$; $b_1 = -0.0014$; $b_2 = -0.001$; $c_1 = 0.0012$; $c_2 = 0.0009$.

Then the variational iteration solutions up to $x_2(t)$ and $y_2(t)$ are as follows:

$$x_1(t) = 3.624e^{0.1t} - 0.224e^{0.2t} + 0.6e^{0.18t}, \quad (19)$$

$$y_1(t) = 10.89e^{0.08t} + 0.36e^{0.18t} - 1.25e^{0.16t}, \quad (20)$$

$$\begin{aligned} x_2(t) = 3.609515785e^{0.1t} &- 0.183867264e^{0.2t} + 0.5919804e^{0.18t} \\ &- 0.04138879997e^{0.28t} + 0.0009983999997e^{0.38t} + 0.01503e^{0.26t} \\ &- 0.00375e^{0.34t} + 0.0003507692307e^{0.36t} - 0.0002341546667e^{0.4t} \\ &+ 0.011364864e^{0.3t}, \end{aligned} \quad (21)$$

$$\begin{aligned} y_2(t) = 11.00045852e^{0.08t} &+ 0.35518824e^{0.18t} - 1.48240125e^{0.16t} \\ &+ 0.17015625e^{0.24t} - 0.006510416667e^{0.32t} - 0.00510624e^{0.28t} \\ &- 0.00024192e^{0.38t} - 0.03354000001e^{0.26t} \\ &+ 0.0008653846153e^{0.34t} + 0.001131428571e^{0.36t}. \end{aligned} \quad (22)$$

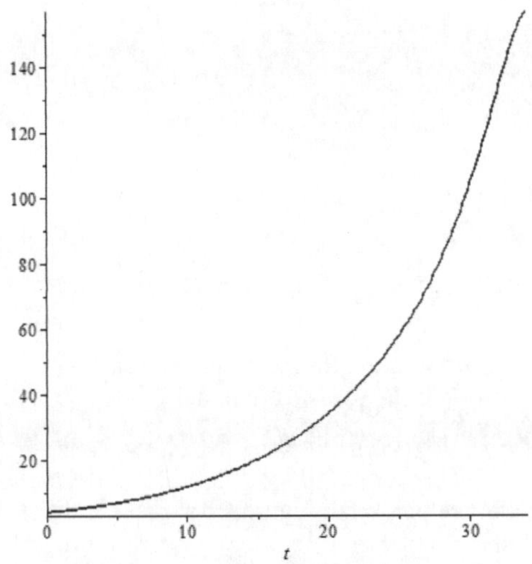

(a). Solution for the first species of the mutualism model.

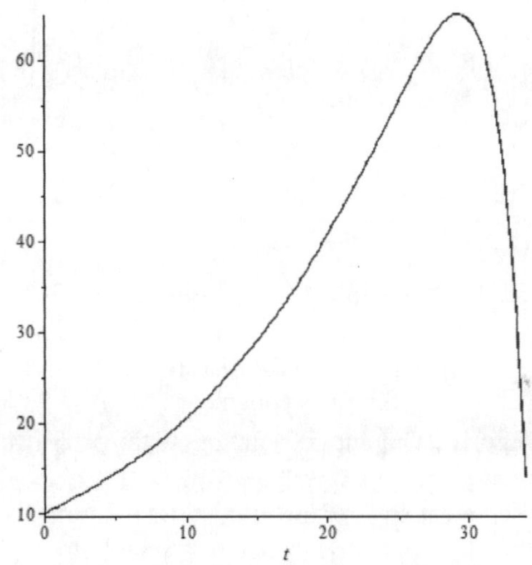

(b). Solution for the second species of the mutualism model.

Figure 1. Graphics of solutions to the mutualism model: (a). $x_3(t)$, (b). $y_3(t)$ with $0 \leq t \leq 34$.

Representatives of the solutions for the mutualism model are plotted in Figure 1 for $x_3(t)$ and $y_3(t)$. We do not write $x_3(t)$ and $y_3(t)$ in this paper, because the formulas are too long. In Figure 1, we observe that due to mutualism, populations of both species increase with respect to time at initial stages of the interaction.

4.2. Parasitism model

Below are given the solution of the system (1)-(2) for parasitism model using the variational iteration method. We assume that $a_1 = 0.1$; $a_2 = 0.08$; $b_1 = -0.0014$; $b_2 = -0.001$; $c_1 = 0.0012$; $c_2 = -0.0009$.

Representatives of the series of variational iteration solutions are:

$$x_1(t) = 3.624e^{0.1t} - 0.224e^{0.2t} + 0.6e^{0.18t}, \quad (23)$$

$$y_1(t) = 11.61e^{0.08t} - 0.36e^{0.18t} - 1.25e^{0.16t}, \quad (24)$$

$$x_2(t) = 3.586909632e^{0.1t} - 0.183867264e^{0.2t} + 0.6311196e^{0.18t} - 0.001643076923e^{0.36t} - 0.0598592e^{0.28t} + 0.0016896e^{0.38t} + 0.01827e^{0.26t} - 0.00375e^{0.34t} - 0.0002341546667e^{0.4t} + 0.01136486399e^{0.3t}, \quad (25)$$

$$y_2(t) = 11.83861929e^{0.08t} - 0.37867176e^{0.18t} - 1.68490125e^{0.16t} + 0.18140625e^{0.24t} - 0.006510416667e^{0.32t} + 0.0006685714285e^{0.36t} + 0.01757376e^{0.28t} - 0.00024192e^{0.38t} + 0.03425999999e^{0.26t} + 0.0008653846153e^{0.34t}. \quad (26)$$

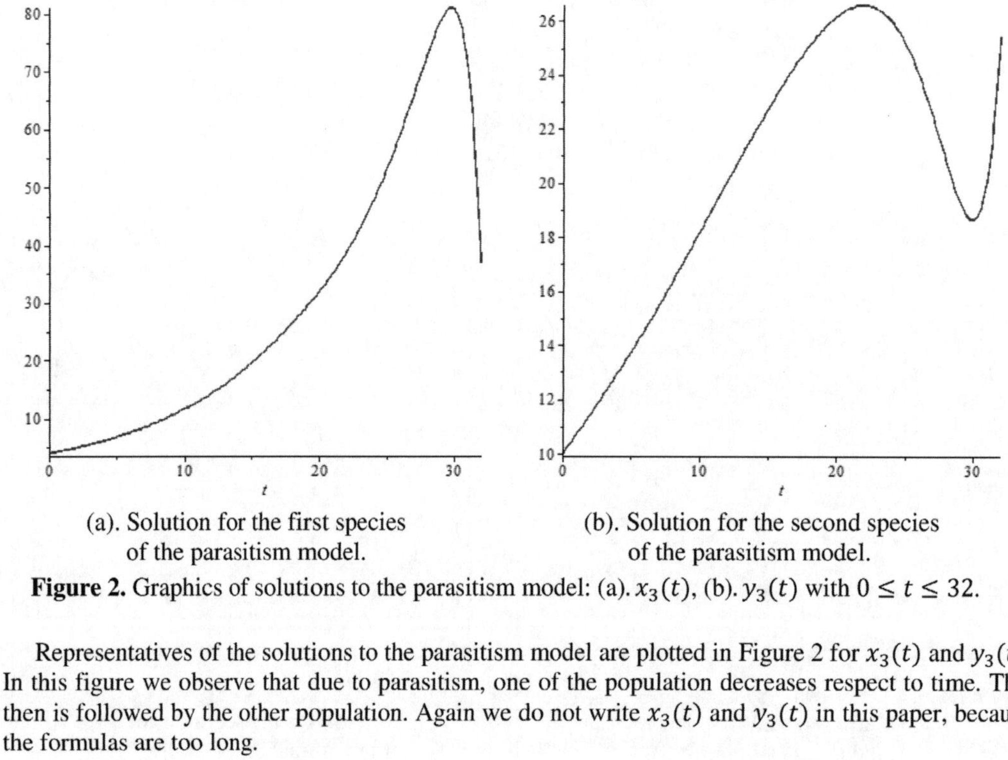

(a). Solution for the first species of the parasitism model.

(b). Solution for the second species of the parasitism model.

Figure 2. Graphics of solutions to the parasitism model: (a). $x_3(t)$, (b). $y_3(t)$ with $0 \leq t \leq 32$.

Representatives of the solutions to the parasitism model are plotted in Figure 2 for $x_3(t)$ and $y_3(t)$. In this figure we observe that due to parasitism, one of the population decreases respect to time. This then is followed by the other population. Again we do not write $x_3(t)$ and $y_3(t)$ in this paper, because the formulas are too long.

4.3. Competition model

Below are given the solution to the system (1)-(2) for the competition model using the variational iteration method. We assume that $a_1 = 0.1$; $a_2 = 0.08$; $b_1 = -0.0014$; $b_2 = -0.001$; $c_1 = -0.0012$; $c_2 = -0.0009$.

Representatives of the series of the variational iteration solutions are:

$$x_1(t) = 4.824e^{0.1t} - 0.224e^{0.2t} - 0.6e^{0.18t}, \quad (27)$$

$$y_1(t) = 11.61e^{0.08t} - 0.36e^{0.18t} - 1.25e^{0.16t}, \quad (28)$$

$$x_2(t) = 4.989257447e^{0.1t} - 0.325793664e^{0.2t} - 0.8400995999e^{0.18t} - 0.00375e^{0.34t} - 0.004227692307e^{0.36t} + 0.07393919997e^{0.28t} - 0.0016896e^{0.38t} + 0.015128064e^{0.3t} - 0.0002341546667e^{0.4t} + 0.09747e^{0.26t}, \quad (29)$$

$$\begin{aligned} y_2(t) = {} & 11.89148417e^{0.08t} - 0.50405976e^{0.18t} - 1.68490125e^{0.16t} \\ & + 0.18140625e^{0.24t} - 0.006510416667e^{0.32t} - 0.006057692307e^{0.34t} \\ & - 0.002057142857e^{0.36t} + 0.01951776e^{0.28t} - 0.00024192e^{0.38t} \\ & + 0.11142e^{0.26t}. \end{aligned} \tag{30}$$

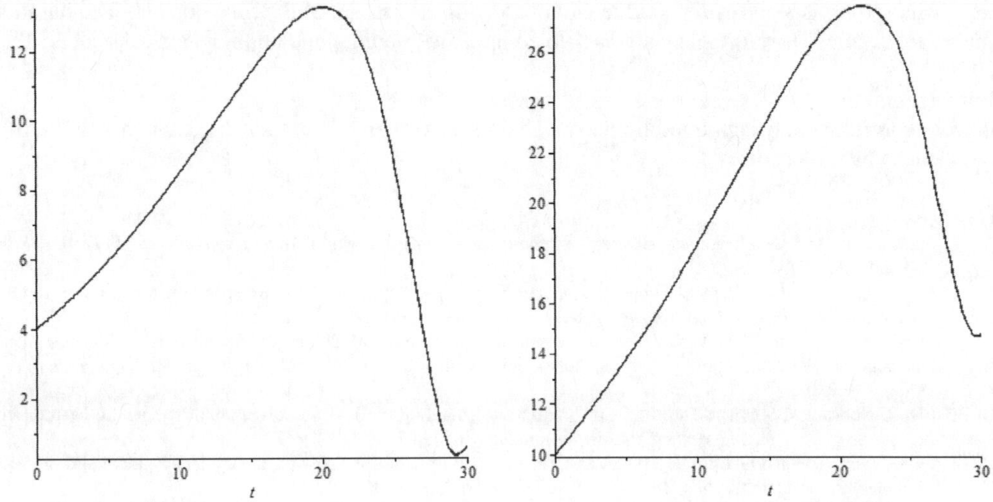

(a). Solution for the first species of the competition model.

(b). Solution for the second species of the competition model.

Figure 3. Graphics of solutions to the competition model: (a). $x_3(t)$, (b). $y_3(t)$ with $0 \le t \le 30$.

Representatives of the solutions to the competition model are plotted in Figure 3 for $x_3(t)$ and $y_3(t)$. Once again, we do not write $x_3(t)$ and $y_3(t)$ in this paper, because the formulas are too long. For small time, both populations increase with respect to time.

Numerical computations for all three cases (mutualism, parasitism, and competition) are given in Table 1. We have compared with these results with the second order Runge-Kutta numerical method. These results are very accurate, as the discrepancy is less than 10^{-6}.

Table 1. Numerical results from the variational iteration method based on the given examples.

t	Mutualism model		Parasitism model		Competition model	
	$x_2(t)$	$y_2(t)$	$x_2(t)$	$y_2(t)$	$x_2(t)$	$y_2(t)$
0.0	4.000000000	10.00000000	4.000000000	10.00000000	4.000000000	10.00000000
0.1	4.042793029	10.07385374	4.042791267	10.06656813	4.033070759	10.06657252
0.2	4.086055587	10.14821700	4.086048406	10.13347257	4.066363553	10.13349044
0.3	4.129792919	10.22309301	4.129776450	10.20071348	4.099879151	10.20075432
0.4	4.174010333	10.29848492	4.173980487	10.26829093	4.133618314	10.26836475
0.5	4.218713190	10.37439597	4.218665655	10.33620501	4.167581809	10.33632224
0.6	4.263906918	10.45082936	4.263837141	10.40445571	4.201770394	10.40462736
0.7	4.309596990	10.52778838	4.309500171	10.47304310	4.236184811	10.47328059
0.8	4.355788954	10.60527627	4.355660046	10.54196712	4.270825824	10.54228248
0.9	4.402488408	10.68329631	4.402322086	10.61122779	4.305694169	10.61163354
1.0	4.449701012	10.76185184	4.449491688	10.68082503	4.340790588	10.68133427

5. Conclusion

Based on the research that has been done, it can be concluded that the variational iteration method can be used to find the solution of the population dynamics model of two species accurately. The population dynamics model of two species being researched is a system of non-linear ordinary differential equations of the first order with initial values. The variational iteration method gives approximate solutions at every time value without any discretisation of the time domain. The iteration formulas are simple. Therefore, they are easy to compute in solving population dynamics models.

Acknowledgment

This work was financially supported by Sanata Dharma University. The financial support is gratefully acknowledged by both authors.

References

[1] Haberman R 1998 *Mathematical Models Mechanical Vibrations, Population Dynamics, and Traffic Flow* (Philadelphia: SIAM)
[2] Chaibi S and Laskri M T 2014 Efficient data selection approach in projected feature space for fast training support vector machines *International Journal of Business Intelligence and Data Mining* **9** 179
[3] Montoya O L Q, Villa L F, Muñoz S, Arenas A C R and Bastidas M 2015 Information retrieval on documents methodology based on entropy filtering methodologies *International Journal of Business Intelligence and Data Mining* **10** 280
[4] Batiha B, Noorani M S M and Hashim I 2007 Variational iteration method for solving multispecies Lotka–Volterra equations *Computers and Mathematics with Applications* **54** 903
[5] Rafei M, Daniali H and Ganji D D 2007 Variational iteration method for solving the epidemic model and the prey and predator problem *Applied Mathematics and Computation* **186** 1701
[6] Shakeri F and Dehghan M 2007 Numerical solution of a biological population model using He's variational iteration method *Computers and Mathematics with Applications* **54** 1197
[7] Mungkasi S and Wiryanto L H 2016 On the relevance of a variational iteration method for solving the shallow water equations *AIP Conference Proceedings* **1707** 050010
[8] Setianingrum P S and Mungkasi S 2016 Variational iteration method used to solve steady state problems of shallow water flows *AIP Conference Proceedings* **1746** 020057
[9] Setianingrum P S and Mungkasi S 2016 Variational iteration method used to solve the one-dimensional acoustics equations *Journal of Physics: Conference Series* accepted
[10] Dispini M and Mungkasi S 2016 Adomian decomposition method used to solve the shallow water equations *AIP Conference Proceedings* **1746** 020055
[11] Dispini M and Mungkasi S 2016 Adomian decomposition method used to solve the one-dimensional acoustics equations *Journal of Physics: Conference Series* accepted
[12] Mungkasi S and Dheno M F S 2017 Adomian decomposition method used to solve the gravity wave equations *AIP Conference Proceedings* accepted
[13] Sharma S and Samanta G P 2015 Analysis of a predator-prey population model *International Journal of Ecological Economics and Statistics* **36** 18
[14] Wazwaz A M 2009 *Partial Differential Equations and Solitary Waves Theory* (New York: Springer)

Adomian decomposition method for solving the population dynamics model of two species

Yulius Wahyu Putranto[1] and Sudi Mungkasi[2]

[1]Postgraduate Program in Mathematics Education, Faculty of Teacher Training and Education, Sanata Dharma University, Yogyakarta, Indonesia
[2]Department of Mathematics, Faculty of Science and Technology, Sanata Dharma University, Yogyakarta, Indonesia

E-mail: juliogoza808@gmail.com, sudi@usd.ac.id

Abstract. Adomian decomposition method has been a powerful method to solve differential equations. In this paper, we propose the method to solve the population dynamics model of two species for mutualism, parasitism, and competition. These three scenarios are considered for the completion of our research. Adomian decomposition method uses initial values of the unknowns and provides series of approximate solutions to the problem. We obtain that the Adomian decomposition method provides fast computation for the solution.

1. Introduction

Population dynamics model has been developed in the field of mathematical biology. Populations of species increase or decrease over time depending on a number of factors. A population certainly interacts with other populations, causing a system dynamics [1].

Two populations interact with the properties of mutualism, parasitism, and competition. These interaction can be formed from a general model (system) of differential equations. Interaction type in both populations can vary depending on the value of each given parameter in the general model. Finding the solution to the model using the aid of computer nowdays is often desired [2-5]. This is because fast and accurate results are important to solve real problems.

In this paper, we solve the population dynamics model of two species using the Adomian decomposition method. The method is chosen because it is meshless, is an analytical approach, possess a fast convergence, and can be implemented on a computer easily [6-11]. It is often related to the variational iteration method [12-16].

The paper is organised as follows. Section 2 contains the problem formulation that we want to solve. Section 3 present the Adomian decomposition method for the population dynamics model. Computational results are provided in Section 4. We conclude the paper in Section 5.

2. Problem formulation

We consider the nonlinear system of the form [12]

$$\begin{cases} \dfrac{dx}{dt} = x(a_1 + b_1 x + c_1 y), \\ \dfrac{dy}{dt} = y(a_2 + b_2 y + c_2 x). \end{cases} \tag{1}$$

Here $x = x(t)$ and $y = y(t)$ are populations of the first and second species at time t, respectively. In addition, $a_1, b_1, c_1, a_2, b_2, c_2$ are constants described as follows:

- x is the population of the first species,
- a_1 denotes the growth rate of the population of the first species,
- b_1 denotes the carrying capacity of the population of the first species,
- c_1 denotes the interacting constant of the population of the first species with the second one,
- y is the population of the second species,
- a_2 denotes the growth rate of the population of the second species,
- b_2 denotes the carrying capacity of the population of the second species,
- c_2 denotes the interacting constant of the population of the second species with the first one.

This model (1) governs mutualism, parasitism, and competition interactions.

3. Adomian decomposition method

We derive the Adomian decomposition method to solve the model following Batiha *et al.* [12].

Model (1) can be rewritten as

$$\begin{cases} \dfrac{dx}{dt} = a_1x + b_1x^2 + c_1xy\,, \\ \dfrac{dy}{dt} = a_2y + b_2y^2 + c_2xy\,. \end{cases} \tag{2}$$

Introducing the derivative operator $L = \frac{d}{dt}$, we obtain

$$\begin{cases} L_x = a_1x + b_1x^2 + c_1xy\,, \\ L_y = a_2y + b_2y^2 + c_2xy\,. \end{cases} \tag{3}$$

Applying $L^{-1} = \int_0^t (.)dt$ to both sides of the nonlinear system (3) gives

$$\begin{cases} L^{-1}L_x = L^{-1}a_1x + L^{-1}b_1x^2 + L^{-1}c_1xy\,, \\ L^{-1}L_y = L^{-1}a_2y + L^{-1}b_2y^2 + L^{-1}c_2xy\,. \end{cases} \tag{4}$$

The Adomian decomposition method admits the decomposition of x and y into an infinite series components

$$x(t) = \sum_{n=0}^{\infty} x_n \qquad\qquad y(t) = \sum_{n=0}^{\infty} y_n \tag{5}$$

and the nonlinear terms x^2, y^2 and xy are assumed to be in the following forms:

$$x^2 = \sum_{n=0}^{\infty} A_n\,, \qquad y^2 = \sum_{n=0}^{\infty} B_n\,, \qquad xy = \sum_{n=0}^{\infty} D_n\,. \tag{6}$$

We define A_n, B_n and D_n as

$$A_n = \sum_{k=0}^{n} x_k x_{n-k}\,, \qquad B_n = \sum_{k=0}^{n} y_k y_{n-k}\,, \qquad D_n = \sum_{k=0}^{n} x_k y_{n-k}\,. \tag{7}$$

We have Adomian polynomials for A_n, B_n and D_n:

$$\begin{aligned} A_0 &= x_0x_0 \\ A_1 &= x_0x_1 + x_1x_0 \\ A_2 &= x_0x_2 + x_1x_1 + x_2x_0 \\ A_3 &= x_0x_3 + x_1x_2 + x_2x_1 + x_3x_0 \\ &\vdots \end{aligned} \tag{8}$$

$$
\begin{aligned}
B_0 &= y_0 y_0 \\
B_1 &= y_0 y_1 + y_1 y_0 \\
B_2 &= y_0 y_2 + y_1 y_1 + y_2 y_0 \\
B_3 &= y_0 y_3 + y_1 y_2 + y_2 y_1 + y_3 y_0 \\
&\vdots
\end{aligned} \tag{9}
$$

$$
\begin{aligned}
D_0 &= x_0 y_0 \\
D_1 &= x_0 y_1 + x_1 y_0 \\
D_2 &= x_0 y_2 + x_1 y_1 + x_2 y_0 \\
D_3 &= x_0 y_3 + x_1 y_2 + x_2 y_1 + x_3 y_0 \\
&\vdots
\end{aligned} \tag{10}
$$

The system of nonlinear differential equations from (4) can be expressed as:

$$
\begin{cases}
x(t) - x(0) = L^{-1} a_1 \sum_{n=0}^{\infty} x_n + L^{-1} b_1 \sum_{n=0}^{\infty} A_n + L^{-1} c_1 \sum_{n=0}^{\infty} D_n \\
y(t) - y(0) = L^{-1} a_2 \sum_{n=0}^{\infty} y_n + L^{-1} b_2 \sum_{n=0}^{\infty} B_n + L^{-1} c_2 \sum_{n=0}^{\infty} D_n
\end{cases} \tag{11}
$$

$$
\begin{cases}
\sum_{n=0}^{\infty} x_n = x(0) + L^{-1} a_1 \sum_{n=0}^{\infty} x_n + L^{-1} b_1 \sum_{n=0}^{\infty} A_n + L^{-1} c_1 \sum_{n=0}^{\infty} D_n \\
\sum_{n=0}^{\infty} y_n = y(0) + L^{-1} a_2 \sum_{n=0}^{\infty} y_n + L^{-1} b_2 \sum_{n=0}^{\infty} B_n + L^{-1} c_2 \sum_{n=0}^{\infty} D_n
\end{cases} \tag{12}
$$

With initial values $x(0) = x_0$, $y(0) = y_0$, we can find the solution of the system. The iterations are determined by the following recursive formulas:

$$
\begin{aligned}
x_0 &= x(0) = x_0, \\
x_{n+1} &= L^{-1} a_1 x_n + L^{-1} b_1 A_n + L^{-1} c_1 D_n,
\end{aligned} \tag{13}
$$

$$
\begin{aligned}
y_0 &= y(0) = y_0, \\
y_{n+1} &= L^{-1} a_1 y_n + L^{-1} b_1 B_n + L^{-1} c_1 D_n.
\end{aligned} \tag{14}
$$

In this paper, we use 7-term approximations to find the solution to the system. The solution is defined by X and Y as

$$X = x_0 + x_1 + x_2 + x_3 + x_4 + x_4 + x_6 + x_7, \tag{15}$$

$$Y = y_0 + y_1 + y_2 + y_3 + y_4 + y_5 + y_6 + y_7. \tag{16}$$

4. Numerical results

For computational experiments in this section, we assume to have initial values $x(0) = x_0 = 4$ and $y(0) = y_0 = 10$. We can determine the type of interaction between both populations from the signs of parameter values. We compute components of 7-term approximations using the computer algebra package Maple. The interaction between populations can be mutualism, parasitism, and competition.

4.1. Mutualism interaction

Given that $a_1 = 0.1$; $a_2 = 0.08$; $b_1 = -0.0014$; $b_2 = -0.001$; $c_1 = 0.0012$; $c_2 = 0.0009$; and $x_0 = 4$ and $y_0 = 10$.

We obtain:

$$
\begin{aligned}
x_1 &= 0.4256000000t, \\
y_1 &= 0.7360000000t, \\
x_2 &= 0.02321664000t^2, \\
y_2 &= 0.02532000000t^2, \\
x_3 &= 0.0008613579093t^3,
\end{aligned} \tag{17}
$$

$y_3 = 0.0005198410667t^3,$
$x_4 = 0.00002377241892t^4,$
$y_4 = 0.00000715508705t^4,$
$x_5 = 4.762365706 * 10^\wedge(-7)t^5,$
$y_5 = 1.122902100 * 10^\wedge(-7)t^5,$
$x_6 = 4.92081361 * 10^\wedge(-9)t^6,$
$y_6 = 3.924828700 * 10^\wedge(-9)t^6,$
$x_7 = -9.99735783 * 10^\wedge(-11)t^7,$
$y_7 = 1.355344556 * 10^\wedge(-10)t^7.$

4.2. Parasitism interaction
Given that $a_1 = 0.1$; $a_2 = 0.08$; $b_1 = -0.0014$; $b_2 = -0.001$; $c_1 = 0.0012$; $c_2 = -0.0009$; and $x_0 = 4$ and $y_0 = 10$.

We obtain:

$$
\begin{aligned}
x_1 &= 0.4256000000t, \\
y_1 &= 0.6640000000t, \\
x_2 &= 0.02304384000t^2, \\
y_2 &= 0.01680960000t^2, \\
x_3 &= 0.0008296779093t^3, \\
y_3 &= 0.0000151441067t^3, \\
x_4 &= 0.00002079741808t^4, \\
y_4 &= -0.00001228646758t^4, \\
x_5 &= 2.877838352 * 10^\wedge(-7)t^5, \\
y_5 &= -4.066091675 * 10^\wedge(-7)t^5, \\
x_6 &= -3.968167659 * 10^\wedge(-9)t^6, \\
y_6 &= -5.050637136 * 10^\wedge(-9)t^6, \\
x_7 &= -4.223308270 * 10^\wedge(-10)t^7, \\
y_7 &= 8.804344521 * 10^\wedge(-11)t^7.
\end{aligned} \tag{18}
$$

4.3. Competition interaction
Given that $a_1 = 0.1$; $a_2 = 0.08$; $b_1 = -0.0014$; $b_2 = -0.001$; $c_1 = -0.0012$; $c_2 = -0.0009$; and $x_0 = 4$ and $y_0 = 10$.

We obtain:

$$
\begin{aligned}
x_1 &= 0.3296000000t, \\
y_1 &= 0.6640000000t, \\
x_2 &= 0.01106304000t^2, \\
y_2 &= 0.01724160000t^2, \\
x_3 &= 0.0001173886293t^3, \\
y_3 &= 0.0000783313067t^3, \\
x_4 &= -0.000004301207450t^4, \\
y_4 &= -0.000007815319423t^4, \\
x_5 &= -1.851824568 * 10^\wedge(-7)t^5, \\
y_5 &= -2.136855195 * 10^\wedge(-7)t^5, \\
x_6 &= -1.635576185 * 10^\wedge(-9)t^6, \\
y_6 &= -7.0058057 * 10^\wedge(-9)t^6, \\
x_7 &= 8.190722735 * 10^\wedge(-11)t^7, \\
y_7 &= 1.240343622 * 10^\wedge(-10)t^7.
\end{aligned} \tag{19}
$$

We find the solution to the system for each initials conditions. Approximate solutions for both populations are given by:

$$X = x_0 + x_1 + x_2 + x_3 + x_4 + x_4 + x_6 + x_7\,,$$
$$Y = y_0 + y_1 + y_2 + y_3 + y_4 + y_5 + y_6 + y_7\,.$$

Approximate solutions for mutualism interaction are given by:

$$X1 = 4 + 0.4256000000 * t + 0.02321664000 * t^2 + 0.0008613579093 * t^3 + 0.00002377241892 * t^4 + 4.762365706 * 10^{-7} * t^5 + 4.92081361 * 10^{-9} * t^6 - 9.99735783 * 10^{-11} * t^7,$$
$$Y1 = 10 + 0.7360000000 * t + 0.02532000000 * t^2 + 0.0005198410667 * t^3 + 0.00000715508705 * t^4 + 1.122902100 * 10^{-7} * t^5 + 3.924828700 * 10^{-9} * t^6 + 1.355344556 * 10^{-10} * t^7.$$

Approximate solutions for parasitism interaction are given by:

$$X2 = 4 + 0.4256000000 * t + 0.02304384000 * t^2 + 0.0008296779093 * t^3 + 0.00002079741808 * t^4 + 2.877838352 * 10^{-7} * t^5 - 3.968167659 * 10^{-9} * t^6 - 4.223308270 * 10^{-10} * t^7,$$
$$Y2 = 10 + 0.6640000000 * t + 0.01680960000 * t^2 + 0.0000151441067 * t^3 - 0.00001228646758 * t^4 - 4.066091675 * 10^{-7} * t^5 - 5.050637136 * 10^{-9} * t^6 + 8.804344521 * 10^{-11} * t^7.$$

Approximate solutions for competition interaction are given by:

$$X3 = 4 + 0.3296000000 * t + 0.01106304000 * t^2 + 0.0001173886293 * t^3 - 0.000004301207450 * t^4 - 1.851824568 * 10^{-7} * t^5 - 1.635576185 * 10^{-9} * t^6 + 8.190722735 * 10^{-11} * t^7,$$
$$Y3 = 10 + 0.6640000000 * t + 0.01724160000 * t^2 + 0.0000783313067 * t^3 - 0.000007815319423 * t^4 - 2.136855195 * 10^{-7} * t^5 - 7.0058057 * 10^{-11} * t^6 + 1.240343622 * 10^{-10} * t^7.$$

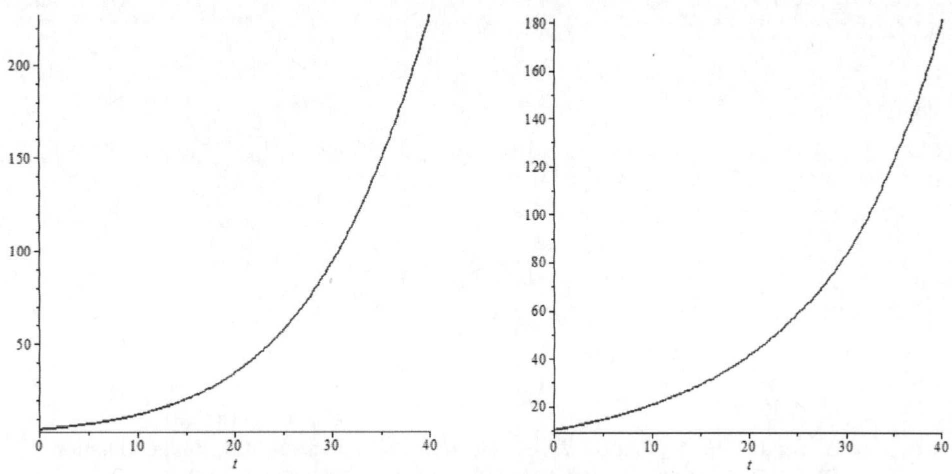

(a). Solution for the first population $X1$. (b). Solution for the second population $Y1$.

Figure 1. Solutions for both populations with mutualism interaction.

IOP Conf. Series: Journal of Physics: Conf. Series **795** (2017) 012045 doi:10.1088/1742-6596/795/1/012045

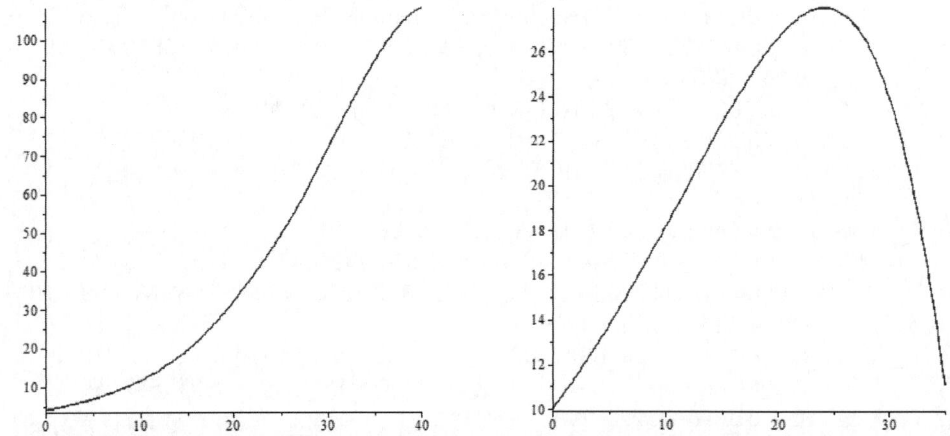

(a). Solution for the first population $X2$. (b). Solution for the second population $Y2$.

Figure 2. Solutions for both populations with parasitism interaction.

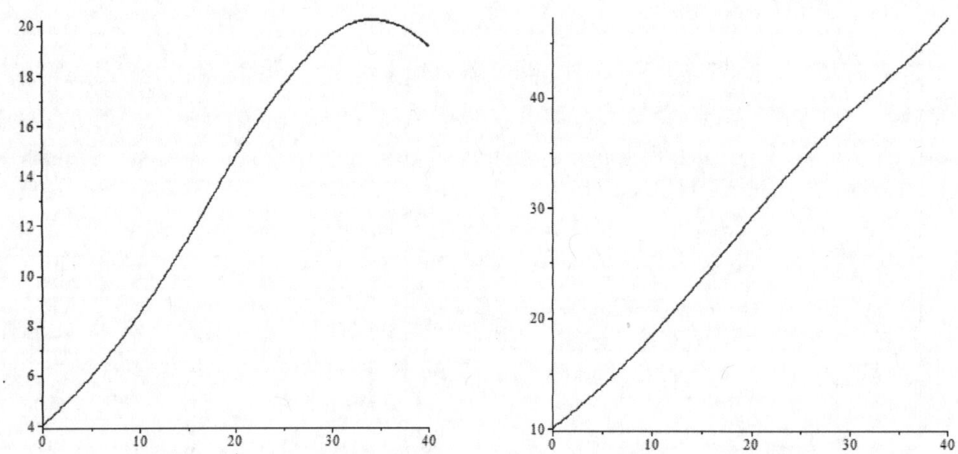

(a). Solution for the first population $X3$. (b). Solution for the second population $Y3$.

Figure 3. Solutions for both populations with competition interaction.

Table 1. Population dynamics for both species obtained using the Adomian decomposition method.

t	Mutualism interaction		Parasitism interaction		Competition interaction	
	$X1$	$Y1$	$X2$	$Y2$	$X3$	$Y3$
0	4.000000	10.00000	4.000000	10.00000	4.000000	10.00000
0.1	4.042793	10.07385	4.042791	10.06657	4.033071	10.06657
0.2	4.086056	10.14822	4.086048	10.13347	4.066363	10.13349
0.3	4.129793	10.22309	4.129777	10.20071	4.099879	10.20075
0.4	4.174010	10.29848	4.173981	10.26829	4.133617	10.26836
0.5	4.218713	10.37440	4.218666	10.33620	4.167580	10.33632
0.6	4.263907	10.45083	4.263838	10.40445	4.201767	10.40462
0.7	4.309597	10.52779	4.309501	10.47304	4.236180	10.47327
0.8	4.355790	10.60527	4.355661	10.54196	4.270819	10.54227
0.9	4.402489	10.68329	4.402324	10.61122	4.305684	10.61162
1	4.449702	10.76185	4.449495	10.68081	4.340776	10.68131

Illustration of the solutions for all three cases are shown in Figure 1, Figure 2, and Figure 3. Numerical results are representatively given in Table 1. From these results, the Adomian decomposition method is successful in solving the population dynamics model. These results compared with the Runge-Kutta numerical solutions lead to the discrepancy of order lower than 10^{-6}. This means that solutions obtained using the Adomian decomposition method are very accurate.

5. Conclusion

We have solved the population dynamics model and for three different sets of parameters for mutualism, parasitism, and competition. The Adomian decomposition method is meshless, so we can obtain approximate solution simply from the explicit solutions. It gives approximate solutions at every time value without any discretisation of the time domain. This is an advantage of using the Adomian decomposition method to solve the problem.

Acknowledgment

This work was financially supported by Sanata Dharma University. The financial support is gratefully acknowledged by both authors.

References

[1] Haberman R 1998 *Mathematical Models Mechanical Vibrations, Population Dynamics, and Traffic Flow* (Philadelphia: SIAM)

[2] Bouakkaz M, Loudcher S and Ouinten Y 2016 OLAP textual aggregation approach using the Google similarity distance *International Journal of Business Intelligence and Data Mining* **11** 31

[3] Giuffrida S, Ferluga G and Valenti A 2015 Capitalisation rates and 'real estate semantic chains': an application of clustering analysis *International Journal of Business Intelligence and Data Mining* **10** 174

[4] Masri H, Krichen S and Guitouni A 2014 A metaheuristic for solving the unsplittable multicommodity flow problem: the maritime surveillance case *International Journal of Business Intelligence and Data Mining* **9** 254

[5] Borruso G 2013 A nodal regional analysis of air passenger transport in Europe *International Journal of Business Intelligence and Data Mining* **8** 377

[6] Biazar J and Montazeri R 2005 A computational method for solution of the prey and predator problem *Applied Mathematics and Computation* **163** 841

[7] Biazar J 2006 Solution of the epidemic model by Adomian decomposition method *Applied Mathematics and Computation* **173** 1101

[8] Rao C V G 2011 A study on series solutions of two species Lotka-Volterra equations by Adomian decomposition and homotopy perturbation method *General Mathematics Notes* **3** 13

[9] Dispini M and Mungkasi S 2016 Adomian decomposition method used to solve the shallow water equations *AIP Conference Proceedings* **1746** 020055

[10] Dispini M and Mungkasi S 2016 Adomian decomposition method used to solve the one-dimensional acoustics equations *Journal of Physics: Conference Series* accepted
[11] Mungkasi S and Dheno M F S 2017 Adomian decomposition method used to solve the gravity wave equations *AIP Conference Proceedings* accepted
[12] Batiha B, Noorani M S M and Hashim I 2007 Variational iteration method for solving multispecies Lotka-Voltera equations *Computers and Mathematics with Aplications* **54** 903
[13] Mungkasi S and Wiryanto L H 2016 On the relevance of a variational iteration method for solving the shallow water equations *AIP Conference Proceedings* **1707** 050010
[14] Setianingrum P S and Mungkasi S 2016 Variational iteration method used to solve steady state problems of shallow water flows *AIP Conference Proceedings* **1746** 020057
[15] Setianingrum P S and Mungkasi S 2016 Variational iteration method used to solve the one-dimensional acoustics equations *Journal of Physics: Conference Series* accepted
[16] Wazwaz A M 2009 *Partial Differential Equations and Solitary Waves Theory* (New York: Springer)

IOP Conf. Series: Journal of Physics: Conf. Series **795** (2017) 012046 doi:10.1088/1742-6596/795/1/012046

Statistical t Analysis for the Solution of Prediction Trash Management in Dusun Tanjung Sari Kec. Ngaglik Kab Sleman, Yogyakarta

Salmahaminati[1], Atina Husnaqilati[2] and Amri Yahya[3]

[1,3]Chemistry Departement, Faculty of Mathematics and Natural Sciences, Islamic University of Indonesia, Yogyakarta, 55581, INDONESIA

[2]Matematics Departement, Faculty of Mathematics and Natural Sciences, Gadjah Mada University, Yogyakarta, 55281, INDONESIA

E-mail: salmahaminati@uii.ac.id

Abstract. Trash management is one of the society participation to have a good hygiene for each area or nationally. Trash is known as the remainder of regular consumption that should be disposed to do waste processing which will be beneficial and improve the hygiene. The way to do is by sorting plastic which is processed into goods in accordance with the waste. In this study, we will know what are the factors that affect the desire of citizens to process the waste. The factors would have the identity and the state of being of each resident, having known of these factors will be the education about waste management, so it can be compared how the results of the extension by using preliminary data prior to the extension and the final data after extension. The analysis uses multiple logistic regression is the identify factors that influence people's to desire the waste while the comparison results using t analysis. Data is derived from statistical instrument in the form of a questionnaire.

1. Introduction

The number of environmental researchers began thinking about the long-term evolution of technology, economic systems, and the environment in the 1980s and 1990s. They proposed a transformation in the management of products, materials, and wastes by looking for a manifestation of sustainable development. Ayres defined the technological evolution of industrial society toward the increasingly efficient use of materials and products [1]. Graedel and Allen using an industrial ecosystem concept in which wastes from one sector would be used as inputs in another sector, as a developed of recycling materials [2]. Braungart argued that all products should be biodegradable or recycled, with shopping and "deshopping" managed [3]. Making these ideas real remains a challenge. Collecting, sorting, and managing the huge array from the consumer products are difficult [4-6].

Padukuhan Tanjungsari, Sleman, Yogyakarta is an area of considerable potential in the management of plastic and paper wastes. There is no optimal utilization of garbage. For these problem, trash management with statistical t analysis proposed to be a solution. Trash management is about colleting, sorting and managing the plastics and papers waste [7-9]. Plastic and paper waste could be recycled or made into other useful items such as the manufacture of bags, and other equipments according to our own creativity.

In this paper, The discussion of this study proposes the analysis using multiple logistic regression is the identify factors that influence people's to desire the waste while the comparison results using t analysis.

2. Methods

The method for this research used multiple logistic regression and t-test [10] to have different mean between two dependent populations. Multiple logistic regression is a popular and widely used for analysis as the linear regression without the dichotomous outcome (e.g., success/failure or yes/no or died/lived). The outcome of the data 1 is dichotomous (yes or no). Model for this method is

$$\log \frac{\pi_i}{1-\pi_i} = \beta_0 + \beta_1 x_{1i} + \beta_2 x_{2i} + \cdots + \beta_p x_{pi}$$

where x_i is covariates or independent variables.

t-test for different mean between two dependent population uses t distribution where the data has amount samples. The mean hipotesis of t-test is nothing difference between 2 data that have same population.The formula is

$$t = \frac{\overline{X}_D - \mu_0}{\frac{s_D}{\sqrt{n}}}.$$

where η_i is zero if two sample don't have difference

3. Results and Discussion

3.1. *Analysis of first survey*

The first survey is the motivation and the score of trash management. It has been done in Padukuhan Tanjung Sari Kec. Ngaglik Kab Sleman, Yogyakarta. Motivation to trash management could be defined in two options. They are yes and no. The defined yes if the person has high motivation to collecting, sorting, and managing the trash product. The defined no if the person does not have motivation to collecting, sorting, and managing the trash product. The motivation is independent variable and the parameter of motivation is dependent variable. The parameter of motivation is about identity (age and education), self-opinion, and economics. Logistic regression analysis used to know about the main factor that influence the motivations with the outcome of two defined options (yes or no) by multiple.

The logistic regression analysis is used by R software. From this analysis, it will show the main factor which influence the dependent variable. Alpha that use for this analysis is 0.05 with confident interval 0.95. The output for this analysis is in table 1

Table 1. P value of independent variable

Variable	P value
Age	0.5627
Education	0.0279
Self-opinion	0.39
Economic	0.5095
Intercept	0.1358

H_0 for this analysis variable is not significant and H_1 variable is significant. H_0 is rejected if P value less than alpha. P value of education is less than 0.05 so the most significant variable in this analysis is only education. Using entering data from this analysis, each dependent variables taken out depend of P value. The variables that taken out are age, self-opinion, and economic. Final output is in table 2.

Table 2. P value of Education and intercept of influence motivation trash management

Variable	P_value
Education	0.0208
Intercept	0.0489

Education is the main factor that influence the motivation for trash management. From this research, It could be concluded that socialization for trash management would be focus in education problem.

3.2. Analysis of the second survey
The second survey in this research is similar with the first survey about management trash that has been done in Dusun Tanjung Sari Kec. Ngaglik Kab Sleman. Score for this part depend on the people who has or no motivation. if score is decline, It define the people in this village have well trash management and if score is increase, it define the people in this village do not have well trash management.

Analysis statistical t is used to know about the differences between two surveys. It is t-analysis for two dependent population. Alpha that use for this analysis is 0.05 with confident interval 0.95. H_0 define if there is no different between two populations and H_1 define if there is different between two populations from the surveys. H_0 is rejected when it less than alpha. P value of this analysis is 0.7131. From this results, It shows that H_0 is not rejected because P_value is higher than alpha.So it can be concluded from the two surveys that socialization of trash management do not have significant influence for the people.

4. Conclusion
Analysis statistical t is used to know about the differences between two surveys. It is t-analysis for two dependent population. Alpha that use for this analysis is 0.05 with confident interval 0.95. H0 define if there is no different between two populations and H_1 define if there is different between two populations from the surveys. H_0 is rejected when it less than alpha. P value of this analysis is 0.7131. From this results, It shows that H_0 is not rejected because P_value is higher than alpha.So it can be concluded from the two surveys that socialization of trash management do not have significant influence for the people.

References
[1] Ayres RU 1989 Industrial metabolism. *In Technology and environment,* edited by J. H. Ausubel and H. E. Sladovich. Washington DC: National Academy Press

[2] Braungart M 1994 Product lifecycle management to replace waste management. In *Industrial ecology and global change,* edited by R. Socolow, C. Andrews, F. Berkhout, and V. Thomas. Cambridge, UK: Cambridge University Press

[3] Graedel TE and Allenby BR 1995 *Industrial ecology* Englewood Cliffs, NJ: Prentice Hall

[4] Gershenfeld N 1999 *When things start to think.* New York: Holt

[5] ISWA (International Solid Waste Association) 1999 *Solid waste on-board weighing systems.* Copenhagen: International Solid Waste Association.

[6] Klausner MW, Grimm, and C. Hendrickson 1998 Reuse of electric motors in consumer products: Design and analysis of an electronic data log. *Journal of Industrial Ecology* 2(2): 89

[7] Wagner SH, Gleskova JC, Sturm, and Suo 2000 Novel processing technology for Macroelectronics. In *Technology and applications of hydrogenated amorphous silicon,* edited by R. A. Street. Berlin: Springer.

[8] Lee Y 1999 *MicroID 13.56 MHz RFID system design guide, AN707 MCRF 355/360 Applications.* Chandler AZ: Microchip Technology

[9] Menell P 1990 Beyond the throwaway society: An incentive approach to regulating municipal

solid waste. *Ecology Law Quarterly* 17: 655
[10] Walpole RE 1995 *Introduction to Statistics 3rd edition* Addison Wasley New York

IOP Conf. Series: Journal of Physics: Conf. Series **795** (2017) 012047 doi:10.1088/1742-6596/795/1/012047

The Stochastic Modelling of Endemic Diseases

Kurnia Susvitasari

School of Math. Science, University of Nottingham, The University Park, NG7 2RD, UK

E-mail: susvitasari@icloud.com

Titin Siswantining

Department Mathematic, University of Indonesia, Kampus UI Depok, 16424, Indonesia

E-mail: titin@sci.ui.ac.id

Abstract. A study about epidemic has been conducted since a long time ago, but genuine progress was hardly forthcoming until the end of the 19th century (Bailey, 1975). Both deterministic and stochastic models were used to describe these. Then, from 1927 to 1939 Kermack and McKendrick introduced a generality of this model, including some variables to consider such as rate of infection and recovery. The purpose of this project is to investigate the behaviour of the models when we set the basic reproduction number, R_0. This quantity is defined as the expected number of contacts made by a typical infective to susceptibles in the population. According to the epidemic threshold theory, when $R_0 \leq 1$, minor epidemic occurs with probability one in both approaches, but when $R_0 > 1$, the deterministic and stochastic models have different interpretation. In the deterministic approach, major epidemic occurs with probability one when $R_0 > 1$ and predicts that the disease will settle down to an endemic equilibrium. Stochastic models, on the other hand, identify that the minor epidemic can possibly occur. If it does, then the epidemic will die out quickly. Moreover, if we let the population size be large and the major epidemic occurs, then it will take off and then reach the endemic level and move randomly around the deterministic's equilibrium.

1. Introduction

Threshold theorem is the most important occurrence in the development of the mathematical theory of epidemic. It states that an epidemic can only occur if the initial number of susceptibles is larger than some critical value which depends on the parameters of the model under consideration (Ball, 1983). Nåsell (2002) also stated that the proportion of infectives in the deterministic model would converge to some equilibrium point as time approached infinity, for which we say that the epidemic enters endemic-equilibrium level or goes extinct. But the stochastic model predicted that once the epidemic is in equilibrium, it would become extinct. Both deterministic and stochastic models were important in any sense to describe this process behaviour. The deterministic model, in fact, is also an approximation of the stochastic model when the population size is sufficiently large (Nåsell, 2002). So in the other words, the stochastic model will have the same threshold value as the deterministic one, but the interpretation of this behaviour is different.

In this paper, we will investigate the behaviour of certain epidemic model, called SIS (*susceptible* $\rightarrow$ *infective* $\rightarrow$ *susceptible*) without demography, through its threshold. The threshold behaviour is usually expressed in terms of epidemic basic reproduction number, R_0. This quantity is usually defined as the expected number of contacts made by a typical infective to susceptibles in the population. It is important to note that in general epidemic modelling, both stochastic and deterministic models have similar threshold value, which is attained at $R_0 = 1$. It then turns out that the models identify two parameter regions, $R_0 \leq 1$ and $R_0 > 1$, with qualitatively different behaviour. We will find out in later section why the theshold value is attained at $R_0 = 1$.

Some simulations will be presented to understand the behaviour of the model priorly under various values of R_0's and then we will provide the mathematical explanation as to why the model behaves as it is.

2. The SIS Epidemic Model without Demography

Suppose that we have a closed mixing population with initially n individuals and m $(m << n)$ infectives. We define SIS model as follows.

At $t = 0$, there are $n - m$ susceptibles who are vulnerable to the diseases. Each infective contacts each susceptible according to independent Poisson processes with rate λ/n and if a susceptible is contacted, it immediately becomes an infective. The infectives also have an infectious period which follow iid exponential distributions with rate γ, and then removed, but once the infectives infectious period is over, the individual will immediately become vulnerable to the disease. Finally, the epidemic ends when no infective is left in the population.

Since we assume the population is closed, the SIS model without demography can be modelled using a univariate process. As we know, the SIS epidemic had two processes: infection and removal with rates $\frac{\lambda}{n}(n-i)i$ and γi, respectively. We can see in Figure 3 that the stochastic processes behaved similarly to deterministic ones. One could interpret the behaviour of the model by simply using deterministic approach, but some processes behave differently as shown in Figure 1, 2, and 4. It means that there is a certain random factor in the stochastic approach that can not be accounted for by deterministic approach.

Before we proceed, we will discuss the threshold behaviour of the model by deterministic approach. Suppose that we define a CTMC (continuous–time Markov chain), $\{Y_n(t) : t > 0\}$ in state space $S = \{0, 1, \ldots, n\}$ and transition rates $\Delta = \{q_{i,i+1}, q_{i,i-1}\}$, where $q_{i,i+1} = \frac{\lambda}{n}(n-i)i$ and $q_{i,i-1} = \gamma i$. If we allow $Y_n(t)$ to be non–integer, we can interpret the transition scheme as $\frac{dY_n(t)}{dt} = \frac{\lambda}{n} Y_n(t)\,[n - Y_n(t)] - \gamma Y_n(t)$.

Let $y(t) = \frac{Y_n(t)}{n}$ represent the proportion of infective in the population at time t, then

$$\frac{dy(t)}{dt} = \lambda y(t) \cdot (1 - y(t)) - \gamma y(t). \tag{1}$$

Obviously, according to (1), the major epidemic occurs if and only if $\frac{dy}{dt} > 0$ and otherwise. At some time t, the increment of $y(t)$ will reach stationary and the process will hit the equilibrium point, as shown in Figure 1, 2, 3, and 4. So, it is possible to determine the equilibrium of $y(t)$ by letting $\frac{dy}{dt} = 0$, which are attained when $y(t) = 0$ or $y(t) = 1 - \frac{\gamma}{\lambda}$.

Now, if $R_0 = \frac{\lambda}{\gamma} \leq 1$, the only possible equilibrium point is at $y(t) = 0$, otherwise, if $R_0 > 1$, the equilibrium is reached at $y(t) = 1 - \frac{\gamma}{\lambda} = 1 - R_0^{-1}$. This is not surprising at all because if the infection rate is negative, the equilibrium is attained when the epidemic dies out. Similar explanation is applied if the infection rate is positive. The point $y(t) = 1 - R_0^{-1}$ is then called the endemic–equilibrium level.

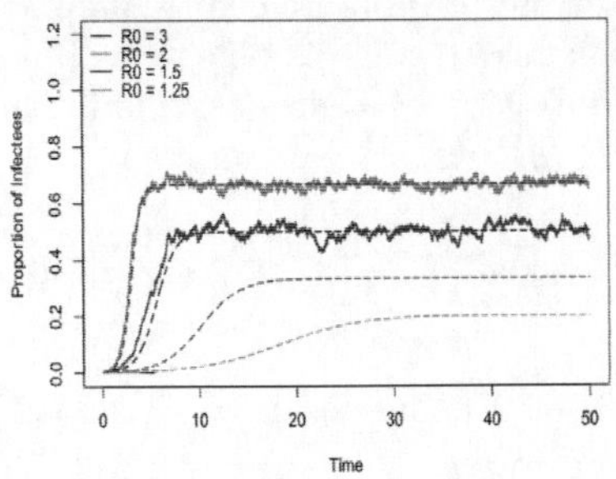

Figure 1: Initial susceptibles = 998

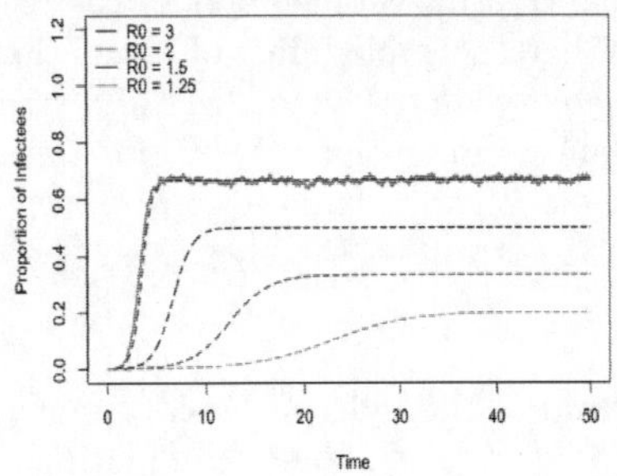

Figure 2: Initial susceptibles = 2998

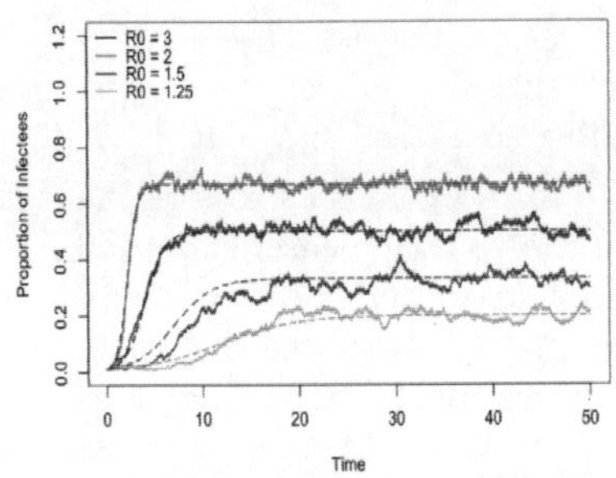

Figure 3: $(S_0, I_0) = (990, 10)$

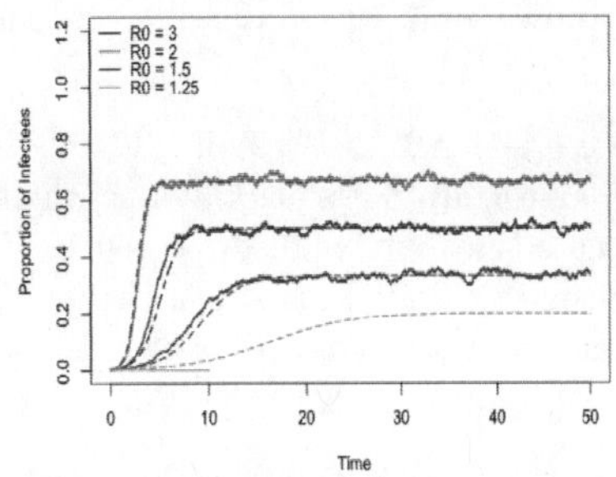

Figure 4: $(S_0, I_0) = (2990, 10)$

In the later section, we will show that the SIS model approached by stochastic process can be approximated by deterministic process if we let $n \to \infty$ in special case.

3. Branching Process

Suppose that in a closed population, just by the end of its lifetime, each individual has independently produced $j \geq 0$ offsprings with probability P_j. Let X_n denote the size of n-th generation, then the process $\{X_n : n = 0, 1, 2, \cdots\}$ is a Markov chain having non-negative integer state space, where state 0 is a recurrent state and others are transient. Since any finite transient state will be visited finitely often, this leads to the conclusion that if $P_j > 0$, then the population will either die out or take off.

Let μ be the mean number of offspring from a single individual, then

$$E(X_n) = E\left(E\left(\sum_{j=1}^{X_{n-1}} \zeta_j \mid X_{n-1}\right)\right), \quad \zeta_j \text{ is number of } j\text{th individual's offspring}$$
$$= \mu^2 E(X_{n-2}) = \cdots = \mu^n E(X_0) = \mu^n, \quad \text{given } X_0 = 1.$$

Suppose that $\pi_0 = \lim_{n\to\infty} P(X_n = 0|X_0 = 1)$ denotes the probability that the population dies out. Note that if $\mu < 1$, then $P(X_n \geq 1) \to 0$ almost surely as $n \to \infty$. This concept is very useful in epidemic modelling because when the expected number of infection is less than 1, then the epidemic will obviously die out. But, if $\mu > 1$, by conditioning on the X_1, $\pi_0 = \sum_{j=0}^{\infty} P(\text{population dies out}|X_1 = j)P_j$.

Note that given $X_1 = j$, the population will eventually die out if and only if each of the j families started by the members of the first generation die out. This statement is supported by the fact that the probability of extinction occuring in the early stage is non-zero. Moreover, the probability that the members of a typical family to die out is π_0. Thus, assuming that each individual acts independently, $P(\text{population dies out}|X_1 = j) = \pi_0^j$.

Hence, π_0 must satisfy

$$\pi_0 = \sum_{j=0}^{\infty} \pi_0^j P_j = E\left(\pi_0^{X_1}\right) = f(\pi_0) \tag{2}$$

where $f(\pi_0)$ is the pgf (probability generating function) of X_1, in such a way that π_0 is the smallest root in $[0,1]$ satisfying equation (2).

4. Density Dependent Population Process

Suppose that $\{\boldsymbol{X}_n(t) : t \geq 0\}$ is the CTMC process defined on d-dimensional integer lattice $\mathbb{Z}^d$ with finite number of possible transitions, given by $q_{i,i+l} = n\beta_l\left(\frac{k}{n}\right)$ where we define $\beta_l : \mathbb{Z}^d \to [0,\infty), l \in \Delta \subseteq \mathbb{Z}^d, \sum_l \beta_l(\boldsymbol{x}) < \infty$ for all $\boldsymbol{x} \in \mathbb{Z}^d$. In this context, we interpret n as the area of region occupied by a certain population. According to Ethier and Kurtz (1986), if the the size of a certain population was k, then the density of that population was $\frac{k}{n}$ with intensities approximated proportionally to its size. For example, in the epidemic case, suppose that there are i-susceptible individuals who are vulnerable to the disease and j-infectious individuals. We assume that the disease transmission has constant rate λ/n and the diseased-individual recovers with rate γ. An infection occurs when an infective makes contact to a susceptible. Therefore, the contact intensity and recovery intensity are $q_{j,j+1} = \frac{\lambda}{n}ij = n\lambda\left(\frac{i}{n}\right)\left(\frac{j}{n}\right)$ and $q_{j,j-1} = \gamma j = n\gamma\left(\frac{j}{n}\right)$, respectively. Both intensities are density-dependent functions.

Now, suppose that Y_l is a independent standard Poisson process, defined for each of possible transition $l \in \Delta$. Therefore, by letting $\boldsymbol{X}_n(0)$ be non-random,

$$\boldsymbol{X}_n(t) = \boldsymbol{X}_n(0) + \sum_l lY_l\left(n\int_0^t \beta_l\left(\frac{\boldsymbol{X}_n(s)}{n}\right)ds\right). \tag{3}$$

Note that the process Y_l is a function of density-dependent. Therefore, the process $\{\boldsymbol{X}_n(t) : t \geq 0\}$ equals to the summation of its initial value and the total increment given by the standard Poisson process during time interval $[0,t]$.

Now let $\tilde{Y}_l(u) = Y_l(u) - u$ be a centered Poisson process at its expectation and $\bar{\boldsymbol{X}}_n(t) = \frac{\boldsymbol{X}_n(t)}{n}$. Following equation (3) and letting $F(\boldsymbol{x}) = \sum_{l\in\Delta} l\beta_l(\boldsymbol{x})$, we can write equation (3) into

$$\bar{\boldsymbol{X}}_n(t) = \bar{\boldsymbol{X}}_n(0) + \frac{1}{n}\sum_{l\in\Delta} l\tilde{Y}_l\left(n\int_0^t \beta_l\left(\bar{\boldsymbol{X}}_n(s)\right)ds\right) + \int_0^t F(\bar{\boldsymbol{X}}_n(s))ds. \tag{4}$$

We will show in a later section that the stochastic process in equation (4) can be approximated by non-random quantity when we let the population size become sufficiently large.

5. The Fundamental Result of Poisson Processes

To start this section, the readers are expected to have prior knowledge of martingale and stopping theorem.

The theorem introduced in this section is called fundamental result of Poisson process. But before that, consider the followings

Theorem 5.1. Let $\{X_n; n \geq 0\}$ be a submartingale with respect to itself for which $X_n \geq 0$ for all $n \geq 0$. Then, for any $\lambda > 0$,

$$\lambda P\left(\max_{0\leq k\leq n} X_k > \lambda\right) \leq E(X_n).$$

(Ball, 2016)

Lemma 5.2. (Borel-Cantelli). Suppose that $\{E_n : n \geq 1\}$ denotes a sequence of events in probability space Ω. If

$$\sum_{n=1}^{\infty} P(E_n) < \infty,$$

then $P\{E_n$ occurs infinitely often for some $n\} = 0$.

Finally, the fundamental result of the Poisson process will be stated formally in the following theorem.

Theorem 5.3. If $\{Y(t), t \geq 0\}$ is a standard Poisson process, then for all $s \leq t$,

$$\lim_{n\to\infty} \sup_{s\leq t} \mid n^{-1}Y(ns) - s \mid = 0$$

almost surely.

Proof. Note that $Y(ns) \sim$ Poisson(ns), for $s \geq 0$ since it is a Poisson process. We can show easily that $(Y(ns) - ns)$ is a martingale. Further, $(Y(ns) - ns)^4$ is a submartingale. Therefore, by applying Theorem 5.1, for any $\varepsilon > 0$

$$P\left(\sup_{0\leq s\leq t} |Y(ns) - ns| \geq \varepsilon n\right) = P\left(\sup_{0\leq s\leq t} (Y(ns) - ns)^4 \geq \varepsilon^4 n^4\right)$$
$$\leq \frac{E(Y(nt) - nt)^4}{\varepsilon^4 n^4} \leq \frac{C(\varepsilon, t)}{n}. \quad (5)$$

Consider equation (5). If we let $n \to \infty$, then $P\left(\sup_{0\leq s\leq t} |Y(ns) - ns| \geq \varepsilon n\right) \to 0$. It implies that $\sum_{n=1}^{\infty} P(|Y(ns) - ns|) < \infty$ since the probability of its supremum tends to zero as n goes to infinity. Therefore, according to Lemma 5.2, $P\left(\sup_{0\leq s\leq t} \left|n^{-1}Y(s) - s\right| \to 0 \text{ as } n \to \infty\right) = 1$. Hence, $\lim_{n\to\infty} \sup_{s\leq t} \mid n^{-1}Y(ns) - s \mid = 0$ almost surely. □

Note that the second term of (4) will converge to 0 if we let n be large according to Theorem 5.3. This suggests that $\bar{\boldsymbol{X}}_n(t)$ will be non-random as $n \to \infty$. So, the process $\bar{\boldsymbol{X}}_n(t)$, given $\bar{\boldsymbol{X}}_n(0)$ is non-random, can be approximated by quantity $\left(\bar{\boldsymbol{X}}_n(0) + \int_0^t F(\bar{\boldsymbol{X}}_n(s))ds\right)$ as $n \to \infty$. Before proving this statement, consider the following lemma.

Lemma 5.4. (Gronwall's inequality). Suppose that $f : \mathbb{R} \to \mathbb{R}$ satisfying $0 < f(t) \leq a + b\int_0^t f(u)du$ for some $a, b > 0$ and $\forall t \geq 0$. Then $\forall t \geq 0$, $f(t) \leq ae^{bt}$.

Now, we will prove the previous statement by proving the below theorem.

Theorem 5.5. Let $\lim_{n\to\infty} \bar{X}_n(0) = x_0$ and $x(t) = x_0 + \int_0^t F(x(u))du$ be the deterministic version of equation (4). Suppose that for each compact $K \subset \mathbb{R}^d, \exists M_K > 0$ such that $\forall x, y \in K, \quad | F(x) - F(y) | \leq M_K | x - y |$. Then, for $0 \leq s \leq t$,

$$\lim_{n\to\infty} \sup_{s\leq t} | \bar{X}_n(s) - x(s) | = 0$$

almost surely.

Proof. Given $x(t) = x_0 + \int_0^t F(x(u))du$, thus

$$| \bar{X}_n(t) - x(t) | \leq | \bar{X}_n(0) - x(0) | + \frac{1}{n} \sum_{l\in\Delta} | l | \sup_{s\leq t} | \tilde{Y}_l \left(n\beta_l \left(\bar{X}_n(s)\right)\right) | +$$

$$\int_0^t M_k | \bar{X}_n(s) - x(s) | \, ds. \tag{6}$$

Now consider the second term of the right hand side of equation (6).

$$\frac{1}{n} \sum_{l\in\Delta} | l | \cdot \sup_{s\leq t} | \tilde{Y}_l \left(n\beta_l \left(\bar{X}_n(s)\right)\right) | = \sum_{l\in\Delta} | l | \cdot \sup_{s\leq t} | n^{-1} Y_l \left(n\beta_l \left(\bar{X}_n(t)\right)\right) - \beta_l \left(\bar{X}_n(t)\right) | .$$

But, according to Theorem 5.3, if we let $n \to \infty$, then $| n^{-1} Y_l \left(n\beta_l \left(\bar{X}_n(t)\right)\right) - \beta_l \left(\bar{X}_n(t)\right) | \to 0$ almost surely. Therefore, equation (6) now becomes

$$| \bar{X}_n(t) - x(t) | = | \bar{X}_n(0) - x(0) | + \int_0^t M_k | \bar{X}_n(s) - x(s) | \, ds.$$

as $n \to \infty$. By applying Lemma 5.4,$| \bar{X}_n(t) - x(t) | \leq | \bar{X}_n(0) - x(0) | \cdot e^{M_K t} = 0$ since $\lim_{n\to\infty} \bar{X}_n(0) = x(0)$. Therefore, $0 \leq \lim_{n\to\infty} \sup_{s\leq t} | \bar{X}_n(s) - x(s) | \leq \lim_{n\to\infty} | \bar{X}_n(t) - x(t) | = 0$. Hence, by sanwich property, $\lim_{n\to\infty} \sup_{s\leq t} | \bar{X}_n(s) - x(s) | = 0$. □

Theorem 5.5 suggests that as $n \to \infty$, the process $\{\bar{\boldsymbol{X}}_n(t), t \geq 0\}$ will resemble the deterministic version $\boldsymbol{x}(t)$. It means that for large n, the process will settle down and randomly move around its mean. We can use this analogy to approximate the stochastic epidemic model when it reaches endemic-equilibrium.

Now, see Figure 1, 2, and 4. In stochastic approximation, even though $R_0 > 1$, there is a non-zero probability that the epidemic will die out. This fact violates our deterministic approximation. Therefore, we will investigate why such scenario occured. Note that in the stochastic version of the SIS model, if the number of susceptibles n is large, the process of infectives can be approximated by a branching process. Following the definition of the SIS model with each shape of an epidemic with few initial infectives, we know that contacts made by a typical infective to susceptibles follow a Poisson process of rate $\beta = \lambda/n$. Suppose that T_i is infection period of i-th typical infective, which is assumed to follow independent exponential distributions with intensity γ and let R be the total number of contacts made by a typical infective in the epidemic model. Then,

$$R \mid T = t \sim \text{Poisson}(\beta t) \tag{7a}$$

$$T \sim \text{exponential}(\gamma). \tag{7b}$$

IOP Conf. Series: Journal of Physics: Conf. Series **795** (2017) 012047 doi:10.1088/1742-6596/795/1/012047

Table 1: Comparison table of probability the epidemics to die out approximated by branching process and empirical with $m = 1$

R_0	Population Size	Branching Process	Empiral (10000 runs)
≤ 1	$N = 100$	1	1
	$N = 1000$	1	1
3.0	$N = 100$	0.3333	0.3420
	$N = 1000$	0.3333	0.3331
5.0	$N = 100$	0.2	0.1997
	$N = 1000$	0.2	0.1996
8.0	$N = 100$	0.125	0.1273
	$N = 1000$	0.125	0.1282

Now, suppose that $f(s), s \in [0, 1]$ is the pgf of R and $g(t) \quad (t \geq 0)$ is the pdf of T. Then,

$$f(s) = \sum_{k=0}^{\infty} s^k \int_0^{\infty} P(R = k \mid T = t)g(t)dt = \frac{\gamma}{\beta + \gamma} \sum_{k=0}^{\infty} \left(\frac{s\beta}{\beta + \gamma} \right)^k .$$

Note that according to Section 3, we find s, such that $f(s) = s$. Therefore, $s = 1$ or $s = \frac{\gamma}{\beta}$. If $\gamma \geq \beta$, then the smallest root is 1, but if $\gamma < \beta$, then the smallest root is $\frac{\gamma}{\beta}$. Hence, suppose there are m initial infectives in the population and assuming that the contacts made are mutually independent, then

$$P(\text{epidemic dies out}) \approx \begin{cases} 1 & \text{if } \gamma \geq \beta \\ \left(\frac{\gamma}{\beta}\right)^m & \text{if } \gamma < \beta. \end{cases}$$

Once again, the above result explains threshold behaviour of the process. All figures confirm the interpretation of threshold behaviour of the SIS deterministic model, that if $R_0 > 1$, major epidemic will occur despite of its initial values and population size. But in the stochastic model (which are displayed in jagged plots), some plots do not mimic the deterministic paths. For example, in Figure 1, the epidemic with $R_0 = 1.5$ and $R_0 = 1.25$ quickly die in a very short time. If we apply branching process approximation, the probability that an epidemic dies out in Figure 1 are $0.11, 0.25, 0.44$, and 0.64, respectively. It seems that as R_0 approaches 1 from the right ($R_0 \downarrow 1$), the greater the probability of an epidemic to die out. But look at Figure 3. Despite similar population size and R_0's, the last two epidemic models (indicated by green and orange colours) do not die out. It turns out that by using the branching process, the probability that the last two epidemics in Figure 3 to die out decreases to 0.017 and 0.107, respectively. So, we could safely claim that when $R_0 > 1$, the stochastic model of the SIS epidemic to die out quickly depends upon the proportion of its initial infectives since intuitively, if there are few initial infectives present in the population, according to the branching process theory, there is non-zero probability that their infectious periods will be over before they make contact with any susceptibles. Another simulation is provided in Table 1, where we compare the probability of the epidemics to die out approximated by brancing process and empirical.

6. Conclusion

We have successfully showed that both deterministic and stochastic models performed similar results when $R_0 \leq 1$. That is, the disease-free stage in the epidemic. But when $R_0 > 1$, the deterministic and stochastic approaches had different interpretations. In the deterministic

IOP Conf. Series: Journal of Physics: Conf. Series **795** (2017) 012047 doi:10.1088/1742-6596/795/1/012047

models, both the SIS and SIR models showed an outbreak of the disease and after some time t, the disease persisted and reached endemic-equilibrium stage. The stochastic models, on the other hands, had different interpretations. If we let the population size be sufficiently large, the epidemic might die out or survive. There were essentially two stages to this model. First, the infection might die out in the first cycle. If it did, then it would happen very quickly, just like the branching process theory described. Second, if it survived the first cycle, the outbreak was likely to occur, but after some time t, it would reach equilibrium just like the deterministic version. In fact, the stochastic models would mimic the deterministic's paths and be scattered randomly around their equilibrium point.

IOP Conf. Series: Journal of Physics: Conf. Series **795** (2017) 012048 doi:10.1088/1742-6596/795/1/012048

A survey on keeler's theorem and application of symmetric group for swapping game

Yohanssen Pratama[1,*] and Yohenry Prakasa[2]

[1]Faculty of Informatics and Electrical Engineering, Del Institute of Technology, Jl. Sisingamangaraja Sitoluama Laguboti Toba Samosir 22381, INDONESIA

[2]Faculty of Mathematics and Natural Sciences, Bandung Institute of Technology

* yohanssen.pratama@del.ac.id

Abstract. An episode of *Futurama* features two-body mind-switching machine which will not work more than once on the same pair of bodies. The problem is *" Can the switching be undone so as to restore all minds to their original bodies?"* Ken Keeler found an algorithm that undoes any mind-scrambling permutation, and Lihua Huang found the refinement of it. We look on the process how the puzzle can be modeled in terms group theory and using symmetric group to solve it and find the most efficient way of it. After that we will try to build the algorithm to implement it into the computer program and see the effect of the transposition notion into the algorithm complexity. The number of steps that given by the algorithm will be different and one of algorithms will have the advantage in terms of efficiency. We compare Ken Keeler and Lihua Huang algorithms to see is there any difference if we run it in the computer program, although the complexity could be remain the same.

1. Introduction

"The Prisoner of Benda", is the episode from the animated television series "*Futurama*" which explain the mechanism of two-body mind-switching machine. Any pair of person (two people) can enter the machine to swap minds, but there is some limitation that should be obeyed. The limitation is the machine will not do the swapping on the same pair of bodies. After all minds has been swapped then it appears a problem that if we want to return al off the swapping minds into its original bodies, we must search the optimum ways to return it. The television show provides an answer using a "Keeler's theorem" [1]. We will use the solution that provides by "Keeler's theorem" and an enhancement of "Keeler's theorem" that propose by Lihua Huang [1] to solve the similar case which is we usually found in the swapping game and try to build some algorithms. The algorithms will be analyzed to see its complexity and efficiency.

The problem to return all the swapping minds into the original body can be modeled using group theory. If there n bodies that involved in the experiment, then it could be write as $\{1,2,...,n\}$. The symmetric group S_n will consist of the $n!$ permutations of $\{1,2,....n\}$. Because the machine switching 2 people mind each time, so we could call a permutation. A permutation which switch the mind's a into b and vice versa is called transposition, it represents 2-cycle (ab). The m-cycle $(a_1...a_m)$ is the permutation which switches a_1's mind to a_2, a_2's mind to a_3, ..., and a_m's mind to a_1 [1]. To divide the products in S_n into transposition (2-cycle), we compute it from right side into left side. For example, $(345) = (34)(45) = (35)(34) = (45)(35)$

The machine has limitation that it won't work more than once on the same pair of bodies, it means each transposition in S_n must be distinct. Let a product P of distinct transposition in S_n, it's represent the successive swapping of minds. Because P is a product, we can also view P as permutation, assumed that this permutation is non-trivial, otherwise it is clear that no one enter the machine. Example, suppose that 4 switches minds with 5, and then 4 switches mind with 3, this mean the product $P = (34)(45)$, or we can say in general, the mind-swapping permutation is $P = (345)$.

We must find a product α of distinct transposition such that permutation $\alpha P = I$, *because it was need* to restore all minds to their original bodies. Such transposition factors in the product α must be distinct with the product of P, because the limitation of the machine (It said that α undo P).

To find a product α that is undo the mind-scrambling permutation $P \in S_n$, we need to consider about which sequence of transposition in S_n that effected P, because they need to be distinct. Keeler's theorem provide such a product $\alpha \in S_{n+2}$. Each factor in Keeler's α contains at least one entry in the set $\{a,b\}$, where

$$a := n+1 \ \text{ and } \ b := n+2 \ ;$$

It is clear that transposition factors in σ are distinct from whatever transpositions that effected P. We can say that a and b as an *outsiders* who never enter the machine during the switching process. a and b is a persons who willing as volunteers to bear another persons mind. They need in order to help another persons to restore their minds into their each original body.

P can be viewed as permutation that was expressed uniquely as the product $P = E_1....E_r$ of non-trivial disjoint cycles $E_1....E_r$ in S_n. For each $i = 1, \ldots, r$, let m_i denote the length of cycle E_i. For discussing Keeler's theorem, we assume that [1]

$$k_1 + k_2 + \ldots + k_r = n$$

2. Keeler's Method

We now describe Keeler's method for constructing a product $\sigma \in S_{n+2}$ which undoes $P = E_1....E_r$. For easier to read the notation, we write $m = m_1$, so that E_1 is a m-cycle $(u_1 \ldots u_m)$ with each $u_i \in \{1, 2, \ldots, n\}$. We can directly check that $\alpha_1 E_1 = (ab)$, where α_1 is the product of $m+2$ transposisitons given by [1]

$$\alpha_1 = (au_1)(au_2)\ldots(au_{m-1})\cdot(bu_m)(bu_m)(bu_1). \tag{1}$$

For each E_i define the similar products α_i of m_{i+2} transpositions, which satisfying

$$\alpha_i E_i = (ab) \ \text{ for } \ i = 1, \ldots, r.$$

Notice that each transposisiton in α_i has the form (az) or (bz) for some entry $z \in E_i$.

Example 1. for $E_1 = (253)$, then by (1) we have $\alpha_1 = (2a)(5a)(3b)(3a)(2b)$, which is

$$\alpha_1 E_1 = (2a)(5a)(3b)(3a)(2b)\cdot(253) = (2)(5)(3)(ab) = (ab)$$

We know that disjoint cycles are commute, so (ab) commutes with every transposition in S_n, then

$$\varphi := \alpha_r \ldots \alpha_2 \alpha_1$$

is a product of distinct transpositions for which $\varphi P = (ab)^r$. So we can conclude as,

IOP Conf. Series: Journal of Physics: Conf. Series **795** (2017) 012048 doi:10.1088/1742-6596/795/1/012048

$$\alpha = \begin{cases} (ab)\varphi\,, & \text{if } r \text{ is odd} \\ \varphi \quad , & \text{if } r \text{ is even} \end{cases} \tag{2}$$

Clearly that α undoes P and α is a product of distinct transpositions in S_{n+2}, each containing at least one entry in a , b, as desired .

From equation (1) and (2), we can see that the number of factors α will be either $n+2r+1$ or $n+2r$. It depend on r, whether it was odd or even. However, for each $r > 2$, the number of factors that needed to undo P can be reduced. For example, for $r = 3$, $P = (12)(34)(56)$ is undone by Keeler's product of 13 transpositions

$$\alpha = (ab)(5a)(6b)(6a)(5b)(3a)(4b)(4a)(3b)(1a)(2b)(2a)(1b),$$

but in reality P can be undone by the product of 11 transpositions,

$$(5a)(6b)(6a)(5b)(1a)(2a)(3a)(4b)(4b)(3b)(1b).$$

After we see the brief of description that become a core of Keeler's algorithm itself, now we will try to build some computer algorithm from it. For the Keeler's algorithm we tried to define the pseudocode as below:

```
BEGIN
  swap(source, destiny, array) {
  if swap = 1, return -1;
    swap source with destiny;
  if all mind already restored to original body, return 1;
    for (source = 0, source = array.length -1,i++)
        for (destiny = 0, destiny = array.length-1, j++)
        {
          if ((source != destiny) && (swap(source, destiny) == 1);
          {
              print swap;
            return 1;
        }
    }
}
  END
```

If we see the pseudocode above, the body number will be stored in array and if the swap was already tried it will return an invalid value. If the source and destination already swapped the records will be updated so it can't be swapped again. When everyone mind already comeback to original body the search will be end. In the next section, we will look at LiHua Huang proposed theorem that is the refinement from the Keeler's theorem. The theorem showing that P can be undone only by $n+ r +2$ distinct transpositions that each containing at least one entries a, b which is better than n + 2r.

3. Analysis of optimal refinement of Keeler's method

Keeler's algorithm was designed to undo every mind-scrambling permutation $P = E_1 \ldots E_r$ that is affected by an unknown sequence of mind swaps. In this section, we will see another algorithm to return all the scramble minds into it original bodies by Lihua Huang.

Theorem 1. Let $P = E_1 \ldots E_r$ be a product of disjoint m_i-cycles $E_i \in S_n$, with $m_i \geq 2$ and $n = m_1 + \ldots + m_r$. Now we define $a = n+1$ and $b = n+2$. Then P can be undone by a product γ of $n+r+2$ distinct transpositions in S_{n+2}, each containing at least one entry in a, b.

Proof. Write $m = m_1$, so that E_1 is a m-cycle $(u_1 \ldots u_m)$. We could define the product that corresponding to the E_1 cycle

$$G_1(a) = (u_1 a)(u_2 a)\ldots(u_m a)$$

which the most left factor is

$$F_1(a) = (u_1 a)$$

Corresponding to each cycle E_i for $i = 1, \ldots, r$, we could define $G_i(a)$ and $F_i(b)$ similarly and set,

$$\dot{\gamma} = (ab) \cdot G_r(a) \ldots G_2(a) \cdot (u_m a) G_1(b)(u_1 a) \cdot F_2(b) \ldots F_r(b)$$

It is quite easy to check that γ undoes P and that γ is a product of $n+r+2$ distinct transpositions in S_{n+2}, each containing at least one entry in $\{a,b\}$.

Example 2. Let $P = (12)(345)(67) \in S_7$. Find $\sigma \in S_9$, a product of distinct transpositions such that each transposition contains at least one of the entries a, b and that α undoes P. Keeler's method:

$$\alpha_1 = (ab)(a1)(b2)(a2)(b1)(a3)(a4)(b5)(a5)(b3)(a6)(b7)(a7)(b6).$$

There are $n+2r+1 = 7 + 6 + 1 = 14$ transpositions in α_1, which undoes P.
LiHua Huang refinement algorithm [1]:

$$\alpha_2 = (ab)(6a)(7a)(3a)(4a)(5a)(2a)(1b)(2b)(1a)(3b)(6b).$$

Notice that there are only $n+r+2 = 7+3+2 = 12 < 14$ transpositions in α_2, which undoes P.

Example 3. Let $P = (12) \in S_2$. Undo P with $\sigma \in S_4$. Keeler's method:

$$\alpha_1 = (ab)(a1)(b2)(a2)(b1).$$

There are $n + 2r + 1 = 2 + 2 + 1 = 5$ transpositions in α_1, which undoes P. LiHua Huang refinement algorithm:

$$\alpha_2 = (ab)(2a)(1b)(2a)(1b).$$

There are $n + r + 2 = 2 + 1 + 2 = 5$ transpositions in α_2, which also undoes P. (Observe that if one switches the disjoint transpositions $(a1)$ and $(b2)$ in α_1 and then renames a, b as b, a, respectively, one can see that α_1 becomes α_2).

From example above, we can conclude that when $r = 1, 2$, both Keeler's method and LiHua Huang algorithm are best possible. However, when r is large, LiHua Huang said have showed that $n + r + 2$ is much less than $n+2r$.

If we look the pseudocode again, then algorithm the complexity will be:

$$\sum_{i=0}^{t-1}\sum_{j=0}^{t-1} 1 = t.t = t^2$$

The algorithm contains two nested loops (first for and second for) which the segment inside loops are executed consecutively t-1 times (t is the total numbers of the input). The complexity of the algorithm will be O(t^2) where t is the length of array. Because both Keeler's and LiHua Huang algorithm has same n instead r and the leading constant in 2r may be ignored, the complexity still leads to O(t^2).

4. Conclusion

From our survey, we can conclude that from simple swapping game, we can search the optimal way to solve the problem. Although the complexity are same for both Keeler's and Lihua Huang algorithm that is O(t^2), but the process that given by the Lihua Huang algorithm is more efficient. For future works, we plans to find another way to utilize the advantages that owned by the Lihua Huang algorithm to improve the running time in the computer program. We hope science can help us solve our daily problem especially on programming or computation problem.

References

[1] Evans R and Huang L 2012 *Keeler's theorem and products of distinct transpositions*
[2] Isaacs M 1994 *Algebra: A Graduate Course,* Graduate Studies in Mathematics vol 100 Amer. Math. Soc., Providence, RI
[3] Bassil Y A 2012 Comparative Study on the Performance of Permutation Algorithms *Journal of Computer Science and Research* **1** 7 -19
[4] Levine A 2010 The Futurama of physics with David X. Cohen. *Amer. Physical Soc. News 19* **3**
[5] Shanker S 2016 Time Complexity of Matrix Transpose Algorithm using Identity Matrix as Reference Matrix *International Journal of Computer Science and Information Technologies* **5** 2347-48
[6] Cormen T Leiserson C Rivest R and Stein C 2009 *Introduction to Algorithms 3rd ed* (MIT Press)
[7] Levitin A 2006 *Introduction to the Design and Analysis of Algorithms 2nd ed* (Addison Wesley)
[8] Donald K 1998 *The Art of Computer Programming (Sorting and Searching)* vol 3
[9] Bona M 2004 *Combinatorics of Permutations* (Chapman Hall-CRC)
[10] Oliver R 2008 On the parity of permutation *Amer. Math. Monthly* **118** 734-35
[11] Neuenschwander D 2001 On the representation of permutations as products of transpositions *Elem. Math.* **56** 1-3
[12] Grime J 2010 *Futurama and Keeler's Theorem : original edit, video exposition of Futurama's episode "The Prisoner of Benda"* available at http://www.youtube.com/watch?v=8M4dUj7vZJc

Iris Segmentation and Normalization Algorithm Based on *Zigzag Collarette*

M. Rizky Faundra[1] and Dwi Ratna Sulistyaningrum[2]

Mathematics Department of Post Graduate Program
Sepuluh Nopember Institut of Technology,
Campus ITS-Sukolilo, Surabaya 60111 INDONESIA

E-mail: rizkyfaundra@mhs.matematika.its.ac.id

Abstract. In this paper, we proposed iris segmentation and normalization algorithm based on the *zigzag collarette*. First of all, iris images are processed by using Canny Edge Detection to detect pupil edge, then finding the center and the radius of the pupil with the Hough Transform Circle. Next, isolate important part in iris based *zigzag collarette* area. Finally, Daugman Rubber Sheet Model applied to get the fixed dimensions or normalization iris by transforming cartesian into polar format and thresholding technique to remove eyelid and eyelash. This experiment will be conducted with a grayscale eye image data taken from a database of iris-Chinese Academy of Sciences Institute of Automation (CASIA). Data iris taken is the data reliable and widely used to study the iris biometrics. The result show that specific *threshold* level is 0.3 have better accuracy than other, so the present algorithm can be used to segmentation and normalization *zigzag collarette* with accuracy is 98.88%

1. Introduction

Biometrics technology has grown rapidly this decade, a technology that uses human characteristics can be used as a self-replacing passwords or ATM card at a signing at the important place that requires a fairly high level of security [1]. Some biometric technologies such as fingerprint recognition and voice has been developed and attract attention but the required technology is sufficiently advanced so that the security system for the better. One biometric technology that is reliable and still needs to be developed is individual recognition through the iris texture [2]. Texture human iris are many and complex consists of two layers, namely ephitelium layer and the stromal layer, the combination of that layer to form a color pattern that is very diverse and naturally occurring three months of birth [3], in addition to the iris pattern is unique even iris left and right eyes on each individual has a different pattern. When compared with other textures such as face, fingerprint or voice, iris texture is more stable due to the location of the iris in the body so that it can be used as a reliable individual recognition [4].

The method of segmentation and normalization in iris has role important to the next process, selection of part of the iris is very important because it determines the percentage of the introduction of the iris recognition system. [5] only select the $(175^0, 225^0)$ and $(310^0, 360^0)$ regions to detect iris outer boundary by using gradient operator for vastly reducing the obsecured by eyelashes and eyelids.

[1] Post Graduate Student at Mathematics Department Sepuluh Nopember Institut of Technology
[2] Senior Lecturer at Mathematics Department Sepuluh Nopember Institut of Technology

[6] only select interval $(0,45^0) \cup (105^0,225^0) \cup (315^0,360^0)$ to avoid noise by eyelid, in this paper, zigzag collarette area of the iris is selected for iris feature extraction because it captures the most important areas of iris complex pattern and concentric with the pupil. The radius of zigzag collarette areas is restricted in a certain range, in Figure 1 describes part of eye image

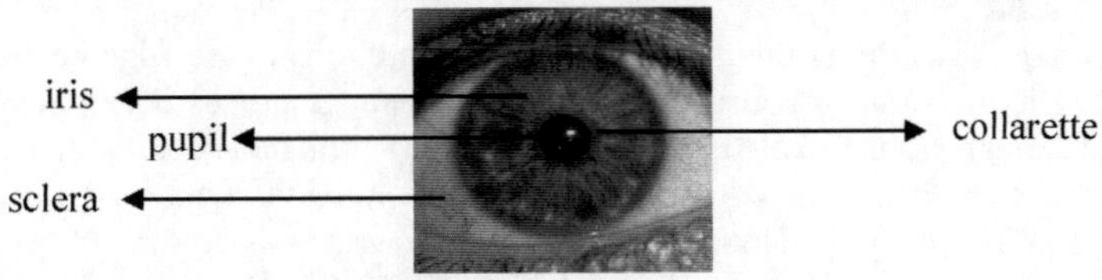

Figure 1. Part of Eye image

2. Iris Segmentation

2.1. Edge Pupil Detection

The pupil is a brown area that is in the middle of the eye, pupil color is very dominant and uniform so that it can be seen the differences between pupils and iris. Edge detection of the pupil is required to seek the boundary between pupil and iris, the method used in this study is the Canny edge detection with initial *threshold*. Canny edge detection has the advantage that the resulting edge as the edge of the actual picture [7]. Here is the Canny edge detection algorithm:

a. Convert image to binary image based on *threshold* with specific level
b. Removal of noise in the image so that it looks the difference between the edge and not edge with a Gaussian filter 5×5
c. Calculate the gradient intensity and direction of the edge with Sobel operator
d. Grouping toward the edges at an angle $0^0, 45^0, 90^0, 135^0$
e. Eliminate pixels which is not a maximum value in the direction of the edge by comparing two neighboring pixels
f. Eliminate lines disconnected as the edges of objects using two threshold values are T1 and T2. Mathematically as follows:

$$f(x,y) = \begin{cases} 0 & ,f(x,y) < T_1 \\ 128, & T_1 \le f(x,y) \le T_2 \\ 225 & ,f(x,y) > T_2 \end{cases} \quad (1)$$

In this study, the value T1 = 0,5 and T2 = 0,5
Perform testing to get the condition in the Figure 2, as follows:

255	255	255
255	128	255
255	255	255

→

255	255	255
255	255	255
255	255	255

Figure 2. Testing change the value 128 to 255

In this research Canny algorithm successfully detects the edge of the pupil in Figure 3.

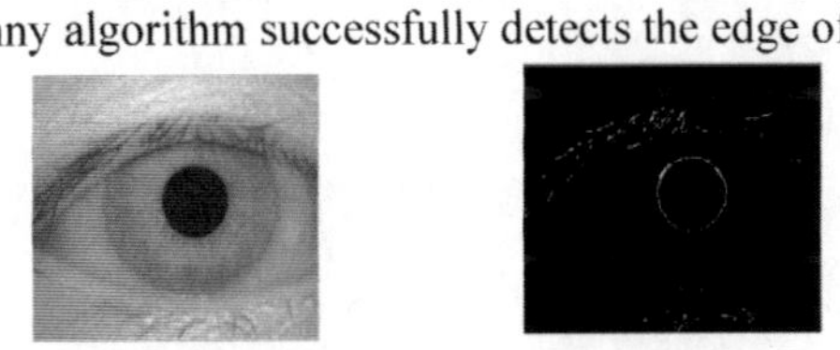

(a) Iris Image **(b) Canny Edge Detection**

Figure 3. Results Canny Edge Detection of eye image CASIA

Figure (3). (a) an original image of gray of the database eye image CASIA seen that there is a difference of color contrast between the pupil with the iris, (b) the result of Canny edge detection visible edges boundary the pupil and iris.

2.2. Finding Circle Pupil

Pupil in the eye shape resembling the geometric shape of a circle so the edge between pupil and iris can be represented in the form of a circle. One method of image processing to detect the shape of a circle is a circle Hough transform [8]. This method requires a binary edge image as an input to describe artificial circles, then circles of artificial sought the maximum cut-off point to find the center point. For all edge points (x_i, y_i), a Hough transform can be written as

$$H(x_c, y_c, r) = \sum_{i=1}^{n} h(x_i, y_i, x_c, y_c, r) \quad (2)$$

where

$$h(x_i, y_i, x_c, y_c, r) = \begin{cases} 1, & if\ g(x_i, y_i, x_c, y_c, r) = 0 \\ 0, & otherwise \end{cases} \quad (3)$$

where $g(x_i, y_i, x_c, y_c, r) = (x_i - x_c)^2 + (y_i - y_c)^2 - r^2$. The three coordinate (x_c, y_c, r) for which $H(x_c, y_c, r)$ is highest will become the coordinate of centre and radius of the circle. In this experiment, the limit radius of between 28-50 pixels for CASIA database [9]. Circle Hough transform algorithm managed to find the coordinates of the center of the pupil and the pupil radius, as in Table 1 below.

Table 1. Finding centre of pupil and radius

Number Eye	Eye Image	Centre of Pupil	Radius
xxx.bmp		135,183	37
xxx.bmp	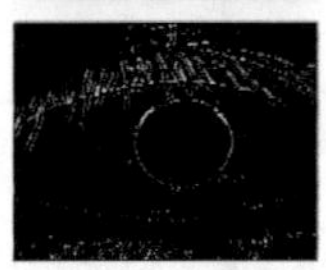	141,183	50

By using the parameters of the center and the radius of the pupil, this research is described a circular shape on the original image with the pixel value $f(x, y) = 255$ where

$$x = A + r\cos\theta\ , 0 \leq \theta \leq 360 \quad (4)$$

$$y = B - r\sin\theta , 0 \leq \theta \leq 360 \quad (5)$$

In the image coordinates can be illustrated form above equation as follows

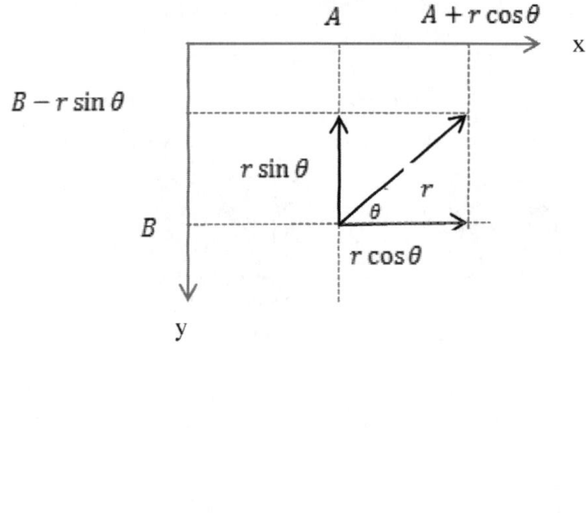

Figure 4. Draw a circle in the image coordinates

This experiment has been successfully depicts a circle on the original image, the following are the results in Figure 5.

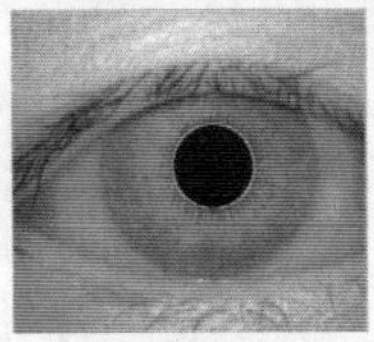

Figure 5. Draw a circle in the original image

2.3. *Zigzag Collarette Isolation*

The next processing is to isolate the area of the iris which is the area zigzag collarette. The area is selected for capture critical information from complex patterns of the iris, but it has little impact on noise such as eyelids and eyelashes. This area is generally concentric and close to the pupil so that this area used to detect the pupil center information and radius limited to a certain distance. In a study conducted by Rai concluded that the zigzag collarette is within 24 pixels of the pupil [10]. In this research zigzag collarette isolation area there are still noise around the pupil resulting geometric shape of the pupil is not completely circular, there are still the eyelashes and eyelids were a little caught up in the area so we need to make noise processing. The following steps are taken to isolate the area zigzag collarette:

a. Draw a circle with a distance of 24 pixels from the pupil
b. Closing pixels outside the zigzag collarette far as 40 pixels by 140 pixels values
c. Limiting the only region with a zigzag collarette equation

$$f(x,y) = g(x+x', y+y') \tag{6}$$

where $x' = [A - r + 24, A + r + 24]$ and

4. Close the pupil and eyelashes with a threshold value

$$f(x,y) = \begin{cases} 140, & f(x,y) \le 100 \\ f(x,y), & other \end{cases} \tag{7}$$

5. Close the eyelid with a threshold value

$$f(x,y) = \begin{cases} 140, & 200 < f(x,y) \le 255 \\ f(x,y), & other \end{cases} \tag{8}$$

In Table 2 below is the result of processing to isolate areas zigzag collarette. The results obtained show that the area has been isolated zigzag collarette well.

Table 2. Isolation zigzag collarette process

Draw zigzag boundary	Close outer zigzag	Limiting the only region with a zigzag collarette	Close the pupil and eyelashes	Close the eyelid
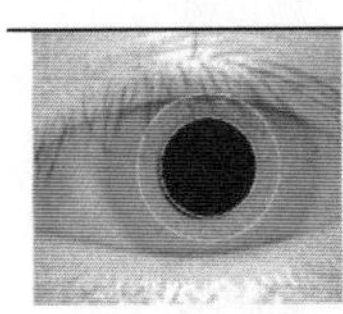	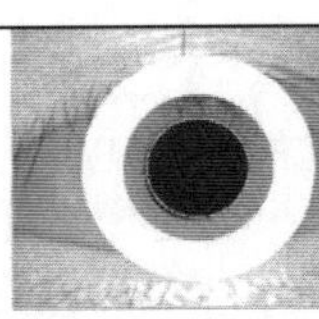	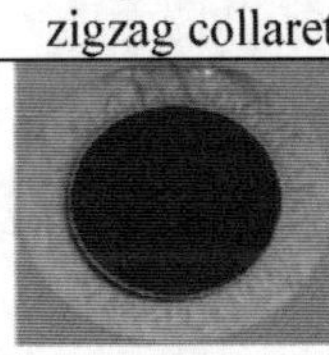	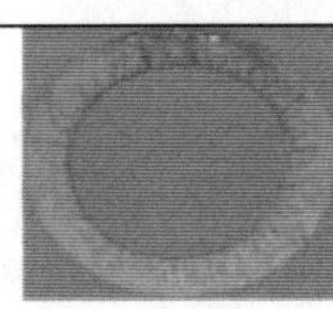	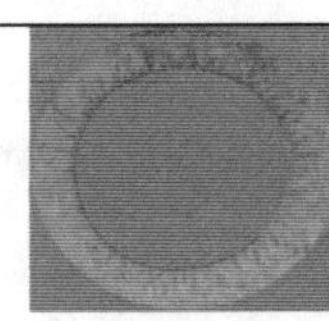

3. Iris Normalization

Further processing is to change the iris shape which is represented by zigzag collarette circular area into a rectangular shape. [4] proposed normalization iris to reduce inconsistencies due to improper iris acquisition by converting to polar form, the method used is called the Daugman Rubber Sheet Model. In this experiment used the formulation Daugman Rubber Sheet Model to map the zigzag collarette into polar form as follows:

$$I(x(r,\theta),y(r,\theta)) \rightarrow I(r,\theta) \quad (9)$$

with

$$x(r,\theta) = (1-r)x_p(\theta) + rx_z(\theta) \quad (10)$$

$$y(r,\theta) = (1-r)y_p(\theta) + ry_z(\theta) \quad (11)$$

where

$x_p(\theta)$ is the coordinates of the x-axis pupil boundary

$y_p(\theta)$ is the coordinates of the y-axis pupil boundary

$x_z(\theta)$ is the coordinates of the x-axis *zigzag collarette* boundary

$y_z(\theta)$ is the coordinates of the y-axis *zigzag collarette* boundary

The equation above interval radius is [0,1] and interval angle is $[0,2\pi]$. This implies that the radius of zigzag collarette has a length of 1 unit while the radius pupil converted into units acquired

$$\hat{r_p} = \frac{r_p}{r_p+24} \quad (12)$$

where r_p is the radius of the pupil in pixels

The next step is to determine the length and width of a pixel on polar coordinates, the research conducted by Rai (2014) obtained the value of the length is 512 pixels and wide 64 pixels to obtain information zigzag collarette. Here are the steps in normalizing zigzag collarette:

a. Determine the length and width of a pixel on polar coordinates that is the width of M pixels and a length of N pixels
b. Calculate the difference of the radius of formula
$$\Delta r = \frac{\hat{r_z}-\hat{r_p}}{M-1} \quad (13)$$
c. Calculate the difference of angle of formula
$$\Delta\theta = \frac{2\pi}{N} \quad (14)$$
d. Remapping of each pixel in accordance with the size of radius $\hat{r_p} + \Delta r$ and angle $\Delta\theta$

The results obtained show that the normalization process goes well and can map each pixel zigzag collarette to a rectangular shape in Figure 6.

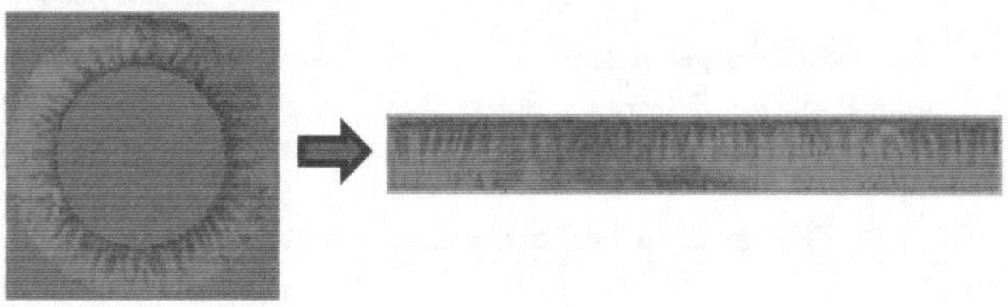

Figure 6. Normalized iris image

4. Experiment Result

Our experiment are performed in MATLAB on a PC AMD A6 1.8 GHz and RAM 2 Gb. The images used in our experiments come from CASIA image database, in our experiments we used 90 images (3 images per person), the proposed iris segmentation and normalisation algorithm has been calculated and accuracy with specific *threshold* level. The experimental result is shown in Table 3 below:

Table 3. Accuracy of Specific *Threshold* Level

Threshold Level	Accuracy
0.2	88.88%
0.3	98.88%
0.4	97.78%
0.5	73.33%

Table 3 illustrates maximum accuracy rate of each level *threshold*. The result show that specific *threshold* level is 0.3 have better accuracy than other, so the present algorithm can be used to segmentation and normalization *zigzag collarette* with accuracy is 98.88%

5. Conclusion

In this paper, we present iris segmentation and normalization algorithm based on *zigzag collarette*. The iris image segmented based on *zigzag collarette*. Canny edge detection and Hough transform can be localized pupil near *zigzag* area very well then Daugman Rubber Sheet Model can represent isolation *zigzag collarette*. The experiment result show that algorithm high performance of 98.88 percent with *specific* threshold level is 0.3.

References

[1] Anil K. Jain, Sharath Pankanti and Salil Prabhakar, "Biometric Recognition: Security and Privacy Concerns", *IEEE Security & Privacy*, 2003.

[2] Li Ma, Yunhong Wang, Tieniu Tan, "Iris Recognition Based on Multichannel Gabor Filter", *ACCV2002: The 5th Asian Conference on Computer Vision*, 2002.

[3] Adler F. H., *"Physiology of the Eye: Clinical Application"*, The C.V. Mosby Company, 1965.

[4] John Daugman, "New Methods in Iris Recognition", *IEEE Transactions On Systems, Man, And Cybernetics—Part B: Cybernetics*, Vol. 37, No. 5, 2007

[5] Lijun Zhou, Yide Ma, Jing Lian and Zhaobin Wang, *"A New Effective Algorithm for Iris Location"*, Proceeding of the IEEE International Conference on Robotics and Biometrics (ROBIO), Shenzhen, China, 2013.

[6] Todor Kazakov, *"Iris Detection and Normalization"*, University of Birmingham Thesis, 2011.

[7] John Canny, *"A Computational Approach to Edge Detection"*, Pattern Analysis and Machine Intelligence, IEEE Transactions on, PAMI-8(6):679–698, 1986.

[8] P.V.C. Hough, *"Method and Means for Recognizing Complex Patterns"*, U.S. Patent 3069654, 1962.

[9] CASIA iris image database.<http: //www.cbsr.ia.ac.cn/IrisDatabase>.

[10] Himansu Rai, *"Iris Recognition Using Combined Support Vector Machine and Hamming Distance Approach"*. Expert Systems with Applications 41 (2014) 588–593, 2014.

IOP Conf. Series: Journal of Physics: Conf. Series **795** (2017) 012050 doi:10.1088/1742-6596/795/1/012050

Patterns relationships of student's creativity with its indicators in learning optical instrument

Sukarmin, Dhian T E V, Nonoh S A, and Delisma W A [1]

Science Education Departement Program Sebelas Maret University, Jl. Ir. SUtami 36A Kentingan Jebres Surakarta 57126, INDONESIA

E-mail: sukarmin67@staff.uns.ac.id

Abstract. This study aims to identify patterns relationships of student's creativity with its indicators in Learning Optical Instrument. The study was conducted at SMPN 2 Sawo. SMPN 1 Jetis, SMPIT Darut Taqwa, SMPN 1 Dander, Bojonegoro and SMPN 3 Plus Al-Fatima. Data analysis used descriptive analysis using the Confirmatory Factor Analysis. Creativity test instruments used have been tested parameters. Creativity indicators used are personal (self-confidence, perseverance), press (spirit, unyielding), process (preparation, incubation illumination, verification) and the product (knowledge, skills). Research Result shows that perseverance and incubation are the highest capabilities and verification capabilities of the lowest. All indicators on student creativity can still be improved. The relationship between creativity with the indicators grouped into a strong, moderate, weak and no relation. Indicators that have a strong relationship ($r \geq 0.50$), namely are personal (self-confidence, perseverance), process (illumination). Indicators that have a connection was ($0.3 \leq r \leq 0.49$) are press (spirit), process (verification). Indicators which have a very low correlation ($r \leq 0.1 \leq 0.29$) are press (unyielding), process (preparation), process (incubation), product (skills) as shown in Figure 1. Indicators that do not have a relationship between the creativity of the students with the indicator that is, product (knowledge).

1. Introduction

Thinking consists of critical thinking skills and creative thinking skills both are interrelated, and are part of the performance assessment. Performance Assessment puts students in real contexts so that learning is easier for an authentic destination. Performance behavior demonstrated work habits, industrious, caring for the timeliness and quality of the work, responsibility, able to work together in groups, task management, systematic, attention to safety, all components can be assessed using observational format. Performance assessment is used for testing the performance of students about critical and creative skills. Both of these skills necessary for the development of the student in the future, therefore, a second instrument need to be trained in the skills of teachers and prospective teachers.

Creativity is the result of interaction between the individual and his environment [1]. Someone affect and are affected by the environment in which it is located. Thus changes in the individual or the environment can support or hinder the creative efforts. The implication is that creative ability can be done through education. Creativity is the ability to create/produce something new [1]. Indicators of

[1] To whom any correspondence should be addressed.

creativity consists of creative personalities include confidence and perseverance, press or urge include giving spirit and unyielding, the creative process includes the preparation, incubation, luminescence and verification, creative products include knowledge and skills.

According to NACCCE (National Advisory Committee on Creative and Cultural Education on) Creativity is an imaginative activity that produces something new and valuable [2]. Creativity is a form of human self-actualization as the most essential, therein involves the ability of rational, emotional or feeling ability, special talents, the ability of imagination, intuition, and fantasy. Based on these descriptions, creativity means something that comes from human activity to create something new and valuable as a result of interaction with the environment that involves the ability of rational, emotional ability or feelings, special talents, the ability of imagination, intuition, and fantasy [3].

The aims of developing creativity: First, to create a child can manifest itself, self-realization is one of the basic human needs. Second, the ability to think creatively can see a wide range resolution of a problem, expressing thoughts that are different from others without being limited and able to generate many ideas. Third, creativity providing satisfaction to the individual, it is important to note because a person's level of achievement of satisfaction will influence the development of social emotions. Fourth, allow human creativity to improve the quality of life, new ideas as pieces of creative thinking is needed to deal with future challenges [1].

The function of the development of creativity in children, children have the opportunity to meet the needs of expression in its own way, creating something else and new. Activities that produce something this can foster an attitude to continue to busy themselves with creative activity will spur the development of cognitive or thinking skills. In mental health, the development of creativity has a therapeutic value for the expression activity of children can channel the feelings that can cause strain on him, among others, feeling better, disappointed, concerned, who may not be able to say. In the development of aesthetics, in addition to activities that are creative expression, children are also socialized and trained to appreciate the variety of beauty among others the beauty of nature, the painting dances, music.

Creativity is necessary to study the physics [4] [5]. Students will easily capture and apply physics concepts to problem-solving. Many studies suggest that the ability of creativity should be owned by every student and can be developed for each student [6] [7] [8]. To measure the ability of the creativity of students have developed an indicator of the ability of creativity [8]. The indicator has translated into an instrument of creative thinking ability tests [9] [10]. In this initial study to investigate how much the relationship with the students' creativity indicator? The benefits of research that, knowing the most powerful indicator of students' creativity.

Research conducted examines the optical instruments. Optical instrument is a tool that works based on the optical properties, such as reflection, refraction, diffraction, interference, and polarization. Optical instrument divided into two optical instrument natural and manmade optical equipment. Ie an optical instrument natural and artificial optical devices are tools of human vision to observe objects that can not be seen clearly by the eye (Kristanta, 2011). To see small objects and objects farther away, the eyes can no longer see clearly, as well as to the eye is defective or damaged. It is necessary for visual aids, such as glasses, magnifying glass, cameras, microscopes and binoculars, and a periscope. The eyes and the visual aids are called optical instruments.

To see small objects and objects farther away, the eyes can no longer see clearly, as well as to the eye is defective or damaged. It is necessary for visual aids, such as glasses, magnifying glass, cameras, microscopes and binoculars, and a periscope. The eyes and the visual aids are called optical instruments. A discussion of optical instrument, requires a lot of ability, among others, describe an optical instrument, describes the process of reflection on optical devices, so that the shadow can be seen and observed, mathematical ability to calculate the magnification and shadow distance to the lens, and therefore the creativity of students is indispensable.

Optical instruments that were examined, namely, the eye and its parts as well as the formation of shadows on the eyes. The camera, the camera has a working principle resembles the human eye. Magnifying glass or a magnifying glass, the magnifier in the form of a convex lens used to observe small objects to make it look bigger and clearer. Microscope, microscope is an optical instrument used for measuring microscopic objects (very small), such as bacteria and viruses. The simplest microscope consists of two positive lenses. Lens close to the object is called the objective lens, while being close to the eye are called the ocular lens. The focal length of the eyepiece is larger than the focal length of the objective lens. So that observation does not tire the eyes, then the eyepiece can be set so that the eye does not need. Binoculars, binoculars or a telescope is an optical instrument used to see objects very far to look more closely and clearly. There are two main types of binoculars, namely binoculars bias, composed of several lenses and binoculars reflection, composed of several mirrors and binoculars lensa.Pada bias, as used objective lenses refract light function. There are two kinds of binoculars bias, ie telescope and binoculars earth. Periscope, periscope is an optical tool that serves to observe objects over long distances or be in a certain angle. Simple shape, in the form of a tube equipped with a mirror or prism at the ends. This prism will reflect light that is parallel to her flat, then arranged so that an angle of 45 ° to the axis of the tube. Periscope used on tanks and submarines. The navigator on a submarine periscope utilize to observe the movements that occur at sea level. Getting a good shadow and clear, demanding high precision and accuracy, so that the creativity of its users is indispensable.

2. Research Methods

The study was conducted at SMPN 2 Sawo. SMPN 1 Jetis, SMPIT Darut Taqwa, SMPN 1 Dander, Bojonegoro and SMPN 3 Plus Al-Fatima. Data analysis is used descriptive analysis using the Confirmatory Factor Analysis [11] [12]. Creativity test instruments used have been tested parameters. Creativity indicators used are 4P: personal (self-confidence, perseverance), press (spirit, unyielding), process (preparation, incubation, illumination, verification) and the product (knowledge, skills). Measurement of student creativity is part of a performance assessment [13].

3. Results and Discussion

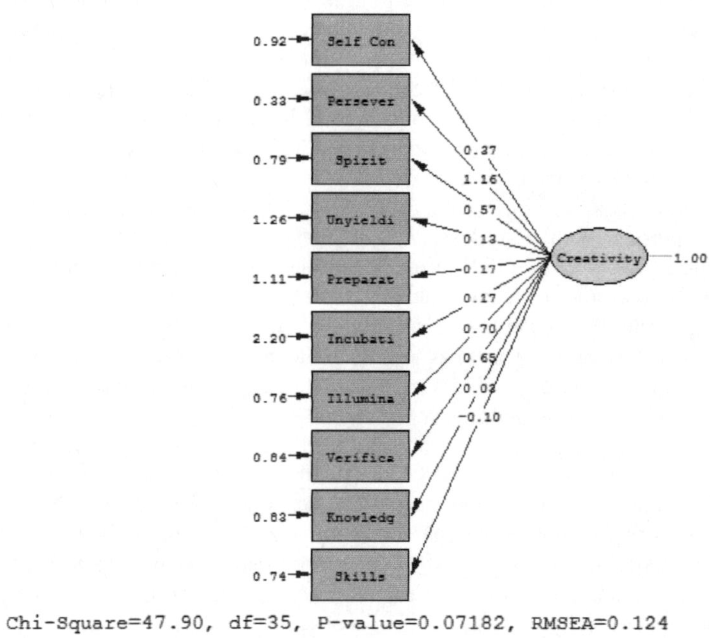

Figure 1. Relationship with Student Creativity indicator

Results in Figure 1 shows that perseverance and incubation are the highest capability and verification capabilities of the lowest. All indicators on student creativity can still be improved. The relationship between creativity with the indicators grouped into a strong, moderate, weak and no relation. Indicators that have a strong relationship ($r \geq 0.50$), personal (self-confidence, perseverance), process (illumination). Indicators that have a connection was ($0.3 \leq r \leq 0.49$), press (spirit), process (verification). Indicators which have a very low correlation ($r \leq 0.1 \leq 0.29$), press (unyielding), process (preparation), process (incubation), product (skills). Indicators that have no connection Indicators that have no connection between the creativity of the students with the indicator that is, product (knowledge) [12].

Components personal indicators, that creative is the "expression" of the uniqueness of the individual in interaction with the environment. Creative expression reflects the originality of the individual. Of unique personal expression, that is expected to create new ideas and innovative products. Therefore, educators should respect personal uniqueness and the talents of their students (do not expect everyone to do or produce the same or have the same interests). Teachers should help students find talent and treasure. Drivers Indicators (press), the creative talents of students will be realized if there are encouragement and support of the environment, or if there is a push in itself (internal motivation) to produce something. Creative talents can thrive in a supportive environment, but can also be inhibited in environments that do not support. In the family, at school, in the work environment, and in society should be recognition and support of the attitudes and behavior of creative individuals or groups of individuals[14].

Of the 10 indicators kretivitas students are the most powerful indicator in the accounts for the ability of students to learn physics krativitas is perseverance. Perseverance in learning physics has an important role in improving students' creativity seen from the correlation value of 1.16. Perseverance is a component of personality. In measuring components used krativitas namely diligence questionnaire completed student. Some of the questions given to students are: I am taking seriously the matter physics complete a task; when doing physics experiments, I always try though chances of success are small. Other indicators that have a strong contribution in shaping the creativity of students for student learning is incubation. Incubation allows students to think in advance in an attempt to resolve the problem. Questions concerning incubation indicators are: In the face of a problem in learning physics, I think to look for solutions; I was discussing with a friend first before doing physics experiments; I pay attention to the stages / steps work first before doing physics experiments.

Process (Process), to develop creative, children need to be given the opportunity to be busy themselves actively. Educators can stimulate involve themselves in creative activities, which helped arrange the necessary facilities and infrastructure. In this case what is important is to give freedom to the child to express himself actively with the requirements do not harm others or the environment. The process implicated themselves creatively without the need to constantly or too fast demanding creative product generates meaningful. It will come by itself in a climate of support, accept, and appreciate. It should also be remembered that the school curriculum is too crowded resulted in no opportunities for creative activities. This type of work is monotonous, does not support the students to express themselves creatively. Products (Product), a condition that allows someone to create a product creative Meaningful personal conditions and environmental conditions. The extent to which both push ("press") person to involve themselves in the process (activity, activity) creative. With its talent and creative personal traits, and with the encouragement (internal or external) to be implicated themselves creatively, then the significant creative products will arise. Educators should respect children's creativity products and communicate it to others [15]. For example by showing or display of children's work. It will arouse the interest of children to be creative.

4. Conclusions and Recommendation

Creativity in the dimension of the person is an attempt to define creativity that focuses on an individual or person of individuals who may be called creative, creativity in the dimensions of process is the creativity that focuses on the thought process that led to unique ideas or creative, creativity in the dimensions of the press is the creativity that emphasizes the press or boost factors, both internal drive yourself a desire and passion to create or implicated themselves creatively, as well as external encouragement of social and psychological environment. Regarding the "press" of the environment, there is an environment that appreciates the imagination and fantasy, and emphasizing creativity and innovation. creativity in product dimension is an attempt of creativity that focuses on the product or what is produced by individuals either something new/original or an elaboration / merging innovative creativity that focuses on creative product emphasis on originality. The 4P indicators developed into 10 indicators of creativity that self-confidence, perseverance, spirit, unyielding, preparation, incubation, illumination, verification, knowledge and skills.

References

[1] Munandar U 2012 *Pengembangan kreativitas anak berbakat* (Jakarta: Rineka Cipta)
[2] 1999 NACCCE report 20-22
[3] Semiawan R C 2009 *Penerapan Pembelajaran Pada Anak* (Jakarta : PT. Indeks)
[4] Arends R 2009 *Learning to Teach* (New York: Mc Graw Hill)
[5] Longair M 2004 *Theoretical Concepts in Physics* (Cambridge: The United Kingdom at the University Press)
[6] Maria K 2004 Developing creativity of students in a computer-assisted learning process *Eur. J. Phys.* 25 279
[7] K R Podolak and M J Walters 2016 *Phys. Educ.* 51 025012
[8] Valentin Voroshilov 2015 *Phys. Educ.* 50 694
[9] Djemari M 2004 *Penyusunan tes hasil belajar* (Yogyakarta: Postgraduate Program of Yogyakarta State University)
[10] Djemari M 2008 *Teknik penyusunan instrumen tes dan non tes* (Yogyakarta: Mitra Cendikia Press)
[11] Embretson and Reise 2000 *Item response theory for psychologists* (London: Lawrence Erlbaum associates publishers)
[12] Hambleton R K, Swaminathan H, and Rogers H J 1991 *Fundamental of item response theory* (Newbury Park, CA: Sage Publication. New Inc.)
[13] M Hibbard, r K 1986 *Performance Assessment in The Science Classroom* (New York: McGraw-Hill)
[14] Kolen M J 2004 *Applied Psychological Measurement* 28 219 – 226.
[15] Master N and John P 1999 *Advances in measurement in educational research and assessment.* (Tokyo: Pergamon)

Students' conceptual understanding consistency of heat and temperature

Indah Slamet Budiarti[1], Suparmi[2], Sarwanto[3], Harjana[4]
Graduate Student's Sebelas Maret University[1],

Lecturers of Sebelas Maret University[2,3,4]
Jl. Ir. SUtami 36A Kentingan Jebres Surakarta 57126, INDONESIA

E-mail: indah_budiarti@student.uns.ac.id, indah_budiarti@yahoo.com

Abstract. The aims of the research were to explore and to describe the consistency of students' understanding of heat and temperature concept. The sample that was taken using purposive random sampling technique consisted of 99 high school students from 3 senior high schools in Jayapura city. The descriptive qualitative method was employed in this study. The data were collected using tests and interviews regarding the subject matters of Heat and Temperature. Based on the results of data analysis, it was concluded that 3.03% of the students was the consistency of right answer, 79.80% of the students was consistency but wrong answer and 17.17% of the students was inconsistency.

1. Introduction

The consistency of students' response in comprehending concepts in Physics demands deeper understanding on some of students to see the equality of physics problems presented in various ways [1]. A deeper understanding of physics concept makes a student consistent about what he or she perceives to be true. An indirect effect of such a consistency is examined using representative consistency in that what he or she perceives to be true or may be contrary to science. However, this is not the case if equivalence is considered from different representation perspectives since such an equivalence requires considering various representations of Physics concepts [2]. Students' consistency will reflect that there are a higher level of understanding in perceiving Physics concepts represented in various problems.

Heat and temperature concept is prevalent throughout the science curriculum at all levels of elementary, secondary, and graduate education [3]. Students have difficulty in the concept of heat transfer using conduction, in a seat belt is made of metal and non-metal [4]. Students have a misconception of heat and temperature. The temperature of an object related to its size [5]. The concept developed by students comes from the interpretation of ideas raised from everyday experience [6]. To understand the concepts of physics, students must be skilled in expressing these concepts in various ways or forms (multiple representations). Representation skills are a must-have capability to interpret and apply the various concepts in solving problems appropriately [7]. Multiple representations are representations of concepts in various way such as verbal, graph, diagram, etc.

A student's ability to succeed at learning is determined, among others, by his or her thinking skills. This is the key point in solving problems during the learning process. By applying thinking skills students can train and develop their cognitive intelligence as well as relate various facts or information to previously acquired knowledge to make a prediction on result or outcome which is then formulated. Teaching-learning activities should involve explicit thinking ability, thereby facilitating categorizing thinking ability based on the available framework [8].

Based on the theoretical and case studies the researchers have conducted studies to investigate the students' conceptual understanding consistency of Heat and Temperature. The research question proposed in this study was How Consistent is the Students' conceptual understanding consistency of Heat and Temperature?

2. Research Method

The research methodology employed in this study was descriptive in nature to identify students' understanding consistency of Heat and Temperature. The samples were taken using purposive random sampling consisting of students of three Senior High Schools in Jayapura City. The testing instruments were adapted from HTCE (Heat and Temperature Concept Evaluation) [9].The data were collected using multiple-choice test items with reasoning as well as interviews. Twenty multiple-choice test items consisting of two items regarding temperature, eight items regarding heat and the other ten regarding heat transfer were administered. The interviews were conducted to collect data on the rationale provided by the students.

3. Result and Discussion

The percentage of the students' understanding consistency based on their answers can be seen in Table 1 below.

Table 1. Percentage of the students' conceptual understanding consistency

Consistency Level	Percentage (%)			
	SMAN 1 Jayapura	SMA Katada	SMA Pembangunan V Yappis	Average
Correctly consistent	4.08	3.33	0	3.03
Consistently with the wrong answer	75.51	80	90	79.80
Inconsistent	20.41	16.67	10	17.17

Based on Table 1 above, it can be seen that the percentage of SMA Pembangunan V Yapis students who consistently gave right answers was 0 % . This suggested that students do not understand the concept of Heat and Temperature. In other words, the students had difficulties understanding equivalent problems presented in various ways of representation. A deeper understanding is achieved when a student is consistent about what he or she perceived to be true. A representation of a problem out of the various forms of representation demands a scientific perception on the part of the students. However, inconsistency in scientific perception resulting from a taken perspective based merely on a single representation rendered the students having a tendency toward consistently giving wrong answers. Based on consistency theory, a learner is brought to a higher level of understanding when exposed to various Physics concepts presented in different problems.

About the data on the percentage of Consistency Level obtained from the students of the three senior high schools taken as samples in this study, the students' conceptual understanding consistency of Heat and Temperature is elaborated as follows:

Temperature concept, featured in the test items could have been answered correctly by the students. "The questions asked the students in the items number 5 and 6 refer to two cups, A and B, of the same material, each containing 150 grams of water. The glasses are in the room temperature of 25 °C. In the beginning, the temperature of the water in glass A was 60 °C and in the cup, B was 75 °C." Which glass reaches the final temperature earlier? Cup A or cup B? The answer yielded in the written test was confirmed by the interview result with a student as follows: "The student states that the temperature reached by an object depends on its size." Such a concept in the student's mind might stem from the student's false understanding of "the heat of an object positively correlates with its mass". It, however, should have been "the heat released from an object positively correlates with its mass." Such students might not have discussed the matter deeply, resulting in the false understanding being internalized. Thus the expression of "the heat of an object positively correlates with its mass is consequently wrongly perceived as "any big object always has a higher temperature than smaller ones. When faced with the problem of multivariable, people generally reduces complexity by either ignoring some variable or variables merge into a relationship of one variable [10]. If time is taken into account, the student tended to think that the temperature would remain the same for a long period, which means that the temperature remains the same at any given time. This can be seen in the student's choice on the graph showing the relationship between temperature and time (Figure 1) below.

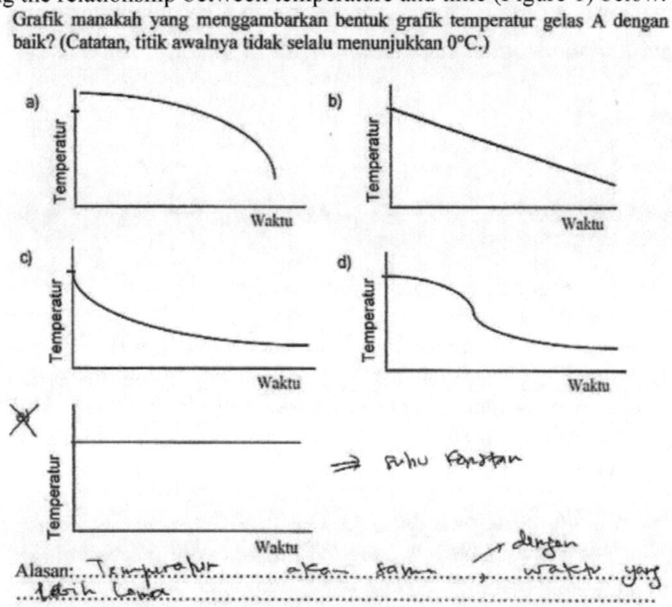

Figure 1. A student's choice on the relationship between Temperature and Time

Heat Concept featured in the items numbers 3 and 4 in the conceptual understanding consistency of Heat revealed that once concept regarding Heat is correctly perceived. Heat is energy released from an object with a higher temperature to another with a lower one. The one releasing heat will have a lower temperature whereas the one receiving it will end up with a higher one. In this case, learners showed a good understanding. A caveat should be noted here that this is true as long as phase transition does not take place. The data showed how Black's Principle is at work as measured by a calorimeter. This particular learning activity taught a valuable lesson to most of the students that they internalized a correct perception. However, about a phase transition, the students were distracted by the problems presented, as evident in a student's choice in Figure2.

Berdasarkan hasil pengukuran temperatur dan kalor yang ditambahkan dalam air, diperoleh grafik sebagai berikut.

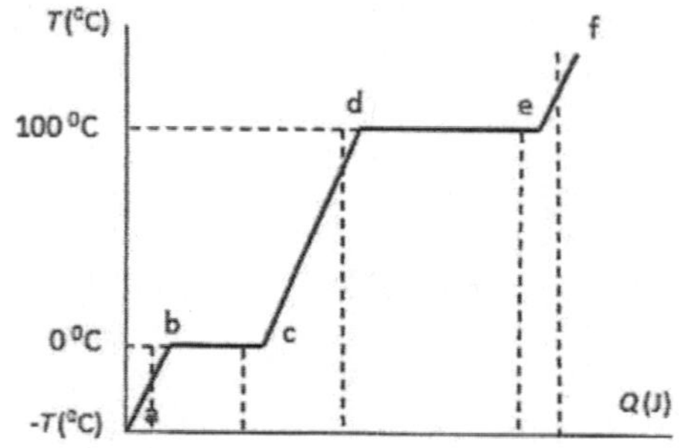

Pada grafik manakah yang menunjukkan semua air menjadi uap.

a. Grafik a-b
b. Grafik c-d
c. Grafik d-e
X. Grafik e-f
e. Grafik c-d

Alasan : Saat D - E air baru proses mendidih, uap terjadi setelah air melewati titik F

Figure 2. A student's answer regarding Heat

Heat Transfer Concept. Based on the data, the students in Level 1 response quality group (Consistently right answer) did give a concrete explanation on the microscopic mechanism about conduction process taking place in metals. They did respond by stating that conduction mainly takes place in solid objects. However, the majority of the students could not yield a satisfying reason why that is the case. This can be seen in Figure 3 below.

Wellem mengambil dua penggaris, berbahan logam dan kayu yang berada di tempat yang sama. Dia menyatakan bahwa penggaris logam terasa lebih dingin daripada penggaris kayu. Apakah penjelasan yang tepat mengenai situasi tersebut?

a. Energi kalor merambat dari tangan ke logam lebih cepat daripada kayu
b. Kayu secara alami merupakan bahan yang lebih panas daripada logam
(c) Penggaris kayu mengandung lebih banyak kalor daripada penggaris logam
d. Logam merupakan radiator panas yang baik daripada kayu
e. Suhu logam lebih rendah dari pada suhu kayu

Alasan: Karena besi lebih dingin

Figure 3. A student's answer of Heat Transfer Concept

Based on the answers of the students, it appears that the students had not understood the problems it faces. At issue, did not mention the type of metal ruler. This proves that the students only knew that only iron is a metal. Supposedly students understand the concept of the rate of heat and energy so that students choose a selection that heat energy travels from hand to metal faster than wood.

4. Conclusion

Based on the result and discussion, it can be concluded that students' conceptual understanding consistency of Heat and Temperature was low. In addition to that, another factor at work was the tendency toward learning Physics by means of memorizing without fully comprehending the subject matter concepts. Furthermore, the day-to-day Physics learning activities in the country still offer students how to solve problems without a fully detailed understanding them. As such, the low mastery of Physics subject matters renders our students only partly understanding Physics concepts.

Acknowledgement
This research was supported by Mandatory grand UNS with contract number 632/UN27.21/LT/2016

References

[1] Ainsworth, S 2006 Learning and Instruction 16 183
[2] Nieminen P, Savinainen A, and Viiri J 2010 Phys. Rev. St Phys. Educ. Res. 6 020109
[3] Gonen S, and Kocakaya S 2010 Eurasian J. Phys. Chem. Educ. 2 1
[4] Mustafa B 2006 EJMSTE 2 2
[5] Mustafa S 2003 Bogazici University Journal of Education 20 1
[6] Luera, G. R., Otto, C. A. & Zitzewitz, P. W 2005 J. Phys. Tchr. Educ 2 43
[7] P. B. Kohl and N. D. Finkelstein. 2005 AIP Conf. Proc. 790 121
[8] Tee et.al. 2012 JRPPTTE 2 212
[9] Pornrat. Et.al. 2013 International Journal of Innovation in Science and Mathematics Education 2 29
[10] Rozier, S., & Viennot, L 1991. IJSE 13 159

Development of two tier test to assess conceptual understanding in heat and temperature

Winarti[1], Cari[2], Suparmi[3], Widha Sunarno[4], Edi Istiyono[5]

[1]Graduate Students, Sebelas Maret University

[2,3]Physics Graduate of Doctorate Program, Sebelas Maret University

[4]Doctorate Program on Science Education. Sebelas Maret University
Jl. Ir. Sutami 36A Kentingan Jebres Surakarta 57126, INDONESIA

[5]Physics Education Department, faculty of Mathematics and Natural Sciences
Yogyakarta State University

E-mail[1]:winarti@student.uns.ac.id, [2]cari@staff.uns.ac.id, [4]edi_istiyono@uny.ac.id.

Abstract. Heat and temperature is a concept that has been learnt from primary school to undergraduate levels. One problem about heat and temperature is that they are presented abstractly, theoretical concept. A student conceptual frameworks develop from their daily experiences. The purpose of this research was to develop a two-tier test of heat and temperature concept and measure conceptual understanding of heat and temperature of the student. This study consist of two method is qualitative and quantitative method. The two-tier test was developed using procedures defined by Borg and Gall. The two-tier test consisted of 20 question and was tested for 137 students for collecting data. The result of the study showed that the two-tier test was effective in determining the students' conceptual understanding and also it might be used as an alternative for assessment and evaluation of students' achievement

1. Introduction

Heat and temperature concepts are found throughout science curricula in Indonesian, at both the pre-college (elementary school and high school) and college levels. A concept related to head and temperature is directly related to physical environment of living organism. Hence heat and temperature are not directly observable quantities. Concepts developed by students originated from the interpretation of idea gained from everyday experiences [7].

Difficulty understanding of heat and temperature concepts have been investigating by many researchers educational. Many students have not understood even misconceptions about the of temperature and heat concept [10]. Students come to class with an understanding that is not empty, the students came up with a variety of knowledge from everyday life. According to [2], students' understanding of the concept of temperature and heat comes from the experience that they get from everyday life. The concept is related to the temperature and heat are directly related to the environment and daily life. Therefore, the temperature and heat can not be directly observed. The result by Gonen research is student conception and misconceptions were acquired and stored, they occurred without ostensible link between everyday life and school experiences. Gonen found that 41

% students had misconceptions on heat and temperature [4]. Based on student responses to all question Gonen found a general misunderstanding of ideas related to thermal equilibrium, heat transfer and temperature concept. Gonen study finding indicated that even though some student has an accurate understanding of Physical processes, their knowledge of heat and temperature concept should be greater, As a matter of fact, the present study reveals those students misconceptions on related mass and heat, and distinction between temperature and heat and those differences in temperature can lead to transfer of heat energy [4]. In another study on the concept of the Black principle, the students have difficulties in distinguishing between $Q_{deliver}$ and $Q_{receive}$ [11]. Many students confused about the concepts of heat and temperature are the same thing, perception of temperature is only about hot and cold, temperature can be transferred. The student memorized this concept and was not able to make a connection between their knowledge and physics phenomena in everyday life [12].

Every day, children are exposed to colloquial term “heat” as a noun, verb, adverb and adjective and these multiple uses may lead to confusion [3]. A student in high school has also great difficulty with energy concept, and the distinction between heat and temperature [5].

According to Meltzer, physics can also be translated into four representations as follows: verbal representation, diagram or images representation, mathematical representation or mathematical symbols, and graphic representation. Thus, solving the physics problems can be conducted by representing the questions of physics in various forms: verbal, graphic, images or pictorial, and math in the form of formulas [8].

This study has two objectives, Firstly, we developed a two-tier test heat and temperature based higher order thinking skill. Secondly, we used the two-tier test instrument to diagnose conceptual understanding heat and temperature student

2. Method

The method of research is the qualitative and quantitative methods. This research was conducted with 2 phase is development and measurements phases. The first phase of this research is to develop open-ended multiple-choice with reasoning instrument test that will be used to measure students' understanding of the concept. According to Borg & Gall, research and development is a process used to develop or validate the products that are used in education and learning. A step-by-step method using the Research and Development (R & D) by Borg & Gall.

Finally, the instrument two-tier test consisted of 20 items. This instrument is a matter of the type open-ended question or multiple-choice with the reasons. In this instrument, the first tier of each item consist of a content question of four choices, the second part is a reasoning for justification for the answer to the first part. To determine the feasibility of this instrument to be used in measuring the students' understanding of the concept of validity and reliability analysis. Logical validity is done by experts determine with V aiken and to determine the validity of empirically used Quest 1 PL. As for the analysis of reliability of the instruments used Cronbach alpha equation

The Second phase is a phase of measurement. This stage involves 137 students of Madrasah Aliyah. In the analysis at this phase is multiple choice answers and reasons answers written by the students. Description of student understanding of concepts derived from the analysis of the answers, reasons, and explanations are given learners. The students' answers are then analyzed based on the classification of understanding of the concept

3. Result and Discussion

In the first step, we develope two-tier test. In the second step, the effects of the two-tier test were discussed. The result of this study showed that the two-tier test was efficient to determine conceptual understanding heat ant temperature of the student.

Results of the analysis showed the instrument developed is effective to use. The Cronbach alpha reliability coefficient was found 0.73 for two-tier test. The instrument consists of 20 two-tier question with multiple choice and reasoning. Some general findings of heat and temperature understanding learned from research are:

Table 1. Description of student conception of heat and temperature

•	**Student Conception**
•	**Student Conception of temperature**
1.	Temperature obtained by measure
2.	Touch and skin can measured temperature
3.	Perception of temperature is only about hot and cold
4.	Temperature can be transferred
•	**Student Conception of Heat**
1.	Substance with big masses and big volume is more heat
2.	Which substance is hotter have more heat
3.	Heat and temperature its same
4.	Heat is not energy
5.	Cold is not heate (ex; ice)

Table 1 shows some of the students' perception of differences of temperature and heat concept. This study aims to identify students' understanding of heat and temperature concepts, particularly for competence on the difference between heat and temperature. In the early part of this chapter of heat and temperature that is how students perceive their temperature. Based students' answers and interviews found that students consider temperatures obtained through measurement. Touch and skin can measure their temperature.

Perceptions of students so that later dug up again farther it until the present study found some misconception. The concept of temperature is not only expressed in the word "cold" or "hot", but rather through the numbers not rely on taste or touch anyone. The more heat, the heat level is something greater and it can be said the higher temperature. Likewise, the cold something then the temperature will be lower. The human sense of touch can only feel hot and cold but can't determine the degree of hotness or coldness of the object. In this concept, the students did not experience an error. Just need a more in-depth explanation so that students can understand the meaning of hot and cold in the concept that not only acquired the sense of taste alone. Such perceptions will bring improper understanding in students. In accordance with the opinion of Ericson (2000) that the hot word is often used in a variety of functions that affect students' concept and understanding.

The next finding is Heat and temperature is same. According to the student who is the heat flow or temperature instead of energy. This is contrary to the concept of heat. Heat is actually the energy transferred due to temperature differences in the two objects. Based on interviews with students stating that the heat can not be directly measured, but if we mix water from two glasses with different temperature, the water temperature will be warm and they conclusion the transfer is the temperature. Specific case for answer the student like a figure 1.

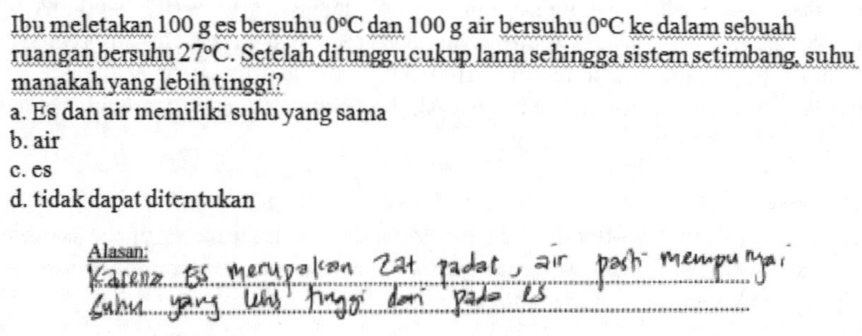
Ibu meletakan 100 g es bersuhu 0°C dan 100 g air bersuhu 0°C ke dalam sebuah ruangan bersuhu 27°C. Setelah ditunggu cukup lama sehingga sistem setimbang, suhu manakah yang lebih tinggi?

a. Es dan air memiliki suhu yang sama

b. air

c. es

d. tidak dapat ditentukan

Alasan:
Karena Es merupakan zat padat, air pasti mempunyai suhu yang lebih tinggi dari pada es

Figure 1. Error representation of temperature concept

The question number 1 its measure students' understanding of the concept by giving an example of a phenomenon when two substances included in a system at equilibrium state. These items require students to classify the different concepts of temperature and heat to restate their respective concepts. Results of the analysis showed some students can work on the problems with both (4% Understanding concept and 19% Partial Understanding, 70% of Specific Misconception. On average students answered that the water temperature will be higher than the temperature of the ice as the water directly increase the temperature while ice must undergo a process of melt prior to temperature increases. Most of the students did not review that the final condition of water, ice, and the rooms are in balance so that all students are the process of temperature rise. In the process, it is water temperature increases more rapidly than ice, who must undergo smelting process first. But once the room reached thermal equilibrium with the water and ice will have the same temperature to room temperature.

The findings of this study are most of the students have alternative conceptions about heat and temperature. Many students were confused about the concept of heat and temperature and could not explain the difference between heat and temperature. Some student still regards that the words heat and temperature are the same. The finding was likely similar to the work by Alwan (2011) and Kesidou and Duit (1993). The response suggests that the student does not always distinguish between heat and temperature. They say heat and temperature as a synonymous word that a single concept. We call this concept "hotness". It has some of the qualities that physicists associated with both heat and temperature. In the reason student, many students have opinion that the hotter object must contain heat

The second research question is if we have two substance, if their masses is different which initial temperature are equally put in the same room that has higher temperature and waiting enough time, which substance heat more?

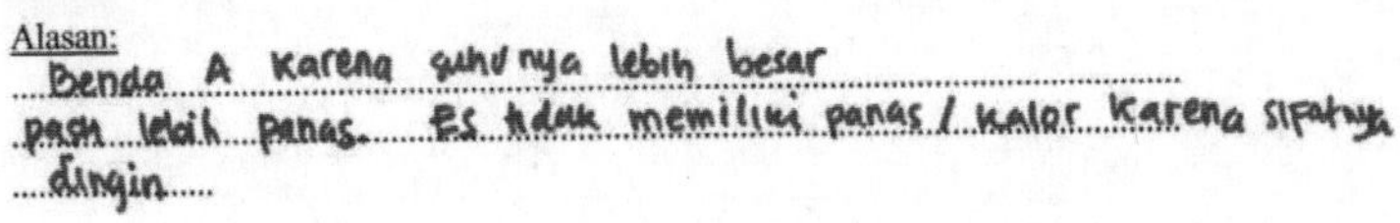

Figure 2. Error representation of heat concept

Based (fig 2) it can be concluded that 54,7 % student is not understanding and 2.98% of the students answered correctly. The initial student assumption that heat of a matter with small mass should be higher, because the volume is small. Some student gives reasons that the heat of substance with small mass should be higher because the volume is small and substance with big mass or big volume takes more heat and substance is hotter its heat is more. Ericson (2000), Gonen (2010) and Alwan (2011) found many student misconceptions about that and the temperature of the body was thought to be related to its size. This study's findings are consistent with Ericson's; Gonen and Alwan.

4. Conclusion

From the analysis and discussion, it can be concluded that the two-tier test can use to determine a conceptual understanding heat and temperature. Two-tier test in this study can help the teacher teach and student learn better because not only measure mathematic skill but its can measure fundamental and the basic concept of heat and temperature, so this instrument can promoting higher order thinking skills.

Many students confused about the concepts of heat and temperature are the same thing. The student memorized this concept and was not able to make a connection between their knowledge and physics phenomena in everyday life.

Acknowledgement

We thank to Sebelas Maret University (UNS). This research was supported by Mandatory Grand UNS with contract number 632/UN27.21/LT/2016.

Reference

[1] Alwan Almahdi Ali. 2011.International Conference on Education and Educational psychology. 12600-614
[2] Arnold, M. And R. Millar.1994. Inter J. of Scie. Edu.. 16 131144.
[3] Erickson, G.L. 1979. Journal of Science Education. 6 322221230
[4] Gonen. Selahatin.,Kocakaya. Serhat. 2010. Eurasian Journal Phys Chem. Education. 21 115.
[5] Kesidou,S,. & Duit, R. 1993.Journal of Research in Science Teaching. 30 1 85106
[6] Lee, O. 2007. Journal Science Teacher Education , 15 733756.
[7] Luera, G. R., Otto, C. A. & Zitzewitz, P. W. 2005. Journal Physics Teacher Education 2438.
[8] Meltzer, D.E. 2004.American Journal of Physics, 72 111432-1446.
[9] Prince. J., Vigeant, Katharyn. 2011. Journal Energy AmericanSociety for Engineering Education. 17 27882
[10] Wiser, M, &Carey, S. 1988. 24 117
[11] Winarti, Cari, Widha S, Edi I,. 2015. Prosiding Seminar Nasional Pendidikan Sains UNS. 19 November 2015. ISSN: 24074659
[12] Winarti, Cari, Suparmi,Widha S, Edi I,. 2016. Procceding Asean Education Symposium. UPI

Uniform circular motion concept attainment through circle share learning model using real media

Ponimin [1], Suparmi [2], Sarwanto [2] and W Sunarno [2]

[1]Science Education Graduate Program, Sebelas Maret University, Surakarta, Indonesia

[2]Lecturer Science Education Graduatel Program, Sebelas Maret University, Surakarata, Indonesia

ponimin@student.uns.ac.id

Abstract. Uniform circular motion is an important concept and has many applications in life. Student's concept understanding of uniform circular motion is not optimal because the teaching learning is not carried out properly in accordance with the characteristics of the concept. To improve student learning outcomes required better teaching learning which is match with the characteristics of uniform circular motion. The purpose of the study is to determine the effect of real media and circle share model to the understanding of the uniform circular motion concept. The real media was used to visualize of uniform circular motion concept. The real media consists of toy car, round table and spring balance. Circle share model is a learning model through discussion sequentially and programmed. Each group must evaluate the worksheets of another group in a circular position. The first group evaluates worksheets the second group, the second group evaluates worksheets third group, and the end group evaluates the worksheets of the first group. Assessment of learning outcomes includes experiment worksheets and post-test of students. Based on data analysis we obtained some findings. First, students can explain the understanding of uniform circular motion whose angular velocity and speed is constant correctly. Second, students can distinguish the angular velocity and linear velocity correctly. Third, students can explain the direction of the linear velocity vector and the direction of the centripetal force vector. Fourth, the student can explain the influence of the mass, radius, and velocity toward the centripetal force. Fifth, students can explain the principle of combined of wheels. Sixth, teaching learning used circle share, can increase student activity, experimental results and efficiency of discussion time.

1. Introduction

Circular motion is an important concept and has many applications in life. Application of uniform circular motion in daily life such as the motion of vehicle wheels, the motion of the planets, the atomic theory, and satellite motion. Understanding of students on uniform circular motion concept is not optimal due to several problems. Students have difficulty on understanding of the velocity direction [1], the direction of the centripetal force [2, 3, 4, 5], acceleration [6, 7, 8, 9] and wheel combination. The difficulties do to the use the lecture method, mathematical approach, limited of media [10], the concept is abstract, and social system of student was not supporting to collaboration. To overcome the problem the real media was developed and used. It was used to visualize of uniform circular motion process. The real media consists of toy car, string, stopwatch, spring balance round table and wheel combination. The real media used to obtain uniform circular motion concept. First, students can explain the understanding of uniform circular motion whose angular velocity and linier velocity is constant. Second, students can distinguish the angular velocity and linear velocity. Third, students can explain the direction of the linear velocity vector and the direction of the centripetal force vector. Fourth, the student can explain the influence of the mass, radius, and velocity toward the centripetal force. Fifth, students can explain the principle of combined wheels. The use of media can increase the student motivation can build the procedural and conceptual understanding, collaborative [11], increase high-level thinking skills [12].

Except the using of media, it is also used circle share model. Implementation of the circle share requires several stages. At first, the students discuss in their respective groups, after that they discuss between groups with circle-share models. The worksheet of each group evaluated by other groups. The first group evaluates the second group, the second group evaluates the third group, n group evaluate the first group (Figure 1).

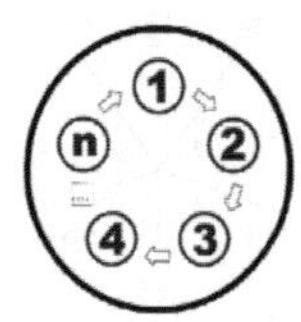

Figure 1. Circle Share Scheme

Circle share model aims to enable students to interact more programmatically and social system that occurs more widely. Interactions between groups can add information, stimulate new ideas and improve intellectual students [13].

2. Methods

This research is a qualitative descriptive study, using a sample of 40 students. The research was conducted through several steps. First, the students determine the working groups. Second, students conduct experiments in real media to obtain uniform circular motion concept, each group evaluated other group through circle-share models. The third, after a circle-share one of the group presents to get a conclusion the concept of uniform circular motion. Fourth, the assessment is to assess students' worksheets and post-test.

3. Result and Discuss

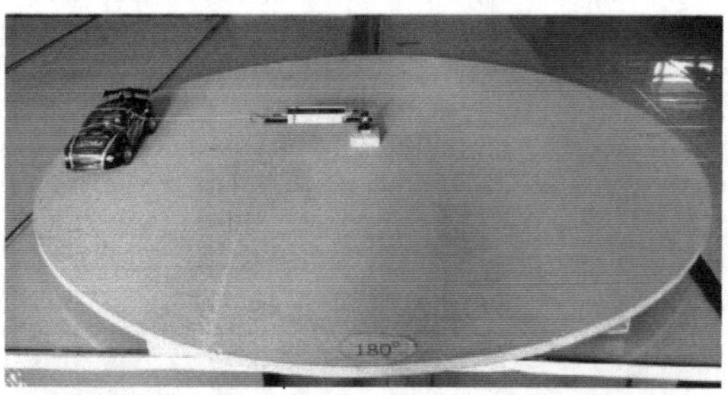

Figure 2. Toy car moving in the round table

The first experiment uses toy car move in a circle to get the concept of circular motion on the path and period. The second experiment, with the same tools, used to demonstrate that the circular motion angular velocity and speed is constant (Figure 2). It takes at an angle of 90°, 180°, 270° and 360° recorded. Next, calculate the angular velocity and linear velocity. The first and second experiment students get the concept that uniform circular motion has circular path, angular velocity and the linear velocity is constant (Table 1).

Table 1. Relationships of time, angular velocity, and linier

Point	Time (s)	Angular Velocity (rad.s^{-1})	Linear velocity (m.s^{-1})
A s/d B	4	0,5 π	0,2π
B s/d C	4	0,5 π	0,2π
C s/d D	4	0,5 π	0,2π
D s/d A	4	0,5 π	0,2π

The third experiment uses string ties toy car given point A, B and C. Then record the time and calculate the angular velocity and linear speed of points A, B, and C. From experiments the students get the concept that the circular motion angular velocity is constant and the linear velocity is influenced by the radius (Table 2).

Table 2. The relationship between the radius, angular velocity, and linear velocity

Point	Radius (m)	Time (s)	Angular Velocity (rad.s^{-1})	Linear velocity (m.s^{-1})
1. A	0, 80	18 s	0,11 π	0.089π
2. B	1,20	18 s	0,11 π	0.130π
3. C	1,60	18 s	0,11 π	0.178π

Figure 3. Out of water hose according to tangent

Figure 4. Direction of v vector and F vector

The fourth experiment uses toy cars equipped with two arrows. The first arrow describes the linear velocity vector. It's the direction of car motion. Second arrow is toward to the center circle to describe the centripetal force. To determine the direction of the linear velocity can be implemented in other ways. First, cut the rope when the toy car is moving and the students observe the direction of motion of the car as the linear velocity direction. Second, use circular flowing water in hose, the water that comes out of the hose is not circular but straight. From the experiments, the students get the concept that the direction of the linear velocity of the tangent in the circle. The direction of the centripetal force is towards the center of the circle (Figure 3 and Figure 4).

The fifth experiment uses tool that consists of three wheels (Figure 3), R_1 concentric with R_2, while R_3 which is connected by rope. Wheel 1 is rotated once with a time of 4 seconds, then the wheels 1 and 2 wheel rotates once and the third wheel rotates twice, and then calculated the angular velocity and linear velocity of each wheel. From the experimental results, the students get the concept that the concentric wheel is same angular velocity but the linear velocity is different (Table 3). The wheels connected by a rope have different angular velocity but they have same linear velocity (Table 4).

Table 3. Relationship angular velocity and linear velocity on concentric wheels

Wheel	Angular Velocity (rad.s^{-1})	Linear velocity (m.s^{-1})
R1	0,25π	0,375
R2	0,25π	0,050

Table 4. Relationship angular velocity and linear speed on wheels connected ropes

Wheel	Angular Velocity (rad.s^{-1})	Linear velocity (m.s^{-1})
R 2	0, 25 π	0,05π
R 3	0, 50 π	0,05π

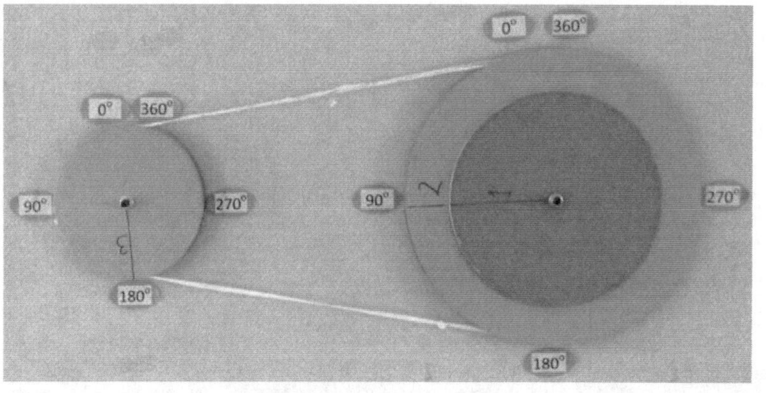

Figure 5. Wheel combination

Table 5. Radius relationship with centripetal force

No	Radius (m)	Force (N)
1	0,80 m	1,5 N
2	1,20 m	1,1 N
3	1,60 m	0,7 N

Table 6. Massa relationship with centripetal force

No	Massa (kg)	Force (N)
1	0,700	0,7 N
2	0,900	1,0 N
3	1,100	1,3 N

Table 7. Velocity relationship with centripetal force

No	Velocity (m.s^{-1})	Force (N)
1	4,5	0,7
2	5,5	1,1

The sixth experiment uses spring balance between the car and the center. Spring balance is used to measure the centripetal force. This tool (Figure 4) is used to determine the influence of the radius, mass and velocity toward the centripetal force qualitatively. From these experiments the students get the concept; that the centripetal force is influenced by radius (Table 5), mass (Table 6) and velocity (Table7).

Figure 6. Linier velocity and centripetal force before circle share

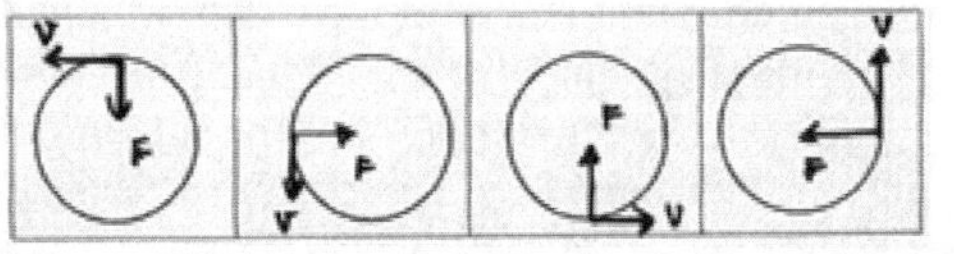

Figure 7. Linier velocity and centripetal force after circle share

Figure 6 and Figure 7 show worksheet of student. Figure 6 is the example of worksheet of leaners on drawing the direction of vector before circle share. The centripetal force vector drawing is correct, the direction toward to the center of the circle, but velocity vector drawing is wrong. Vector velocity straight and touch in the circle, but the vector drawing is curved. Figure 7 is the example result worksheet of leaners on drawing the vector direction after circle share. The drawing of centripetal force vector is correct, the direction toward the center of the circle. The drawing of linear velocity vector is correct too, it is straight and touches the circle. Another student worksheet can be found in Appendix 1 and Appendix 2.

Student achievement focus on the data writing, data analysis and conclusions score before and after circle share. Furthermore, before and after the circle share score compared.

Table 8. Score of experiment

No	activity	before circle share	after circle share	Increase	Percentage
1	data recorded	9,54	10,00	0,45	4,80 %
2	data analysis	7,10	9,55	2,45	34,51 %
3	conclusions	7,60	9,45	1,85	24,34 %

Table 9 is student score. The average score before circle of data record is 9.54 and the score after circle share is 10.00, an increase of score is 0.46 or 4.80%. The average score of data analysis before the circle share is 7.10 and after the circle share is 9.55, an increase 2.45 or 34.51%. The average score on conclude before circle share is 7.60, after circle share is 9.45, an increasing 1.85 or 24.34%. Thus the circle share can improve students learning score in the experimental results; recorded data, data analysis and conclusions.

The end outcome of students consists of experimental results and post-test. All of 40 students, maximum score is 93, an average of 82, a minimum of 64. The student's score of 35 students are more than 70 (87.5%). The score of five students are less than of 70 (12.5%), according to the records when activity turns these students are less active in conducting experiments and discussions so that the concept obtained is not optimal.

Using real media makes it is easier for students to accept the concept. Students receive information through the senses; visual, hear, and touch [14]. Students accepted concept then stored in long-term memory, so it is not easily lost and easy to be called back [15, 16]. Learning uniform circular motion using real media can overcome difficulty of concept understanding of uniform circular motion. This media for six experiments, five experiments worked well. One experiment is the centripetal force is not optimal due to the frictional forces against the boards' toy cars [17]. Circle-share model application can improve learning outcome. Circle-share is a model of discussion to collaborate and cooperative each student group. Collaborative and cooperative learning can improve social skills and academic achievement [12]. Circle share model is discussion forms the application of a two-steps approach can reduce the cognitive load of students [18]. Interaction between groups is a widespread form of collaboration that can stimulate new ideas and improve student's intellectual [13].

4. Conclusion

Learning of uniform circular motion using real media and circle share models obtain some of the findings. The first finding is the use of real media that can visualize process of the uniform circular motion. The results showed that media is acceptable to teaching learning uniform circular motion. Teaching learning is well done, the students are active and motivated, student learning outcomes are better. Students obtain concept of uniform circular motion correctly. First, students can explain the meaning of uniform circular motion that has circular path with a constant angular velocity and speed. Second, the students obtain the concept that the angular velocity is constant but linier velocity being influenced by radius of the circle. Third, the direction of the linear velocity is the direction of the tangent of the circle and the direction of the centripetal force is always towards the center of the circle. Fourth, the students obtain concept that the concentric wheel has same angular velocity but linear velocity is different. The wheels are connected by a rope has different angular velocity but the linear velocity is same. Fifth, that centripetal force is influenced by the radius, the larger radius causes smaller centripetal force; if the mass is greater so the centripetal force is greater too; If the speed is greater, so the centripetal force is greater too.

The second finding is that circle-share model can increase the experimental results of students. It can add new information from another groups, work in the group is not yet optimal. Circle-share model is a way to create more extensive social system than the social system within the group. Each group programmatically and systematically evaluates the worksheet of other groups. Evaluates to other groups can add the new information in each group. The response of groups to one another is different when it gets the results of his evaluation. The attitude of the group responding to the evaluation result is different; it makes different increasing one group with another group.

Acknowledgements
The author would like to thank Prof. Suparmi who helped write this article. Thanks also to Beta N P, Dewa, Indah S B, Siska, Winarti and Sumarsih who all help finishing this article.

Appendix 1. Student's worksheet before circle-share

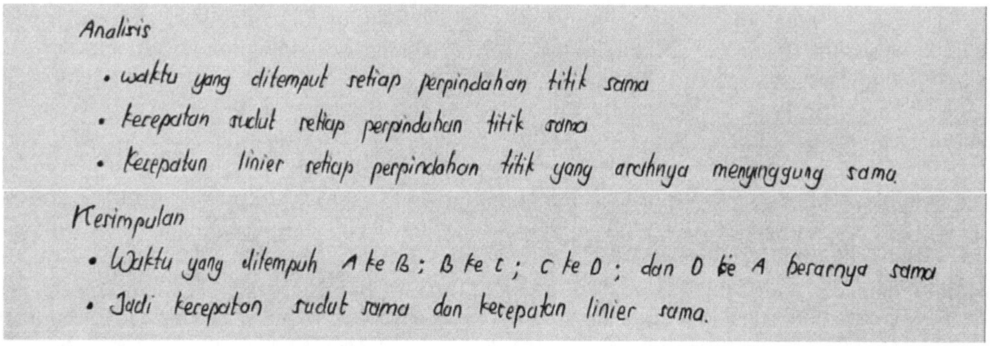
Analisis
• waktu yang ditempuh setiap perpindahan titik sama
• kecepatan sudut setiap perpindahan titik sama
• Kecepatan linier setiap perpindahan titik yang arahnya menyinggung sama
Kesimpulan
• Waktu yang ditempuh A ke B ; B ke C ; C ke D ; dan D ke A besarnya sama
• Jadi kecepatan sudut sama dan kecepatan linier sama.

Appendix 2. Student's worksheet after circle-share

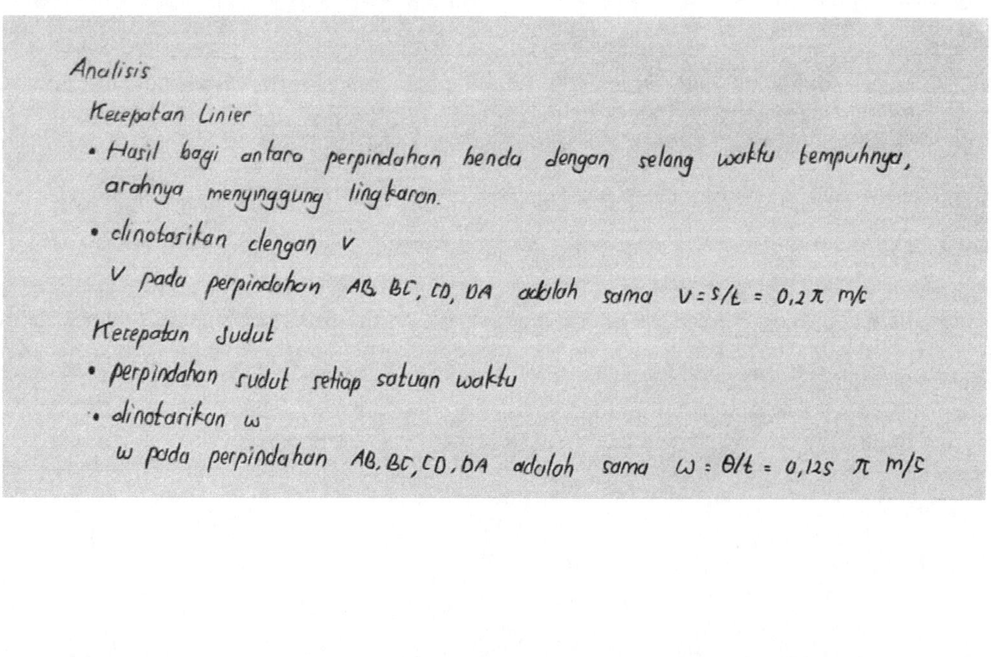
Analisis
Kecepatan Linier
• Hasil bagi antara perpindahan benda dengan selang waktu tempuhnya, arahnya menyinggung lingkaran.
• dinotasikan dengan v
v pada perpindahan AB, BC, CD, DA adalah sama v = S/t = 0,2 π m/s
Kecepatan Sudut
• perpindahan sudut setiap satuan waktu
• dinotasikan ω
ω pada perpindahan AB, BC, CD, DA adalah sama ω = θ/t = 0,125 π m/s

IOP Conf. Series: Journal of Physics: Conf. Series **795** (2017) 012053 doi:10.1088/1742-6596/795/1/012053

Kesimpulan

- dari percobaan tersebut disimpulkan bahwa kecepatan linier (v) dan kecepatan sudut (ω) adalah konstan.

References

[1] Ken z *Physics Teacher* 2006 p 207.
[2] Jonathan M 2005 *The Physics Teacher* **43** p 91.
[3] John LM 2000 *The Physics Teacher* **38** p 354.
[4] Koyuncu K, Kemal Y and Hakam 2013 *Physics Education* **48** (2).
[5] David C 2006 *Physics Teaher* **44** p 619.
[6] Isaac N and Richard C H 2000 *American Journal of Physics* **68** p 637.
[7] Harvey S L 2002 *American Journal of Physics* **70** p 490.
[8] John K 2012 *Physics Teacher* p 16.
[9] David W K 1995 *The Physics Teacher* **33** p 565.
[10] Shaona Z, Jing H, Nathaniel P, Xiaojun W, Liangyu P, Hua X, and Lei 2011 *European Journal of Physics* **32** 1597–1606.
[11] Kukliansky I and Eshach H 2014 *Journal of Science Education and Technology* **23** p 108-15.
[12] Barak M, Ben C D, Zoller U 2007 *Research Science Education* **37** p 353-9.
[13] Manabo S 2008 *E-Journal of All India Association for Educational Research* **20** (3 & 4).
[14] Sharon E S, Deborah L L and James D R 2011 *Instructional Technology and Media for Learning* Pearson Education.
[15] Schunk D H 2009 *Learning Theories an Educational Perspective* Person Education International.
[16] Smith E E and Kosslyn S M 2007 *Cognitive Psychology Main and Brain, 1st ed* Person Education.
[17] Molina-Bolivar and 2Cabrerizo-vilches 2014 *Physics Education* **49** (4) p 400-5.
[18] Franz-Josef S and Franz X B 2010 *International Journal of Science Education* 32 (6).

Multi-perspective views of students' difficulties with one-dimensional vector and two-dimensional vector

Ahmad Fauzi, Kunthi Ratna Kawuri, Retno Pratiwi

Physics Education Department of SebelasMaret University,
Jl. Ir. Sutami 36A Kentingan Jebres Surakarta 57126, INDONESIA

E-mail: fauziuns@gmail.com

Abstract.Researchers of students' conceptual change usually collects data from written tests and interviews. Moreover, reports of conceptual change often simply refer to changes in concepts, such as on a test, without any identification of the learning processes that have taken place. Research has shown that students have difficulties with vectors in university introductory physics courses and high school physics courses. In this study, we intended to explore students' understanding of one-dimensional and two-dimensional vector in multi perspective views. In this research, we explore students' understanding through test perspective and interviews perspective. Our research study adopted the mixed-methodology design. The participants of this research were sixty students of third semester of physics education department. The data of this research were collected by testand interviews. In this study, we divided the students' understanding of one-dimensional vector and two-dimensional vector in two categories, namely vector skills of the addition of one-dimensionaland two-dimensional vector and the relation between vector skills and conceptual understanding. From the investigation, only 44% of students provided correct answer for vector skills of the addition of one-dimensional and two-dimensional vector and only 27% students provided correct answer for the relation between vector skills and conceptual understanding.

1. Introduction

In order to improve physics education quality, teacher needs to understand what is happening in his classroom and what are students doing and learning when they learn. According to many research, students are engaged in learning or constructing the concepts of physics and changing their prior conceptions about how the world works. Much research has been done on how students change their ideas. It is suggested that learning physics involves some rearrangement/refinement of existing concepts or maybe the elimination of old concepts (prior conceptions) and the addition of new concepts in the mind of the student. Since 1980s, Physics education research has devoted a great deal of energy and attention to the question of how to help students learn/construct the concepts of physics faster/better and the difficulties that students have in changing their prior conceptions.

Vectors are component of mathematics which is an essential language for physics[1]. Usually, senior high school students learn vectors as the first topic when they start learning their first physics course. Even in some introductory university physics textbooks, vector concepts are set as the first topic. In Indonesia, senior high school students and university students learn vectors after they have studied measurement. In university levels, especially in physics education, students learn vectors twice. In the first semester they must study about vector in fundamental physics course and in third semester they must study vector in classical mechanics course. In classical mechanics students learn vector when they start learning their first classical mechanics course. Generally, the vector concepts consist of vector magnitude, addition, multiplication, subtraction, and component of vector. Aguirre argued that lacking of vectors understanding might cause serious difficulties when student's must study physics concepts that included vector concepts [3,7]. Garcia et al said that students' difficulty in learning physics is caused by a lack of skills and understanding about mathematics, especially vectors. Thus, understanding vector quantity is necessary in success of learning physics[2]. However, many physics education researchers found that students still hold misconception about vectors although they have studied it[4,6].

Shaffer and McDermott found that introductory students, pre-college teachers and physics graduate students have some serious conceptual and reasoning difficulties to treat velocity and acceleration as vectors in one and two dimensions [4]. Umporn and Narumon said that students still hold misunderstanding about vector addition after a traditional teaching. The students have some difficulties to add vectors graphically without any physical context. They also still confused about the "tip-to-tail" strategy add vectors graphically[7]. Barniol and Zavala said that Students who completed a calculus-based type course of mechanics have serious difficulties sketching the unit vector in the direction of a vector in the Cartesian coordinate plane[5]. Students also have difficulties with mathematical operations of vectors in which sometimes they forget the rules of the operations or they carry their difficulties with algebra into the vector operations. Barniol and Zavala also said that students face difficulties with the term of magnitude and direction or dot product with crossproduct consequently they difficult to interprete the operations [8]. Barniol and Zavalafound that, even after three introductory physics courses, students still have difficulties with vector components, in particular choosing answers with incorrect magnitudes. Some students think that the magnitude of a component is equal to the magnitude of the vector, and others know the "rule" that components are shorter than the vector, but have problems identifying the magnitude of the components graphically[9]. Barniol and Zavala found that no significant difference between the medians obtained students' understanding of vector concepts in problems with no physical context, and with three mechanics contexts: force, velocity, and work [12].

Some researchers have suggested steps to reduce misunderstanding about vector. Umporn and Narumon suggested to probe students' understanding of vector concepts even after the lesson. Active learning and teaching combined with careful instruction give much promise for helping students significantly improve their conceptual understanding [4,7]. Ostermeier et al argued that the constructivist conceptual change approaches have positive effects of students' interest and motivation as well as competencies in science and mathematics[10]. Duit and Treagust argued that conceptual change strategies may only be efficient if they are embedded in a conceptual change supporting learning environment that includes many additional features such as specially organized instruction based on models of teaching [11].

So far, the data of research about vector misunderstanding collected by written tests. However, the reports of conceptual change often simply refer to changes in concepts, such as on a test, without any identification of the learning processes that have taken place. In addition, students' interview is needed to judge conceptual change. The interviews are needed to interpret the data to avoid ambiguous decision. We argue the importance of research to explore students' understanding of one-dimensional vector and two-dimensional vector in multi perspective views in order to address to judge conceptual change with minimum ambiguous decision using test and interviews perspective.

2. Methodology

Our research study adopted the mixed-methodology design. The participants of this research were sixty students of third semester of physics education department. The data of this research were collected by test and interviews. We adopted the instrument test to asses students' understanding of one-dimensional and two-dimensional vector developed by McDermott and Shaffer (2005) and Umporn and Narumon (2011). In this study, we divided the students' understanding of one-dimensional and two-dimensional vector in two categories, namely vector skills of the addition one-

IOP Conf. Series: Journal of Physics: Conf. Series **795** (2017) 012054 doi:10.1088/1742-6596/795/1/012054

dimensional and two-dimensional vector and the relation between vector skills and conceptual understanding. The data of this research consist of two data namely essay test as quantitative data and interviews as qualitative data. Based on the incorrect answer, we choose ten students as the interviews participants.

3. Result and Discusion

3.1. Result

The essay test of vector addition one-dimensional and two-dimensional vector consist of five problems and the essay test of relation between vector skills and conceptual understanding consist of three problems.

3.1.1. Vector Addition

The problems of vector addition one-dimensional and two-dimensional presented below.
Problems 1, 2 and 3. Draw the resultant these vectors!

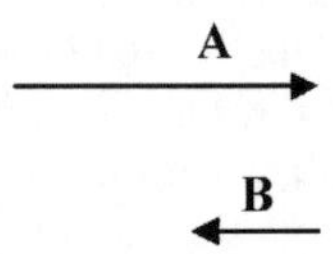

1)

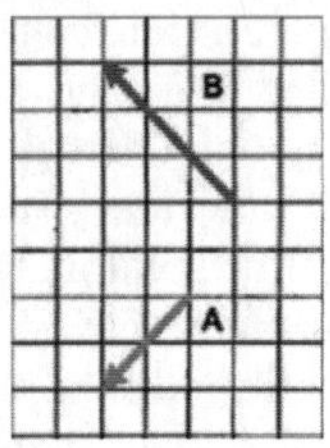

2)

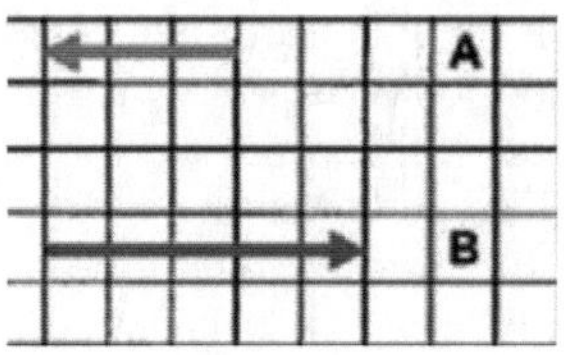

3)

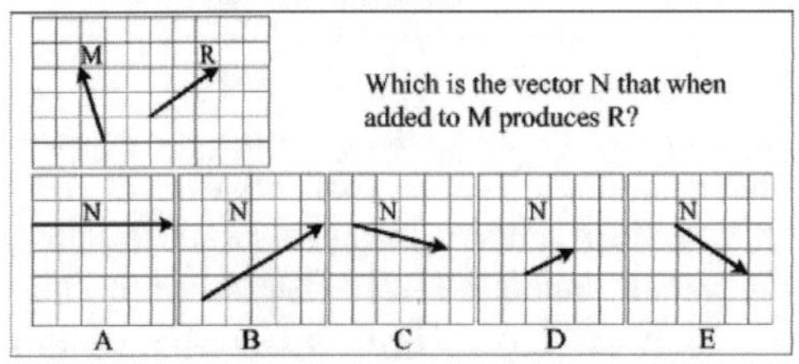

4)

Problem 5) Generally, in vector addition **C**= **A** + **B** the magnitude **C** should just equal to magnitude A dan plus the magnitude B. Is this conclusion always correct? if vector **A** and **B** parallel, is the magnitude **C**= **A** + **B** equal to the sum the magnitude **A** and **B**? if vector **A** and **B** antiparallel, the magnitude of **C** equals the difference of the magnitude **A** and **B**?
The correct answer of students about one and two-dimensional vector addition problems presented by table below.

Table 1. The correct answer of students about one and two-dimensional vector addition problems

Problems Code	Description	Result (%)
1)	one dimensional vector addition	50
2)	two dimensional vector addition	90
3)	one dimensional vector addition	40
4)	two dimensional vector addition	20
5)	One and two dimensional vector addition	20

IOP Conf. Series: Journal of Physics: Conf. Series **795** (2017) 012054 doi:10.1088/1742-6596/795/1/012054

According to table 1, we concluded that the average of the correct answer of students about one and two-dimensional vector addition problems only 44 %. The majority of students (90%) answered the problem number 2 correctly and only 20% of students answered the problems number 4 and 5 correctly. Almost all of the students with the incorrect answered gave statement that they confused about the problems as the reason. The statement of students that they do not familiar to problem about one dimensional vector addition is shown in figure 1.

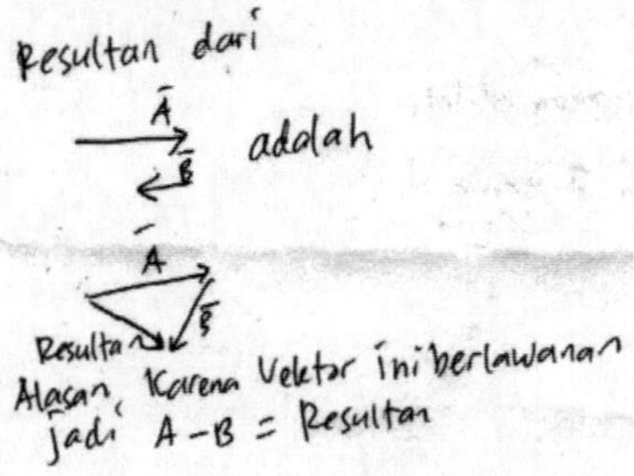

Figure 1. Screenshot of problem number 1

Most of students who answered the problems number 1 and number 3 incorrectly said that they have never known about the problem especially about parallel and antiparallel vectors in one-dimensional vector problems. Most of students who answered the problems number 1 and number 3 correctly said that they have known about one dimensional vector addition from the text book. According to the students, in classroom the lecture never teach about one dimensional vector addition. We also found that some students were still confused using the "tip-to-tail" strategy add vectors graphically. The students said that they're confused using the "tip-to-tail" strategy add vectors graphically when the magnitude and direction of vectors are different.

Based on the students' interviews, we found that their work was consistent to their statement. In spite of the problems number 2 and 4 is identical, we found that students added the vectors by creating a wrong triangle because they just adding two vectors without knowing how to do it properly. The majority of students gave the incorrect answer to the problem of two-dimensional vector additions because they forgot that vectors can be moved to make a connection.

Most of students (80%) gave incorrect answer for problem number 5. Based on the test and the interview we concluded that most of students still confused about parallel and antiparallel vectors. The prove that support our conclusion is shown in figure 2.

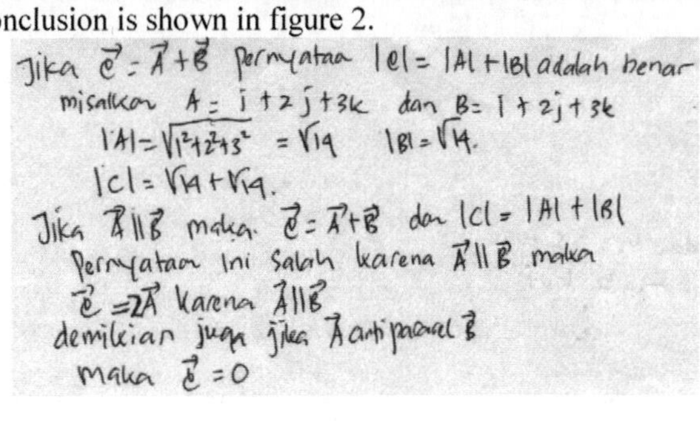

Figure 2. Screenshot of problem number 5

Based on the interview, we found that most of students that gave incorrect answer just considered the magnitude of each vectors without considering the angle between the vectors. For the problems about parallel and antiparallel vectors we found that students gave incorrect answer because they were not careful to distinguish between scalar and vector quantities.

3.1.2. Vector Skills and Conceptual Understanding

The problems of vector skills and conceptual understanding presented below.

Problem 6)

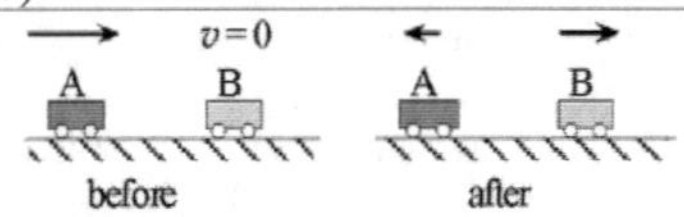

Draw vectors to show the direction of the acceleration of each cart during the collision and compare the relative magnitudes of the average accelerations.

Problem 7)

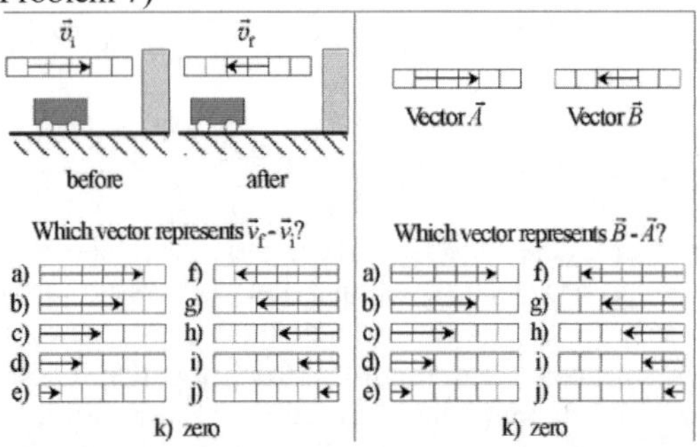

Problem 8) Choose the correct response to make the sentence true: A component of a vector is (a) always, (b) never, or (c) sometimes larger than the magnitude of the vector.

The correct answer of students about vector skills and conceptual understanding problems presented by table below.

Table 2 The correct answer of students about vector skills and conceptual understanding problems

Problems Code	Description	Result (%)
6)	one dimensional collision of two carts	20
7)	one dimensional collision of cart	30
8)	components of vector	30

According to table 2, we concluded that the average of the correct answer of students about the relation between vector skills and conceptual understanding problems only 27%. The minority of students (20%) answered the problem number 6 correctly and average 30% of students answered the problems number 7 and 8 correctly. Based on the test and interview, we found that almost all students who answered incorrectly gave a statement that they confused about the problems as their reason. The example incorrect written test answer is shown in figure 3.

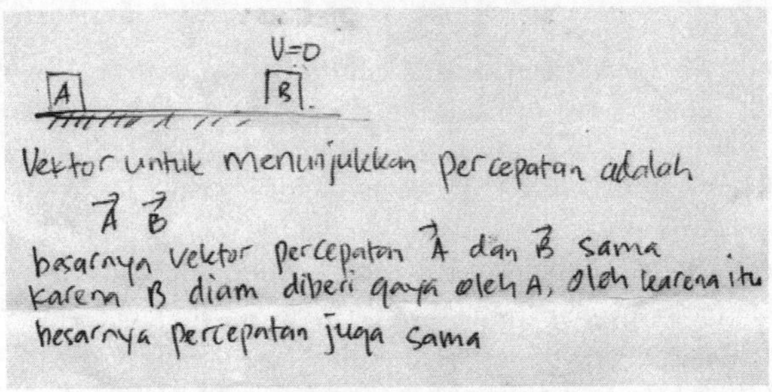

Figure 3. Screenshot of problem number 6

Problem number 6 described the relation between vector skills and conceptual understanding. Most of students were unable to draw the vector in order to show the acceleration each cart during collision. According to students, the direction of acceleration during collision is same as the direction of the velocity vector before collision. The cart B has same direction to cart A because received force from cart A. Most of students could not identify the acceleration as the change of the velocities that can be found by subtracting the initial from the final velocity each cart. Based on interview to students who gave correct answer for problem 6 we concluded that they understand that velocity equivalent to acceleration, so the average accelerations is the same as for the changes in velocity, consequently cart A has the larger acceleration, which is opposite to the direction of the acceleration of cart B.

Problem number 7 and 8 also described relation between vector skills and conceptual understanding. According to the test and interview we found that most of students have difficulties about manipulation vector included physics concept or without include physics concept. The student`s difficulties of vector manipulation is shown in figure 4.

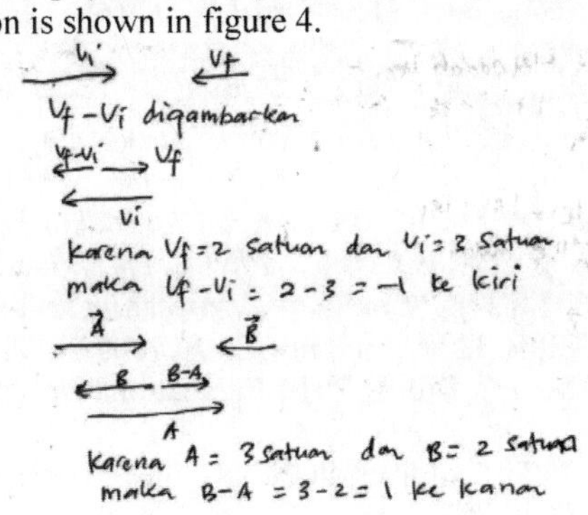

Figure 4. Screenshot of problem number 7

Problem number 7 is similar to problems number 1 and 3. We also found that the answer of students for the problems number 7, 1, and 3 were consistent. Based on the interview, we concluded that the students have difficulties to correlate the vector manipulation and the physics concept. We concluded that most of students still confused about parallel and antiparallel vectors. We found that most of students ignored the magnitude of vector to solve problems number 7 and 8. We found that the students just drawn the vector without considering the magnitude of vector accurately. The students who answered the problem number 8 correctly could use the relation between vector skills and conceptual understanding correctly. They said that based on Pythagorean theorem, the magnitude of a vector is always larger than the absolute value of each component.

3.2. Discusion

From our investigation, we found that after a conventional teaching about vector, some students ignored the direction of vector. Most of students still confused about parallel and antiparallel vectors in one-dimensional vector problems. Many students were still confused to use the "tip-to-tail" strategy add vectors graphically. We found that the students added the vectors by creating a wrong triangle because they just adding two vectors without knowing how to do it properly. The majority of students who answered the problem two-dimensional vector additions incorrectly because they forgot that vectors can be moved to make a connection although they have already taken fundamental physics and classical mechanics course. We also found that the students just drawn the vector without considering the magnitude of vector accurately. Our findings are consistent to the previous research [4,6,7, 9,12]. Based on the test and interviews, we argued the importance of improving the quality of science instruction include instructional methods.

4. Conclusion

The results of this research become complements the study on how to investigate the students' difficulties of one-dimensional and two-dimensional vector. By using multi-perspective views (test perspective and interviews perspective), we found that 44% of students provided correct answer for vector addition of one-dimensional and two-dimensional vector and only 27% students provided correct answer for the relation between vector skills and conceptual understanding. Based on the test perspective and interviews perspective we found that our finding is consistent.

References

[1] Knight RD 1995 *The Physics Teacher*, 33(2), 74-77
[2] García FS, Alfaro-Avena LL, and Dena-Ornelas, O 2008 *Revista Mexicana de Física E* 54(1) 7-14.
[3] Aguirre JM1988*Physics Teacher* 26 212.
[4] Shaffer PS and McDermtt LC 2005 American Journal of Physics, 10.1119/1.2000976
[5] Barniol P and Zavala G 2012 *AIP Conference Proceedings* 1413 115
[6] Flores S, Kanim SEand Kautz CH 2003*American Journal of Physics,* 10.1119/1.1648686.
[7] Wutchana U and Emarat N 2011 Eurasian *J. Phys Chem. Edu*3(2):102-111
[8] Barniol P and Zavala G 2012 *AIP Conference Proceedings* 4403-8
[9] Barniol P and Zavala G 2010*AIP Conference Proceedings* 1289, 341
[10] Ostermeier C, Prenzel M, and Duit R 2010 International Journal of Science Education, 32(3), 303–327.
[11] Duit R and Treagust DF 2012 Issues and Challenges in Science Education Research: Moving Forward, 10.1007/978-94-007-3980-2_4,
[12] Barniol P and Zavala G 2014 Physics Education Research 10.1103/PhysRevSTPER.10.020115

Consistency argued students of fluid

Viyanti[a], Cari[b], Suparmi[b], Winarti[a], Indah Slamet Budiarti[a], Jeffry Handika[a], Fatma Widyastuti[a]

[a]Postgraduate student of UNS Science Education, Jl. Ir. Sutami 36A Jebres Surakarta;

[b]Lecture of UNS, Jl. Ir. Sutami 36A Jebres Surakarta

Email: viyanti@student.uns.ac.id

Abstract. Problem solving for physics concepts through consistency arguments can improve thinking skills of students and it is an important thing in science. The study aims to assess the consistency of the material Fluid student argmentation. The population of this study are College students PGRI Madiun, UIN Sunan Kalijaga Yogyakarta and Lampung University. Samples using cluster random sampling, 145 samples obtained by the number of students. The study used a descriptive survey method. Data obtained through multiple-choice test and interview reasoned. Problem fluid modified from [9] and [1]. The results of the study gained an average consistency argmentation for the right consistency, consistency is wrong, and inconsistent respectively 4.85%; 29.93%; and 65.23%. Data from the study have an impact on the lack of understanding of the fluid material which is ideally in full consistency argued affect the expansion of understanding of the concept. The results of the study as a reference in making improvements in future studies is to obtain a positive change in the consistency of argumentations.

1. INTRODUCTION

The ability of students to integrate knowledge aspects in isolation is a major factor in cognitive development. This capability allows students to focus and increase the understanding of the physics content element. The elements of physics content built by the construction of the development of knowledge based on understanding the situation of argumentations. [7] revealed that skills can argue to accompany students to understand the epistemology of scientific thinking and to uncover phenomena and proof of concept using thought. Reinforced the opinion [10], science not only find or memorize facts; but also how to build argumentations and consider how a phenomenon described. General direction used in construction to develop arguing students using analytical framework [14]. The main elements of a framework of skills Toulmin argued the claim, data, warrant, backing, and rebuttal. Toulmin Model framework adopted by many other researchers to characterize skill argue in science [6].

The analytical framework developed by loading argued skills to some ideas and facts how the student can reveal phenomena of physics in understanding the science of different contexts. Analysis of the context of science can be seen from the viewpoint of various interpretation and focus on the problem thinking activities. However, developments in the field have not been skilled for students to integrate different scientific contexts. It is pointed out that the students focus on providing solutions mathematically completion of the give physical meaning of the problem is given. Reaffirmed the results of the study [2] teachers spend more time to decrease in mathematical formulas. Students tend to focus on solving problems in mathematical physics indicate less inconsistency in the conception of a student on a concept. Related to this, students should be able to utilize mathematical physics problems by linking theory with the given problem, so it looks justification argumentations although students use mathematical language. Justification argumentation contains supporting theory can be visualized in consistency argumentation. According to [3], the consistency of the response of students to understand the concept of physics requires a deeper understanding to see equality of physics problems which is outlined in a various ways. Consistency argues student of physics problems faced bring students a better understanding of the level.

The importance of facilitating learning constructing consistency in argumenting not to rule obtaining core skills of students. Core skills acquired through activities, 1) discussion, 2) experimentation, 3) observation, 4) scientific verifications of law through experiment, 5) to formulate and test hypotheses, 6) presentation and communicate the procedures and results of the study, 7) documented scientific activities, 8) identify and summarize information. Core skills with a learning environment that empowers consistency argued allegedly encouraging students to think independently. Based on the study of theory and problems of field, researchers have conducted studies to assess the consistency of the material Fluid argue student. The formulation of the problem in this research is How the consistency of the material Fluid Students Argumentation?

2. RESEARCH METHODS

This study is a preliminary study using a descriptive survey method. The data were obtained using a multiple choice test and interview based on the aim of seeing consistency student argued. Item test consists of eight items, each item, is displayed in the form of verbal, images and tables. The materials tested are fluid physics.

Categorizing consistency argued refer to students categorized as inconsistent if more than two pairs of test and students categorized consistently one completely consistent if all the test questions were given consistently and correctly. Categorizing argue consistency are assessed using a rubric quality arguments modified from [13], modified criteria of consistency of production argumentation [5]. Data were obtained from three universities by the number of consecutive samples: IKIP PGRI Madiun 25 students; UIN Sunan Kalijaga 52 students; and the University of Lampung 68 students.

3. RESULTS AND DISCUSSION

Items used in the study have been analyzed for validity and reliability. The trial items to 40 students at the University of Lampung are out of the main sample. Related to that obtained an average of 3.01 with ttabel tcount 2.02 indicates that the items were valid. Reliability of items in the category enough with probability value of 0.43 and a significance level of 0.05. The percentage of students based on the consistency of the study argue against the answers given are presented in Table 1.

Table 1. Percentage of Students Argue Consistency

Consistency Level	Percentage (%)		
	IKIP PGRI Madiun	UIN Sunan Kalijaga	Univ.Lampung
Consistency Right	2,78	7,17	4,61
Consistency with Right answer	20,7	30,7	38,4
Not Consistent	76,5	62,2	57

Table 1 presents the percentage level of consistency argues, generally students have absolutely lowest consistency 2.78%. This reflects a failure to understand the students' matter of understanding the concept. Related to that, do interviews to some students with the lowest level categories to dig problems happened. The summary of the interview that "students are accustomed to solve about a matter of using the mathematical equation because it is easier to finish than about the shape of understanding". In addition, students also have low response capability in the pouring concepts into written form by giving the argumentations on the answers given.

Consistency argue demanding a greater understanding of students to see equality between the grounds answer choice answers. The tendency of students to guess very high in this study, it can be seen, none can provide the argumentation according to a key sentence in the given problem. According to [3], "the consistency of response of the students in understanding the concept of physics requires a deeper understanding of the students to see the equality of physics problems outlined in a various way". With a deep understanding will make students consistent with what was understood and believed to be true and to bring the level of understanding is better in view of various concepts of physics as outlined in various problems.

Consistency response Argumentations Correct Answer Student

Exposure results in consistency student response argues correct answers confirm that there needs to be a paradigm shift by planting a concept which integrated to provide students matter contains the basic concepts of physics to practice the skills of consistency argumentations. The data description of examples of student answers correct to argue the consistency of students is presented in Figure 1 and 2.

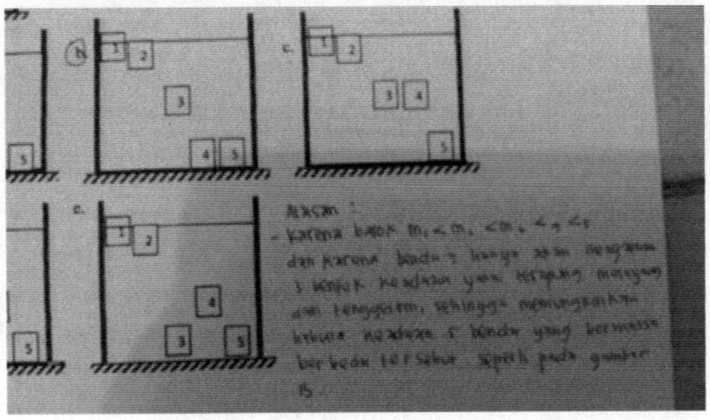

Figure 1. Example of correct answers students consistency (Source: Loveredu, et al, 2005)

Problems are given in Figure 1, the students asked to draw the final position of objects 1, 3, and 4 (the position of objects 2 and 5 have been determined). This problem requires an answer and right reason. Based on the average data of students chose answer b, while the written reasons have given argumentations according to the keywords which implicitly presented on the matter, the object has a mass of "varied". Figure 1 represents the answer and the reason for the student, "because the beam m1 <m2 <m3 <m4, <m5 and because the objects were just having three forms of state are floating, floating and sinking, making it possible that the state 5 objects have masses of different such as in the picture B ". The argumentation given still looks weak, the student should provide argumentations from the viewpoint of the density of objects and liquids. This is confirmed by the statement [8], that the sequence analysis of the most significant argumentation if it follows the pattern of claims, counter-claims responsive, and combines previous ideas. This is supported by the statement [4], presents the basic argumentation in the real-world applications with logical statements, it is very important for the development of science.

Related to this, arguing consistency students will be categorized above level 1 if students begin the analysis from the perspective of two objects a predetermined position and has the same density with liquid. Students can assume that the first thing definitely has a smaller density of objects 2 and 3 objects greater density of objects 2 so that the beam 3 will sink and the beam 1 will float part. Associated with the mass of the object "varied", the student should have a density to know 3 objects larger than the object 2, so that students can conclude that the object 3 sinks. Students are able to put forward three additional criteria of an argumentation declaration indicates that the student has a good scientific literacy. In line with the [15] revealed that solving through consistency argumentation can increase thinking skills that are important ideas in science.

Student Response Not Consistency and Wrong

Weak consistency of the students indicated that learning is not a visualization field of physics, this has resulted in a fluid concept only refers to merely memorizing. Students have the tendency prioritize how to solve problems with mathematical equations without understanding the physical meaning of each of the equations used to solve problems. This resulted in the lack of understanding of the mastery of materials physics.

Consistency and student response does not indicate that any of the argumentations presented are not supported by the data to support the claim. The data description of a sample of students did not answer the consistency of arguing and one presented in Figure 2.

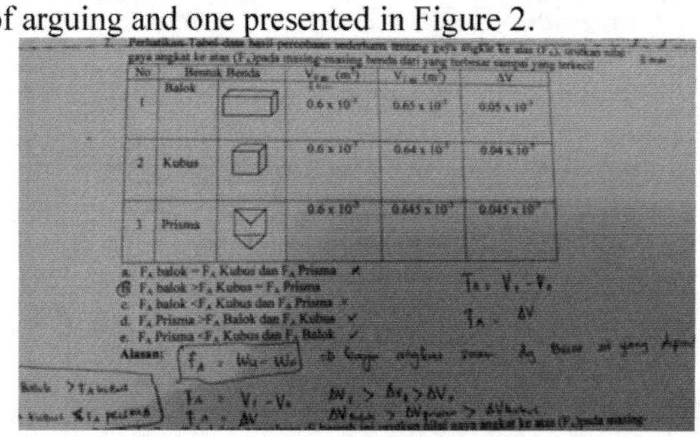

Figure 2. Example of a student answers incorrectly categorized as inconsistent and (Source: [1]).

The problems that are presented in Figure 2 relates to the upward lifting force (FA). Figure 2 is an example answer students chose answer b by writing reason only contains the mathematical equation without giving any physical meaning. Answers and the reasons given by students tends to guess and not consistent. This implies that student hardship reveal physical force from the equation. Supposedly written mathematical equation that can relate problems with the theory that implicitly disclosed in the matter. Students can start writing argumentations phases: 1) the students write the equation FA = ρgV, with expressing meaning that the FA on each body area affected touchpad means the material to the water affects the shapes of objects FA experienced objects; 2) related to the equation FA, FA values consecutive beams 0:49 N; 0:39 cube N; and prism 0:04 N; 3) The student must be able to remember that the volume of the submerged object is obtained by reducing the volume of water before it dipped to the volume of water after it is dipped; 4) students should be able to identify that the volume, type of material and assumed the same types of objects Mass large, so if the mass of the same type, the third volume of the object assumed to be equal.

The ability to analyze measures indirectly about the completion of the student's consistency argue convincingly give reasons for the answer selected. Consistency requires a theoretical argumentation in justifying the interpretation presented in the matter. [12] revealed a positive effect on the consistency argue student conceptual change. The consistency argued stages working requires clear and measurable student to consider the logical structure. Low ability in understanding problem in the matter resulted in the lack of understanding of students to a physics concept.

[11] argues, students will not be involved in building the skills to argue unless they have a fundamental scientific literacy. In developing consistency argumentation contrasts with the scientific concept that will produce proper discourse. Basically, the students had the idea to try to solve problem but always to no avail, and sought to assert their own argumentations with weak statement. [17] reveal that when students are given directives explicitly in the argumentation, the opportunity to practice the science content, allowing students to more easily analyze the claim argumentation. Therefore, students can not be involved in the concept and theory argues consistency if you do not have the background

knowledge. This statement is in line with the results of the preliminary study [16] if the students only focus on empowering skills of argumentation (elements claim) without empowering other elements of argumentation skills, this affects the weak understanding of the concept of students. This means that students not only do not understand aspects of the task level of abstraction, but also the students have to struggle to understand quickly the construction of a task so that it can consistently argued.

4. CONCLUSION

Based on research data on average consistency argued for the right consistency, consistent with incorrect and inconsistent answers, respectively, are 4.85%, 29.93% and, 65.23%. The data obtained in this study illustrates the undeveloped construction of student knowledge. Should, consistency argued encourage students skillfully argued using Toulmin's argument pattern. Toulmin's argument pattern enables students can identify key statements as the basis for the level of argumentation Toulmin next element. That is, students need to understand more deeply the content if the questions given to them are abstractions. This has significant implications on learning how to promote the development of consistency argumentations.

References

[1] Abidin, Zainal D.K. Dina, Rif'ati. Yushardi. 2013. Pengaruh Bentuk Benda dan Kedalaman Terhadap Gaya Angkat ke Atas (Fa) pada Fluida Statis. Tidak Diterbitkan. library.unej.ac.id/client/search/asset/571
[2] Alfiani. 2015. Prosiding Seminar Nasional Fisika (E-Journal) SNF 2015 http://snf- unj.ac.id/kumpulan-prosiding/snf2015/ Volume IV, oktober 2015
[3] Ainsworth. 2006. [online].tersedia: http://www.nottingham.ac.uk/search aspx?q=ainsworth+considering+learning+with+multiple+representations. [08 juli 2016]
[4] Aristotle. 1991. On rhetoric: A theory of civic discourse. Oxford, UK: Oxford University Press.
[5] Chien-Hsien Chen and Hsiao-Ching She. 2012. Educational Technology & Society, 15 (1), 197–210.
[6] Erduran, S., Simon, S., & Osborne, J. 2004. Science Education, 88, 915-933.
[7] Howe, C., Mcwilliams, D., & Cross, G. 2005. British journal of psychology, 97, 67-93.
[8] Leitao, S. 2003. Written Communication. 20, 269-306.
[9] Loverude, Michael E., Kautz, Christian H., & Heron, Paula R. L. 2003. American Journal of Physics 71, 1178. 2003; doi: 10.1119/1.1607335
[10] McNeill, K. & Krajcik, J. 2008. Journal of Research in Science Teaching. 45(1). 53-78.
[11] Norris, S., & Phillips, l. 2003. Science Education, 87, 224-240.
[12] Nussbaum, E. M., & Sinatra, G. M. 2003. Contemporary Educational Psychology, 28(3), 384 – 395.
[13] Stapleton, Paul., Wu, Yanming. 2015. Journal of english for academic purposes 17 (2015) 12-23
[14] Toulmin, S. 1958. The Uses of Argument. Cambridge: Cambridge University Press.
[15] Wolfe, C. 2011. Written Communication. 28(2). 193-219.
[16] Viyanti, Cari, Widha Sunarno, Zuhdan Kun P. 2016. JP2F Volume 7 April 2016.
[17] Zohar, A., & Nemet, F. 2002. Journal of Research in Science Teaching, 39(1), 35 – 62.

A model of scientific attitudes assessment by observation in physics learning based scientific approach: case study of dynamic fluid topic in high school

Elvin Yusliana Ekawati[1]

1 Physics Departement of Sebelas Maret University, Surakarta Indonesia

E-mail: elvin.pfisika@gmail.com

Abstract. This study aimed to produce a model of scientific attitude assessment in terms of the observations for physics learning based scientific approach (case study of dynamic fluid topic in high school). Development of instruments in this study adaptation of the Plomp model, the procedure includes the initial investigation, design, construction, testing, evaluation and revision. The test is done in Surakarta, so that the data obtained are analyzed using Aiken formula to determine the validity of the content of the instrument, Cronbach's alpha to determine the reliability of the instrument, and construct validity using confirmatory factor analysis with LISREL 8.50 program. The results of this research were conceptual models, instruments and guidelines on scientific attitudes assessment by observation. The construct assessment instruments include components of curiosity, objectivity, suspended judgment, open-mindedness, honesty and perseverance. The construct validity of instruments has been qualified (rated load factor > 0.3). The reliability of the model is quite good with the Alpha value 0.899 (> 0.7). The test showed that the model fits the theoretical models are supported by empirical data, namely p-value 0.315 (≥ 0.05), RMSEA 0.027 (≤ 0.08)

1. Introduction

Physics is a branch of science that is obtained through scientific methods to unravel the mysteries of matter and its interactions with other objects in the universe. For teaching physics, scientific activity is a learning activity that can restore the spirit of the application of scientific methods to find concepts, laws, principles, theories and other physical products. On the other hand, there are some topics in physics that physics teacher assumed as an abstract matter, because the teacher presents difficulties for scientific activity in the topic. One of them, the topic of dynamic fluid on learning physics in high school. To organize scientific activities through experiments, preparation for carrying out scientific activities quite time-consuming, in addition to the school laboratory facilities are often considered inadequate. For example in the scientific activities with the aim to apply the principles of Bernoulli, the teacher can not perform using venturi meter as in figure 1, because the tool is not owned by the school.

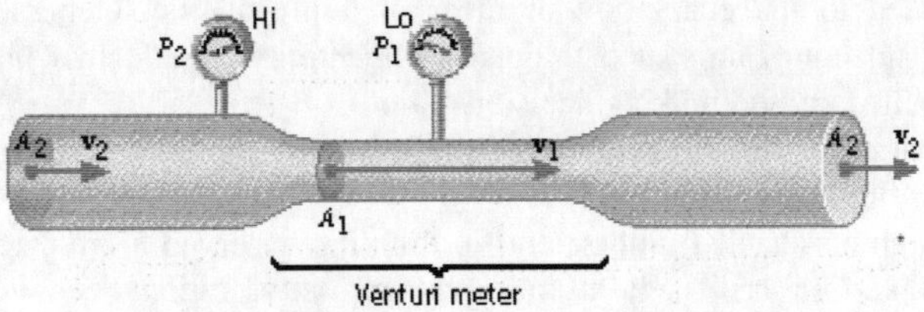

Figure 1. Venturi meter [1]

Based on interviews with teachers, materials related to the application of Bernoulli's principle is rarely taught using trial, more likely is presented in the form of lectures and discussions were accompanied by exercises, individual work or group work. On the other hand, teachers found to meet the assumptions: the flow is steady and laminar, the fluid is incompressible, and the viscosity is small enough to be ignoring [1], rather difficult met. Consequently, the results of data collection, data analysis and conclusions practicum students often do not correspond to the Bernoulli principle.

The situation is causing high order thinking skills and scientific attitude of students is not well developed. To overcome these problems, a teacher must prepare more creative learning by making simple tools and apply a scientific approach. One model of learning that applies scientific approach is Guided Inquiry Learning Laboratory (GIL). The model is in accordance with the teaching of physics to students in high school, because it can improve scientific thinking skills or develop a scientific attitude. GIL is one of the activities the students to identify problems based on phenomena such as formulating problems and hypotheses, designing an experiment with the guidance of teachers, conduct experiments, draw conclusions trial, presented the results of observations of the experiment and work on the problems associated with materials such trials, teachers comment that the ongoing investigation[2]. GIL consists of 5 stages of learning, namely: observation, manipulation, generalization, verification and application[3]. These stages are very relevant to the steps of the scientific method.

The application of scientific approaches assumed to be used as an attempt to develop a scientific attitude of students. Scientific attitude becomes important to be developed, because without scientific attitude the students will lose spirit in the study of physics, especially in high school. Baumel & Berger in his research report provides some suggestions:

- the scientific attitude needs to be examined for all levels of ability,
- science teacher (physics) not only evaluate the ability of knowledge, but the development of scientific attitude of students as well,
- Teachers should provide guidance to students that the scientific attitude is an integrated part of the behavior / habits of learners in everyday life, which includes: intelectual honesty, willingness to admit error, listening to others' ideas, and dealing with facts in an unbiased way makes a favorable and lasting impression upon the pupils [4].

Scientific attitude is a different meaning to the attitude towards science and the ability to carry out scientific procedures. Gardner referred to 'attitudes towards science' (including aspects such as attitudes towards scientists and interest in science) and 'scientific attitudes' (such as openmindedness, Scepticism etc) [5]. Attitudes have emotional content and vary in intensity and generality according to the range of objects or situations over they've apply. Mostly attitude to learn are difficult, extinguishing from affective attributes of personality such as interest, appreciation, likes, Opinion values, ideals and character traits [4]. In further examining this latter construct Gauld and Hukins concluded that 'the components of the scientific attitude Mentioned in the literature seem to fall into three broad groups'. Reviews these groups are:

- general attitude towards ideas and information, such as curiosity, openmindedness, Scepticism;

- attitudes related to the evaluation of ideas and information, Generally labeled as critical mindedness and containing such things as Objectivity, intellectual honesty and caution in drawing Conclusions and making decisions; and
- a commitment to a particular scientific beliefs such as loyalty to truth, cause and effect relationships, etc [6].

The habit of thought associated with scientific thinking deserved more careful consideration. To be scientific mean that one has such attitudes as curiosity, rationality, willingness to suspend judgment, open mindedness, critical mindedness, objectivity, honesty and humility etc. attitude regulate behavior that is directed towards or away from some object or situation group of objects or situations [7]. In other hand, Jones & Butts developing scales for measuring measuring students' scientific attitudes along four dimensions, concern for new evidence relating to currently held ideas, belief in causation, honesty in reporting and scepticism (involving suspension of judgement, concern for evidence and use of many sources of information) [8].

The presence of scientific attitude assessment instruments become part needs to be prepared by the teacher, to monitor the progress of scientific attitude of students. Emphasis on the development of the scientific attitudes was pointed out by Heiss [9] who stated that the development of the scientific attitudes and the ability to use the methods of science are major goals of science instruction. It was further pointed out that considerable attention has been given to ways of teaching the scientific method, whereas less attention has been given to the development of scientific attitudes. In his suggestions for developing the scientific attitudes, Heiss asserted that the main handicap in this area was the lack of an appropriate instrument for measuring these attitudes. There are several forms of assessment instruments that can be used to determine the scientific attitude of students, and assessment instruments in the form of observation is a model that assumed matches assessment for the scientific approach, especially with models Guided Inquiry Laboratory. The results of the Kozlow & Nay provided the inspiration to develop forms of assessment is not only using the techniques of inventory. In line with the development of educational measurement, has so far not been found studies that develop assessment instruments scientific attitude in learning high school physics which use techniques other than inventory, such as the valuation techniques of observation and assessment colleagues [10].

Based on the above issues, this study was focused on the preparation of the models of scientific attitude assessment for teaching physics in high school. The valuation model is expected to help teachers to foster the development of a scientific attitude of students in the organization of learning physics that apply scientific approach, with case studies on the topic of dynamic fluid.

2. Method

This study uses a model of research and development of Plomp which uses five phases, namely the preliminary investigation phase, design phase, realization/construction phase, test, evaluation, revision phase, and implementation phase [11]. This research was conducted in Surakarta, using schools representing of high, medium and low achievements in 2016. The determination used these schools using purposive sampling technique. The schools name were SMA N 1 Surakarta, Batik 2 Surakarta SMA and SMA Al Islam Surakarta, with the number of subjects was 122 students. Data related to school achievement accessed from the new admissions system online in Surakarta on http://www.ppdbsolo.net. These schools have laboratory facilities and supporting scientific activities in teaching physics.

The quantitative data obtained from observation sheet with rating scale and semi-questionare with likert scale. Qualitative data obtained from focus group discussions, interviews, and observations. Data analysis technique performed quantitatively and qualitatively. Analysis of quantitative data used to obtain evidence of the validity of the content by using a formula Aiken [12]. Analysis to obtain evidence that a valid instrument construct's, using confirmatory factor analysis of order 2nd by using a version of the lisrel 8.50 program, while for reliability using Cronbach Alpha formula. Qualitative data analysis is used to describe the comments of validator and students to the developed product.

3. Result and Discussion

The instruments developed in this study to assess the attitudes of students, with the scientific observation by teachers in teaching physics. Scientific activity is a prerequisite assessment and trials carried out on the topic Fluid Dynamic. In the study, scientific activities carried out by the scientific approach, especially with models Guided Inquiry Laboratory (GIL). Components scientific attitude assumed to be observed in every phase of the learning model GIL, with a guide in Table 1. Teachers can use in the assessment of these guidelines, but it may be possible scientific attitude to be expected in the indicator is not according to the guidelines.

Table 2. Syntax of GIL and Observed Components of Scientific Attitude

GIL Phase	Student Activity	Scientific Attitudes Observed
Phase 1 *Observation*	Students observe natural phenomena that exist. Students describe in detail what they saw. Questions in the beginning used to help students in the process of discovery.	Curiousity,
Phase 2 *Manipulation*	Students provide suggestions and ideas for the investigation of finding answers to the problems. Students design and conduct experimental procedure designed	Curiousity, Open-mindedness
Phase 3 *Generalization*	Students make observations on the results of the experiment, collect data and establish principles or rules of appropriate data. Students discuss the results of observations logically based on data findings	Objectivity, honesty
Phase 4 *Verification*	Students present findings that other groups interested in his findings as well as to prove the results of experiments on each group.	Objectivity, Open-mindedness
Phase 5 *Application*	Students conclude the findings. Students are guided to solve the problem of applications in everyday life by administering a matter or issue	Willingness to suspend judgment, Perseverence

In GIL learning model, each student is grouped with each group consists of a maximum of 5 students. One sheet of assessment instruments used to assess the scientific attitude of the group. The instrument contains statements related to student activities should be observed, selection assessment score that consists of the numbers 1, 2, 3, 4, 5 and rubric to help teachers in providing assessment scores. Table 2 presented the examples of instruments and sections.

Tabel 2. Sample Instrument of Scientific Attitudes Assessment

No	Observed activity	Student Name:			
		1	2	3	4
1	Before starting eksperimen activity, teacher ask to student for learning the task sheet and student disposed as				

Instructions for giving a score:
Score 1: quite
Score 2: reading the task sheet, but lean quite
Score 3: reading the task sheet, and ask to her / his friend
Score 4: reading the task sheet, and ask to her / his teacher

IOP Conf. Series: Journal of Physics: Conf. Series **795** (2017) 012056 doi:10.1088/1742-6596/795/1/012056

Instruments model of scientific attitude assessment in this study was observations form, which include grilles, observation sheets (instrument), assessment guidelines and scoring rubric. Constructs instruments of scientific attitude assessment developed, consists of 6 components and 13 indicators. Item 2 for indicator 2 was bad items that eliminated, so the result of development of instruments received 12 indicators with 12 items, each item is represented by two items. In Table 3 are presented the components and indicators are reported as a result of research.

Table 3. Component and Indicator of Scientific Attitude Assessment in Physics Learning at High School

Component of Scientific Attitudes	Indicator	Item Code	No Item
Curiousity	1.1. Prioritize asked to face the new situation	CR 1	1
	1.2. Having a desire to update knowledge	CR 2	2
Objectivity	2.1. Observing and recording the facts without the influence of personal pride	OB 1	3
	2.2. In interpreting the results of observations are not influenced by others	OB 2	4
Wilingness to Suspend Judgment	3.1. Do not rush to draw conclusions before enough evidence collected	SS 1	5
	3.2. Willingness to accept the facts as supported by convincing evidence	SS 2	6
Open-mindedness	4.1. Willingness to renew their opinions and conclusions	OP 1	7
	4.2. Not necessarily reject the opinions of others	OP 2	8
Honesty	5.1. Report the observing result despite contradict with hypotesis	HN 1	9
	5.2. Assume all available information/data when establish conclusion	HN 2	10
Peserverence	6.1. Complete all scientific activities until end	PS 1	11
	6.2. Not giving up when doing practicum	PS2	12

Results of reliability analysis indicate that the instrument has been qualified reliability with Cronbach Alpha coefficient is 0.899 (≥ 0.70). CFA 2nd order analysis results show the proposed empirical measurement model has a good model fit. Figure 2 shows the fulfillment of criteria, ie, p-value 0.315 (≥ 0:05), RMSEA = 0.027 (≤ 0:08). Rated load factor standard of all indicators in the model of instruments of scientific attitude assessment have met the requirements in the moderate category (> 0.3).

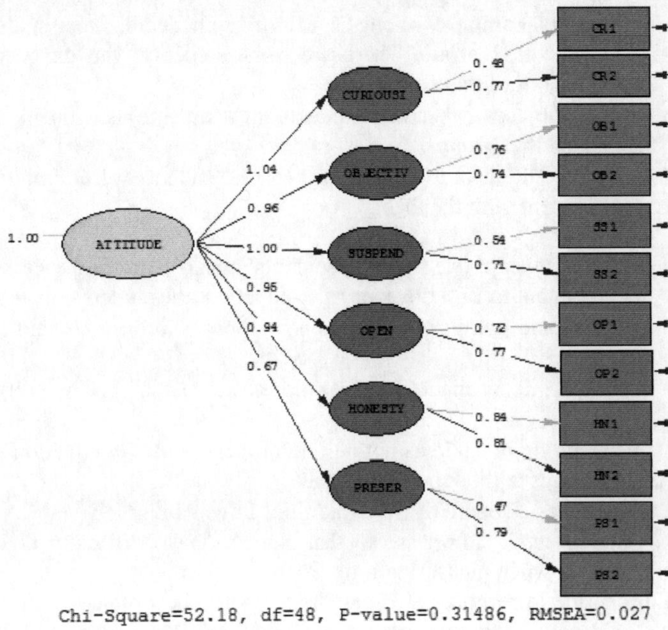

Figure 2. Empiric Model of Instrumen of Assessment Scientific by 2nd order CFA

Decisions of analysis result related goodness of fit, by comparing the index with a cut-off criteria[13] which is presented in Table 4.

Tabel 4. The result of goodness of fit analysis the model of instrumen of scientific attitude assessment

No	Goodness of Fit Index Criteria	Cut of value	Result	Conclusion
1	*p-value*	> 0.05	0.31486	Good fit
2	RMSEA	≤ 0.08	0.027	Good fit
3	RMR	< 0.05	0.025	Good fit
4	NFI	≥ 0.90	0.97	Good fit
5	NNFI	≥ 0.90	0.99	Good fit
6	CFI	≥ 0.90	1.00	Good fit
7	IFI	≥ 0.90	1.00	Good fit
8	RFI	≥ 0.90	0.95	Good fit
9	GFI	≥ 0.90	0.93	Good fit
10	AGFI	≥ 0.90	0.90	Good fit

Empirically, instruments of scientific attitude assessment can be declared to have content validity, construct validity and reliability that meet the criteria. In addition, valuation models with this observation allow teachers to monitor the progress of scientific attitude of students. In general, three components of scientific attitude that needs more attention, namely perseverance, honesty and

IOP Conf. Series: Journal of Physics: Conf. Series **795** (2017) 012056 doi:10.1088/1742-6596/795/1/012056

objectct. Based on observations with qualitative approach, records obtained during the learning physics model GIL, namely:

- Students were motivated to complete scientific activity will decrease as students in other groups have completed the task group. There are some students who gave up or did not complete the task when the hour lesson ends.
- The data collected is not objective, because students must assume as a theory in a book or observations of students in other groups.
- Some students did not write the data to be honest, because students were afraid the results of observations are not consistent with the theory.

Scientific attitude identified the weak, due to some students' mistakes in the process of scientific activity. For example, in an experiment to find the continuity equation, the students make mistakes:

- drafting tool trial (with a hand pump) which does not support the flow of fluid to move with constant
- do not perform step experiments made steady fluid flow, tend to be less serious conduct experiments
- fluid flow is unstable resulting in students not getting proper continuity equation. But students still report what their findings, although aberrant theory
- on tools that use tap water as a source of fluid, the fluid flow is relatively stable, but the water that comes out of the faucet is colorless, so that students have difficulty determining the beginning and end position when measuring time.

Consequently, the results of the experiment the students are not in accordance with the principles continuity. In the example of a table on the observation and analysis of data obtained by the students in figure 3, and figure 4, it can be shown that the experimental results deviate from theory students, but the students made their conclusion in accordance with the theory (shown in figure 5)

The Observation Table Continuity of Principles

No	Hose 1 d_1 = 1,5 . 10^{-2} m A_1 = 1,766 . 10^{-4} m^2				Hose 2 d_2 = 0,7 10^{-2} m A_2 = 0, 385 10^{-4} m^2			
	s_1 (m)	t_1 (s)	v_1 (m/s)	$A_1 v_1$	s_2 (m)	t_2 (s)	v_2 (m/s)	$A_2 v_2$
1	0,7	3	2,1	3,7086 . 10^{-4}	0,7	4	2,8	1,078 . 10^{-4}
2	0,7	3	2,1	3,7086 . 10^{-4}	0,7	5	3,5	1,3475 . 10^{-4}
3	0,7	5	3,5	6,1810 . 10^{-4}	0,7	5	3,5	1,3475 . 10^{-4}
4	0,7	4	2,8	4,9448 . 10^{-4}	0,7	5	4,2	1,617 . 10^{-4}
5	0,7	3	2,1	3,7086 . 10^{-4}	0,7	5	3,5	1,3475 . 10^{-4}

Figure 3. Sample of Student's Observation Table

Data Analysis

Based on the observation table Principle of Continuity, value A_1 v_1 ....>......... A_2 v_2

(fill in the blank with <, = or >)

Figure 4. Sample of Student's Data Analysis

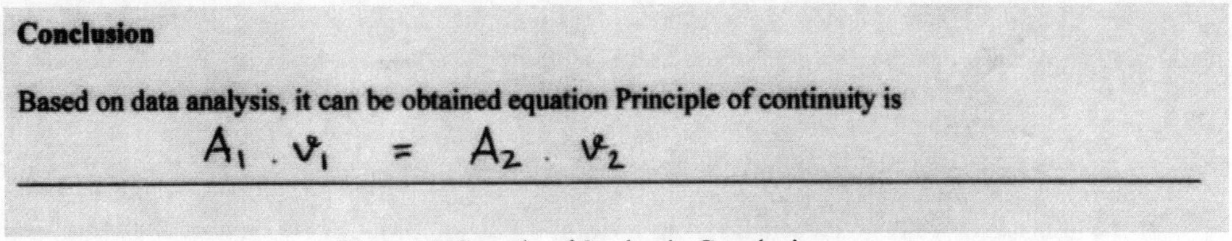
Conclusion

Based on data analysis, it can be obtained equation Principle of continuity is

$A_1 . v_1 = A_2 . v_2$

Figure 5. Sample of Student's Conclusion

In this case, the student does not use the data as a source of information in formulating conclusions. Students indicated making conclusions related to the principle of continuity as written in the reference book.

Overall, the model assessment conducted scientific attitude in learning physics, especially with models GIL for dinamics fluid topic, can foster positive scientific attitude. Students looked eager to follow the teaching of physics. On the other hand, teachers can guide students to discover the physics concepts with the scientific method

4. Conclusion

Model scientific attitude assessment in learning physics that has been generated, it used to improve the quality of learning physics-based scientific approach. One of them was implemented through the learning model GIL. Assessment will not be obtained by optimal when learning of physics do not organize scientific activities. Not all components of the scientific attitude can be observed in teaching physics in high school. Components scientific attitude can be assessed using instruments developed in this study, namely: curiousity, objectively, willingness to suspended judgment, open-mindedness, honesty and preserverence. That component, is important component of scientific attitude to be developed and can be observed relative ease in learning physics in high school.

This research has produced the instruments of scientific attitude assessment that meet both criteria, and match the learning model GIL. In addition, the instrument fit for use by teachers in teaching physics in high school, especially on the topic of dynamic fluid.

References

[1] Giancoli, D. C, 2005. Physics: principles with applications (6th ed.). USA: Pearson Prentice Hall, pp 274

[2] Wenning, C. J. 2010. Levels of inquiry: Using inquiry spectrum learning sequences to teach science. *Journal Physics Teacher of Education* 5(4) 11-19.

[3] Wenning, C. J. 2011. The Levels of Inquiry Model of Science Teaching. *Journal Physics Teacher of Education*, 6 (2) 9-12

[4] Baumel H B & Berger J J 1965An attempt to measure scientific attitude *Science Education* **49** (3) 267-269

[5] Gardner P L 1975 Attitudes to science: A review *Studies in Science Education* **2** 1 - 41Dachlan, U. 2014. Panduan Lengkap Structural Equation Modeling: Tingkat Dasar, Semarang: Lentera Ilmu. pp 182

[6] Gauld C F & Hukins A A 1980 Scientific attitudes: a review. *Studies in Science Education* **7** 129-161

[7] Pitafi A I & Farooq M 2012 Measurement of scientific attitude of secondary School students in Pakistan. *Academic Research International*, **2** (2) 379 -392

[8] Jones B & Butts B 1983 Development of a set of scales to measure selected scientific attitudes. *Research in Science Education* **13**, 133 – 140

[9] Heiss ED1958 Helping Students Develop a Scientific Attitude *The Science Teacher* **25** 371-373

[10] Kozlow M J & Nay M A 1976 An approach to measuring scientific attitudes. *Science Education* **60** (2) 147-172. USA: John Wiley & Sons, Inc.

IOP Conf. Series: Journal of Physics: Conf. Series **795** (2017) 012056 doi:10.1088/1742-6596/795/1/012056

[11] Plomp T J 1997 Educational design: introduction. From Plomp (eds.). Educational & training system design: *introduction. Design of education and training (in Dutch).* Utrecht (the Netherlands): Lemma. Netherland. Faculty of Educational Science and Technology, University of Twente.
[12] Aiken, L. R. (1985). Three coeficients for analyzing the reliability and validity of ratings. *Educational and Psychological Measurement*, 45, 131 -142
[13] Dachlan U 2014 Panduan Lengkap Structural Equation Modeling: Tingkat Dasar Semarang: Lentera Ilmu. pp 182

Students' representation about Newton law: consequences of "zero intuition"

Jeffry Handhika[1], C. Cari[2], A. Suparmi[3]

[1]Graduate Students, SebelasMaret University, IKIP PGRI MADIUN,
Jl. Setiabudi No. 85, INDONESIA,

[2,3]Physics Department of Post Graduate Program SebelasMaret University,
Jl. Ir. Sutami 36A KentinganJebres Surakarta 57126, INDONESIA

e-mail: jeffry.handhika@yahoo.com

Abstract. Newton's laws can represent in the language of verbal, mathematical, physical, and visual. Students who understood concept would express the concepts in various representations consistently. In this research, a mathematical presentation used to reveal the student's concept understanding about Newton first law. The results showed that 21.87% of the students changed the mathematical presentation of Newton's first law ($\sum \vec{F}=0$) into verbal representation incorrectly. Changing the mathematical form of Newton's first law into the form of ($0=\sum \vec{F}$) caused the percentage of students who did not respond increased, further concluded that "zero intuition" in the equation of Newton first law caused misconceptions.

1. Introduction

Understanding of Newton's laws has been discussed intensively by physicists and physics educators [1, 2, 3, 4, 5, 6]. Understanding of physics language, mathematics, visual, and intuition is the key to understanding physics concept [7]. Misconceptions and incorrect conception can occur if one language is not understood, or only one language that dominates over the others when taking conclusions. For example, if the mathematical language is dominant in explaining the concept, it causes incorrect conception [4, 5]. This condition will be reinforced in this research when equalization of the "zero" in the mathematical representation of Newton's law with "nothing" such that causes different verbal representations.

In fact, there are three types of the meaning of "zero" in the equation of Newton firs law. Zero as (1) the "zero intuitive", which means "nothing", (2) a "zero number" which used to represent numbers and (3) a "mathematical zero" according to modern mathematics [8]. "Zero intuitive" has extraordinary consequences in physics. Newton first law, for example, students can interpret that no forces are acting on the object [2]. This concept could certainly cause misconceptions on the student. In mathematics, Zero is a member of the set count numbers, integers, real numbers, and complex numbers, zero is a separator between the positive numbers and negative numbers, zero is not said to be positive or negative, zero is a number that does not have a sign [8]. So, the mathematician does not define zero as "nothing."

In the language of physics, the zero appears because there are physical quantities are mutually eliminated, especially for vector addition. In Newton's first law ($\Sigma \vec{F}$=0), "zero" appears because the algebraic operations of the net force acting on the object are zero. In mathematics, the "zero" can also obtain from the operation of algebra, variables such as x multiplied by y produces "zero" if both x and y are equal to zero, and x or y is equal to zero [5]. In Newton's second law ($\Sigma \vec{F}= m.\vec{a}$), $\vec{a} = 0$, then the equations and statements going back to the Newton first law. If m = 0, although mathematically produce a value of zero, Newton's laws do not apply, because the mass (m) is not the result of algebraic operations, mass is physical quantities. Newton's third law is not the same as Newton's first law, and most important is not another form of Newton's first law [9]. The Newton's first law can be another form of Newton Second Law [10], with consequences mass of the object should not be equal to zero.

Understanding of the language of mathematics and physics should be understood comprehensively if one language is more dominant over the other lead to occur incorrect conception. As a result, inconsistencies concept can raise cognitive conflict. Cognitive conflict can potentially cause misconceptions [7] and can cause anxiety if it not managed properly. Students Verbal representations have investigated by using mathematical presentations ($\Sigma \vec{F}$=0), cognitive conflicts raised by providing other forms of mathematical presentation (0=$\Sigma \vec{F}$).

2. Methods

This research method is descriptive qualitative. Subjects in this study were students at the IKIP PGRI Madiun who take a course in mechanics (32 students). Data collected through the test, where the students have to represent mathematical equation into verbal representation. In turn Verbal representations obtained are retested by giving a mathematical presentation in another form. Then finally unstructured interviews on students conducted to clarify the student answer (reasoning of his/her representation) [11, 12].

3. Results and Discussion

From the 32 students, there are 71.87% of students could represent the concept verbally and correctly, it found that there are two types of incorrect verbal representations.

Table 1. Percentage of verbal representations for $\Sigma \vec{F}$=0

Presentation	Verbal Representation	Percentage (%)
$\Sigma \vec{F}$=0	**Correct Representation**	
	Net force equal to zero, forces are acting on the object with the resultant forces are equal to zero	71.87
	Incorrect Representation	
	Type 1: Force equal to zero, there is no force acting on the object	6.25
	Type 2: The net force is equal to zero, meaning there is no force acting on the object.	15.62
	No Response	
	No response	6.25

Incorrect Representation type 1 caused by the students who are not reading the mathematical symbol (sigma) and has a "zero intuition." In type 2, the students understand the mathematics language partially, and they still have "zero intuition." Another mathematical representation is given to evaluate verbal representations which reported by students (Table 2).

Table 2. Percentage of verbal representation for $0=\sum \vec{F}$.

Presentation	Verbal Representation		Percentage (%)
	Correct Representation		
$0 = \sum \vec{F}$	"Zero" equal to the net force, forces are acting on the object with the resultant of forces are equal to zero		31.25
	Incorrect Representation		
	Type 1.	Zero is equal to the force; there is no force acting on the object	6.25
	Type 2.	a. There is no force acting on an object is equal to net force b. The object stopped because there is no force, if there is a moving object, the object has force.	9.37
	Type 3.	Mathematically, the equation is the same shape	3,12
	No Response		
	No response		50,00

The percentage of correct representation has decreased from 71.87% to 31.25% due to changes in the form of a mathematical presentation $0=\sum \vec{F}$. (Table 2); based on the interview, there was information that the students are confused in representing the mathematical equation. An example of the interview described below:

Lecturer : Why not give a response?
Student : I'm confused, sir
Lecturer : Try to express what you're confusing?
Student : Zero may not be the same as net force
Lecturer : Why? Can you give an explanation?
Student : Only this time I found zero equal to net force, this looks unusual. In physics, books not described the mathematical form like this.
Lecturer : What do you think? Can you express your argument?
Student : wait a minute, sir, (students write down their arguments) (Fig. 1).

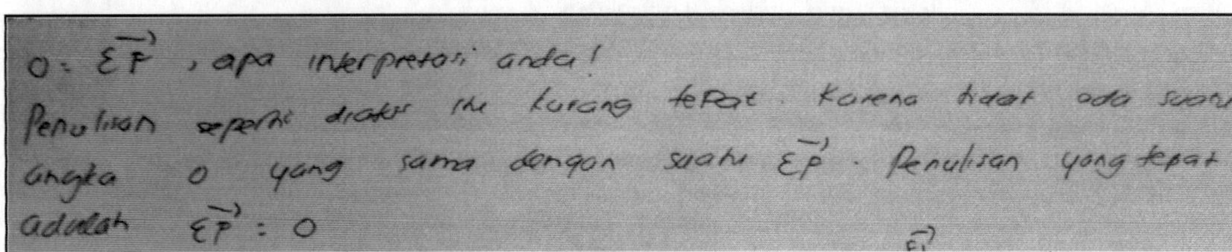

Writing as above it ($0 = \sum \vec{F}$) was not quite right because there is no "zero" equal $\sum \vec{F}$. Appropriate writing is $\sum \vec{F}=0$.

Figure 1. The students write arguments

On the argument of type 1 (Table 2), students have misconceptions, because intuition is more dominant than others to get the conclusion. Besides that, The students also do not understand the mathematical language well. In type 2, intuition still dominates, although the mathematical language also appears to explain the verbal representation of Sigma. The dominance intuition, in this case, may cause a misconception. Intuition comes from the students themselves, simultaneously together with the conception that possessed by students, and able to answer problems that arise in him at that time. Type 3, mathematical language is more dominant than any other language.

Physics language more dominant than mathematical language to make decisions of presentation (Fig.1), the student not confident that the presentation of $0 = \sum \vec{F}$ is correct in physics because the student's never discovered a form of the equation $0 = \sum \vec{F}$ in the physics book. Understanding mathematics and physics language comprehensively help the student to get incorrect concepts. Changes in the mathematical presentation caused a significant impact on student representation. Cognitive conflict arises because there is the unusual presentation ($0 = \sum \vec{F}$). Students who have an understanding of mathematical language better than physics language argued that the form of presentation $\sum \vec{F}=0$ is equal to $0 = \sum \vec{F}$ be the same equation mathematically, without detailed explanation.

Physics can represent in four languages, (1) intuition, (2) mathematical (symbols and graphics), (3) communication, (4) physics language. All must be understood comprehensively. The misconception may occur if the language of intuition is more dominant than the other three languages. Incorrect conception occurs when the communication and mathematical language is more dominant. The dominance of using physics language can cause the student cannot provide a detailed explanation, domination of mathematical language causing physical limitations of the delivery of verbal language.

4. Conclusion

Zero intuition leads the students experiencing misconceptions. In Newton's laws, zero intuition can appear on Newton first and second law. In Newton's first law, zero intuition led to misconceptions on the student. The mathematical presentation can be used to identify students who experienced zero intuition. Changes the form of mathematical representation led to cognitive conflict. Students who experience zero intuition will argue with intuition languages. Introducing mathematics, physics, and intuition language comprehensively is the one way to prevent zero intuition.

References

[1] Anderson J L 1990 Am. J. Phys. **58** 1192
[2] Hsu L 2001 The Physics Teacher **39** 206
[3] Viennot L 2012 Physics Education **47** 5
[4] Handhika J, Cari C, Soeparmi A, and SunarnoW 2016 *AIP Conf. Proc.* 1708070005
[5] Handhika J, Cari C, Soeparmi A, and Sunarno W 2015 *ICMSE 2015* PE 34 37
[6] Hestenes D, Wells M, and Swackhamer G 1992. *Phys. Teach.* 30 141
[7] Cari C, Suparmi A, Handhika J 2016 *ICOPIA 2016*
[8] Boucenna A 2007 https://arxiv.org/ftp/arxiv/papers/0707/0707.3579.pdf
[9] Lee C 2011 *Found Phys.* 41 1810
[10] Lenzen V F 1937 *Isis* 27 2 258
[11] Waldrip B, Prain V, and Sellings P 2013 *Instr Sci* 41 165
[12] Lee H S and Park J 2012 *International Journal of Science and Mathematics Education* 11 1391

IOP Conf. Series: Journal of Physics: Conf. Series **795** (2017) 012058 doi:10.1088/1742-6596/795/1/012058

Using word-pictorial presentation model to simplify understanding concept test of Newton's law

Jeffry Handhika[1], C. Cari[2], A. Suparmi[3], W. Sunarno[4]

[1]Graduate Student's, Sebelas Maret University, IKIP PGRI MADIUN, Jl. Setiabudi No. 85, INDONESIA,

[2,3,4]Physics Department of Post Graduate Program Sebelas Maret University, Jl. Ir. Sutami 36A Kentingan Jebres Surakarta 57126, INDONESIA

e-mail: jeffry.handhika@yahoo.com

Abstract. The information in the presentation of the test is a factor that affects student's response; it can be the correct or incorrect answer. Assume that students already know the information for an example speed. If the questions related to kinematics, then the chances of students answered correctly are greater than if it related to the dynamics. This potential is used principally to Simplify Student Understanding Concept Test of Newton's Law. FCI (Force Concept Inventory) numbers 13 and 14 used as a comparison of Word- Pictorial WPP model. WPP model applied to the students who take the course of mechanics (22 students). Obtained a positive correlation between WPP and FCI numbers 13 and 14, with value of r_{13} = 0, 72, r_{14} = 0, 88. It can conclude that the WPP model could be used as an alternative assessment to assess student understanding concept of Newton's Law

1. Introduction

Not all questions of FCI [1] can be understood by the student well, caused not all questions were answered correctly by the students. Question FCI contains WPP that presents information about the physical quantities of material content directly or indirectly. Assume the amount of speed already known by students, rationally; the opportunities students can answer questions related to kinematics greater than dynamics. It would be Easier for students to deciding the correct answer If the students already know the information, but if the presentation that requires the integration concept, it is difficult, for students to selecting the correct answer.

The information can be shaped as presentation words, verbal, images, graphics, and mathematical symbols. Furthermore, information will be filtered and included in the working memory [2]. Integration of it and prior knowledge raises student information which used to deciding the answer (Fig. 1). The process of integration of verbal information modeling, Picture, and prior knowledge enables intuition involvement in generating new information. If it is received is different from the initial knowledge possessed by students, it will bring cognitive conflict. The negative impact of cognitive conflict is the emergence of anxiety that caused the student has difficulty in deciding the answer or may not answer.

IOP Conf. Series: Journal of Physics: Conf. Series **795** (2017) 012058 doi:10.1088/1742-6596/795/1/012058

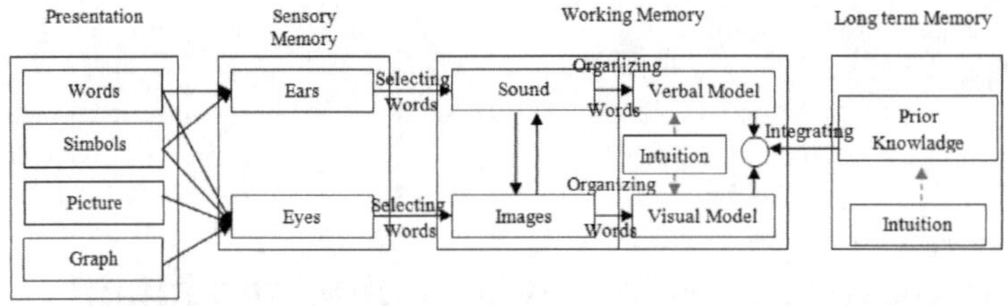

Figure 1. Cognitive Theory of Presentation, adapted from [2]

Prior knowledge, information model, and intuition are the three essential factors in the decision answers to the questions given. FCI is well used to evaluating of understanding force concept [3] and conceptual coherence [4] but has the disadvantage, the number of the questions (28) would require more time when implemented in learning. Development of an evaluation tool based on the presentation contained in the FCI is needed.

The use of the image has a stronger effect than the words. Memory stored in the form of images is more durable than the word [5]. Students will further develop the meaning behind the pictorial presentation more than words [6], and auditory stimuli are more accurately recalled than visual [7]. Presentation in the form of images and words has the advantages of each. On this test model, it displayed overall.

2. Methods

This research is a basic stage to develop the instrument more efficient than the FCI. The first step is analyzing the parts of a question on FCI. The parts that exist in the FCI profiled as a reference to develop a model of WPP. The final step is testing the model of WPP to 22 students who take a course in mechanics. FCI score numbers 13, 14 are used to validate WPP. Comparing FCI test with other tests ever conducted by other researchers [8, 9]. FCI numbers 13 and 14 are Newton's Third law which Newton first law as a distractor. Both problems have been due to the findings of previous studies [10], that students have the conception that Newton's first law is an another form of the Newton third law, so they concluded that Newton first law is equal to Newton third law.

3. Results and Discussion

3.1. First Step, analyze the parts of a question on FCI

FCI question number 13 (Fig.2) [1] relating to Newton's third law, the information presentation physical quantity "speed" in a "word" and "image" (interaction cars). To answer this question, students must understand the relationship between speeding up and force.

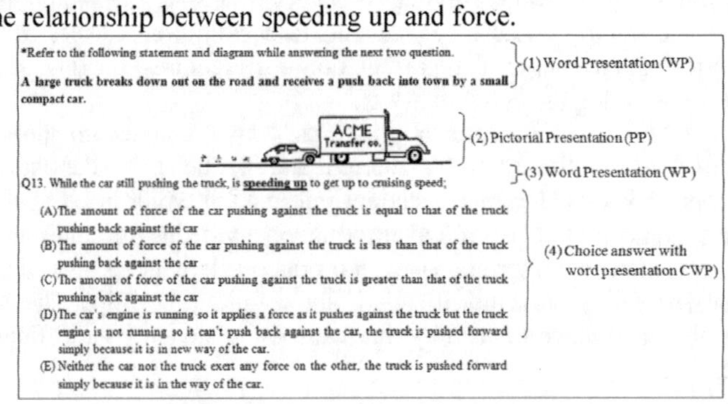

Figure 2. Question FCI Number 13

There are four parts in question FCI (Fig. 2). Part (1) is the information presented through WP that describes the command to complete the test and the events of the PP. Part (2) is a PP which describes the events which were described by WP. Presentation (1) and (2) are mutually reinforcing so that the students more easily understand the events. Part (3) a WP that provides physical quantity information "Speeding up" and "speed," were associated with kinematics, and then "pushing" was related to dynamics. PP shows an object interaction. Based on the information, the students have opportunities to conclude that this case is Newton's third law, but it would be different if the multi-presentation presented cause anxious, students would be unable to integrate all presentations. In FCI number 14, reinforced with a WP which presented physical quantities "constant speed" as a distractor (Fig. 3) [1].

Q. 14. After the person in the car, while pushing the truck, reaches the cruising speed at which he/she wishes to continue to travel at a constant speed; →WP as a distractor

(A) The amount of force of the car pushing against the truck is equal to that of the truck pushing back against the car
(B) The amount of force of the car pushing against the truck is less than that of the truck pushing back against the car
(C) The amount of force of the car pushing against the truck is greater than that of the truck pushing back against the car
(D) The car's engine is running so it applies a force as it pushes against the truck but the truck engine is not running so it can't push back against the car, the truck is pushed forward simply because it is in new way of the car.
(E) Neither the car nor the truck exert any force on the other, the truck is pushed forward simply because it is in the way of the car.

Figure 3. Question FCI Number 16 with WP as a distractor

WP as a distractor makes students confused in choosing an answer. For students who have misconceptions and believe that in the Newton's first law "no forces are acting on the object," then they will choose the answer (E) because it makes more sense for them. For students who have a conception that mass influences forces of action-reaction, then they will choose answer (B), while the answer (C) selected because they have the intuition to assume that high speed will produce a larger force. Students who not understand the concept of physics will choose Answer (D). For students who understand physics concepts well, the questions which submitted by the various presentations must have answered correctly [11]. For students who understand the concept partially will bring cognitive conflict and make it difficult for them to decide the answer [12]. Opportunities to get an incorrect answer, for the students who understand the concept partially [11].

3.2. Second Step, create WPP model which considering the parts of FCI questions

In the first step, it has explained that the FCI has four parts, (1) WP, (2) (PP), and (3) (CWP). In exceptional cases, WP may act as distractors. WP and PP presented simultaneously based on the research [4, 5, 6]. CWP is not used to develop this model because the student's conception should not be restricted. WPP model results as follows:

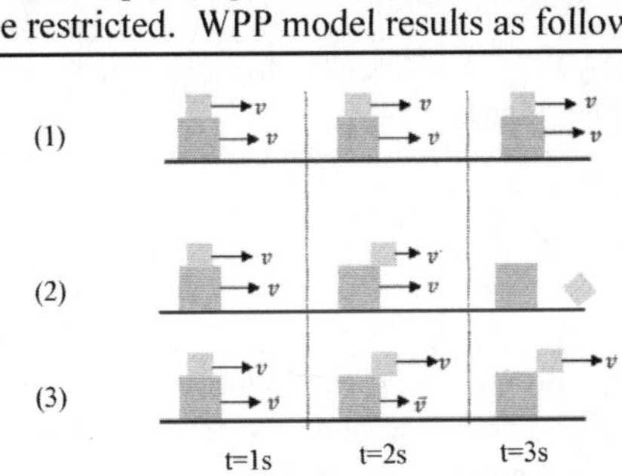

Figure a.

* Small and large box move with the speed and time, according to the conditions (1), (2), and (3), Which v is speed of box, arrows indicate the direction and magnitude of speed (Fig. a). Base on this condition, determine which condition represent Newton's first law? Give your arguments!

Figure 4. WPP Model

3.3. Testing the model of WPP

Model of WPP tested to 22 students who take a course in mechanics. The result of the test (Table. 1a, 1b) shows that the student's partially understanding about Newton's law.

Table 1a. Students answer about Newton's Law, before Discussion

	Choice case						
	1	2	3	1 & 2	1 & 3	2 & 3	1, 2, & 3
Answer (%)	0.09	0.18	0.09	0.18	0.14	0.32	0

Table 1b. Students answer about Newton's Law, after Discussion with peers

	Choice case						
	1	2	3	1 & 2	1 & 3	2 & 3	1, 2, & 3
Answer (%)	0.14	0.09	0.14	0.23	0.41	0.14	0

WPP positively correlated to the FCI test (Table 1). FCI score numbers 13, 14 are used to validate WPP. This numbers will be answered correctly by the students who understand Newton's first and third law well. Problem WPP requires the ability to understanding of Newton's first and third law. FCI Question number 13 gives the correlation value smaller (0.72) than question number 14 (0.88). The information is presented in number 13 more detail, not caused student's anxiety so that they can get the correct answer easily. In the model of WPP, understanding the concept of the students can be identified more detail, because students are required to provide arguments of their options. The argument that has been presented by the students can be used to identify students' understanding of the concept (Figure 5a, 5b).

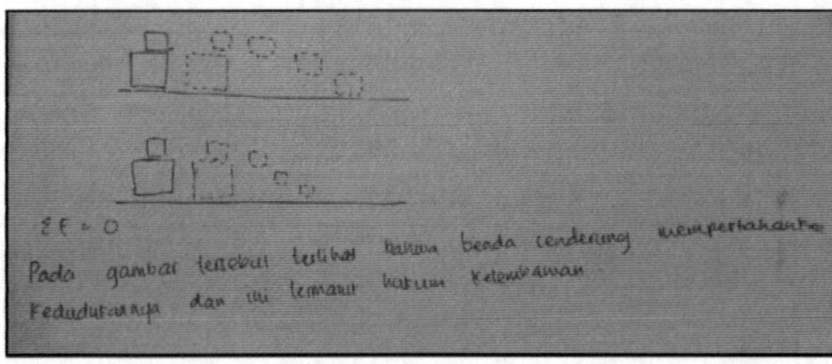

In this figure shows that the object tends to maintain its position and this includes the law of inertia.

Figure 5a Example 1. The argument by student

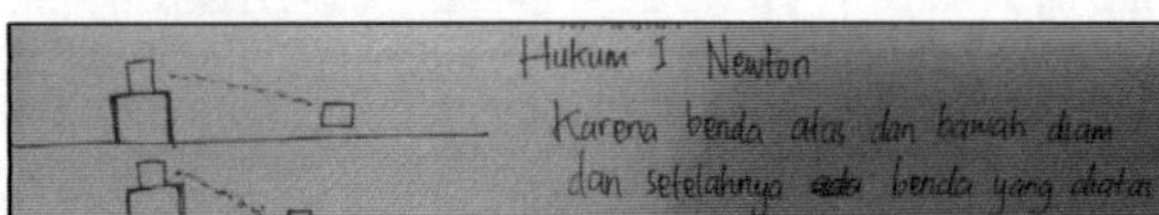

Because the objects that are in the position of the top and the bottom at rest, then the object positioned at the top is moving and then stopped.

Figure 5b Example 2. The argument by student

From the arguments had presented, many students still choose case 2 and 3 (0.32%) (Table 1a). Case 2 and 3 chosen because according to the textbooks information and examples are given by the teacher while they were in high school level. Many teachers cited Newton's first law as "a car braking process" so that many students conclude that Newton's first law on the choice of b and c. The students assume inertia as an object that maintains its position (Fig. 5.a) and the objects at rest, then moving and stopped (Fig 5.b). After being given the time to discuss with peers, with problems to distinguish between case 2 and 3, caused student changed the answer choices, selection of case 1 and three is more dominant (Table 1b). Arguments students (Fig. 6) provide information that the discussion helps students to find the correct concept.

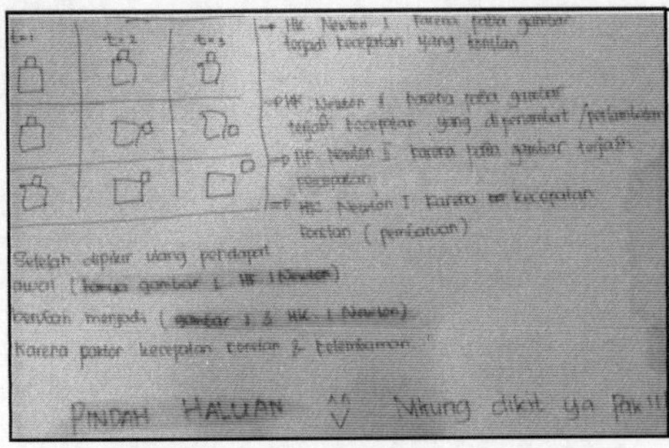

After the discussion, which was initially only chosen case (1) as Newton's first law, transformed into (1) and (3), because they get information about the constant speed and definition of inertia. Case (3), initially answered Newton's second law, and then converted into Newton's first law when viewed from the small box.

Figure 6. Example Answer Student after discussions with peers

Still found that the students who did not select an answer, and convinced the case (2) as Newton's first law. Not easy to change the conception of students who already stored in its memory. Car braking events which visualized in the textbooks are information that goes through the dual channel stored in long term memory. Cognitive conflict through discussions has been conducted to examine and clarify the prior knowledge of the students. They can change the answer choices. In the case of (1) rarely exemplified in the learning process in high school, and not found in everyday life, so it can be difficult for students to understand Newton's first law. Case (1) are found in textbooks and presented in textual presentation, and delivered verbally by most teachers, but not exemplified the event in detail. These events can only exemplify in the laboratory scale. Also found students who answered the case (3), arguing that the case (3) be the third Newton's laws. Action and reaction can occur in a small box and a large box, but the arguments have presented by the student is not appropriate.

4. . Conclusions

WPP can use as alternative tools to measure student understanding concept of the Newton's law. The arguments submitted by students is one of the strengths of this model. Found information that is giving the example of Newton's laws for "a car braking event" is represented differently by the student. Students represent the case as Newton's first law without considering the condition and which objects observed. Information delivered from textbooks, and teachers or lecturers are not examined more deeply by the students. Recommended for teachers and lecturers to submit an example of Newton's laws in detail and clearly and maximize the learning activities at the laboratory to describe the ideal conditions of physics.

Acknowledgement

This research supported by Mandatory grant UNS with contract number 632/UN27.21/LT/2016 and IKIP PGRI Madiun

References

[1] Hestenes D, Wells M, and Swackhamer G 1992 Phys. Teach. **30** 141
[2] Mayer R E 2009 *Multimedia Learning, Second Edition* (New York: Cambridge University Press)
[3] Huffman D and Heller P 1995 Phys. Teach **33** 138
[4] Savinainen A and Viiri J 2008 International Journal of Science and Mathematics Education **6** 719
[5] Tulving E and Craik F I M 2000 The *Oxford Handbook of Memory* (New York: University Press)
[6] Craik F I M and Lockhart R S 1972 Journal of Verbal Learning and Verbal Behavior **11** 671
[7] Leo D C 2014 Journal of Undergraduate Psychological Research **9** 18

IOP Conf. Series: Journal of Physics: Conf. Series **795** (2017) 012058 doi:10.1088/1742-6596/795/1/012058

[8] Thornton R K, Kuhl D, Cummings K, and Marx J 2009 Physical Review Special Topics - Physics Education Research **5** 010105
[9] Yasuda J, Taniguchi M 2013 Physical Review Special Topics - Physics Education Research **9** 010113
[10] Jeffry Handhika, C. Cari, A. Soeparmi and Widha Sunarno 2016 AIP Conf. Proc. 1708 070005
[11] Handhika J, C Cari, Soeparmi A, and Sunarno W 2015 ICMSE 2015 PE 34 37
[12] Cari C, Suparmi A, Handhika J 2016 ICOPIA

Student representation of magnetic field concepts in learning by guided inquiry

Siska Desy Fatmaryanti[1,2], Suparmi[1], Sarwanto[1] and Ashadi[1]

[1] Doctorate Program on Science Education, School of Postgraduate, Sebelas Maret University, Jl. Ir. Sutami 36A Kentingan Jebres Surakarta 57126, INDONESIA

[2] Department of Physics Education, Faculty of Teacher Training and Education, Muhammadiyah University of Purworejo, Jl. KH. Ahmad Dahlan No 3 Purworejo 54211, INDONESIA

E-mail: sd_fatmaryanti@yahoo.com

Abstract. The purpose of this study was to determine the change of student's representation after the intervention of learning by guided inquiry. The population in this research were all students who took a fundamental physics course, consisted of 28 students academic year 2016, Department of Physics Education, Faculty of Teacher Training and Education, University of Muhammadiyah Purworejo. This study employed a quasi-experimental design with group pre-test and post-test. The result of the research showed that the average of students representation of magnetic field before implementation of guided inquiry was 28,6 % and after implementation was 71,4%. It means that the student's ability of multi-representation increase. Moreover, the number of students who is able to write and draw based on experiment data increased from 10,7% to 21,4 %. It was also showed that the number of student with no answer decreased from 28,5% to 10,7%.

1. Introduction

The teaching of fundamental physics is usually presented only in abstract concepts. Students often tend to memorize physics formula oriented problem-solving strategies [1,2]. As a result, many students do not understand the whole concepts of introductory physics. Physics concepts should be part of learning a physics process [3]. The teaching of an abstract concept in learning physics is presented the physics concepts in the form of mathematics. As a result, students do not understand physics concept. They could not represent a concept with multi-representations [4]. To improve student's understanding of physics concept, teacher should apply learning strategies which engage students in active learning using inquiry.

The concept of a magnetic field is a very important in a physics. This concept is introduced to the students from elementary school until the senior high school and this concept is observed in daily life. In general, the students feel that the concepts of magnetism are as a difficult and abstract concepts [5,6]. One of the statements in Ausubel's theory is that the most important factor that influence learning is what has been known to students [9]. In this research, the real natural phenomena were observed directly by the student, and then they were actively involved in explaining observation facts/data by drawing, formulating mathematics equation, drawing of a pattern of the magnetic field and formulating the definition of concept that relate some variable.

There are many reasons to believe that multiple representations of concepts are corresponding with the ability to construct, interpret and transform between different representations that correspond to the same physical system or process play a positive role in learning physics. First, the experts often use multiple representations as a first step in a problem-solving process [7,4]. Second, students who are taught explicit by using problem-solving strategies emphasizing the use of different representations of knowledge at various stages of problem-solving, construct higher quality and complete representations and perform better than students who learn traditional problem-solving strategies [8]. Third, multiple representations are very useful in translating the initial, mostly a verbal description of a problem into a representation more suitable for mathematical manipulation [8] because the process of constructing a representation of a problem makes it easier to generate appropriate decisions about the solution process. Also, getting students to represent it in different ways helps them to shift their focus from merely manipulating equations toward understanding physics. Some researchers have argued that to understand a concept thoroughly, one needs to be able to recognize and manipulate it in a variety of representations [4].

One of learning model that can maximize the process of learning is the inquiry learning model [9]. Inquiry learning model can be defined as the intentional process of diagnosing problems [10]. In Inquiry learning model, knowledge and classroom experience must be combined with subject matter knowledge in ways that allow students to use scientific reasoning and critical thinking to develop their understanding of science.

Many educators discussed the nature of the inquiry by making use of mostly two concepts such as open inquiry and guided inquiry (Hassard, in [11]). The open inquiry was reported that this learning model did not have a significant effect on improving students' academic achievements and developing their scientific process skills [11, 12]. This is because the different interpretations between a concept by researchers restrict reform works about the scientific inquiry with a concept that is understood by the teachers [13]. According to Furtak [14], teaching with scientific method are the boundaries of the traditional method, where certain answers by the teachers are transferred to the students and the open inquiry model, the students tried to construct their own problems and discuss problem solutions. In guided inquiry model, teachers and learners play a crucial role in asking questions, developing answers and structuring of materials and cases. The usage of guided inquiry method is very important in the transition from lecturing method to other teaching methods which are less and more clearly structured for alternative solutions.

Learning activities in guided inquiry model help students to develop their individual responsibility, cognitive levels, report making, problem-solving and understanding skills. According to National Research Council [15], guided inquiry can facilitate focusing on learning and the developing certain scientific concepts. In this Learning, students still in the teachers' guidance to focus their attention on the content, because sometimes they have less suitable means for discovering scientific thinking processes [12].

The purpose of this study was to determine the student's representation after the intervention using a learning by guided inquiry and the effectiveness of a learning by guided inquiry in building student's representation. Intervention from this learning model done with involving the magnetic field phenomena on the mathematics, illustration (pattern of the magnetic field and magnetic force), and verbal representations, so hopefully will be able to bring up representations of the student with the category of "good representations".

2. Method

2.1. Research Design

In this study used the quasi-experimental design. Quasi-experiment is a research design involving an experimental approach but where random assignment to treatment and comparison groups has not been used. The design in this stage is group pre-test and post-test design

2.2. Participants

The population in this research consists of all students in the 2016 year of fundamental physics course, Department of Physics Education, Faculty of Teacher Training and Education, University of Muhammadiyah Purworejo. The number of students from whom the samples were taken is 28 students.

2.3. Procedure

The students' representation is confirmed through the procurement of essay test questions encourage them to think of magnetic field phenomena. The representation test consists of 4 questions for the definition of the magnetic field and 6 questions for the direction of the magnetic field. The questions are selected to encourage representation processing among students by transforming visual representation into verbal representation, picture and mathematical representation. The intervention in this study is done by using steps of inquiry learning.

3. Results and Discussion

The Questions regarding students' representation of the magnitude and direction of the magnetic field required the students to use their representation in transforming magnetic field phenomena to verbal and from verbal to mathematical form (calculations). Furthermore, the students are also asked to describe it as well as to perform the transformation from a verbal statement into a picture by drawing the direction.

In Figures 1 and 2 stated an analysis of the student's answer to two questions in the diagram apparent.

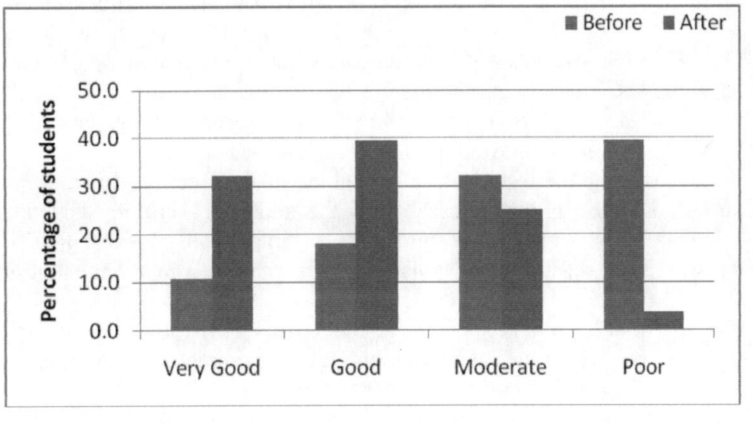

Figure 1. Graph of Student's representation on definition of the magnetic field

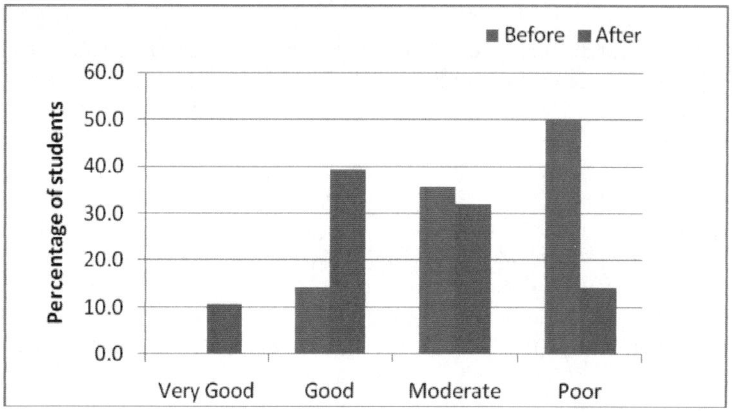

Figure 2. Graph of Student's representation on direction of the magnetic field

Based on the analysis of student's representations of the charts above, it seems like the students' representation of the magnetic field with various questions from interpretation down to verbal-to-mathematics and verbal to- picture transformations and vice versa indicate significant differences before and after the implementation of guided inquiry.

In Figure 1 and Figure 2, students who were initially unable to write the equation of magnetic field magnitude by directly translating picture from the experiment. For question 1(figure 1), the percentage of students who are able to accurately reply, interpret and transform the representation is 25% (very good), 35,7% (good) and 39,3% (moderate), whereas the remaining percentage of students still have low capabilities to interpret and transform the representation. For question 2 (Figure 2), the percentage of students with very good, good, and moderate representations are 10,7%, 39,3%, and 32,1% respectively, and the remaining students have bad representation.

Difficulties in transforming these students are the ability to interpret the direction of magnetic field in a picture; it does not appear in picture created by the students. Therefore, students need guidance and training in making representations imagination to build good understanding. Based on these results, students still have difficulty in transforming verbal to visual, although the study was conducted.

The results can also be seen from the percentage of students' ability to create images magnitude and direction of magnetic field, and write magnetic field equations directly from the experiment and the balanced equation. In Table 1 below contained more results.

Table 1. Percentage of students who can answer questions before and after the learning

No	Representation	Percentage of students (%)			
		Question 1		Question 2	
		Before	After	Before	After
1	Writing a magnitude of magnetic field equation based on fact of experiment	14,3	25	7,1	14,3
2	Drawing a direction of magnetic field based on fact of experiment	25	35,7	21,4	35,7
3	Incorrectly writing /drawing	35,7	39,3	39,3	39,3
4	No Answer	25	0	32,1	10,7

Table 1 shows that the students have been able to draw a direction of the magnetic field based on fact of the experiment, as many as 35,7% (question 1 and question 2) and these results better than before the application of guided inquiry learning, although not maximized. Not the maximal percentage of student who is can draw because students do not realize that a direction of the magnetic field or magnetic force was always perpendicular. In this case, the student just writes the equation with a direct view without settling into vector roles. These are some of the student's answer.

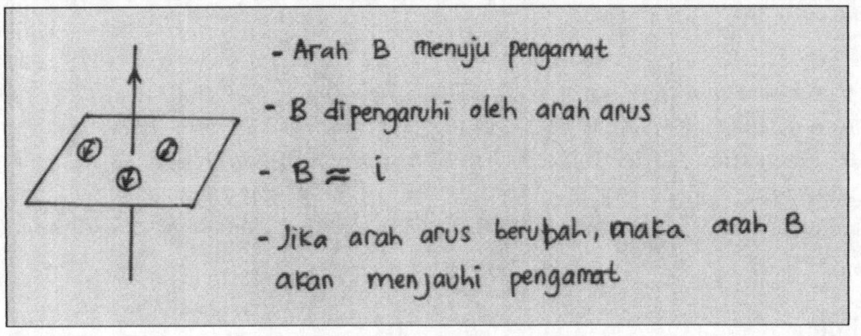

Figure 3. Sample of student's answer about direction of magnetic field with moderate category

In Figure 3 shows that the students' answers to the category of "moderate" only involve verbal representations and drawings. on the representation of the image, they have difficulty interpreting the pattern and direction of the magnetic field with the direction of the compass needle deviation. While on their verbal representation can only be concluded that the magnitude and direction of the magnetic field are only influenced by the strong electric current. in this category of students have not reached the make decisions with mathematical representations and graphics.

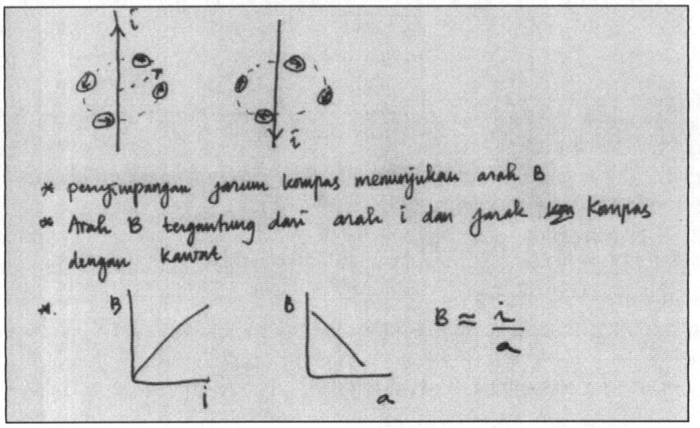

Figure 4. Sample of student's answer about direction of magnetic field with good category

In Figure 4 shows that the students' answers to the category of "very good" involving verbal representations, drawings, graphs and mathematical. the representation of their images was able to create an image with its own representation based on observations on deviation compass needle. While on their verbal representations can be concluded that the magnitude and direction of the magnetic field are not only influenced by the strong electric current, but also within the compass of the wire. Although simple, but students have also been able to represent their observations into a mathematical relationship and chart form between B, i and a.

4. Conclusion

Learning physics based on inquiry is performed with the purpose of increasing the students' analogical thinking capacity on the magnetic field by establishing a student's representation. The appearance of a student's representation reflected about his/her ability to interpret all physics phenomenon, as seen from the students' answers in verbal, mathematical or symbolic forms and picture representation. The result of this research shows that the students' representation has been well formed. In this case, before the implementation using inquiry, the students' representation was generally in the category of "poor," but after the implementation, it improved to "moderate" and "good". The establishment of students representation shows that there has been an improvement in their ability to understand the concept, as well the ability interpret and transform from the fact of an experiment to multi-representations [6,4].

This research finding shows that the learning by inquiry can be used to train the students in transforming representation. In learning the magnetic concept, the student is not only able to learn using mathematics, but also understand the direction of magnetic field and magnetic force through picture representation. Physics learning which only focuses on mathematics will result in a shallow understanding [6,16].

5. Acknowledgment

This research was supported by the Muhammadiyah University of Purworejo for facilities. The researcher thank to all institution that has been supporting this research and to the reviewers.

References

[1] Heller J and Reif F 1984 Prescribing effective human problem-solving processes: problem description in physics *Cogn. Instruct* **1(2)** 177-216.

[2] Henderson E C, Yerushalmi, Kuo V H, Heller P and Heller K 2004 Grading student problem solutions: The challenge of sending a consistent message *Am. J. Phys.* **72** 164-169

[3] Chabalengula, Mumba and Mbewe 2011 How Preservice Teachers Understand and Perform Science Process Skills *Eurasia Journal of Mathematics, Science & Technology Education* **8(3)** 167-176

[4] Maries A 2011 *Role of Multiple Representations in Physics Problem Solving* Dissertation (University of Pittsburgh)

[5] Saarelaeen M 2011 *Teaching and Learning of Electric and Magnetic Fields at University level* Dissertation in Forestry and Natural Science (Finland: University of Eastern)

[6] Fatmaryanti, S.D, Suparmi, Sarwanto, and Ashadi 2015 *Analysis of Magnetism Problems based on concept required and science generic skills* International conference on science and science education UKSW

[7] Zhang J and Norman D 1994 Representations in Distributed Cognitive Tasks *Cog. Sci.* **18(1)** 87-122

[8] Heuvelen V A 1991 Learning to think like a physicist: A review of research-based instructional strategies. *Am. J. Phys*. **59(10)** 891-897

[9] Arends R I 2013 *Learning to Teach* (Jakarta: Salemba Humanika)

[10] Linn M C, Davis E A, and Bell P 2004 Inquiry and technology. In M C Linn, E A Davis, and P. Bell (Eds.), *Internet Environments for Science Education (pp. 3-28).* Mahwah, NJ: Lawrence Erlbaum Associates.

[11] Berg C A R, Bergendahl C V, Lundberg K S and Tibell L A E 2003 Benefiting from an open-ended experiment? A comparison of attitudes to, and outcomes of, an expository versus an open-inquiry version of the same experiment. *International Journal of Science Education,* **25**: *1-22.*

[12] Kai H W and Krajcik J S 2006 Inscriptional practices in two inquiry-based classrooms: A case study of seventh graders' use of data tables and graphs *Journal of Research in Science Teaching,* **43** (1) 63-95.

[13] Wallace C S and Kang N H 2004 An investigation of experienced secondary science teachers' beliefs about inquiry: An examination of competing for belief sets *Journal of Research in Science Teaching* **41(9)** 905-935
[14] Furtak M E 2006 The problem with answers: An exploration of guided scientific inquiry teaching. *Science Education* **90(3)** 453-467.
[15] National Research Council 2001 *Inquiry and the National Science Education Standards* (Washington, DC: National Academy Press)
[16] Albe V, Venturini P, and Lascours J 2001 Electromagnetic Concepts in Mathematical Representation of Physics *Journal of Science Education and Technology* **10** 197 -203
[17] Saglam M and Millar R 2006 Upper High school Students Understanding of electromagnetism. *International Journal of Science Education* **28 (5)** 543-566

Preliminary study of acceleration based sensor to record nile tilapia (*Oreochromis niloticus*) feeding behavior at water surface

Aji Subakti, Zarah F. Khotimah and Fajar M. Darozat
Department of Innovation PT. Multidaya Teknologi Nusantara
Jl. Cikutra Baru V No.18 Bandung 40124, INDONESIA

E-mail: fajar@efishery.com

Abstract. In this preliminary study, the acceleration based sensor was developed to monitor the activity of Nile tilapia (*Oreochromis niloticus*) feeding behavior at the water surface. This study was conducted for three weeks in a fish pond with 40 m^2 in size, stocked with 850 fingerlings of Nile tilapia strain Nirwana-2 (average biomass of 13 g, fed four times a day at 8 am, 12 pm, 4 pm, and 8 pm). The acceleration sensor system was installed floating in the pond and was designed in a way so that the xz plane of the sensor will be parallel with water surface, while the y-axis will be pointing downward. By sensing the acceleration caused by the surface wave, the activities of fish near surface water could be monitored. The result showed that there were three distinctive patterns could be observed which was related to the feeding activity of fish. Generally, it can be concluded that this acceleration based sensor system can be integrated with automatic feeder machine, in particular by analyzing the recorded patter, it is possible to monitor when the fish stop eating, and so the right amount of feed could be given to the fish.

1. Introduction

Aquaculture is known as the fastest-growing food production system with the average level of annual growth between 1970 and 2008 at 8.3%, showed impressive growth as the production of capture fisheries has been static since the late 1980s. Aquaculture has expanded, intensified and diversified to bridge the gap between global supply and demand of food as world human population keeps increasing. As it was projected by United Nations (2013) that human population will reach 9.6 billion by 2050 and 10.9 billion by 2100, the increase in aquaculture production is required [1]. Feeding management is one of the important aspects of aquaculture business as feeding cost contributes to 30-70% of total operating expenses [2, 3]. Feeding management efficiency regarding feeding rate, frequency and timing need to be adjusted to the required level and follows the feeding behavior of reared fish to avoid both over- and underfeeding. Overfeeding, in particular, can give a detrimental effect to the growth of fish due to the overload of stomach and intestines which further leads the decrease of digestive efficiency and feed utilization [4, 5]. Overfeeding can even cause deterioration in water quality and further affect the health of fish by weakening the immune system and increasing susceptibility to infection [6, 7]. Thus, feeding management efficiency prevents these problems from happening.

There are two strategies to increase feeding management efficiency, first by hiring experienced employee (which is harder to find) and second by implementing automatic fish feeder technology. The later can be conducted by integrating it with a sensor system to detect and monitor fish behavior during feeding and to determine the amount of feed required in reaching fish satiation. Even though manual feeding is the most used feeding method in aquaculture, the intensification which usually includes a more complex system, a larger production unit and a greater quantity of supplemental feed can be optimized by the use of automatic feeder. In addition to that, automatic feeder supports more frequent feeding as well as night feeding [7].

Nile Tilapia (*Oreochromis niloticus*) has been known as one of commercially important fish species for aquaculture, ranked as the second most important fish species after the carp due to their high growth rates, tolerance to environmental stress, high market demand, etc. [8, 9]. Several factors affect feed requirement of Nile tilapia, including biological factors (size, age, condition, and reproductive state of the fish), environmental factors (temperature, dissolved oxygen, water quality, and photoperiod) as well as feeding management (in term of the amount and frequency of feeding). The important point needs to be understood related to feeding management in Nile tilapia is its natural feeding habit. Tilapia is daytime and surface feeder which means that they eat at the water surface and commonly known eat more during the day [10].

Accelerometer has been used to monitor very small movement such as microtremors [12] and landslide movement direction [13]. The same device could be used to measure wave properties (amplitude, wavelength) at water surface generated by fish movements.

2. Methods

This study was conducted in three weeks, in which 850 animals of mixed-sex Nile Tilapia fingerlings strain Nirwana-2 (average biomass of 13 g) were stocked in one 40 m^2 earthen ponds located in Bandung, West Java province. Fish were fed commercial pellets with 30-33% crude protein content four times a day at 8 am, 12 pm, 4 pm, and 8 pm. Feed was applied at 6% of estimated biomass during the study. The pond continuously flowed with freshwater from the river with an average flow rate around 10 liters per minute.

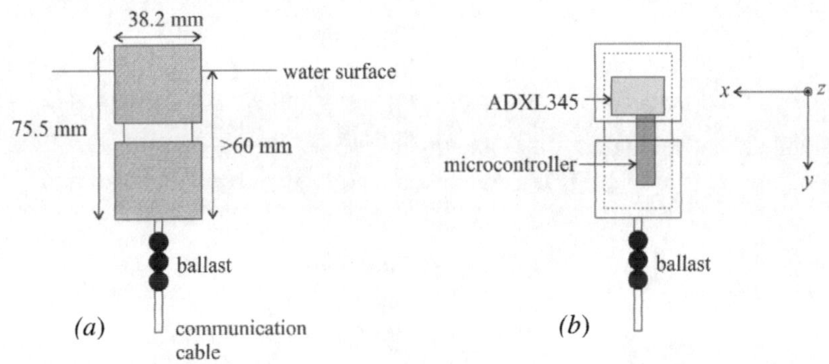

Figure 1 *(a)* The illustration of sensor tube when floating at the water surface. More than 80% of the tube submerged underwater. *(b)* The illustration of ADXL345 and microcontroller position inside the tube. The direction of each axis is also shown in the picture.

Acceleration sensor ADXL345 was used to record pond ripple. The sensor was placed inside a PVC tube with inner diameter of 28.7 mm, outer diameter of 38.2 mm, and height of 75.5 mm. Some weights were added to the tube to make sure it floated vertically at the water surface with more than 80% of its body submerged as illustrated in Figure 1. At this position, X and Z axis of the sensor would be parallel with water surface, while Y-axis would be pointing downward. Sensor configured at sensitivity of $3.77\times10^{-3}\,g$ with a maximum measurement range of $\pm2g$ [11]. Sensor captured data for each axis simultaneously every 100 ms, followed by sending them to the receiver which was placed outside the pond through a communication cable. The communication cable also acted as an anchor so the sensor would not drift to the edge of the pond by wind or water current.

3. Results and Discussion

The results showed that there was no fish mortality during the study period. According to the study, this variety of fish showed well adaptation with the rearing environment. After arrival, several critical points checked were fish movement, oxygen, and water condition in the bag, feeding respond and if there is any abnormality on the body. Water quality parameters were checked prior stocking and were in the average optimum value for grow-out tilapia: 7.0-11.1 mg/L, 7.65-8.96, 25.9-27.7°C and 0 mg/L for dissolved oxygen (DO), pH, temperature, and ammonia respectively. Good water quality parameter supported good feeding respond and intake, thus would be advantageous for gathering data from the sensor.

There were three distinctive patterns of acceleration recorded by the sensor as a function of time. The first pattern was an idle pattern as shown in Figure 2. This pattern occurred because almost no ripple presented at the water surface. This pattern can be interpreted as no significant activity performed by the fish at the surface. The X and Z acceleration recorded by the sensor was not zero because the real sensor axis was slightly tilted from expected axis due to soldering process and center of mass position which was not exactly at the center of the tube. The slow change of acceleration at the X and Z axis can be seen in Figure 2*(a)*, which was caused by the wind. In the Figure 2*(b)*, it is shown there was some little yet fast change acceleration at X and Z axis which was caused by fish activity below the surface, especially the fish activities near the communication cable. Meanwhile, at the Y-axis, the acceleration recorded by the sensor relatively constant at around 1055 mg for both graphs, which is equal to the addition of 1000 mg caused by gravity and 55 mg caused by sensor offset and noises.

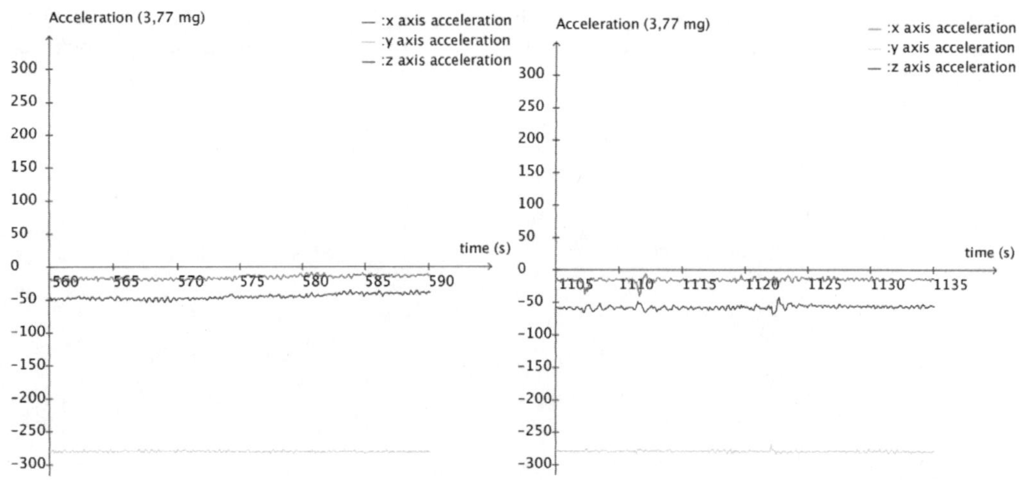

(a) *(b)*

Figure 2 *(a)* Idle pattern recorded by the sensor at 6 September 2016 from 2:09:20 to 02:09:50 local time zone. *(b)* Idle pattern recorded by the sensor at 6 September 2016 from 2:09:20 to 02:09:50 local time zone.

The second pattern was spike pattern as shown in Figure 3, which occurred when a very short physical impulse given to the sensor. The source of this impulse was mostly from the fish. This pattern was occasionally shown up. Almost all spike captured by the X and Z axis, which means most of the time the physical impulse from fish given to the sensor horizontally. The Y-axis of the sensor is relatively immune from fish disturbances. Sometimes spike pattern also occurred at night, which is mean that fish activities not completely stopped even after the sun set.

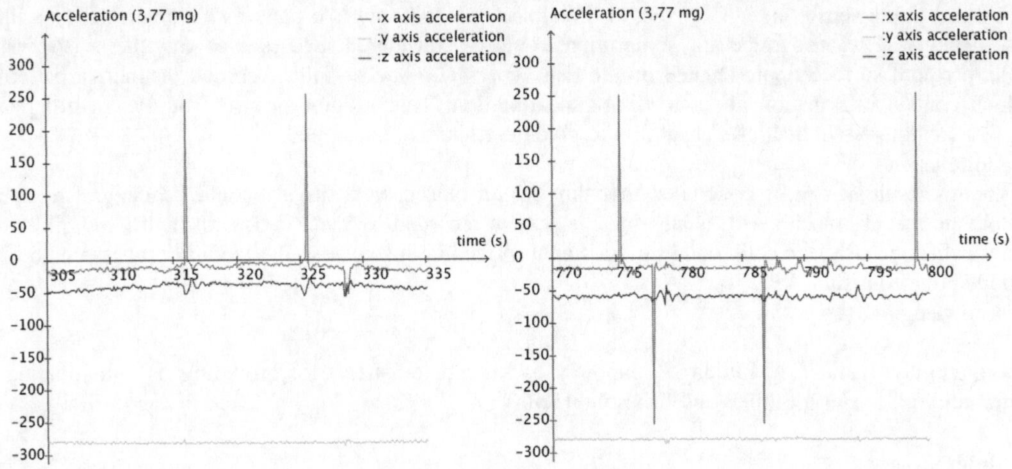

(*a*) (*b*)

Figure 3 *(a)* Two spike patterns recorded by the sensor at 6 September 2016 from 2:05:05 to 02:05:35 local time zone. Each spike caused by a physical impulse that directed to the +X axis and slightly to the –Y axis. *(b)* Four-spike patterns recorded by the sensor at 6 September 2016 from 12:12:50 to 12:13:20 local time zone.

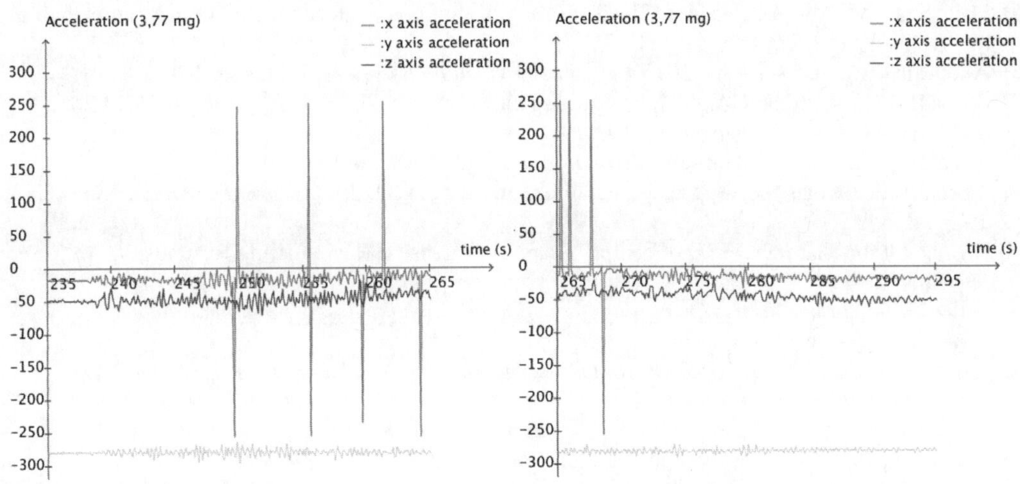

(*a*) (*b*)

Figure 4 Feeding pattern recorded by the sensor at 6 September 2016 from 08:03:55 to 08:04:55 local time zone.

The third pattern was feeding pattern as shown in Figure 4, which only showed up when feed is given to the fish. The acceleration recorded by each axis changed rapidly. The change of value was significant enough to be differentiated from noises. This rapid change of acceleration was caused by surface ripple generated by fish movement in their attempt to snatch the floating feed. Over the time, the value of acceleration change became smaller and smaller, as fewer and fewer fish went to the surface. By looking at the Y-axis data, the pattern started around 08:03:58 and ended around 08:04:45. It seemed to be a short duration, but it was normal since the fish only three weeks old. Several spike patterns were also recorded at Z-axis as the fish unintentionally collided with the sensor.

Although the sensor could only record three patterns, and only two patterns have correlation with fish behavior at water surface, this experiment could be considered success. Not only the sensor was able to record surface ripple created by the fish, it was also successfully recorded distinctive pattern which contain information about feeding time duration. This is important information in order to design control system that have capability to give the right amount of feed.

4. Conclusion

With this result, it can be concluded that this design of sensor positively could be integrated with automatic feeding machine. By analyzing the pattern recorded by the Y-axis sensor, it is possible to know when the fish stop eating and send a signal to feeding machine so the feeding process could be stopped immediately.

Acknowledgments

We gratefully thank PT. Multidaya Teknologi Nusantara (eFishery) for providing us with funding, automatic fish feeder machines and fish pond facilities.

5. References

[1] FAO 2016 *The State of World Fisheries and Aquaculture* Rome 200 pp
[2] Nekoubin H and Sudagar M 2012 *World Appl. Sci. J.,* 17 (8): 1001-1004
[3] Rana KJ, Siriwardena S and Hasan MR 2009 *FAO Fisheries and Aquaculture Technical Paper* No. 521 63p
[4] Ajani F, Dawodu MO and Bello OA 2011 *Afr. J. Agric. Res* 6 (2): 318-322
[5] Daudpota AM, Abbas G, Kalhoro IB, Shah SSA, Kalhoro H, Hafeeez-ur-Rehman M and Ghaffar A 2016 Pakistan J. Zool 48 (1): 171-177
[6] Abou-Zied RM and Ali AAA 2015 *Egyptian J. Nutrition and Feeds* 18 (2): 451-459
[7] Sousa RMR, Agostinho CA, Oliveira FA, Argentim D, Novelli PK and Agostinho SMM 2012 *Aqr. Bras. Med. Vet. Zootec* 64 (1): 192-197
[8] Ronald N, Gladys B and Gasper E 2014 *J. Aquac Res Development* 5 (2): 222
[9] Yakubu AF, Nwogu NA, Apochi JO, Olaji ED and Adams TE 2014 *Journal of Aquatic Science* 2(1): 1-4
[10] Riche M and Garling D 2003 Feeding Tilapia in Intensive Recirculating Systems *NCRAC Fact Sheet Series #114*
[11] ADXL345 datasheet http://www.analog.com/media/en/technical-documentation/data-sheets/ADXL345.pdf
[12] Dai, K. and Li, X. 2014 *Surface Wave Survey Method and Accelerometer-Based Field Testing Cases. Innovative and Sustainable Use of Geomaterials and Geosystems.* Geo-Hubei 2014: pp. 79-86.
[13] Lemke, J., Hagy, M., and Brittsan, D. 2008 *Continuous Real-Time Monitoring of an Active Landslide Using a High-Density Accelerometer Network.* GeoCongress 2008: pp. 500-508.

IOP Conf. Series: Journal of Physics: Conf. Series **795** (2017) 012061 doi:10.1088/1742-6596/795/1/012061

Processing of micro-nano bacterial cellulose with hydrolysis method as a reinforcing bioplastic

Maryam Maryam[1], Rahmad Dedy[2] and Yunizurwan Yunizurwan[1]

[1]Agro-Industrial Engineering, Polytechnic ATI Padang

[2]Chemical Engineering of Agricultural Product, Polytechnic ATI Padang
Jl. Bungo Pasang, Tabing, Padang 25171, West Sumatra, INDONESIA

e-mail: maryam.atip@gmail.com

Abstract. Nanotechnology is the ability to create and manipulate atoms and molecules on the smallest of scales. Their size allows them to exhibit novel and significantly improved physical, chemical, biological properties, phenomena, and processes because of their size. The purpose of this research is obtaining micro-nano bacterial cellulose as reinforcing bioplastics. Bacterial cellulose (BC) was made from coconut water for two weeks. BC was dried and grinded. Bacterial cellulose was given purification process with NaOH 5% for 6 hours. Making the micro-nano bacterial cellulose with hydrolysis method. Hydrolysis process with hydrochloric acid (HCl) at the conditions 3,5M, 55°C, 6 hours. Drying process used spray dryer. The hydrolysis process was obtained bacterial cellulose with ±7 μm. The addition 2% micro-nano bacterial cellulose as reinforcing in bioplastics composite can improve the physical characteristics.

1. INTRODUCTION

Nanotechnology is the ability to create and manipulate atoms and molecules on the smallest of scales. Their size allows them to exhibit novel and significantly improved physical, chemical, biological properties, phenomena, and processes because of their size. Nanotechnology can be defined as everything related materials, systems and processes that are operated on a scale of 1-100 nanometers (nm). Nanotechnology offers great opportunities for the development of innovative products and applicable to various fields such as packaging, biomedical, electronics, optics and agriculture. The physical characteristics and new phenomena would arise if materials on the nanoscale due to changes in the functional properties of the material, related to the changing nature of the dispersion.

Coconut water as waste has great potential as a raw material of bacterial cellulose. Bacterial cellulose as a result fermentation of *Acetobacter xylinum* bacterial in coconut water medium. Bacterial cellulose has several advantages as a source of cellulose compared to wood or non-wood resources, among others, it can be harvested in a shorter time, easily cultivated, and reduce the exploitation of forests to meet the needs of cellulose for the industry. Furthermore, not like other cellulose, bacterial cellulose has no lignin, pectin, and hemicelluloses which are impurities in cellulose which mean fewer steps in production. Bacterial cellulose has a high content of cellulose fibers and has a high mechanical characteristic. The physical characteristics of bacterial cellulose are purity, crystallinity, high mechanical strength, and low-cost polymers. The study of microbial cellulose for various areas of application urgently needed to increase the added value the products of bacterial cellulose and is not limited to use as a food product.

Nanotechnology, every material can be lighter, more stable and increase in functionality. Nanotechnology offers many advantages and opportunities in the field of product innovation and development of bioplastic. Moreover, the strength, flexibility, and stability are considered the most important characteristic on producing bioplastic. Using filler which may increase those properties is an alternative solution for that problem [1].

There are several methods for the synthesis nano cellulose, the method of mechanical, chemical, and biological. Several previous research that use a chemical method by means of a strong acid hydrolysis. Wang et al (2008) using sulfuric acid, hydrochloric acid and distilled water in the ratio 3: 1: 6 (v / v) were accompanied by stirring 50 Hz (3000 rpm) for 10 hours, resulting nano cellulose measuring 20-90 nm [2]. Rosa et al (2010) using the base material pieces of coconut delignification in the previous stage, which then results delignification coconut pieces hydrolyzed with sulfuric acid, resulting nano cellulose measuring 5-6 x 58-515 nm[3]. Sadegifar et al (2011) using hydrogen bromide (HBr) with a concentration of 2.5 M temperature is 80°C for 3 hours, then filtered using Whattman paper. The resulting Nano cellulose measuring 100-400 x 7-8 nm [4].

Ieolovich (2012) made nano cellulose was made with variations of the reaction temperature and ratio of acid to cellulose, produced nano cellulose 1150-200 x 10-20 nm size [5]. Xiong et al (2012) with waste raw cotton which is hydrolyzed using sulfuric acid 63% by weight total of 33 mL at a temperature of 44°C with stirring and ultrasonikasi at 50 Hz for 3 hours. The adding water distillation which aims to stop the process of hydrolysis, water results distillation is added as much as 5 times the volume of the mixture. The resulting nano cellulose measuring 10-65 nm [6]. Bernando et al (2012) using a base material of bamboo fiber, bamboo fiber before hydrolysis stage soaked in NaOH 2% w / w for 4 hours and bleaching at a temperature of 80 ° C for 3 hours. Furthermore pulp already bleaching hydrolyzed using sulfuric acid 64% by weight and the resulting nano cellulose measuring 100-130 x 5-8 nm [7].

Hydrolysis of cellulose that is commonly used is strong acid. Strong acids can eliminate the amorphous part of a chain of cellulose so that the insulation on the crystalline cellulose can be done [8]. The mechanism of cellulose hydrolysis with strong acids shown in Figure 1.

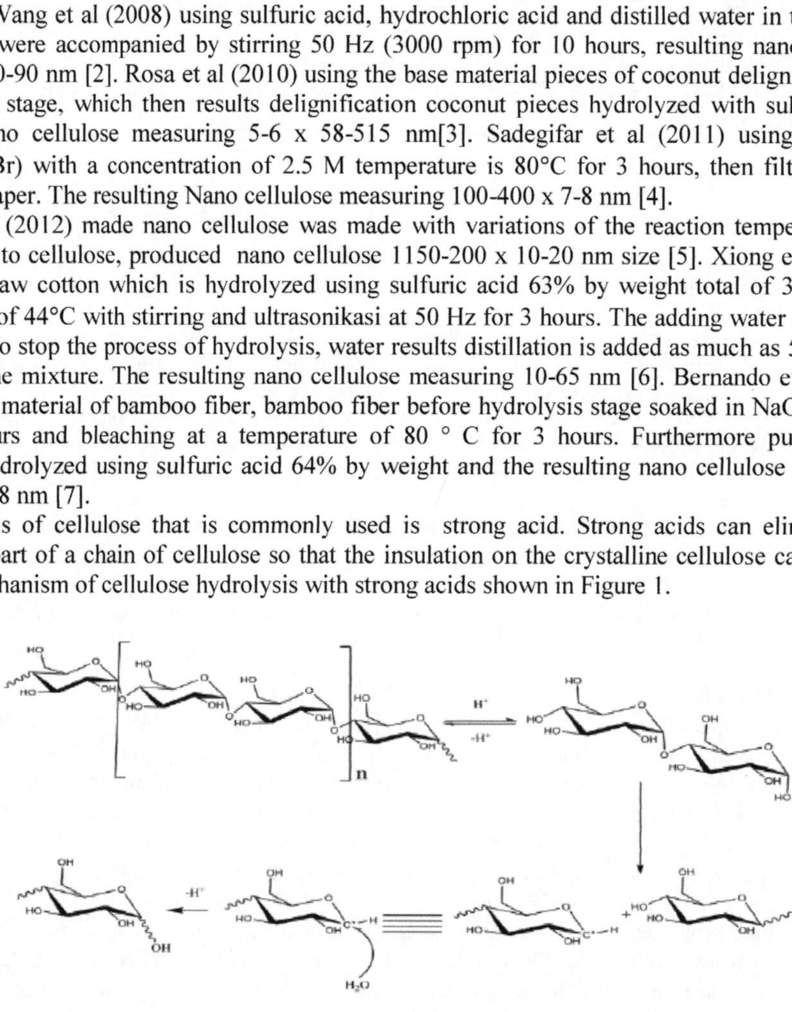

Figure 1. Mechanism of Acid Hydrolysis [9]

Acid hydrolysis is main process used in producing cellulose nanocrystal, where the arrangement of small blocks separated from the cellulose fibers. Cellulose is composed of amorphous regions and crystalline regions. The amorphous area has a lower density than crystalline regions, so that when the cellulose is given treatment using strong acids, the amorphous regions will break and release the crystalline regions. The nature of cellulose nanocrystal depends on various factors, such as a source of cellulose, reaction time, temperature, and type of acid used for the hydrolysis process. Sulfuric acid and hydrochloric acid is often used in the production of nanocrystals of cellulose. Dispersabilitas of nanocrystal cellulose derived from both acid type is different, because of the abundance of sulfate groups on the surface. The nanocrystal cellulose derived from hydrolysis using sulfuric acid can be dispersed easily in water while nanocrystal cellulose derived from hydrolysis using hydrochloric acid does not disperse and the suspension solution tends to flocculate [10]. The purpose of this research is obtaining micro-nano bacterial cellulose as reinforcing bioplastics and knowing the characteristics of bioplastics.

2. MATERIAL AND METHOD

2.1. Materials and tools

Materials were used coconut water, urea, sugar, acetic acid, hydrochloric acid, sodium hydroxide, sodium hypochlorite, sago starch, and glycerin, and distilled water. Tools were used hotplate stirrer, beaker glass, 41 Whatman filter papers, plastic mold, and Buchi B 290 spray dryer.

2.2. Methodology

This research was conducted in three stages: the first is a production of bacterial cellulose, the second is making of micro-nano from bacterial cellulose that is produced in the first stage, and final stage is manufacturing bioplastics. The method of making micro-nano cellulose bacteria is acid hydrolysis process. Firstly, bacterial cellulose powder was heated in NaOH 5%, 100 ^{o}C for 4 hours. After that washed and filtered and then blended. Hydrolysis process with hydrochloric acid (HCl) at the conditions 3,5M, 55°C, 12 hours. The results, micro-nano cellulose bacteria from this step, was bleached, neutralized, and then dried with a spray dryer. Casting method was used to manufacture the bioplastics. Treatment of adding nano cellulose (0%, 2%, 4%) into 100 ml of distilled water plus 3 ml of glycerol and 10 g of sago starch and then stirred for 1 hour at 80-85 ^{o}C. After 1 hour, the solution then printed in the mold.

3. RESULT AND DISCUSSION

3.1. Bacterial Cellulose

Bacterial cellulose (BC) was made from coconut water for two weeks (Figure 2a). Base on analysis this cellulose has 0,9-1,0% of cellulose and 380, 56 Mpa in the tensile strength test. The more layers formed, the higher tensile strength number. Modulus young approximately 18 Gpa, elongation 11,3 % and density around 1, 15 g/cm^{3}. Bacterial cellulose has high porosity, hydrophilic and 700% water absorption. Bacterial Cellulose was dried and grinded as the Figure 2b and 2c. SEM testing results of bacterial cellulose can be seen in figure 3. Bacterial cellulose was given purification process with NaOH 5% for 6 hours.

2a 2b 2c

Figure 2. Bacterial Cellulose

Figure 3. SEM test of bacterial cellulose

3.2. Nano bacterial cellulose

Hydrolysis process using hydrochloric acid (HCl) with 3.5M concentration conditions, 55°C temperature for 6 hours. This process has been work out in reducing the size of the bacterial cellulose fibers to ± 7 μm. SEM testing results against bacterial cellulose can be seen in Figure 4. To the extent, this has not been obtained yet cellulose particle size with nanometer size.

Figure 4. SEM test of micro-nano bacterial cellulose

Preparations of micro-nano bacterial cellulose can be stored in a wet form can be seen in Figure 5a. Drying is done by spray drying to obtain micro-nano bacterial cellulose in a powder, as in Figure 5b, 5c.

5a 5b 5c

Figure 5. Micro-nano bacterial cellulose

3.3. Physical characteristic of bioplastic

Micro-nano bacterial cellulose produced from hydrolysis at 3.5M concentration 55°C temperature for 6 hours selected as a filler in the manufacture of bioplastics. Distribution of micro-nano bacterial cellulose in bioplastics can be seen from the test results of SEM in Figure 6. Micro-nano bacterial cellulose distributed within the matrix of starch, this means that the process of adding nanofiller bacterial cellulose to the matrix goes well. Bioplastics surface that looks less flat due to the molding process bioplastics with the casting method manually.

Figure 6. SEM test of bioplastic (treatment 2%)

Test FTIR (Fourier Transform Infrared Spectroscopy) as an analysis conducted to determine the functional groups on the polymer. FTIR spectra generated from testing bioplastics are presented in Figure 7. Based on the image seen on the identification of functional groups in bioplastic films formed on IR absorption spectrum is a combination of the specific functional group contained in each of the constituent components of bioplastics. So it can be seen clearly that the bioplastics obtained a bioplastic produced through a process of mixing (blending), it can be seen because no new functional groups. Also be concluded that the plastic is formed while still having hydrophilic properties such as its constituent components. The existence of functional groups such as CH (frequency 2900-3010 cm-1), C-NO_2 (frequency 1300-1500 cm-1), -N = N (frequency from 1575 to 1650 cm-1) and - (CH_2) n (frequency 750 cm-1), shows the plastic film can be degraded properly.

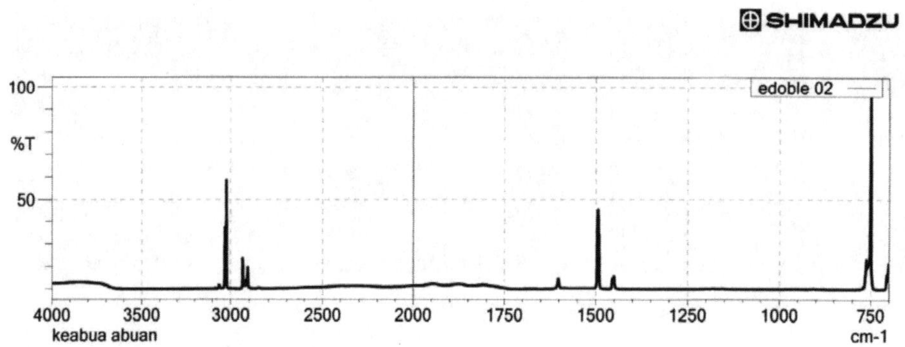

Figure 7. FTIR test of bioplastic (treatment 2%)

The addition of micro-nano bacterial cellulose as filler bioplastics significant effect increases the value of tensile strength and tensile modulus. The highest tensile strength at treatment addition of micro-nano cellulose 2% of 25.62 Mpa. This value is higher than the tensile strength in bioplastics breadfruit starch-chitosan-sorbitol with a value of 16.34 Mpa [11]. The resulting rigid bioplastic indicated by the tensile modulus values high despite already plus plasticizer glycerol. The tensile modulus value was higher when compared to the starch bioplastics breadfruit-chitosan-sorbitol with a value of 2.72 Mpa [11]. According to Bourtoom (2006) To increase the elasticity of the bioplastic can be done by the addition of plasticizers [12]. The tensile strength value meets into standard tensile strength values, with the value of biodegradable tensile strength of 10-100 Mpa and tensile strength values seen from polypropylene (conventional plastic) value of 24.7 which was close enough [13]. The mechanical properties of bioplastics with sago starch matrix and filler micro-nano bacterial cellulose when compared to conventional plastics and biodegradable plastics produces mechanical properties are not much different. Values for tensile strength bioplastic is now approaching the standard, but the young's modulus and elongation still not meet the standards.

4. CONCLUSION

Micro-nano-sized bacterial cellulose which was produced is ± 7 μm. The addition of micro-nano bacterial cellulose as filler bioplastics significant effect increases the value of tensile strength and tensile modulus. The physical characteristics of bioplastics by adding 2% of micro-nano bacterial cellulose which are tensile strength of 25,65 Mpa, Young modulus of 3.73 Mpa, elongation of 4.98% and density of 0.32 g / cm3. The tensile strength value has fulfilled the standard of bioplastics.

ACKNOWLEDGEMENT

This work is supported by Centre of Training and Education Ministry of Industry Indonesia. The analysis was done at Polytechnic ATI Padang.

REFERENCES

[1] Wicaksono R, Syamsu K, Yuliasih I, Nasir M, 2013. Karakteristik Nanoserat Selulosa dari Ampas Tapioka dan Aplikasinya Sebagai Penguat Film Tapioka. *J Tek Ind Pertanian*. 23 (1) : 38-45.

[2] Wang, N.; Enyong, D.; Rongshi, C., 2008. Preparation and Liquid Crystalline Properties of Spherical Cellulose Nanocrystals. *Langmuir*, Vol. 24, 5-8

[3] Rosa, M. F.; Medeiros, E. S.; Malmonge, J. A.; Gregorski K. S.; Wood, D. F.; Mattoso, L. H. C.; Glenn, G.; Orts, W. J.; Imam, S. H., 2010. Cellulose nanowhiskers from coconut husk fibers: Effect of preparation conditions on their thermal and morphological behavior. *Carbohydrate Polymers*, 81, 83-92

[4] Sadeghifar, H.; Ilari, F.; Sarah, P. C.; Dermot F. B.; Dimitris S. A., 2011. Production of

cellulose nanocrystals using hydrobromic acid and click reactions on their surface. Springer. *Journal Material Science*.
[5] Ioelovich Michael. 2012. Optimal Conditions for Isolation of Nanocrystalline Cellulose Particles. *Journal of Nanoscience and Nanotechnology*, 2(2): 9-13
[6] Xiong, R.; Xinxing, Z.; Dong, T.; Zehang, Z.; Canhui, L., 2012. Comparing microcrystalline with spherical nanocrystalline cellulose from waste cotton fabrics. *Cellulose*, 19, 1189–1198.
[7] Bernardo, S. L. B.; Fabiano V. P.; Jean, L. P.; Bruno J., 2012. Preparation morphology and structure of cellulose nanocrystals from bamboo fibers. *Cellulose*, 19, 1527–1536.
[8] Isdin O. *Nanoscience in nature: cellulose nanocrystals*. Surg, 2010, 3(2)
[9] Yue, Y., A. 2007. Comparative Study of Cellulose I and II Fibers and Nanocrystals. *Thesis*. Heilongjiang Institute of Science and Technology.
[10] Peng, B.C., Dhar, N., Liu, H.L., dan Tam, K.C. 2011. Chemistry and Applications of Nanocrystalline Cellulose and Its Derivatives: A Nanotechnology Perspective. *The Canadian Journal of Chemical Engineering*. 89: 1191-1205.
[11] Setiani, W., T. Sudiarti dan L. Rahmindar. 2013. Preparasi dan Karakterisasi Edible Film dari Poliblend Pati Sukun-Kitosan. *Jurnal Kimia Valensi* 3(2) : 100-109 . Jurusan Kimia Fakultas Sains dan Teknologi UIN Sunan Gunung Djati. Bandung
[12] Bourtoom, T. 2006. Plasticizer Effect on the Propertes of Biodegradable Blend Film From Rice Starch-Chitosan. *Songklanakarin Journal of Science and Technology*.30 (Suppl.1), 149-155.
[13] Krochta, J.M and M. Johnston.1997. Edible and Biodegradable Polymer Film. Challenges and Opportunities. *Food Tech.* 51(2):61-74

Explorative study of tropical pitcher plants (*Nepenthes sp.*) types and insects that trapped inside in Sebangau National Park Palangka Raya Central Kalimantan

Nanik Lestariningsih and Denik Setyaningsih

Biology Department Of IAIN Palangka Raya
Jl. G.Obos Kompleks Islamic Centre Palangka Raya, Central Kalimantan 73112

Email : nanik.lestariningsih@iain-palangkaraya.ac.id

Abstract. Pitcher plants (*Nepenthes sp.*) is a plant with unique shape either of shades of colors, pouch shape and its capability in catching insects. Pitcher plant (*Nepenthes sp.*) is one of the plants that protected under Law Number. 5 of 1990 about Conservation of Biological Resources and Ecosystem and Government Regulation Number 7/1999 about Preservation of Plants and Animals. Sebangau National Park is one of representative of peat swamp ecosystem and one of some types of pitcher plants (*Nepenthes sp.)* habitat. This study aimed to determine the types and diversity levels of pitcher plants (*Nepenthes sp.*) and the trapped insects inside in Sebangau National and to determine the differences of diversity levels of pitcher plants (*Nepenthes sp.)* and the trapped insects inside in the opened and closed forest in Sebangau National Park. The research type was conducted descriptive qualitative research. The method used survey method with purpossive sampling technique.The result of the study the number of pitcher plants (*Nepenthes sp.)* obtained in opened forest were three types consist of *Nepenthes mirabilis, Nepenthes rafflesiana* and *Nepenthes gracilis* with two types insects trapped inside those were Diptera ordo and Hymenoptera ordo. While the number of pitcher plants (*Nepenthes sp.*) obtained in closed forest as many as two types consist of *Nepenthes ampullaria* and *Nepenthes rafflesiana* with two type insects trapped inside those were Diptera ordo and Hymenoptera ordo. The results of the analysis calculation pitcher plants (*Nepenthes sp.)* species diversity index in opened and closed forest showed lower category. The diversity in row were 1 and 0,45 with H' criteria ≤ 1 low diversity. The results of the study of insects trapped inside of pitcher plants (*Nepenthes sp.)* obtained in opened and closed forest showed low category. The diversity in a row were 0,63 and 0,52 with the criteria of H' ≤ 1 low diversity.

1. Introduction

Pitcher plants (*Nepenthes sp.)* is one of the plants that is protected under Law No. 5 of 1990 on Conservation of Biological Resources and Ecosystem and Government Regulation No. 7/1999 about Preservation of Plants and Animals. This is in line with the regulations of the Convention on International Trade in Endangered Species (CITES), of the 103 species of pitcher plants in the world that have been published, two types: N. rajah and N. khasiana are listed on Appendix-1. The rest are on Appendix-2. This means that forms of trading activity is very ristricted. Population *Nepenthes sp.* predicted to continue to decline from year to year in Indonesia[1]. If this declanation continues to happen

[1] Muhammad Mansur, *Nepenthes Kantong Semar yang Unik.*

then it will not be found again in the diversity of pitcher plants types (*Nepenthes sp.)* In their natural habitat Pitcher plants (*Nepenthes sp.)* Is unique in terms of shape, size, and color hue diverse pockets as a result of a modification of the tip of the leaves change shape and function to trap insects or other small animals [1,2,3,4,5]. Therefore *Nepenthes sp.* species diversity and insects trapped in it is worthed to be investigated.

2. Methods

The research is carried out a qualitative descriptive study was a study that aims to create a picture of the situation, data collection instruments based on research and analysis of quantitative data / statistics and survey method[2]. The steps in the research of this description was by direct observation, document, identify, classify, describe, and calculating the diversity index. Data collection research used *purposive sampling* techniques[3]. The data collection technique involved determining the location of research, mapping and mapping plot transect lines, until the retrieval, recording descriptions of morphological characteristics *Nepenthes sp.* and the characteristics of the body structure of insects trapped in it, conduct an inventory and identification, and the manufacture of wet preserved insect specimens. Sebangau National Park has an area 214.809.000 m^2. Consists of three areas, namely Habaring Hurung resort area of Palangka Raya, Pulang Pisau district resort, and Katingan district resort. Research sites in the region of 1 resort Habaring Hurung Palangka Raya. The research location was open forest and closed forest in Sebangau National Park Resort Habaring Hurung. Plot observations made discontinuous [7,8], mapping the plot was done by creating a transect line measures 5 x 50 m by 5 transects with a distance of 5 m. Each transect was divided into 10 subplots with a size of 5 x 5 m. The number of transects in the field that is 5 with a length of each transect is 50 m. The photo locations of the survey figure 1.

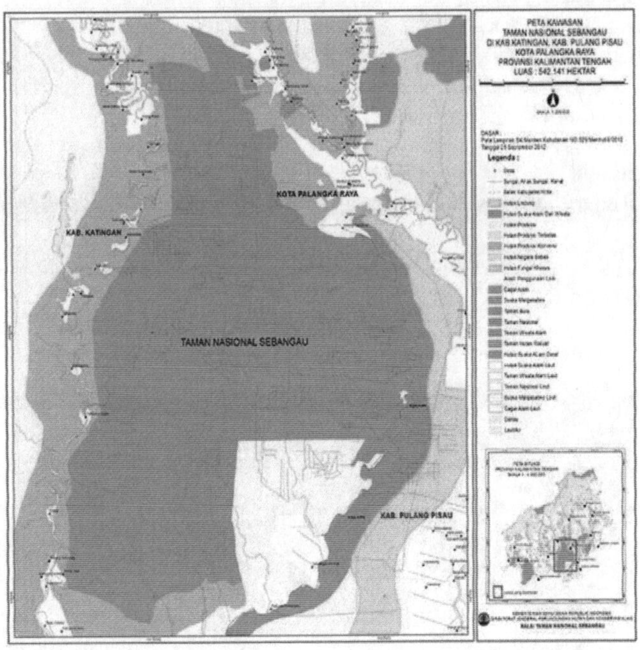

The photo of the forest in Sebangau National Park Resort Habaring Hurung figure 2.

---

[2] Sugiono, *Metode Penelitian Pendidikan kuantitatif, Kualitatif dan R & D.*
[3] Fachrul, M. F, *Metode Sampling Bioekologi.*

Figure 2. Forest in Sebangau National Park Resort Habaring Hurung

The habitat of *Nepenthes sp.* is in the peat forest or peat soil texture and tolerant to high moist condition. Soil pH ranging between 4 to 6.5. While soil moisture ranges between 27-29^{0}C.
After collecting the data then analyzing the quantitative descriptive statistic. Calculating the diversity index, evenness, richness and density of *Nepenthes sp.* and insects trapped in it. Diversity was analyzed using the Shannon-Wiener formula.

3. Results and Discussion

From the steps or procedures discussed in method section the endangered plants identified as follows:

3.1 Type *Nepenthes sp.* type was identified and found in Open Forest in Sebangau National Park Resort Habaring Hurung

The photo of the *Nepenthes sp.* types are shown in figure 3
Specimen 1

Figure 3. The Bag *Nepenthes gracilis* Korth.

Nepenthes gracilis Korth. growing climbing with stem length ≤ 5 m, diameter ≤ 5 mm, has a triangular shape, and tems color from green to reddish brown. It has a thick leaves, lancet-shaped, green or red, leaf length ≤ 16 cm, width ≤ 3 cm, tendrils length of ≤ 26 cm. Bag top and bottom have cylindrical shape, height ≤ 17 cm, width of mouth ≤ 5 cm, width ≤ 5 cm cover, green, brown or reddish.

Spesimen 2

Figure 4. Pockets *Nepenthes mirabilis* (Lour.)

Figure 4 shows the photos of the bag *Nepenthes mirabilis* grows climbing rod length ≤ 5 m, diameter ≤ 5 mm, cylindrical rod shape, green or red or green pink. Has a thin leaves with an elongated shape or a lancet, is green with a slippery surface, ≤ 15 cm leaf length, leaf width ≤ 8 cm, long tendrils of ≤ 23 cm. Shaped pockets are on the waist, height ≤ 20 cm, width of mouth ≤ 4 cm. Color bags rangs from green, red, green or red striated. *Nepenthes mirabilis* has shaped flower bunches, length ≤ 45 cm, sepals round to elliptical, red.

Spesimen 3

Figure 5. The bag Nepenthes rafflesiana

Figure 5 shows the photos of the bag *Nepenthes rafflesiana* grows to the height ≤ 4 m, diameter ≤ 1 cm, have round rods, green, red or brown. Leaves are thick, smooth, oblong-shaped, green or red on the bottom green, red bone mother, length ≤ 50 cm, 9 cm and a width ≤ ≤ 75 cm long tendrils. Bags are under the oval-shaped, winged, height ≤ 24 cm, width of the mouth of the bag ≤ 6 cm, white with red striped, red striped maroon with green, green with brown striated. Pockets are on a funnel-shaped, wingless, height ≤ 30 cm, width of mouth ≤ 7 cm, green with brown striated. Shaped flower bunches, length ≤ 50 cm, sepal's elliptical, red maroon.

3.2 *Type Nepenthes sp.* obtained in Closed Forest in Sebangau National Park Resort Habaring Hurung
Spesimen 1

Figure 6. Bag of *Nepenthes ampullaria* Jack.

Figure 6 shows the photos of the bag *Nepenthes ampullaria* grows climbing and surface soil, trunked cylindrical, brownish green, length ≤ 10 m, diameter ≤ 5 mm. It has thick leaves, green, lancet-shaped, length ≤ 33 cm, width ≤ 8 cm, and long tendrils of ≤ 51 cm. It has Bag-shaped jars and clustered appear on the surface ground or hanging on the trunk that grows upright, having wings, elliptical shaped cover, height ≤ 17 cm, width of mouth ≤ 6 cm. Colors rangs from green bag, green with red mouth and red striated green, reddish or brown.
Spesimen 2

Figure 7. Bottom Bags *Nepenthes rafflesiana* Korth.

Figure 7 shows the photos of the bag *Nepenthes rafflesiana* has a cylindrical rod, green or brownish red color, height ≤ 2 m, diameter ≤ 1 cm. it has green leaf or red at the top and bottom, thick, smooth, oblong shape, length ≤ 54 cm, width ≤ 10 cm, and long tendrils of ≤ 67 cm. Bags are under the oval and wingless, funnel-shaped bag over and wingless, height ≤ 22 cm, width of mouth ≤ 7 cm, red maroon premises striated striated green and white with red on the muut.

Type Nepenthes (*Nepenthes sp.)* Are found in open forests Sebangau National Park Resort Habaring Hurung totaling 3 species are *Nepenthes mirabilis*, *Nepenthes rafflesiana* and *Nepenthes gracilis*. While the type of pitcher plants (*Nepenthes sp.*) Found in the closed forests in Sebangau National Park Resort Habaring Hurung totaling 2 species are *Nepenthes ampullaria* and *Nepenthes rafflesiana*.

3.3 Insects trapped in *Nepenthes sp.* in the forest open and closed in Sebangau National Park Resort Habaring Hurung

Spesimen 1. Ordo Diptera

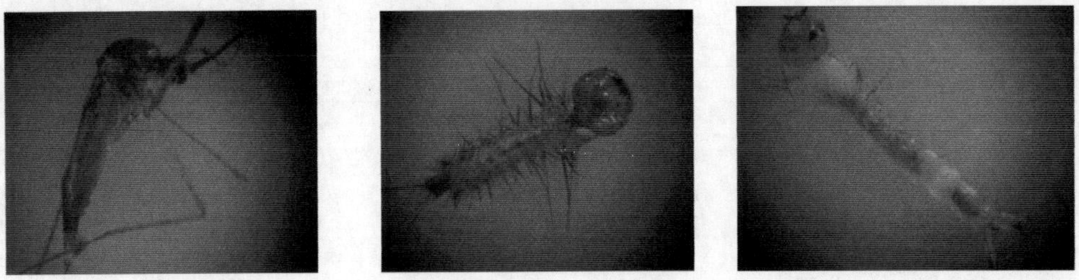

Figure 8. Insects ordo Diptera

Specimen 2. **Ordo Hymenoptera**

Figure 9. Ordo Hymenoptera

Insects trapped inside pitcher plants (*Nepenthes sp.*) are found in open forests and closed forest ç in Sebangau National Park Resort Habaring Hurung consists of two orders is the order Diptera and the order Hymenoptera.

3.4 Number of *Nepenthes sp.* and insects trapped in it in the open and closed forests in Sebangau National Park Resort Habaring Hurung.

From the observation, the insects trapped in *Nepenthes sp.* that grows at the open and closed forest of Sebangau National Park Resort Habaring Hurung are listed in the table 1.

Table 1. Number of *Nepenthes sp.* and insects trapped in it in the open and closed forests in Sebangau National Park Resort Habaring Hurung

No.	Open Forest				Closed Forest			
	Species	Total	Insects	Total	Species	Total	Insects	Total
1	*Nepenthes gracilis*	20	Ordo Hymenoptera	72	*Nepenthes rafflesiana*	1	Ordo Diptera	4
			Ordo Diptera	42				
2	*Nepenthes mirabilis*	22	Ordo Hymenoptera	54	*Nepenthes ampullaria*	5	Ordo Hymenoptera	12
			Ordo Diptera	18			Ordo Diptera	41
3	*Nepenthes rafflesiana*	5	Ordo Hymenoptera	17				
			Ordo Diptera	9				

3.5 Diversity of *Nepenthes sp.* Sebangau National Park Resort Habaring Hurung

The results of the analysis are listed in table 2.

Table 2. Diversity index of *Nepenthes sp.* in the open and closed forests in Sebangau National Park Resort Habaring Hurung

Open Forest				Closed Forest		
NO	Species	H'	KR%	Species	H'	KR%
1	*Nepenthes gracilis*	-0,36	47%	*Nepenthes rafflesiana*	-0,15	17%
2	*Nepenthes mirabilis*	-0,37	43%	*Nepenthes ampullaria*	-0, 30	83%
3	*Nepenthes rafflesiana*	-0,24	11%			
Diversity Index = 1				Diversity Index = 0.45		
Evenness Index = 0,87				Evenness Index = 0.41		
Wealth index = 0,43				Wealth index = 0.82		

The highest relative density type of pitcher plants (*Nepenthes sp.)* Obtained in open forest in Sebangau National Park Resort Habaring Hurung is *Nepenthes mirabilis* with a value of 47% of individuals/m^2. While the lowest relative density is *Nepenthes rafflesiana* with a value of 11% of individuals/m^2. The highest relative density type of pitcher plants (*Nepenthes sp.)* Obtained in closed forest in Sebangau National Park Resort Habaring Hurung is *Nepenthes ampullaria* with a value of 83% of individuals per m^2 and the lowest relative density is *Nepenthes gracilis* with a value of 17% of individuals per m^2.

3.6 Diversiy of insects trapped inside *Nepenthes sp.* Sebangau National Park Resort Habaring Hurung

The results of the analysis are listed in table 2.

Table 3. Diversity index of insects trapped in *Nepenthes sp.* in the open and closed forests in Sebangau National Park Resort Habaring Hurung.

Open Forest				Closed Forest		
NO	Ordo	H'	KR%	Ordo	H'	KR%
1	Hymenoptera	-0,33	21%	Hymenoptera	-0,27	67%
2	Diptera	-0,19	79%	Diptera	-0,37	33%
Diversity Index = 0,52				Diversity Index = 0.63		
Evenness Index = 0,74				Evenness Index = 0.91		
Wealth index = 0,26				Wealth index = 0.14		

The highest relative density of insects trapped inside pitcher plants (*Nepenthes sp.)* Obtained in closed forest in Sebangau National Park Resort Habaring Hurung is the order Hymenoptera for 67% while the lowest relative density is the order Diptera with 33%. The highest relative density of insects trapped inside pitcher plants (*Nepenthes sp.)* Obtained in open forest in Sebangau National Park Resort Habaring Hurung is the order Diptera with a value of 79% while the lowest relative density is the order Hymenoptera with a value of 21%.

References

[1] Akhriadi, P., Hernawati, Primaldhi, A., and Hambali, M. 2009. *Nepenthes naga*, a New Species of Nepenthaceae from Bukit Barisan of Sumatra. *Reinwartia*. 12(5): 339.

[2] Damayanti, F., Mansur, M., and Roostika, I. 2011. Diversity of *Nepenthes* sp. in West Kalimantan. *International Journal of Biodiversity and Conservation*. 3(13): 705.

[3] Clarke, C., Lee, C.C. & McPherson, S. 2006. *Nepenthes chaniana* (*Nepenthaceae)*, a new

species from north-western Borneo. *Sabah Parks Nat. Jour.* 7: 53 – 66.

[4] Clarke, C., Moran, J. A., and Chin, L. 2010. Mutualism Between Tree Shrews and Pitcher Plants. *Plant Signaling and Behavior*. 5:10: 1-3.

[5] Bauer, U and Federle, W. 2009. The Insect-Trapping Rim of Nepenthes Pitchers Surface Structure and Function. Plant Signaling & behavior. 4:11

[6] Cheek, M. and Jebb, M. 2013. Recircumscription of the *Nepenthes alata* Group (Caryophyllales: Nepenthaceae), in the Philippines, with Four New Species. *European Journal of Taxonomy*. 69:1.

[7] Fachrul, M. F. 2012. Metode Sampling Bioekologi. Penerbit Bumi Aksara. Jakarta.

[8] Natalia, D., Umar, H., and Sustri. 2014. Pola Penyebaran Kantong Semar (*Nepenthes tentaculata* Hook.F) di Gunung Rorekautimbu Kawasan Taman Nasional Lore Lindu. Warta Rimba2 (1) : 35-44

[9] Grafe, U. T., Schoner, C. R., Junaidi, A., and Schoner, M. G. 2011. A Novel Resource Service Mutualism Between Bats and Pitcher Plant. *Evolutionary Biology*. 7:436-439.

[10] Hernawati. 2001. A Preminilary Research to Conserve *Nepenthes* spp. in West Sumatra. Final Report *Nepenthes* Project 2001. Padang: Supported by BP Conservation. *Nepenthes* Team.

[11] Mansur,M., *Nepenthes Kantong Semar yang Unik.* Jakarta: Penebar Swadaya, 2006.

[12] Moran, J. A., Merbach, M., Livingston, J. N., Clarke, C. M., and Booth, W. E. 2001. Termite prey specialization in the pitcher plant *Nepenthes albomarginata* evidence from stable isotope analysis. *Annals of Botany*. 8: 307-311.

[13] Sugiono. "Metode Penelitian Pendidikan Kuantitatif, Kualitatif dan R & D." Bandung: Alfabeta Bandung. 2009.

Ecofriendly Synthesis of nano Zero Valent Iron from Banana Peel Extract

Sunardi[1], Ashadi[2], Sentot Budi Rahardjo[3] and Inayati[4]

[1]Chemical Analyst, Setia Budi University,
Jl. Let. Jend. Sutoyo, Mojosongo Surakarta 57127, INDONESIA
Environmental Science of Post Graduate Program Sebelas Maret University, Jl. Ir. Sutami 36A Kentingan Jebres Surakarta 57126, INDONESIA
[2]Departement of Chemistry Education, Sebelas Maret University, Surakarta, Indonesia
[3]Departement of Chemistry, Sebelas Maret University, Surakarta, Indonesia
[4]Chemical Engeneering, Sebelas Maret University, Surakarta, Indonesia

E-mail: nardi_usb@yahoo.co.id

Abstract. In this study, nano Zero Valent Iron (nZVI) were synthesized from banana peel extract (BPE) and ferrous sulfate. During the synthesis of nZVI both the precursor and the reducing agent were mixed in a clean sterilized flask in 1:1 proportion. For the reduction of Fe ions, 5 ml of filtered BPE was mixed to 5 ml of freshly prepared 0.001 M – 0.005 M aqueous of $FeSO_4$ solution with constant stirring at room temperature. Within a particular time change in colour from brown to black color obtained by nanoparticles synthesis. A systematic characterization of nZVI was performed using UV-Vis. UV–visible absorption is used to investigate SPR. Characteristic surface plasmon absorption band was observed at 210 nm for the black colored nZVI synthesized from 0.001-0.005 M ferrous sulfate with BPE concentration 5 ml. It has been found that the optimum concentration for the synthesis of nZVI is 0.001M Fe^{2+} ions. There is small decrease in the intensity of SPR band from 0.001 to 0.005 M. The characterization size of nZVI was performed using TEM. The result shows that formation of particles size of nZVI was more 100 nm.

1. Introduction

The term nano comes from the Greek word meaning very small. Nano-size shows one billionth of a meter, or 10^{-9} [1-3]. Nanotechnology is defined as the manipulation of the material through a particular chemical and / or physical processes to synthesize materials with specific characteristics, which can be used in particular applications. Nanotechnology is the art of how to create and manipulate materials at the nanoscale (1-100 nm). Nanotechnology is a field of science that is growing rapidly with a variety of applications in science and technology[9].

1

Nanotechnology relates to the synthesis of nanoparticles about size, shape, chemical composition and dispersity aimed for human benefits. At this time, chemical and physical methods have been successfully produce pure, nanoparticles are well defined, but it is expensive and harmful to the environment. Utilization of biological organisms such as plant extracts, biomass plants and microorganisms could be an alternative to chemical and physical methods for the production of environmentally friendly nanoparticles[2]. Nanobiotechnology is a related field between biology and nanotechnology. It is an alternative for the synthesis of nanoparticles that are environmentally friendly by using biological resources such as plants and microorganisms[10-11].

Green synthesis nanoparticle has been achieved by using plant extracts for reducing and capping agents. Many research has been done in the biosynthesis of silver nanoparticles with plant parts such as Punica granatum peels [13], Citrus sinensis peel [14], lemon leaves [15], Myrica esculenta leaf [16], Wrightia tinctoria leaves [17] and mango peel [18].

In this study, we have reported for ecofriendly synthesis of nano Zero Valent Iron (nZVI) using the peels extract of the banana plant. Aqueous ferrous sulfate solution reacts with banana peel extract, causing rapid formation of very stable, crystalline nZVI. The rate of synthesis of the nanoparticles very quickly, which justifies the use of crop residues in the biosynthesis of the iron nanoparticles through ecofriendly methods and more secure. In the next section we have described the synthesis of nZVI by discoloration, changes in absorbance and particle size are formed after reduction.

2. Numerical Methods

2.1. Preparation of Banana peel extract
Banana peel extract was used as a reducing agent for the synthesis of nZVI. The fresh banana peels were washed repeatedly with distilled water to remove dust and dirt in it. Approximately 25 g of peel was taken in a glass of 250 ml containing 75 ml of double-distilled water and then boiled peel at 80 ° C for 10 minutes and filtered through Whatman No 1 filter paper. The resulting filtrate is stored at 4 ° C and is used as a reducing and stabilizing agents.

2.2. Synthesis and Characterization nZVI
In this study, the nZVI synthesis of both the precursor and the reducing agent is mixed into clean sterilized flask in a 1: 1 proportion. For the reduction of Fe ion, 5 ml filtered BPE mixed with 5 ml of freshly prepared 0.001 M - 0.005 M aqueous solution of $FeSO_4$ with constant stirring at room temperature. In particular time changes the color changes from brown to black shows the synthesis of nanoparticles. The formation of nanoparticles can be observed by UV-Visible spectroscopy at a wavelength of 150-550 nm. Observation of ultraviolet-visible spectra obtained using Shimadzu UV-1650pc Spectrophotometer. The images of nZVI were analyzed with a Philips EM 400T Transmission Electron Microscopy (TEM) operated at 100 kV.

3. Results and Discussion

Green synthesis of nano Zero Valent Iron and characterization
The detailed study of the green synthesis of nZVI was performed using BPE. The formation of colloids nZVI investigated with observation of the color change of the solution. The emergence of black colour in the reaction showed the formation of nZVI. Figure 1 shows the image discoloration of synthesis reaction nZVI: Tube A contains ferrous sulfate, tube B contains banana peel extract and tube C contains nZVI in colloidal form. Banana peel contains polyphenols which can reduce Fe ions into Fe^0. Polyphenols are biology components that interact with the metal salts through functional groups, OH and mediate their reduction to nanoparticles [19].

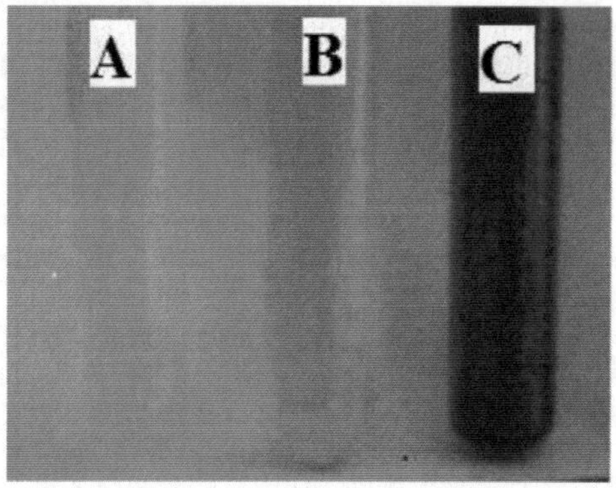

Figure 1. Tube A contains ferrous sulfate, tube B contains banana peel extract and tube C contains colloidal nZVI

UV–Visible analysis
Early detection is done by visual observation of the reaction solution changes color nZVI. These changes were associated with excitation of surface plasmon resonance (SPR) in nZVI. To observe the SPR, usually using a UV-visible. The observation of the characteristics of the surface plasmon absorption band at a wavelength of at 210 nm for the black colored nZVI from BPE concentration (5 ml) with 0,001-0,005 M ferrous sulfate [3, 12]. The optimum concentration for synthesis nZVI is 0,001 M of Fe^{2+} ions. There was a decrease in the intensity of the SPR band from 0,001 to 0,005 M. There is a decrease in the intensity of the SPR band when the concentration is increased further. The decrease in the intensity of the SPR band [Fig. 2 (a-e)] is due to the formation of less nZVI despite high initial concentration of Fe^{2+} ion

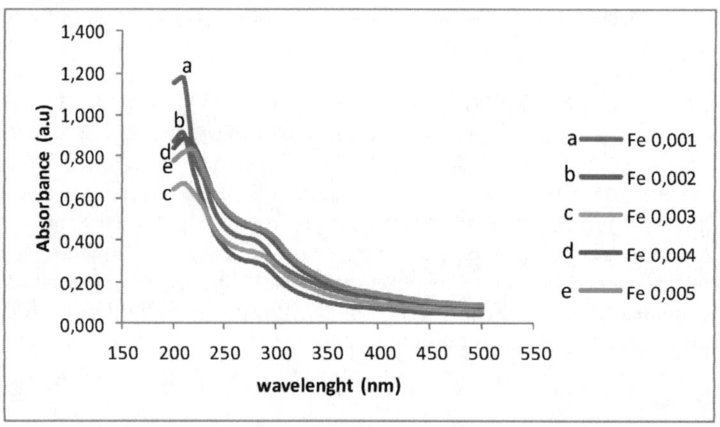

Figure 2. UV Vis spectra for the nZVI prepared with 5 mL Banana peel extract with various Fe^{2+} aqueous concentration

TEM analysis
According Hoag et al. (2009), the reaction between Fe ions and extract containing polyphenols produce nZVI [20]. Samples were characterized using transmission electron microscopy (TEM). TEM analysis is presented in Figure 3. From the image confirm that the density grid elements in each zone is related with the intensity shading. The size similar nZVI can be seen in Figure 3. However, due to the agglomeration causes the particle size nZVI becomes larger 100 nm [21]. These results require further study and research.

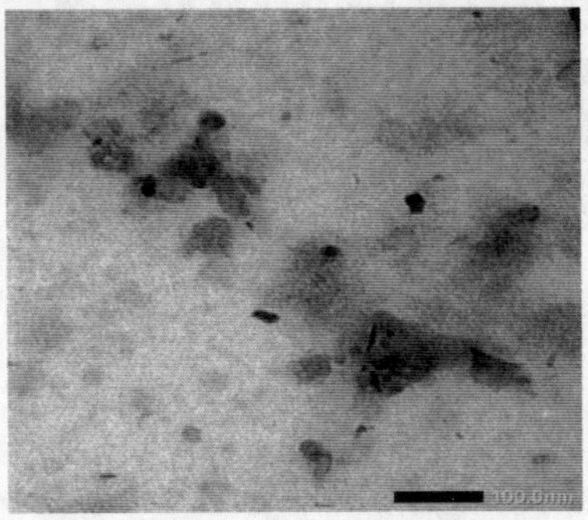

Figure 3. TEM image nZVI synthesized using BPE

Conclusion

In conclusion, nano Zero Valent Iron can synthesized directly betwen ferrous sulfate with banana peel extract in aqueous media without the addition of chemicals (capping agent). So this process is environmentally friendly nanoparticle synthesis. However, the size of particles nZVI was more 100 nm.

References

[1] Varahalarao V., D.S.V.G.K. Kaladhar, Mohan behara, G Kishore Naidu and B Sujatha 2013. *Chemijournal.* 1(4) 22
[2] Monalisa Pattanayak, Debabrata Mohapatra and P.L. Nayak 2013 *Middle-East Journal of Scientific Research.* 18(5) 623
[3] Monalisa Pattanayak and P.L. Nayak 2012 *World Journal of Nano Science & Technology.* 2(1) 6
[4] R. Yuvakkumar, V. Elango, V. Rajendran, and N. Kannan 2011 *Digest Journal of Nanomaterials and Biostructures.* 6(4) 1771
[5] Mahendra Rai and Alka Yadav 2013 *IET Nanobiotechnol.* 7(3) 117
[6] Parashar, V., Parashar, R., Sharma, B., and Pandey, A.C. 2009 *Dig. J. Nanomater. Biostruct.* 4(1) 45
[7] Parashar, U.K., Saxena, P.S., and Shrivastava, A. 2009 *Digest J. Nanomater. Biostruct.* 4 (1) 159
[8] Bankar, A., Joshi, B., Kumar, A.R., Zinjarde, S. 2010 *Colloids Surf. A, Physicochem. Eng. Asp.*, 368 58
[9] Rai, M. and Ingle, A. 2012 *Appl. Microbiol. Biotechnol.* 94(2) 287
[10] Renugadevi, K. and Aswini, R.V. 2012 *Int. J. Nanomater. Biostruct.* 2(2) 5
[11] Albrecht MA, Evan CW, and Raston CL, 2006 *Green Chem* 8 417
[12] T. Kokila, P. S. Ramesh and D. Geetha, 2015. *Appl Nanosci.* DOI 10.1007/s13204-015-0401-2.
[13] Guangquan L, Dan H, Yongqing Q, Buyuan G, Song G, and Yan C 2012 *Int J Mol Sci* 13 466
[14] Ahmad N, Sharma S, Alam MK, Singh VN, Shamsi SF, Mehta BR and Fatma A, 2010 *Colloids Surf B* 81(1) 81
[15] Kaviya S, Santhanalakshmi J, Viswanathan B, Muthumar J, and Srinivasan K, 2011 *Spectrochim Acta Part A Mol Biomol Spectrosc* 7 594
[16] Padma S and Shukla D, 2012 *Appl Nanosci* 2 163
[17] Phanjom P, Zoremi E, Mazumder J, Saha M, and Baruah SB. 2012 *Int J NanoSci Nanotechnol* 3 73
[18] Bharani M, Karpagam T, Varalakshmi B, Gayathiri G, and Priya LK. 2012 *Int J Appl Biol Pharm Technol* 3 58
[19] Bar H, Bhui DK, Sahoo GP, Sarkar P, Pyne S, and Misra A 2009 *Colloids Surf A Physic- ochem*

Eng Asp 348 212
[20] Hoag, GE., Collins, JB., Holcomb, JL., Hoag, JR., Nadagouda, MN., and Varma, RS. 2009 *J. Mater. Chem.* 19 8671
[21] S Machado , S.L. Pinto, J.P. Grosso, H.P.A. Nouws, J.T. Albergaria, and C. Delerue-Matos. 2013 / *Science of the Total Environment* 445–446 1

IOP Conf. Series: Journal of Physics: Conf. Series **795** (2017) 012064 doi:10.1088/1742-6596/795/1/012064

Preparation of TiO_2 photocatalyst with the matrix of palm wood (*Arenga pinnata*) waste in the photodegradation of *batik* wastewater

Dian Kresnadipayana[1], Endang Tri Wahyuni[2], Sri Juari Santosa[2], Mudasir[2]

[1] Health Science Faculty, Setia Budi University, Jl. Letjen Sutoyo, Mojosongo, Surakarta 57127, INDONESIA

[2] Chemistry Department, Mathematic and Natural Science Faculty, Gajah Mada University, Sekip Utara, Bulaksumur, Yogyakarta 55281, INDONESIA.

E-mail: diankresna@setiabudi.ac.id cc: dian.kresnadipayana@gmail.com

Abstract. The study aimed to the preparation of TiO_2 photocatalyst with the matrix from palm wood waste whose has lignin and cellulose content. TiO_2 photocatalyst with the matrix from the wastewater of palm wood waste (TiO_2/pww) was used as photocatalyst in photodegradation of *batik* wastewater. TiO_2 solid was dissolved in ethanol and aquadest, added with the powder of wood palm waste and stirred with a magnetic stirrer for 16 hours. Then separation was carried out using buchner and filtrate and residue were obtained. The filtrate was disposed and the residue was calcined with various temperatures for 3 hours. The temperatures in this research were 100 °C (TiO_2/pww-100); 200°C (TiO_2/pww-200); 300°C (TiO_2/pww-300). Analysis and characterization of TiO_2/wwp were conducted using X-ray diffraction (XRD) and spectrophotometer Fourier Transform Infra Red (FTIR) methods. Photocalalytic TiO_2/wwp use the batch system in a reactor with UV light 40 watts, 220 volts and length wave 360 nm the plate magnetic stirrer. Liquid waste batik adds TiO_2/wwp with time variation. At XRD analysis showed that the preparation of TiO_2/pww could be done on the heating TiO_2/pww temperature of 100°C and 200°C. At the temperature of 300°C, it was indicated that the lignocelluloses in palm wood waste were burned, meaning that few lignocelluloses remained. The result of FTIR analysis showed clearly that at the temperature of 300°C, a few spectrum of lignocelluloses remained in palm wood waste, while at a temperature of 100°C and 200°C, spectra of lignocelluloses of palm wood waste remained. The result of photocatalysis test indicated that TiO_2/pww could reduce 40%, 72%, 81% and 64% COD for TiO_2 (control), TiO_2/pww-100, TiO_2/pww-200 and TiO_2/pww-300, respectively.

1. Introduction

TiO_2 photocatalyst has some advantages; it is non-toxic agent with high photoactivities, stability, affordable price, and capability to work under ultraviolet light of sun light spectrum. Photodegradation of organic compounds, like acetaldehyde, dimethyl sulfide, dimethyl disulfide and methyl mercaptan use TiO_2 photocatalyst [1]; new methylene blue [2]; lignin [3]. The use of TiO_2 photocatalyst in the form of powder to process wastewater is generally practical; however, this powder will form suspension in wastewater solution. This cause solution to become turbid and therefore light absorption by substrate is less perfect. As a consequence, photicatalyst is getting less effective. Moreover, the separation of photocatalyst solid from the solution, for recovery or regeneration, is difficult to be done.

One of efforts to improve TiO_2 activity as photocatalyst is by creating thin layer of TiO_2. Modification has been done by putting TiO_2 with various metal oxide, active carbon [4] [5]; and chitosan [6]. Other innovative modification has been using organic compounds, like cellulose [7] [8]; and lignocellulose [9]. Coloring agents that are widely used in textile industry are black, red, and golden yellow remazols. In staining process, only 5% of these compounds are used, while the rest 95% will be disposed as wastewater. These compounds are stable enough, and hence, they are very difficult to be degraded in the nature. Furthermore, they are harmful for environment, and particularly, in high concentration, they can increase COD (Chemical Oxygen Demand). This triggers researchers to conduct this research, which aims at synthesizing TiO_2 by using the matrix of palm wood waste which contains lignin and cellulose. TiO_2 with the matrix of palm wood waste is utilized as photocatalyst in the photodegradation of *batik* wastewater. In this research, study on TiO_2 synthesize with the matrix of palm wood waste, the effect of calcination temperature in the synthesis, and the effect of time variation in the application of *batik* wastewater management.

2. Numerical Methods

Approximately 50 mL TiO_2 solid was dissolved in ethanol and aquadest, and 10 g of palm wood waste powder was added and then stirred using magnetic stirrer in 16 hours. Afterwards, separation was carried out using buchner which later produced filtrate and residue. Filtrate was disposed and residue was calcined with various temperatures in 3 hours. The temperature variations in this study were 100°C, 200°C; and 300°C. Analysis and characterization were conducted using X-ray diffraction (XRD) and spectrophotometer Fourier Transform Infra Red (FTIR) methods. Photocatalysis test of TiO_2/pww was done using batch system in reactor which was equipped with 40 watt and 220 volt lamp at 340-390 nm wavelength, and magnetic stirrer plate. *Batik* wastewater was added with TiO_2/pww by considering time variation. After TiO_2/pww with the greatest *batik* waste water degradation activity was obtained, time optimization was carried out with time variations of lighting duration in UV reactor, i.e. 6, 12, 24, 36 and 48 hours, respectively.

3. Results and Discussion

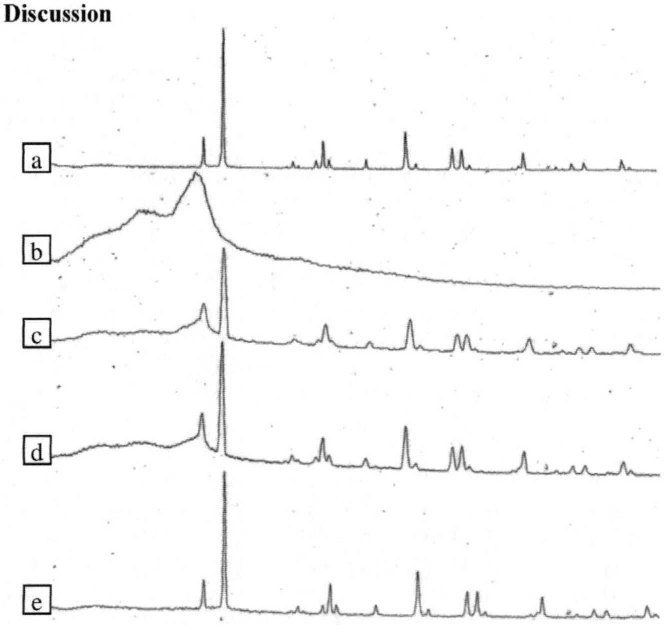

Figure 1. XRD. (a) TiO_2; (b) palm wood waste (pww) (c) TiO_2/pww-100; (d) TiO_2/pww-200; (e) TiO_2/pww-300

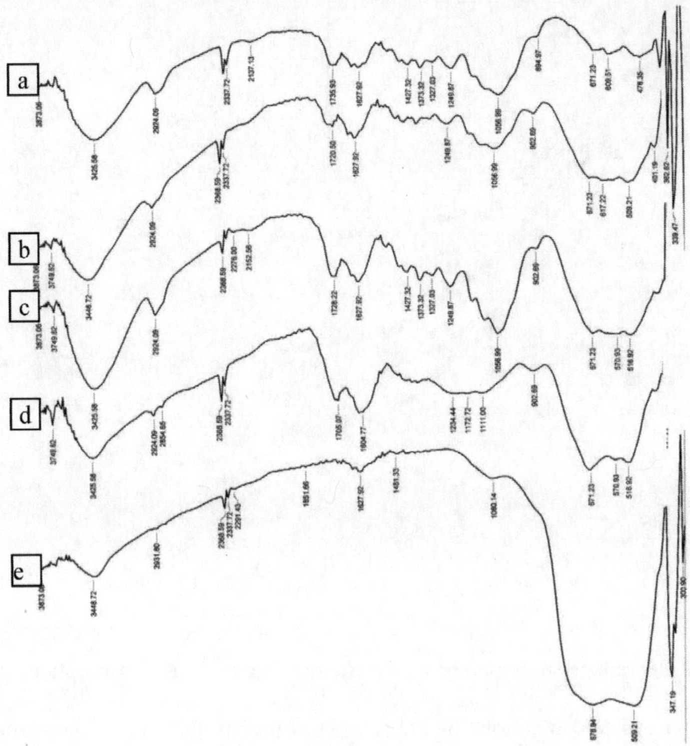

Figure 2. FTIR. (a) palm wood waste (pww); (b) TiO_2/pww-100; (c) TiO_2/pww-200; (d) TiO_2/pww-300; (e) TiO_2

Photocatalyst was made by dissolving TiO_2 in ethanol and water, and adding several grams of palm wood waste (lignin-cellulose). This process yielded solid which was later heated with various temperatures in 3 hours. Photocatalyst produced was later characterized using XRD and FTIR. In the structure model of lignin and cellulose, –CH_2OH, –OH, –C–O– and –OCH_3 group in aromatic chain appeared. Oxygen in this group was rich with free electrons, and therefore, it could interact with metal cations. Moreover, lignin heating would form pores which could increase the surface area as an attachment space for the other materials.

XRD analysis indicated that preparation of TiO_2/pww could be carried out in the heating of TiO_2/pww at the temperatures of 100°C and 200°C. At the temperature of 300°C, lignocelluloses in palm wood waste were totally burned or small amount of lignocelluloses remained (Figure 1). X-ray diffraction at 300°C dehydration showed crystal phase of microsphere composite TiO_2 in TiO_2/cellulose with X-ray diffraction pattern from cellulose component which disappeared [7]. In the FTIR analysis demonstrated clearly that small number of lignocelluloses spectra in palm wood waste remained at the temperature of 300°C and lignocelluloses spectra in palm wood waste also remained at the temperature of 100°C and 200°C (Figure 2). Compared to cellulosic/SiO_2 nanocomposites, OH stretching peak of 3418-3460 cm^{-1} [10] and 3421 cm^{-1} of lignin [11].

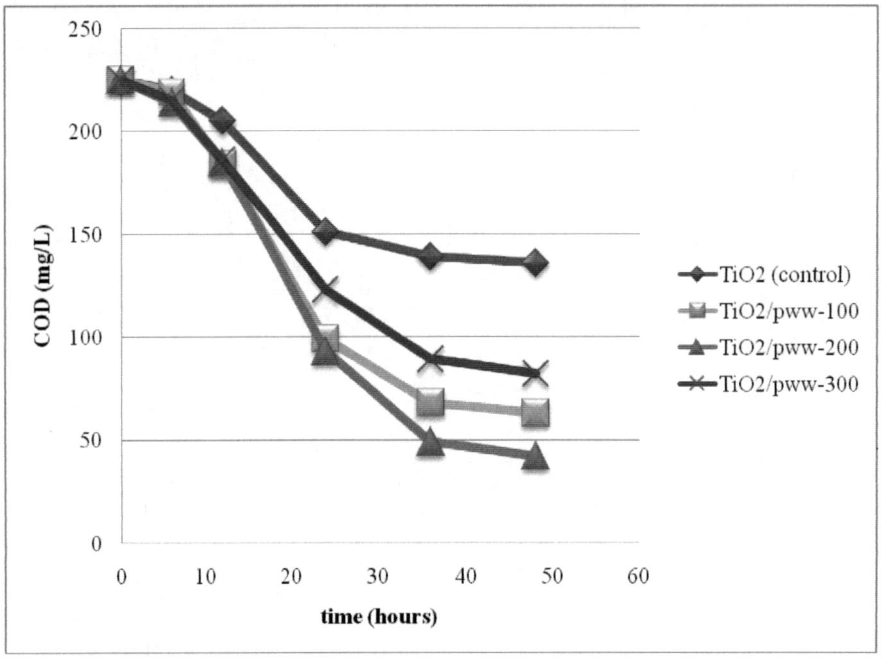

Figure 3. COD reduction in batik liquid waste on TiO_2 and TiO2/pww photocatalyst.

TiO_2/pww photoactivity testing was done by applying it as photocatalyst in photodegradation process of *batik* wastewater. The photodegradation was performed using batch system in a closed reactor equipped with UV lamp and magnetic stirrer plate. The process was accomplished by lighting the mixture which consisted of *batik* wastewater and TiO_2/pww powder, and stirring in certain duration. In photocatalyst test, TiO_2/pww was found to be able to reduce COD, i.e. approximately 40%, 72%, 81%, and 64%, respectively, for TiO_2 (control), TiO_2/pww-100, TiO_2/pww-200 and TiO_2/pww-300 (Figure 3). Acetaldehyde in cellulose/TiO_2 composite with UV radiation was reduced into 45%. Cellulose TiO_2 nanoparticles indicated high photocatalytic activities in MO degradation high concentration under weak UV light radiation. [9].

Conclusion

Preparation of TiO_2 photocatalyst with the matrix of palm wood waste can be used as TiO_2 photocatalyst which can significantly reduce COD in *batik* wastewater. The most effective TiO_2 photocatalyst in the photodegradation of *batik* wastewater is the preparation of TiO_2 with calcination temperature of 200°C.

Acknowledgments

Thankful is acknowledged to Directorate of Research and Community Service, Ministry of Research, Technology and Higher Education, Republic of Indonesia through Setia Budi University for the Doctoral Dissertation Research Grant under the contract numbers; between the Directorate General of Research Strengthening and Development and Kopertis Region VI, 644/M/KP/XII/2012, 23 December 2015, and between and Kopertis Region VI and Setia Budi University, 020/K6/KM/SP2H/ PENGEMBANGANKAPASITAS/2016, 4 May 2016. very grateful to Prof. Dr. Endang Tri Wahyuni, MS as dissertation supervisor for her helpful information and suggestion throughout this research.

References

[1] Tanizaki T, Murakami Y, Hanada Y, Ishikawa S, Suzuki M, Shinohara R 2007 *Journal of Health Science*, **53** (5) 514–519
[2] Ali R and Siew OB 2006 *Jurnal Teknologi*, **45** (F) 31 – 42
[3] Machado AEH, Furuyama AM, Falone SZ, Ruggiero R, Perez DS, Castellan A 2000 Chemosphere **40** 115-124
[4] Janus M, Kusiak E, Morawski W 2009 *Catal Lett* **131**:506–511
[5] Park Y, Kim W, Park H, Tachikawa T, Majima T, Choi W 2009 *Applied Catalysis B: Environmental* **91** 355–361
[6] Hui L.K., Zainal Z, Hussein MZ, Abdullah AH, Hamadneh LR 2009 *Journal of Hazardous Materials* **164** (138–145
[7] Nagaoka S, Hamasaki Y, Ishihara S-I, Nagata M, Iio K, Nagasawa C, Ihara H 2002 *Journal of Molecular Catalysis A: Chemical* **177** 256
[8] Morawski AW, Kusiak-Nejman E, Przepio´rski J, Kordala R, Pernak J 2013 *Cellulose* **20**:1293–1300
[9] Liu S, Tao D, Bay H, Liu X 2011 *Journal of Applied Polymer Science* **126** E281
[10] Pang SC, Chin SF, Yih V 2011 *Adv. Mat. Lett* **2**(2) 119
[11] Xu F, Yu J, Tesso T, Dowell F, Wang D 2013 *Applied Energy* **104** 801–809

IOP Conf. Series: Journal of Physics: Conf. Series **795** (2017) 012065 doi:10.1088/1742-6596/795/1/012065

Synthesis propanol by esterification and reduction reaction

Salmahaminati[1] **and Jumina**[2]
[1]Chemistry Departement, Faculty of Mathematics and Natural Sciences, Islamic University of Indonesia, Yogyakarta, 55581, INDONESIA

[2]Chemistry Departement, Faculty of Mathematics and Natural Sciences, Gadjah Mada University, Yogyakarta, 55281, INDONESIA

E-mail: salmahaminati@uii.ac.id

Abstract. Synthesis of propanol from propanoic acid had been done. Propanol was synthesized via two steps. They are; esterification of propanoic acid and methanol in the presence of the sulfuric acid catalyst with the mole ratio of 4:3 to produce methyl propanoate, and reduction of methyl propanoate with sodium using ethylene glycol as the solvent to yield propanol. Structural characterizations of methyl propanoate and propanol were done using IR, ^{1}H-NMR, and GC spectrometers. The results show that esterification of propanoic acid with methanol produced methyl propanoate in 75% yield. Reduction of methyl propanoate using ethylene glycol as a solvent produced propanol in yield of 77%.

1. Introduction

Diesel fuel from plant oils such as rapeseed, soybean, and palm could be producing by enhancing technology. It has largely centered on transesterification of oils with methanol to produce fatty acid methyl esters [1]. Life cycle assessments (LCAs) were published by comparing biodiesel to conventional petroleum diesel [2].

Biodiesel as a substitute for petroleum diesel helps reduce GHG emissions; there are quality issues associated with its widespread use [3,4]. Even though it has been documented that the methyl esters have many desirable fuel qualities such as good cetane number and lubricity, the amount of biodiesel added to a petroleum diesel blend is most often limited to 5% or less due to poor storage stability, marginal cold flow properties, excessive solvency and engine compatibility issues [5-8]. Furthermore to these quality issues, an overall energy balance that includes the life cycle of the raw materials used in biodiesel production indicates that petroleum-derived diesel is still more energy efficient [9].

In this paper, propanol is synthesized. The discussion of this study proposes that substituting propanol for petroleum fuel will reduce in both fossil fuel consumption as well as green house gas emissions. The esterification reaction and external electron donor in reduction reaction are studied in detail.

2. Materials and Characterization

The materials used are: propanoic acid (100%), ethanol, H_2SO_4 (98%), sodium bicarbonate ($NaHCO_3$), sodium sulfate anhydrous (Na_2SO_4), sodium (Na), ethylene glycol, HCl (36%), For characterization, we used Infrared spectrometer (IR, Shimadzu Prestige-21), Gas Chromatography (GC-Hewlett Packard 5890 series II) and Gas Chromatography-Mass Spectrometer (GC-MS, Shimadzu QP-2010S).

3. Methods

Methyl propanoate synthesis

Propanoic acid 50 mL (0.67 mole), methanol 20 mL (0.5 mole) and two mL sulfuric acids were added in the reflux system. The mixture was refluxed and stirred with a magnetic stirer for 14 h. The product was distilled at a temperature of 70-80 °C. Then, it was extracted with ten mL of 10% $NaHCO_3$. The bottom layer was separated, and the top layer is dried with anhydrous Na_2SO_4. The product was weighed and analyzed by GC, IR and ^{1}H-NMR Yield: 33 g (0.4 mole)

Propanol synthesis

Methyl propanoate 5 mL (0.045 mol) was added in a 100 mL three neck flask with 10 g (0.45 mol) of natrium. The mixture is stirred with a magnetic stirrer and heated with an oil bath. After sodium melting, we added ethylene glycol as a solvent. (Added as soon without removing the reflux system). The mixture was refluxed for one hour. The product was distilled at the temperature of 90-100 °C. The product was weighed and analyzed by GC. Yield: 2.2 g (0.035 mol).

4. Results and Discussion

Synthesis of methyl propanoate

An esterification is a reaction of alcohols, carboxylic acids and catalyzed by strong acid to produce an ester. The mol ratio of propanoic acid and methanol using H_2SO_4 as a catalyst in this paper is 4:3. The synthesis was done with the addition of the reactants and refluxed for 14 h [10]. Reaction mechanisms of the esterification are in Figure 1 [11].

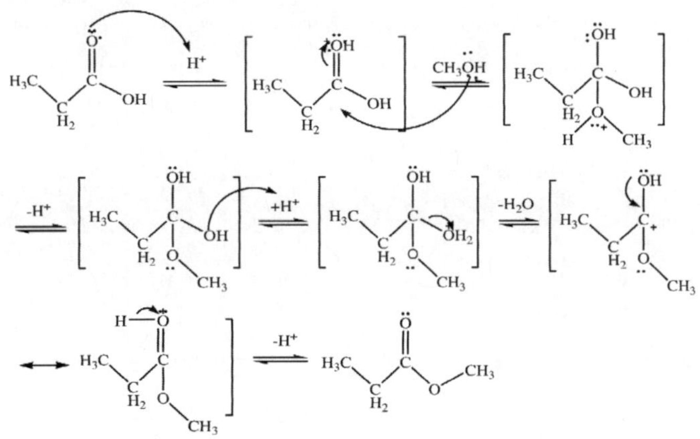

Figure 1. Mechanism reaction of esterification

The esterification reaction of propanoic acid and methanol with the sulfuric acid catalyst will produce methyl propanoate. The results of GC analysis product compared with chromatograms of the esterification results that have been added by spiking of propanoic acid and methanol. The chromatograms are presented in Figure 2.

IOP Conf. Series: Journal of Physics: Conf. Series **795** (2017) 012065 doi:10.1088/1742-6596/795/1/012065

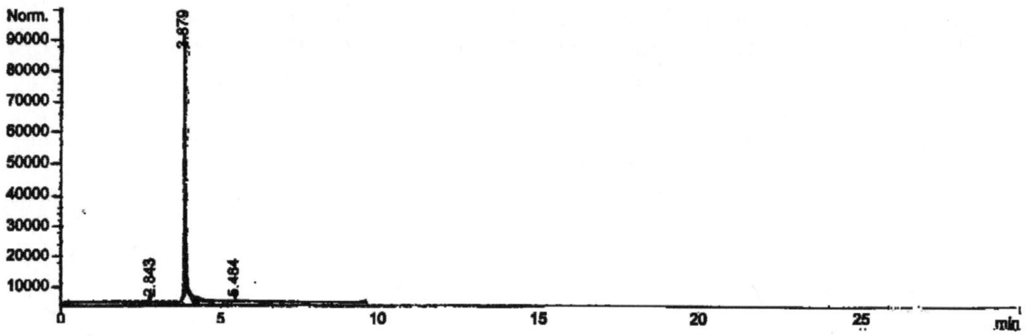

(a)

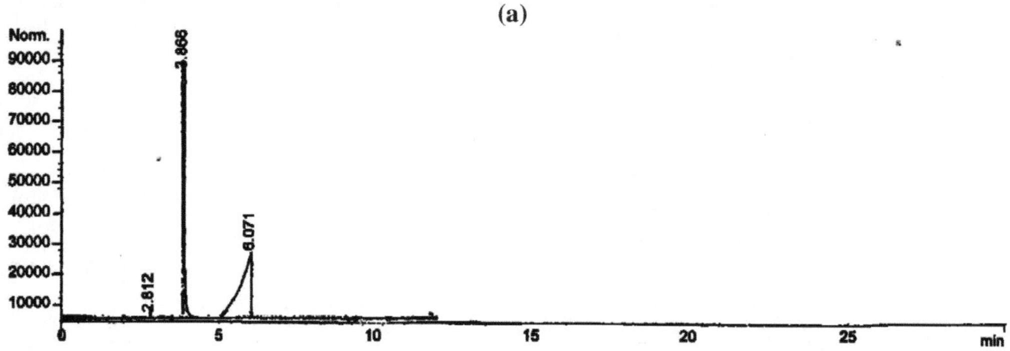

(b)

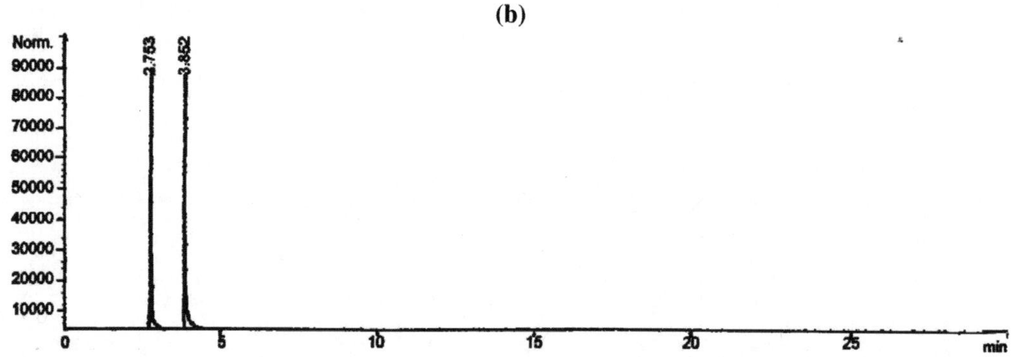

(c)

Figure 2. Chromatogram (a) methyl propanoate product, (b) spiking propanoic acid on the methyl propanoate product, and (c) spiking methanol on the methyl propanoate product

Based on Figure 2 (a), the peak at a retention time (t_R) 3.852 minutes with a percentage of 99% is expected to be the product, methyl propanoate. The peak has appeared at the retention time (t_R) 6.071 minutes after spiking with propanoic acid (b) and the peak at a retention time (t_R) 2,812 minutes has increased after spiking with methanol (c). Thus, it estimated that there is still a little methanol and propanoic acid in the product.

The results of analysis using IR spectrometer provides a spectrum in Figure 3, and the data analysis is in Table 1.

IOP Conf. Series: Journal of Physics: Conf. Series **795** (2017) 012065 doi:10.1088/1742-6596/795/1/012065

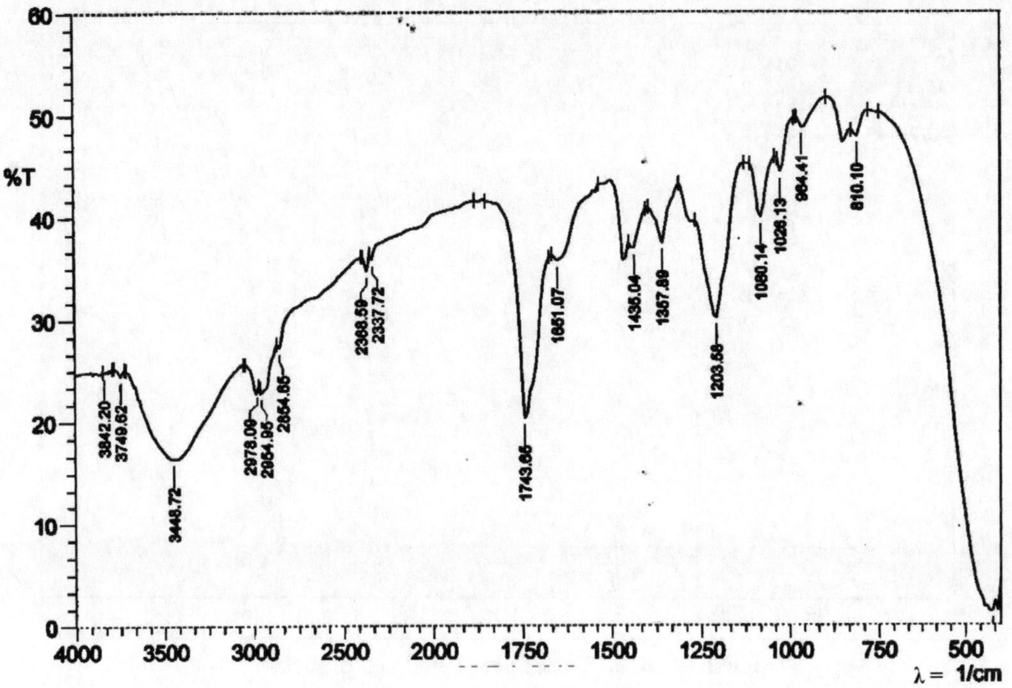

Figure 3. Infrared spectrum of methyl propanoate

At 3448 cm^{-1} region shows a vibration range of hydroxyl (OH) methanol remaining in the fractions of methyl propanoate. Uptake observe for a residual alcohol resulting in the range of the C-O at 1080 cm^{-1} region. Uptake sharply with strong intensity in the area of 1743 cm^{-1} is the vibration of the carbonyl group (-C = O) ester and reinforced by their absorption at 1203 cm^{-1} that indicates the range -C-O-C- ester. Absorption at wave numbers 2978 and 2854 cm^{-1} is the stretching of absorption Csp^3-H in the alkyl group. Uptake in the area in 1357 and 1435 cm^{-1} are indicated the presence of C-H vibration of the bends methylene group (-CH_2) and vibration of C-H bend a methyl group (-CH_3)

Table 1. Infrared spectrum data of methyl propanoate

Wavenumber (cm^{-1})	Functional group
3448	–OH alcohol
1080	–C–O alcohol
1743	–C=O carbonyl
1203	–C–O–C–
2854 and 2978	Csp^3-H
1435	–CH_2–
1357	–CH_3

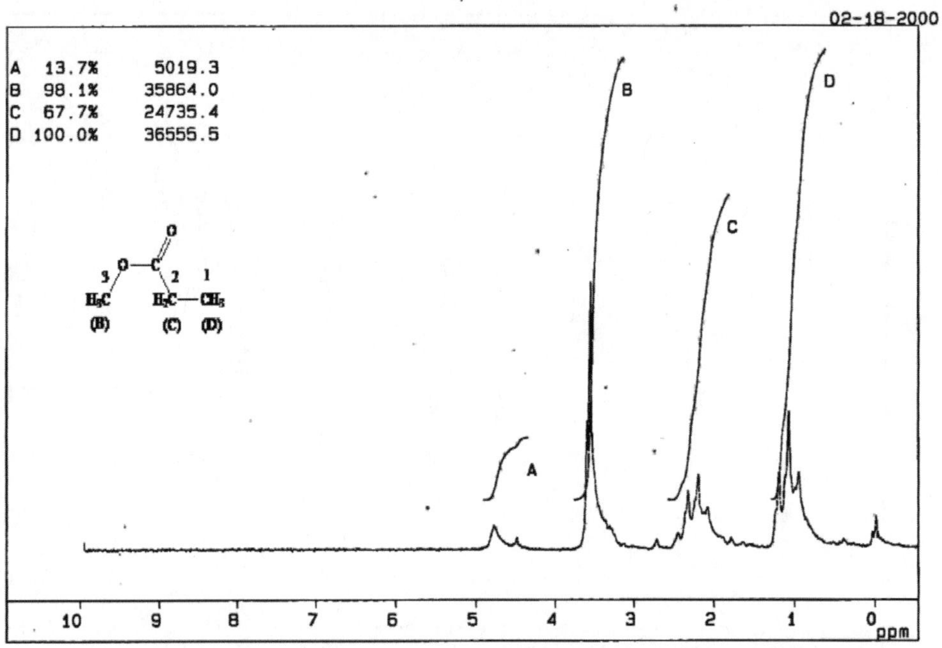

Figure 4. ^{1}H-NMR spectrum of methyl propanoate

Analysis product using ^{1}H-NMR spectrometer, Figure 4, shows the four areas that describe the chemical shift of four protons with different chemical environments. A chemical shift of 4.80 to 4.60 ppm is a singlet appearance, equivalent to one hydrogen atom of the first proton from the -OH in methanol remaining in the fractions of methyl propanoate. Singlet appearance is caused by the influence of an electronegative of O atom. Chemical shift area B (δ 3.70 to 3.40 ppm, triplet, 3H) estimated from proton on C3 ($-OCH_3$). Chemical shift area C (δ 2.40 to 2.00 ppm, quartet, 2H) estimated from proton ester $-CH_2-C=O$, equivalent to three H atoms, is estimated from the three protons at C1. Chemical shift area D (1.30 to 1.10 ppm, triplet, 3H) estimated from the group $-CH_3$ protons at C1, triplet appearance, equivalent to 2 H atom, is estimated from the two protons at C2.

According to the analysis by GC, IR and ^{1}H-NMR spectrometers, it shows that the propanoic acid could be esterification with methanol (4:3) using concentrated sulfuric acid catalyst to produce methyl propanoate as a main component with 75 % percent of the results.

Biopropanol synthesis; reduction of methyl propanoate with Na in ethylene glycol

The reduction reaction of methyl propanoate would produce methanol and propanol. The GC analysis products from the reduction of methyl propanoate are shown in Figure 5.

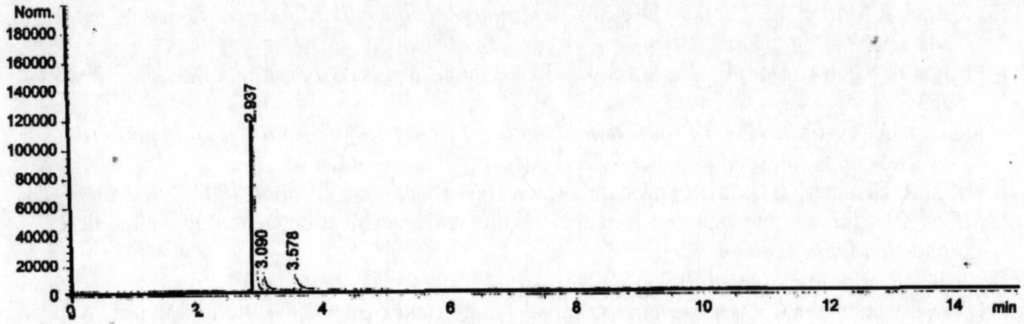

(a)

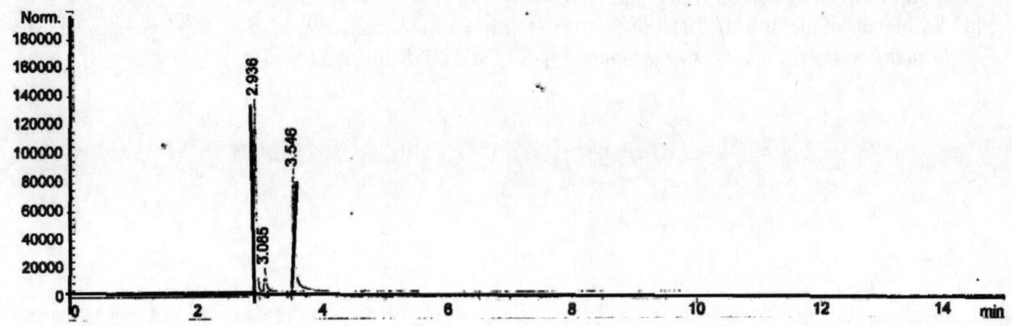

(b)

Figure 5. Chromatogram (a) the product, (b) spiking 1-propanol of the product

Based on Figure 5 (a), the peak at a retention time (t_R) 3.578 minutes is expected to be the product, propanol because it has increased after spiking with 1-propanol (b) but it estimated that there is still an impurity on the products because we found two peaks at a retention time 2.936 and 3.085 minutes. From this results, it shows that the sodium metal could reduce methyl propanoate to be methanol and propanol.

5. Conclusion

Analysis by GC, IR and ^{1}H-NMR spectrometers shows that the propanoic acid could be esterification with methanol (4:3) using sulfuric acid catalyst to produce methyl propanoate as a main component with a yield of 75 %. From these results, it can be ascertained that the sodium metal can reduce methyl propanoate into methanol and 1-propanol. 1-propanol product obtained has a yield of 77% of the results.

6. References

[1] Radich A 2004 *Biodiesel Performance Costs, and Use Energy Information Administration* http://www.eia.doe.gov/oiaf/analysispaper/biodiesel/
[2] Hill J, Nelson E, Tilman D, Polasky S, Tiffany D 2006 Environmental, economic and energetic costs and benefits of biodiesel and ethanol biofuels *Proceedings of the National Academy of Sciences.* 103 11206

[3] Kalnes T, Marker T, Shonnard DR 2007 *International Journal of Chemical Reactor Engineering* 5 1542
[4] Edgar L, Yijun L, Dora E, Kaewta S, David AB, and James GG *Ind. Eng. Chem. Res.* **2005** *44* 5353
[5] Kinast JA, Tyson KS 2003 *Production of biodiesel from multiple feedstocks and properties of biodiesel and biodiesel/diesel blend Final report*; NREL: Golden, CO
[6] Ma FR, Hanna MA, Biodiesel production: a review *Bioresour. Technol.* 1999 *70* 1
[7] Haas MJ, The interplay between feedstock quality and esterification technology in biodiesel production *Lipid Technol.* 2004 *16* 7
[8] Talley P, Biodiesel *Render* 2004 (Sept)
[9] Edwards R, Mahieu V, Griesemann J, Larivé J, and Rickeard DJ 2004 Well-to-Wheels Analysis of Future Automotive Fuels and Powertrains in the European Context *SAE Technical Paper* 1 1924
[10] Smith PWG, Furnish BS, Hannaford AJ and Tatchell AR 1989 *Vogel's Textbook of practical organic chemistry* John Wiley and Sons New York
[11] Salmahaminati, Jumina 2016 *3rd International Conference on Research Implementation, and Education of Mathematics and Science* C-89 ISBN 978-602-73159-1-4

IOP Conf. Series: Journal of Physics: Conf. Series **795** (2017) 012066 doi:10.1088/1742-6596/795/1/012066

Understanding valence-shell electron-pair repulsion (VSEPR) theory using origami molecular models

Teguh Endah Saraswati[1,2*], Sulistyo Saputro[2], Murni Ramli[2], Danar Praseptiangga[2], Nurul Khasanah[3], and Sri Marwati[4]

[1]Department of Chemistry, Mathematic and Natural Science Faculty, Sebelas Maret University, Jl. Ir. Sutami 36A Surakarta 57126, INDONESIA

[2]Center for Japanese Studies, Sebelas Maret University, Jl Ir. Sutarmi 36 A Surakarta 57126, INDONESIA

[3]Madrasah Aliyah Negeri 1 Surakarta, Jl. Sumpah Pemuda No.25 Banjarsari Surakarta 57136, INDONESIA

[4]Madrasah Aliyah Negeri 2 Surakarta, Jl. Brigjen Slamet Riyadi No.308 Laweyan Surakarta 57141, INDONESIA

*E-mail: teguh@mipa.uns.ac.id

Abstract. Valence-shell electron-pair repulsion (VSEPR) theory is conventionally used to predict molecular geometry. However, it is difficult to explore the full implications of this theory by simply drawing chemical structures. Here, we introduce origami modelling as a more accessible approach for exploration of the VSEPR theory. Our technique is simple, readily accessible and inexpensive compared with other sophisticated methods such as computer simulation or commercial three-dimensional modelling kits. This method can be implemented in chemistry education at both the high school and university levels. We discuss the example of a simple molecular structure prediction for ammonia (NH_3). Using the origami model, both molecular shape and the scientific justification can be visualized easily. This 'hands-on' approach to building molecules will help promote understanding of VSEPR theory.

1. Introduction

Molecular geometry governs many chemical properties, including reactivity, polarity, phase of matter, colour, magnetism, and biological activity. A molecule's shape can be predicted by valence-shell electron-pair repulsion (VSEPR) theory, which explains that electron pairs will position themselves as far from each other as possible to minimize repulsion [1-3]. The arrangement of the atoms in a molecule can be studied using two-dimensional (2D) or three-dimensional (3D) models. Conventionally, students determine the shapes of molecules using the valence-shell electron-pair repulsion (VSEPR) theory based on the 2D Lewis electron dot structure.

Interactive models are particularly effective for teaching students to recognize molecular structures. To be truly interactive, a model should be 3D; students often have difficulty imagining a real molecular structure when it is represented by a 2D drawing. For example, in the case of a molecule with four electron groups, a student might draw a square planar molecule with 90-degree angles between the bonds. A 3D model, conversely, is more intuitive.

The commercial 3D models available (e.g., Molymod, 3D simulation software) are generally of limited accessibility due to high cost and the requirement that users possess specialized computer skills even though it was proved could be used to improve the spatial ability of the high-school chemistry students [4]. Alternatively, 3D molecular models can be made by hand using plasticine, straws and toothpicks or matchsticks. Although the materials involved are inexpensive, the relative fragility of these models restricts their applicability.

Here we introduce origami modelling for visualising and exploring molecular shape. Origami, the Japanese art of paper folding, transforms a flat sheet of paper into intricate 3D designs. Origami has applications beyond its traditional aesthetic value; for instance, the technique has been used in teaching mathematics [5]. In chemistry, origami models can be used to approximate 3D molecular structures and explore molecular function [6].

Our origami model is adopted from Robert M. Hanson's *Molecular origami: precision scale models from paper* [7]. Here we explain how a 3D origami model shows differences in molecular geometry that cannot be depicted using the traditional 2D drawing method.

2. Two-dimensional (2D) and three-dimensional (3D) molecular models

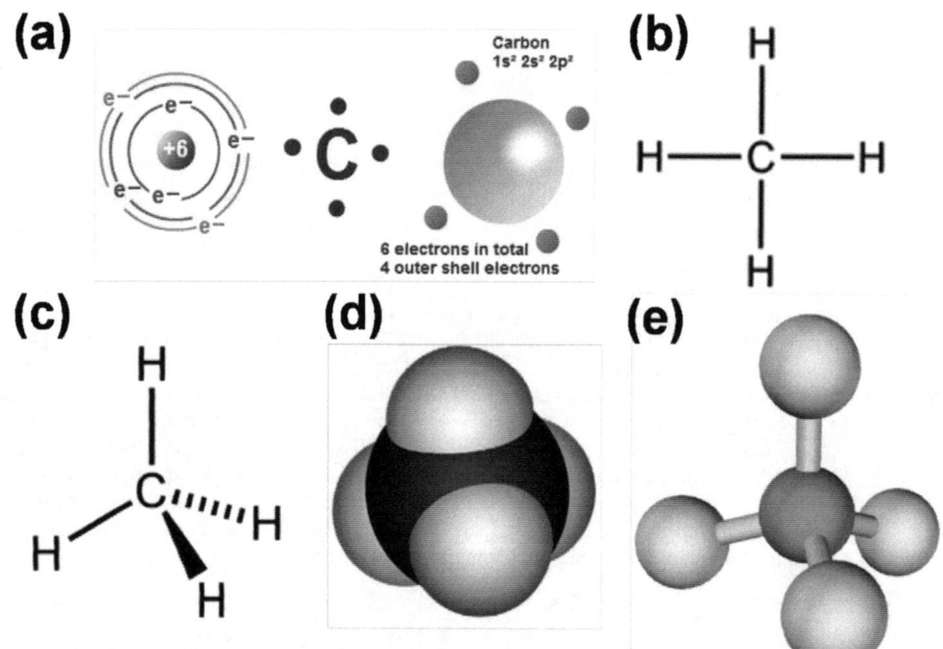

Figure 1. The valence shell electrons of the carbon atom (a) and the tetrahedral methane molecule (b), with 2D and 3D drawings by line (c), ball (d) and stick and ball (e).

To represent molecular configurations on a 2D surface (e.g., paper, blackboard or screen), we often use perspective drawings in which the direction of a bond is specified by the type of line connecting the bonded atoms. Figure 1 shows 2D and 3D perspective drawings of methane (CH_4), which has a tetrahedral molecular geometry. Conventionally the central carbon atom is placed in the plane of the drawing surface. Bonds to this atom that lie in the same plane are represented by a line of normal thickness, whereas bonds lying out of this plane are represented differently. Bonds directed in front of the plane are shown as wedges, with an atom or group at the thicker end of the wedge interpreted as nearer to the viewer than the carbon atom at the narrower end of the wedge. Bonds directed behind the plane of the carbon atom are hashed or broken.

The perspective drawings in Figure 1 show whether atoms and bonds are in plane with the screen surface, behind it or in front of it. However, these representations do not accurately convey the bond angles and therefore provide only limited insight into the molecular properties of the compound.

3. Theoretical background: molecular shape and valence-shell electron-pair repulsion (VSEPR) theory

Molecular geometry is determined by the relative three-dimensional positions of the atomic nuclei. A molecule's shape is dependent on the preferred spatial orientation of covalent bonds to atoms having two or more bonding partners. The three-dimensional configuration of a molecule determines properties including its reactivity, polarity, phase of matter, colour, magnetism, and biological activity.

Molecular geometries can be explained using Lewis structures, VSEPR theory, valence bond theory, hybridization theory and molecular orbital theory; here we use primarily VSEPR theory. To determine molecular geometry, it is important to understand the difference between bonding electron pairs and nonbonding electron pairs. Bonding pairs are electrons shared by the central atom and another atom, while nonbonding pairs are electrons on an individual atom that are not shared with another atom. The term 'electron-pair geometry' refers to the bonding and nonbonding electron domains; 'molecular geometry' describes the resulting bond angles between the central and terminal atoms, which together dictate the shape of a molecule [8].

The general procedure for determining the molecular geometry involves first drawing the Lewis structure, which shows the 2D distribution of atoms and electrons and can be used to determine the number of bonding and nonbonding electron pairs on the central atom. The molecular geometry can then be predicted using the VSEPR theory.

The VSEPR (pronounced "VES-per") theory predicts the shapes of molecules and ions by assuming the valence-shell electron pairs are as far apart as possible, thus minimizing electron pair repulsions. Electron pairs or groups of electron pairs in the valence shell of an atom are known as the electron domain. The electron domain is divided into bonding domains, which include electrons in bonds, and nonbonding domains, which include unshared electrons [8, 9]. A bonding domain contains the electrons that are shared between two atoms, as a single, double or triple bond, while a nonbonding domain (also known as a lone pair) contains valence electrons that are associated with a single atom. Table 1 summarizes the molecular and electron pair geometries for different combinations of bonding and nonbonding electron domains [8, 9].

Table 1. Molecular geometries predicted by VSEPR theory [8, 9]

Electron pairs			Arrangement of pairs (A: central atom; X: surrounding atom; E: lone pair electrons)		Molecular geometry
Total	Bonding (X)	Lone (E)			
2	2	0	Linear	Linear AX_2	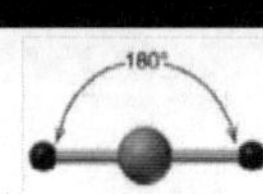
3	3	0	Trigonal planar	Trigonal planar AX_3	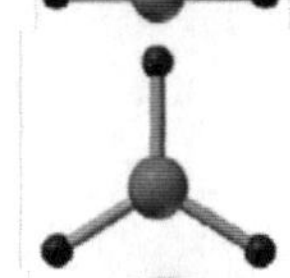
	2	1		Bent (or angular) AX_2E_1	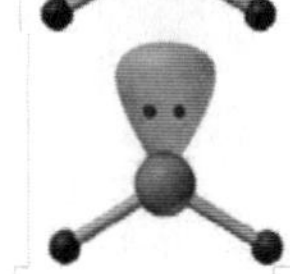

IOP Conf. Series: Journal of Physics: Conf. Series **795** (2017) 012066 doi:10.1088/1742-6596/795/1/012066

Table 1. Molecular geometries … (cont.)

Electron pairs			Arrangement of pairs (A: central atom; X: surrounding atom; E: lone pair electrons)		Molecular geometry
Total	Bonding (X)	Lone (E)			
4	4	0	Tetrahedral	Tetrahedral AX_4	
	3	1		Trigonal pyramidal AX_3E_1	
	2	2		Bent (or angular) AX_2E_2	
5	5	0	Trigonal bipyramidal	Trigonal bipyramidal AX_5	
	4	1		Seesaw (or distorted tetrahedron) AX_4E_1	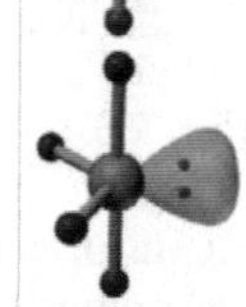
	3	2		T-shaped AX_3E_2	
	2	3		Linear AX_2E_3	
6	6	0	Octahedral	Octahedral AX_6	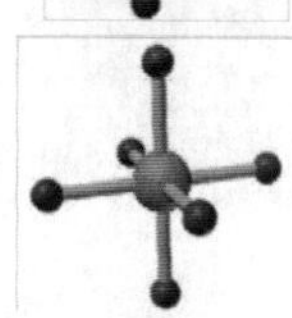
	5	1		Square pyramidal AX_5E_1	
	4	2		Square planar AX_4E_2	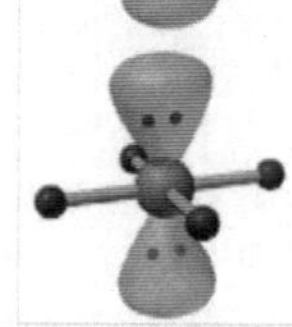

If a molecule has two bonding electron domains, its molecular geometry is predicted to be linear. If a molecule has four bonding domains, its molecular geometry is predicted to be tetrahedral. However, when the molecule has one or more nonbonding domains the molecular geometry may change. For example, a water molecule has two bonding domains and two lone pair electrons. A water molecule's geometry is predicted to be neither linear nor tetrahedral, but rather a bent structure. According to VSEPR theory, the lone pairs are positioned adjacent to each other to minimize repulsion between the electron domains.

4. Three-dimensional (3D) modelling of molecular geometry using origami based on VSEPR theory

Origami modelling provides a 'hands-on' approach for students to practise visualising 3D molecular structures. Flinn Scientific has developed origami patterns for several molecular geometries, such as trigonal pyramidal, tetrahedral, seesaw, square pyramidal, trigonal bipyramidal and octahedral [10]. Another molecular origami kit is provided by Hanson [6]; see Figure 2 for one example of trigonal pyramidal patterns. The origami patterns are drawn with solid lines to represent 'mountain' folds, which fold away from the viewer, and dashed lines to represent 'valley' folds, which fold toward the viewer. The shaded regions can be removed or simply folded back and tucked into the hidden recesses of the model.

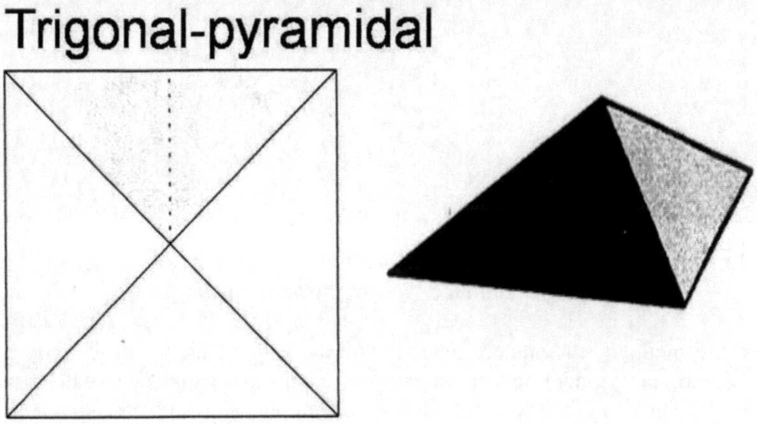

Figure 2. Origami patterns for trigonal pyramidal molecular geometries [6].

To illustrate the advantages of origami modelling over conventional methods, consider as an example the simple chemical structure of ammonia (NH_3). Traditionally, to determine the molecular geometry one must first draw its Lewis structure. The number of valence electrons on each atom can be determined from the electron configurations of the elements. The electron configurations of hydrogen and nitrogen are shown below:

H: $1s^1$

N: [He] $2s^2\ 2p^3$

Thus, hydrogen and nitrogen have valences of one and five electrons, respectively. Using these valence numbers, the electrons can be combined to form covalent bonds to draw a Lewis structure in which all of the elements except the hydrogen atoms have an octet of valence electrons, as shown in Figure 3a.

The Lewis structures do not accurately represent the molecular geometry of ammonia because its shape is not flat as described by the octet rule. Even though the molecule has three bonding electron pairs, according to VSEPR theory it is not trigonal planar because it also has a lone pair electron.

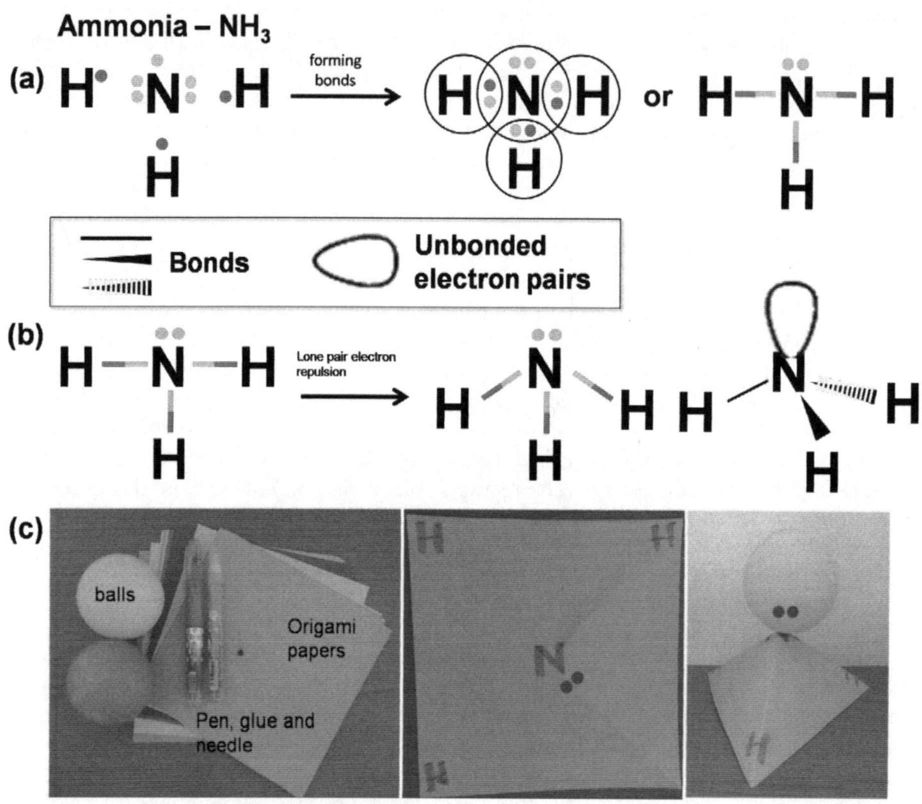

Figure 3. The Lewis structure (a), conventional drawing of 3D molecular geometry (b) and origami molecular model (c) of the ammonia molecule.

Using the origami method, students are able to explore VSEPR theory by directly manipulating physical models. The origami model only requires paper, pen, glue, balls and needles. Before folding the trigonal pyramidal pattern shown in Figure 3c (centre), the nitrogen and hydrogen atoms are planar. However, the lone pair electron on the nitrogen atom (represented by a ball attached to the top corner of the nitrogen) causes the molecular geometry to change. As VSEPR theory predicts, electron repulsions occur not only among the bonding pair electrons but also between the bonding electrons and the lone pair electron [1-3]; that is, three N-H bonds are repelled by the lone pair electron on the nitrogen atom. The lone pair electron will reduce the likelihood of a planar geometry because the repulsion between bonding pair electrons and the lone pair electron is greater than the repulsion among the bonding pair electrons [1-3]. Finally, using this 'hands-on' approach, the student is able to construct a molecule's geometry while considering the scientific reasoning behind it.

5. Implementation of the origami model for teaching VSEPR theory in high school chemistry

Next the origami model for molecular geometry was introduced in a high school chemistry class. The students had already received traditional instruction (without the 3D molecular geometry model) on the topics of electron configuration, valence electrons, Lewis structures, bonding and nonbonding electron pairs within the electron domain, the octet rule, and VSEPR theory.

At the beginning of the study, the students were given three pre-test questions to gauge their initial understanding of the material discussed in class. After the origami model was used to teach VSEPR theory, the students were asked to answer post-test questions. The results from the pre- and post-tests were analysed to determine whether the students understood VSEPR theory more fully after using the origami model.

One student's pre- and post-test responses are shown in Figures 4 and 5. Because the tests were administered in Indonesian (the national language where the study was conducted), these examples have been translated to English.

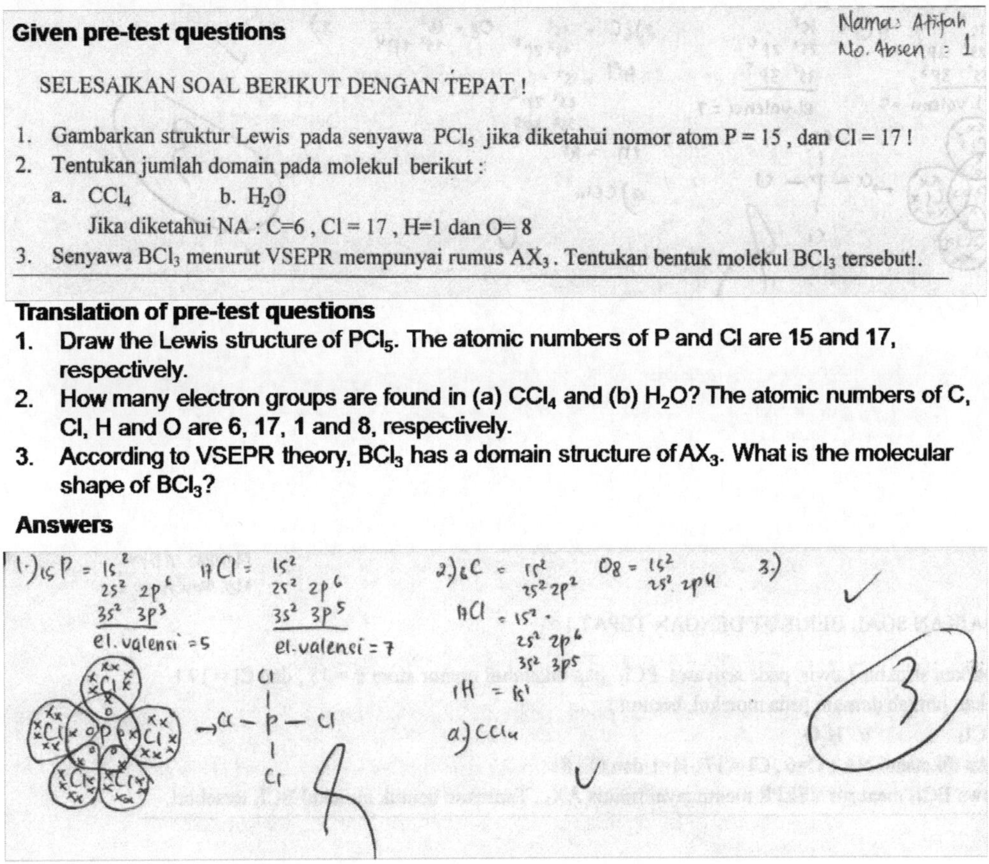

Given pre-test questions

Nama: Atifah
No. Absen : 1

SELESAIKAN SOAL BERIKUT DENGAN TEPAT !

1. Gambarkan struktur Lewis pada senyawa PCl_5 jika diketahui nomor atom P = 15 , dan Cl = 17 !
2. Tentukan jumlah domain pada molekul berikut :
   a. CCl_4 b. H_2O
   Jika diketahui NA : C=6 , Cl = 17 , H=1 dan O= 8
3. Senyawa BCl_3 menurut VSEPR mempunyai rumus AX_3 . Tentukan bentuk molekul BCl_3 tersebut!.

Translation of pre-test questions

1. **Draw the Lewis structure of PCl_5. The atomic numbers of P and Cl are 15 and 17, respectively.**
2. **How many electron groups are found in (a) CCl_4 and (b) H_2O? The atomic numbers of C, Cl, H and O are 6, 17, 1 and 8, respectively.**
3. **According to VSEPR theory, BCl_3 has a domain structure of AX_3. What is the molecular shape of BCl_3?**

Answers

Figure 4. Pre-test questions and example answers.

Although the pre- and post-tests were not identical, the questions focused on the same concepts. For example, question 3 of the pre-test (Figure 4) and question 1 of the post-test (Figure 5) both were intended to assess students' understanding of the trigonal planar molecular shape. The post-tests also contained related questions about V-shaped geometry (questions 2 and 4) and trigonal pyramidal geometry (questions 3 and 5). These similar questions were prepared to compare students' comprehension of particular topics before and after instruction using the origami model.

As shown in Figure 4, one student correctly answered the pre-test question about the Lewis structure (question 1) but not the questions 2 and 3 about the number of bonding and nonbonding electron pairs, even though she had received traditional instruction on all this material. Perhaps she knew how to draw the Lewis structure but struggled with the theoretical background and its use in predicting molecular geometry.

Given post-test questions

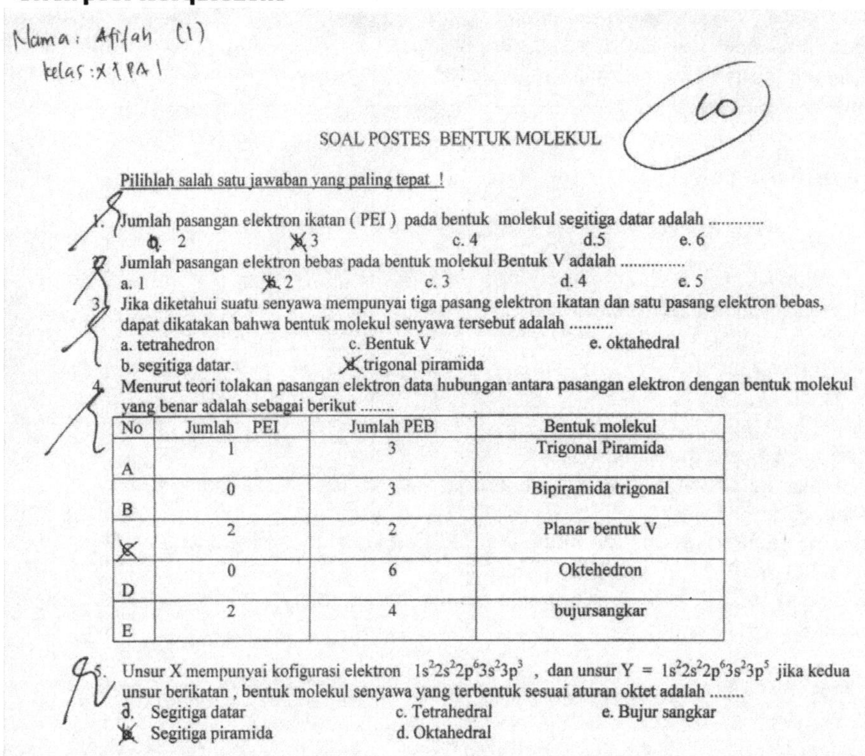

Nama: Afifah (1)
kelas: X IPA 1

60

SOAL POSTES BENTUK MOLEKUL

Pilihlah salah satu jawaban yang paling tepat !

1. Jumlah pasangan elektron ikatan (PEI) pada bentuk molekul segitiga datar adalah ....
   a. 2 b. 3 c. 4 d.5 e. 6
2. Jumlah pasangan elektron bebas pada bentuk molekul Bentuk V adalah ....
   a. 1 b. 2 c. 3 d. 4 e. 5
3. Jika diketahui suatu senyawa mempunyai tiga pasang elektron ikatan dan satu pasang elektron bebas, dapat dikatakan bahwa bentuk molekul senyawa tersebut adalah ....
   a. tetrahedron c. Bentuk V e. oktahedral
   b. segitiga datar. d. trigonal piramida
4. Menurut teori tolakan pasangan elektron data hubungan antara pasangan elektron dengan bentuk molekul yang benar adalah sebagai berikut ....

No	Jumlah PEI	Jumlah PEB	Bentuk molekul
A	1	3	Trigonal Piramida
B	0	3	Bipiramida trigonal
C	2	2	Planar bentuk V
D	0	6	Oktehedron
E	2	4	bujursangkar

5. Unsur X mempunyai kofigurasi elektron $1s^2 2s^2 2p^6 3s^2 3p^3$, dan unsur Y = $1s^2 2s^2 2p^6 3s^2 3p^5$ jika kedua unsur berikatan , bentuk molekul senyawa yang terbentuk sesuai aturan oktet adalah ....
   a. Segitiga datar c. Tetrahedral e. Bujur sangkar
   b. Segitiga piramida d. Oktahedral

Translation of post-test questions:

1. How many bonding electron pairs are in a trigonal planar molecule?
   a. 2 b. 3 c. 4 d. 5 e. 6
2. How many nonbonding electron pairs are in a V-shaped molecule?
   a. 1 b. 2 c. 3 d. 4 e. 5
3. A compound with three bonding electron pairs and one lone pair electron has which of the following molecular shapes?
   a. tetrahedral b. trigonal planar c. V shape d. trigonal pyramidal
   e. octahedral
4. Assuming the repulsion of lone pair electrons, choose the correct combination of data from a-e below:

	Number of bonding electron pair	Number of lone pair electron	Molecular shape
a	1	3	Trigonal pyramidal
b	0	3	Trigonal bipyramidal
c	2	2	V shape (bent)
d	0	6	Octahedral
e	2	4	Square planar

5. An element X has the electron configuration $1s^2 2s^2 2p^6 3s^2 3p^3$, while an element Y has the electron configuration $1s^2 2s^2 2p^6 3s^2 3p^5$. If elements X and Y formed a bond, the molecular shape in accordance with the octet rule would be which of the following?
   a. trigonal planar b. trigonal pyramidal c. tetrahedral d. octahedral
   e. square planar

Figure 5. Post-test questions and example answers.

As shown in Figure 5, the same student was able to answer all of the post-test questions correctly. After the origami model lesson, she knew how many bonding electron pairs are found in the trigonal planar molecular shape (question 1), whereas she could not answer the relevant question in the pre-test.

Moreover, the student confidently gave the correct answers in the post-test for both related questions 2 and 4 which asked about V-shape geometry, and for other related questions 3 and 5 which asked about trigonal pyramidal. Correct answers to these recurring questions indicate an improvement in the student's understanding of molecular geometry according to VSEPR theory after using origami models.

6. Conclusions

Origami molecular modelling is simple, readily accessible and inexpensive compared with other sophisticated methods such as computer simulation and commercial modelling kits. Using this 'hands-on' approach, students can predict the molecular shape and understand the justification according to VSEPR theory.

Acknowledgement

This work was supported in part by Grants of PNBP 2016 Sebelas Maret University provided by Minister of Research, Technology and Higher Education, the Republic of Indonesia under project No. 624/UN27.11/PL/2015 and 632/UN27.21/LT/2016.

References

[1] Rayner-Canham G and Overton T 2010 *Descriptive Inorganic Chemistry* (New York, USA: W. H. Freeman and Company)

[2] Housecroft C E and Sharpe A G 2005 *Inorganic Chemistry* (London: Pearson)

[3] Muller U 2006 *Inorganic Structural Chemistry* (England: John Wiley and Sons., Lyd)

[4] Barnea N and Dori Y J 1999 High-School Chemistry Students' Performance and Gender Differences in a Computerized Molecular Modeling Learning Environment *Journal of Science Education and Technology* **8** 257-71

[5] Mitchell D 1997 *Mathematical origami: geometrical shapes by paper folding*: Tarquin Pubns)

[6] Davis J, Leslie R, Billington S and Slater P R 2010 Origami: a versatile modeling system for visualising chemical structure and exploring molecular function *Chemistry Education Research and Practice* **11** 43-7

[7] Hanson R M 1995 *Molecular origami: precision scale models from paper*: University science books)

[8] Jespersen N D, Brady J E and Hyslop A 2012 *Chemistry: The Molecular Nature of Matter* (New York, USA: John Wiley and Sons, Inc.)

[9] Ebbing D D and Gammon S D 2009 *General Chemistry* (New York, USA: Houghton Mifflin Company)

[10] FlinnScientific VSEPR Origami - Student Activity Kit.

Energy efficiency evaluation of hospital building office

Indah Fitriani, Senot Sangadji and S.A. Kristiawan[1]

Civil Engineering Post graduate Program Sebelas Maret University
Jl. Ir. Sutami 36A Kentingan Jebres Surakarta 57126, INDONESIA

E-mail: indahfmku@gmail.com,s.sangadji@ft.uns.ac.id, s.a.kristiawan@ft.uns.ac.id

Abstract. One of the strategy employed in building design is reducing energy consumption while maintaining the best comfort zone in building indoor climate. The first step to improve office buildings energy performance by evaluating its existing energy usage using energy consumption intensity (Intensitas Konsumsi Energi, IKE) index. Energy evaluation of office building for hospital dr. Sayidiman at Kabupaten Magetan has been carried out in the initial investigation. The office building is operated with active cooling (air conditioning, AC) and use limited daylighting which consumes 14.61 kWh/m2 /month. This IKE value is attributed into a slightly inefficient category. Further investigation was carried out by modeling and simulating thermal energy load and room lighting in every building zone using of Ecotect from Autodesk. Three scenarios of building energy and lighting retrofit have been performed simulating representing energy efficiency using cross ventilation, room openings, and passive cooling. The results of the numerical simulation indicate that the third scenario by employing additional windows, reflector media and skylight exhibit the best result and in accordance with SNI 03-6575-2001 lighting standard. Total thermal load of the existing building which includes fabric gains, indirect solar gains, direct solar gains, ventilation fans, internal gains, inter-zonal gains and cooling load were 162,145.40 kWh. Based on the three scenarios, the thermal load value (kWh) obtained was lowest achieved scenario 2 with the thermal value of 117,539.08 kWh.The final results are interpreted from the total energy emissions evaluated using the Ecotect software, the heating and cooling demand value and specific design of the windows are important factors to determine the energy efficiency of the buildings.

1. Introduction

The global contribution from buildings towards energy consumption, both residential and commercial, has steadily increased, reaching figures of 20% to 40% in developed countries [1,2]. Office buildings are types of commercial buildings with large amounts of energy consumption. Energy saving has always been one of the ultimate objectives for efficient building design is that be able in sustainable development. One of the strategies employed in building design, reducing energy consumption while maintaining the best comfort zone in building indoor climate. In tropical countries such as Indonesia, the major energy consumption in buildings is for air-conditioning and lighting which should be properly managed to avoid high cost. Lighting consumption depends on the purpose of the building, the use of daylight, illumination levels for certain areas and the hours of usage. A well-designed lighting system in buildings would not only significantly reduce long-term energy utilization, but it would also reduce long terms cost.

The study of lighting performance in the building tends to concentrate on aspects relating to energy consumption and involves building design including that of fenestrations and lighting level, whether daylighting and artificial lighting [3].

IOP Conf. Series: Journal of Physics: Conf. Series **795** (2017) 012067 doi:10.1088/1742-6596/795/1/012067

In this study, the object of research is a hospital building office at Kabupaten Magetan it is required that 30% of working floor area has natural light with 300 lux [8]. Based on preliminary observations, the existing of natural conditions in the space of building visible dark in the north and west sides of the building, but in the south side its visible over bright. This research is intended to develop an accurate computer model of office buildings in dr. Sayidiman Hospital at Kabupaten Magetan about daylight and thermal (a) to ensure that the model is accessible and adaptable for future users, (b) to investigate the lighting and thermal performance within the office model using simulation models, (c) to concentrate on the energy saving through daylighting and thermal control.

2. Methodology

The study of daylighting and thermal performance of buildings can be achieved by experimental measurements and computer modeling. As the experimental measurements are costly and time-consuming, computer modeling has recently attracted great attention for their simplicity and reability. Computational simulation is the best method for this study because it can predict the lighting and thermal performance in a building. Computer-based building energy simulations have proven a useful tool in analyzing lighting and thermal schemes.[4,5].

Magetan city is located in East Java, Indonesia. It is located at 7⁰38'30" north latitude and 111⁰25'45" – 111⁰20'30" east longitude. Typical year weather data of Surabaya city where located in East Java is adopted, which is very close to Magetan. Then relevant data is put into weather tool of Ecotect which could automatically output the analysis results of Magetan climate.

2.1. Energy Consumption Intensity

The energy consumption intensity (Intensitas Konsumsi Energy, IKE) is a term to determine amount of energy use in buildings and has been applied in various countries (ASEAN, APEC), expressed in units of kWh / m2 per month [6]. Table 1 presents the standard IKE buildings in Indonesia [7].

Tabel 1. Category level of energy consumption per month based on IKE

IKE (kWh/m²/month)	Category
4,17 - 7,92	Extremely efficient
7,92 - 12,08	Efficient
12,08 - 14,58	Slightly efficient
14,58 - 19,17	Slightly inefficient
19,17 - 23,73	Inefficient
23,75 - 37,75	Extremely inefficient

The energy consumption intensity in each room can be calculated using the following equation:

$$IKE = \frac{P_K}{A} \quad (1)$$

where IKE = energy consumption intensity (kWh/m^2/month), P_K = total electricity consumption (kWh), A = area (m²). It should be emphasized that as-built drawing in each room of the building should complete. Criteria such as the air-conditioned room or not air-conditioned room and spacious rooms should be evident and accurate. The data type of equipment, the amount of equipment, the power supply of any type of electrical appliances as well as the area for each room also needs to be included in the report energy consumption.

2.2. Lighting

Natural light sources can be categorized into direct (in the form of direct light or diffuse light from the sky) or indirect (reflection light or natural light that has been modified from its main source) [8]. SNI (Indonesian national standard) does not take into account solar intensity, therefore lighting level data were taken as a uniform, which is 10.000 lux. Table 2 shows that standard illumination index to achieve visual comfort for the occupant in the building. The standard is based on the function and effectiveness of the lighting space [9].

Table 2. Standard illumination (Indonesian national standard, SNI) index

Type of Building	Rooms	The recommended illumination values (Lux)
Office	President room	350
	Workspace room	350
	Meeting room	300
	Computer room	350
	Active archive room	300
	Archive room	150

2.3. *Thermal*

The physical aspects of thermal comfort depends on six main factors which serve as an interrelated system is influenced by psychological factors; (1) ambient air temperature, (2) mean radiant temperature, (3) relative humidity (4) air movement, (5) clothing insulation, (6) metabolic heat rate [10]. The method of cooling load temperature difference is needed to cope with the external heat load caused by the heat entering through conduction (walls, ceilings, glass, partitions, floors), radiation (glass), and convection (ventilation and infiltration). Internal heat load caused by the heat arising persons/occupants, lights and equipment/machinery [11].

To determine the internal heat gains due to electrical appliances within the building, a stastic model based on the load profile or the power density of electrical appliances has been applied in most of building energy simulation tools. These gains are usually used to calculate the heating and cooling energy demand of a building and to design an adequate HVAC system for building.

3. Results and Discussion

3.1. *Intensity of Energy Consumption*

To calculate the IKE air conditioning in a room, the necessary data collection on the type of air-conditioning installed in the room below its installed power capacity (Watt). Observation and data collection on the building of the electrical appliances shown in Table 3.

Table 3. Type of equipment and power capacity (Watt) on the office building

Type of Load	Electrical Equipment	Capacity (Watt)	Quantity (Unit)	Total	Sub Total
1 Light	DL PL-C Lamp	18	53	954	1,918
	RM 2xTL5 Lamp	28	33	924	
	Baret TL Lamp	20	2	40	
2 HVAC	AC 2,5 PK	1,865	15	27,975	47,371
	AC 2 PK	1,492	10	14,920	
	AC 1,5 PK	1,119	4	4,476	
3 Others	Exhaust fan	50	2	100	7,685
	Computer	125	16	2,000	
	Laptop	20	10	200	
	Printer 1	40	4	160	
	Printer 2	20	9	180	
	LED TV 32"	100	2	200	
	LED TV 29"	70	8	560	
	Foto Copy	175	2	350	
	Refrigerator	200	1	200	
	Telephone	20	11	220	
	Finger print	10	1	10	

Type of Load	Electrical Equipment	Capacity (Watt)	Quantity (Unit)	Total	Sub Total
	Dispenser	100	2	200	
	Water heater	300	2	600	
	Wifi server	150	1	150	
	Camera CCTV	120	6	720	
	Integrated sound system	835	1	835	
	Water pump	1000	1	1,000	
				Total	56,974

It is assumed that the number of hours worked (operational activities) in one day for eight hours, starting from 07.30 am to 15:30 pm and factor load requirements for six working days which is 24 days in a month. The consumption of electric energy per month is 10.939 kWh/day. The total areas offices of the selected were 748.85 m2. Thus IKE air-conditioned 14.61 kWh/m2/month. This value when associated with the IKE standards for yhe air-conditioned room for an office building is included in a slightly inefficient category.

3.2. Energy Simulation

3.2.1. Lighting.

This research was carried out during a sunny day on July 9th, 2016 at 12.00 pm using lux-meter placed 75 cm above the floor at several points assuming the average height of working table when people were sitting in the room. The values obtained were used as the benchmark. In figure 1 shows the location point in the building, respectively.

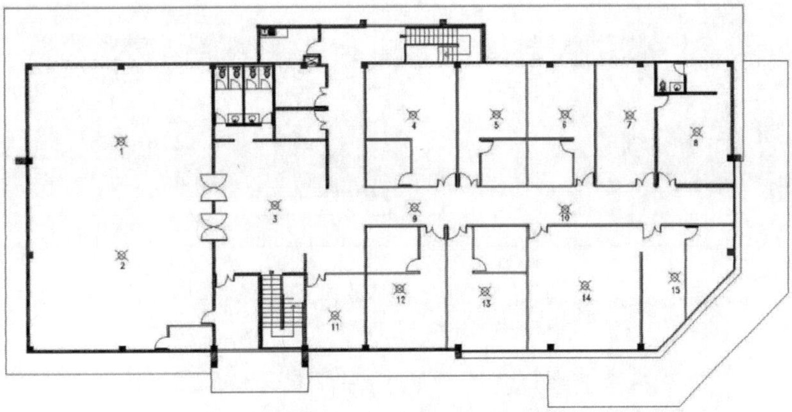

Figure 1. Location points

The average of illuminance level is about under 350 lux. Stated that an office should have a design of 350 lux for illuminance level illustrated in Table 4. So, the interior illuminance level in office areas is less than the recommended value, based on the SNI.

a. Lighting Analysis

The results of the simulation using Ecotect programming of natural lighting in the northern and eastern side of the building is the color of blue and purple means look less good lighting quality while on the south side have the quality of lighting the excess is marked in yellow. From the measurement results obtained that natural and artificial lighting in a space not meet its standard of 350 lux for the boardroom, a work room and a computer room, 300 lux to 100 lux and a meeting room for the corridor. Simulation with Ecotect of the existing condition of the building, given three experiments conducted, the experiment can be seen as in Figure 4.

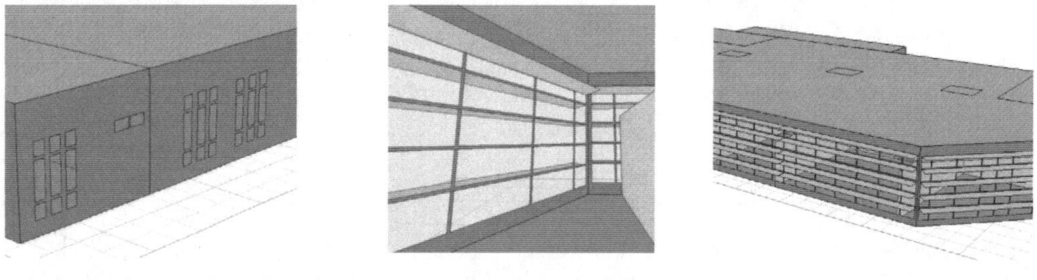

Figure 4. (a) Scenario 1 (b) Scenario 2 (c) Scenario 3

In the first scenario of the existing building is added with the opening of the window on the north side, the blue color is an added window. Scenario 2 Each window openings both existing as well as additional on the north, south, and west in the given additional media self-light reflector is placed at a position above the human eye. In addition to the south side added shading to reduce incoming sunlight, so not too much glare and heat in the room. While in scenario 3 Any opening additional windows either existing or in the north, south, and west, which has been given additional self-light and shading, then added with a skylight in the roof of the building.

b. Simulation Analysis

There are three results of the following simulation have analyzed by Ecotect simulation, Consist of scenario 1, scenario 2, and scenario 3.

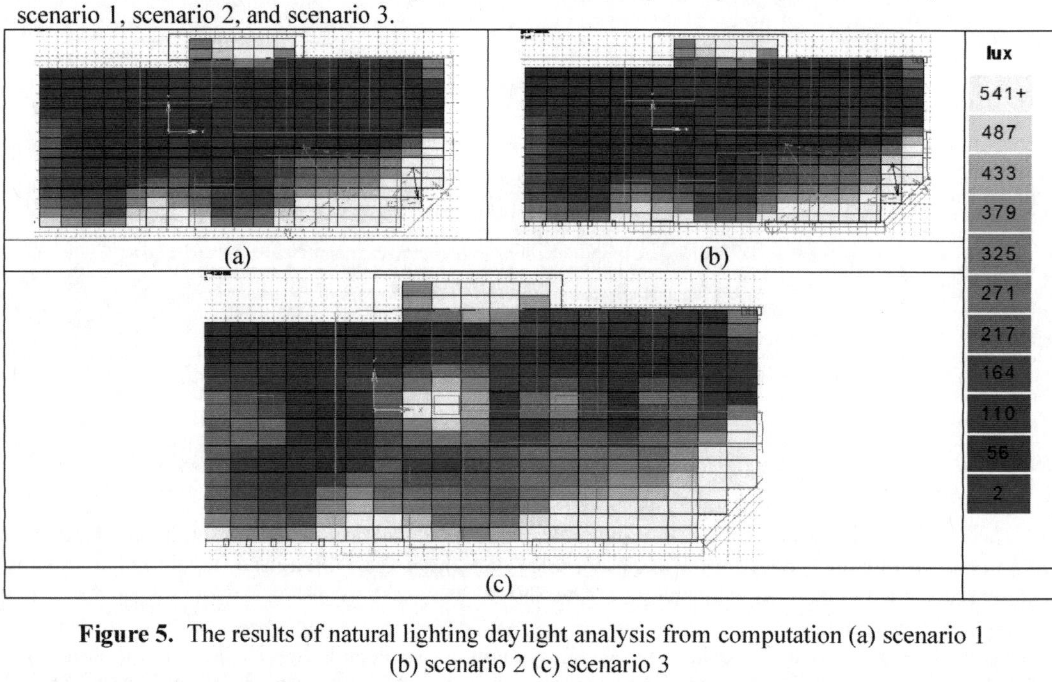

Figure 5. The results of natural lighting daylight analysis from computation (a) scenario 1 (b) scenario 2 (c) scenario 3

Daylight analysis results as seen in Figure 5 consist of first scenario (a) with the addition of light windows of the north side of the lux level is increasing although not up to the middle of the room, while the south side is still too strong. Second scenario (b) visible in the middle of the room lux level is increasing because it is assisted by a self-light from windows north and south sides, while on the south side looks lux levels declining as by shading of windows on the south side. The third scenario (c) with an additional four roof sky the center of the room appear brighter than the other scenarios. The comparison data between existing lux levels and three scenario are summarized in Table 4.

Table 4. Lux value of the comparison rooms each scenario without artificial light

No	Rooms	Points	Lux Existing	Lux Sim. 1	Lux Sim. 2	Lux Sim. 3
1	Big meeting room 1	1	83	148	195	305
2	Big meeting room 1	2	139	250	275	350
3	Free Function	3	174	250	280	300
4	General affair room	4	47	255	275	265
5	Officialdom room	5	53	260	280	265
6	Program & planning room	6	27	245	265	265
7	Small meeting room	7	42	250	280	345
8	President room	8	37	245	275	300
9	Corridor 1	9	47	97	250	280
10	Corridor 2	10	91	125	265	275
11	Head of administration room	11	187	187	257	300
12	Service room	12	243	243	275	310
13	Support room	13	583	583	350	350
14	Finance room	14	500	500	350	350
15	Head of finance room	15	600	600	350	350

Shown in Table 5 comparison lux levels at any points of space, and the scenario three which has a lux level approaching 03-6575-2001 and SNI standard 6197: 2011. Graph of simulation results also be displayed in Figure 6.

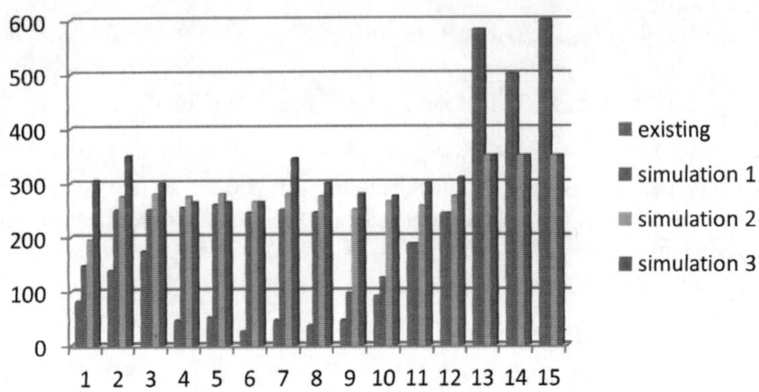

Figure 6. lux comparison graph

3.2.2. Thermal Simulation

The average annual of thermal reveal that the hottest day is on November 2nd where it reached 37,4°C and the coldest day is on Agust 3rd where it down to 18,6°C.

Table 5. Comparison result of thermal load and cooling load each scenario

IOP Conf. Series: Journal of Physics: Conf. Series **795** (2017) 012067 doi:10.1088/1742-6596/795/1/012067

No	Type of Load		Scenario 1 (kWh)	Scenario 2 (kWh)	Scenario 3 (kWh)
1	Fabric gains	2,873.77	2,773.77	2,465.07	2,463.81
2	Indirect Solar Gains	638.64	600.64	538.24	537.67
3	Direct Solar Gains	1,026.61	1,026.61	646.06	646.06
4	Ventilation Gains	1,243.29	1,243.29	1,242.01	1,242.01
5	Internal Gains	1,331.36	1,331.36	1,332.02	1,332.02
6	Inter-zonal Gains	130.73	130.73	125.27	88.38
7	Cooling Load	154,901.00	153,901.00	111,190.42	112,148.51
	Total	162,145.40	161,007.40	117,539.08	118,458.45

Illustrated in Table 5 kWh comparison to the existing level of the building and each scenario, and which has the lowest load total is in scenario two, as expressed in Figure 7 is a graph of each scenario.

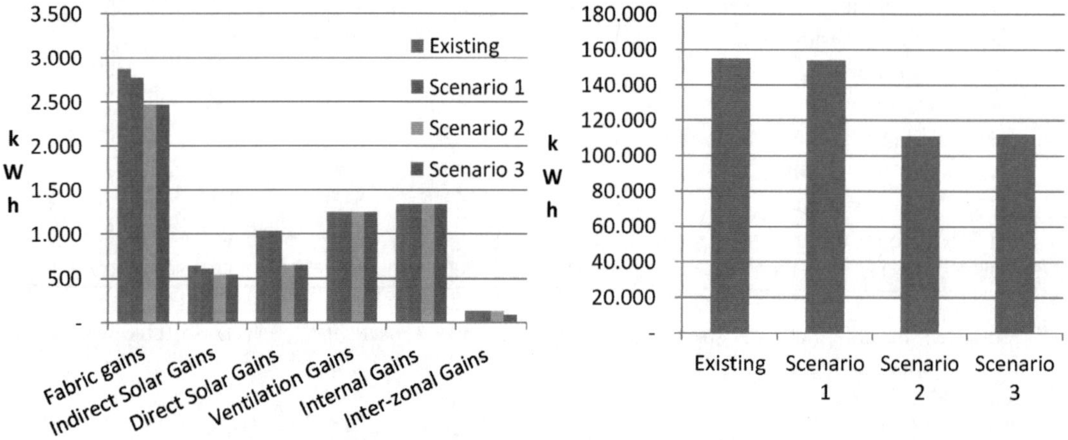

(a) (b)

Figure 7. (a) heating load (b) cooling load

By the meteorological analysis of Ecotect, the design strategies are achieved. Through the simulation analysis and comparison of building natural ventilation. So what have been done is that ecological energy-saving planning will save a lot of time and energy and can create much more pleasant and comfortable space.

4. Conclusion

For the data analysis, dr. Sayidiman Hospital the value of IKE air-conditioned space is at 14.61 kWh/m2/month. This energy consumption intensity (intensintas konsumsi energi, IKE) index value is attributed into a slightly inefficient.

Three scenarios of building energy and lighting retrofit have been performed simulating representing energy efficiency using cross ventilation, room openings, and passive cooling. The results of the numerical simulation indicate that the third scenario by employing additional windows, reflector media, and skylight exhibit the best result and in accordance with SNI 03-6575-2001 lighting standard.

Total thermal load of the existing building which includes fabric gains, indirect solar gains, direct solar gains, ventilation fans, internal gains, inter-zonal gains and cooling load reached 162,145.40

kWh. Based on the three scenarios, the thermal load value (kWh) obtained was lowest achieved scenario 2 with the thermal value of 117,539.08 kWh. The final results are interpreted from the total energy emissions evaluated using the Ecotect software, the heating and cooling demand value and specific design of the windows which are some of the important factors to determine the energy efficiency of the buildings. However, as more accurate results and the shorter time step of simulations are required, the gains have to be modeled in more detail.

References

[1] E.H. Borgstein, R. Lamberts, Developing energy consumption benchmarks for buildings: bank branches in Brazil, Build, 82 (0) (2014) 82-91.

[2] L.Perez-Lombard, J.Ortiz, C.Pout, A review on buildings energy consumption information, Energy Build. 40 (3) (2008) 394-398.

[3] E. Shen, J. Hu, M. Patel, Energy and visual comfort analysis of lighting and daylight control strategies, Build. Environ. 78 (0) (2014) 155-177.

[4] GREENSHIP, Technical guide green building assessment tools for new buildings 1.2. GBC Indonesia, 2013.(in Indonesia).

[5] A.S. Choi, Practical applications of form factor computation in lighting calculation, Build, Environ. 37 (11) (2002) 1107-1115.

[6] R. Johnson, R. Sullivan, S.Selkowitz, S. Nozaki, C. Conners. Glazing energy performance and design optimization with daylighting, Energy Build. 6 (4) (1984) 305-317.

[7] Indonesian National Standard SNI 03-6169-2000, Procedures energy audits in buildings, National Standardization Agency, Jakarta, Indonesia, 2000 (in Indonesia).

[8] Ministry of National Education, Guidelines for the implementation of energy conservation and control in the Ministry of National Education, Jakarta,Indonesia 2004, (in Indonesia).

[9] Indonesian National Standard SNI 03-6197-2000, Energy Conservation in Buildings Lighting Systems, National Standardization Agency, Jakarta, Indonesia.2000. (in Indonesia).

[10] Boutet, Terry S, Controlling air movement, A manual for architects and builders, McGraw-Hill. ISBN 0070067139, 1987.

[11] ASHRAE Handbook: Fundamentals, American Society of Heating Refrigerating and Air Conditioning Engineers, Inc.Atlanta, 1993.

The simulation of cropping pattern to improve the performance of irrigation network in Cau irrigation area.

Retno Wahyuningsih, RR Rintis Hadiani and Sobriyah

Civil Engineering Department of Post Graduate Program Sebelas Maret University, Jl. Ir. Sutami 36A Kentingan Jebres Surakarta 57126, INDONESIA

E-mail: naoki.noq2@gmail.com, rintish@gmail.com, sobriyah@ft.uns.ac.id

Abstract. Cau irrigation area located in Madiun district, East Java Province, irrigates 1.232 Ha of land which covers Cau primary channel irrigation network, Wungu Secondary channel irrigation network, and Grape secondary channel irrigation network. The problems in Cau irrigation area are limited availability of water especially during the dry season (planting season II and III) and non-compliance to cropping patterns. The evaluation of irrigation system performance of Cau irrigation area needs to be done in order to know how far the irrigation system performance is, especially based on planting productivity aspect. The improvement of irrigation network performance through cropping pattern optimization is based on the increase of water necessity fulfillment (k factor), the realization of planting area and rice productivity. The research method of irrigation system performance is by analyzing the secondary data based on the Regulation of Ministry of Public Work and State Minister for Public Housing Number: 12/PRT/M/2015. The analysis of water necessity fulfillment (k factor) uses Public Work Plan Criteria Method. The performance level of planting productivity aspect in existing condition is 87.10%, alternative 1 is 93.90% dan alternative 2 is 96.90%. It means that the performance of the irrigation network from productivity aspect increases 6.80% for alternative 1 and 9.80% for alternative 2.

1. Introduction

The development of an area which is followed by the growth of the population requires the increase of water necessity and food necessity fulfillment. The increase of population along with the increase of land necessity reduces the very limited agriculture area, especially in Java Island. Agriculture is the main user of the world's water resources. The contribution of irrigated agriculture to food production is important. Therefore, sustainability of irrigated agriculture would demand the efficient management of the available finite water resources under the existing constraints [1]. The indicator of the performance is based on the peasants' satisfaction of available and reliable irrigation water in order to fulfill irrigation water necessity. Besides, it is needed to have cooperation between government and peasants about the procedures of how to use water well [2]. Water shortage is the main limiting factor for agricultural productivity in many countries and improving water use efficiency in agriculture has been the focus of numerous studies. The usual approach to limit water consumption in agriculture is to apply water quotas and in such a situation farmers should use an irrigation schedule that maximizes the yield [3]. The Government's budget is insufficient to fund all of the necessity needed every year. The budget includes rehabilitation, operation, maintenance and handling after the natural disaster [4].

Some researches have been done to improve the fulfillment of irrigation water necessity and the production result. The non-linear optimization model for irrigation and cropping pattern optimization are suggested in order to maximize the result in the limitation of available water resources [1]. The uncertain and functional interval methods are introduced in the research about the uncertain water production function of the plant itself and the optimal allocation of water resources in irrigation area. This approach has important value in theoretical and practical for optimal irrigation schedule and has a broad prospect for researchers and the development of related agriculture water management [5]. Other researchers present two mathematical models that consider the physical and chemical site-specific management zones within the parcels. The first model is for the crop planning problem which assesses the chemical and physical management zones to determine the optimal crop pattern for maximizing the farmers' expected profit. The second model is a real-time irrigation method that takes the solution t the crop planning problem as an input [6]. A random-fuzzy-variable is developed for crop area optimization in response to such complexities in more efficient and sustainable ways. The developed model is helpful for managers in gaining insight into the trade-offs between the system benefit and the constraint violation risk [7].

The benefits of irrigation modernization include increased water efficiency and productivity, improved operation and management of irrigation systems and working conditions of farmers, but increased energy demands and investment amount.[8]. Many studies in recent years have focused on developing methodologies to quantify water and energy efficiency with improved irrigation systems. Lecina et al. (2010) analyzed modernization in the Alto Aragón irrigable area. They highlighted some advantages and disadvantages of migrating from the surface to sprinkler irrigation systems and quantified water savings and some of the associated economic and social impacts [9]. Abadía et al. (2010) developed a comparative analysis of energy efficiency in more than 30 collective irrigation networks in Murcia and Castilla-La Mancha, proposing some measures to improve energy efficiency and the main problems with these infrastructures [10]. Rodríguez Díaz et al. (2012) analyzed some modernized areas in Andalucía that migrated from surface irrigation systems to sprinkler and drip irrigation systems, highlighting some strengths and weaknesses of this modernization[11].

The regulation of Ministry of Public Work and State minister for Public Housing Number 12/PRT/M/2015 explains irrigation as an effort to provide, regulate and drain irrigation water in order to support some kinds of farming including surface irrigation, swamp irrigation , underground irrigation, pump irrigation and embankment irrigation. Irrigation Network is the channel, building and complementary building which is one unity needed to provide, divide, give, use and drain irrigation water [12]. Cau irrigation area in Madiun district, East Java province irrigates 1.232 Ha of land covering Cau Primary channel irrigation network, Wungu secondary channel irrigation network and Grape secondary channel irrigation network. The problems in Cau irrigation area are the limited water availability especially during dry season (planting season II and III) and the cropping pattern in the dry season is not obeyed. The evaluation of irrigation system performance in Cau irrigation area needs to be done to see how far the performance of irrigation system is so that it can be improved. The evaluation of the performance of irrigation system is to know the condition of an irrigation system including physical infrastructure aspect, planting productivity aspect, supporting means aspect, personnel organization aspect, documentation aspect, and the institutional condition aspect. The improvement of irrigation network performance through physical infrastructure aspect has great value but is very dependent on the budget provision. Because of that, this study is sought to improve the performance of irrigation network through planting productivity aspect which covers three things such as irrigation water necessity fulfillment ("k" factor), the realization of planting area, and rice productivity. The simulation of some cropping pattern alternatives is needed to get the best "k" factor value (=1).

2. 2. The Necessity of Water Explanation

The analysis of water necessity fulfillment (k factor) refers to Public Work Plan Criteria (KP-01). It used secondary data including rainfall data, climatology data (temperature, air humidity, wind velocity and sunlight), water availability data (debit), data of plant condition report and rice production data. The stages done in the cropping pattern simulation are : counting water necessity in the rice field, counting plant water necessity, counting water availability in the intake, and the cropping pattern simulation.

2.1. Water Necessity in the Rice Field

The amount of water needed by plant in a paddy field terrace is represented in this equation:

$$NFR = Etc + P + WLR - Re \quad (1)$$

with:

NFR = water necessity in the farm (mm/day),
ETc = plant water necessity (mm/day),
WLR = water layer replacement (mm/day),
P = percolation (mm/day),
Re = effective rainfall

2.2. Plant Water Necessity

Plant water necessity is the amount of water needed to replace the losing water as the result of vaporization. The amount of water plant necessity (consumptive use) is counted based on this formula:

$$ETc = Kc \, . \, ETo \quad (2)$$

with:

Etc = plant evapotranspiration (mm/day),
ETo = plant reference evapotranspiration (mm/day),
Kc = plant coefficient.

Water necessity during land preparation is counted based on the formula V.D Goor-Zijlstra as follow:

$$IR = \frac{M.e^k}{e^k - 1} \quad (3)$$

$$M = Eo + P \quad (4)$$

$$K = M.T/S \quad (5)$$

with:

IR = irrigation water necessity (mm/day),
M = water necessity to replace the losing as the result of evaporation and percolation in the rice field which has been saturated,
Eo = open water evaporation is taken 1.1 Eto during land preparation period (mm/day),
P = percolation (mm/day),
T = the duration of land preparation,
S = the water needed for saturation is added by 50 mm.

Irrigation Water Necessity for rice:

$$IR = NFR / e \quad (5)$$

Irrigation Water Necessity for soy:

$$IR = (Etc - Re) / e \quad (6)$$

dengan:

NFR = water necessity for irrigation,
Etc = evaptranspiration (mm),
P = percolation (mm/day),
Re = effective rainfall (mm/day),
WLR = water layer replacement (mm),
e = overall irrigation efficiency.

2.3. Water Availability

The amount of water availability is counted based on the mainstay debit which is predicted to exist continuously in certain amount and certain period of time. The mainstay debit is minimum debit for the possibility of standard fulfillment which has been determined and can be used for irrigation. The possibility of fulfillment determined 80% means the possibility of river debit is lower than mainstay debit which is 20%. Mainstay debit analysis using statistic possibility method with normal distribution will give a good result if available debit observation data is more than 20 years (Directorate General of Irrigation, 1986). The formula used is:

$$Q80 = \bar{x} - (0{,}821.Sd) \quad (7)$$

dengan:

Q80 = mainstay debit with possibility 80% (m3/sec),

$\bar{x}$ = debit ½ average monthly,

Sd = standard deviation.

2.4. Effective Rainfall

Effective rain is the amount of rainfall which is used to fulfill plant necessity. One of the ways used to count effective rain is with the R80 method. The calculation of R80 rain by using formula as follow:

$$R_{80} = \bar{x} + k.Sd \quad (8)$$

with:

R_{80} = the amount of rainfall in certain reset period,

$\bar{x}$ = monthly rainfall average (mm),

K = frequency factor (-0,842 for 20% dry)

Sd = standard deviation.

2.5. K Factor

K factor is counted in order to show the stability between the necessity and availability of water. The formula used to count the amount of k factor is as follow:

$$k = \frac{Q_t}{Q_b} \quad (9)$$

dengan:

Q_t = available outflow debit.

Q_b = outflow debit needed.

The value of k factor is about 0 up to 1 in which the worst condition is shown by 0 value which means the water necessity is not fulfilled at all and the best condition is shown by k value = 1 which means the water necessity can be wholly sufficed.

3. Result and Discussion

The method used to calculate potential evaporation (Eto) by using modified-Penman method is shown in Table 1. The result of R80 rain calculation with normal distribution approach by using equation formula (8) is shown in Table 2.

Table 1. Evapotranspiration Calculation Result with Penman Method

Month	Eto
January	1,97
February	1,65
March	1,75
April	1,59
May	1,45
June	1,46
July	1,56
August	1,86
September	2,04
October	2,23
November	2,07
December	1,86

Table 2. 80% Rain Calculation Result (R_{80})

Month	R_{80} (Daily)
January	6,09
February	6,55
March	4,75
April	5,21
May	1,70
June	0,00
July	0,00
August	0,00
September	0,00
October	0,00
November	2,69
December	7,59

Based on the Irrigation Plan Criteria (KP-01) which is issued by Directorate General of Irrigation and Department of Public Work 2013, the value of effective rain for rice is calculated as 70% of mainstay rain which is 80%. Moreover, it can also be calculated based on the monthly minimum rainfall (R80) with 5 year-reset periods multiplied by rain coefficient and rotation system. Meanwhile, for soy, the amount of effective rainfall is determined by a biweekly method which is related to monthly average rainfall (fulfilled 50%) and monthly average plant evapotranspiration.

Water necessity for preparing the land determines the maximum necessity of irrigation water. The amount of water necessity for land preparation depends on the amount of land saturation, the length of land cultivation and the amount of evaporation and percolation. The calculation of water necessity during land preparation for rice is shown in Table 3.

Table 3. Water Necessity during Land Preparation for Rice

Month	Jan	Feb	Mar	Apr	May	Jun	Jul	Aug	Sep	Oct	Nov	Dec			
Eo	2,17	1,81	1,94	1,75	1,60	1,61	1,71	2,04	2,25	2,45	2,28	2,05	Eo	1,1	Eto
M=Eo+P	3,67	3,31	3,42	3,25	3,10	3,11	3,21	3,54	3,75	3,95	3,78	3,55	P	1,5	mm/day
K=MT/S	0,55	0,50	0,51	0,49	0,46	0,47	0,48	0,53	0,56	0,59	0,57	0,53	T	30	day
e^k	1,73	1,64	1,67	1,63	1,59	1,59	1,62	1,70	1,75	1,81	1,76	1,70	S	200	Mm
$IR=Me^k/(e^k-1)$	8,67	8,46	8,52	8,43	8,34	8,34	8,40	8,59	8,72	8,84	8,74	8,60	Mm/day		

Water necessity for Soy is needed to cultivate the land which is going to be planted and create a humid condition for newly growing seedbed. The amount of needed water is between 50-100mm. The calculation of water necessity during land preparation for rice is shown in Table 4.

Table 4. Water Necessity during Land Cultivation for Soy

Month	Jan	Feb	Mar	Apr	May	Jun	Jul	Aug	Sep	Oct	Nov	Dec			
Eo	2,17	1,81	1,94	1,75	1,60	1,61	1,71	2,04	2,25	2,45	2,28	2,05	Eo	1,1	Eto
M=Eo+P	3,67	3,31	3,42	3,25	3,10	3,11	3,21	3,54	3,75	3,95	3,78	3,55	P	1,5	mm/day
K=MT/S	2,20	1,99	2,05	1,95	1,86	1,86	1,93	2,13	2,25	2,37	2,27	2,13	T	30	day
e^k	9,02	7,30	7,79	7,05	6,42	6,45	6,88	8,28	9,48	10,71	9,67	8,41	S	50	Mm
$IR=Me^k/(e^k-1)$	4,12	3,84	3,93	3,79	3,67	3,68	3,76	4,02	4,19	4,36	4,22	4,03	Mm/day		

In improving the performance of irrigation network based on the plant productivity aspect refers to three things such as water necessity fulfillment analysis (k factor), the realization of planting area and rice productivity. For the calculation of the water balance, the taking necessity which is produced for used cropping pattern will be compared to mainstay debit. If the river debit is abundant, the water necessity of irrigation area will be fulfilled. If the river debit is not abundant or even sometimes lacked debit, there will be three choices which must be considered (KP-01, 2013). Those choices are the width of irrigation area reduction, cropping pattern modification and technical rotation. In this study, the fulfillment of the necessity is by modifying the cropping pattern. From the simulation of cropping pattern alternative, it is expected to get optimal k factor value (=1), so that it can improve the value of irrigation network performance. The existing cropping pattern dan some cropping pattern alternatives are shown in Table 5.

Table 5. The Existing Cropping Pattern and Alternative Cropping Pattern

Condition	Plant Season	Crop Pattern	Planting Area (ha) Rice	Soy
Existing	I	Paddy	1.111	-
	II	Paddy	1.111	-
	III	Paddy - Palawija	221	214
Alternative 1	I	Paddy	1.232	-
	II	Paddy - Palawija	830	281
	III	Paddy - Palawija	350	574
Alternative 2	I	Paddy	1.232	-
	II	Paddy - Palawija	649	462
	III	Paddy - Palawija	350	574

From the existing cropping pattern and alternative cropping pattern, it can be got the value of irrigation water necessity. The stability of water balance can be obtained by comparing the value of water necessity and the value of water availability (k factor). Next, the value of physical condition of k factor analysis can be seen and in the form of a percentage. The calculation of physical condition of k factor analysis can be seen in Table 6.

Table 6. The Value of k Factor Analysis Physiscal Condition

Condition	Planting Season	"k" factor	Information	The Value of the Physical Condition (%)
(1)	(2)	(3)	(4)	(5)
Existing	I	1,00	Fulfilled	91,00
	II	0,73	Not Fulfilled	
	III	1,00	Fulfilled	
Alternative 1	I	1,00	Fulfilled	95,00
	II	0,85	Not Fulfilled	
	III	1,00	Fulfilled	
Alternative 2	I	1,00	Fulfilled	100,00
	II	1,00	Fulfilled	
	III	1,00	Fulfilled	

The value of irrigation network performance from the plant productivity aspect, appropriate to the Regulation of Ministry of Public Work and Public Housing Number:12/PRT/M/2015, has value as much as 15,00. That value includes 9.00% of factor analysis, 4.00% of planting area realization, and 2,00% of rice productivity. The value of physical condition which has been found out and multiplied by the value of weight condition will gain the value of weight condition. The calculation of the value of weight condition is explained in Table 7.

Table 7. The Value of The Weight Condition of Plant productivity Aspect

planting productivity aspect	Weight Value (%)	The Value of the Physical Condition (%)			Weight Condition Value (%)		
		Existing	Alternative 1	Alternative 2	Existing	Alternative 1	Alternative 2
(1)	(2)	(3)			(4)=(2)*(3)/100		
"k" factor	9,00	91,00	95,00	100,00	8,19	8,55	9,00
realization of planting area	4,00	71,89	88,39	88,39	2,88	3,54	3,54
paddy productivity	2,00	100,00	100,00	100,00	2,00	2,00	2,00
Amount	15,00				13,07	14,09	14,54

The value of irrigation network performance from plant productivity aspect can be obtained by comparing the value of weight condition and the expected value of weight condition. The performance value obtained is in the form of a percentage (%). The calculation of the performance value of irrigation network from plant productivity aspect can be seen in Table 8.

Table 8. The Value of Irrigation Network Performance from Plant Productivity Aspect

Condition	Weight Value (%)	Weight Condition Value (%)	the value of plant productivity performance
(1)	(2)	(3)	(4) = (3)/(2)*100
Existing		13,07	87,10
Alternative 1	15,00	14,09	93,90
Alternative 2		14,54	96,90

The result of the calculation value of plant productivity performance shows that the existing condition is 85.10%; the first cropping pattern alternative simulation is 88,10%, and the second alternative cropping pattern simulation is 92,50%. It means the first alternative simulation has 3.00% of performance value improvement compared to the existing condition. Meanwhile, the second alternative simulation has 7.4% of performance value improvement. The graph comparing the value of plant productivity performance of the existing condition, alternative 1 and alternative 2 is shown in Figure 1.

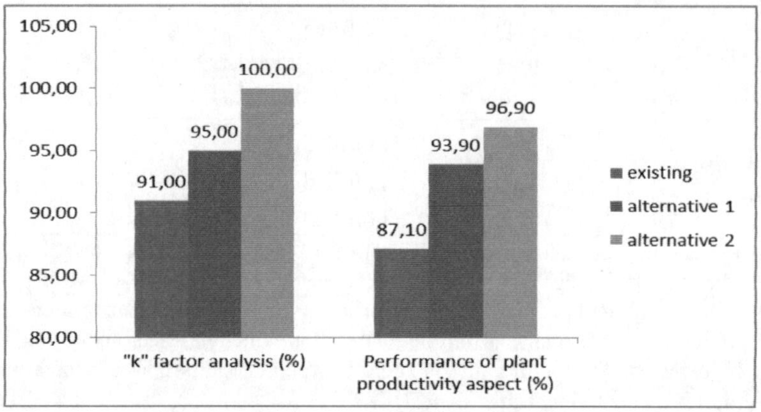

Figure 1. The comparison between k Factor value and the value of Plant Productivity Performance

From the cropping pattern alternative simulation, it can be concluded that cropping pattern alternative 2 is chosen.

4. Conclusion

From the analysis result on the existing condition in Cau Irrigation area, it shows that the water availability is still sufficient for planting season I and III. Yet, it is insufficient in the planting season II. The simulation of cropping pattern is done by determining some alternatives of the possible cropping pattern. From the analysis result, the simulation of cropping pattern alternative 2 can improve the plant productivity performance with rice cropping pattern with 1.232 Ha of width during planting season I; Rice with 649 Ha of width and Soy with 462 Ha of width during planting season II; Rice with 350 Ha of width and Soy with 574 Ha during planting season III. With Public Work Plan Criteria Method (KP-01), the value of k factor in cropping pattern alternative 2 is considered fulfilling in water necessity fulfillment. Based on the regulation of Ministry of Public Work and Housing Number 12/PRT/M/2015, it can improve the performance of irrigation network in plant productivity aspect which is improving 9.80%. It means the performance of irrigation network can be improved from plant productivity aspect apart from the other aspects as listed in the regulation of Ministry of Public Work and Housing Number 12/PRT/M/2015. The improvement of plant productivity is meant to be other alternatives in improving irrigation network performance without including budget like physic infrastructure aspect.

5. Reference

[1] Garg, N. K., and Sushmita M. Dadhich. "Integrated non-linear model for optimal cropping pattern and irrigation scheduling under deficit irrigation." Agricultural Water Management, 2014.

[2] Mattamana, Binoy Aliyas, Shiney Varghese, and Kichu Paul. "Irrigation System Assessment-Farmer's and Manager's view." International Journal of Engineering Science and Innovative Technology (IJESIT), 2013.

[3] Linker, Raphael, et al. "Optimal model-based deficit irrigation scheduling using AquaCrop: A simulation study with cotton, potato, and tomato." Agricultural Water Management, 2016.

[4] Agus Dharma. "The development of Water Resources Policy and Its Effect on the Irrigation Management" (In Indonesia), Jakarta, 2013.

[5] Tong, Fangfang, & Ping Guo. "Simulation and optimization for crop water allocation based on crop water production functions and climate factor under uncertainty." Applied Mathematical Modelling, 2013.

[6] Cid-Garcia, Nestor M., Angel G. Bravo-Lozano, and Yasmin A. Rios-Solis. "A crop planning and real-time irrigation method based on site-specific management zones and linear programming." Computers and Electronics in Agriculture, 2014.

[7] Li, Mo, and Ping Guo. "A coupled random fuzzy two-stage programming model for crop area optimization."—A case study of the middle Heihe River basin, China. Agricultural Water Management, 2015.

[8] Tarjuelo, José M., et al. "Efficient water and energy use in irrigation modernization: Lessons from Spanish case studies." Agricultural Water Management, 2015.

[9] Lecina, Sergio, et al. "Irrigation modernization and water conservation in Spain: The case of Riegos del Alto Aragón." Agricultural Water Management, 2010.

[10] Abadía, R., et al. "Comparative analysis of energy efficiency in water users associations." Spanish Journal of Agricultural Research, 2010.

[11] Díaz, J. A., et al. "Modernizing water distribution networks: Lessons from the Bembézar MD irrigation district, Spain." Outlook on AGRICULTURE, 2012.

[12] Ministry of Public Work and Housing, "Exploitation and Maintenance of Irrigation.", Regulation of the Minister of Public Works Number: 12/PRT/M/2015, Jakarta, 2015.

[13] Ministry of Public Works of the Republic of Indonesia, Directorate General of Water Resources, "Public Work Plan Criteria, Irrigation Network Planning (KP-01)", Jakarta, 2013.

Analysis of slope slip surface case study landslide road segment Purwantoro-Nawangan/Bts Jatim Km 89+400

Joko Sidik Purnomo[1], Yusep Muslih Purwana and Niken Silmi Surjandari

[1]Civil Engineering Department of Post Graduate Program Sebelas Maret University
Jl. Ir. Sutami 36A Kentingan Jebres Surakarta Postal Code 57126, INDONESIA

E-mail: jokosidikp@yahoo.com ,ymuslih@yahoo.com,
nikensilmisurjandari@gmail.com

Abstrack. Wonogiri is a region of south eastern part of Central Java province which borders with East Java and Yogyakarta Province. In Physiographic its mostly undulating hills so that the frequent occurrence of landslides, especially during the rainy season. Landslide disaster that just happened that on the road segment Purwantoro-Nawangan / Bts Jatim Km 89 + 400 were included in the authority of the Highways Department of Central Java Province. During this time, Error analysis of slope stability is not caused by a lot of presumption shape of slip surface, but by an error in determining the location of the critical slip surface. This study aims to find the shape and location slip surface landslide on segment Purwantoro - Nawangan Km 89 + 400 with the interpretation of soil test results. This research method is with the interpretation of CPT test and Bore Hole as well as modeling use limit equilibrium method and finite element method. Processing contours of the slopes in the landslide area resulted in three cross section that slopes A-A, B-B and C-C which will be modeling the slopes. Modeling slopes with dry and wet conditions at the third cross section slope. It was found that the form of the slope slip surface are known to be composite depth 1.5-2 m with safety factor values more than 1.2 (stable) when conditions are dry slopes. But its became failure with factor of safety < 0.44 when conditions are wet slopes.

1. Introduction

Wonogiri is a region of south eastern part of Central Java province which borders with East Java Province and Yogyakarta Province. In Physiographic region of Wonogiri mostly undulating hills so that the frequent occurrence of landslides, especially during the rainy season. Landslide disaster that just happened that on the road Purwantoro-Nawangan/Bts Jatim Km 89 + 400 were included in the authority of the Highways Department of Central Java Province. The location of a provincial road that connects Wonogiri Regency of Ponorogo, so that the existence of these roads is vital to support the transportation flow of goods and services for both regions. The main problem in the stability analysis is to determine the slip surface landslides [1]. During this time, error analysis of slope stability is not caused by a lot of presumption shape slip surface, but by an error in determining the location of the critical slip surface [2]. This paper aims to find the shape and location slip surface landslide on segment Purwantoro - Nawangan Km 89 + 400 with the interpretation of soil test results as well as modeling use limit equilibrium method and finite element method.

2. Slip surface location with CPT test

To determination of the positions of landslides in the field is not conducted directly, but attributed to determining the shear stress was drained soil based on the correlation value of the cone (q_c) from CPT testing whose value is directly proportional. The collapse of the slope can be caused by the disruption of stability, when the shear stress is greater than the allowable shear strength in the ground, then the process of land mass movement will occur. Based on test results of CPT at some point in parallel directions landslide, slope pieces obtained by positioning of points conical low value. When these points are connected will be seen a field that is a collection point or a weak point called the slip surface. Slip surface of depth was predicted by the test results of CPT in study area according to the method Suryolelono is modeled as interface element [3]. In addition with this method the potential for further sliding to predict in case of disturbances on the slope. Slopes will adjust to the new shape of the slope with the slope angle is smaller than the natural slope angle of the slope forming soil type.

3. Layout and location slip surface

Landslide is one type of soil or rock mass movement that generally occurs on a slope of 20°-40 ° to the moving masses in the form of residual soil, sediment colluvial and weathered volcanic rocks. Colluvial residual soil and soil are generally a loose and can store water. As a result, the shear strength is relatively weak, especially if the water contains more saturated and pressing. Increased water saturation can occur if the land is gone in the top layer of soil or rock that is more cohesive and impermeable. So that the water that seeped into the soil difficult to penetrate a layer of soil or rock beneath it, and only accumulate in the soil is relatively loose. Contact between layers of soil or rock that is impermeable to the mass of soil on top is often a slipe surface of ground movement.

Slip surface can also be the zone that is the boundary the level difference of rock weathering, field discontinuity rock, and the rock layers such as clay stone, silt stone, shale and tuff. Masses of soil and rock that does not move the soil or bedrock that is more compact and more massive example andesite and andesite breccia rocks. The emergence of permeation or springs on the slopes generally occurs at the contact zone between the water impermeable rock masses or layers of soil/rock that passes water. This contact zone is often a slip surface movement.

Landslide on residual soil, especially on steep slopes, slip surface not like the type of landslide in the circular shape. On residual soil slip surface of a relatively shallow depth, often with a slightly curved or nearly planar (Figure 1). Nevertheless, the volume of landslide material is still very large [4].

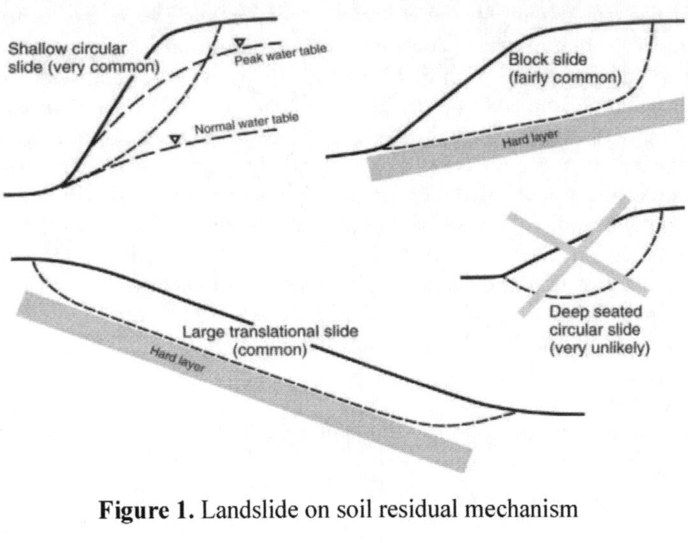

Figure 1. Landslide on soil residual mechanism

The mechanism of landslides through higher groundwater levels often occur in residual soil slopes and colluvial. Colluvial residual soil layer or serves as a free aquifer and aquifer were hanging (perched aquifer) with the condition of the ground water level fluctuates depending on the amount of infiltration of rainwater. Free aquifer (phreatic aquifer) is the flow of water in the saturated condition that occurs in the aquifer that has a water level that is not pressurized. While hanging aquifer is a layer saturated with water that is above a layer of soil or rock that is less watertight than the main aquifer.

The presence of a layer of relatively less impervious causes the water caught in it and form a layer of water-saturated. The aquifer that hanging is not too wide and only contains water only during the rainy season. Besides resulting rise in groundwater levels, the pervasive rain water into the slope can also lead to a) increase the weight volume of soil and rock, b) reduction in pore water pressure is negative (suction) in the zone is not saturated (unsaturated), c) an increase in the positive pore water pressure, d) internal erosion and e) changes in the mineral content of constituent masses of soil and rock on the slope. The effectiveness of rain in triggering landslide depends on the amount of rainfall and the long rains, the graduation rate of water in the soil and water saturation in the slope before the rain [5].

4. Slip surface location process

In the plane strain problem, the slip surface is assumed to be a circular arc in many existing LEM. However, numerical analysis has revealed that the shape of the slip surface might be a combination of a straight line and a circular arc or a combination of a straight line and a logarithmic spiral. Optimisation techniques have been shown to be the most efficient means of locating non circular slip surfaces [6–10].

In the SRM, the critical slip surface is found automatically from the increase in the shear strain due to the reduction of the shear strength. Typically, the critical slip surface is approximated using technical measures, mesh deformation plots [11], velocity fields [12], etc. Cheng et al. [13] defined the critical slip surface as a wavy line that connects the maximum shear strain increment. Zheng et al. [14] similarly defined the critical slip surface as a wavy line that connects the maximum equivalent plastic strain, where the wavy line is smoothed using the least squares method.

5. Research methods

In this study, the data used is secondary data such as bore log data, soil properties, the data shear strength of soil investigation soil mechanics laboratory Faculty of Engineering, Sebelas Maret University. Photo documentation and the results on location surveys in the form of coordinates of ground cracks are taken directly from location. Land contour data downloaded free from website *www.tanahair.indonesia.go.id* with form of *shp*. Land contour is processed with the help of software and performed to obtain the land cracks plotting area landslide. Landslide areas are plotted coordinates of any point soil cracks, and then determine the 3 cross-sectional area slopes that represent landslide. Interpretation of the results on the CPT test 4 point parallel directions landslide, cross section obtained by positioning points conical low value. When these points are connected will be seen a field that is a collection of weak points called slip surface. Processing contours of the slopes in the landslide area produce three cross section that slopes A-A, B-B and C-C which will be modeling. This research method is with the interpretation of CPT test and bore hole combined with a cross section of slope and modeling shape of slip surface using limit equilibrium method and finite element method. Modeling slopes with dry and wet conditions at the third cross section slope.

6. Result and discussion

Soil samples will drill test results conducted laboratory tests to determine the physical properties of soil, one of which is the shear strength.

IOP Conf. Series: Journal of Physics: Conf. Series **795** (2017) 012069 doi:10.1088/1742-6596/795/1/012069

Table 1. Soil test result

No.	Sample	type of investigation										
		w	Gs	grainsize		Atteberg		γ_B	direct shear		Classification	
		%		%		%		gr/cm^3	c (kg/cm^2)	ϕ (°)		
1	BH 1 04.00 - 04.50 M (UDS)	45,15	2,64	Gravel	0,18	LL	48,60	1,722	0,459	10,84	SC = Sandy Clay	Layer 1
				Sand	69,25	PL	37,75					
				Silt & Clay	30,57	PI	10,85					
2	BH 1 6.50 - 7.00 M (UDS)	42,96	2,66	Gravel	0,85	LL	46,98	1,652	0,262	12,83	SC = Sandy Clay	
				Sand	53,57	PL	38,65					
				Silt & Clay	45,58	PI	8,33					
3	BH 1 14.50 - 15.00 M (DS)		2,67	Gravel	2,87	LL	42,42	-	-	-	Compact rocks	Layer 2
				Sand	35,42	PL	35,37					
				Silt & Clay	61,72	PI	7,05					

From processing and plotting the contours of soil cracks, then the resulting landslide areas. The next step is to perform the the interpretation of the data results CPT test in the third cross section slope so is known composition of soil forming slopes.

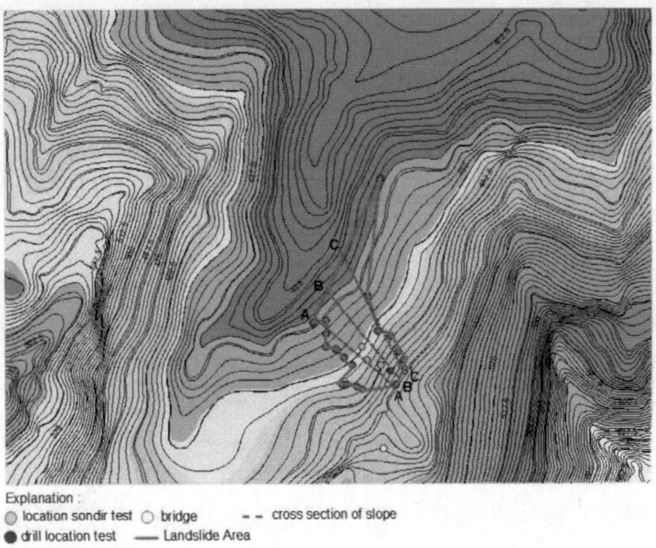

Figure 2. Landslide area to determine the cross section

In figure 3-5 that the results CPT on S1, S2, S3, and S4 show that the hard soil (q_c = 250 kg/cm^2) is located at a depth beetwen 3-9 m. On the soil drill test results have N-SPT value greater than 50 (N-SPT> 50) is located at a depth of over 7 m (hard soil). The test results CPT at 4 point in parallel directions landslide, cross section obtained by positioning points conical low value. Based on this, a cross-section of A-A, B-B and C-C are synchronized to CPT results by connecting the weakest points will form a slip surface symbolized by dashed lines. Prediction calculation depth slip surface as figure 3-5 showed that depth which occurred betwen 1.5-2 m.

IOP Conf. Series: Journal of Physics: Conf. Series **795** (2017) 012069 doi:10.1088/1742-6596/795/1/012069

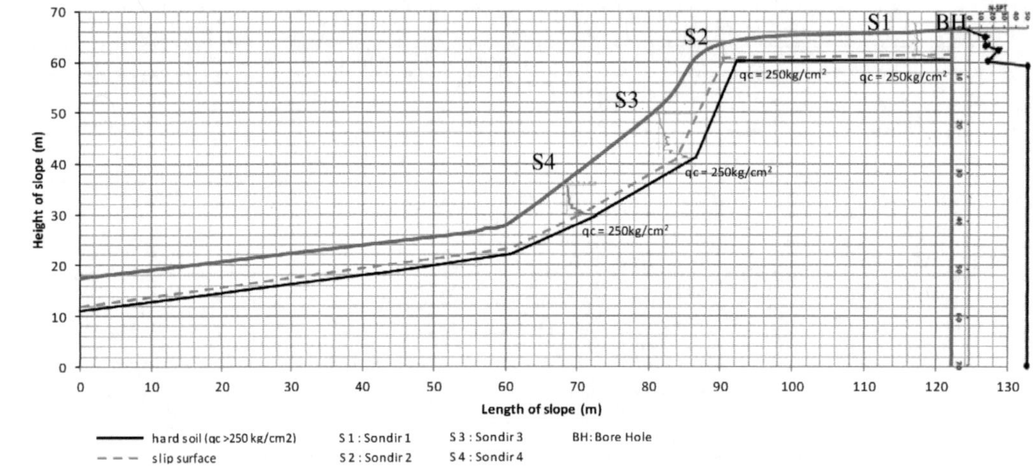

Figure 3. Composition of the soil forming slopes A-A

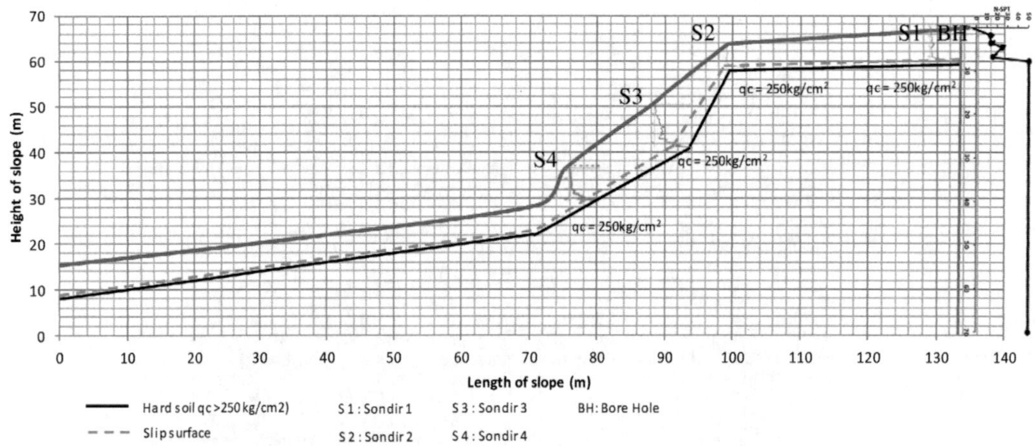

Figure 4. Composition of the soil forming slopes B-B

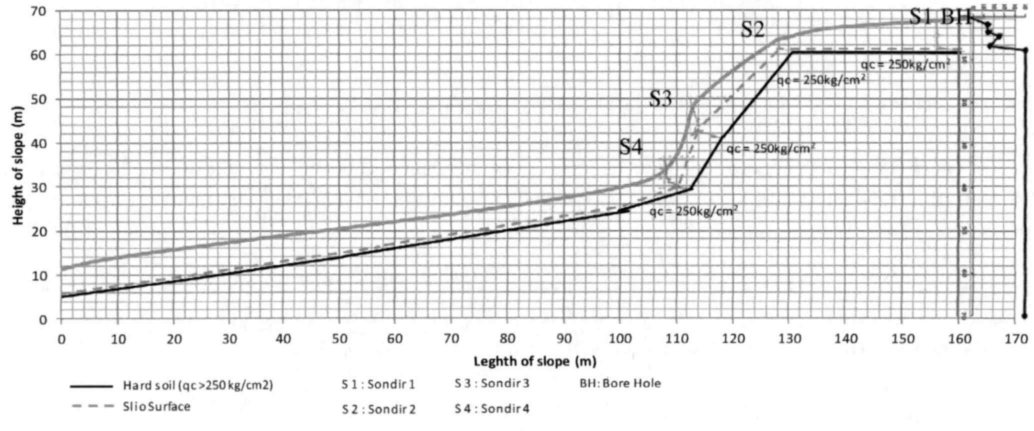

Figure 5. Composition of the soil forming slopes C-C

IOP Conf. Series: Journal of Physics: Conf. Series **795** (2017) 012069 doi:10.1088/1742-6596/795/1/012069

Modeling slopes with LE and FE methods produces that slopes in wet conditions make the slope becomes unstable with a safety factor of less than 0.44 with LE methods and less than 0.33 with FE methods. Shape critical slip surface at the time of wet conditions is shaped composite. By knowing the shape and location of slip surface proper, then handling the strengthening of the slope can be done appropriate. Comparison shape slip surface and safety factor of slope are shown in the figure 6-8.

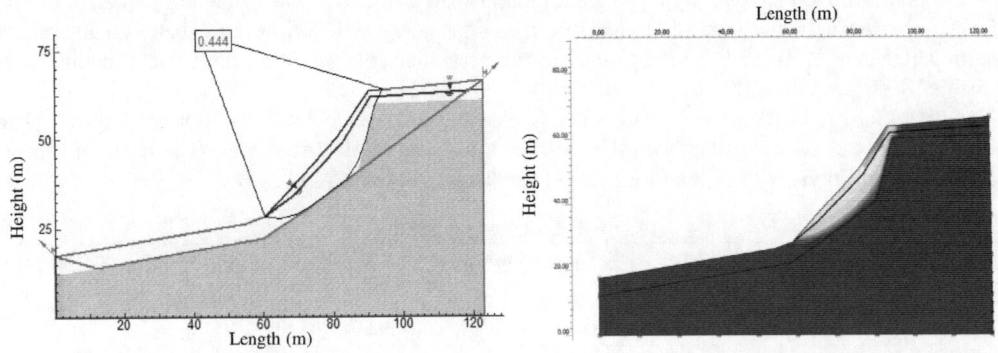

(a) (b)

Figure 6. Slip surface slope A-A (a). Limit Equilibrium (b). Finite Element

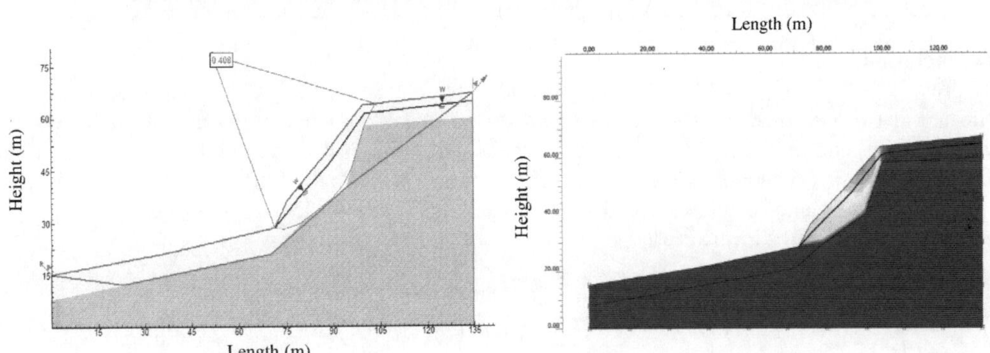

(a) (b)

Figure 7. Slip surface slope B-B (a). Limit Equilibrium (b). Finite Element

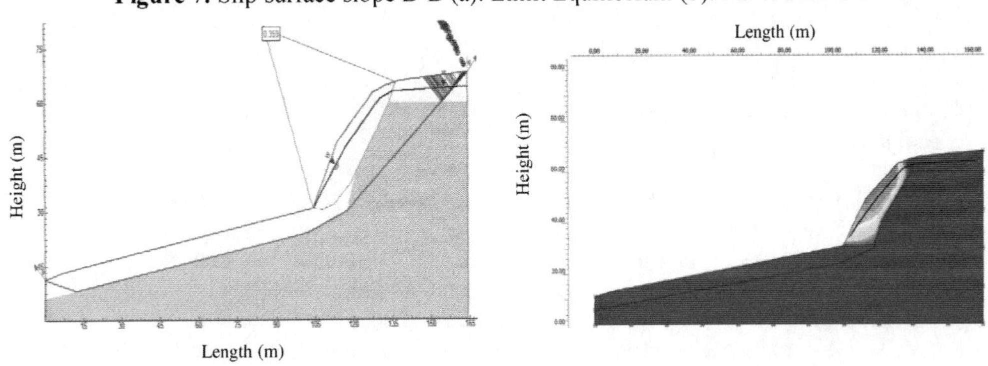

(a) (b)

Figure 8. Slip surface slope C-C (a). Limit Equilibrium (b). Finite Element

The comparison shape of slip surface and safety factor between limit equilibrium and finite equilibrium method can be known, that the shape resulting from both methods have a slip surface which shape almost the same. Figure 4-6 (a) search the slope slip surface using the LE method where this method uses the principle of equilibrium forces and moments. Shape slip surface in wet conditions produces composite shape (combination of circular and flat shape). The occurrence collapse of the slope in figure 4-6 (b) using strength reduce method (SRM). In SRM, the strength parameters of soil will be decreased until the slope fails and the factor of safety will be the ratio between the actual strength parameters of the soil and the critical parameters, therefore on the finite element method does not recognize shapes slip surface.

The definition of factor of safety in SRM is exactly same as in limit equilibrium method [15]. Safety factor in wet condition between the two methods have small differences. The results factor of safety in wet conditions with LE and FE methods are shown in table 2.

Table 2. Factor of safety

Cross Section			Limit Method	Finite Method	Description
Slope	A-A	Dry	3.812	-	Safety
		Wet	0.444	0.294	Failure
Slope	B-B	Dry	3.135	-	Safety
		Wet	0.408	0.290	Failure
Slope	C-C	Dry	2.957	-	Safety
		Wet	0.359	0.336	Failure

7. Conclusion

Interpretation of the results of soil tests on roads Purwantoro-Nawangan/Bts Jatim Km 89 + 400 indicated that slip surface consist of more than two flat surfaces connecting points of the lowest cone (weakest point) and the shape of slip surface with a depth 1.5-2 m. In interpretation of the slip surface with two-dimensional modeling using limit equilibrium method and finite element method in wet conditions on the slopes of a cross section A-A, B-B and C-C produce a composite (combination of circular and flat shape) slip surface.

References

[1] Bardet J P and M M Kapuskar 1989 *Computers and Geotechnics Journal* **Vol. 8 No. 4** pp 329 – 348
[2] Bowles E 1984 *Physical and geotechnial properties of soil* (New York: McGrawHill)
[3] Suryolelono K B 1993 *Civil Engineering Forum* No. 11/1 August 1993.
[4] Wesley L D 2010 *Geotechnical engineering in residual soils* (New Jersey: John Wiley & Sons, Inc)
[5] Karnawati D 2005 *Land Mass Movement of Natural Disasters in Indonesia and Handling Efforts*. (Yogjakarta: Geology Engineering Gadjah Mada University)
[6] Arai K and Tagyo K 1985 *Soils.Found.***25(1)** 43-51
[7] Greco V.1996.*J Geotech Eng* **122(7)** 517-25
[8] Malkawi A,Hasan W and Sarma S 2001 *J. Geotech.Geoenviron.Eng* **127(8)** 688-98
[9] Zolfaghari AR, Heath AC and Mccombie PF 2005 *Comput.Geotech* **32(3)** 139-52
[10] Cheng YM, Li , Chi S C and Wei WB 2007 *Comput.Geotech* **34(2)** 92-103
[11] Kim J Y and Lee S R 1997 *Comput.Geotech* **21(4)** 295-313
[12] Dawson E M,Roth W H and Drescher A 1999 *Geotechnique* **49(6)** 835-40
[13] Cheng Y M, Lansivaara T and Wei W B 2007 *Computers and Geotechnics* **34(3)** 137-150
[14] Zheng H, Sun G and Liu D 2009 *Comput. Geotech.2009* **36(1-2)** 1-5
[15] Griffiths D and Lane P 1999 *Geotechnique* **49(3)** 387-403

Determination of the bridge maintenance and rehabilitation priority scale in kabupaten Pinrang

Asrul Nurdin[1], S.A. Kristiawan[2] and Dewi Handayani[2]
Civil Engineering of Post Graduate Program Sebelas Maret University
Jl. Ir. Sutami 36A Kentingan Jebres Surakarta 57126, INDONESIA
Email: asruldirga@yahoo.com

Abstract. This study attempts to get the support system to help stakeholders and the decision makers in the determination of priorities the maintenance and rehabilitation of the Bridge .The location of the study is done in Kabupaten Pinrang. Bridge damage value is specified using assessment standard Bridge Management System (BMS). There are 13 Bridges by details as follows: good condition with the condition 1 in 6 Bridge, damaged with the condition two as many as five fruit Bridges and there are two Bridges by value condition of three which means need handling rehabilitation. From our observation examination Bridges and it has been obtained the results of then afterward determined the priority scale by using the method Analytic Hierarchy Process. The discussion or interview of 27 respondents in completion question to determines the priorities of handling bridges by three (3) criteria and three sub-criteria with weights each: a condition of damage with weights 0,491 or amounting to 49,1 % followed by criteria policy with weights 0,324 % then criteria volume traffic weight of 0,185 or 18.5 %. Based on the results of this can be concluded that the implementation of a method of Analytic Hierarchy Processing is effective in determining the scale of priorities for the stakeholders. Having acquired value bridge condition and weights of the assessment results of the, then is needed to Analysis of spatial use application ArcGIS to obtain information condition Bridge Map.

1. Introduction

Bridge such an important part in a system of the road network, because the effect would who mean that if the bridge collapse or did not function properly, collapse the bridge will reduce or restrain traffic, which means disturb the flow of the transportation of people and goods. For that reason should this infrastructure maintained well so its performance can be increased or defended

Obstacles faced by office works Kabupaten Pinrang that is they did not the implementation of the maintenance Bridges like program routine maintenance and maintenance intervals. Which routine maintenance performed to repair damage bridge on the damage level of light. Depart from this problem hence writers interested to draw up a system the determination of to judge priority maintenance bridge approach in Analitical Hierarchy Process and the device system information that can be showing the combination of technical data bridges and information supporting analysing data with on the Arc Info

Examination and maintenance Bridge is a series of process of Bridge Management System which includes routine maintenance intervals for the purpose to give a monitoring system and control maintenance better [1]. Maintenance done be gave a hope that all the in Indonesia will be effective and the purpose of an examination this Bridge was to convince that bridges are working safely and the need for held a certain action to the maintenance and repair of periodically of maintenance in bridge any very important to therefore condition bridge is staying on its function, Huang, y.h [2]. Assessment of damages bridge do with some components, of the components next divided into such element .Elements bridge itself formed of elements at the level of smaller as constituent. Each the destruction on element starting from elements smallest given code for uniformity understanding and ease in insert data .Started at the level of 1 structure bridge in A whole , the level of 2 is part of the river , building up and building bottom .While the level of elements 3,4,5 is the smallest of the level of 2. Damage-damage that has been identified then given judgment whereby a judgment this condition is done based on the categories the value of the structure , the level of damage, the quality of damage, their influence on the functions of an element concerned and on other element. From scoring the condition of the bridge it will be in a kind of handling know as on a table in the following.

Table 1. Types Of Handling

Value Of The Condition	Handling
0-2	Routine Maintenance
3	Rehabilitation
4-5	replacement

1.1. Analitycal Hierarchy Process (AHP)

AHP a model supporting a decree being developed by Thomas l.Saaty.Supporting model decision will disassemble problems multi-factors or multi the criteria a complex being a hierarchy, according to saaty [3], Analytical Hierarchy Process the had the capability to solving problems that multi the criteria based on a comparison preperensi of any element and their hierarchy [4]

Tabel 3. Scale Of Comparison

Intensity interests	Description
1	Second as important element
3	One element a little more important
5	One element more important
7	single element obviously very important
9	One element absolute far more important
2,4,6,8	values between two values taking into account adjacent

Saaty (1993) determined scale quantitative 1 (one) to 9 (nine) to assess the comparative degree the interests of a elements against another .As the table above .The weighting for the judgment of gregarious expressed with find the mean of geometric (geometric mean from a judgment rendered by all the members of the group .Value geometric is formulated with equation

$$GM = \sqrt[n]{X_1 \, x \, X_2 \, x \, \ldots\ldots x \, X_n}$$

1.2. Do multiplication elements in a row and in nth root as shown in equation this:

$$w_i = \sqrt[n]{a_{11} \, x \, a_{12} \, x \, \ldots\ldots x \, a_{1n}}$$

1.3. Counting a vector priority or a vector eigen and the magnitude of the weighting of each element can be obtained by equation this:

$$X_i = \frac{w_i}{\Sigma w_i}$$

1.4. Counting value eigen maximum (λ_{maks}), by means of multiplying matrix resiprokal with weights obtained , the result of a sum operation matrix is the value of eigen maximum (λ_{maks}) with equation the following

$$\lambda_{maks} = \Sigma a_{ij} \, x \, X_i$$

1.5. Calculation index consistency
This calculation in mean to know consistency answer that it would affect to the truth the results.Calculation index consistency can be done by using equation

$$CI = \frac{\lambda_{maks} - n}{n-1}$$

To know ci good enough or not , needs to be know consistency ratio (CR) .The ratio the consistency of a is parameter to check whether comparison in pairs has done with consequent or not using equation Under this

$$CR = \frac{CI}{RI}$$

Value Random Indices (RI) depending on the size of matrix as shown in table 2.8 . The following

Tabel 4. Random Indeks

Ordo matrik	1	2	3	4	5	6	7	8	9	10
RI	0	0	0,58	0,9	1,12	1,24	1,32	1,41	1,45	1,49

1.6. Geographic information system (GIS)
Geographic Information System (GIS) Is a information systems that designed to work with the data referenced spatial or coordinat geography or in other words a sig is a system a database with a special ability to handle the referenced data into the room (spatial) at the same time as a set of operation work. The use of Geographical Information System aimed at building a system information a bridge that can be simplified and accelerate in the provision of information condition and the proposed priority for a bridge [5] major role sig is help in the planning activities and management. The one generated by sig is input in the process of planning and management. Geographical information system were divided into two groups the system manual (analogous), and automatic system (based digital a computer). The difference most basic located on how to funds. There is every indication that the use of computers to handle geographic information will continue to increase dramatically in the future and that the user base will continue to broaden. But trends in the software industry and in society more generally suggest that the nature of GIS will continue to evolve [9]

2. Research Methods
Methods used in the research by system bridge management system (bms and qualitative based on an analysis weighting elements and criteria analitical hierachy process (ahp).In this method be a assessment of the interests of criteria with spread kusioner With involves several stake holders and decision maker , where in research by involving three factors criteria::

- Factors Condition Bridge
- Factors Volume Traffic
- Factors Policy

Of each above criteria next in the form of some sub part to form a hierarchy Matrix comparison processed using equation. Weighting received in matrix comparison carried out consistently, in measuring based on value CR . Comparison received if CR<0,1. When CR $\geq$ 0,1 so comparison changed to meet the criteria output of process of analysis is a CR<0,1 database maintenance and rehabilitation bridge based Geographical Information System in the city Pinrang. In detail process of analysis geospatial is as follows:

1. Make layer shapefile are still in arc.map.
2. Check updating the map.
3. Processing maps and attribute-nya in ArcGIS.
4. Editing a map with use toolbar editing as append is, dissolve, merge and split tool/add vertec to make location a bridge to map.

5. Spatial namely the analysis and classification symbology display a map arc. A binder with 3 category into good, enough and broken
6. At the end of work or hard work in format digital (Soft Copy and casts Hard Copy) covered report, mold map basic, map condition bridge, and table attribute.

3. Results and Discussion

Of the survey conducted then done inventory 13 bridge carried on two district city are Sawitto and district Paleteang obtained data as can be seen in table below.

Table 5. Data – data Bridge

Number bridge	Name Bridge	Size		Structure over the building			NK	Indication	Type handling
		P	L	G	LT	SDR			
7315301	Lompobattang	17	6	GK	BA	Concrete	3	Damaged	Rehabilitation
7315302	BauMassepe	32	7	GB	BB	Concrete	2	Minor Damage	Maintenance
7315303	Paleteang	16	5	GB	BB	Concrete	2	Minor Damage	Maintenance
7315304	Pacongang	24	6	GB	BA	Concrete	1	Good	Maintenance
7315305	Kamp. Jaya	10	7	GB	BB	Brick	1	Good	Maintenance
7315306	Pawelloi	9,5	7	GK	BA	Brick	1	Good	Maintenance
7315307	Langnga	22	7	GB	BA	Concrete	2	Minor Damage	Maintenance
7315308	Beruang	21	7	GB	BA	Concrete	2	Minor Damage	Maintenance
7315309	Lerang-lerang	11	6	GB	BA	Concrete	1	Good	Maintenance
7315310	Maccorawalie	25	5	GB	BA	Concrete	2	Minor Damage	Maintenance
7315311	Corawali	12	5	GB	BA	Concrete	1	Good	Maintenance
7315312	Kamp. Tengah	8,5	4,5	GB	BA	Concrete	3	Damaged	Maintenance
7315313	Jampue	4	7	GB	BA	Brick	1	Damaged	Maintenance

Nb :
GK = Girder Composite
GB = Girder Concrete
BA = Concrete Asphalt
BB = Reinforced Concrete

From the data we can see value damaged bridge, who into the the run would obtained 3 (three) the damage. Group damaged bridge in accordance technical criteria bridge, that is good - damaged, heavily damaged and critical / collapse .In accordance with a system handling bridges in this study, for the good s /d damaged and heavily damaged next Are sorted the priority scale Handling indicative conducted in bridge that has been in the survey and considered is 2 bridge to the management of heavily damaged, 5 bridge to the management of maintenance periodical and 6 of the bridge only need handling routine maintenance. While the destruction at the bridge is on element couples (a rift, the rupture, on element concrete (loss of concrete, crack, rupture/a partial loss of concrete, on element the flow of the river and mud sediment, garbage, scouring, on element a surface coating (rough, crack, wavy), on element parapet (rusts, the rupture/missing, damage components, components missing , the decline in the quality of paint) , elements pipe cucuran (of the elements missing, components is congested)

Traffic

The volume of traffic flow the conversion equivalence a passenger car (emp) fluctuating daily and times in the discussion next calculation, the motor vehicle that passes through segments it consists of: Light Vehicle (LV), Heavy Vehicles (HV) Motorcycle (MC).

Tabel 7. Value Traffic

Segments		**wide**	**LHR**
1	Bau Massepe	4	1164,5 smp/jam
2	Teuku Umar	3	2284,5 smp/jam
3	Bulu Paleteang	4	4441,5 smp/jam
4	Lompo Battang	4	2277,5 smp/jam
5	Latahere	4	1158,5 smp/jam
6	Beruang	4	4432,5 smp/jam
7	Serigala	4	2276,5 smp/jam
8	Briptu Suherman	7	4468,8 smp/jam
9	Wahidin Sudirohusodo	4	3367,5 smp/jam
10	Kamp. Tengah	3	311,5 smp/jam

Value

The weight with the methods AHP as outlined in sub. Chapter mentioned above applied on the implementation of the determination of the priority scale maintenance and rehabilitation bridge in Kabupaten Pinrang by using a condition of damage, traffic and policy. The weight in table above according to the analysis and next can been concluded in figure below.

Weight		
Criteria	**Sub-Criteria**	**Indication**
Condition (0,491)	**River Flow** (0,411)	**Minor Damage** (0,170)
		Damaged (0,830)
	Under Construction (0,414)	**Minor Damage** (0,170)
		damaged (0,830)
	Building Upon (0, 145)	**Minor Damage** (0,170)
		damaged (0,830)
Traffic (0,185)	< **50** (0,072)	
	51 – 200 (0,102)	
	201 – 500 (0,296)	
	> **500** (0,530)	
Policy (0,324)	**Work Plan (0,237)**	
	Developmen Plan (0, 763)	

Figure 3. weight a hierarchy the priority scale maintenance and rehabilitation bridge

The result of all the process start in the provision of value condition bridge , the calculation of traffic and weighting by means of a method of Analitical Hierarchy Processing in pour or displayed in a system the

Geography Information System (GIS) viem several species of map as in stated in figure below determination of priority maintenance and rehabilitation bridge.

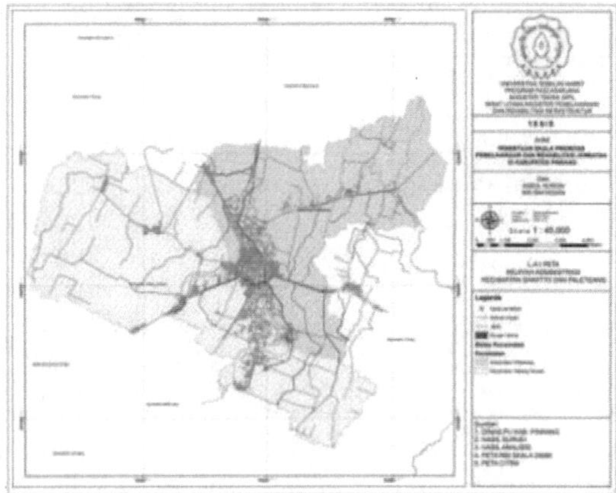

Figure 4. Map damaged bridge that is described caprice or condition of the damage to research this

4. Conclusion

Based on the results of the analysis and discussion in chapter before , so a conclusion can be drawn as follows

From the weighting of three (3) criteria through the results of kusioner from 27 respondents obtained weight each for criteria a condition of damage of 0,491 or (49,1 %), volume traffic with weights 0,185 or (18.5 %) whereas for criteria policy of 0,324 or 32,4 %.Of the result concluded that the criteria a condition of damage have the importance higher than other criteria.

Of the 13 locations bridge who are the objects in this research the scale of priority there are two (2) the point bridge that requires handling rehabilitation, for a bridge who with a condition of damage while only need handling maintenance intervals as many as six fruit bridge while 5 bridge does not require handling serious it still in good condition.

References

[1] The directorate general for marga department of public works of the republic of indonesia 1993 .A guide maintenance and rehabilitation bridge.

[2] Huang Y. H 2012. *A Model for Concurrent Maintenance Of Bridge Elements.* Automation in Construction 21 pages 74-80. Univ.of Science and Technology. Taiwan

[3] Saaty, T, 2008. *Decision Making with the Analytic Hierarchy Process* International Journal Sciences Services Volume 1 Number 1. Pitsburgh

[4] Junaidi et al 2006. Proceeding seminar on national management technology .Priority handling of road improvement project on segments roads in kabupaten kapuas by the method of ahp .Technology institute of ten november

[5] Subagio G 2008. Forum Teknik Sipil No.XVIII/3-September 2008, *Sistem Informasi Manajemen Jembatan Berbasis Web Dengan Metode Bridge Condition Rating*. Universitas Gajah Mada

[6] Abdul Rahaman.S Prioritization of Sub Watershed Based on Morhometric Characteristics Using Fuzzy Analytical Hierarchy Process and Geographical Information System-A Study Of Kallar Watershed, Tamil Nadu. International Conference on water Resources, Coastal And Ocean Engineering (ICWRCOE 2015), University Tiruchirappali, India

[7] Rahmawaty 2014. Integrated Geographic Information System and Global Positioning System for Mapping of Forest Plants in Supporting Natural Resources Protection. 2nd Humboldt Kolleg in conjuction with International Conference on Natural Sciences, HK-ICONS 2014.

[8] Giovanni Ottomano Palmisano 2015. Greenways for rural sustainable development: An integration between geographic information systems and group analytic hierarchy process. L and Use Policy 50 (2016) 429-440

[9] Michael F Goodchild 2015, Geographic informasi system (GIS). University of California, Santa Barbara, CA, USA

[10] Alejandro Ruiz-Padillo 2016. Application of the fuzzy analytic hierarchy process in multi – criteria decion in noise action plans ; Prioritizing road stretches. Environmental Modelling & Software 45-55. Industrial Engineering and Transportation Departement.

[11] Saptarshi Sasmal 2008, Condition evaluation of exixting reinforced concrete bridges using fuzzy based analytic hierarchy approach. Expert Systems with Applications 35 (2008) 1430-1443. Structural Engieering Research Centre, CSIR Campus Taramani, Chennai 600 113, India.

[12] Krishna P. Paudel 2009, Geographic informasi system (GIS) based model of dairy manure transportation and application with environmental quality consideration. Waste Management 29 (1634-16430. Department of Agricultural Economics and Agribusines, Lousiana State University, Baton Rouge United States.

[13] Ayaho Miyamoto 2001. Development of bridge management system for existing bridges. Advances in Engineering Sofware 32 821-833. Department of Computer and system Engineering. Yamaguchi University, Tokiwadai, Ube, Yamaguchi, 755-8611, Japan.

The effect of numbered heads together (NHT) cooperative learning model on the cognitive achievement of students with different academic ability

Marleny Leasa[1], Aloysius Duran Corebima[2]

[1]Faculty of Teachers Training and Education, Pattimura University Jl. Ir. M. Putuhena, Unpatti Campus Poka Ambon 97223, INDONESIA
[2] Faculty of Mathematics and Sciences, State University of Malang, Jl. Semarang No. 5, 65145 East Java-INDONESIA

Abstract. Learning models and academic ability may affect students' achievement in science. This study, thus aimed to investigate the effect of *numbered heads together* (NHT) cooperative learning model on elementary students' cognitive achievement in natural science. This study employed a quasi-experimental design with *pretest-posttest non-equivalent control group* with 2 x 2 factorial. There were two learning models compared NHT and the conventional, and two academic ability high and low. The results of ana Cova test confirmed the difference in the students' cognitive achievement based on learning models and general academic ability. However, the interaction between learning models and academic ability did not affect the students' cognitive achievement. In conclusion, teachers are strongly recommended to be more creative in designing learning using other types of cooperative learning models. Also, schools are required to create a better learning environment which is more cooperative to avoid unfair competition among students in the classroom and as a result improve the students' academic ability. Further research needs to be conducted to explore the contribution of other aspects in cooperative learning toward cognitive achievement of students with different academic ability.

1. Introduction

One of the aims of science education to elementary students is to prepare them before encountering real environment. Science learning has helped develop students science literacy as they also take a part in the society who are faced with various problems related to science such as pollution, flood, global warming, diseases, and recycling. In achieving the goal, the students are necessarily taught how to develop a higher order thinking which covers the analysis, evaluation, and creations domains. As as result, the students are more aware of and able to deal with the problems. They can be engaged in a learning activity which improves their sientific performance as well as their ability to define the issues. Cooperative learning is an especially effective method to use with any problem-solving task because it encourages people to express divergent points of view.

In Indonesia learning framework, cognitive achievement still becomes the main focus for teachers and schools, as well as students' main target. A cognitive achievement is the representation of students' intelligence quotient. Several education policies were done to improve students' learning outcome to obtain the standard. Therefore, the learning plan, practice, and evaluation must involve cognitive aspects holistically. In fact, the learning pattern in Ambon-Indonesia, especially in science subjec tends to focus on the low cognitive. It is because many teachers think that elementary school students cannot perform high order thinking skill. Thus the learning pattern is constructed as simple as possible, as long as students know. The evaluation instrument is in the form of multiple choice or essay test which is dominated by C1-C3 cognitive level. This condition reflects that the orientation of the learning pattern in classroom focuses on the low level of cognitive achievement.

The researchers found the teachers' difficulties when they are involved in a social activity, particularly for the teachers in KKG (*Teacher Work Group*) IV and VI of Sirimau in July-September 2016. The lack of understanding and best practices of the implementation of various student-centered learning models makes the teachers cannot design a relevant activity. The reality reveals that it is necessary to provide learning models and materials which are suitable for the development of elementary school students' characteristics. Based on Piaget's development level, elementary school students are at the level of concrete operational which needs a social interaction including with teachers and peers to understand the concept learned, especially in science learning. The cooperative model is a constructivism and colaborative which encourages students to interact with other people [1]. Cooperative helps students to develop their understanding when doing social interaction in a group and presentation, one of this model is Numbered Heads Together (NHT), which considered relevant for elementary school students. NHT emphasizes members' responsibility to do their tasks based on their number. Students are to show their capability and use any strategies to reveal their responsibility. Students are also trained to understand every task comprehensively. Hence, the students' responsibility is not merely on the individual project, but also as a member of a particular group. This process encourages students' ability to solve problems and students comprehensiveness. Thus, NHT can improve students' cognitive achievement. Besides, as a part of cooperative learning models, NHT contributes students' social skill improvement, when they interact with teachers and peers during the learning [2].

The class is a diversity miniature of students. One of the diversity aspects is unavoidably academic ability. In a cooperative learning atmosphere, heterogeneity is a valuable. Besides promoting students to the real world, heterogeneity also provides students to build a good relation or social competition [3]. Students' different academic ability can be accommodated within NHT learning pattern. Learning process and students' motivation might improve if teachers can conduct an appropriate learning that is suitable with students' various need and ability [4]. Compared to low academic ability students, those with high academic ability tend to pursue more challenging objectives, have more self-motivation, more idealism to be a success, and less hesitation to their academic performances [5].

Many previous studies reveal the effect of NHT [6-14]. A beneficial process in cooperative learning especially in the cognitive achievement is the social interaction. Vygotzky's (1978) theories emphasize the contribution of social interaction to cognitive development which may impact to academic achievement. Educators construct an education through individual interaction within society and learning is considered as the result of the internalization of social interaction. An interaction fosters an individual to cope with the diversity of life styles, experiences, attitudes, ideas, and opinions [15]. Also, Slavin [16] asserts that cognitive perspective focuses on the interaction among students in groups. The interaction contributes to the improvement of the learning process as well as encourages students' academic. It is in line with Pham's [17] statement that cooperative learning model enables high and low ability students to help and motivate each other to improve their academic performances.

The purpose of this study was to examine the comparison of students' achievement based on learning model and different academic ability. There are three hypotheses of the study: 1) There is a different cognitive achievement of students who experience NHT cooperative learning and students who experience conventional models. 2) There is a different cognitive achievement of high academic ability students and low academic ability students. 3) There is a different cognitive achievement in learning model interaction and academic ability. This information are important in encouraging the constructivism learning which is mostly omitted during the learning activity.

2. Method

This study was a comparative study, which was conducted in the form of quasi-experimental with pretest-posttest non-equivalent control group design. This study used 2 x 2 factorial pattern, including two variables with two levels. The first variable is the learning models, involving NHT cooperative learning and conventional learning model. The second variable is students' academic ability which is divided into high and low academic ability.

Four core schools in Sirimau District area were selected as the population of the study. The researchers chose the schools randomly. The samples of the study were 100 fifth graders who enrolled in the academic year 2016-2017. Their academic ability comprised two elements including high and low performance based on the students' academic record scores in the previous semester and an interview. The students who attained scores in the interval top 33,3% were considered as high academic ability students, while the students who obtained scores in the interval bottom 33,3% were considered as low academic ability students.

The cognitive achievement test which was constructed by the researcher referred to the domain from level C1 to C6. The instrument has been going through construct validation, and empirical try out in the form of essay test. The cognitive achievement test was administered twice, in pretest and posttest. The data of the study were analyzed descriptively and statistically using ana Cova in SPSS 17.00 computer program for windows to test the hypotheses of the research.

3. Results and Discussion

The research results describe the students' cognitive achievement and the hypotheses. The data description of the cognitive achievement is in the form of the mean score of pretest and posttest based on the learning models and the students' academic ability. Figure 1 presents these results as follows.

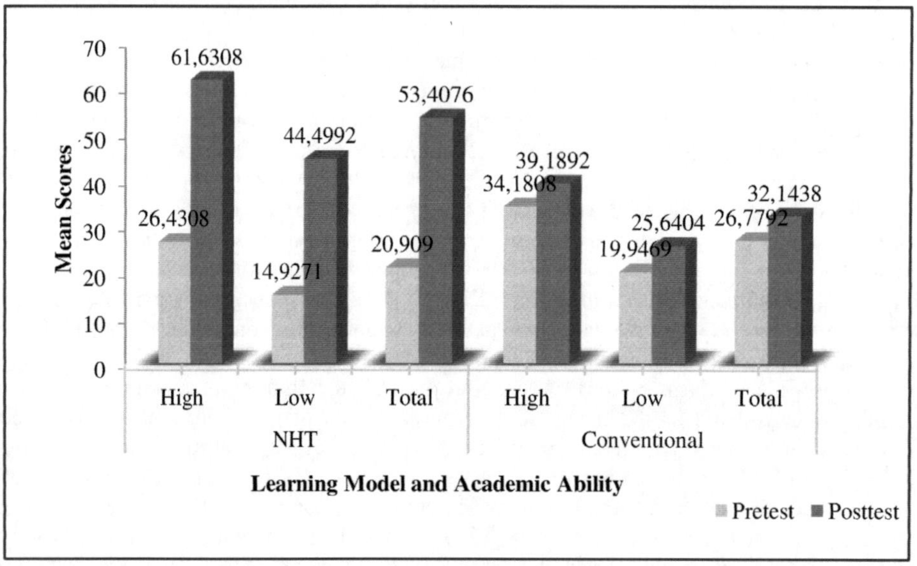

Figure 1. The comparison graphic of the students' cognitive achievement based on learning models and academic ability

Figure 1 shows that: 1) in low academic groups, the mean pretest score of the students who experienced conventional learning is 33.63% higher than the mean pretest score of the students who experienced NHT. 2) In high academic groups, the mean pretest score of the students who experienced conventional learning is 29.32% higher than the mean pretest score of the students who experienced NHT. 3) In low academic ability groups, the mean posttest score of the students who experienced NHT is 73.55% higher than the mean pretest score of the students who experienced conventional. 4) In high academic groups, the mean posttest score of the students who experienced NHT is 57.26% higher than the mean pretest score of the students who experienced conventional. Table 1 depicts the results of the ana Cova test.

Table 1. The ana Cova test result of the effect of learning model and academic ability to cognitive achievement

Source	Type III Sum Of Squares	Df	Mean Square	F	Sig.
corrected model	17844,845[a]	4	4461,211	74,662	,000
intercept	10878,018	1	10878,018	182,052	,000
pre test	587,390	1	587,390	9,830	,002
learning model	10896,453	1	10896,453	182,361	,000
academic	1782,420	1	1782,420	29,830	,000
model * academic	123,656	1	123,656	2,069	,154
error	5676,461	95	59,752		
Total	206497,357	100			
corrected total	23521,306	99			

Table 1 illustrates some information as follows.

3.1 The difference of the students' cognitive achievement based on learning model

The F observed of the learning model is 182.361, p-value 0.000 < α (0.05). Hence, H0 stating there is no cognitive achievement difference between students who experienced NHT and conventional learning model can be rejected. It means, there is an effect of learning model to students' cognitive achievement. Therefore, it is not necessary to conduct Least Significant Difference (LSD) test for the present study only consisted of two groups. Table 2 provides information on students' diverse learning achievement in every model.

Table 2. The mean of the cognitive achievement of each learning model group

No.	Learning Model	Pretest	Posttest	Different
1	Conventional	26.779	32.144	5.365
2	NHT	20.909	53.408	32.499

Based on Table 2, it can be seen that the group which was treated using NHT cooperative learning has a higher mean than the group which was treated using conventional. The cognitive achievement of the students who experienced conventional improves 20.03%, while the students who experienced NHT improve 155.43%. It means the learning outcomes of the students who experienced NHT is 72.45% higher than those who experienced conventional model.

The cognitive achievement mean score of the students who experienced NHT improved higher than mean score of the students who experienced conventional model. Cooperative learning model encourages the students to cooperate in a group so well that creates meaningful learning process and enhances the students' cognitive achievement. This process might help students to develop academic performances [18].

There are several supporting aspects in NHT cooperative learning activity. They are openness, critics/suggestions, assistance, dynamic, and tolerance [19]. In a multi-dimension cooperative class, a group dynamics is an inevitable condition. Willingness to obtain group objectives needs all of the above-mentioned aspects. Openness means that students and teachers need to be sincere to accept suggestions during the learning activity. Teachers may know some things that students do not know, vice versa. Therefore, both of them can complement one another during the learning activity. In cooperative class, each person can deliver opinion or critics, to explain some misconceptions of some things. Critics are addressed to construct knowledge appropriately. They encourage people to develop a mindset that every person is in the right track. Some assist will be provided to encourage the similarity point of view. A good thing in cooperative learning class is dynamic.When an individual admits his weakness, he has to be ready to change. The dynamic aspect appears in cooperative learning. Another aspect is tolerance, in which students need to tolerate persistent differences.

Based on the analysis of the participants' responses and interview results, there were some differences found in their answers. To answer question number 2c "Explain the reasons of why smoking is not good for health?". The answers is presented in figure 2 and 3 below.

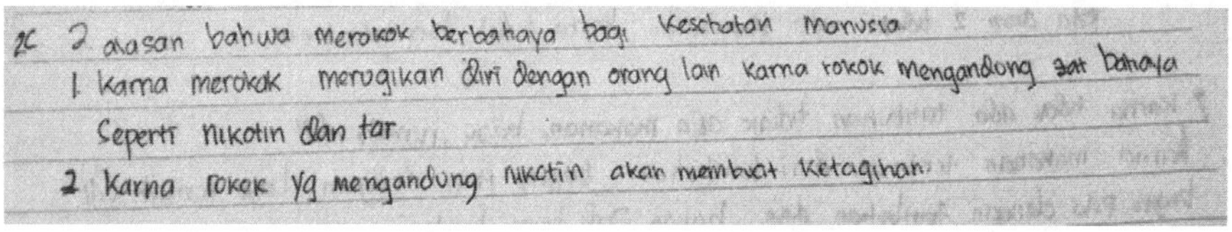
2c 2 alasan bahwa merokok berbahaya bagi kesehatan manusia
1. Karna merokok merugikan diri dengan orang lain karna rokok mengandung zat bahaya
seperti nikotin dan tar
2 Karna rokok yg mengandung nikotin akan membuat ketagihan

Figure 2. Student answers for the problem number 2c

Students who learned using NHT model provided responses as follows. Smoking is not good for health because; 1) a cigarette contains chemicals which are harmful such as nicotine and tar. 2) nicotine inside the cigarette is addictive. The students explained the reasons of why smoking can be disadvantageous; because it contains chemicals such as nicotine and tar, which can cause addiction. The students can express their opinions and associate them with the impacts of smoking based on their own life experiences.

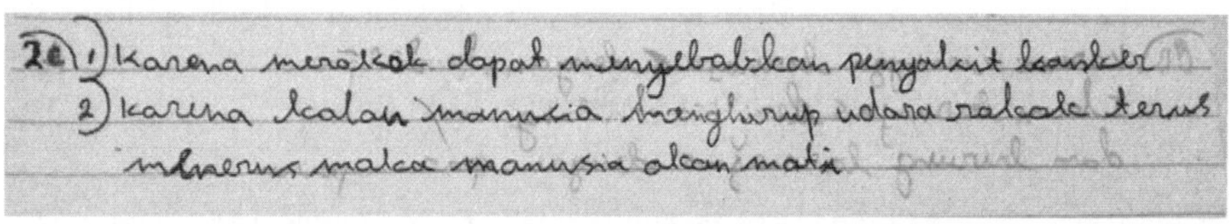
2c 1) karena merokok dapat menyebabkan penyakit kanker
2) karena kalau manusia menghirup udara rokok terus
menerus maka manusia akan mati

Figure 3. Student answers for the problem number 2c

Students who were taught with conventional models provided reasons related to the effect of smoking on human's health that is cause diseases. They answered: 1) smoking habit can result in cancer, 2) humans risk their lives by inhaling the smoke produced from cigarettes.

The students who experienced learning with NHT were used to building a concept to think deeper or to think at the higher order level in facing problems. They are trained to argue and explain rationales of a problem. First , they will solve a problem individually and then discuss it in a group while communicating, sharing experiences with other students and resolve problems together. On the other hand, the students who were accustomed to learning with conventional models were exposed to memorization of a concept from textbooks. They were likely to have little interaction with other students because they could not work cooperatively.

Cognitive achievement provides information related to achievement or index which is valuable and can be used to examine the cooperative activity in a group [20]. The students who were involved showed better cognitive achievement than the students who learned individually learning as the consequence of transfer knowledge activity from group to individual. In cooperative learning, the member shared information partially, then fused, corrected, and revised so that their knowledge and understanding improved. Also, students comprehend the collective knowledge of the group. The NHT cooperative learning group test score should have smaller standard deviation than the conventional learning group test score.

NHT emphasized on the students' responsibility. This learning model fostered the students to learn the material given as well as to examine their comprehension. NHT provided times for students to think and find appropriate solutions. Interaction among students also helped the students to strengthen their understanding and long-term memories to the solutions. NHT stressed the students' participation in learning especially in answering each question so that the students had to be ready anytime. The students shared knowledge until they were able to find the appropriate solution. This model was able to provide opportunities for the students to think deeper, adjust every solution given by the group member, provide the best solution, and make a responsible decision.

3.2 The effect of academic ability to cognitive achievement

The F observed of the academic ability is 29.830, p-value 0.000 < α (0.05). Hence, H0 stating there is no cognitive achievement difference between high and low academic ability students cannot be accepted. It means, there is an effect of the academic ability to students' cognitive achievement. Therefore, it is not necessary to conduct LSD test for the present study only consisted of two groups. Table 3 presents differences in students' performance from each academic group.

Table 3. The mean of the cognitive achievement of each academic ability group

No	Academic Ability	Pretest	Posttest	Different
1	Low	17.537	34.693	17.155
2	high	30.151	50.859	20.708

Table 3 indicates that students with high academic performance had higher mean scores compared to those coming from low academic performance. The cognitive achievement of the low academic ability students improves 97.82%, while the cognitive achievement of the high academic ability students improves 160.23%. It means the cognitive achievement of the high academic ability students is 33.03% higher than the low academic ability.

The ana Cova test result shows that there is a cognitive achievement difference between high academic ability students and low academic ability students. Learning outcomes of students with low academic ability have significantly improved from pretest to posttest. The interaction between high academic ability students and low academic ability students in the group influenced their learning process. The process contributes to the individual behavior changes which cause achievement improvement [21]. The low academic ability students tended to face some difficulties and thus they needed help or supervision during peer interaction, especially with high academic ability students. During the interaction activity, the students clarified and shared ideas between group members. The interaction contributes to the achievement improvement. Interaction may assist students in finding solutions to problems, as well as motivate them to resolve the problems by themselves.

Toward question 3a "Fish use their gills to breathe, whales and dolphins breathe though their lungs, and insects use trachea to take in oxygen. What can you infer from this statement?. The students provided various answers are presented in figure 4 and 5 below.

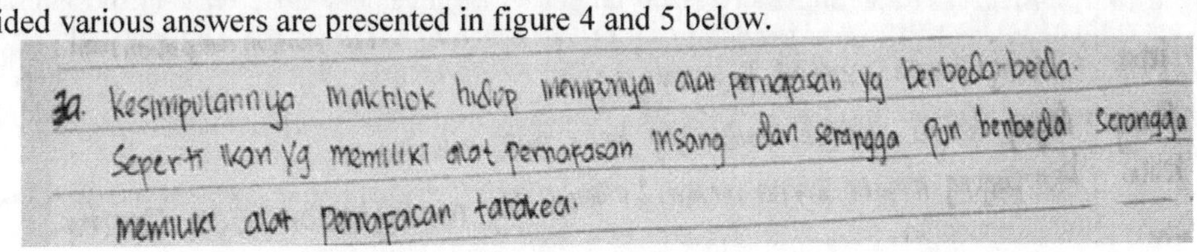

Figure 4. Student answers for problem number 3a

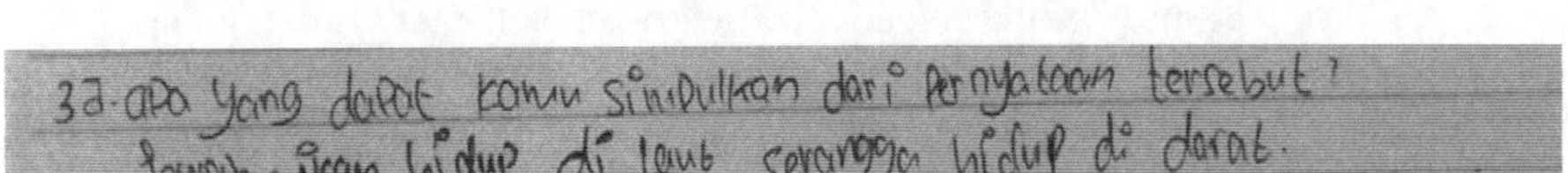

Figure 5. Student Answers for Problem Number 3a

Students with higher academic ability answered: it can be concluded that living creatures use different organs to breathe. For example, fish have gills and insects have trachea. Students with low academic ability answered: fish live in the sea, insects live on the land. These results indicated that the students with higher academic ability can carefully draw a conclusion based on the statement. The students were well informed that fish, whales, dolphins, and insects represent the animal groups while gills, lungs, and trachea are organs of respiratory system. These students had an ability to examine facts and use their logic to draw a conclusion. Meanwhile, the low ability students associated their thoughts with animals' habitats. They tended to make a wong inference since they were not able to synchronize their logic. This probably happens because they forgot or even had no idea about animals' respiratory system organs such as gills, lungs, and trachea.

Different academic ability emphasizes different experiences of students. Students construct new experiences during the learning process. There is a study that compares students' experiences in cooperative learning and In a big group conducted by Mulryan [22]. In her study, Mulryan observed fifth and sixth-grade students since they were involved in small and big groups in mathematics subject. Holistically, she found that the students tended to participate in the small groups rather than in big groups. However, their performance to the tasks given depended on their academic level. The high academic ability students were significantly involved in the small groups than in big groups, while the low academic ability students in the small groups was not significantly different than their involvement in big groups. This fact shows not only the importance of comparing students' experiences based on instructional models but also considering the grouping and managing students who can influence various students.

High academic ability students tend to show prominent performance compared to low academic ability students. They tend to be superior in group discussion activity or presentation, due to their self-confidence. After conducting ana Cova test analysis, it is showed that early academic ability might influence students' cognitive achievement. The students with high early ability tend to have a high academic ability. Hence, the high academic ability students can improve their achievement by attaining information during learning process and keeping the information as new knowledge.

Teachers are expected to supervise and guide students in a group activity. It enables low academic ability students to be involved in doing group tasks. The result of this study shows that the improvement of the achievement also occurred on low academic ability students. It is necessary to highlight that low academic ability students were doing good performances in constructing their knowledge. Students are more motivated when they are allowed to interact with each other in a group.

3.3. The difference of cognitive achievement based on the interaction between learning model and academic ability

The F observed of the interaction between learning model, and academic ability is 2.069, p-value 0.154 > α (0.05). Hence, H0 stating there is no cognitive achievement difference in the interaction between achievement and different academic ability is accepted. Hence, the alternative hypothesis stating there is no different cognitive achievement in the interaction between learning model and academic ability can be rejected. It means, there is no effect on the interaction between learning model and academic ability towards students' cognitive achievement. The ana Cova test result shows that there is no effect on the interaction between learning model and academic ability towards students' cognitive achievement. Therefore, it is necessary to conduct LSD test to examine the interaction position. Table 4 shows the results of the LSD test.

Table 4. The mean of the cognitive achievement on the interaction group

No.	Interaction	Pretest	Posttest	Different	Mean	Notation			
1	conventional - low	19,947	25,640	5,693	26,907	a			
2	conventional - high	34,181	39,189	5,008	35,830		b		
3	NHT - low	14,927	44,499	29,572	47,397			c	
4	NHT - high	26,431	61,631	35,200	60,790				d

The results provide the following information: 1) The students' cognitive achievement in conventional-low academic ability is significantly the lowest that the other interaction. 2) The students' cognitive achievement achievement in NHT-high academic ability is significantly the highest among the other three interaction. The cognitive achievement in the conventional-high academic ability group improves 14.65%, in the conventional-low academic ability improves 28.54%, in the NHT-high academic ability improves 133.18%, and in the NHT-low academic ability is 198.11%.

The ana Cova test result shows that there is no effect on the interaction between learning model and academic ability towards students' cognitive achievement achievement. The interaction between learning strategy and academic ability does not influence students' satisfaction in learning [23]. For students who have a moderate academic ability, cooperative or conventional learning makes their process easier. While for low academic ability students, either cooperative or in conventional model class might still make them difficult to learn.

The cognitive achievement between NHT-high academic ability and NHT-low academic ability is not significantly different, likewise conventional-high and low academic ability. As a consequence, there are no cognitive achievement differences in the interaction between learning model and academic ability. Thus, all interaction causes the same effect. The NHT provides the same opportunity to think and memorizes each solution [24]. Besides, NHT also facilitates the students to have self-preparation in justifying the group performances, so that the low academic ability students tried their best. In the conventional learning, the real competition is unavoidable. Both the high and low tried to get the maximum result. While in fact, the improvement of the low academic ability students was not significant.

4. Conclusions

The findings provide the following conclusions: 1) NHT is 72.45% more potential to improve cognitive achievement rather than conventional learning model. 2) The high academic ability students' cognitive achievement is 30.03% different compared to the low academic ability students. 3) There is no effect on the interaction between learning model and academic ability. However, based on the LSD test, it is obtained that the mean of the cognitive achievement in the interaction between conventional-low academic ability is significantly different from the cognitive achievement in the interaction between conventional-high academic ability, NHT-low academic ability, and NHT-high academic ability. Moreover, the other interaction groups show the significant mean of the cognitive achievement.

References

[1] Senior C, Howard C, Reddy P, Clark R and Lim M 2012 *Stu. High. Educ.* **37** 957
[2] Ibrahim A 2012 *Inter. J. Sci. and Math. Educ.* **10** 293
[3] Bullard J and Bullock JR 2004 *J. Early Child. T. Educ.* **25** 39
[4] Alfassi M 2004 *Learn. Envi. Res.* **7** 1
[5] Bouffard SM and Hill NE 2005 *S. Psy. Educ.* **8** 441
[6] Maheady L 2001 *J. Behav. Edu.* **15** 25
[7] Haydon T, Maheady L and Hunter W 2010 *J. Behav. Educ.* **19** 222
[8] Bawn S 2007 *Thesis* The Evergreen State College
[9] Bertucci A, Johnson DW, Johnson RT and Conte S 2012 *J. Educ. Res.* **105** 329
[10] Thurston A, Topping KJ, Tolmie A, Christie D, Karagiannidou E and Murray P 2010. *In. J. Sci. Educ.* **32** 551
[11] Malanga PR and Sweeney WJ 2008 *J. Beh. Educ.* **17** 187
[12] Miaz Y 2010 *Res. J. Soc. Sci.* **8** 40
[13] Munawaroh 2016 IOSR J. Res. Method. Educ. **5** 24
[14] Yamarik S 2007 *J. Eco. Educ.* **38** 259
[15] Ari RB and Friedrich KP 1996 *Eu. J. Inter. Stu.* **7** 17
[16] Slavin RE 2015 *Inter. Encyc. Soc. Beh. Sci.* **4** 881
[17] Pham THT 2013. *Higher Educ. R. & Dev.* **32** 993
[18] Zhong YB, Lei YM, Wang LXB, Yang L and Lin GH 2014 *Inf. & Eng. and Ope. Res. Man.* 525
[19] Noaparast KB and Khosravi Z 2010 *Inter. Educ.* **21** 281

[20] Hsiung CM 2010 *J. Eng. Educ.* 45
[21] Gillies RM and Ashman AF 2000 *J. Spec. Educ.* **34** 19
[22] Mulryan CM 1995 *Elem. Sch. J.* **95** 297
[23] Lee C, Ng Maureen and Phang R 2002 *A. Pac. J.Educ.* **22** 3
[24] Hunter WC, Dieker LA, and Whitney T 2016 *J. Educ. Psy. Cons.* **26** 186

Air Pollution Simulation based on different seasons

Muhaimin[1]

Chemistry Education Department of Mathematics and Natural Science Faculty, Islamic University of Indonesia, Yogyakarta

Email: muhaimin@uii.ac.id

Abstract. Simulation distribution of pollutants (SOx and NOx) emitted from Cirebon power plant activities have been carried out. Gaussian models and scenarios are used to predict the concentration of pollutant gasses. The purposes of this study were to determine the distribution of the flue gas from the power plant activity and differences pollutant gas concentrations in the wet and dry seasons. The result showed that the concentration of pollutant gasses in the dry season was higher than the wet season. The difference of pollutant concentration because of wind speed, gas flow rate, and temperature of the gas that flows out of the chimney. The maximum concentration of pollutant gasses in wet season for SOx is 30.14 $\mu g/m^3$, while NOx is 26.35 $\mu g/m^3$. Then, The simulation of air pollution in the dry season for SOx is 42.38 $\mu g/m^3$, while NOx is 34.78 $\mu g/m^3$.

1. Introduction

Qualitative aspects of the theory of dispersion of pollutants in the air is to describe the state of emissions in the atmosphere from pollution sources which in assumed to be point source. Qualitatively, this dispersion theory can approximate pollutants concentration in the ambient air, one of the models of dispersion of pollutants in the air are Gaussian Plume [1]. Gaussian Plume Model is an approach used to study the contaminants in the air because of the turbulent diffusion and advection caused by the wind [2]. Gaussian Plume Models are often used to predict the spread of continuous air pollution originating from the surface or plateau. It is assumed that the spread of air pollution has a gaussian distribution or have a normal probability distribution. This model assumes that the atmosphere had a stagnant and homogenous state, the concentration of pollutants will be distributed normally. Input for this model is very simple which include wind speed, distance distribution of pollutant gasses from sources and distribution coefficient of pollutant gasses [3]. The equations of the model for the Gaussian Plume dispersion of pollutants in the air can be expressed as follows [4]:

$$C\,(x,y,z) = \frac{Q}{2\pi u \sigma_y \sigma_z} exp\left(\frac{-y^2}{2\sigma_y^2}\right)\left(exp\left(\frac{-(z+H)^2}{2\sigma_z^2}\right) + exp\left(\frac{-(z-H)^2}{2\sigma_z^2}\right)\right)$$

where,

$C(x,y,z)$: Gas concentration in x,y,z coordinate (g/m^3),

[1] To whom any correspondence should be addressed.

Q	: Gas emission rates from (g/s),
U	: Wind speed average (m/s),
σ_y	: Plume standard deviation in line with y-axis (m),
σ_z	: Plume standard deviation in line with z-axis (m),
H	: Effective stack height (m)

The Gaussian Model is the most mathematical model to predict air pollution dispersion. It is widely used, easy to be understood and applied, and the assumption of the model that the air pollutant will be dispersed based on the normal statistical distribution [5,6]. Dispersion simulation uses equation to predict the concentration of air pollutant that emitted by a point source. Meteorology and emissions data are used to predict the dispersion of pollutans in the atmosphere [7,8].

The purposes of this study were to determine the distribution of the flue gas from the power plant activity, differences of pollutant gas concentrations in the wet and dry seasons.

2. Experimental methods

Parameters of air pollution in this research were SO_x and NO_x. In Indonesia has two seasons, they are the dry season (April - September) and the wet seasons (October - March). In this paper will discuss the simulation of air pollution based on the different season. This research used quantitative analysis method using Gaussian Plume equation. The data used in this study is secondary data. The secondary data on climatology used was in year 2012-2013. Then, the data of pollutant gas was taken from Cirebon Coal Fired Power Plant. The distance dispersion of pollutants was observed in this research is in line with the x-axis of the sources of pollutants (chimney) is 20,000 meters, while the distance is in line with the y-axis is 1,500 meters.

3. Results and Discussion

To determine the direction of dispersion of pollutants in the air as a result of the activity of the power plant, modeling the wind direction is done by using data of wind speed and direction obtained from air monitoring of Cirebon power plant in West Java, during one year from October 2012 until September 2013 using WRPLOT. In the wet season, the dominant wind direction was blowing from Northeast and southwest at speeds ranging from 2,05 m/s, shown in Figure 1(a). Then, in the dry season showed that the dominant wind direction was blowing from north and south at speeds ranging from 1,21 m/s, shown in Figure 1(b).

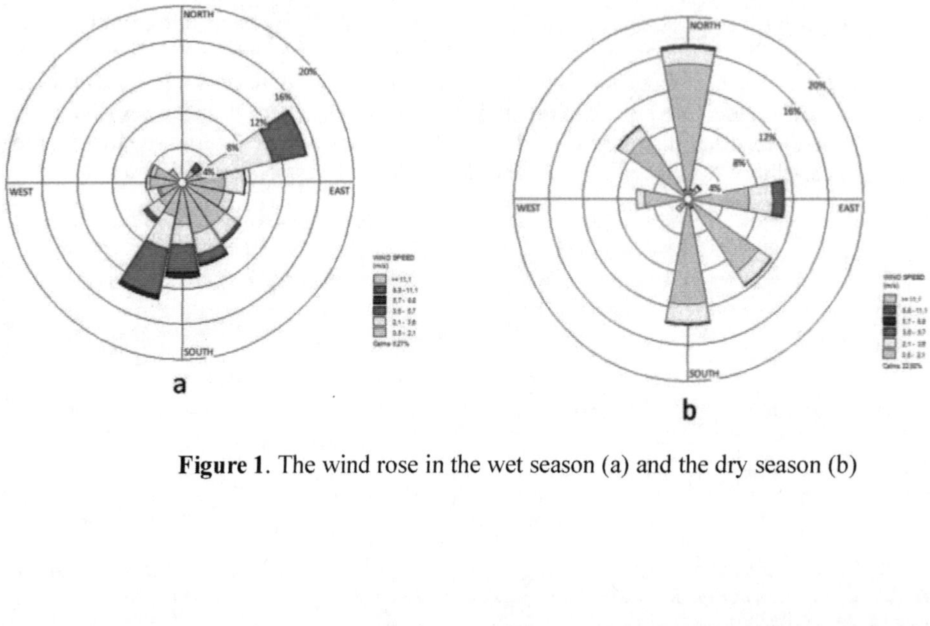

Figure 1. The wind rose in the wet season (a) and the dry season (b)

The simulation results of gas dispersion of pollutants into the atmosphere was done by using Gaussian Plume equation taking into account with the meteorological and environmental monitoring data which conducted by Cirebon power plant. They have been carried out in April 2012 to March 2013. The specification of secondary data used in the simulation of pollutant dispersion around Cirebon power plant is shown in Table 1.

Tabel 1. The Secondary Data Used in Simulation

Parameter	Value		Unit
Inside Diameter of Stack	6,85		m
Physical Stack Height	215		m
	Dry Season	Wet Season	
Stack Gas Exit Temperature	406.407	404.822	K
Temperature of Ambient Air	302.410	301.221	K
Stack Gas Exit Velocity	0.165	0.207	m/s
Wind Speed	1.235	1.958	m/s
Emission Rate:			
• SOx	18.713	18.683	g/s
• NOx	15.356	16.331	g/s
Distance Distribution Of Pollutant Gasses	100 - 20000		m

3.1. Dispersion model of SO_x

Combustion of fuels containing sulfur, such as coal and oil are used as fuel in power plant such that resulted sulfur dioxide (SO_2). It can cause health problems in humans and the environment [9,10] and react to the other chemical in the ambient air to create sulfate particles [14]. The sulfur compounds formed in the atmosphere are two kinds of SO_2 and SO_3, both called by SO_x and from the burning of fossil fuels around 75% of the total amount of emissions of SO_x [13].

Figure 2 showed that the maximum concentration of SO_x in the dry season was 42.38 μg /m^3 (about 780.78 meters from the source in the direction of the x-axis), while the wet season is 30.14 μg/m^3 (about 740.74 meters from the source direction of the axis x). In the region, there are no settlements, only rice fields. The difference is caused by the concentration factor of water present in the atmosphere, the water content in the rainy season will be higher than in the dry season.

The high rainfall during the wet season will dissolve compounds of pollutants in ambient air so that it will experience wet deposition. So that the wet deposition can lower the concentration of pollutants in ambient air, or in other words, the presence of rainwater, pollutants in the atmosphere will experience a dilution that resulted in a decrease in the concentration of pollutants [16].

The removal of SO_2 in the atmosphere at a wet deposition process is influenced by some factors namely rainfall, the length of the rain, frequency of rain, and the relative amounts of SO_2 and sulfate particle size distribution [11].

In the dry season does not happen sulfate dilution process to pollutants in ambient air, because the water content in the atmosphere is less common than in the wet season. High sulfate concentrations indicate that oxidation and changing of SO_2 to sulfate rely on the reaction of photochemical oxidation, the sulfate concentration in the day is higher than at night, this is caused by the presence of sunlight [12]

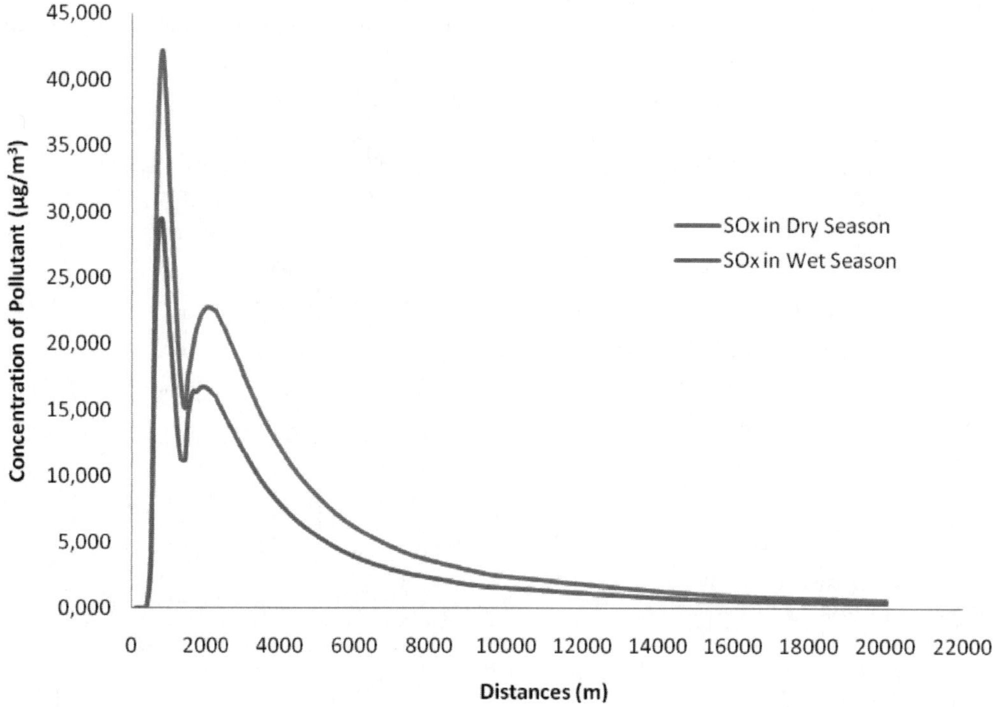

Figure 2. Dispersion of pollutant SO_x in the wet and the dry seasons.

3.2. Dispersion model of NO_x

NOx is a definition used to explain the amount of nitrogen oxide (NO), Nitrogen Dioxide (NO_2), and nitrogen oxides, and several others. It is a pollutant that can result if the acid rain reacts with volatile organic compounds through complex reactions, assisted by sunlight and water vapor in the atmosphere. The main source of NO_x emissions derived from the activities of electricity generation and motor vehicles [14]. The impact of NO_x gasses are: increasing the concentration of ozone, cause respiratory disease and cancer in humans, increasing global climate change, reducing agricultural productivity result from the deposition of ozone, and decrease biodiversity ecosystems [13].

One of the efforts to control NO_x concentrations in ambient air is the air dispersion modeling. Accordng to the Figure 3 showed that the maximum concentration of NO_x in the wet season is at 26.35 $\mu g/m^3$ which is located at a distance of 740.741 meters from the source (chimney). Whereas in the dry season, the maximum concentration of NO_x was 34.78 $\mu g/m^3$ at a distance of 780.781 meters from the source (chimney). This difference can be caused by ambient air temperature, wind speed, rainfall, and others. In the dry season more diffuse plume of NO_x distributed than in the dry season, it can be seen from the reach of the lowest concentration that are at different distances. NO_x concentration of the lowest in the wet season is at a maximum distance of about 18 km while the dry season is at a distance of over 20 km.

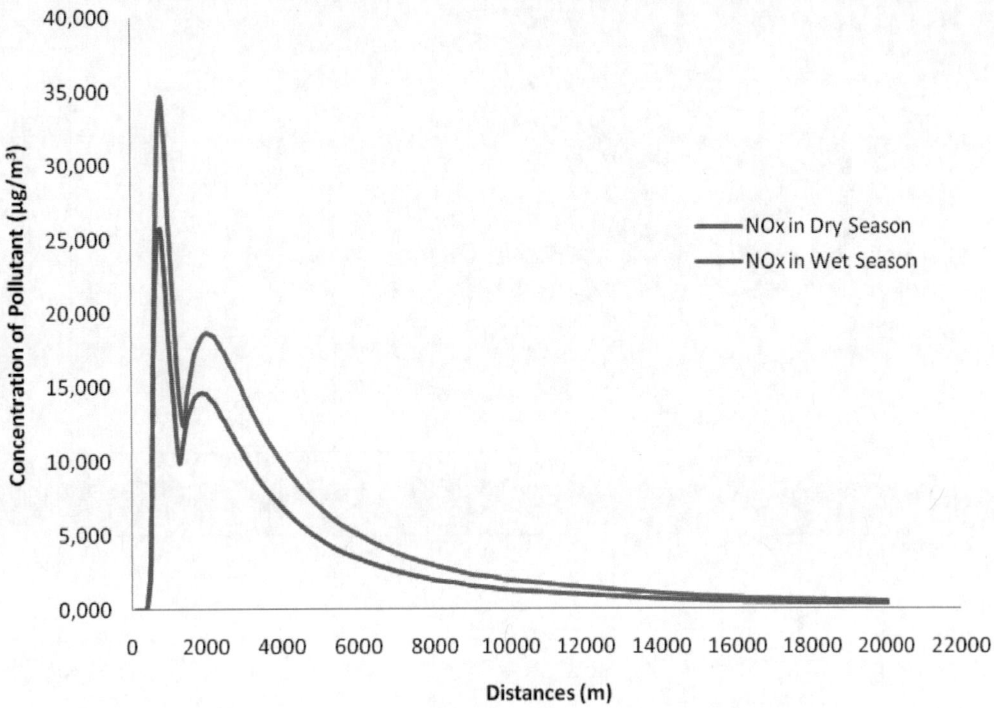

Figure 3. Dispersion of pollutants NO_x in the wet and the dry seasons.

As well as SO_x, NO_x in the atmosphere will also experience deposition both wet and dry deposition. In the wet season NOx will experience wet deposition, whereas in the dry season NO_x pollutants will experience dry deposition. In the process of wet deposition compound ammonium (NH_4^+), nitrate (NO_3) will be up to the surface through rain water, while in the dry deposition of nitrogen compounds will eliminate gasses NH_3, NO_2, HNO_3 (with particles NH_4 $^+$, NO_3, organic compounds), organic acids, and aerosols containing nitrogen. Release of NOx compounds from the combustion reaction is a major component in the formation of nitrogen oxides (NO). Zeldovich reaction showed that the nitrogen oxide (NO) is generated from the oxygen supply of free air around 200,000 ppm in the atmosphere at temperatures above 1300°C. at a temperature below 760°C, Concentration of NO compounds is very low even not be produced [13]. NO production is a function of the fuel to air ratio [15]. The distribution of pollutant gasses SOx and NOx in the dry and the wet season are shown in Figure 4.

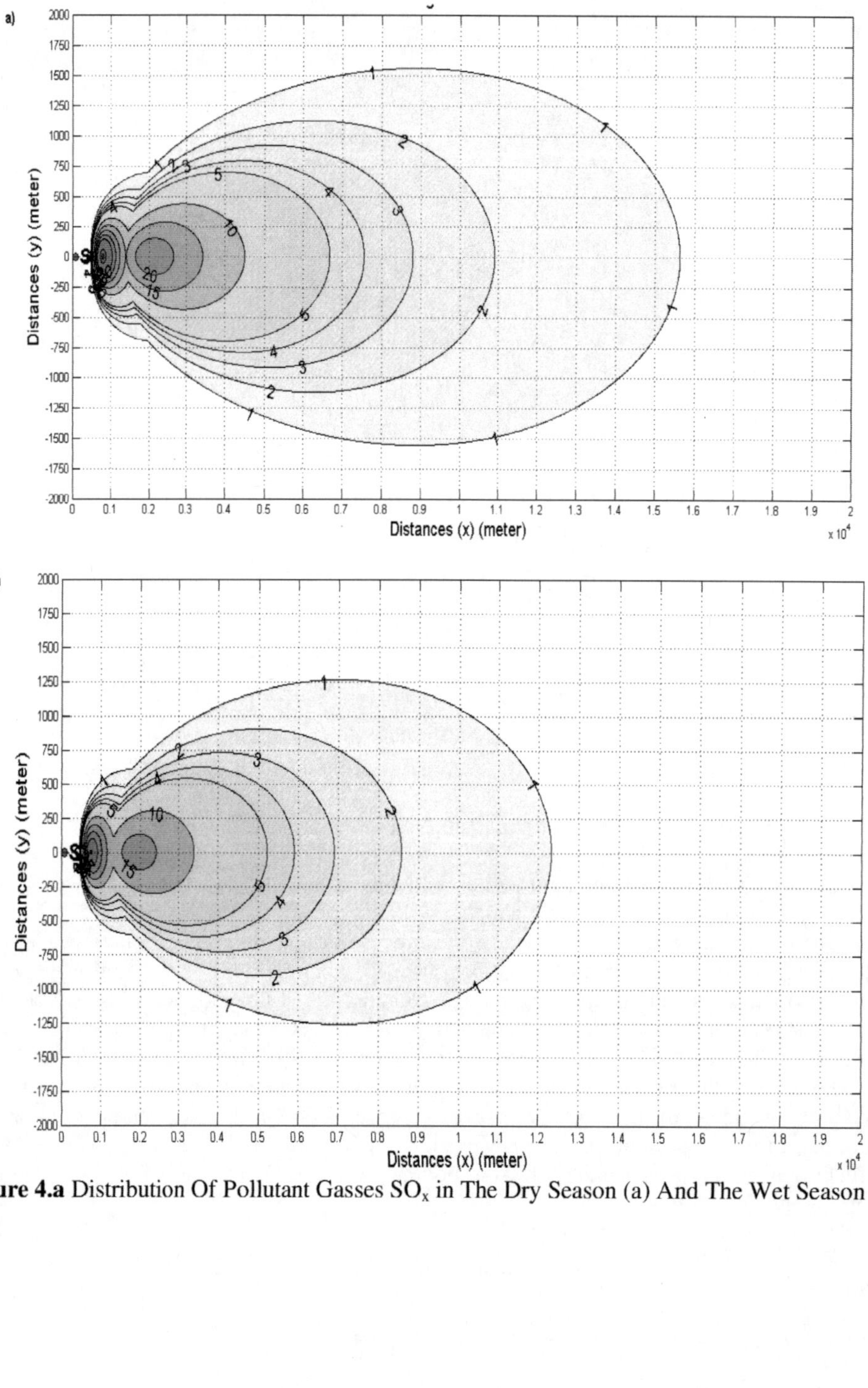

Figure 4.a Distribution Of Pollutant Gasses SO_x in The Dry Season (a) And The Wet Season (b)

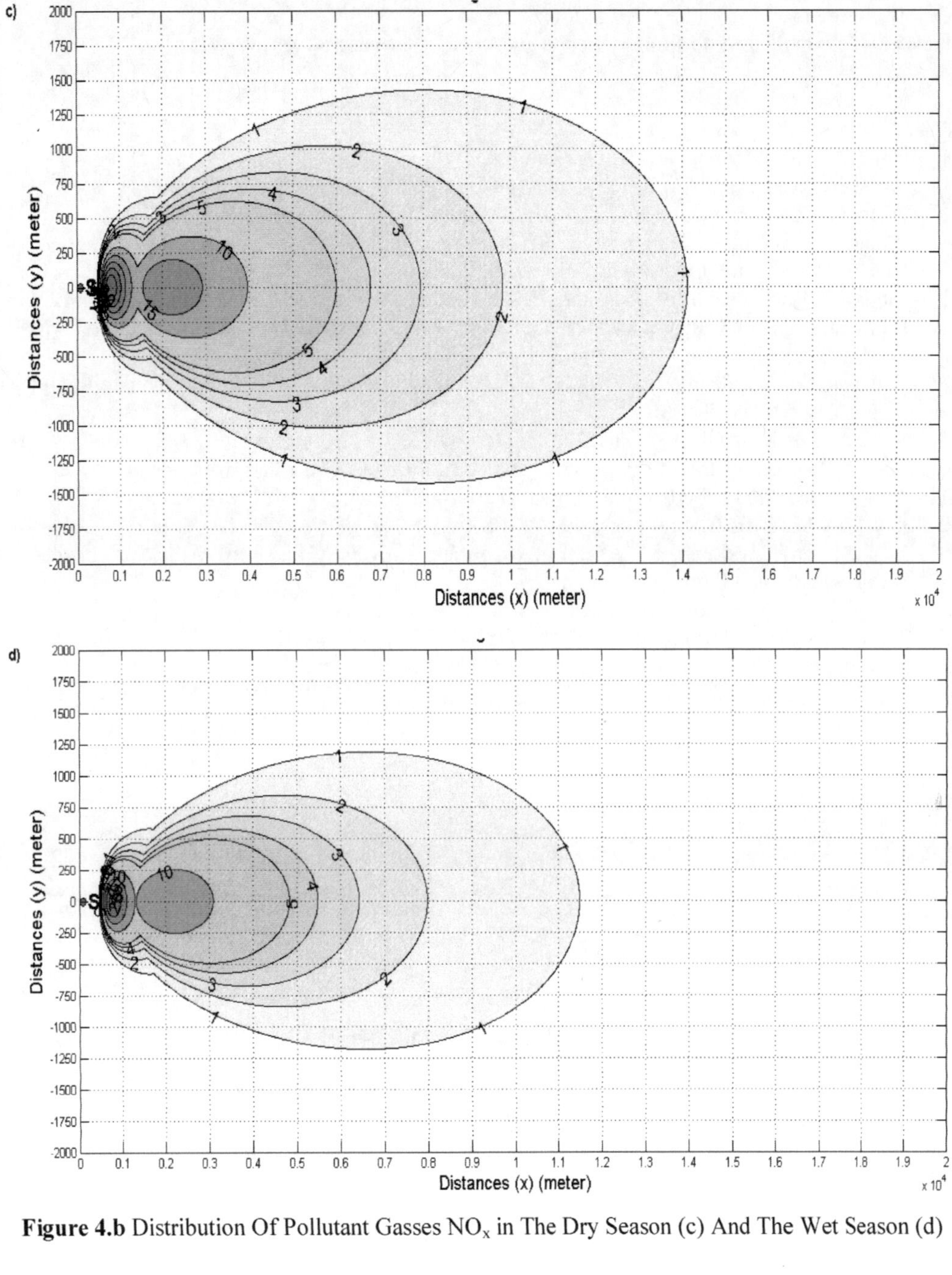

Figure 4.b Distribution Of Pollutant Gasses NO_x in The Dry Season (c) And The Wet Season (d)

4. Conclusion

In conclusion, the concentration of pollutant gasses in the dry season was higher than during the wet season. The maximum concentration of pollutant gasses in wet season for SO_x is 30.14 $\mu g/m^3$, while NO_x is 26.35 $\mu g/m^3$. Then, The simulation of air pollution in dry season for SO_x is 42.38 $\mu g/m^3$, while NO_x is 34.78 $\mu g/m^3$.

Acknowledgment
Thankful is acknowledged to PLTU Cirebon and PSLH UGM as partner research.

References

[1] Coll, J., 2002, London, New Fetter Lane.
[2] Stockie, J.M., 2011 SIAM Review 53 2 349–372.
[3] Ukaigwe, S.A., and Osoka, E.C., 2013 IJSR, 2, 9.
[4] Weiner and Matthews, 2003 Elsevier Science, 4.
[5] Ukaigwe, S.A., and Osoka, E.C., 2013 IJSR, 2, 9.
[6] Modi, M., Ramachandra,V.P., Ahmed S.K.L., and Hussain, Z., 2013, IJPCBS 3 4 1224-1230.
[7] Bijaksana, A.M.A., Sjahrul, M., Nadjamuddin Harun, and Rudy Djamaluddin, 2012 IJCEE-IJENS, 03 12.
[8] Grigoras, G., Cuculeanu, V., Ene, G., Mocioaca, G., Deneanu, A., 2012 Romanian Reports In Physics, 64(1), 173–186.
[9] Srivastava, R.K., 2000 U.S. EPA, Washington, D.C.
[10] Flagan, R.C., and Seinfeld, J.H., 1988 Prentice Hall Englewood Cliffs, New Jersey.
[11] Anonymous, 2003 Alberta Environment, California.
[12] Khoder, M.I., 2002 Chemosphere 49 675–684.
[13] Lagzi, I., and Mészáros, R., 2013 Eötvös Loránd University Facultity of Science Institute of Geography and Earth Science.
[14] Ma, S., 2010 Energy and Power Engineering 2 190-195.
[15] Varma, S.A.K., and Srimurali, M., 2013 IOSR-JESTFT 6 3 08-11.
[16] Venkatram, A., 2004 Atmos. Environ. 38 16 2439-2446.

Detection of Atrial Fibrillation Using Artifical Neural Network with Power Spectrum Density of RR Interval of Electrocardiogram

Adfal Afdala, Nuryani Nuryani and Anto Satrio Nugroho
Physics Department of Post Graduate Program Sebelas Maret University,
Jl. Ir. Sutami 36A Kentingan Jebres Surakarta 57126, INDONESIA

E-mail: nuryani@mipa.uns.ac.id

Abstract. Atrial fibrillation (AF) is a disorder of the heart with fairly high mortality in adults. AF is a common heart arrythmia which is characterized by a missing or irregular contraction of atria. Therefore, finding a method to detect atrial fibrillation is necessary. In this article a system to detect atrial fibrillation has been proposed. Detection system utilized backpropagation artifical neural network. Data input in this method includes power spectrum density of R-peaks interval of electrocardiogram which is selected by wrapping method. This research uses parameter learning rate, momentum, epoch and hidden layer. System produces good performance with accuracy, sensitivity, and specificity of 83.55%, 86.72 % and 81.47 %, respectively.

1. Introduction

Atrial fibrillation (AF) is a serious heart condition where the atria contracts irregularly in rapid succession so that blood can not completely pump into ventricle [1]. Blood is not pumped completely become one of the causes blockage of blood to the brain and increases the risk of stroke [2]. Atrial fibrillation is a heart disorder that is difficult to detect, this is because the appearance that cannot be predicted certainly.

One way to detect atrial fibrillation is to studying the morphology of electrocardiogram. Atrial fibrillation has morphological characteristics in the electrocardiogram that is missed of P wave, and irregularity of heart rate which is represented by R peaks [3]. Figure 1 shows the appearance of the electrocardiogram in patients with normal (NSR) and atrial fibrillation (AF).

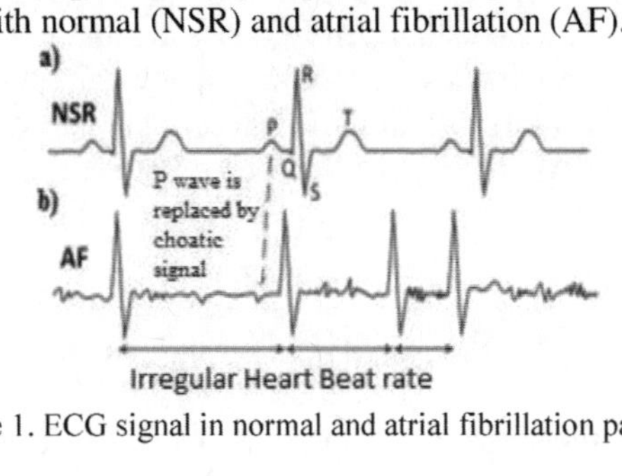

Figure 1. ECG signal in normal and atrial fibrillation patient

Characteristics of atrial fibrillation can be used as a marker of the emergence of atrial fibrillation, therefore, the features used in this study is the culmination of irregularity information R. Several studies have been done using this feature including the detection of atrial fibrillation by using a fuzzy inference system with the features of the P wave on the electrocardiogram [5], automatic detection by using support vector mechine [6]. The classification used is back propagation neural network.

2. Numerical Methods

There are three general stages in this study. Figure 2 shows each of the stages and they will be explained as well. The Data were taken from the database of physionet atrial fibrillation with a sample frequency of 250 Hz, it was taken with a record time span of one hour.

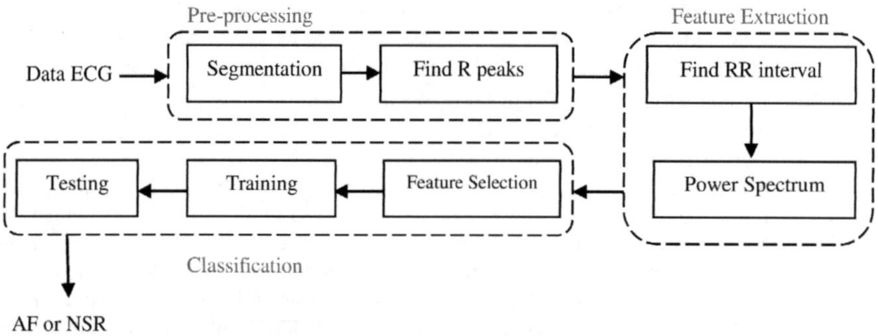

Figure 2. Flow chart for the research

a. Preprocessing

This is the stage of the data preparation. This phase consists of several stages inside namely segmentation and positioning of R peak. Segmentation is the process of data sharing wherein each segment consists of 10000 of data, if it is converted into units of time with data sampling frequency of 250 Hz, then each of segment represents 40 seconds of ECG recording. This process is intended to make the appearance of irregular heart rate can be monitored easily.The next stage is the stage of determining the position of the R peak, where R peak position can be searched using the algorithm Pan-Tomkin with an accuracy of 99.3% [7].

b. Feature Extraction

Extraction feature has two stages in it. They are named the stages of finding R peak intervals in each segment and the power spectrum as well.R peak intervals are the characteristic of atrial fibrillation, so it can be used as a feature to detect atrial fibrillation of the heart. The interval distance between the peak of R (RR interval) can be calculated by using equation 1.

$$RRI(n) = R(n+1) - R(n) \tag{1}$$

The next stage is the stage of finding the power spectrum in each segment. Power spectrum is one characteristic of a signal [8].

c. Classification

Classification stage starts from selecting the best features. The best feature is a feature that provides a lot of information to group and normal atrial fibrillation. Feature selection using wrapping techniques that run at softwear weka by using the power spectrum as attributes.Next Stage is training of a neural network by using the feature selection that is resulted from the previous stage. Neural network used is named back propagation neural network. In the training phase, the main objective is to find the optimal weights to detect atrial fibrillation. Optimal weight is represented by good training performance. Training performance can be seen by analyzing the mean square error with the increasing of repetitions. The parameter used is the learning rate, momentum, hidden layers and the number of epochs. The next stage is the testing phase. This phase is to test the weight performed on test data and produce the output of accuracy, specifications, and sensitivity.

3. Results and Discussion

Parameters at the training stage is set with number of learning rate = 0.05 momentum = 0.9, epoch = 40000, and the hidden layer = 5 with outuput 1.
The selection feature stage produces 5 selected features. The results of the performance of each feature can be seen in Figure 3.

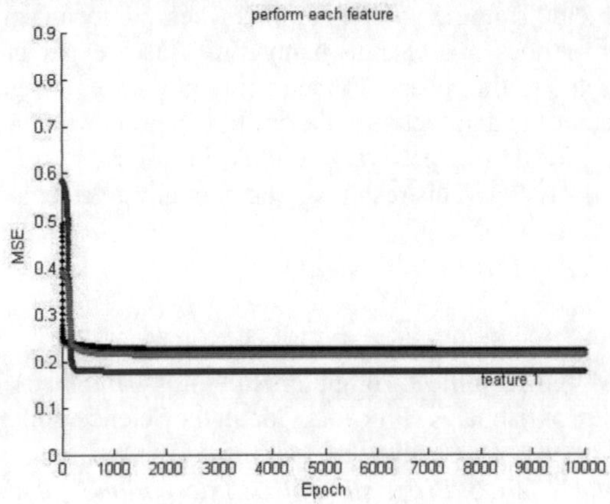

Figure 3. Perform each feature

Figure 3 shows the performance of each feature that has been carried out through training of artificial neural networks. The blue line is the feature that has the best performance of the other features. This performance can be improved by combining all the features. A good feature has small and convergent mean square error. Figure 4 shows the results of the performance of some combined features.

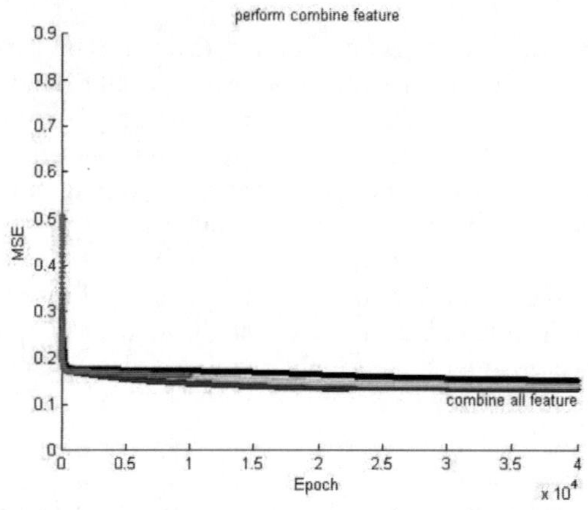

Figure 4. Perform combine feature

Figure 4 shows the blue line that is a combination of the five features with smallest mean square error that is 0:13 and it is convergent. Training will be stopped when the mean square error is convergent in which the mean square error does not change many times as we get epoch.The training process produces useful weights for the testing phase. In the testing phase, the weights will be used to test the system's ability to sort whether the data belong to atrial fibrillation or not. The decision is determined by the threshold used. This research respectively resulted in accuracy, sensitivity, and specifications that are 83.55%, 86.72% and 81.47%. This result is good enough to detect atrial fibrillation.

4. References

[1] Shadnaz A, Alireza M, & Maryam M.Computers in biology and medicine.**60**,132-142 (2015)
[2] Padmavathi, K and Ramakrishna, K. Procedia Computer Science.**46**,53-59 (2015).
[3] Camm, A.J. Europan Society Of Cardiology.**14**,1385-1413 (2012).
[4] Z. Huang, Z. Chen, and Y. Zhao, *Proceeding of 1st International Conference on Bioinformatics and Biomedical Engineering*,Wuhan,2007, pp 625-628.
[5] Anderson, L., Bailey, B., and Bolkhovsky, J.,Thesis, Worcester Polytechnic Institute,West Midlands,2011.
[6] Clifford, G D. "Advanced Methods for ECG Analysis".(Artech House, United Kingdom,2006).p.56
[7] Pan J dan Tomkins W. IEEE Transactions on Biomedical Engineering.**32**.230-235 (1985).
[8] Moody G.B and Mark R.G. IEEE Engineering in Medicine and Biology Magazine.**20**.45-50 (2001).

Rheological properties of carbon nanotubes-reinforced magnetorheological elastomer

S A A Aziz[1], SA Mazlan[1] ,N I Nik Ismail[2], Ubaidillah[3], MHA Khairi[1], NA Yunus[1]

[1]VSE Research Laboratory, Malaysia – Japan International Institute of Technology, Universiti Teknologi Malaysia, Jalan Sultan Yahya Petra, Kuala Lumpur, Malaysia, 54100
[2]Advanced Processing & Product Technology Centre, R&D Centre of Excellence (COE), Malaysian Rubber Board, Experimental Station, 47000 Sg. Buloh, Selangor, Malaysia
[3]Mechanical Engineering Department, Faculty of Engineering, Universitas Sebelas Maret, Jl. Ir. Sutami 36A, 14 Kentingan, Surakarta, 57126, Central Java, Surakarta

E-mail: amri.kl@utm.my

Abstract. Magnetorheological elastomer (MRE) based on the natural rubber with different types of multiwall carbon nanotubes (MWCNT) as additives were synthesized. MRE with pristine MWCNTs was prepared as a control and the carboxylated (MWCNT-COOH), as well as hydroxylated (MWCNT-OH) were introduced as new additives in MRE. Their rheological properties under different magnetic field were evaluated by using the rheometer (MCR 302, AntonPaar, Austria) equipped with the electromagnetic device. The dependency of MREs towards excitation frequencies under different magnetic field was investigated. It is shown that the storage modulus and loss factor of MRE with functionalized MWCNTs exhibited noticeable increment in MR performance compared to control parallel with the frequencies increment.

1. Introduction

Magnetorheological (MR) materials have attracted interest in academic and industrial areas due to their mechanical and rheological properties can be changed and altered rapidly and reversibly by applying magnetic field [1,2]. Among MR materials family, MR elastomer (MRE) have been extensively studied due to their obvious advantage; MREs has a stable performance since the magnetic particle cannot settle with time, and also their unique mechanical performance in which their field dependent modulus compared to the other MR materials.

MREs mainly consists of non-magnetic solid-phase material such as rubber, micrometer size soft magnetic particles, and additives. There are two kinds of dispersion of soft magnetic particles in MREs namely: isotropic and anisotropic. In the stage of designing the MRE, material selection plays important roles in better performances. Few types of rubber have been used by other researchers like Silicon rubber [3,4], natural rubber [5,6] and other types of rubber [7,8]. Carbon black is frequently used as filler on fabricating the MRE. However, recently many other ingredients and additives have been explored by another researcher to get the optimum performance of MR effect for variable applications. Nanomagnetic materials have attracted the remarkable attention of researchers due to their strong interconnected three-dimensional networks at relatively lower filler loading [9].

Nayak et al., fabricated silicon rubber with dispersed 7% nano-sized carbon black and tested in the oscillation mode. It has been found that Young modulus and rheological properties of MRE with nano-sized carbon black increased with 56% and 22% compare to MRE without carbon black [10]. Li et al.

[11,12], investigated dynamic mechanical behavior silicone rubber based MRE using 1 and 3.5 wt% of multiwall carbon nanotube as additives. By conducting both shear and compression investigations, at all dynamic, MRE with MWCNT exhibit higher zero-field stiffness and damping and the dynamic mechanical properties also increased. However, more study has to be continued further to investigate the contribution of MWCNT in MRE properties. Also, functionalized MWCNTs, carboxylated (MWCNT-COOH) and hydroxylated (MWCNT-OH) also have been introduced, and their rheological properties such as storage modulus, and loss factor are investigated in details.

2. Materials and Fabrication

2.1 Raw Material

Natural rubber (NR) manufactured by Malaysian Rubber Board was selected as the matrix. Soft magnetic carbonyl iron particles (type OM) with an average diameter of 6μm purchased from BASF, Germany were used. Three types MWCNT i.e. Pristine MWCNT, carboxylated (COOH-MWCNT) and hydroxylated (OH-MWCNT) were used as an additive which obtained from Shenzhen Nanotech Port Co. Ltd, China. The outer diameter of 10-20nm and particle length in between 10-30 μm with a purity level of MWCNT used is 95%. The vulcanizing agent and the plasticizer were sulfur and epoxidized palm oil (EPO), respectively. Meanwhile, the zinc oxide (ZnO) and stearic acid were used as activators. The N Cyclohexylbenzothiazole-2-sulphenamide (CBS) used as an accelerator. Santoflex 13 used as antiozonant and as a synthetic polymer stabilizer.

2.2 MRE Fabrication

In this study, three groups of MRE samples were prepared with different types of MWCNT; MWCNT, MWCNT-COOH, and MWCNT-OH (0.1 wt%). For comparison, the reference samples consist of NR and CIP was also prepared. The fabrication process of MREs requires three steps: sonication of MWCNTs using horn type ultrasonicator, mixing the rubber and other additives via double roll mill, which ensures all components were dispersed homogeneously in the NR matrix and then vulcanized to ensure the structures were locked into place in the matrix. First, 0.1 wt% of MWCNTs were sonicated in the epoxidized palm oil (Rovpro 5300, Rovski Industries Sdn. Bhd) for 15 minutes with 10 seconds off pulse by using the horn type ultrasonicator (FB 705, Fisher Scientific (M) Sdn Bhd, Malaysia). Secondly, 100 phr rubber, 10 phr epoxidized palm oil, 30 wt% of carbonyl iron particles (CIP), 0.1 wt% of MWCNT and additives (5 phr zinc oxide, 2 phr stearic acid, 1.5 phr santoflex, 2.5 phr sulfur, and 1 phr cyclohexyl-2-benzothiazolesulfenamide were mixed via double roll mill. Finally, the MRE samples were vulcanized at 150°C for 60 minutes under high pressure (MDR 2000).

2.3 Rheological Properties

The rheological properties of MRE samples were measured by using an oscillation magnetorheological device (MRD), Rheometer (Physica MCR 302, Anton Paar, Germany) under the variable magnetic fields. The circular sample with the 1mm thickness was used and placed in the cavity base plate. Each sample has been tested in the oscillatory mode to investigate the dynamic rheological properties. The electric currents were varied at 0 and 4 A. The magnetic flux density calculated which corresponding to the 0, and 4 A were 0, and 643 mT, respectively. MR effect formula can be calculated as below;

$$\frac{G'_{\max} - G_0}{G_0} \times 100\% \qquad (1)$$

The samples were tested under the frequency sweep test. The shear strain was fixed at 0.6%, and the frequency was varied from 0.1 to 100 Hz.

3. Results and Discussion

The variation of storage modulus and loss factor of the MRE samples are shown in Figure 1 (a)-(b) as a function of frequency. The linear viscoelastic response of MRE samples has been studied using constant strain amplitude of 0.6% and frequency sweep from 0.1 to 100 Hz. Figure 1(a-b) hows the storage modulus increase with the increasing of the frequency with and without magnetic field. In the case of MRE samples without MWCNT, the initial and maximum storage modulus is found to be considerably lower than the MRE samples with carboxylated and hydroxylated MWCNTs. Upon addition of 0.1 wt% of carboxylated and hydroxylated MWCNT, both storage modulus showed noticeable increment, nevertheless the pristine MWCNT stay lower trend in the region of lower and highest frequencies. The distribution of CIP and MWCNTs in the matrix was random [13] , which lead to larger moveable of CIP and MWCNTs in the matrix, resulting in higher storage modulus. In addition, the high specific areas and essential modulus of MWCNT enhance the stiffness of rubber. This phenomena also will lead to increment in the storage modulus [14].

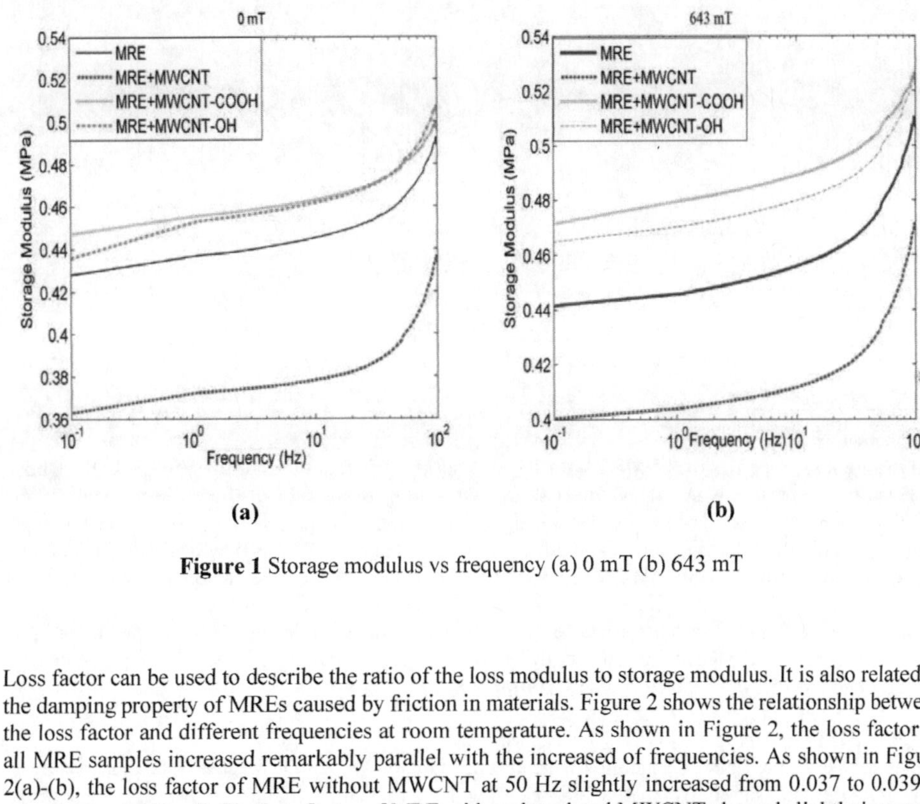

Figure 1 Storage modulus vs frequency (a) 0 mT (b) 643 mT

Loss factor can be used to describe the ratio of the loss modulus to storage modulus. It is also related to the damping property of MREs caused by friction in materials. Figure 2 shows the relationship between the loss factor and different frequencies at room temperature. As shown in Figure 2, the loss factor of all MRE samples increased remarkably parallel with the increased of frequencies. As shown in Figure 2(a)-(b), the loss factor of MRE without MWCNT at 50 Hz slightly increased from 0.037 to 0.039 at zero field and 643 mT. The loss factor of MRE with carboxylated MWCNT showed slightly increased compared to MRE with pristine and hydroxylated MWCNTs. However at frequencies above 60 Hz, all the MRE samples with and without MWCNTs, decreased rapidly with and without magnetic field. Moreover, MRE with MWCNTs show lower loss factor as compared to MRE without MWCNT. It is due to the decreasing in the slipping displacement as a results from the strong interaction force between matrix and particles which lead to decreament in energy dissipation [15]. In addition, this result is consistent with the results of Jiang et. al [16] which natural rubber has better filler compatibility compared to other rubbers which lead to better filler matrix interaction [17] . Besides, this finding is in agreement with Ramana et al. whereby the functional groups of MWCNTs attached to the surface of MWCNTs will react with the MRE during curing process. Subsequently the covalent linkage would built which contribute to better interaction of the MRE [18]. At higher magnetic field, all the MRE samples with the MWCNTs showed almost the same trend at higher magnetic field of 643 mT.

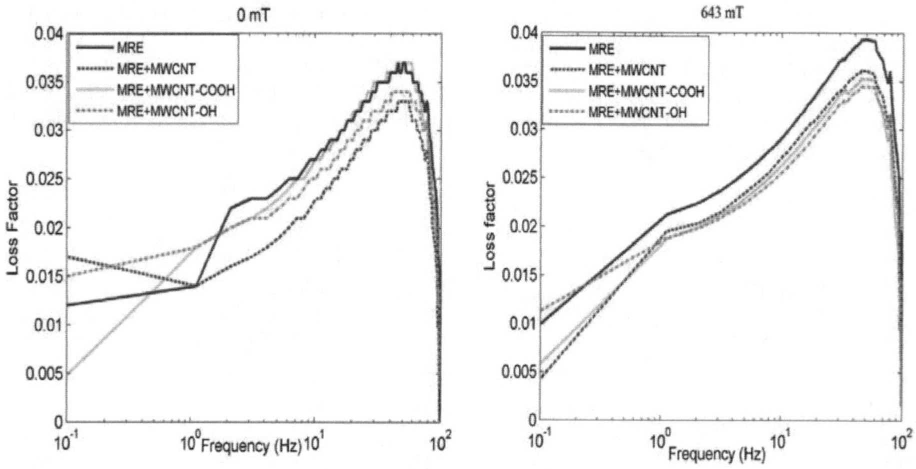

(a) (b)

Figure 2 Loss factor vs frequency (a) 0 mT (b) 643 mT

4. Conclusion

The rheological properties of the MRE with and without the different types of the MWCNTs have been analyzed in the present work. It has been shown that the incorporation of the different types of the MWCNTs affected towards the MR performance. The change brought by the study is tremendously important to interpret the function of the different types of the MWCNT as new additives in the MRE materials. The addition of the MWCNTs into the matrix leads to lower loss factor ratio and improved storage modulus. The MRE without MWCNT has shown the highest loss factor in both zero field modulus and at higher current induced. Rheological studies indicate that the interaction and compatibility between MRE with CIP and MWCNTs play an important role for the improvement of modulus and changes in loss factor of MRE properties.

5. Acknowledgements

The author gratefully acknowledge the financial funded by the Ministry of Higher Education, Malaysia PRGS Grant (Vote No: 4L667), Universiti Teknologi Malaysia under GUP (Vote No: 13H55) and also Malaysian Rubber Board for their technical advices and facilities.

6. References

[1] Zhang W, Gong X, Xuan S, and Jiang W, 2011 *Ind. Eng. Chem.* Res. **50**, 6704.
[2] Carlson J.D. and Jolly M.R. 2000 *Mechatronics* **10**, 555 (2000).
[3] Li W.H. and Zhang X.Z. 2010 *Smart Mater. Struct.* **19**, 035002 (2010).

[4] Hu Y, Wang Y.L, Gong X.L, Gong X.Q, Zhang X.Z, Jiang W.Q, Zhang P.Q, and Chen Z.Y. 2005 *Polym. Test*. **24**, 324.
[5] Chen L, Gong X.L, and Li W.H., 2007 *Smart Mater. Struct*. **16**, 2645.
[6] Ju B, Tang R, Zhang D, Yang B, Yu M, and Liao C, 2015 *J. Magn. Magn. Mater*. **374**, 283.
[7] Ubaidillah, Sutrisno J, Purwanto A, and Mazlan S.A, 2015 *Adv. Eng. Mater*. **17**, 563.
[8] Wei B, Gong X, and Jiang W, 2009 *J. Appl. Polym. Sci*. **116**.
[9] Mordina B, Tiwari R.K, Setua D.K, and Sharma A, 2014 *J. Phys. Chem. C* **118**, 25684.
[10] Nayak B, Dwivedy S.K, and Murthy K.S, 2015 *J. Intell. Mater. Syst. Struct*. **26**, 830.
[11] Li R. and Sun L.Z, 2013 *J. Eng. Mater. Technol*. **135**, 021008.
[12] Li R and Sun L.Z, 2011 *Interact. Multiscale Mech*. **4**, 239.
[13] Aziz S.A.A., Mazlan S.A, Ismail N.I.N, Ubaidillah U, Choi S.-B., Khairi M.H.A, and Yunus N.A, 2016 *Smart Mater. Struct*. **25**, 077001.
[14] Ponnamma D, Sadasivuni K.K, Grohens Y, Guo Q, and Thomas S, 2014 *J. Mater. Chem.* C **2**, 8446.
[15] Chen L, Gong X.L, and Li W.H, 2008 *Polym. Test*. **27**, 340.
[16] Jiang H.-X, Ni Q.-Q, and Natsuki T, 2011 *Polym. Compos*. **32**, 236 (2011).
[17] Chen S, Chen J, Xu K, Pan R, Peng Z., Ma L., and Wu S, 2015 *Polym. Compos*. **16**.
[18] Venkata Ramana G., Padya B., Kumar R.N., Prabhakar P., and Jain P.K, 2010 *Indian J. Eng. Mater. Sci*. **17**, 331.

Performance of polymer nano composite membrane electrode assembly using Alginate as a dopant in polymer electrolyte membrane fuel cell

S. Mulijani [1]
Department of Chemistry Bogor Agriculture University,
Darmaga Campus, INDONESIA

E-mail: srimulijani@ipb.ac.id

Abstract. Polymer membrane and composite polymer for membrane electrode assembly (MEAs) are synthesized and studied for usage in direct methanol fuel cell (DMFC). In this study, we prepared 3 type of MEAs, polystyrene (PS), sulfonated polystyrene (SPS) and composite polymer SPS-alginat membrane via catalyst hot pressed method. The performance and properties of prepared MEAs were evaluated and analyzed by impedance spectrometry and scanning electron microscopy (SEM). The result showed that, water up take of MEA composite polymer SPS-alginate was obtained higher than that in SPS and PS. The proton conductivity of MEA-SPS-alginate was also higher than that PS and PSS. SEM characterization revealed that the intimate contact between the carbon catalyst layers (CL) and the membranes, and the uniformly porous structure correlate positively with the MEAs prepared by hot pressed method, exhibiting high performances for DMFC.

1. Introduction

Polymer electrolyte membrane fuel cell (PEMFC) has great potential for movable and stationary power supply application wherein the fuel cell must be capable of operating at high power density with low content of catalyst to reduce its cost. In order to reduce the prime cost of PEMFC, the performance improvement of membrane electrode assembly (MEA) becomes a key point. MEA typically consists of proton exchange membrane (PEM), catalyst layers and gas diffusion layers (GDLs), in which the electrolyte, catalytic particles and reactants have a close contact with each other. The GDLs facilitate transport of reactants and products between fuel cell flow in the field and the ionomer electrolytes-catalysts interface. The state-of the art MEA consists of perfluorosulfonic acid (PFSA) polymer membranes such as Nafion® and Pt/C electrocatalysts. In spite of good physical and chemical stability and high proton conductivity, PFSA suffers from limited operation temperature, high cost and environmental concerns [1-2]. Therefore, significant effort has been made to develop nonfluorinated hydrocarbon polymers based on aromatic plastics, such as sulfonated

[1] To whom any correspondence should be addressed.

polystyrene (SPS), sulfonated polysulfone (SPSf), sulfonated poly(ether ether ketone) (SPEEK), sulfonated polybenziimmidazole (SPBI) and sulfonated poly (aryl ether kethone) (SPAEK) as an alternative proton exchange membranes for PEMFCs. These sulfonated aromatic polymers possess excellent chemical resistance, high thermo-oxidative stability, good mechanical properties and low cost [3-4].

MEA for PEMFCs is generally fabricated by the hot pressing method, wherein a PEM is sandwiched between an anode and a cathode electrode and pressed at a high temperature. However, the MEAs fabricated by the conventional hot pressing method are usually inclined to delamination both at anode and cathode interfaces after long-term operation, leading to degradation in cell performance. Thus, conventional hot-pressing method can only offer physical contact, rather than achieving a true interfacial contact between the PEM and the electrodes during the MEAs fabrication process. Some researchers faced delamination problem in hot-pressing method in case of using liquid methanol as a fuel in a direct methanol fuel cell (DMFC), because swelling of the PEM in methanol solution which can greatly affect the intimate contact between the PEM and electrodes in MEAs. In this context, it is essential to mention some of the recent studies in which effort has been made to improve the interfacial contact in the MEAs for DMFC. Liang *et al.*[5] introduced a simple method for fabricating the MEAs as compared to dry

2. Experimental Methods

Polymer material used in the study was prepared from styrofoam (PS) waste and it's synthesized by sulfonation reaction. Membranes were prepared by dissolving a known amount of sulfonated poly styrene (SPS) in dichloromethane (10%) under constant stirring. Polymer SPS-Na-alginate nanocomposite membranes were prepared by mixing SPS and Na-alginate and dissolving in dichloromethane with ultrsonication process.Three types of proton exchange membranes PS, SPS and SPS-Na-alginate were used as the polymer electrolyte membrane for MEA preparation. Before use, the membranes were pretreated in hydrogen peroxide solution (5 wt.%) at 80℃ for approximately 1 h; after being flushed with deionized water, it was transferred into a 0.5M sulphuric acid solution and boiled at 80℃ for an additional 1 h, then washed with deionized water [8].

The catalyst inks were prepared by dispersing catalysts powder into a mixture of dichloromethane and membrane solution. The catalyst used for anode and cathode layers were Ag/C (20 wt.% Ag/AgCl) and 5% membrane solution. The mixture was ultrasonically dispersed for 30 min before being used. All MEAs used in this study were prepared using hot pressing method at 120-125℃ for approximately 5 min.

Electrochemical impedance was conducted for a single cell by feeding 1.0M methanol at a flow rate of 1mlmin$-^1$ using a high-pressure piston pump and by purging dry oxygen gas into the cathode at a flow rate of 20mlmin$-^1$ at 40°C at a constant dc potential controlled by the potentiostat with applying a small alternating voltage (10mV) and varying the frequency of the alternating voltage from 1×10^4 to $1\times10-^2$ Hz. Scanning electron microscope (SEM) was used to evaluate the cross section of the MEA. The sample with gold sputter coatings was carried out on the desired membrane samples at pressure ranging between 1 and 0.1 Pa. The DMFC performance test was conducted at 40.0℃ by feeding 1.0M methanol at a flow rate of 1mlmin$-^1$ using a high-pressure piston pump and by feeding dry oxygen gas into the cathode at a flow rate of 20mlmin^{-1} at ambient pressure by fixing the load current, which was controlled with an electric load system.

3. Result and Discussion

The DMFC performance was evaluated through by polarization curve, also known as a voltage–current density (V-I) curve, is commonly used to evaluate fuel cell performance. Figure 1 is a representation of typical polarization compares the performance of the single cells with the membrane PS, SPS, SPS-alginate and MEA-PS, MEA-SPS, MEA-SPS alginate with 1.0 M methanol at 50℃. The Figure 1 shows that all of the MEA membranes have better fuel cell performance than that in single membrane. This suggests that

the membrane with this type of membrane electrode assembly can improve fuel cell performance. These results also reveal that in the very low current density region, the cell using MEA-SPS exhibited a slightly lower open circuit voltage (OCV) and performance than the cell with the MEA-PS and MEA-SPS-alginate. This can be attributed to the fact that the rate of methanol cross over for the MEA-SPS is higher than that in the MEA-SPS and MEA-SPS-alginate.
However, in higher current density region particularly above 500 mA/cm2 the cell using MEA-SPS-alginate showed much better performance as compared to the cell with MEA-PS and MEA-SPS. This indicates that the MEA-SPS-alginate has a lower mass transfer resistance than the MEA-PS and MEA-SPS. The lower mass transfer resistance for the MEA-SPS-alginate might be related to the possible formation of hydrophilic pores in the MEA fabrication process. Chaudry and Sengawa [6] have reported increase in hydrophylicity properties and also the conductivity with addition of small inorganic polymer such as monmorilonite, alginate, silica, etc. Earlier reports suggest that sulphonated of polystyrene has ability to increase the hydrophilicity properties of the membrane.

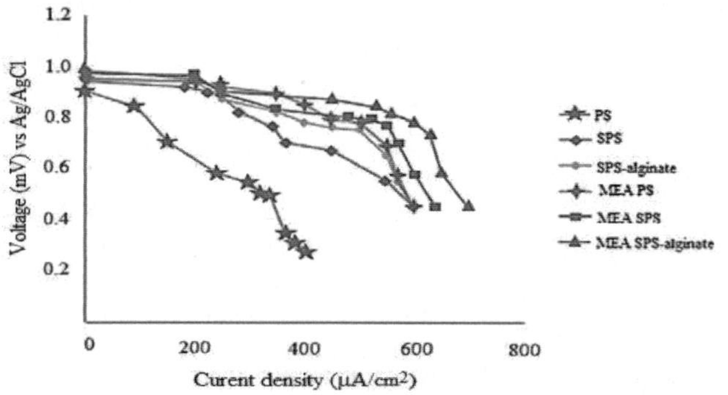

Figure 1. Polarization of comparison of cell performance for membrane

Generally with sulfonated acid membranes, there is no preferential permeation of water or methanol; both decrease with increasing sulfonated content. For PS membranes the total flux of methanol and water increases with increasing temperature and feed methanol concentration. Therefore, sulfonated membranes seems promising and worth investigating for DMFC application. Figure 2 exhibit the result of proton conductivity in different electrode such as carbon, cuprum and iron. In order to provide enough hydration to the membrane and activated the anode catalyst activation of the prepared membranes is necessary. MEA SPS alginate activated membrane appeared to have higher proton conductivity than that the MEA SPS activated membrane. Alginate as a dopants incorporated into the pores of membrane of MEA will affect the high conductivity of membrane. The effect of incorporating alginate as inorganic dopants in the membrane depends on the nature of the dopant in question. If the dopant is not acidic it simply acts as a barrier to both proton and methanol flow. The dopant in this case is non-conducting and acts as a barrier to both vehicular diffusion and Grotthuss diffusion. It obstructs the former by preventing the vehicular transport of hydronium ions, and it also obstructs the latter by disturbing the hydrogen bond network. Apparently this should be particularly effective in blocking ing methanol flow as methanol molecules are much larger than protons, but it also affects proton conductivity adversely.

Another effect of higher conductivity of MEA SPS-alginate membrane is due to its contents of sulfonate group in the SPS. The particle has an SO_4^{2-} group attached which attracts protons and continues the hydrogen bond network. This particle would still block vehicular diffusion but will only slightly reduce Grotthuss diffusion, if at all. Depending on the acidity of the particles, it might even increase the proton exchange sites and Grotthuss diffusion and the overall conductivity.

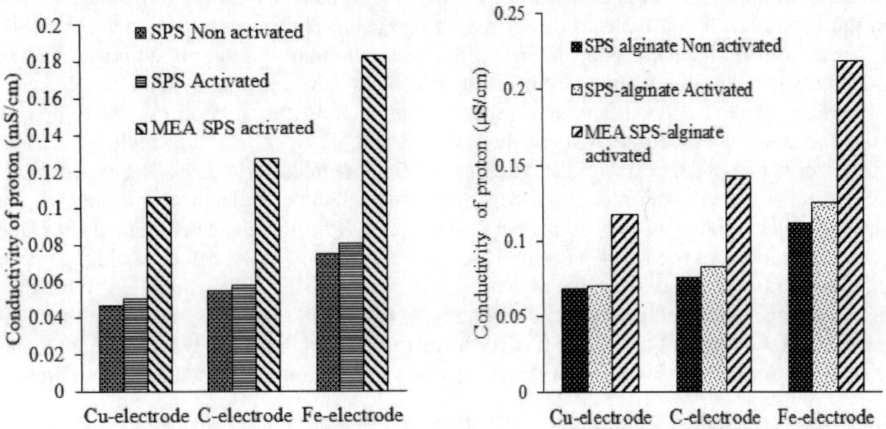

Figure 2. Proton conductivity of SPS and SPS-alginat

Sulfated dopants, such as sulfated zirconia and sulfated silica are especially popular in current research [11] and extensive research has been done with these inorganic particles in PEMFCs. Incorporation of sulfated zirconia particles (ZrO_2/SO_4^{2-}) into Nafion via the sol-gel method has shown increased water uptake and proton conductivity in hydrogen/oxygen fuel cells due to the additional acid sites [12]. However, due to methanol crossover, increased water uptake is undesirable in DMFCs. The acidity of the membrane should be such that overall methanol crossover is reduced while maintaining conductivity. The results revealed that the water up take of MEA SPS (13.68%) was higher compared to SPS (5.15%) and PS (0.69%). The contribution of alginate and sulfonate in the MEA-SPS-alginate membrane would increase proton conductivity and water uptake. While the water up take might increase methanol crossover in the membrane, sulfonate group increases the proton exchange site.

Evaluation of sulfonate group in the sulfonation process of the PS was confirmed qualitatively by FTIR. The process, which occurs exclusively in the phenyl rings should be manifested by the appearance of new vibration characteristic of sulfonic acid groups; by changes in the aromatic C–H out of plane bending region and perhaps by splitting of some bands (e.g. the C–O stretch of the substituted rings), which is due to the reduced symmetry and direct influences of sulfonic acid groups vibration frequencies after the sulfonation process.

Significant differences in FTIR spectra of sulfonated PS were observed compared to pure PS. The peak identified at 1500 cm^{-1} proved the presence of C–C aromatic ring in the PS, which referred to the 2, 4-substitution aromatic ring indicating that only first substitution was involved in this study. As sulfonated PS became water-soluble when the sulfonation degree was approaching 100% the first type of substitution reaction is more favorable. Therefore, the study followed first type substitution. The sulfonation process decreased the intensity of the aromatic C–C vibration and the peak was observed to split to two absorption bands at 1495 and 1475cm^{-1}. These two bands increase with the increasing concentration of sulfonic acid group in the sulfonated PS membranes. Occurrence of a new absorption band at 1028.35 cm^{-1}, 1181.27cm^{-1}, 1275 cm^{-1}and 1069 cm^{-1} upon sulfonation indicated presence of the sulfonic acid groups in PS. In summary, the changes observed in the FTIR spectra indicated the presence of $-SO_3H$ groups.

The cross section of membrane morphology was evaluated to support the performance of DMFC. Figure 3 shows the SEM cross-sectional images of membrane PS, SPS, SPS-alginate and MEA SPS-alginate, respectively. It is clear from the SEM images that for the MEASPS-alginate, by hot pressing, the interfacial contact between the membrane and the electrodes proves to be good. However, for the membrane SPS-alginate, which was activated, alginate seems algomerated in the membrane. Dopant of alginate increased both proton conductivity and the strengthen of hot pressing method to reduce the delamination problem that occurred at the interfaces of the MEAs. Liang *et al* reported that delamination problem seems to be more serious both at the anode and cathode interfaces of the conventional MEA; in fact, this could be the main cause for the low performance of the cell with the MEA after being operated for 4 days. The existence of a better interfacial contact at the cathode interface for the MEA SPS-alginate explains why it yielded more stable performance than that of MEA SPS after operating the cell. The prevalence of a good interfacial contact at the cathode interface of the MEA SPS-alginate can be attributed to the effect of alginate as an inorganic dopant into membrane. In addition, the alginate presence in the MEA SPS-alginate also roughens the surface of the membrane, which enhances the contact region of the MEA-SPS-alginate membrane with the electrodes during the fabrication of the MEA. It is pertinent to mention that the solvents methanol/water present in the membrane dispersion play an important role while fabricating the MEA by the hot pressing method.

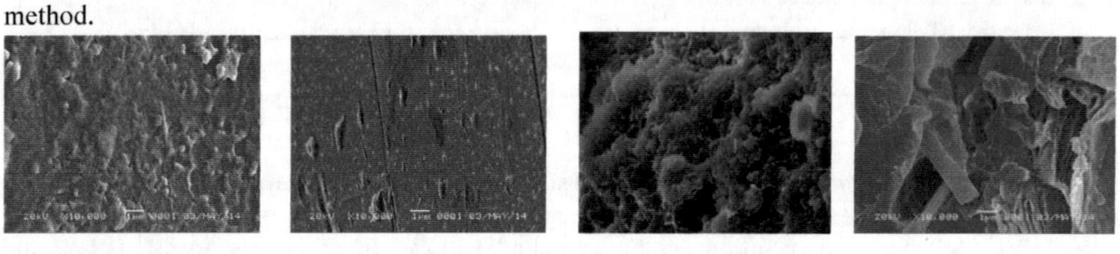

(a) (b) (c) (d)

Figure 3 Cross section morphology of membranes (a) PS, (b) SPS (c) SPS-alginate, dan (d) MEA SPS-alginate

In conclussion, a kind of polymer electrolyte composite inorganic dopant was introduced between the PEM and the electrodes during the MEA fabrication process to achieve a better interfacial contact. The dopant provides a better resistant and stronger binding force between the membrane and the electrodes. The performance of the cell using the MEA-SPS-alginate was found to be more stable. The cross-sectional analysis of the MEAs after the long-term operation revealed that the MEA-SPS-alginate retained a better interfacial contact between the membrane and the electrodes as compared to the MEA-SPS. The proton conductivity results revealed that the MEA-SPS-alginate exhibited a higher value than the MEA-SPS, confirming that the MEA-SPS-alginate had a better interfacial contact between the electrodes and the membranes. The inorganic content needed for the right balance varies. Sulfonate group attached in PS and alginate doped into MEA-SPS-alginate membranes seems to have potential advantages for DMFCs over plain Nafion membranes.

4. References

[1] Woong JA, Yim SD, Choi YW, Sohn YJ, Park SH, Yoon GY, Park GG, Yang TH, Kim KB 2011 *Electrochemica Acta* 56, 7732-7739.

[2] Kendrick I, Smotkin ES 2014 *International Journal of Hydrogen Energy*. 39, 2751-2755.

[3] Mulijani S, Dahlan K, and Wulanawati A, 2014 *IJMMM* 2 36-40.
[4] Lakshmi RTSM, Meier HJ, Schlenstedt K , Choudhary CVV, Varna IK 2005 *J. of Membrane Science.* 26, 27-35
[5] Liang ZX, Zhao TS and Prabhuram J 2006 *Electrochemica Acta* 51, 6412-6418.
[6] Chaudhary S, Sengwa RJ 2010 *IJEMS* 18 147-156.
[7] Cho Y-H *et al* 2012 *International Journal of Hydrogen Energy.* 37, 5884-5890
[8] Huaneng BJ, Bladergroen P, Sivakumar, Linkov V, Shan J 2012 *Int. J. of Elect. Sci.* 7 4223-4234.
[9] Wu QW, Zhao TS, Chen R, Yang WW 2010 *Int. J. of Hydrogen Energy* Vol.35, pp.10547-10555.
[10] Jalani Nikhil 2006 [Report]. - Worcester: Worcester Polytechnic Institute,.
[11] Pyoungho C, Jalani Nikhil H and Ravindra D 2005 *J. of The Elect. Society*, Vol. 152,. - A1548 .